NEUROSCIENCE RESEARCH PROGRESS

REPAIR OF THE HUMAN BRAIN AND SPINAL CORD

NEUROSCIENCE RESEARCH PROGRESS

Additional books in this series can be found on Nova's website under the Series tab.

Additional e-books in this series can be found on Nova's website under the e-book tab.

NEUROSCIENCE RESEARCH PROGRESS

REPAIR OF THE HUMAN BRAIN AND SPINAL CORD

GISELHER SCHALOW

NOVA BIOMEDICAL

New York

Library of Congress Cataloging-in-Publication Data

ISBN: 978-1-62948-314-6
Library of Congress Control Number: 2014958461

Published by Nova Science Publishers, Inc. † New York

Contents

Preface

With the single-nerve fiber action potential recording method and the single-motor unit electromyography the functioning of the human central nervous system (CNS) is analyzed at the single-neuron level under physiologic conditions and following injury, malformation and degeneration. It is shown that the self-organization of the neuronal networks of the human CNS by phase and frequency coordination becomes impaired following all nervous system diseases.

Out of the differences between the functioning of the healthy and pathologically functioning CNS, a repair treatment is developed called Coordination Dynamics Therapy (CDT). This movement-based learning therapy is able to improve almost every nervous system in its functioning by functional and structural repair for all ages including premature born babies and in aging.

The therapy progress with CDT can be quantified by using the System Theory of Pattern Formation. By pattern change, given by a special CDT device, a single value is obtained for the quality of CNS functioning.

Especially the tremor in Parkinson's disease and the urinary bladder repair in spinal cord injury are analyzed in detail by the human neurophysiologic recording methods and clinical assessments.

The repair of the human nervous system is shown in cross-sectional and longitudinal studies in mild and severe traumatic brain and spinal cord injuries. By applying CDT, CNS functioning can also be improved in Parkinson's disease and in aging. The rate of improvement/repair by learning is measured in the healthy and the diseased nervous systems. It is emphasized that the efficacy of movement-based learning therapies may differ by a factor of 100. A powerful tool in learning is the learning transfer. By training integrative movements, including automatisms, not only the trained movements can be improved, but also the vegetative and higher mental functions can be repaired as, for example, continence, speech and cardio-vascular performance.

Giselher Schalow, MD
Dr. Med. Habil., Dr. Rer.Nat., Dipl.Ing.
Untere Kirchmatte 6
CH-6207 Nottwil, Switzerland
g_schalow@hotmail.com

Chapter I

Introduction to Coordination Dynamics Therapy

1. Learning for Repair

One explanation for how multicellular organisms produce adaptive changes without genetic changes may lie in the development of complex physiological and behavioral systems coordinated by the central nervous system.

The nervous system permits rapid adaptation to changing environmental conditions without genetic mutation [76] by coordinating inputs from the internal and external environment (these external influences may be used to construct a formula for the somatic nervous system by using the equation of motion within the system theory of pattern formation) via receptors and directing a complicated physiological response through various effector systems to maintain homeostasis. In many instances, homeostasis is maintained by reflexes, automatisms and fixed action patterns in response to a stimulus. A problem arises when a stimulus or event in the environment does not determinately predict a condition or appropriate response to that condition.

To cope with environmental complexity and ambiguity, an organism requires mechanisms that allow experience to effect relatively long-lasting changes in behavior. Through a nervous system, the organism can accomplish this through a mechanism that has been called learning. Behavior modification is achieved through changes of neuronal network functioning [22].

In the human nervous system, learning is used for repair. By learning the injured nervous system adapts to movements given and dictated by devices and by re-learning of volition old learned movements. When the CNS is injured, the rate of learning is reduced. The rate of repair depends on the learning strategy, that is to say the treatment method, and the severity of the CNS injury. With the treatment method 'Coordination Dynamics Therapy' rates of repair may be measured and compared with the rate of learning of the healthy human CNS.

2. Specific Neural Activity Is Required for Development and Repair

During movement-based learning neuronal networks are specifically activated. The specific activity of α-motoneurons, γ-motoneurons, parasympathetic neurons and the different kinds of skin and urinary bladder receptors at the neuron level under physiologic and pathophysiologic conditions has been analyzed in detail [14]. Some data on human neurophysiology are given below.

Neuronal specific activity seems to be required to regulate development [35] and for the repair of the human CNS. It seems that activity controls neurogenesis, maturing of neurons, specific migration of neurons and the development of their axons and dendrites. The specific complexity of axonal and dendritic arborization and weights of synapses develops through this natural activity, generated by natural movements.

There is an indication that genetic programs initiated at the progenitor stage are modulated during development by activity [35]. A critical period for enhanced synaptic plasticity in newly generated neurons in the adult brain has been identified. Adult-born neurons exhibit the same classic critical period plasticity as neurons in the developing nervous system [23].

3. Epigenetic Mechanisms

If only genomes were important to evolution, humans would not have developed any further than mice. In fact, plants would have the same chance of ruling the world because humans, mice, and Arabidopsis genome share the same number of genes, approximately 25,000. It is the epigenome that has evolved tremendously since the appearance of the first multicellular organisms [21].

Simple organisms, such as bacteria, increase their rate of spontaneous mutations to enable the survival of species in a changing environment. Multicellular organisms use complex mechanisms coordinated by the central nervous system (CNS) to behaviorally adapt to changing environments without having to undergo the difficult process of mutating their genome [22]. Their behavioral adaptation depends on learning, supported by epigenetic mechanisms. What sets the injured human nervous system apart however is that it may be partially repaired by learning [14]?

The rate of hippocampal neurogenesis significantly changes on physiological and pathological influences, such as physical activity, environmental enrichment, stress, and aging. It has been suggested that epigenetic mechanisms could be sensors of environmental changes and fine modulators of adult hippocampal neurogenesis [21]. It is likely that permanent changes in gene expression patterns are achieved through permanent changes of epigenetic mechanisms [22]. Epigenetic mechanisms are likely used to maintain cell memories, how cells maintain their differentiated characteristics, as well as maintain and change neuronal networks to maintain long-term changes in behavior and to repair CNS injury by learning. An important question is how movement-based learning can modulate the epigenetic mechanisms to repair more efficiently.

4. Coordination Dynamics Therapy for Repair

The prevailing belief is that substantial repair of the injured, malformed or degenerating human central nervous system (CNS) is not possible. However, up-to-date human neurophysiological research shows that a partial repair of the injured nervous system is possible through the performance of a special intensive movement-based learning therapy called ***Coordination Dynamics Therapy (CDT)*** [14]. The extent of repair depends on the severity of the injury. The CNS of the patient can learn to repair itself. CDT teaches the CNS how to repair itself by learning. All repair mechanisms are already in the nervous system, they need however, to be activated and optimized by learning.

Coordination Dynamics Therapy (CDT) is largely based on measurements of the human CNS [12, 14] and has been shown to be able to improve CNS functioning after stroke [1], traumatic brain injury [2, 7], spinal cord injury [3, 4, 10, 13], cerebral palsy [6], cerebellar injury [9], in Parkinson's disease [5] and after hypoxic brain injury [8]. Speech had been induced and improved in a patient with severe cerebral palsy [14] and urinary bladder functions were repaired in spinal cord injured patients [13, 14].

5. Rate of Repair

To better understand the repair strategies and to improve the efficiency of repair, this book shall measure and compare rates of repair. It has been shown that the CNS repair may need years [10, 13] of efficient therapy administered at the limits, or 'aggressive therapy'. The rates of improvement in CNS functioning in healthy subjects and those with traumatic brain injury and differing spinal cord injuries shall be contrasted.

6. Stability of Movement Patterns

Repair requires not only the re-enablement of a movement or other pattern, but the pattern also requires certain stability. To cover any appreciable distance, the patient must be able to walk or run for some time without losing the movement pattern. In the vegetative nervous system the pattern stability is even more important. The urinary bladder continence pattern is only able to be achieved in the patient with when this automatic pattern is sufficiently stable so that the patient is continent for a few hours and can control it. Therefore, stability of self-organized neuronal network patterns is essential for normal life. A bird would not live very long if the flying pattern were to become unstable. Such pattern stability may be measured continuously and improved through movement-based therapy in a patient with severe traumatic brain injury when running on a treadmill.

To aid in the understanding of pattern formation and pattern stability, the Method section combines human neurophysiologic measurements on the variability of phase and frequency coordination among single neuron firings with the pattern stability defined at the macroscopic level within the System Theory of Pattern Formation (Figure 1).

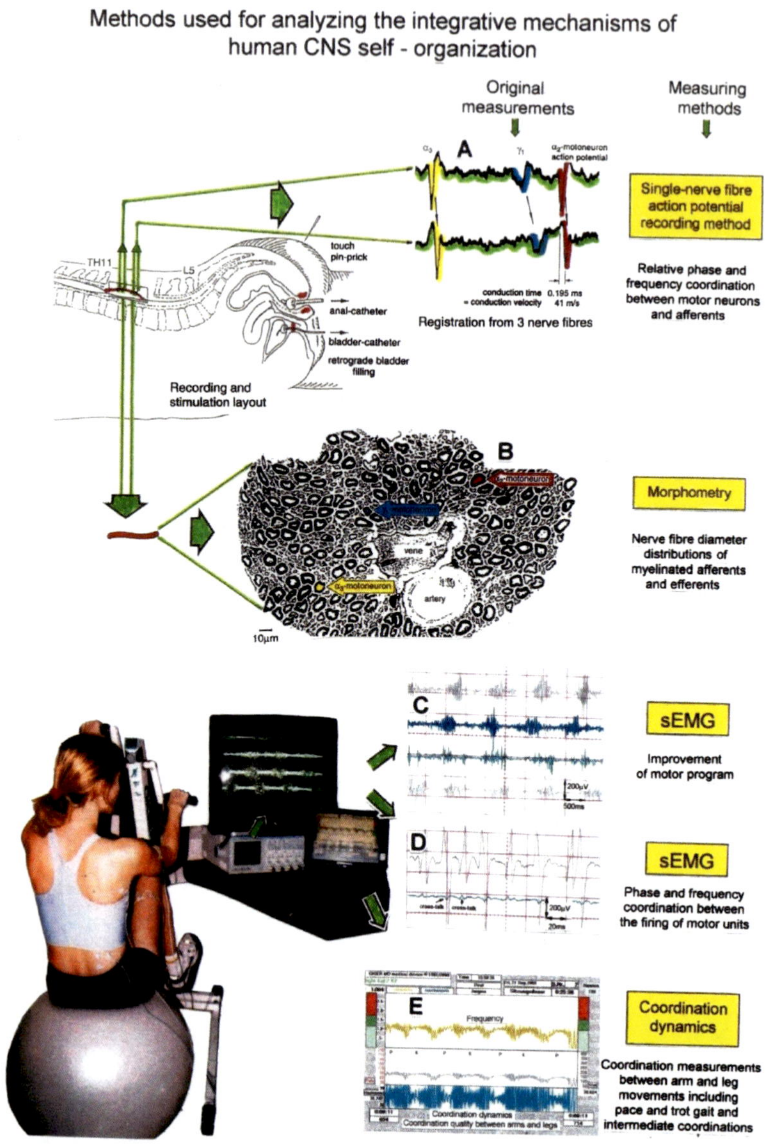

Figure 1. Schematic layout of the measuring methods used to study the self-organization of neuronal networks of the human CNS under physiologic and pathophysiologic conditions.

A, B. By recording with two pairs of platinum wire electrodes from sacral nerve roots containing between 200 and 500 myelinated nerve fibers, records were obtained in which single-nerve fiber action potentials (APs) could be identified from motoneuron axons (main AP phase downwards) and afferents (main AP phase upwards). By measuring the conduction times and with the known electrode pair distance (10 mm) conduction velocities could be calculated. Conduction velocity distribution histograms were constructed in which the myelinated nerve fiber groups larger than 4μm could be characterized by group conduction velocity values. Recording was followed by morphometry. Distributions of nerve fiber diameters were constructed and nerve fiber groups were characterized by the peak values of asymmetrical distributions. By correlating the peak values of the velocity distributions with those of the diameter distributions obtained for the same root, a classification scheme was constructed of the human peripheral nervous system. It became thus possible to record natural impulse patterns simultaneously from identified single afferent and efferent nerve fibers and to analyze self-organizing mechanisms of the human CNS under physiologic and pathologic conditions.

C, D. By performing surface electromyography (sEMG) with up to 4 electrode pairs (2 indifferent electrodes, 1 earth electrode; pre-amplification x1000, 4-channel oscilloscope) the changes of motor programs were measured (C). When recording from appropriate patients, natural activation patterns of several single-motor units were obtained (D) and coordination between motor units firing could be studied.

E. Coordination between arms and legs was quantified by the single integrative parameter, arrhythmicity of turning (df/dt or df/dt/f; f = frequency) during exercising on a special coordination dynamics therapy device.

In this way, the quality of CNS organization can be measured macroscopically and non-invasively via stability changes upon pattern change. When the patient is exercising on a special CDT device, which imposes on the patient continuously changing arm and leg movements, such pattern change can be made visible (Figure 174) and measured by what may be called Coordination Dynamics values.

Chapter II

Method: Scientific Basis for Coordination Dynamics Therapy

1. Repair Strategies in General

1.1. Movements, Coordinated Firing of Nerve Cells and Neural Network Stability, Have to Be Trained for Repair

In nervous system injuries, movements and other patterns become impaired and the coordinated firing of nerve cells and neural network stabilities become impaired [12,14]. To repair nervous system functioning, movements have to be re-learned, the coordinated firing of nerve cells has to be improved, the communication of neural sub-networks improved and the stability of physiologic patterns increased and those of pathologic patterns, such as spasticity, decreased.

Movement-based learning can repair impaired CNS functioning, because this is essentially the method which the CNS uses during development for correcting CNS development 'en route'. Trial and error eliminating processes are used to improve CNS self-organization [14].

CDT uses these movements of the developing child for improving and repairing CNS function, namely creeping, crawling, standing up, walking, running, jumping (Figure 2) and training balance. Additionally, the coordinated firing of nerve cells can be trained when a subject is exercising on the special CDT device (Figures 2A,33), which is precisely calibrated so that the CNS can re-learn the coordinated firing of nerve cells up to within a few milliseconds (Figure 9) from the movement induced afferent input.

Through exercising on the device, physiologic movement patterns improve and their stability will increase and the stability of pathologic patterns will decrease by learning transfer [12, 14] (Figures 18,19). The patient can only train those movements which he can perform independently or with the assistance of a therapist (Figures 2,58).

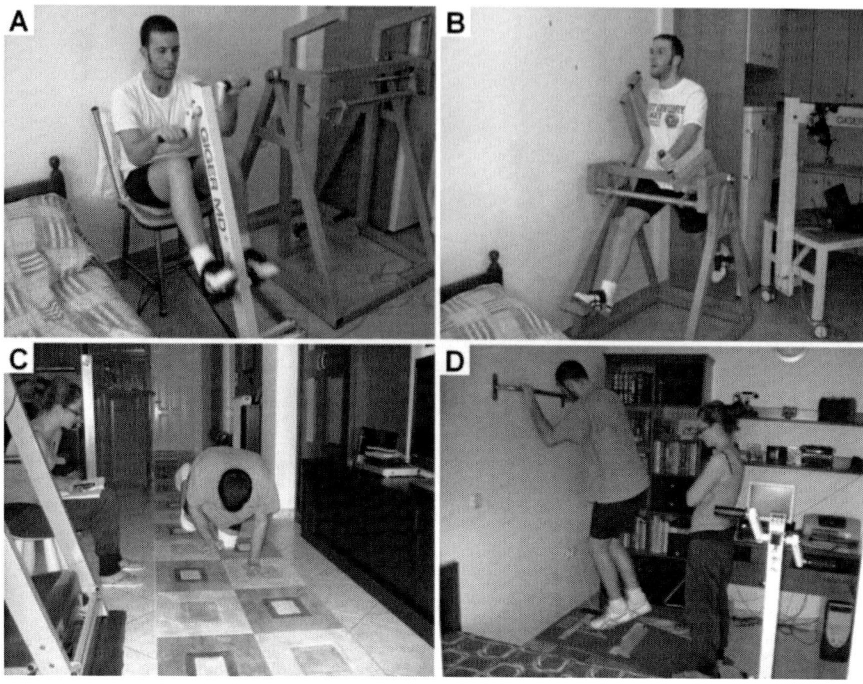

Figure 2. Patient with severe brain injury following a car accident during exercising on the special CDT device, sky-walking, crawling and jumping on springboard.

1.2. Learning Transfer from Movements to Improve Autonomic and Higher Mental Functions

A powerful tool to repair the injured CNS by movement-based learning is the learning transfer. By learning transfer [12, 14] patterns which cannot be trained by movement per se, for example urinary bladder, sexual and higher mental functions can be partly repaired. Through movement-based learning (CDT), urinary bladder functions [13] and speech [14] could be repaired in patients with spinal cord injury and cerebral palsy. The scientific basis for learning transfer is given by the System Theory of Pattern Formation and Human Neurophysiology [12, 14]. Since sub-neural networks of the human CNS can generate different patterns and the CNS must be viewed as one network, one trained pattern will also influence the other, untrained patterns to become more physiologic in functioning.

1.3. Different Repair Strategies for Humans and Animals: Learning for Repairing the Human CNS

Repair strategies of the nervous system in animals and humans are different. In animals, in which the neural networks are more hard-wired, the nerve fiber growing strategy is used. In humans, in whom the neural networks are so complex as to enable speech and reading, the movement-based learning strategy is used; learning is primarily used and not the enhancement of growing of nerve fibers. Animal research is therefore only of limited assistance to the repair of the human nervous system [14].

2. Human Electrophysiology

2.1 Single-Nerve Fiber Action Potential Recording for Analyzing CNS Self-Organization

2.1.1. Single-Nerve Fiber Action Potentials to Measure Phase and Frequency Coordination among Identified Afferent and Efferent Nerve Fibers in Nerve Roots

The development of the single-nerve fiber action potential recording method was possible because of the unique anatomical landscape of the human spinal canal. Because of the ascensus of the spinal cord, the lumbosacral nerve roots have become very long and form the cauda equina (Figure 4). Since the caudal sacral nerve roots are very thin (Figure 4) and nerve roots are only ensheathed by a thin layer of cells (Figure 1), they are ideal for recording single-nerve fiber action potentials (APs) from undissected nerve roots. Since humans have no tail, continence (mainly S2 to S5) and sexual functions are mainly located in the conus medullaris only. Those functions are therefore represented in the lower sacral nerve roots and they are not intermingled with tail functions as they are for example in rats, cats and dogs. A human CNS, dissected by the Author, is shown in Figure 4A.

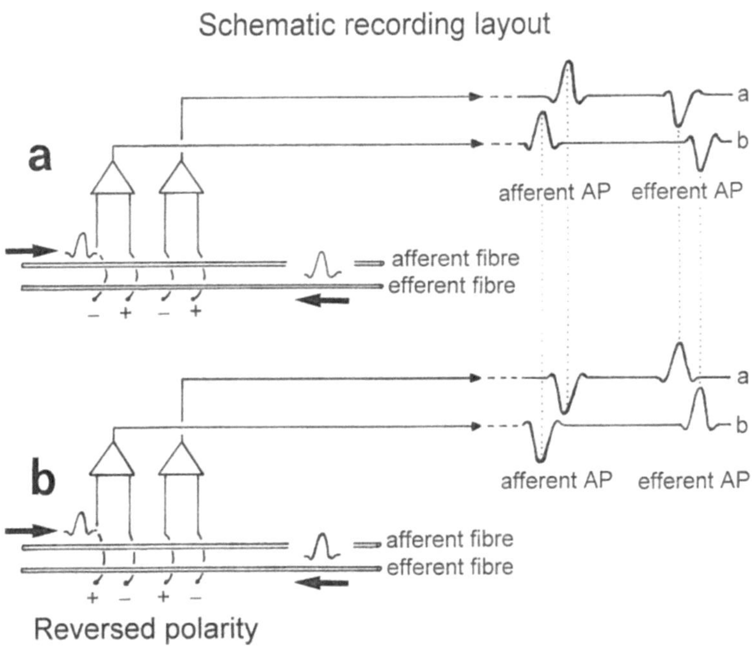

Figure 3. Schematic layout for recording single-nerve fiber action potentials (APs). The reversing of the inputs to both preamplifiers does not change the ability to differentiate between afferent and efferent APs.

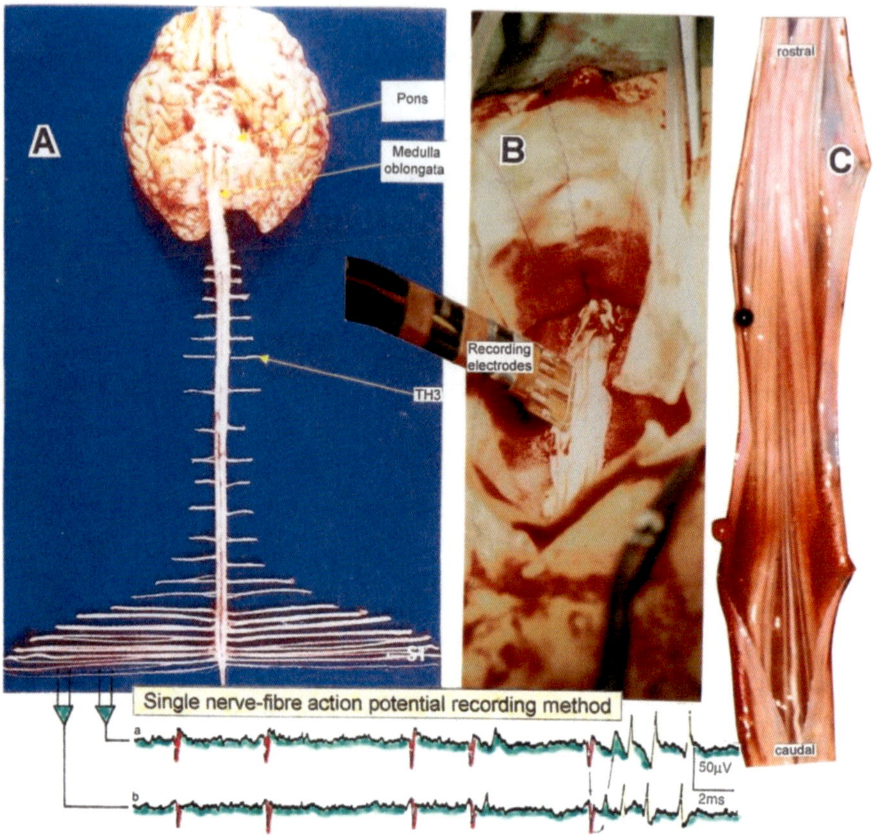

Figure 4. Layout of the recording of single-nerve fiber action potentials to analyze the self-organization of neuronal networks of the human CNS under physiologic and pathophysiologic conditions. A, B, C. By recording with two pairs of platinum wire electrodes (B) from sacral nerve roots (cauda equina, C) containing between 200 and 500 myelinated nerve fibers, records were obtained in which single nerve-fiber action potentials (APs) were identified from motoneurons (main AP phase downwards) and afferents (main AP phase upwards). By measuring the conduction times and with the known electrode pair distance (10 mm), conduction velocity distribution histograms were constructed in which the myelinated nerve fiber groups larger than 4μm could be characterized by group conduction velocity values. After the recording, morphometry was performed. Distributions of nerve fiber diameters were constructed and nerve fiber groups were characterized by the peak values of asymmetrical distributions. By correlating the peak values of the velocity distributions with those of the diameter distributions obtained for the same root, a classification scheme was constructed of the human peripheral nervous system. It became thus possible to record natural impulse patterns simultaneously from identified single afferent and efferent nerve fibers and to analyze self-organizing mechanisms of the human CNS under physiologic and pathologic conditions. A. Human CNS with the schematic illustration of the recording layout and original record of single nerve-fiber action potentials. Note the time calibration of 2ms. B. Intraoperative recording layout (when implanting a bladder stimulator) with two pairs of wire electrodes and one temperature sensor. A thin nerve root is positioned over the platinum wire electrodes. C. Dissection of the human cauda equina; at the caudal end of the filum terminale with thin nerve roots are seen. D. Schematic splitting of the activity of several nerve fibers into simultaneous impulse patterns of single fibers by comparing waveforms, conduction velocities and reoccurring characteristic impulse patterns (rhythmic firing of sphincter motoneurons). The different conduction times and waveforms were recognized on an expanded time scale. Stretch receptor and secondary muscle spindle afferents contribute to the drive of sphincter motoneurons and form, together with other afferents, regulation units.

The schematic recording layout is shown in Figure 3 to record single-nerve fiber action potentials (APs) extracellularly from undissected nerve roots with two pairs of platinum wire electrodes (electrode pair distance 10 mm; electrode distance in each pair 4 mm), pre-amplified (x1000), filtered (RC-filter, passing frequency 100 Hz - 10 kHz), and displayed on a digital storage oscilloscope, and also stored using a PCM-processor and a video recorder. APs from the afferent and the efferent fibers can clearly be distinguished since in the electrode arrangements, the main phase (second phase) from the afferent fiber is upwards and that of the efferent fiber is downwards (Figure 3A). E.g., the AP of a skin afferent fiber reaches a pair of electrodes first as negative and then as positive. According to the electrode setting used, the main phase is upwards. An AP of a motoneuron, coming from the opposite direction, would reach the electrodes in the order positive-negative. The potential changes are therefore opposite and the main triphasic AP will point downwards. An AP in an afferent fiber reaches first the caudal electrode pair and then the rostral pair, whereas an AP of the efferent fiber reaches first the rostral electrode pair and then the caudal one. The conduction times are therefore also opposite. The reversing of the inputs to both preamplifiers does not change the ability to differentiate between afferent and efferent APs (Figure 3B).

A real recording arrangement during an operation is shown in Figure 4B. To obtain natural impulse patterns simultaneously from several single afferent and efferent nerve fibers to analyze CNS functioning, the summed impulse traffic has to be split into the impulse patterns of single fibers. The splitting is achieved by recognizing the APs from certain single fibers on the basis of wave form comparisons on the two recording traces and the conduction time which an AP needs to travel from one electrode pair to the other one (10mm) and selecting these APs out. The splitting of the skin afferent activity upon touch or pin-prick into the natural impulse patterns of the different single touch and receptors is shown in Figure 5. Such messages inform the CNS about changes in the periphery. Similar natural impulse patterns inform the CNS about changes in the urinary bladder or anal canal such as bladder filling or bladder or anal canal catheter pulling (Figure 152). Urinary bladder afferents stimulated by electrical stimulation is shown in Figure 6.

As we can measure the natural impulse patterns, generated by certain single receptors in the periphery, which run into the spinal cord (CNS) and those patterns which leave the cord in ensembles of single fibers simultaneously, it becomes possible to analyze the integrative properties of the largely unchanged CNS of brain-dead humans (HTs) at a cellular level. This also means that the change in function, caused by a CNS injury, can be identified. The use of this new electrophysiological recording technique has brought new understanding to the functioning of the human CNS and has allowed for much greater opportunity in identifying techniques for its repair.

For the analysis of CNS functioning must first identify the kind of nerve fibers from which the recordings are taken. A classification scheme of human peripheral nerve fibers is needed.

Conduction velocities of single nerve fibers were calculated from the conduction distance (electrode pair distance = 10 mm) and the conduction times (the time difference of a certain AP between the traces from two pairs of wire electrodes). Velocity distributions of afferent and efferent fibers were constructed and distribution peaks were correlated to certain nerve fiber groups. The nerve roots from brain dead and surgical patients could be removed, fixated, embedded and stained. Mean nerve fiber diameters could be measured, and nerve fiber diameter distributions constructed (morphometry).

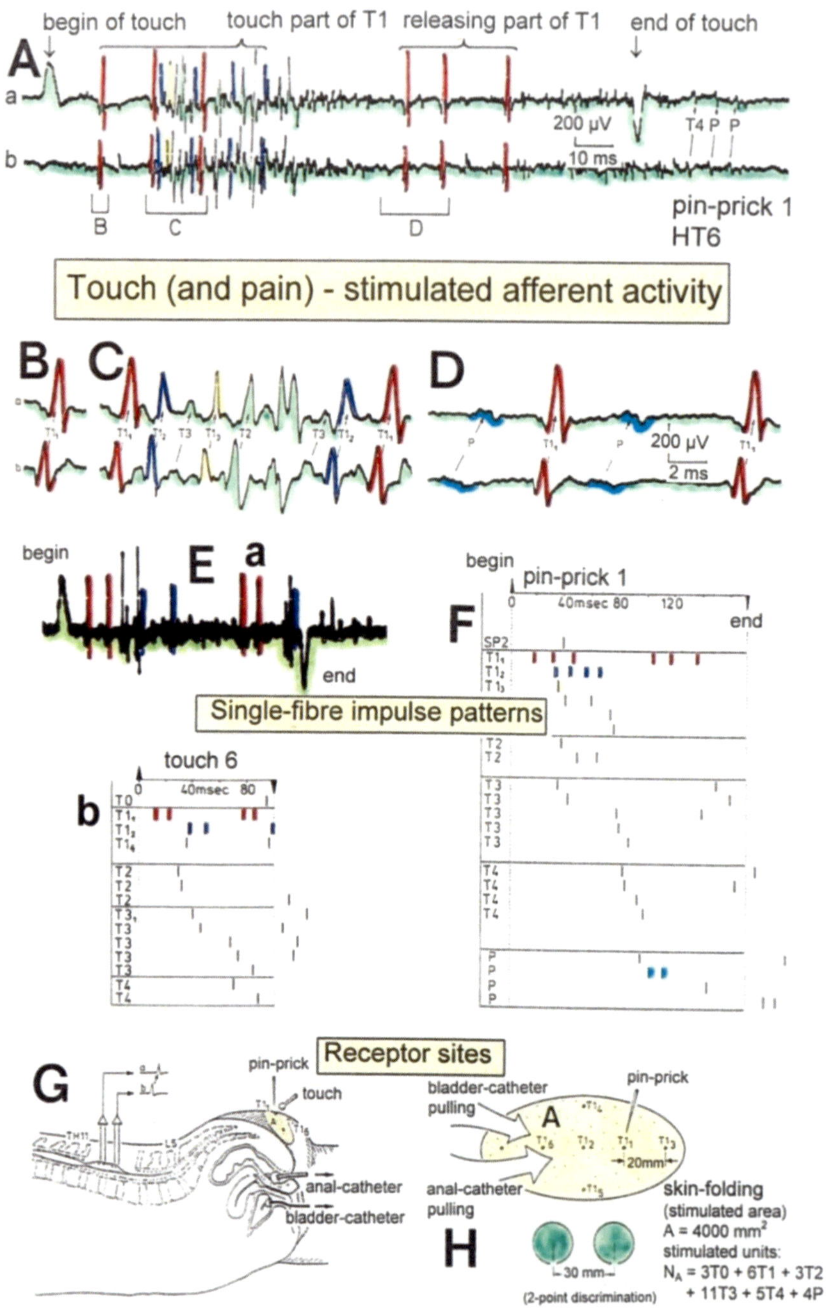

Figure 5. Touch (and pain)-stimulated afferent activity. Touch and pain activity stimulated by pin-pricking (A) and touching (Ea) at S5 or Co dermatomes and recorded extracellularly from a dorsal coccygeal root (brain-dead human HT6). T1, T2, T3, T4, P = mark action potentials (APs) from single touch and pain fibers. Subscripts 1, 2, 3 mark single fibers. A. Whole sweep following pin-prick 1 shown at a slow time base. The large upward artifact on trace 'a' marks electronically the beginning of the pin-prick. The large downward artifact on trace 'a' marks the end of the pin-prick. Note that 2 intervals of high activity of large APs occur, one after the beginning of the pin-prick with 1 AP in front, and a second before the end of the pin-prick; potentials with small amplitude follow potentials of large amplitude. Time intervals B, C and D are shown in a time-expanded form in Figures B, C and D. B, C,

D. Time expanded sweep pieces of A. Identified APs are indicated. Note that the APs from the $T1_1$ touch unit can be safely identified by the waveforms in B, C, D. Eb, F. AP occurrence patterns of single touch and pain fibers following short touch 6 and pin-prick 1. No pain afferents are stimulated upon touch 6. Upon pin-prick 1, the single-fiber AP activity of the different touch and pain groups is identified by the AP waveforms on traces 'a' and 'b', and by the conduction times. The single touch afferents of the T1 group are marked with subscripts. One active secondary muscle spindle afferent fiber (SP2) could always be identified in F. Note that for pin-prick 1, touch and pain afferents are stimulated whereas for touch 6 only touch afferents. G. Recording and stimulation arrangement for simultaneous recording of several single touch and pain units. A = area stimulated by skin folding, drawn in H in more detail. $T1_1$, $T1_6$ = suggested touch points of the $T1_1$ and $T1_6$-units. H. Drawing of the very approximate skin area stimulated by skin folding. $T1_{1\text{-}6}$ = suggested focal T1 touch points. Two-point discrimination indicated for the sake of comparison. N_A = number of stimulated units in the dorsal coccygeal root. Skin tractions evoked by anal and bladder-catheter pulling are indicated by the large open arrows.

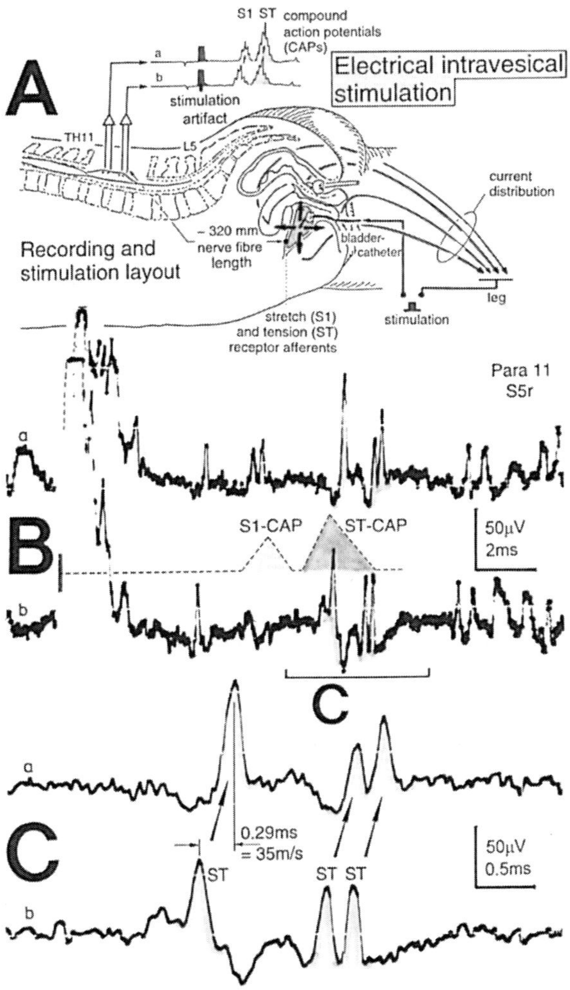

Figure 6. A. Recording and stimulation layout for electrical intravesical stimulation (electro negativity in bladder). B. Recording from an S5 root following electrical stimulation. Compound action potentials (CAPs) of S1 and ST afferents schematically redrawn from Figure 9A. C. Time-stretched sweep piece of the recording in B. Single tension receptor afferent APs (ST) are indicated.

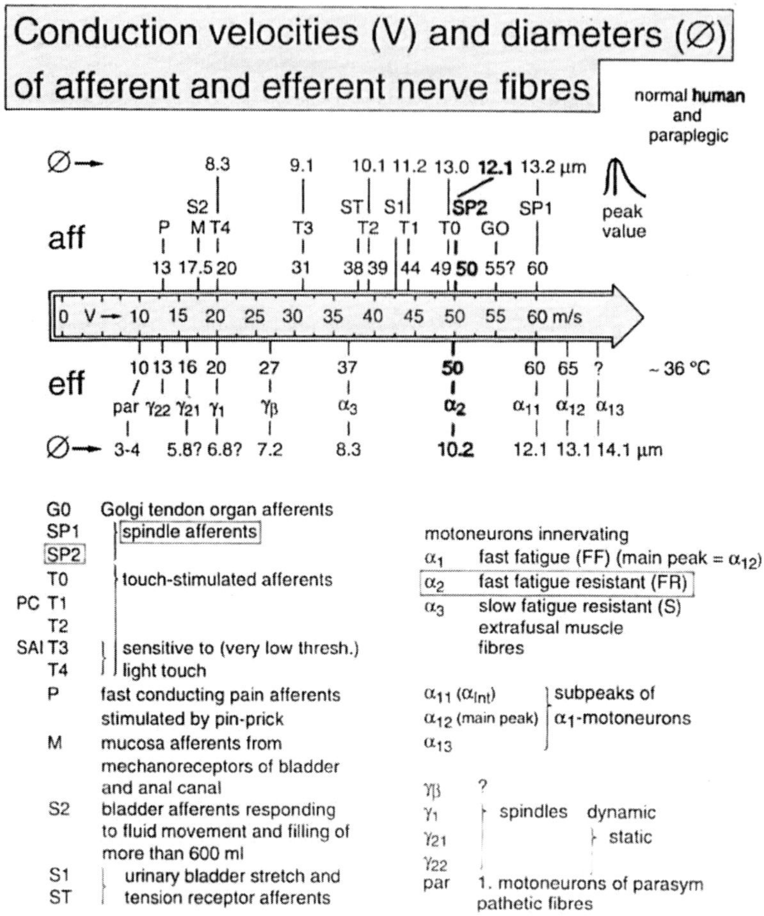

Figure 7. Classification scheme for human peripheral nerve fibers. Conduction velocities (V) and nerve fiber diameters (∅) of afferent and efferent nerve fiber groups in normal humans and in patients with a traumatic spinal cord lesion for 0.5 to 6 years. The splitting of the α_1-motoneurons into the 3 subgroups, α_{11}, α_{12}, α_{13}, has not yet been confirmed.

By correlating the identified conduction velocity peaks with nerve fiber diameter peaks, a classification scheme for the human Peripheral Nervous System (PNS) was constructed, in which individual groups of nerve fibers are characterized by group conduction velocities and group nerve fiber diameters (Figure 7). This classification scheme is still incomplete and only holds for nerve fibers thicker than approx. 3.5µm. The classification schemes for animals do not apply to humans. For example, conduction velocities in rats, cats and dogs (max $\approx$ 120 m/s) are much higher than those in humans (max $\approx$ 70 m/s).

2.1.2. Premotor Spinal Oscillators

Typical firing patterns of motoneurons can be observed when motoneurons are activated with increasing strength of adequate afferent input. With low afferent input, the motoneurons fire occasionally. With increasing input they fire intermittently in an oscillatory manner and then continuously in an oscillatory manner. The demonstration that neurons of the CNS, in

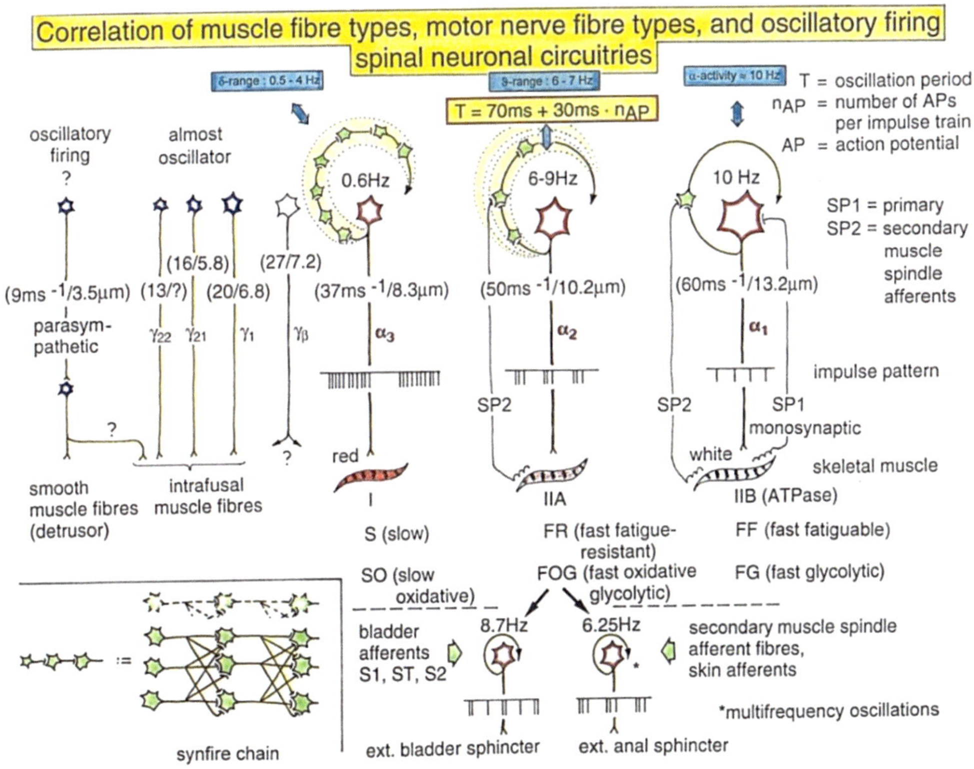

Figure 8. Correlation of muscle fiber types, motor nerve fiber types, and oscillatory firing spinal neuronal networks, based on histochemical, morphological and physiological properties. This figure provides a simplified correlation between muscle fiber, motoneuron and sacral oscillator types. No additional subtypes have been included. The existence of α_1-motoneuron (FF) oscillators firing at 10 Hz has been predicted and they have been identified in paraplegics (unpublished observation). α = motoneuron, γ_1, γ_2 = dynamic and static fusimotors, parasympathetic = parasympathetic preganglionic motoneuron. S1, ST, S2 = stretch, tension and flow receptor afferents.

this case motoneurons, can fire both in an oscillatory manner and non- oscillatory manner is very important for the understanding of the functioning of the human CNS. To describe the functioning of the CNS merely by reflex pathways and loops or coupling of rigid oscillators (of cellular or network origin) is in contradiction to empirical human data, namely that premotor spinal oscillators self-organize as was concluded from measurements of simultaneous natural impulse patterns of afferent and efferent fibers. In what follows, we shall concentrate mainly on the oscillatory firing of motoneurons, which takes place for high activation. In this high activation mode they can also be used as a reference basis when defining phase relations and thus phase and frequency coordination can be measured among neuron firing. For high and rather constant afferent input it was found that α-motoneurons fire repeatedly with impulse trains according to their type (Figure 8). The α_1-motoneurons (FF) fire rhythmically at around 10 Hz (range 8 to 20) with an impulse train consisting of 1 AP; α_2-motoneurons (FR) fire at approx. 6 to 9 Hz with 2 to 5 APs per impulse train, and α_3-motoneurons (S) fire with a frequency in the range of 1 Hz and with long impulse trains consisting of up to 40 APs (and more). The rhythmic firing patterns of α-motoneurons are probably generated by local neuronal networks of the spinal cord since oscillatory firing can be recorded from motoneurons of the disconnected spinal cord. It may be that the motoneuron is a part of the spinal oscillator. The oscillation period (T) is roughly related to the number of action potentials (APs) per impulse train (n_{AP}), and this can be expressed by the formula: $T = 70ms + 30ms \bullet n_{AP}$. A typical premotor α_2-oscillator fires with 3 APs every 160ms ($T = 70ms + 30ms \bullet 3 = 160ms$), and can change its firing pattern to 2 APs every 130ms for less activation or to 4 APs every 190ms for higher activation.

The α_1-oscillators respond very dynamically, but have few oscillator network properties. Their firing is absolutely correlated to the firing of primary spindle afferent fibers. The α_2-oscillators respond less dynamically, have strong oscillatory properties and self-organize by the adequate afferent input patterns from several kinds of receptors including secondary muscle spindle and urinary bladder afferents. The behavior of α_3-motoneurons is more static and their input is polymodal. The dynamics of responding to inputs increases from α_3 to α_2 to α_1-oscillator in accordance with the dynamics of the 3 muscle fiber types the α-motoneurons innervate. The slow (S), medium fast (FR) (fast fatigue-resistant) and fast contracting muscle fibers (FF) (fast fatigable) have their own corresponding premotor networks in the spinal cord, namely that in which the α_1, α_2 and α_3-networks are integrated in (Figure 8).

2.1.3. Phase and Frequency Coordination among Neuron Firing for Human CNS Self-Organization

Relative phase and frequency coordination between the APs of the oscillatory firing α_2-motoneuron O2 and the secondary muscle spindle afferent fiber SP2(1) can directly be seen in the original recordings in Figure 9. The firing of the oscillator and the sweep pieces which are shown time-expanded are indicated at the summary trace. Figure 9B,C shows the AP-impulse train of oscillator O2 in connection with one of its driving spindle afferent AP. Because of the duration of the phase relation of around zero milliseconds between the firing of the driving SP2(1)-fiber (firing mostly every 80ms) and the impulse train of the oscillatory firing motoneuron O2 ($T(O2) \approx 160ms$), the SP2(1)-fiber AP (every second AP) appeared at a similar time as the impulse train. Because the AP of the spindle afferent fiber had a characteristic waveform, it was easy to extract its impulse pattern from the summed impulse traffic of this S4 dorsal root. During touch-induced skin afferent activity (Figure 5), the activities of the motoneuron and the spindle afferent fiber were covered by the skin afferent activity. After the cessation of the skin afferent activity the afferent and efferent APs were found again at their expected time positions of the regular firings. The phase coordination between the firings of the oscillatory firing motoneuron O2 and the secondary muscle spindle afferent fiber SP2(1) at the time when records B,C were taken, was 1.6ms (3ms - 1.4ms, Figure 9B,C). In Figure 9D, the relative frequency coordination between the firings of the SP2(1)-fiber and the impulse train of the oscillator is indicated. For the time period evaluated, the correlation between the firing of the motoneuron and the spindle afferent fiber was in the range of between 3 and 5ms (Figure 9D).

In Figure 9E, considerations concerning the relative frequency coordination are extended to the activity of further afferent fibers and γ-motoneurons of the same root. G of Figure 9e shows sweep pieces of the original recordings; A through F shows the interspike interval distributions of spindle afferents and γ-motoneurons. It can be seen from the overlapping of the oscillator frequency T(O2) and T(O2)/2 distribution ranges and the interspike interval distributions of the afferents that, from the viewpoint of frequency coordination, fiber SP2(1) contributed strongly to the drive of oscillator O2, whereas there was a weaker contribution from other afferents, i.e., less overlapping between the distributions of the afferents and the range of the basic frequency or the first harmonic of the oscillator. Also, γ-motoneurons showed only little frequency correlation at that time period.

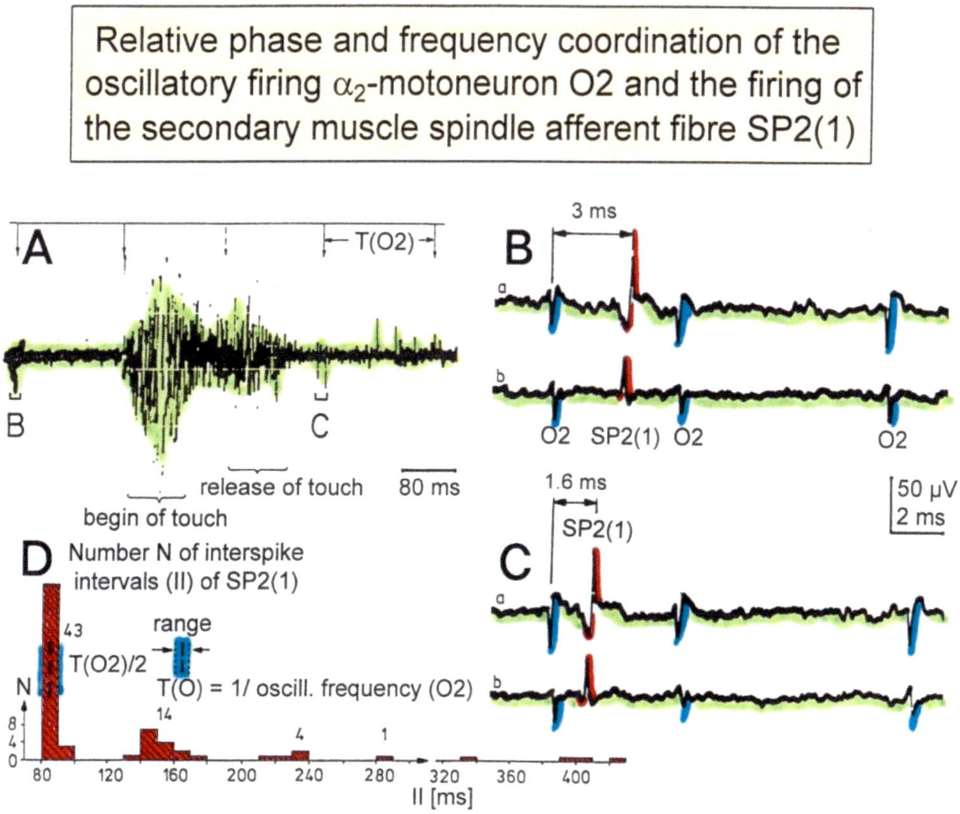

Figure 9 A-D. Time relation between the occurrence of the action potentials (APs) of oscillatory firing α_2-motoneuron O2 and the firing of the secondary muscle spindle afferent fiber SP2(1). HT6. S4 dorsal root recording. A. Overall view of the used sweep piece; only trace "a" shown. Four oscillation cycle periods of motoneuron O2 are indicated (T(O2)). The APs of the impulse trains can be recognized only partly, because of the slow time base and poor digitalization. One impulse train (dashed arrow) is lost in the touch stimulated activity, which consists of a touch (large overall activity) and a release part (lower overall amplitude). B,C. Sweep pieces from A, time stretched. In B, motoneuron impulse train APs is marked O2, spindle afferent APs are marked SP2(1). Note that the APs of the spindle afferent fiber are not time-locked to the first AP of the impulse train of the rhythmically firing motoneuron (relative phase coordination). Digitalization 4 times better than in A, but still rather poor, as can be seen from the low amplitudes of the motoneuron APs on trace "b" in C. D. Occurrence of interspike intervals of the secondary muscle spindle afferent fiber SP2(1). The numbers give the amount of IIs in each distribution peak. The oscillation period of motoneuron O2 (and the range of variation) and the half period are indicated by short dashed lines. Note that the IIs of fiber SP2(1) are very similar to the oscillation period (or the half of it) of α_2-motoneuron O_2 (relative frequency coordination).

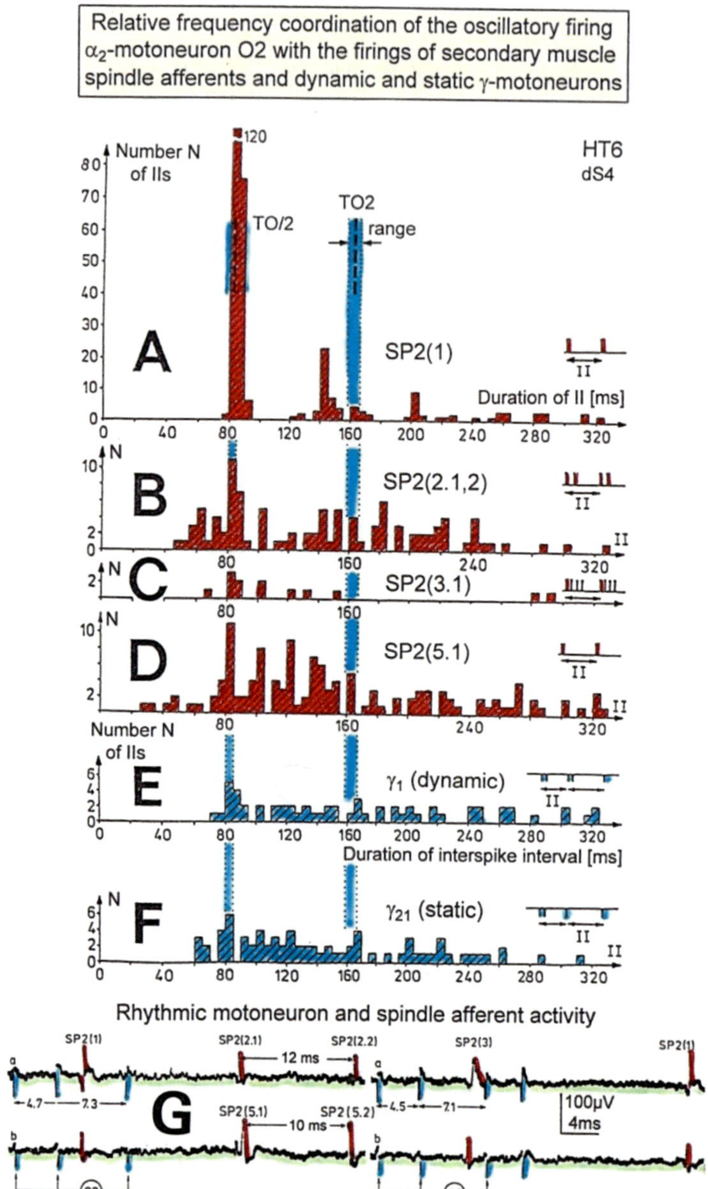

Figure 9E. Interspike interval distributions of single endings of four secondary muscle spindle afferents (SP2) and two γ-motoneurons, recorded simultaneously. In A, the oscillation period TO2 (impulse train length = 3 APs) with its range of simultaneously recorded oscillatory firing α_2-motoneuron O2 (see G) is drawn for comparison; also, the halves of the oscillation period TO2/2 are indicated. Note that the interspike interval distributions of spindle afferents and γ-motoneurons have shortest interspike interval, nearly identical to the half of the oscillation period (relative frequency coordination). The schematic impulse pattern in A to F shows the procedure for measuring the interspike intervals. Original records of the firing patterns of α_2-motoneuron O2 and the secondary muscle spindle afferents SP2(1), SP2(2), SP2(3) and SP2(5) are shown in G. Brain-dead human HT6, dS4 root.

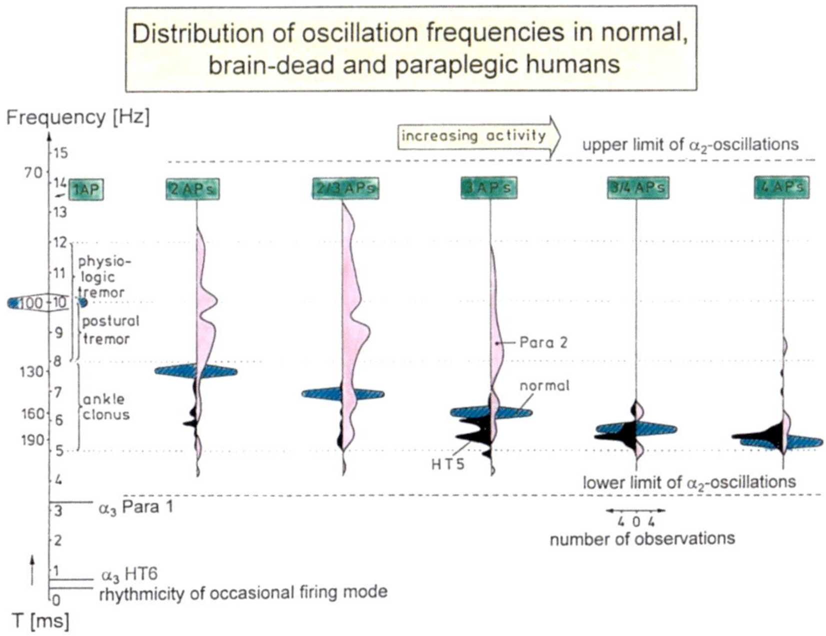

Figure 10. Frequency distributions of oscillation frequencies of continuously oscillatory firing α_2-motoneurons with increasing number of APs per impulse train (increased activity) in paraplegic 2 (open), in brain-dead HT5 (filled), and probably normal human (cross-hatched). Frequencies and rhythmic activity changes in the occasional and oscillatory firing mode are indicated. Ranges of physiologic tremor, postural tremor and ankle clonus are also drawn. Note that frequencies for the brain-dead HT5 are too low, and the oscillation frequencies of the spinal cord isolated for a long time (Para 2) are too high and too spread as compared to the theoretically predicted frequency ranges (cross-hatched). T = oscillation frequency.

Phase and frequency coordination among α_1-motoneuron firings can directly be seen in recordings of single-motor unit surface electromyography (Figures 13-15).

The fact that neurons fire in a relatively coordinated way of up to a few milliseconds is used for re-organizing the injured CNS by re-learning phase and frequency coordination between neuron firings when exercising movements coordinated with an exactness of up to a few milliseconds, using a special coordination dynamic therapy device, i.e., by instrumented supervised phase and frequency re-learning (Figure 150).

2.1.4. Coordinated Firing of Neurons, Their Impairment and Repair by Learning

In every CNS injury, surgery, degeneration, malformation or aging the phase and frequency coordination [15, 16] of neuron firing becomes impaired [17] and disturbs or destroys CNS self-organization. Also the organization of sub-networks and their communication becomes impaired. The impairment of the sub-network premotor spinal oscillator upon injury can be nicely measured by the variation of the oscillation frequency (Figure 10). In rather physiologic cases, such as in *brain-dead* humans, the premotor α_2-oscillators fire rhythmically with a certain Eigen-frequency. Following spinal cord injury, the oscillators lose their Eigen-frequency to some extent and fire at many different frequencies (Figure 10).

The impaired coordination between nerve cells and arm and leg movements following injury [12, 14] can be improved especially by exercising on the special CDT device, which is very precisely manufactured.

The exactness of the device guarantees that the coordinated arm and leg movement induced afferent input (like in Figure 5) can teach the neurons of the CNS to improve their coordinated firing up to within a few milliseconds. Since the neurons work as coincidence and more widely as coordination detectors (Figure 11), this improved coordinated firing improves for example the communication among nerve cells and neural sub-networks (especially between networks across an injury site as for example in spinal cord injury) because the threshold of neuron excitation is reached earlier. The training of phase and frequency coordination via coordinated arm, leg and trunk movements improves not only the self-organization of the corresponding sub-networks but also the functioning of the CNS neural networks in general.

Some motor and other patterns re-appear upon this improved coordination at the neural and movement level. For example, child with cerebral palsy learned to speak following exercise on the special device through learning transfer [14]. Surface Electromyography may be used to achieve correlation between neuronal network functioning and the movements necessary for movement-based learning.

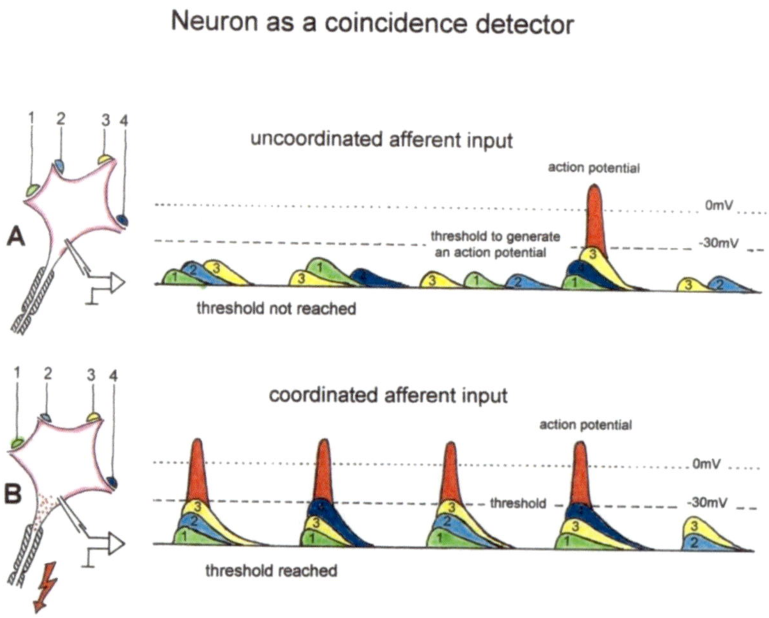

Figure 11. Neuron operating as a coincidence detector. A. Afferent input is reaching rather uncoordinated the cell soma. Only sometimes an action potential is generated, because the threshold of action potential generation is mostly not achieved. B. The action potentials in fibers 1 through 4 are reaching time-coordinated the dendrites or the cell soma. The postsynaptic potentials add up and the threshold is achieved at approximately –30mV, and action potentials are generated time-coordinated at the axon hillock. In the real CNS mostly many more smaller postsynaptic potentials will contribute to the generation of an action potential and passive conduction from the dendrites to the cell soma has to be taken into account. Coordinated afferent input may thus induce or enhance (coordinated) communication between neuronal network parts following CNS injury.

2.2. Surface Electromyography to Record Motor Programs, Oscillatory Firing, and Phase and Frequency Coordination among Motor Units

Another human electro-physiologic tool to measure natural impulse patterns of neurons is the surface electromyography (sEMG). With the same recording system used to record singe-nerve fiber APs, just replacing the wire electrodes with EMG surface electrodes, single-motor unit firing and motor programs can be recorded non-invasively (Figures 13-17). The sEMG recording arrangement is shown in Figure 12 for recording motor programs from an infant.

2.2.1. Oscillatory Firing of Motoneurons and Motor Units

In Figure 17, the different frequency patterns of oscillatory firing of motoneurons are shown. Original records were taken with the single-nerve fiber action potential recording method from motoneuron axons and surface electromyography (sEMG) from single motor units. α_1-Motoneurons innervate FF-type muscle fibers and fire rhythmically with impulse trains consisting of 1 action potential in the order of 10Hz (Figure 8). α_2-Motoneurons innervate FR-type muscle fibers and fire rhythmically with impulse trains consisting of 2 to 5 action potentials in the range of 4 to 7 Hz. The amplitude of the extracellular action potential

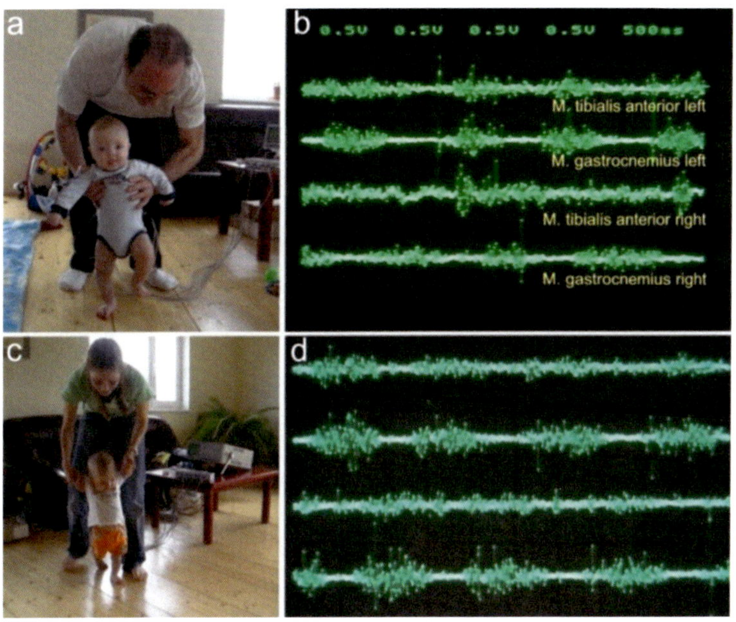

Figure 12. Surface EMG obtained from the healthy 5-months-old (a,b) and 8-months-old old "Jürgen" (c,d) during supported walking. a. Walking resembles automatic stepping, because of the strong lifting of the left knee. The toes of the right foot are plantar flexed, which is not physiologic. b. Surface EMG motor programs of left and right tibialis anterior and gastrocnemius muscles. Note that there is no antagonistic action between the tibialis anterior and gastrocnemius muscles. The right tibialis anterior muscle shows no motor program. c. The walking is more walking like and not so any more much automatic stepping like. d. Better motor programs then 3 months earlier (b). Still there exists no antagonistic action between the tibialis anterior and gastrocnemius muscles. The activation of the right tibialis anterior muscle is a bit better than 3 months ago (b).

of the α_2-motoneurons (axon group diameter = 10.2μm, axon group conduction velocity = 50m/s) is on average smaller than that of the α_1-motoneurons (axon group diameter = 13.1μm, axon group conduction velocity = 65m/s) (Figure 7), depending on the position of the axon in the nerve root with respect to the recording electrodes. FR-type motor unit potentials have much smaller amplitudes than the motor unit potentials of FF-type muscle fibers.

The α_3-motoneurons (axon group diameter = 8.3μm, axon group conduction velocity = 37m/s) innervate S-type muscle fibers and fire oscillatory at a frequency of around 1 Hz with long impulse trains (up to 50 action potentials per impulse train). The motor unit firing of single S-type muscle fiber motor units could not be safely identified by sEMG because their amplitudes are still smaller than those of FR-type motor units and are thus difficult to identify. The impulse patterns of oscillatory firing motoneurons obtained with sEMG are similar or the same as those obtained with the single-nerve fiber action potential recording method (Figure 17). This confirms the accuracy of the single-nerve fiber action potential recording method. Since sEMG is a non-invasive recording method, oscillatory firing can be recorded easily when using appropriate patients.

2.2.2. Motor Program Generation, Oscillatory Firing and Coordination among α_1-Motor Units (FF-Type)

When a healthy person exercises on the special coordination dynamics therapy device, coordinated motor programs can be recorded from the different muscles (Figure 15). The patterns of recruitment of motor units cannot be seen in such a motor program. The number of activated motor units is so high that single motor units cannot be followed. But when only a few motor units can be activated in a certain muscle, then the pattern of activation of the motor units and the coordination between them can be seen. If the CNS of a patient is functioning rather physiologically as a result of a long lasting intensive coordination dynamics therapy, then an analysis of the generation of the motor program becomes possible based on single motor unit firing.

It can be seen from Figures 13 and 14 how the CNS generates a motor program in a muscle, if only a few α_1-motor units can be activated in that muscle. By increasing the load on the special CDT device (in Newton's), more muscle power is required to perform the movements and the motor unit firing rates increase in that the motor units start to fire more rhythmically and with increasing frequency. Motor units coordinate their firing in order to avoid tremor.

In a patient with a spinal cord injury (injury level sub L2 following an accident), the left tibialis anterior muscle (third trace in Figure 13) was already strongly re-innervated after two years of treatment. The overlapping of the motor unit potentials was so strong that single motor unit firing could only be seen occasionally. In the right tibialis anterior muscle (top trace in Figures 13,14) however, oscillatory and coordinated firing of mainly 3 single α_1-motor units could be identified. Figure 13H shows the wave forms of the 3 different motor unit potentials (1, 2, and 3). Motor unit 1 had the lowest threshold and the largest amplitude. In the majority of cases, the motor units can be identified safely, though not always. It is especially difficult when the motor units fire simultaneously or nearly simultaneously. It will now be shown how the firing frequencies and coordination between motor units firing changes for different movements and loads.

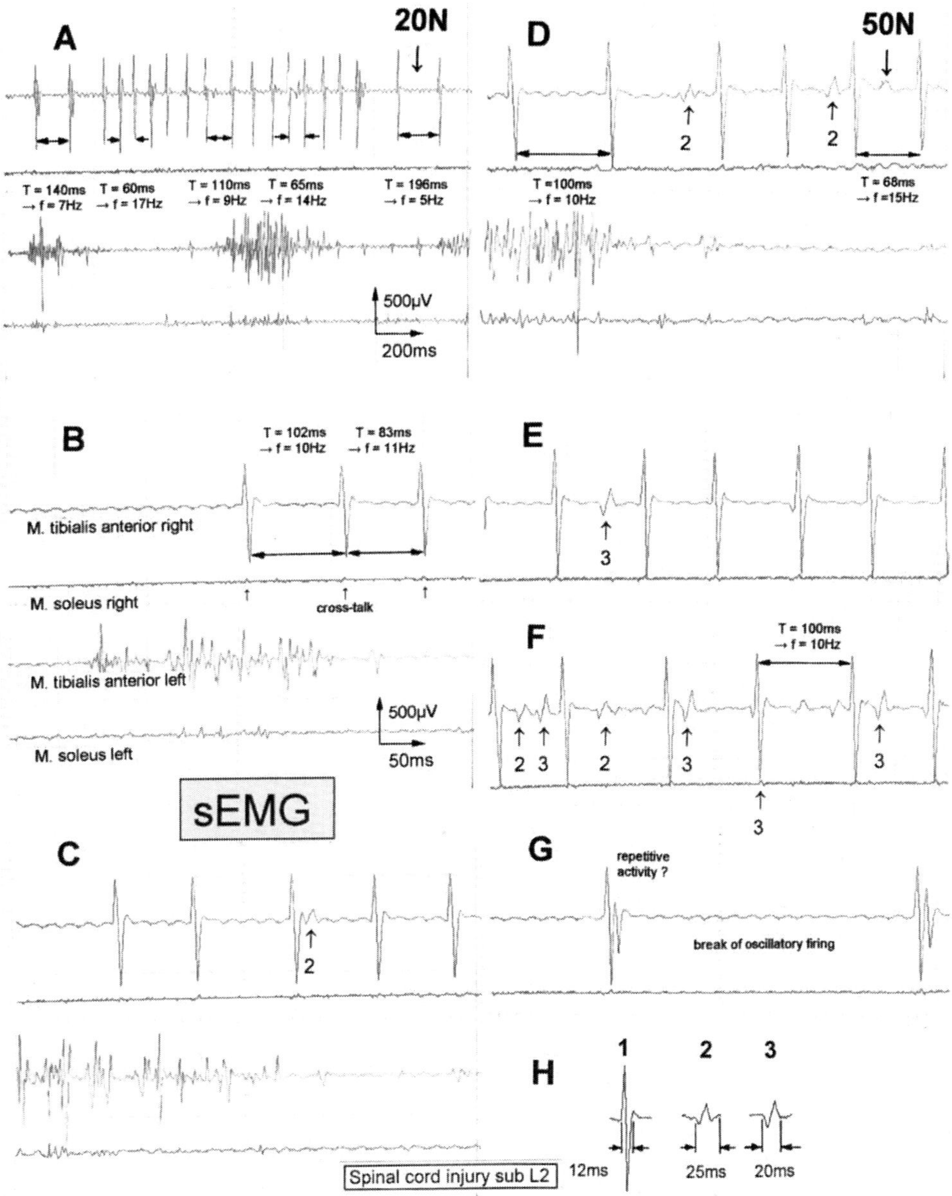

Figure 13. Non-synchronous partly coordinated firing of FF-type motor units of the right tibialis anterior muscle from a patient with spinal cord injury sub L2 suffered in an accident, measured by surface EMG; the firing was activated by exercising on the special coordination dynamics therapy device at the loads of 20 and 50N (Figure 1). Oscillation periods (T [ms]) and oscillation frequencies (f [Hz]) of oscillatory firing motor unit 1 are partly indicated. The waveforms of 3 motor units are identified in the right tibialis anterior muscle (top trace) and are shown in 'H'. Motor units 2 and 3 are marked in 'A' through 'G'; the identification of large amplitude motor unit 1 is obvious. The left tibialis anterior muscle (third trace) shows already a motor program; little activity in the soleus muscles for this low-load activation.

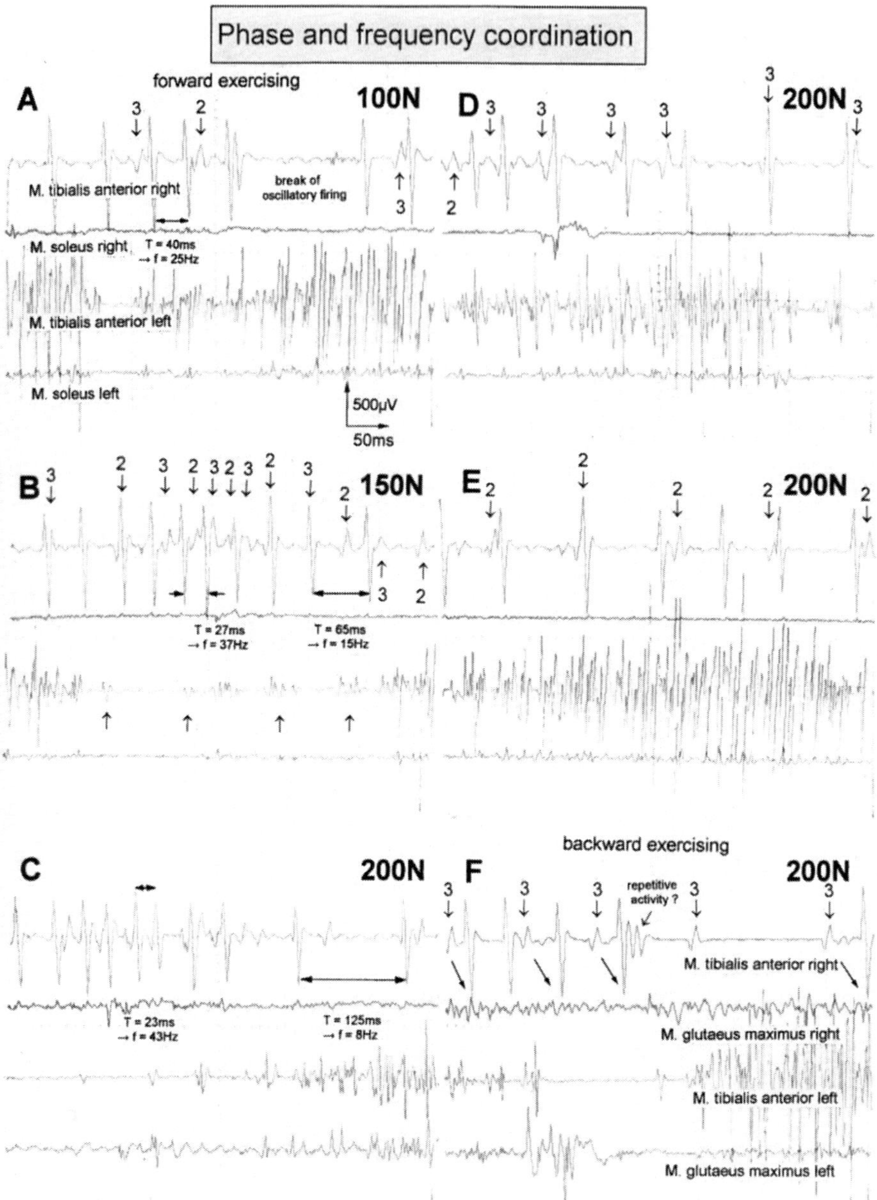

Figure 14. Phase and frequency coordination between oscillatory firing motor units (FF-type) during the generation of a motor program during exercise on the special coordination dynamics therapy device at loads increasing from 100 to 200N. Oscillation periods (T) and oscillation frequencies (f [Hz]) of oscillatory firing motor unit 1 are partly indicated. 'A,B,D,E' same recording situation as in Figure 3; 'C,F' soleus electrodes shifted to gluteus muscles to check early re-innervation upon therapy. The waveforms of the 3 identified FF-type motor unit potentials '1', '2', and '3' are the same as in Figure 3; motor units '2' and '3' are partly marked. In 'F', some coordination's between motor unit '3' and '1' are marked.

2.2.3. Firing Frequency Increases with Increasing Load

In Figure 13A, motor unit 1 fired oscillatory for a certain period of time to generate a motor program when the patient exercised at the load of 20N on the special CDT device. The frequency fluctuated between 5 and 17Hz. Quite often, the frequency of this motor unit 1 was in the order of 10Hz for low load exercising at 20 or 50N (Figure 13). Motor units 2 and 3 fired mostly only occasionally. In Figure 13F, motor unit 3 fired transiently oscillatory at the frequency of approximately 6Hz.

For exercising at higher loads, the firing frequencies of the motor units increased. At 100N, motor unit 1 reached 25Hz (Figure 14A); the corresponding values at 150N and 200N being 37 Hz (Figure 14B) and 43 Hz (Figure 14C). Also, motor units 2 and 3 increased their firing rates. They started to fire also rhythmically with increasing frequencies for increasing load. At 150N (Figure 14B), motor unit 2 and 3 reached the frequency of 22Hz and 25Hz respectively. Exercising at 200N is a rather high load, even for a healthy person. The lowest frequencies measured at 200N were 9Hz and 10.5Hz for motor unit 2 (Figure 14E) and 3 (Figure 14F), respectively. For some time motor unit 1 stopped firing oscillatory (Figure 14F).

2.2.4. Motor Program Development

In the left tibialis anterior muscle (Figures 13, 14, third trace), motor unit activity was recorded when muscle power was needed, and none or nearly none was recorded when no muscle power was needed, in some similarity to the physiologic motor programs (Figure 1C). In the right tibialis anterior muscle (top trace) the activation pattern was different. When the muscle had to be activated the motor units were more activated (Figs.13A, 14B) and when no muscle power was needed, less activity was recorded (Figs.13B,G, 14E,F). The activity of the motor units increased and decreased. For high load exercising, the motor units fired rhythmically with increasing and decreasing frequencies. During low load exercising, the motor units fired rhythmically with low frequencies or stopped firing in an oscillatory manner and fired only occasionally. The activity changes from motor unit firing to no motor unit firing during the motor program were not as pronounced in the right tibialis anterior muscle, where only a few motor units could be activated, as in the left tibialis anterior muscle. It was as if inhibition were missing. Inhibition is necessary to give motor unit firing patters their structure. As measured earlier, the high activity mode of α-motoneuron firing is the oscillatory firing.

2.2.5. Phase and Frequency Coordination between the Firing of the Motor Units

But how did the motor units coordinate their firing? As can be seen from Figures 13 and 14, the motor units did not synchronize their firing. In Figure 13F, motor unit 1 is firing in an oscillatory manner at around 10Hz and motor unit 3 shows a phase lag of 22ms with respect to motor unit 1 (the 2 right potentials). But when motor unit 2 fired additionally (on the left part of the trace), motor unit 3 changed its phase with respect to motor unit 1 so that all 3 motor units fired unequally to give rise to a smooth non-rhythmic muscle contraction. In Figure 14D (200N), motor unit 3 fired approximately at the same time as motor unit 1. The motor unit potentials partly added up, because of partial synchronized firing of the corresponding motoneurons. The distributed firing of motor units was poor at that time interval. In Figure 14E, motor unit 2 fired approximately at the same time as motor unit 1.

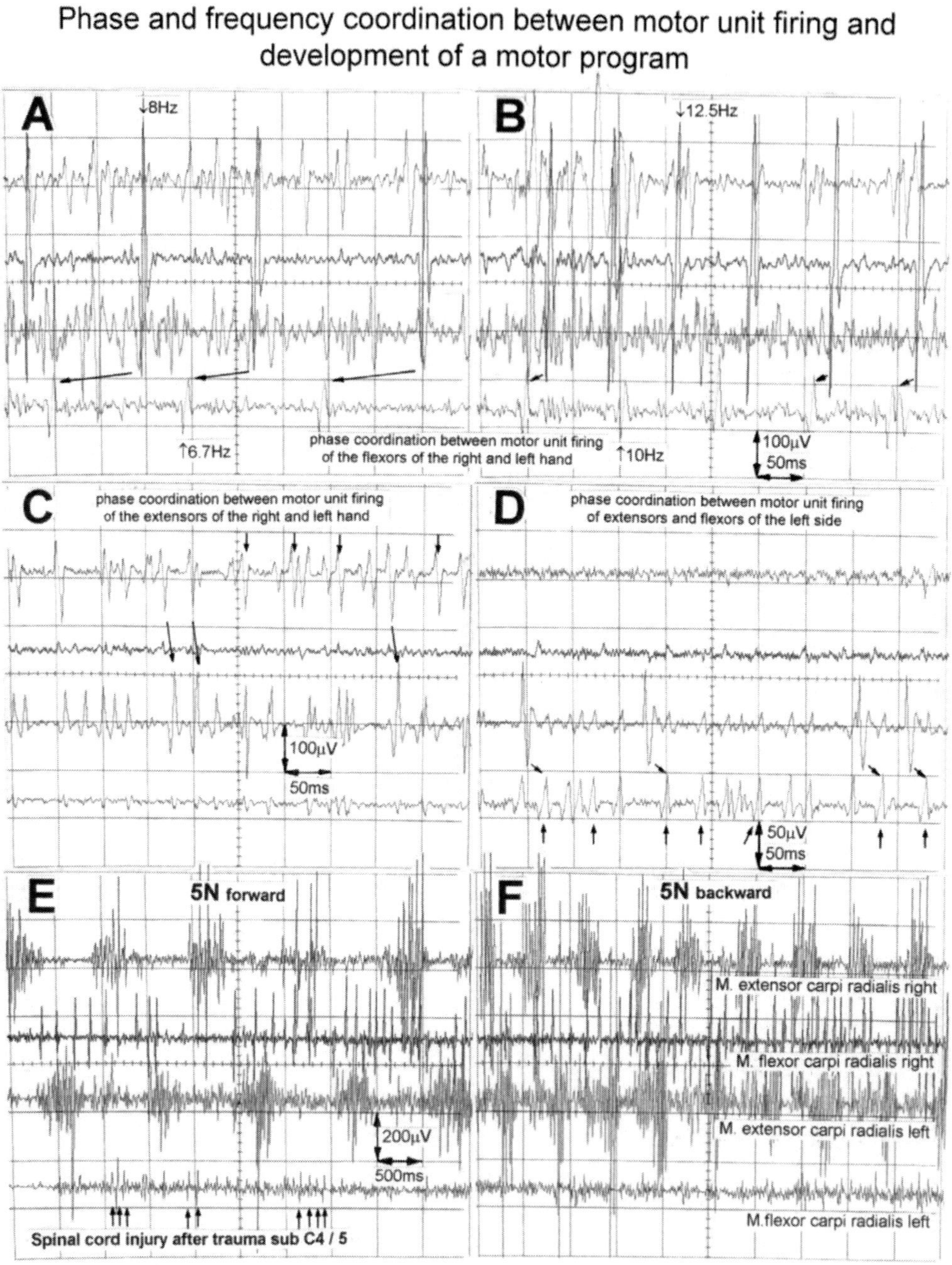

Figure 15. Phase and frequency coordination of single-motor unit firing between different muscles and between different arms. Phase coordination's are indicated by the arrows between motor unit potentials. A-D. The patient with a spinal cord injury was in position at the special coordination dynamics therapy device but activated muscles on volition upon looking onto the oscilloscope screen. E,F. Patient exercising on the special device. Note the beginning of the appearance of a motor program in the right flexor carpi radialis muscle for 5N forward turning (similar to Figure 4E, right tibialis anterior sweep). Recordings for a patient with a complete spinal cord injury sub C4/5.

In Figure 14F, motor unit 3 fired in phase correlation to motor unit 1, but it was not synchronized. When motor unit 1 transiently stopped to fire in an oscillatory manner, motor unit 3 continued firing in an oscillatory manner. When motor unit 1 started to fire in an oscillatory fashion again, the same phase coordination built up to the firing of motor unit 3. For the high activation parts in Figure 14B,C, motor units 1, 2 and 3, and some other unidentified units, fired very close to each other (but not in a synchronized manner) and the potentials overlapped, because the frequencies were very high to generate strong muscle power.

2.2.6. Phase and Frequency Coordination's between Motor Unit Firing of Different Muscles and Different Arms

Figures 13 and 14 show relative phase and frequency coordination for motor unit firings in the same muscle, namely the right tibialis anterior muscle, of a patient with a spinal cord injury sub L2 suffered during an accident. Figure 15 shows phase and frequency coordination among motor units between different muscles and arms in a spinal cord injury patient (sub C4/5). A single FF-type motor unit of the right flexor carpi radialis muscle fired at approximately 8Hz when activated on volition (Figure 15A). In Figure 15B, the same motor unit is seen firing at 12.5Hz. This motor unit fired in relative coordination (marked with the arrows) with another single FF-type motor unit in the left flexor carpi radialis muscle. The phase relations have different values in 'A' and 'B'. However, also note that the motor unit in the flexor changed its oscillatory firing from 6.7Hz to 10Hz, probably to develop more motor unit force. Not every motor unit potential is correlated. There may be correlations to motor units of other muscles which were not recorded from. In Figure 15C, phase relations are indicated with the arrows from the right extensor carpi radialis to the left extensor carpi radialis. In Figure 15D, phase relations are indicated by the arrows from the left extensor carpi radialis to the left flexor carpi radialis. Thus, there is coordinated firing between all motor units activated on volition, when the patient is not exercising on the special coordination dynamics therapy device, but is in position to do so.

During exercise on the special coordination dynamics therapy device at 5N, a motor program was partly generated in the right and left extensor carpi radialis muscles (Figure 15E, F). In the right flexor carpi radialis muscle, the CNS tried to build up a motor program with 2 or 3 innervated motor units, in similarity to that seen in Figure 1C. In the left flexor carpi radialis muscle no motor program was generated, even though small motor unit potentials can be seen (marked by the arrows in Figure 15E). It was measured in Figure 15 that phase and frequency coordination also takes place among motoneurons, innervating different muscles. Above it was shown that there is coordinated firing between afferents and efferents (Figures 9, 9E). It is likely that phase and frequency coordination is a general organization principle of the human CNS.

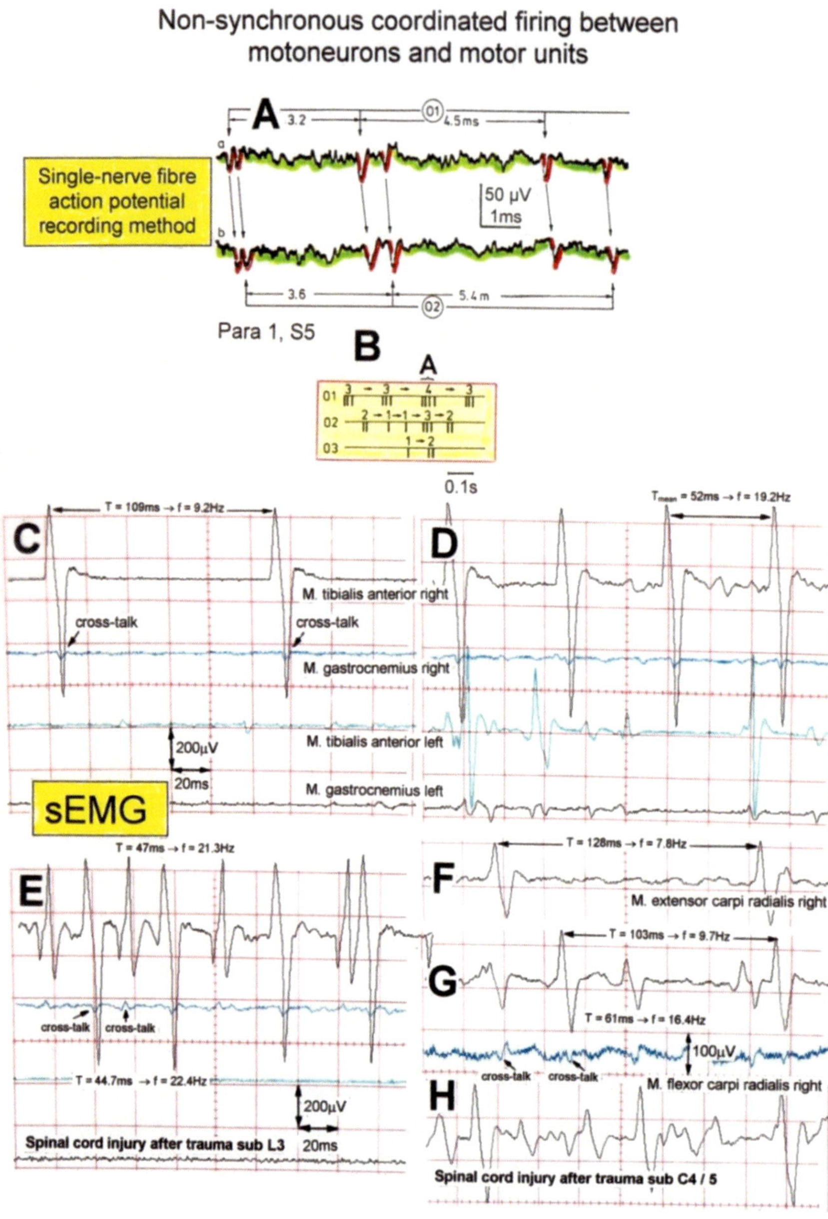

Figure 16. Non-synchronous relative coordinated firing of motoneurons from patients with spinal cord injury, recorded with the single nerve-fiber action potential recording method (A,B) and with sEMG from FF-type motor units (C-H) activated on volition. Oscillation periods (T) and oscillation frequencies (f [Hz]) are indicated. Note that the waveform identification of the motor unit action potentials in C, D, and E is safe. Note further that the α_1 – motoneuron, giving rise to the FF-type motor unit potential of 'C', fired in 'C' oscillatory at 9.2Hz, in 'D' at 19.2Hz, while in 'E' it fired in non-synchronous relative coordination with another α_1 – motoneuron giving rise to another FF-type motor unit potential. Even the cross-talks from M. tibialis anterior to M. gastrocnemius were of such a high quality that they could be used for waveform identification and thus for firing pattern identification (C, D, E). The oscillatory firing patterns of motoneurons, recorded from motoneuron axons (A, B), showed normally no synchronization, as reported earlier.

2.3. Correlation of Natural Efferent Impulse Patterns, Recorded with the Single-Nerve Fiber Action Potential Recording Method and Single-Motor Unit Semg

With the single-nerve fiber action potential recording method natural impulse patterns were recorded simultaneously from several single afferent and efferent nerve fibers and functions of the human CNS analyzed. The disadvantage is that this is an invasive method. With surface EMG natural impulse patterns of several motor units and motor programs were recorded. The disadvantage is that one can obtain only natural impulse patterns from the motoneurons and not from afferent fibers. The advantage, however, is that it is a non-invasive recording method.

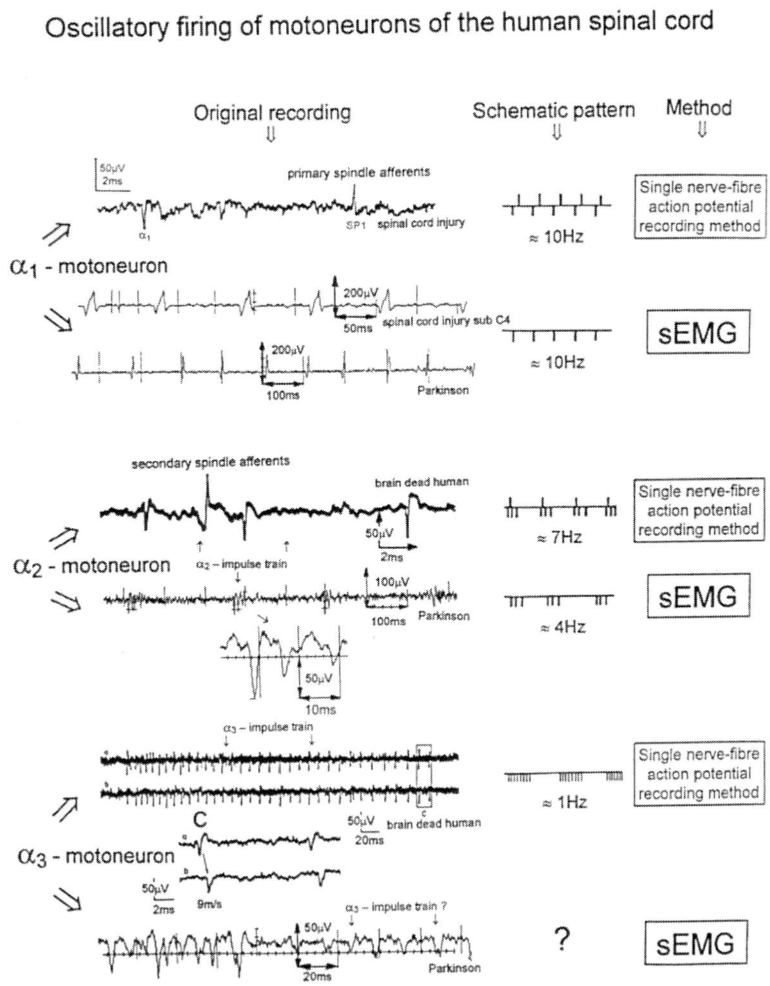

Figure 17. Oscillatory firing patterns of α_1, α_2, and α_3-motoneurons recorded from motoneuron axons with the single-nerve fiber action potential recording method and by surface electromyography (sEMG) from FF, FR, and S-type motor units. The left panel shows original recordings, the middle panel the schematic patterns; the recording methods are indicated on the right side. The recordings were taken from patients with spinal cord injury **and Parkinson's disease and** from brain-dead humans.

By correlating the natural impulse patterns obtained with the single-nerve action potential recording method with those obtained with the surface EMG, one can verify whether the natural impulse patterns are the same or similar and can also verify the accuracy of the measured patterns. Organization can be measured from the neuron level and followed right up to the muscle level.

As Figure 16 shows, the quality of the sEMG recordings can be much better than that of the single-nerve fiber recordings. The amplitude of the single-motor unit APs is in the range of 20 times greater than those of single-nerve fiber APs and the duration of the motor unit potentials is also in the range of 20 times longer (Figure 16). The precise amplitude and duration depends on the kind of fiber (see above). It is very important that the natural firing patterns obtained with both electrophysiological recording methods are the same for the different kinds of motoneurons (Figure 17). Both human electrophysiological methods can therefore be used to analyze human CNS functioning.

Figure 12 shows that the motor program improves during development between the age of 5 to 8 months and from Figure 27 it can be seen that the motor program improves, i.e., repairs following six months of CDT. However, a safe quantification of improvement is not entirely possible.

As we need to measure CNS functioning and learning in patients reliably and non-invasively and want to characterize CNS functioning and learning by a single, authoritative parameter a further human CNS measuring tool is needed. By using the System Theory of Pattern Formation and making the step from the muscle and motor program level to the movement pattern level, it will become possible to measure the quality of CNS organization by one value, namely the coordination dynamics value, and to better understanding of human CNS organization.

3. Integrative Physiology: System Theory of Pattern Formation

3.1. The System Theory of Pattern Formation for understanding Neurotherapy

To understand the on-going changes of movement performance in the patient, the System Theory of Pattern Formation is used. In a complex system like the human CNS, patterns are generated by a nervous system which seeks cooperative stability. Stability is what defines collective states. The system has the tendency to slip into the collective states to which it is attracted. When an infant crawls, its arms and legs are strongly attracted to the 'pace' and 'trot' gait patterns. The attraction is so strong that intermediate crawling patterns seemingly do not exist, as if the patterns are hard-wired. But with the help of the special CDT device the CNS can generate intermediate coordination patterns. A patient with a CNS injury often crawls with intermediate arm and leg coordination patterns and has to re-learn the pace and trot gait coordination's for CNS repair and shifts in this way the attractors for crawling to the pace and trot gait coordination's. Attractive states and attractors of CNS organization can be pictured as a ball in a potential well or more generally in an attractor layout (Figure 18).

Changes in CNS functioning are characterized as continuous stabilization and destabilization, over time, of preferred attractor states.

To reduce for understanding the complexity of human neural networks of the many billions of neurons, *order parameters* or *collective variables* are introduced for certain movements. An equation of motion describes the coordination patterns dynamics. However, coordination patterns are not only determined by the task or biological function. Patterns adjust continuously to requirements from the environment (transmitted by impulse patterns from stimulated receptors in the periphery), memory, intention, and support given by a therapist. The specific requirements are captured by the concept of behavioral information and are made part of a vector field that attracts toward the required patterns. The coordination pattern dynamics, characterized by equations of motion of collective variables (the vector **X**), takes the general following form [18]:

$$d\mathbf{X}/dt = \mathbf{F}_{intr}(\mathbf{X}) + \sum c_{inf}\mathbf{F}_{inf}(\mathbf{X},t) \tag{1}$$

where $\mathbf{F}_{intr}$ designates the *intrinsic dynamics* of the nervous system. These intrinsic dynamics capture the anatomical (neuronal network structure), physiological and pathological states of the CNS and its muscular-skeletal elements.

$\sum c_{inf}\mathbf{F}_{inf}(\mathbf{X},t)$ represents the sum of external influences ($\mathbf{F}_{inf}(\mathbf{X},t)$) with their relative strength (c_{inf}) pertaining to each influence. The so-called *behavioral information* $\mathbf{F}_{inf}(\mathbf{X},t)$ includes cognitive states, emotional states, intentions, motivations, instructions, inter-personal coordination, movement support etc.

While applying therapy to the patient these extrinsic influences become extremely important, because the intrinsic (pattern) dynamics can be changed with these extrinsic influences by altering the equation of motion. By modulating the behavioral information, the intrinsic dynamics of the neuronal networks can be influenced further, that is if CDT is no longer efficient in repairing the injured CNS, requiring the therapy to be updated.

If the behavioral information includes the exercising of extremely coordinated, integrative movements, like exercising on the special CDT device, then the quality of CNS self-organization can be enhanced by improving the exactness of self-organization, namely the precision of phase and frequency coordination between neuron and neural assembly firings. By improving the precision of organization of the intrinsic dynamics, that is the specific variability of the injured networks, certain patterns do eventually re-appear.

3.2. Clinical Implications for Treatment Derived from the Equations of Motion of the Collective Variables (Formula 2)

1. Behavioral requirements $\mathbf{F}_{inf}$ (like intention, support, and instruction) affect the whole coordination dynamics, including stability, rather than only certain coordination patterns. The *change of the whole coordination pattern dynamics* of the CNS by the behavioral information *is one scientific basis for learning transfer* between different patterns and stability changes of patterns (as for example the reduction of spasticity). The other scientific basis for learning transfer is followed from human neurophysiology, namely that nerve cells or neural sub-networks are involved in different neural network organizations [14].

2. Intrinsic coordination tendencies captured by the intrinsic dynamics influence the performed pattern systematically because the degree to which intrinsic tendencies conflict or agree with the required patterns determines the variability of the performed coordination pattern.
3. A reduction in stability of movements and other patterns when intrinsic and informational requirements conflict, may lead to loss of stability and abrupt change while behavioral information is changing smoothly.
4. The intrinsic dynamics $\mathbf{F}_{intr}$ include vegetative and higher mental functions (these are also patterns of the coordination dynamics), which indicate that via exercising coordinated movements with support and instructions ($\mathbf{F}_{inf}$), urinary bladder function, intelligence and speech may be partly repaired following CNS injury or malformation.
5. When in an injured CNS with a certain set of behavioral information ($\sum c_{inf}\mathbf{F}_{inf}$) the intrinsic coordination dynamics ($\mathbf{F}_{intr}$) can no longer be influenced during coordination dynamics therapy, then this set of behavioral information has to be changed (using different $\mathbf{F}_{inf}$), or balanced differently (using different $\mathbf{c}_{inf}$), to further improve CNS organization dynamics.
6. However, the equations of motion of the coordination pattern dynamics (formula 2) provide no information about the specific behavioral information (F_{inf}) and training intensity ($\mathbf{c}_{inf}$) with which the CNS can be efficiently repaired in the patient. We need to have detailed knowledge of the human CNS on the single neuron and neural assembly level [14], as well as the knowledge at the integrative level, to find the specific behavioral information for the repair of the human CNS.

A novel step in coordination dynamics therapy is the inference derived from the formula 2 of the equation of motion. It suggests that the movement training not only improves the performance of that particular movement, but also improves the other non-trainable functions by transfer of learning. These functions include vegetative functions like bladder control, speech and higher mental functions. Furthermore, we have a means by which the stability of physiological network states can be increased (e.g., movements, continence, continuous concentration in performing certain tasks, speech etc.) and simultaneously the stability of pathological network states, like spasticity, decreased. The coordination (pattern) dynamics therapy partly based on the System Theory of Pattern Formation thus offers us an important theoretical basis and a practical tool to diagnose, quantify and repair the malfunctioning human nervous system at the macroscopic level.

3.3. Geographical Landscape of Attractors

The drawback of the equation of motion of the order parameters (formula 2) is that it is normally not possible to find a mathematical solution to it. But by defining a potential function and by picturing the attractive states and attractors by a ball in a potential well or rather by a ball moving in a geographical landscape of attractors (Figure 18), we form a theoretical basis to understand and measure stability of certain coordinated movement patterns (i.e., the deepness of the potential well of an attractor) in patients with CNS injury who receive on-going therapy.

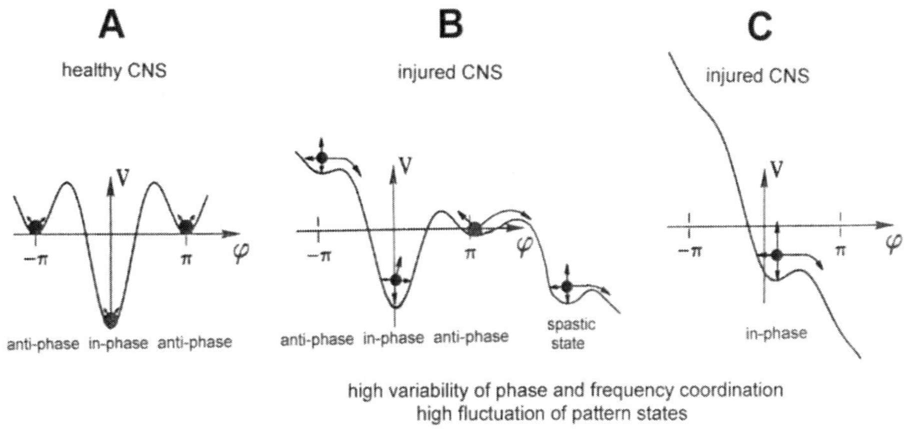

Figure 18. The potential, V(φ), of the coordination dynamics for jumping on springboard of a healthy (A) and injured CNS (B,C). The region around each local minimum acts like a well that weakly traps the system into a coordinated state. Behavioral changes are represented by the over-damped movement of a rolling ball in the potential "landscape". High fluctuations (indicated by long arrows attached to the ball (network state)) in the stable state, due to high variability of phase and frequency coordination (in the injured case), will have a greater probability of "kicking" the system out of the basins of attraction (B,C) than for low fluctuations (short arrows) (A), due to small variability of phase and frequency coordination (in A). In B, only the in-phase jumping is stable, even though the fluctuation is high. In C there is only an attractor basin for the in-phase jumping, but the fluctuation is so high that there is a high probability that the system is kicked out of the basin of attraction. The patient can no longer jump in anti-phase and has difficulty with jumping in-phase. The stability of jumping depends on the motor program (deepness of basin of attraction) and the fluctuation of the pattern state (moving of the ball) caused by the increased variability of phase and frequency coordination due to the injury.

3.4. CNS Repair upon Stability Changes of Physiologic and Patho-Physiologic Patterns: Improvement of Geographical Landscape of Attractors

When jumping on springboard (Figure 2D) the pattern changes can be represented by the over damped movement of a rolling ball in the potential landscape for the physiologic (Figure 18A, little fluctuation of phase and frequency coordination) and the pathologic case (Figure 18B, C, large variability). In the healthy CNS, the phase and frequency variability is small (short arrows of the moving ball) and the jumping in in-phase and in anti-phase is stable (Figure 18A). Following injury, the potential landscape is deformed and the fluctuation of the network states, generating jumping, is high (Figure 18B). The in-phase jumping is still stable in spite of the increased fluctuation (larger fluctuation arrows), because the basin of attraction is deep. The jumping in anti-phase became unstable because the basin of attraction is shallow and the increased fluctuation in the state has a greater probability of "kicking" the system out of the basin. A switch into a spastic state is also possible. In severe CNS injury or malformation, the patient can no longer jump in anti-phase because of the missing of attractors for anti-phase jumping (Figure 18C). The attractor layout is asymmetrical and deformed. Support is needed for anti-phase jumping. Jumping in-phase is still possible but unstable.

Upon performing very exact coordinated movements, imposed by devices, the nervous system of the patient learns to reduce the variability of phase and frequency coordination and achieves in this way a small fluctuation of the network states again as shown in Figure 18A. The progress in treatment (learning) is that the in-phase jumping in Figure 18C and the anti-phase jumping in Figure 18B become stable (Figure 18A) again. Also the potential landscape will change due to the reduction of the phase and frequency variability. The important consequence for treatment is that when exercising on special devices and reducing in this way the variability of phase and frequency coordination, the patient can induce the formation of patterns again, without having trained them (learning transfer). But exercising only on the special device is not enough in itself to improve the attractor layout sufficiently. Regrettably, there is no miracle device. Many different movements have to be trained to repair the CNS.

3.5. Reduction of Spasticity

When performing movements like walking, running, crawling, or exercising on a device which imposes highly coordinated movements on the patient, the coordination dynamics can be changed in the way that the stability of spastic states decreases and the stability of the movement states increases. Such changes of coordination pattern dynamics can be pictured again by means of an attractor layout. An attractor is pictured as a potential well (attractor valley) into which a rolling ball is attracted. The position of the ball represents the momentary state of the system. Figure 19 shows schematically such an attractor layout with the two attractor's *spasticity* and *coordinated movement*. When exercise is commenced (A), the spastic state is very stable (the attractor valley is deep) and the state of the system is attracted towards the attractor state *spasticity*. With exercise, the attractor layout is changing in the short-term memory in the way that the attractor *spasticity* is getting shallower and the attractor *physiologic movement* is getting deeper (B). Because of fluctuation due to variability of phase and frequency coordination, the position of the ball, which represents the momentary state of the system, is switching between the attractor states *spasticity* and *movement*. Spasticity and movements are present simultaneously in the patient. With further exercise, the attractor *movement* becomes deeper (more stable) than the attractor *spasticity*. The patient can now perform the movements with little or no spasticity. The transient reduction of spasticity in the short-term memory, achieved by many hundreds of coordinated movements can last anything from a few minutes up to several hours; this is indicated in Figure19 by the two long arrows. The shorter backward arrow (from right to left) indicates that spasticity has slightly reduced in the long-term memory. The coordination dynamics have changed. Repeated exercising will further reduce the stability of *spasticity* and increase the stability of the coordinated *movement* and will further change intrinsic coordination tendencies in the long-term memory.

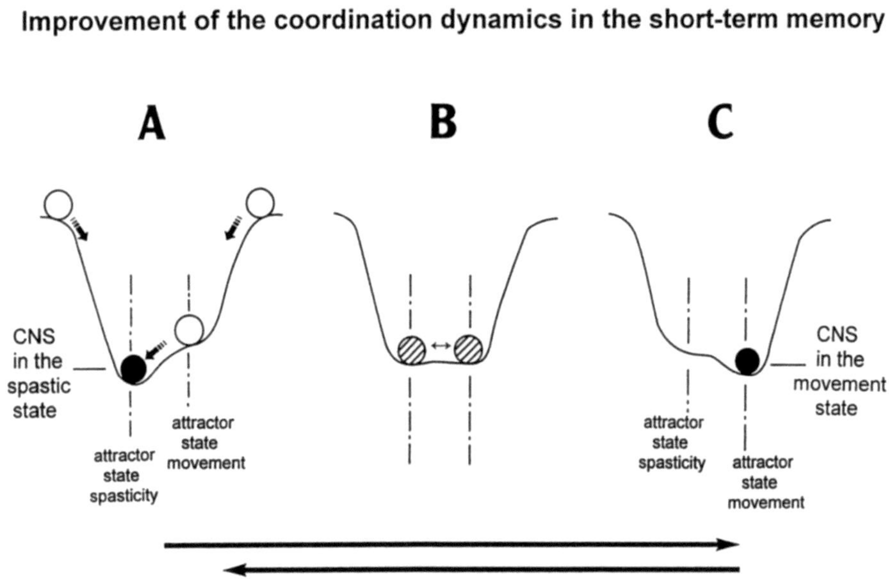

Figure 19. Therapy-induced spasticity reduction in the short-term memory. The position of the ball represents the state of the system and the potential well, the attractor. The ball is attracted to the stable position in the deepness of the whole, called attractor state. The attractor layout, consisting of two attractive states of different stability, is changing upon exercising very coordinated rhythmic movements. Black ball = stable state, open ball = very unstable state, hatched ball = spasticity and movement co-exist.

3.6. Quantifying CNS Functioning by Measuring Pattern Stability upon Pattern Change When Exercising on the Special CDT Device

Experimentally, the underlying dynamics of coordinated movements can be found in the temporal stability of coordination patterns and can be assessed through pattern change. A change of the coordinated movement patterns is generated, when a subject is exercising on the special CDT and recording device, where the coordination between arms and legs, imposed by the device, changes continuously between pace and trot gait and backwards. The stability of the intrinsic coordination tendencies is measured by the deviations and differential stability during the performance of these rhythmic movements.

The plotting of the differential stability over time of the frequency of exercising generates an attractor layout for this special movement (Figure 20) and the mean stability per minute can be measured by the arrhythmia of exercising (df/dt:f, f = frequency; or dv/dt, v = angular velocity). Such value, the so-called coordination dynamics value, quantifies CNS functioning objectively, integratively and non-invasively. The assessment of quality of CNS organization by pattern change is a novel step in CDT.

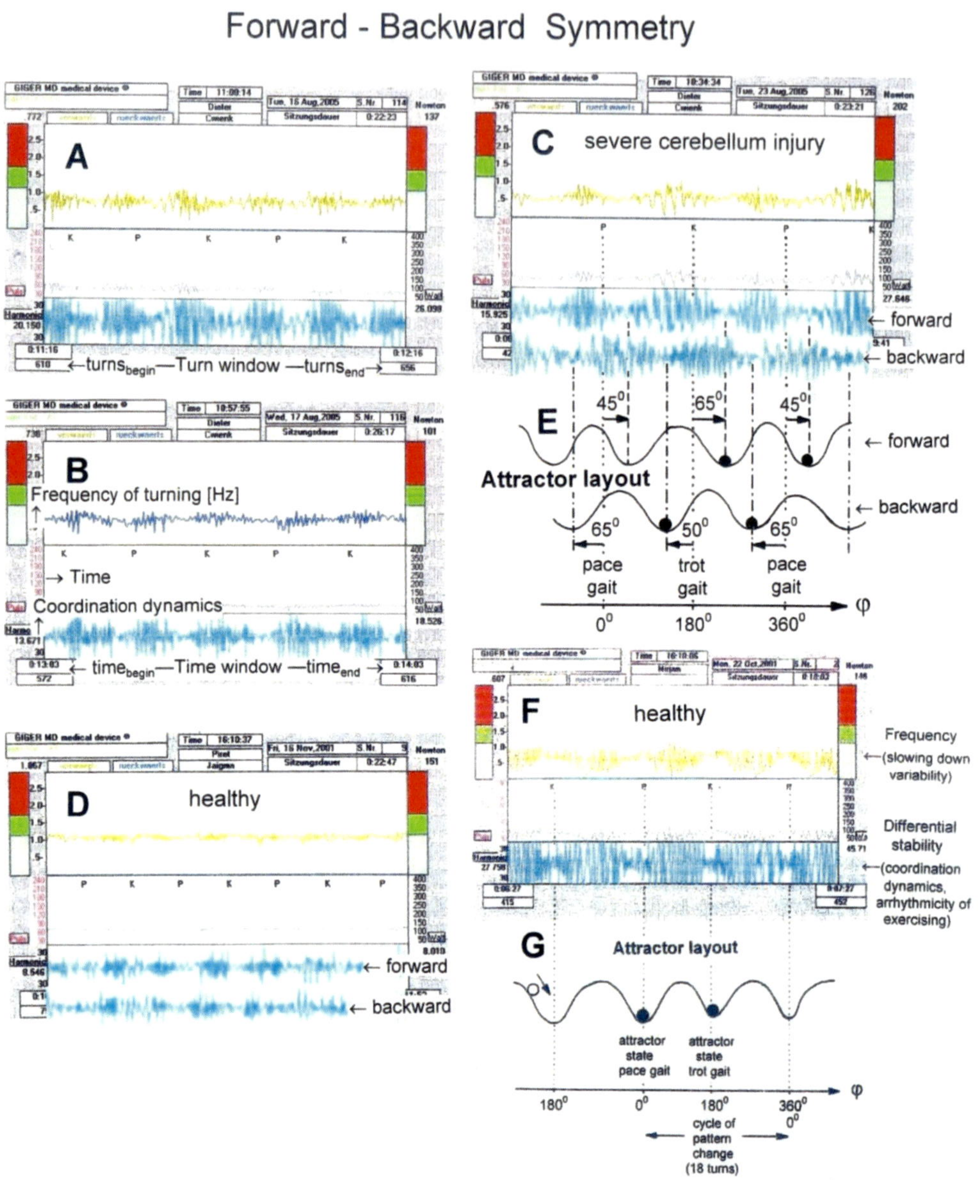

Figure 20. Forward-backward movement symmetry impairment in a patient with severe cerebellum injury for exercising on the special coordination dynamics therapy device (A-C). Coordination dynamics figures of a healthy person are shown for comparison in D and F. In G the attractor layout is constructed for the healthy case. The attractors are the pace and trot gait movements. The state of the system (the white ball) is attracted to the ground of the potential well. Being in the attractor state, the ball is pictured black. In E the attractors of a brain injured patient are not at the pace and trot gait coordination's for forward and backward exercising.

3.7. Forward-Backward Symmetry Impairment

As the coordination dynamics traces in Figure 20 show, the lack of symmetry between forward and backward exercising can be made visible and quantified in a patient with cerebellum and cerebrum injuries by measuring the coordination dynamics (temporal stability) through pattern change. When exercising at a load of between 100 and 200N, the values for the exercising frequency and the coordination dynamics increased and decreased rhythmically both for forward (Figure 20A) and backward turning (Figure 20B). But when turning forward, the arrhythmicity was small to the right side of the pace (P) and trot gait (K) coordination's (Figure 20A) and when turning backwards (Figure 20B), the arrhythmia was small to the left side of the pace (P) and trot gait (K) coordination's. Therefore, there seems to be a mirror picture shift of the arrhythmia between exercising forward and backward with respect to the pace and trot gait coordination's.

In the case of a healthy patient, the lowest arrhythmia of exercising (the highest pattern stability) was positioned at the pace and trot gait coordination's (Figure 20F). To make this mirror image change with respect to forward and backward exercising better visible in the patient, the coordination dynamics traces for forward and backward exercising are arranged together in Figure 20C (lower part). The recordings were taken from one week later to show that patterns of highest stability were at the same place with respect to the pace and trot gait coordination's to show that there was no phase drifting. It can clearly be seen in Figure 20C that the periods of lowest arrhythmia shifted rather symmetrically away from the pace (P) and trot gait (K) coordination positions with respect to forward and backward exercising. When picturing this rather opposite symmetric shift of the arrhythmia of exercising in an attractor layout of pattern formation in the framework of the system theory of pattern formation, the attractor states (patterns with highest stability) shifted away from the pace and trot gait coordination patterns (Figure 20E). In the attractor layout of a healthy, physically active girl (Figure 20G), drawn from the coordination dynamics recording of Figure 20F, the stable movement patterns (when exercising on the special device) are lying at the pace and trot gait coordination's. For the patient with the brain injury, the stable movement patterns (small arrhythmia of exercising) shifted with respect to the cycle of pattern change (lower part of Figure 20G) between $\varphi = 45°$ and $65°$ forwardly for forward exercising, and between $\varphi = 50°$ and $65°$ backwardly for backward exercising (Figure 20E). This rather opposite shift of the attractor states shows similarity to the rather mirror picture change of the antagonicity impairment of the antagonistic muscle activation of the tibialis anterior and gastrocnemius muscles, measured by surface EMG in this patient (Figure 21). We can therefore measure symmetry impairment of CNS organization with respect to forward and backward exercising with the special CDT device. Symmetry diagnostic with the special CDT device can also be performed with respect to right and left and rostral and caudal.

4. Surface Electromyography (sEMG) for Optimizing Therapy

4.1. Motor Pattern Diagnostic by sEMG for Updating Coordination Dynamics Therapy (CDT)

In Figures 13 to 15 sEMG demonstrated coordinated firing of single motor units. But sEMG can also be used to improve or update CDT. In the above coordination pattern dynamics, characterized by equations of motion of collective variables, it was emphasized that the intrinsic dynamics $\mathbf{F}_{intr}(\mathbf{X})$ (CNS functioning) can be modulated by the treatment (behavioral information $\sum c_{inf}\mathbf{F}_{inf}(\mathbf{X},t)$ upon changing the equation of motion, if CDT is no longer efficient in repairing the injured CNS. The system theory of pattern formation, however, is giving no information on how to change the therapy for further improvement of CNS organization. But such information can come from human neurophysiology. It will be demonstrated now how motor patterns, recorded with sEMG, from different movements can show us which movements have the best motor patterns and should be trained most because they offer the highest rate of learning. We shall first analyze recorded motor patterns when the patient with the brain injury is exercising forward and backward on the special CDT device. The coordination pattern dynamics give us information about CNS functioning in general. The sEMG motor patterns show what is wrong in the different motor patterns and gives indications as to what to change in the administered therapy. Natural impulse patterns at the neuron level give us information about CNS functioning at the neural level.

4.2. Symmetry Improvement of Motor Programs of Antagonistic Muscles by Increasing the Integrativity of the Movement

In a patient with cerebrum and cerebellum injury, surface electrodes were placed on the right and left tibialis anterior and the right and left lateral gastrocnemius muscles. When the patient only exercised with the legs, like on a fitness bicycle, the left gastrocnemius muscle was activated too late with respect to the left tibialis anterior muscle (Figure 21A, at $\varphi = 200^0$ instead of 180^0 (antagonistic muscles) for exercising in the forward direction, and it was activated too early for exercising in the backward direction (Figure 21B, at 150^0 instead of 180^0). The antagonicity of the motor programs thus showed pathologic symmetry between exercising in the forward and backward direction, namely $\Delta\varphi_{forward} = 20^0$ against $\Delta\varphi_{backward} = -30^0$. When exercising with arms and legs (Figure 21C,D) this asymmetric coordination did not change in the stronger left leg.

However, an improvement of the symmetry of antagonistic coordination between the gastrocnemius and the tibialis anterior muscles could be observed in the right weaker leg when exercising in the forward or backward direction using arms and legs as compared to only using the legs. Upon exercising in the forward direction only using legs, the right gastrocnemius muscle was activated too late with respect to the right tibialis anterior muscle (Figure 21A, at 230^0 instead of 180^0; $\Delta\varphi_{forward} = 50^0$), and too early for exercising in the backward direction (Figure 21B, at 65^0 instead of 180^0; $\Delta\varphi_{backward} = -115^0$). When exercising in the forward direction with arms and legs, the right gastrocnemius muscle was only activated a little late (Figure 21C, only at 210^0 instead of 180^0; $\Delta\varphi_{forward} = 30^0$). When exercising in the backward direction, the right gastrocnemius muscle was not activated as early with respect to the right tibialis anterior muscle (only at 105^0 instead of 180^0; $\Delta\varphi_{backward} = -75^0$). Thus, it may be stated that the antagonistic coordination between the right tibialis anterior and the right gastrocnemius muscles improved for exercising in the forward and backward direction by 20^0 and 40^0 respectively upon exercising with arms and legs as compared to only using the legs, like on a fitness bicycle. More generally speaking, this antagonistic symmetry between the tibialis anterior and the right gastrocnemius muscles improved by 60^0 for exercising in the forward and backward direction following exercise with both arms and legs. During this more integrative exercising on the special CDT device using both arms and legs, the antagonistic activation of the tibialis anterior and the gastrocnemius muscles improved in the short-term memory. Therefore in this patient the antagonicity impairment could be treated by exercising very coordinated movements with arms and legs on the special device. It is likely that, other, very coordinated four-limb movements could also improve the antagonicity of muscle activation.

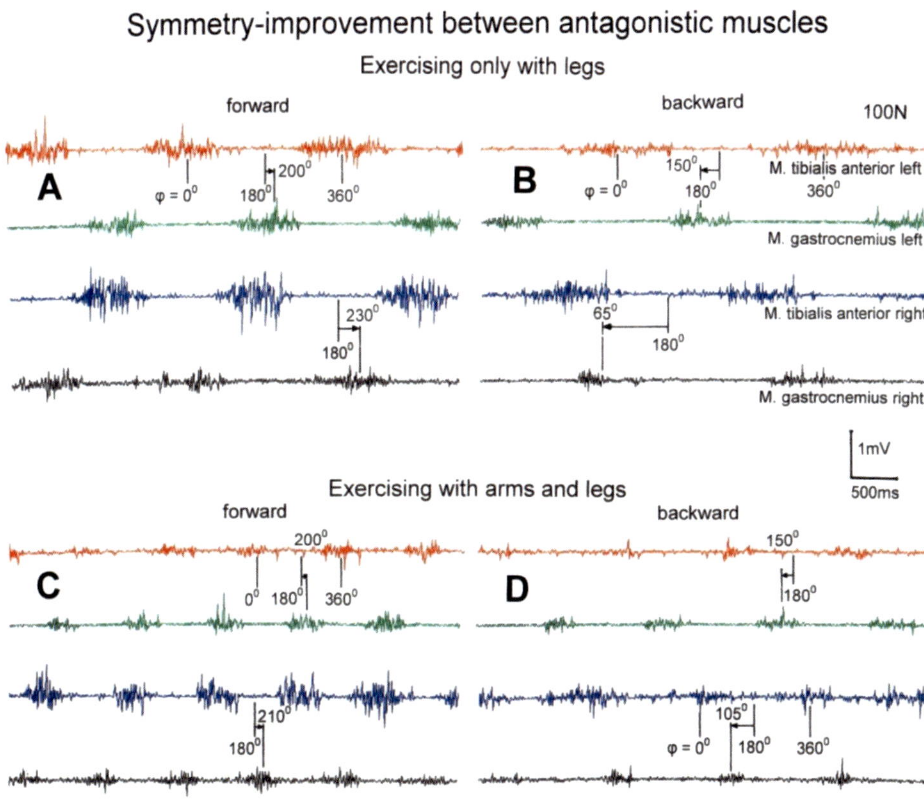

Figure 21. Motor programs of right and left tibialis anterior and gastrocnemius muscles recorded by sEMG for the patient with a severe cerebellar injury during exercise on the special coordination dynamic therapy device in the forward (A,C) and backward direction (B,D) only using legs (A,B) and with arms and legs (C,D). Note that the gastrocnemius muscle is not activated properly antagonistically with respect to the tibialis anterior (at 180 degree): it gets activated earlier or later, depending upon whether movements are performed in the forward or backward direction and upon whether only legs or arms and legs are used.

4.3. Learning in the Short-Term Memory from the Better Opposite Side

The better, in terms of antagonistic activation, left side did not improve in this patient, whereas the poor right side did improve by 36% ($\Delta\varphi_{forward}$ - $\Delta\varphi_{backward}$ reduced from 165^0 to 105^0). With the integrativity increasing from coordinated leg movements to coordinated arm and leg movements, the better left side remained unchanged, whereas the poor right side improved by 36%. This one-sided symmetry improvement can be interpreted in the way that the poor right side learned in the short-term memory from the good left side. This network responsible for this learning effect may be located in the spinocerebellum and/or in the spinal cord and is called co-movement.

4.4. Judgment of Motor Walking Programs in a Patient with Balance Problems

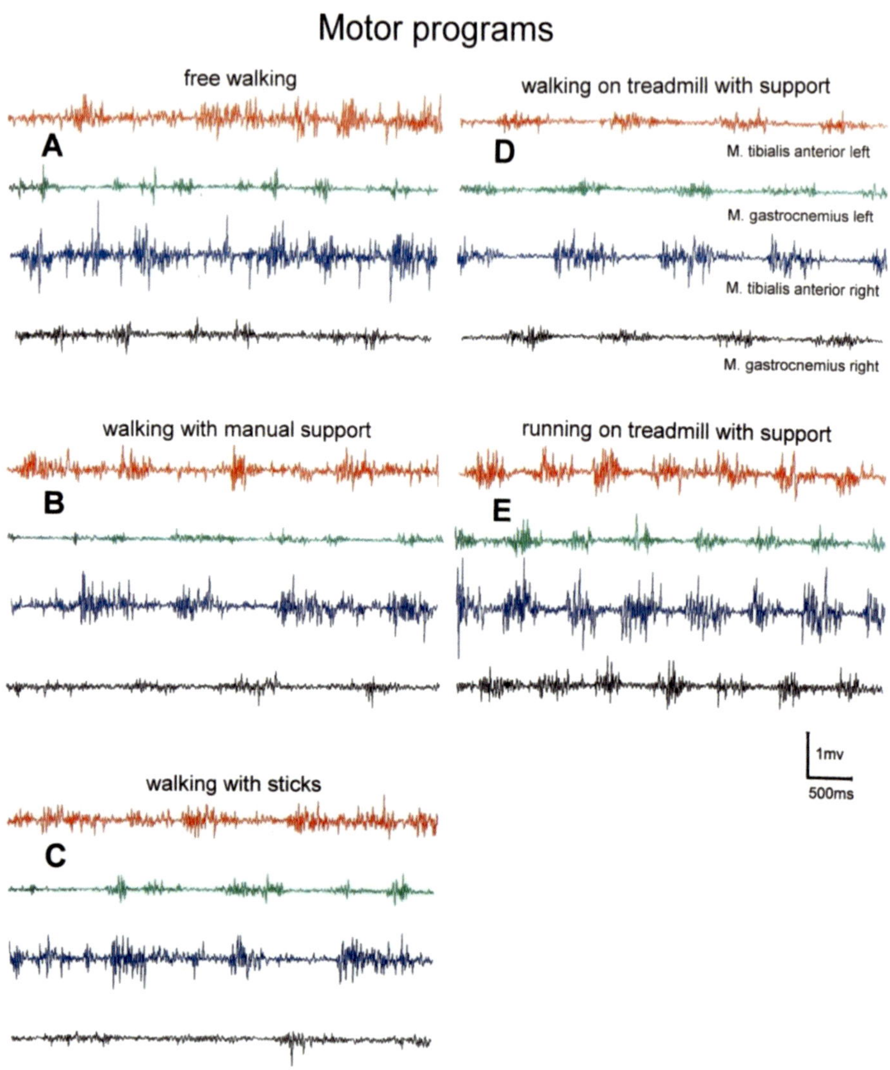

Figure 22. Surface EMG motor program of the right and left tibialis anterior and gastrocnemius muscles for free walking (A), walking with manual support (B), walking with sticks (C), walking (D) and running (E) on the treadmill with support. Note that the motor programs for walking with strong support (D,E) are much better than those for walking with little or no support (A,B,C).

In a patient with cerebellum and cerebrum injury, changes in the walking program will be analyzed when the patient had to balance unaided. Since the vestibulocerebellum was also heavily damaged, increasing motor program deficits for increasing balance needs can be expected. Figure 22 shows sEMG motor programs for free walking (Figure 22A) and for supported walking (manual support) (Figure 22B), walking with sticks (Figure 22C), walking on treadmill (Figure 22E) and running on treadmill (Figure 22E). The motor programs for free walking (A) were worse than those for the different kinds of supported walking (B-E). With the increasing support (from B,C to D,E) the motor programs improved. Obviously, the patient had balance problems: due to the effort of attempting to maintain balance by himself, his motor programs worsened. The interpretation of this worsening of the motor programs with the increasing balance needs is that following free walking, the damaged vestibulocerebellum could no longer sufficiently coordinate balance with walking. The motor program for supported walking or running (B-E) were best for running (E), which indicates that the higher speed improves the performance and the learning as long as the patient can manage the speed.

4.5. Comparison between Motor Programs When Swinging and Jumping On Springboard

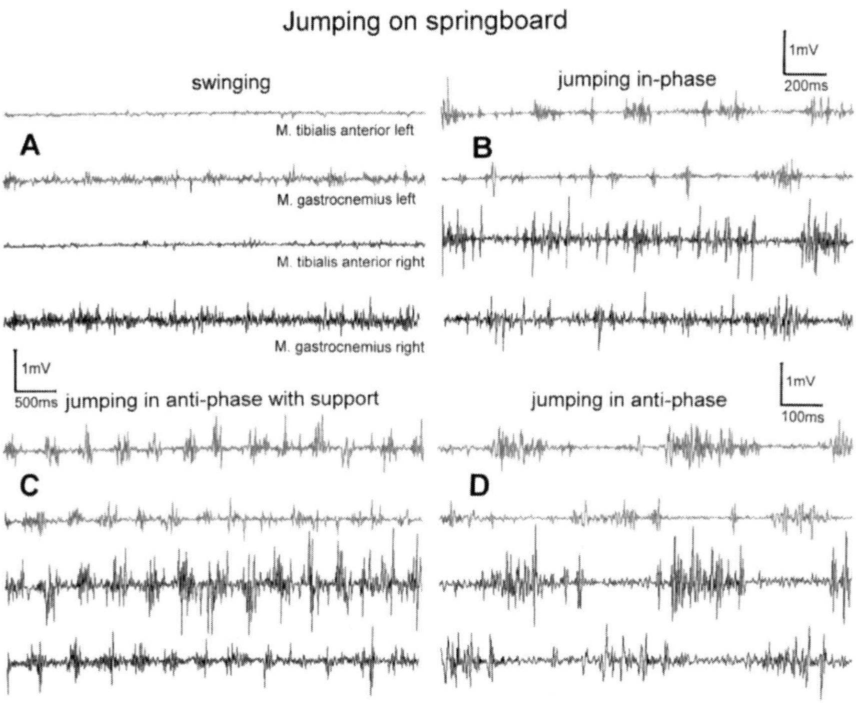

Figure 23. Surface EMG motor program of the right and left tibialis anterior and gastrocnemius muscles for swinging (A) and jumping on springboard in-phase (B), in anti-phase with support (C), and in anti-phase without support (D). Note that the motor programs are better for jumping in anti-phase (C,D) than for jumping in in-phase and swinging. Note further that continuous synchronous oscillatory firing of motor units can be seen on the trace taken from the right tibialis muscle for jumping in in-phase (B).

To discover which motor programs are best for performing rhythmic, dynamic, stereotyped movements for functional repair at the neural ensemble level, sEMG was recorded for the different kinds of springboard movements. Motor programs during swinging and jumping on springboard are shown in Figure 23. For swinging (Figure 23A) the tibialis anterior and gastrocnemius muscles were only slightly activated rhythmically, i.e., no real motor program can be seen. When jumping in in-phase, the leg muscles were more activated and motor programs can be partly identified (Figure 23B). In the right tibialis anterior muscle, synchronized oscillatory firing of motor units (pathologic) can be seen. For jumping in anti-phase with (Figure 23C) and without support by a therapist (Figure 23D) the motor programs were quite good, even though rhythmic firing of motor units can also clearly be identified during the motor bursts (right gastrocnemius in Figure 23D).

The conclusion which may be drawn from this is that jumping on the springboard will train the neuronal networks of the patient more effectively than swinging, since real motor programs could only be identified during jumping. The general conclusion for treatment is that sEMG can help to improve the rate and understanding of repair.

4.6. Symmetry Learning to Enhance the Efficiency of CNS Repair

Learning transfer from one hand movement to the symmetric one has been known to occur [18]. This means that the symmetry counterpart improves without being trained itself. It has been shown (see above) that the antagonicity between the tibialis anterior and gastrocnemius muscles in the poor right leg improved in the short-term memory by 36% when increasing the integrativity of exercising by changing from only leg movements to coordinated arm and leg movements.

Assuming that repeated improvements in CNS functioning in the short-term memory will slowly manifest themselves in the long-term memory would mean that there is learning transfer occurring in relation to the improvement of antagonicity from the left leg to the right leg. A particular movement can be also improved by training its symmetrical counterpart. If, for example, the walking pattern on the right side has been impaired in a stroke patient for a few years, this pathologic walking pattern has become an old-learned movement and is difficult to change. Its counterpart, namely backward walking, has however not become an old-learned movement after injury. Additional training of backward walking will therefore more efficiently improve forward walking by learning transfer. In conclusion, movements and especially automatisms have to be trained with their symmetry counterparts to enhance the efficiency of CNS repair.

5. CNS Functioning Can Be Assessed Non-Invasively by Measuring the Coordination Dynamics

5.1. Plausible Explanation of Measuring CNS Functioning

A prerequisite of the repair of the human CNS is that the quality of its functioning must be measured. This can be achieved when a patient or a healthy subject is exercising on a precisely calibrated device for complicated coordinated arm and leg movements. If the CNS is functioning well, the CNS can cope with the difficult coordination's and the rhythmicity of exercising is good. If the CNS is functioning poorly, the rhythmicity of exercising is poor, because the CNS has difficulty in generating these difficult coordinated movement patterns. This instrument is the special Coordination Dynamic Therapy (CDT) and recording device (Figures 174,209-211). While performing changing coordinated arm, leg and trunk movements, imposed by the device, CNS functioning can be measured non-invasively by the arrhythmicity of exercising with the changing coordination's between arms and legs. Further, through exercising with different combinations of arms and legs in both forward and backward directions, right-left, rostral-caudal, and forward-backward symmetry impairments of CNS movement organization can be measured, trained and repaired.

The pace and trot gait coordination's are mainly generated in the neuronal networks of the spinal cord, because a new-born baby can step automatically (Figure 24). It cannot control the start and the termination of the stepping automatism. The afferent input from the heel strike is one of the stimuli which starts the automatic stepping. With development of the infant, the neural networks of the spinal cord become more and more under the control of supraspinal centers. During early development therefore, the infant has significant difficulty in exercising on the special CDT device. This is not due to the size of the device.

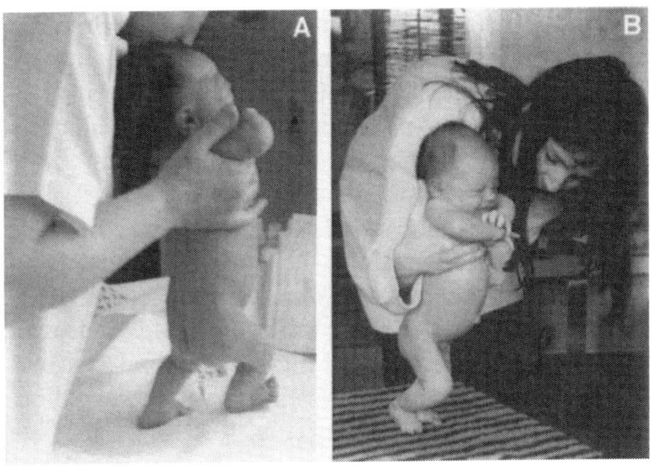

Figure 24. Automatic stepping in a newborn infant. A. A 5-day-old infant, Juliane, performing primary automatic stepping; slight backward posture. The heel of the right foot touched the ground first. B. Infant Juliane, 8-day-old, performing automatic stepping.

For performing the intermediate coordination's between pace and trot gait, the spinal cord's neuronal networks have to be varied by supra-spinal centers to accomplish these difficult coordination's. When the brain-injured patient is performing these movements, with the help of the device, the brain's networks are involved in the generation of these complex patterns.

If the brain is functioning well, the CNS has no problem in generating the difficult coordination's and the arrhythmicity of exercising is slight. If the brain is functioning poorly, because of a brain injury, the patient struggles to generate these intermediate coordination's between pace and trot gait. The arrhythmicity of exercising is very apparent. Therefore, the measuring of the coordination dynamics is really assessing CNS functioning, including brain functioning. With respect to therapy, the training of the intermediate coordination's between pace and trot gait also repairs brain functions.

5.2. Scientific Basis for Measuring CNS Functioning

The means by which CNS functioning may be evaluated with the different measuring methods shall now be discussed (Figure 1).

5.2.1. Relative Phase and Frequency Coordination between the Firings of α and γ-Motoneurons and Secondary Muscle Spindle Afferents Recorded with the Single-Nerve Fiber Action Potential Recording Method

With the single-nerve fiber action potential recording it was shown that the neurons of the human CNS organize themselves by phase and frequency coordination (Figures 9, 9E). Following injury, this organization principle is disrupted [14].

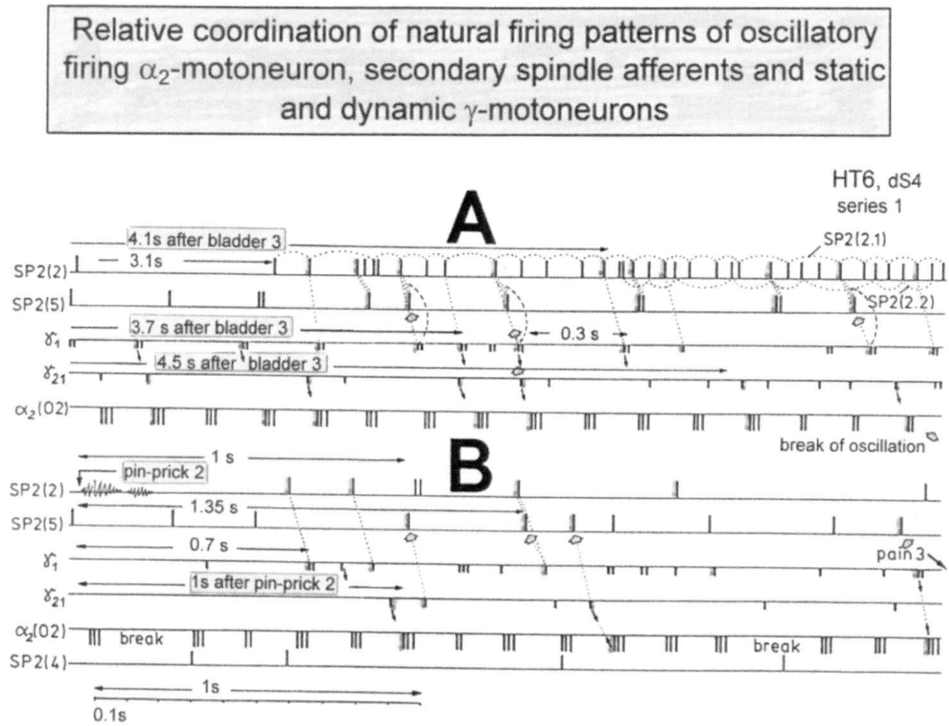

Figure 25. Phase and frequency coordination between the extracellular recorded action potentials of simultaneously recorded γ-motoneurons (γ_1 and γ_{21}), secondary spindle afferent fibers (SP2(2), SP2(4), SP2(5)) and oscillatory firing α_2-motoneuron O2 following bladder catheter pulling (bladder 3) (A) and pin-prick 2 (B). B was recorded before A. In A the impulse patterns of the 2 encoding sites SP2(2.1) and SP2(2.2) of the single parent fiber SP2(2) are indicated by the dotted curves. Times to the activity increases of γ-motoneurons and secondary spindle afferents following stimulation are indicated. Similar time intervals of the occurrence of γ-motoneuron APs and SP2(5) fiber APs (phase coordination) are indicated by the open arrows, and the similar time intervals of γ-motoneuron APs and α-motoneuron APs are indicated by small arrows. Similar time intervals of the APs of fibers SP2(2) and SP2(5) are indicated by the double dotted lines, those of γ_1-APs and the SP2(2) fiber APs by a dotted line, and those of γ_1-APs and the SP2(2)-SP2(5) correlation by a curved dashed line. HT6; dS4-root.

Figure 25 shows schematically a recording from a dorsal S4 root of a brain-dead human. Of the summed afferent and efferent impulse traffic the natural impulse patterns of one α_2-motoneuron, a dynamic (γ_1) and a static γ-motoneuron (γ_{21}) and two to three secondary muscle spindle afferent fibers could be extracted (for classification see Figure 7). The natural stimulations performed were pin-pricking (pain) of sacral dermatomes inside the continence automatism zone (Figure 5) and urinary bladder catheter pulling (Figures 5,153). It can be seen from Figure 25 that the α_2-motoneuron (α_2(O2)) fired in an oscillatory manner with 2 to 3 impulses per impulse train and sometimes there was a break in the oscillatory firing as it is shown in Figure 26 for the firing of the α_1-motor units (FF). The impulse train for α_1-motor units consisted of only one action potential (Figures 8, 17). Phase coordination's between the α_2-motoneuron, the γ-motoneurons and the secondary muscle spindle afferent fibers are indicated by different arrows and the dotted and dashed lines. It can be seen from Figure 25 that there were many coordination's between the different neurons. The relative phase and frequency coordination seems to hold for all neurons and is an integrative mechanism for the self-organization of the neuronal networks of the human CNS.

But the recording from single-nerve fibers is an invasive recording method and cannot help to quantify learning following therapy.

5.2.2. Phase and Frequency Coordination between Motor Unit Firing and Building Up of a Motor Program with Increasing Load Recorded with Single-Motor Unit Semg When Exercising on the Special CDT Device

Figure 26A-D shows how motoneurons coordinate their firing and how the CNS generates a motor program during exercise on the special CDT device. Approximately 5 motor units could be activated in the right tibialis anterior muscle of a spinal cord injury patient. The motor units marked 1 and 2 had a low threshold, and the motor unit 3 a higher threshold. Their waveforms are shown time-stretched in the inserts of Figure 26A,C. For high exertion, when exercising on the special CDT device at 200N, it appears that 2 other motor units could be activated. This section shall on focus on the firing of these 3 motor units, to study their coordination during the exercising on the special CDT device at the load increasing from 20 to 200N.

The patient whose record is shown in Figure 26A was exercising at 20N. Motor unit 1, with a large motor unit potential, fired in an oscillatory manner at around 10Hz and the motor unit can thus be suggested to be innervated by an α_1-motoneuron. Motor unit 2 fired at around 5Hz in between the action potentials of unit 1 to avoid synchronization, but fired in every second interspike interval due to having half of its frequency. Motor unit 2 was likely also innervated by an α_1-motoneuron, even though the frequency was low (5Hz) for the oscillatory firing of an α_1-motoneuron. There was slight inhibition acting on motor unit 1 after the 6th spike since the interspike interval greatly exceeded 100ms. When increasing the load to 50N (Figure 26B), motor unit 1 slightly increased the frequency to increase the activity, and motor unit 2 increased its frequency from to 5 to 10Hz to increase the activity. The higher threshold motor unit 3 started to fire occasionally. After the 4th motor unit 1 potential, motor unit 2 stopped firing in an oscillatory manner. It appears that motor program inhibition worked transiently on the oscillatory firing sub-network (to reduce the frequency of oscillatory firing or to stop the oscillatory firing) and gave rise to a transient break in the oscillatory firing.

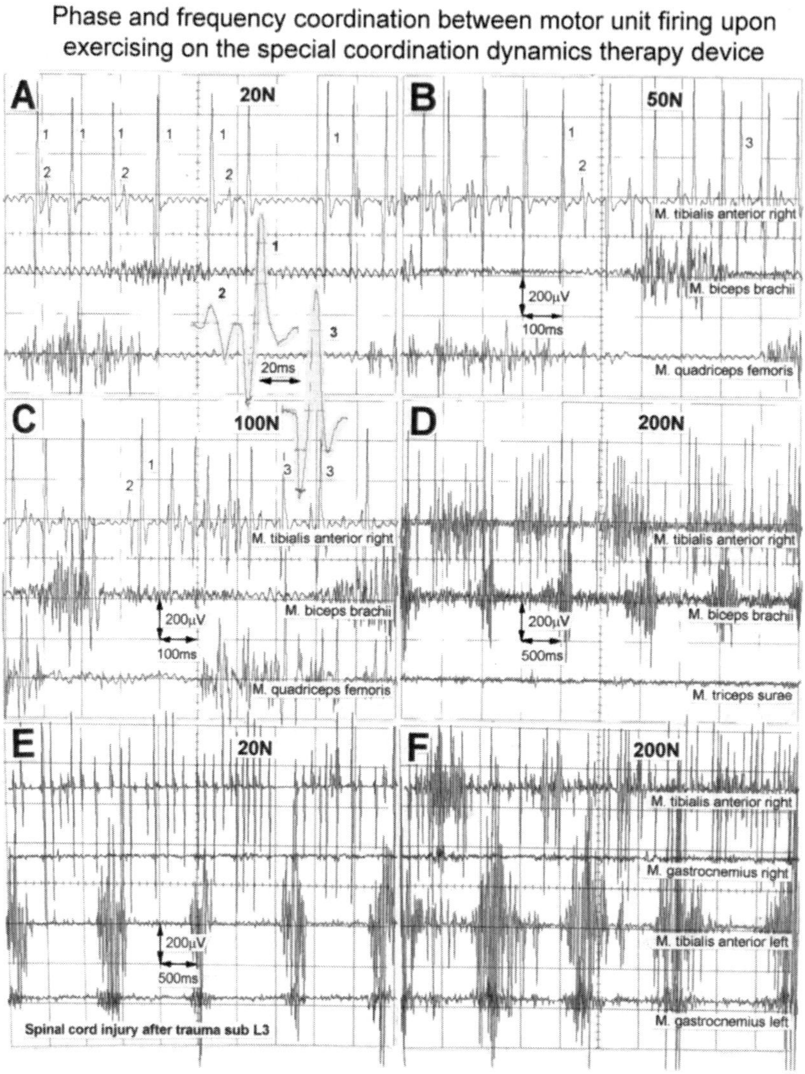

Figure 26. Phase and frequency coordination between motor unit firing (in the same muscle) during the generation of a motor program during exercise on the special coordination dynamics therapy device at loads increasing from 20 to 200Newton. The wave forms of 3 identified motor unit potentials of FF-type '1', '2', and '3' are inserted in 'A,C' and some of them are marked in 'A,B,C'. Note in 'A' that motor unit '1', firing at approx. 9Hz, coordinates its firing with motor unit '2', firing at approx. 4.2Hz, by firing twice as often. In 'B' both motor units fire non-synchronously at around 10Hz with relative phase and frequency coordination. A, B. No motor program can be seen in the right tibialis anterior muscle as in the biceps brachii and quadriceps femoris muscles during exercising at 20 and 50N. C. The oscillatory firing of motor unit 1 and 2 (and 3) was transiently interrupted (inhibited) upon exercising at 100N, thus showing the beginning of a motor program. In the motor program burst, the frequencies of the motor units are increasing and decreasing. D. Upon exercising at 200N, the motor program became more pronounced. Probably two more high-threshold motor units contributed to the motor program burst in addition to the 3 motor units identified in 'A-C'. E, F. Improvement of the motor program from 'E' to 'F' in the right tibialis anterior muscle of the same patient on another day. But due to flu the motor program of the right tibialis anterior muscle was in 'F' not as good as in 'D'. The activity in the left gastrocnemius muscle consists partly of cross-talk. Recording from a patient with a complete spinal cord injury sub L3.

When exercising at 100N (Figure 26C) the way in which the motor program was generated can be seen. When an activity burst is activated, for example in order contract a muscle, motor units are activated to fire in an oscillatory manner for the duration of the burst. The frequency of the oscillation increases and decreases. Other motor units were also activated, but first the ones with the lowest threshold. The activated motor units avoided synchronization. When the motor program burst was supposed to be terminated, the motor unit firing stopped. They were probably inhibited. When exercising at 200N (Figure 26D), a motor program can be seen in the right tibialis anterior muscle (note the time axis change). Four weeks later, when the patient had a flu (Figure 26E,F), the motor program developed in the right tibialis anterior muscle worsened.

The healthy motor program of the biceps brachii (spinal cord injury sub L3) looks completely different (Figure 26A,B). No coordination can be seen and analyzed. The difference is understandable. In the right tibialis anterior muscle, the CNS could only activate 5 motor units and only α_1-motor units (large motor unit potential and only one action potential per impulse train characterizes α_1-motor unit (FF) oscillatory firing (Figure 17)); no α_2 (FR) and α_3-motor units (S) were innervated. In the biceps brachii muscle, all types of motor units could be recruited, each type being in abundance. While exercising at 20N (Figure 26A), mostly only α_2 (FR) and α_3-motor units (S) (identified by the small motor unit action potential amplitude) and maybe a few α_1-motor units (FF) of small size were activated in the biceps brachii muscle. For 50N (Figure 26B), motor units of α_1-type (FF), were activated. These motor units have naturally larger amplitudes. Since the recruitment is according to the size principle in each motoneuron group and the group recruitment depends on the speed of movement, quite a complexity is present. Further, the transiently oscillatory firing of motoneurons will be shorter, especially if the patient is exercising very fast. In the partly impaired motor supply of the quadriceps femoris muscle (injury level sub L3), some motor units can be identified (Figure 26B). In the better left tibialis anterior muscle, no coordination could be extracted from the sum of the motor units activity; too many motor units were functioning to permit identification of single motor unit firing (Figure 26E,F). However, many α_1-motor units were activated in the left tibialis anterior muscle up to a size of 2mV (Figure 26F), whereas in the biceps brachii muscle they were still under 1mV (Figure 26D) for the load of 200N.

In Figure 26 relative phase and frequency coordination is shown for motor units in the same muscle, namely the right tibialis anterior muscle of a patient with a spinal cord injury sub C4/5. In Figure 15 phase and frequency coordination among motor units was shown between different muscles.

By recording single-motor units non-invasively with the sEMG phase and frequency coordination and the building up of a motor program can be identified and followed throughout a program of exercise at increasing loads. However, sEMG is not capable of quantifying the improvement of CNS functioning following therapy in human patients.

5.2.3. From sEMG Motor Programs to High-Load Coordination Dynamics to Evaluate CNS Functioning upon Therapy

5.2.3.1. Pathologic Patterns of Motor Activation

Surface EMG was performed in a patient with an incomplete cervical spinal cord injury sub C5/6 while he was undergoing CDT. Figure 27 shows sEMG recordings of motor patterns while the patient was exercising on the special CDT device at the beginning of therapy (Figure 27A,B) and again after 6 months of therapy (Figure 27C,D). The motor patterns were very largely abnormal at the commencement of therapy (Figure 27A,B). The amplitude of sEMG activity was highest in muscles proximal to the injury, i.e., the biceps brachii, while significantly smaller in the muscles distal to the injury site, i.e., tibialis anterior and gastrocnemius muscles. Note the difference in calibration for displaying the amplitude. The motor patterns in the muscles distal to the injury site C5/6 show obvious improvement after 6 months of CDT (Figure 27C in comparison to Figure 27A,B).

5.2.3.2. Motor Pattern Stability

At the beginning of therapy, the motor patterns varied significantly while turning on the special CDT device, as can be seen in Figure 27A and 27B. This variation persisted for as long as 6 months after starting therapy. In other words, the stability of the motor pattern for exercising on the special device was very low at the beginning of therapy.

This low pattern stability is also reflected in the high CDT values (high arrhythmicity of exercising = low pattern stability) at the beginning of therapy (Figure 50). Clinically this motor pattern variability correlated with the instability of movement performance, observed during walking and running on treadmill at the beginning of the therapy period.

5.2.3.3. Impairment of Reciprocal Relationship of Antagonist Muscles

The natural reciprocal relationship of agonist and antagonist muscles of the leg, namely the tibialis anterior (dorsal flexor) and gastrocnemius muscle (plantar flexor), was reasonably well preserved on the left side (Figure 27B,C) but was significantly impaired on the right side. It persisted for as many as 6 months in spite of therapy (Figure 27D). This impairment of antagonistic muscle action contributed to the patient's abnormal gait.

5.2.3.4. Measurements of Temporal Stability of Movement Patterns by Pattern Change

As explained above, the sEMG motor patterns were generated and recorded when the patient was exercising on the special CDT and recording device. The mean temporal stability of these motor patterns over one minute (= coordination dynamics value = mean arrhythmicity value) was used to quantify improvements of CNS functioning through therapy (see above, Figure 50). To measure differences in temporal stability of different movement patterns, such coordinated patterns, involving arm and leg movements, need to be generated with accuracy. This is achieved by the high precision mechanics of the special device. The arms turn at a slightly higher frequency than the legs. The relationship of revolution between arm leavers and legs paddles is 19:18. This difference in revolution produces the changing of coordination patterns of arm and leg movements. Such pattern change must also be seen in the sEMG motor patterns.

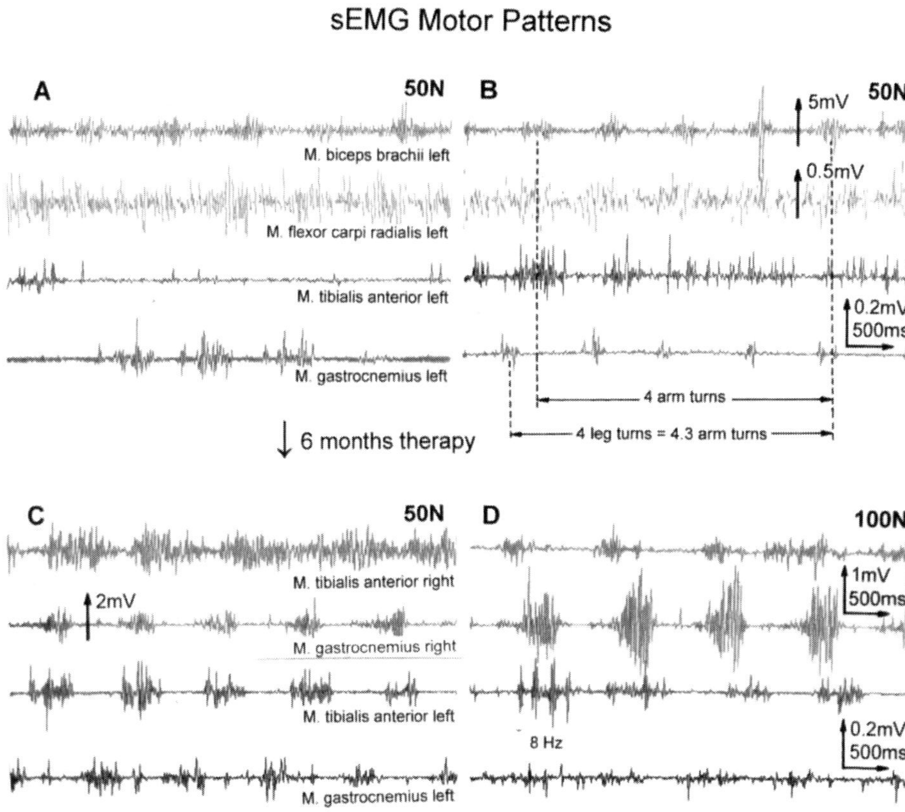

Figure 27. Motor patterns of arm and leg muscles recorded by sEMG for the patient with a 50% spinal cord injury at C5/6 levels at the beginning of the coordination dynamics therapy (A,B) and after 6 months of therapy (C,D). Note that the motor patterns of the left tibialis anterior and gastrocnemius muscles improved during these 6 months of therapy (from A,B to C); the patterns varied quite much from recording (A) to recording (B). The sEMG activity has a large amplitude for the biceps brachii muscle and a low amplitude for the tibialis anterior and gastrocnemius muscles (A,B) (note the different calibrations). The patterns were generated when the patient was exercising on the special coordination dynamics therapy device for turning. To measure stability differences of patterns (system theory of pattern formation), the arms turn at a higher frequency than the legs. The revolution relation between arm leavers and leg pedals is 19:18; this relation generates the changing coordination between arms and legs. The different turning frequencies of arms and legs can be seen in the sEMG patterns (B; 4 leg turns = 4.3 arms turns). Upon increasing the load from 50 (C) to 100N (D) during exercising on the special device for turning, the motor patterns of the left tibialis anterior and gastrocnemius muscles deteriorated (D): specific patterns were lost, antagonicity was lost, pathologic synchronization of premotor spinal α_1-oscillators at 8Hz occurred; this deterioration of motor patterns stability is in accordance with the increase of the coordination dynamics values (= increase of arrhythmicity of exercising) at higher loads.

The way that the muscle activations changed between the left biceps brachii muscle (arm) and left gastrocnemius muscle (leg) can be seen in Figure 27B. During 4 gastrocnemius activations (4 turns), the biceps were activated approximately 4.3 times (4.3 turns). This seemingly simple change of arm and leg movements between pace and trot gait is an extremely difficult pattern for the CNS to generate. All the muscle patterns of arms and legs have to be changed. And if the device is positioned further away from the trunk, then the trunk also has to perform rotating movements in coordination. Since this trunk rotation is changing with the coordination position between arms and legs, all the segmental trunk muscles change their coordination in the rostral and caudal direction. If the CNS is functioning physiologically, then the subject can turn relatively easily and smoothly. But if the CNS is injured, then such ongoing change of coordination's among arm, leg, and trunk muscles becomes a very difficult task for the CNS. Patients often sweat due to the stress which the CNS undergoes in generating these complicated movements as the pattern keeps changing.

As can be seen from Figure 27B, during the time period of the recording, the patient's CNS could manage the change of the coordination between pace and trot gait quite well. Chaotic activation was not observed on either the biceps brachii or gastrocnemius muscle traces during the movement pattern change.

5.2.3.5. Impairment of Pattern Formation Is Revealed with More Integrated CNS Activation at Higher Loads

After increasing the load on the special device from 50 (Figure 27C) to 100N (Figure 27D), the motor patterns of the left tibialis anterior and gastrocnemius muscles partly deteriorated (Figure 27D). The specific patterns and their antagonistic relationship were partly lost. Pathologic synchronization of premotor spinal α_1-oscillators occurred at 8 Hz, as can be seen from the trace of the left tibialis anterior muscle (Figure 27D). The increasing impairment of CNS self-organization with increasing load of exercising, made visible by the sEMG recordings, is in accordance with the increase of the arrhythmicity of exercising on the special CDT device for increasing load (Figure 28, increase of coordination dynamics values at higher loads).

5.2.3.6. Increase in Temporal Instability of Movement Patterns While Exercising Against High Loads

From the above discussion, the increase of the coordination dynamics values, i.e., deterioration of coordination dynamics, when exercising against high loads is understandable (Figure 28). It is because of the reduction in the temporal stability of movement pattern, as shown in sEMG recordings. Kinesiologically, this increased temporal instability of movement pattern formation is apparent when coordination dynamics traces are plotted for one minute against increasing load (Figure 28).

As can be seen from Figure 28, the arrhythmicity of exercising increased with increasing load. Exercising at 20 (A) and 50N (B), in the recording, the arrhythmicity varied without any specific structure. At 100 (C) and 150N (D), however, structure becomes visible in the coordination dynamics traces. Note in Figure 28, that the coordination dynamics values increased with increasing load. When exercising against a load of 200N in the forward direction (E), it can be seen that there is rhythmic increase and decrease of the arrhythmicity of exercising. At times, when the arrhythmicity is small, the arm and leg movement patterns show a high degree of stability. Different patterns of coordinated arm and leg movements have different degrees of stability. Those patterns which are more stable have small amplitude of arrhythmicity; they are the attractor states among different movement patterns. The unstable movement patterns (with significant arrhythmicity in exercising) do not constitute attractor states. Sudden instabilities of movement patterns are perceived by the patient and healthy subjects as sudden increases in the resistance to smooth turning. These are sudden perturbations of CNS organization and are felt as a stutter, which interrupts smooth turning.

5.2.3.7. Exercising at High Loads Reveals Impairment in the Symmetries of CNS Organization

While the patient was exercising against a high load of 200N, rhythmic changes of the arrhythmicity of exercising were observed during forward (Figure 28E) as well as backward exercising (Figure 28F).

Coordination Dynamics in dependence on the load

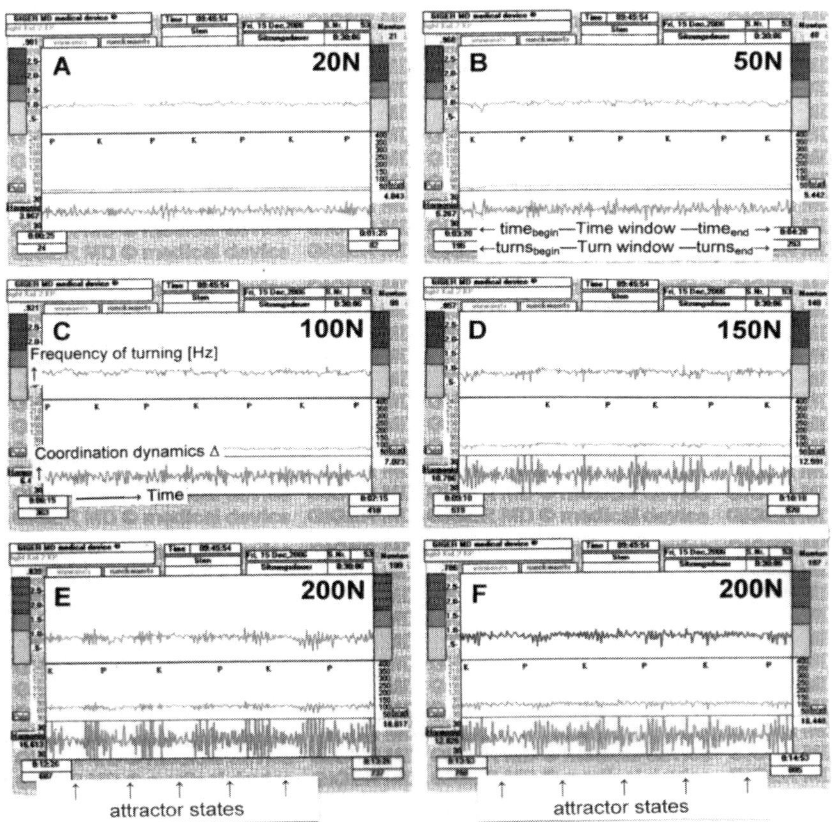

Figure 28. Coordination dynamics measurements in a patient with a 50% spinal cord injury upon exercising on the special coordination dynamics therapy and recording device for forward turning at increasing load from 20 to 200N (A-E). The recording sweeps are 1min long. F. Exercising in the backward direction at 200N. Upper trace = frequency; lower trace = coordination dynamics = arrhythmicity of exercising. P = pace gait coordination, K = trot gait coordination. "P to K" and "K to P" = coordination changes from pace to trot gait and from trot gait to pace gait. Note that for 20N (A) and 50N (B) the coordination dynamics increase is more noise like whereas from 100N (C) to 200N (E) there are rhythmic increases and decreases of the coordination dynamics with the changes of the coordination. Note further that the lowest arrhythmicity of exercising (highest pattern stability = attractor state) lies for forward exercising at 200N (E) to the right side of the pace (P) and trot gait coordination's (K) and for backward exercising (F) to the left side of the pace and trot gait coordination's.

Interestingly, the location of attractor states (patterns of high stability) were seen to be situated at different locations from the calibrated states of 'pace' (P) and 'trot gait' (K), as shown in Figure 28E,F. For forward turning, the attractor is located on right side, while for backward turning, the attractor is located on left side of the pace and trot gait patterns. This impairment of symmetry for forward-backward movements (different attractor states) only became visible with the very integrated activation of the CNS required when exercising against high loads.

5.2.3.8. Improvement of Symmetries of CNS Organization Increased Pattern Stability

As the impairment of symmetry may be diagnosed with the highly integrated activation of the CNS, training of those movements at high loads is required to improve symmetry. As the pattern stability with respect to symmetry increases, movement performance improves (Figure 18). In the case of an asymmetrical injury and attendant asymmetric CNS organization, Figure 18B,C), the movement pattern 'jumping-in-anti-phase' can be so unstable that the patient is unable to perform the movement. In terms of system theory, the potential well for 'jumping-in-anti-phase' in the attractor layout is very shallow or does not exist anymore (Figure 18C). In a patient with incomplete spinal cord injury C5/6, initially, only in-phase jumping was possible without support. The anti-phase jumping needed assistance and effort from the patient. With training, the patient re-learned to jump in anti-phase; the attractor basin for 'jumping-in-anti-phase' became deeper and the phase and frequency variability became smaller so that the movement pattern became stable and thus the movement could be performed independently. At the end of therapy, the patient was even able to jump unassisted with a skipping rope.

5.3. Conclusion

Within the System Theory of Pattern formation, CNS functioning was quantified by pattern stability during pattern change while exercising on the special CDT device. Pattern stability was given by the deepness of the potential well and the variability of phase and frequency coordination among the neurons of the neuronal networks as demonstrated in Figure 18 by the moving (indicated by arrows) of a ball (the pattern state) in a potential well. The change of the coordinated movement patterns is generated, when the subject is exercising on the special coordination dynamics recording device, where the coordination between arms and legs, imposed by the device, changes continuously between pace and trot gait. The change of the coordination's can be seen when the therapist is looking at the position of arms and legs. The change of the arm and leg coordination's can also be made visible by the change of motor patterns (sEMG) as shown in Figure 27B.

The temporal stability of the intrinsic coordination tendencies is measured by the deviations in differential stability during the performance of these rhythmic movements. The plotting of the differential stability over time of the frequency of exercising generates an attractor layout for this special movement (Figure 20) and the mean stability per minute can be measured by the arrhythmicity of exercising (df/dt, f = frequency). This value, the so-called coordination dynamics value (Δ = df/dt:f), quantifies CNS functioning objectively, integratively and non-invasively.

But as shown in Figure 26, the generated motor pattern and therefore the CNS organization is conditioned to some extent by the load under which the patient exercises. For low-load exercising at 20N (26A), only few motor units were activated (tibialis anterior right trace). With increasing load, more and more motor units were activated until they fused (26A to D). Also, the coordination dynamics values depend strongly on the load (Figure 28) as one might expect. These variations will be discussed further later on and used for assessing rates of repair (Figure 50). With increasing load the coordination dynamics values Δ are higher (worse) and the integrativity of CNS organization increases. As will be shown below, the repair of the integrated functions, i.e., the ability coordinate various parts of the body together, need more time (Figure 50). But the repair of the integrated functions is the most meaningful real repair which includes substantial learning transfer, i.e., functions such as continence which cannot be specifically trained can be repaired by learning transfer. The CNS of the patient is routinely assessed by Coordination Dynamics values for low load (20N) and at higher loads. The sum of the coordination dynamics values is then calculated. This is the so called 'high-load' test. Impairment of pattern stability and pattern performance, which can be seen in sEMG recordings, correlate with the impairment of the coordination dynamics (increased coordination dynamics values). The quality and stability of motor pattern organization correlates, on the other hand, with the quality and stability of neuronal network organization, measured with the single-nerve fiber action potential recording method. CNS functioning can therefore be effectively assessed by measuring the coordination dynamics.

Movement-based learning, accomplished by CDT, is successful in repairing the CNS through functional reorganization. However, in severe brain and spinal cord injuries (SCI), structural repair is also required, including the building of new neurons from stem cells. It was shown that new motoneurons could be built in SCI, but only to a very limited extent [11]. Cell physiology and genetics, especially epigenetics, have to be correlated to movement-based learning to further enhance the functional and structural repair of the human CNS.

6. Cell-Physiology and Epigenetics: An Attempt to Use Animal Data on Hippocampus Learning for Repair in Humans

6.1. Limited Neurogenesis in Human from Endogenous Stem Cells

Through CDT, a few new nerve cells (motoneurons) can be built from endogenous stem/progenitor cells in a patient with a cervical spinal cord injury. This resulted in the recovery of some finger functions [11]. Exogenous stem cell therapy, by contrast, is unlikely to be effective. This is because the administered cells have to proliferate and have to be integrated into the injured adult neural networks. But this proliferation and integration does not seem to be possible in humans, because physiological neural activity is required to stimulate the membranes of the administered stem/progenitor cells for proliferation and integration to occur (Chapter VI of [14]). Secondly, the communication distance between nerve cells during development and repair seems to be in the range of 0.001mm, which is difficult to achieve during exogenous stem cell therapy (Chapter 1 of [14]). Administered stem/progenitor cells which are not adequately stimulated for proliferation and integration will die.

6.2. Excitation-Neurogenesis Coupling in Learning: Comparison between Animal and Human

1. Adult neural stem/progenitor cells (NPCs) can intrinsically sense excitatory neural activity and thereby implement a direct coupling between excitation and neurogenesis [24]. This finding in animals is supported by human research; new motoneurons could be built in the human spinal cord, activating finger function [11]. The dorsal flexion of the finger (extension) was generated by dorsal flexion neural networks at or in the SCI's (C5/6) spared neural networks for dorsal flexion of the hand (Figure 29). The excitatory neural activity of the remaining changed network seems to have sensed the repair of the network in which neurogenesis was also involved.

2. The activation of excitation-neurogenesis coupling is sensed in animals via Ca_v 1.2/1.3 (L-type) Ca^{2+} channels and NMDA receptors on the proliferating precursors. The Ca^{2+} channels are opened via excitatory stimuli. Therefore, NPCs themselves can act as the signal detection and processing elements, mediating adult excitation-neurogenesis coupling [24]. The impact of excitation could also be indirectly felt on the surrounding mature hippocampal cells. There is nothing in human research, treatment (Chapter II of [14]) or human neurophysiology (Chapter III of [14]) which sheds light on how neuron production is achieved. If there are no NPCs in the human spinal cord, the NPCs would have to be stimulated in other locations such as the sub-ventricular zone of the lateral ventricle and transported via the liquor to the spinal cord. To induce migration of NPCs and direct them to the site where they are needed, integrative and specific movements would have to be trained.

3. The excitation through Ca^{2+} channels and NMDA receptors act to inhibit expression of the glial fate genes Hes1 and Id2 and increase expression of NeuroD, a positive regulator of neuronal differentiation. The response in phenotype-determining genes occurs soon after the initiation of the excitatory stimulation [24]. There is no information from treatment in humans and human neurophysiology on this matter (Figure 29). The index finger started to work after three years of treatment. The growth of the motoneuron axon from the spinal cord to the muscle activating the index finger requires at least two years. The building of new motoneurons (and interneurons) may need six months to a year in humans.

4. Many newborn neurons die within the first 2 weeks in vivo [24]. If, in humans, CDT is stopped for more than 3 days, progress will halt and may even go into reverse. Recovered functions start to decline. An incomplete SCI patient (C5/6) trained six months and got a bit better. Then he stopped treatment for six months and started therapy again. It seemed that he had lost all recovered functions in the interim.

5. How does neurogenesis allow the network to adapt to different levels of memory storage demands, which in turn are correlated with the activity level in the network? In this view, activity-dependent neurogenesis may affect future memories or the degradation and clearance of previously stored memories [24]. If memory is seen in the form of patterns of neural activity, activity-dependent neurogenesis supports the re-learning of lost pattern

functions and supports clearance of post-injury developed pathologic patterns, like spasticity. Therefore, neurogenesis helps the degradation of pathologic activity patterns. The degradation of pathologic activity patterns includes those of higher mental functions. Following severe brain injury a 25-year-old male patient became aggressive. After 3 months of CDT the aggressive behavior disappeared completely. He even began to display signs of happiness. Because of the short therapy time, the improvement in mood was probably mainly due to a functional repair by CNS reorganization, not substantial neurogenesis. In a 25-year-old patient with cerebral palsy, aggressiveness declined after 6 months of CDT. In this case, neurogenesis may have contributed to the reduction of the aggressiveness. When CDT is applied, movement, vegetative and higher mental functions always move in the direction of physiologic functioning and back to the patient's pre-existing character. Epigenetics seem to have a supervisory role over the repair in such a way as to ensure that no pathologic functioning occurs.

6. (a) Neurogenesis elicited a more rapid loss or clearance of previously stored old memory and (b) the newest memories were recalled at a higher fidelity (conclusions from a model [24]). In patients, new neurons with new dendrite and axonal trees can contribute to the reduction of spasticity which is likely to be mainly due to the network impairment and the loss of neural network complexity due to the injury. This gives an explanation as to why, with CDT, spasticity reduces in the long-term memory. However, it is possible that physiologic patterns may also lose stability.

Physiologic, stereotyped patterns like the trot gait coordination and walking also have to be trained in addition to complicated coordination's in order to enhance the complexity and accuracy of neural networks. After repeated performance of the high load tests on the special CDT device, a healthy pupil in the developmental study (Chapter II of [14]) felt that his trot gait coordination (for walking) transiently weakened. But a pain escape walking for years due to a damaged hip will change more the walking pattern. With new neurons, and in turn with higher network complexity, re-learned patterns can be performed more accurately and the variability of the pattern is higher. In poor network complexity, the rate of movement-based learning is smaller (see below). New neurons will improve the accuracy and variability of re-learned patterns.

7. The newest memories were recalled at a higher fidelity, despite the fact that network size was not allowed to increase in the network model. This can be intuitively understood from the lack of a favorable environment for new neurons in old memories. The synapses of the new neurons can be devoted more fully to the newer memories (patterns), which are then more accurately stored.

This advantage that the new neurons with their new synapses enhance the accuracy of the stored memories (patterns) was dramatically greater in networks that had been more active and had been required to store many memories (patterns) [24].

This finding from a neural network model offers one understanding of why the patient has to train at his limits (more network activity) and has to train different movements as the networks are required to store more memories in the form of patterns. Therefore, training different movements at the limit will increase the quality of the re-learned movement patterns and will in turn give rise to learning transfer, so that other functions (patterns), which cannot be trained such as continence and speech, are repaired as well. New neurons open more possibilities for learning and re-learning and most likely serve to increase the rate of repair.

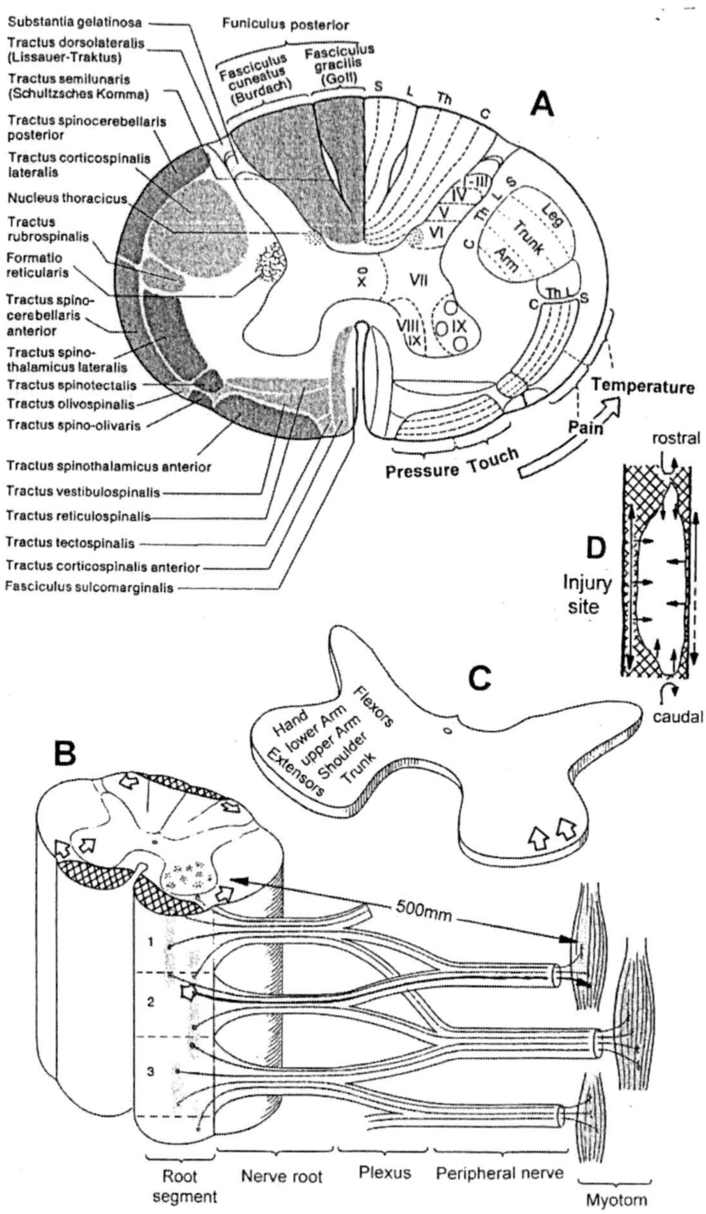

Figure 29. Schematic drawings of the human spinal cord with a cervical injury. A. Spinal cord cross-section with fiber tracts and grey matter. Note the regaining of sensitivity upon CDT from pressure to touch, pain, and temperature is indicated by an open arrow. B. Spinal cord section with suggested spared matter (cross-hatched). Open arrows indicate direction of structural repair. The axonal outgrow of a newly build motoneuron is indicated by a bended and long arrow. C. Motoneuron sites for serving different functions are indicated. Note the extensor motoneuron cell bodies are sited more adjacent to the spared spinal cord matter then the flexors and should be generated first in this case, indicated by open arrows. D. Injury site of the patient. Small short arrows indicate direction of structural activity-dependent repair. The structural repair starts from the activated spared matter into the cavity. The ascending and descending tract activity is indicated by long arrows, and the tract and network activity rostral and caudal to the injury site is indicated by bended arrows.

8. It is of interest to determine the factors that create the conditions for excitation-neurogenesis coupling to occur. Cells with neurogenic potential exist elsewhere in the brain, including the neocortex, but contribute to neurogenesis only in the sub-ventricular zone and the dentate gyrus sub-granular zone. However, when cells from non-neurogenic areas are removed and transplanted into neurogenic areas, neuronal progeny results [24]. There are a number of factors that may contribute to the neurogenic potential to induce competence for excitation-neurogenesis coupling in the resident population of NPCs:

a) Type of (natural) local neuronal activity.
b) Access of the NPCs to the local activity.
c) Ability of the local environment to induce activity-sensing competence in the NPCs.
d) Factors intrinsic to the resident NPCs themselves.

Clinical research offers no direct answer. The naturally activated local neural networks surrounding the injury site and the cells with neurogenic potential will be important for the growing of axonal and dendrite trees and neurogenesis. It is conceivable that progenitor cells are able to identify natural activity related to repair which would trigger neurogenesis and proliferation. The frog experiments of Chapter I of *Human Neurophysiology* [14] on development and repair show how complicated and differentiated the establishment of a seemingly simple innervation patterns can be.

9. Connection of NPCs to the activated networks. The activation of Ca_v 1.2/1.3 channels on adult NPCs potently influenced neurogenesis. The channels open primarily in response to the depolarization provided by excitatory synaptic inputs. But proliferating NPCs in vivo are presumably not synaptically connected [24]. The mechanisms required for the activation of the NPCs are not clear.

There is no answer from movement-based learning therapy regarding how the NPCs are stimulated. It can be deduced from improvements in patients that the repair, including structural repair, is constantly adapting to the needs of the specifically activated circuitry. During treatment, the CNS networks are activated integratively, but specifically those sub-networks are included in the activation which have to be repaired.

6.3. Critical Period Plasticity

Adult-born neurons exhibit the same critical classic period plasticity as neurons in the developing nervous system [23]. When patients have a break in their treatment when training at the limit, their functions deteriorate (see for example Figure 46b). This can be quantified by measuring the coordination dynamics. This experience in patients is in accordance with the transient nature of plasticity. In order not to miss the enhanced synaptic plasticity which occurs during the critical period in newly generated neurons of the adult brain, the treatment should not be stopped for longer than 3 days, especially when we do not know precisely when the critical periods will occur.

6.4. Neural Network and Pattern Stability

An aspect which is overlooked in animal research is that neural network patterns need to be stable. Such network stability becomes impaired if the structure of networks is changed. Kelso stated that the stability property of network functions is 'foreign' in research [18]. Recently, it was stated that 'the transient nature of such enhanced plasticity may provide a fundamental mechanism allowing adult-born neurons within the critical period to serve as major mediators of experience-induced plasticity while maintaining stability of the mature circuitry' [23].

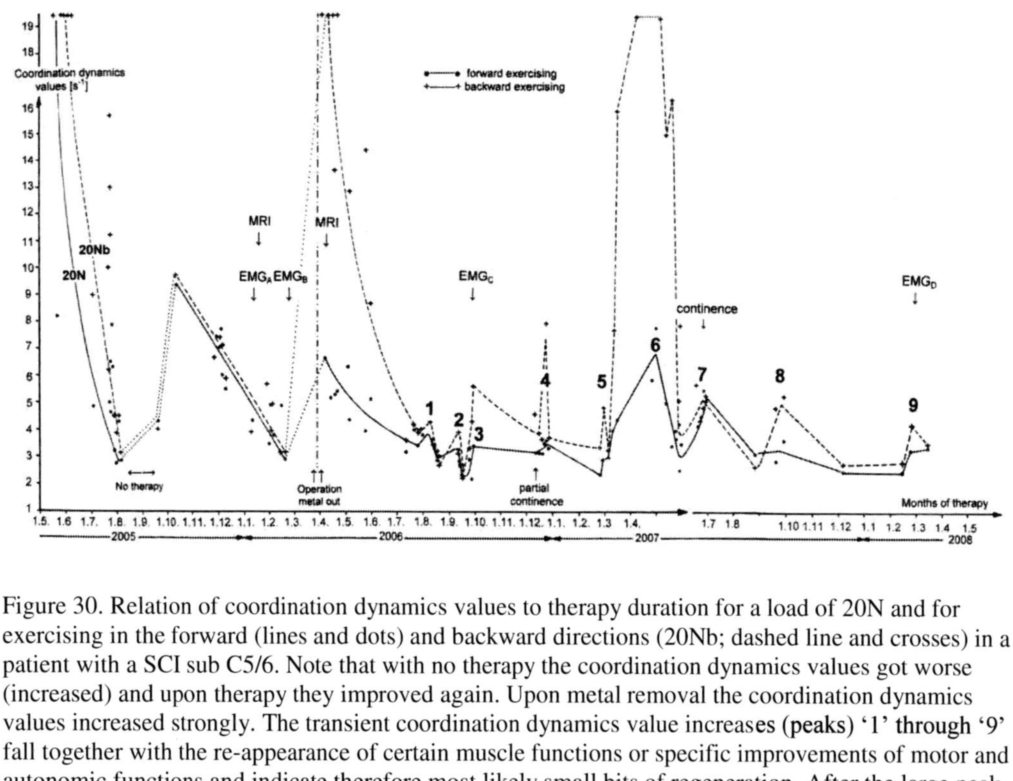

Figure 30. Relation of coordination dynamics values to therapy duration for a load of 20N and for exercising in the forward (lines and dots) and backward directions (20Nb; dashed line and crosses) in a patient with a SCI sub C5/6. Note that with no therapy the coordination dynamics values got worse (increased) and upon therapy they improved again. Upon metal removal the coordination dynamics values increased strongly. The transient coordination dynamics value increases (peaks) '1' through '9' fall together with the re-appearance of certain muscle functions or specific improvements of motor and autonomic functions and indicate therefore most likely small bits of regeneration. After the large peak '6' of transient coordination dynamics value increase urinary bladder functioning was re-learned.

In humans, with every structural change, the stability of CNS self-organization is impaired. This must then be re-established by movement-based learning (Figure 30). Chapter III of [14] showed that with CNS damage, neuronal network deterioration can be best observed in the firing patterns of the premotor spinal oscillators. During repair, instabilities can also occur. With every regained function, instabilities occurred, which were quantified by coordination dynamics values and repaired with the ongoing movement-based learning therapy (Figure 30). The incidence of instabilities with network changes is entirely logical. If one changes an electronic circuit by adding a new connection, the circuitry function will change and instabilities may also occur, by starting to oscillate, for example.

6.5. Repair Connected to Blood Vessels

Proliferating cells and putative progenitors in both sub-granular (SGZ) and sub-ventricular (SVG) are closely associated with the vasculature, indicating that factors released from the blood vessels may have a direct impact on adult progenitors [19,27]. "Infusion of vascular endothelial growth factor (VEGF) promotes cell proliferation in the SVG and SGZ, which can be blocked by a dominant-negative VEGF receptor 2 [20]. In addition, VEGF is required for increased neurogenesis in adult mice exposed to an enriched environment or given the opportunity of voluntary exercise, which are both known to enhance adult neurogenesis.

Following human SCI, autonomic innervation, especially of the skin blood vessels, is often impaired. Patients often suffer from fragile skin and easily get pressure ulcers, which heal slowly because of poor blood supply. Intensive CDT repairs this impaired blood supply within 6 to 9 months so that the skin is no longer vulnerable and the occurrence of pressure ulcer reduces. That means movement-based learning is repairing the blood supply and the microcirculation and may have, therefore, an impact on adult progenitors in human.

6.6. Repair Influences from Distant Excited Networks

Neural progenitors can be indirectly influenced by neurons outside the microenvironment that are connected to neurons within the neurogenic niche through neural circuits. Both local and distal neurons can exert direct influences on neural progenitors through the ambient levels of neurotransmitters in the neurogenic niche, or even through synaptic contact with neural progenitors. Therefore, adult neurogenesis is subject to complex extrinsic regulation [32].

CDT tries to activate the different kinds of neural sub-networks in close proximity to the injury site to start the repair from the microenvironment adjacent to the injury site. But by the training of integrated movements in which also the injured circuits are included in the activation, neurons (networks) outside the injury microenvironment are stimulated to influence the injured networks for repair through neural or neural network connections.

6.7. It Is Learning That Achieves Repair

It is learning and not simply training that elicits the survival effects of new neurons in the hippocampus. Learning appears to promote the survival of newborn neurons in cognitively unimpaired aged rats [25]. Learning elicits different influences on neural precursors at different developmental stages. The regulation of SGZ neurogenesis by hippocampus-dependent learning is complicated and can be affected by factors such as the age of the newborn neurons, the stage of learning and specific learning protocols [32]. When the human patient is exercising on the special CDT device, he should not just turn, but should try to turn more smoothly. He should try to reduce the arrhythmicity of exercising. The patient should try to improve the performance of the movement by learning. During learning, it is essential to concentrate and to be aware of what needs to be corrected. It is also important to evaluate the performance by looking at the coordination dynamics curves on the computer screen and correction of errors in successive attempts. When the patient is well-practiced at exercising smoothly, the skill can be accomplished without conscious effort, much like in walking, swimming, cycling or skiing. Simply exercising will also improve CNS functioning. However, the rate of learning is significantly slower. This is the learning of automatic movements, in which the process is subconscious. A problem in some patients, for example in severe cerebral palsy, is that the cognitive functions are so impaired that they cannot understand that they have to learn to improve their nervous system functioning. The hope in such cases is that the simple training improves their cognitive functions to a point that they can understand that they have to improve the performance of the trained movements to improve their CNS functioning.

6.8. Microenvironment (Neurogenic Niche) Permissive for the Differentiation and Integration of New Neurons

In Chapter I of [14] it was shown that the development and the repair were very similar when the two kinds of motoneurons innervated two kinds of target cells (muscle fibers). In adult frogs a slow muscle fiber can be innervated by two motoneurons. By cutting one axon, the membrane changes its properties in the denervated part. It can now generate action potentials which cannot be generated in the still innervated part of the muscle fiber. By partial denervation [28], therefore, we obtained innervation and denervation membrane properties in the same cell; but with more than one cell nucleus. In the developing frog, the muscle fiber has approximately a length of 1mm. The distance of action of the neurotrophin is shorter than 0.5mm. In order not to influence neighboring muscle fibers (Ø ≈ 15μm), the distance of action of the neurotrophin secreted from the motoneuron must be shorter than 15μm for building up the specific physiologic innervation pattern. It is likely that the distance of action of the neurotrophin is in the range of 1μm, which is governed by the distance between the neuron and the muscle fiber (Chapter I of [14]).

One may assume that a similar detailed neurogenic landscape is present in the neural networks of the human CNS. It is difficult to see how such microenvironment can be generated during the administration of neural stem/progenitor cells for treating patients with SCI or Parkinson's disease. Before treating human patients the microenvironment should be clarified in humans and grounded in a scientific basis. To induce competence for excitation-neurogenesis and excitation-repair coupling the following should be explored in humans:

a) What kind of local neural activity is needed?
b) How can the access of the NPCs to the local activity be achieved?
c) What is the ability of the local environment to induce activity-sensing competence in the NPCs?
d) What is the intrinsic state of the transplanted NPCs after injection?
e) Proliferating cells and putative neural progenitors in both subgranular zone of the dentate gyrus and subventricular zone of the lateral ventricles are closely associated with the vasculature, indicating that factors released from the blood vessels may have a direct impact on adult neural progenitors [19]. How can blood vessel supply be achieved for the administrated stem/progenitor cells?

f) Astroglia is supposed to induce neurogenesis from adult neural stem cells [34].

During stem cell therapy in spinal cord injury, the astrocytes will not contribute to the stimulation of neurogenesis, because outside the spinal cord matter there is no astroglia. How can the potential contribution from the astrocytes be simulated in a physiological context?

6.9. The Necessity of Adequate Activation of Networks to the Repair of the Human CNS

New neurons can be built in the animal and human CNS (Chapter II of [14]). Electrical activity has been shown to regulate development in a variety of species and in various structures [37], including the spinal cord and cortex. Within the mammalian cortex specifically, the development of dendrites and commissural axons in pyramidal cells is activity-dependent [38]. Excitatory stimuli act directly on adult hippocampal neural stem/progenitor cells to favor neuron production [24].

In Chapter I of [14] it was shown that there was a close similarity between development and repair in the frog peripheral nervous system. The case reports and group studies of Chapter II of [14] indicate, on the other hand, that there is only some similarity between development and repair in the human CNS.

Assuming that data concerning the animal CNS development, learning and memory formation in the animal hippocampus partly hold in human development and repair, it may be understood how the natural impulse patterns can change the neural networks for repair through movement-based learning.

If it is only the overloaded networks, which induce CNS repair, then those networks which must be repaired and those which will take function over must be sufficiently stimulated. The load in neural networks is increased by going to the limits of exercising and by improving the coordinated firing of neurons through improving the coordination of arm and leg movements, since neurons work as coincidence and more generally as coordination detectors (Figure 11). Further, by training complicated coordinated movements, neural network patterns can be reached, which lie deeply in the complexity of CNS organization. For functional repair especially in spinal cord injury the function of tracts has to be changed. By training different movements, different kinds of tract will become overloaded when going to the limits of these coordinated movements. By training impaired functions (with support), the corresponding damaged circuits and the functionally connected networks will become stressed and receive the stimulus for repair. As the building of new motoneurons in a human patient indicates, stressed networks in particular seem to also be the stimulus for neurogenesis. The building of new neurons is a powerful repair strategy, especially when certain kinds of neurons are missing. But as the human research indicates, the neurogenesis is very limited in the human CNS and the benefit from such repair strategy needs more than a year of intensive training to contribute to functional recovery.

6.10. Selective Requirement for Natural Activity in Specific Neurogenesis and in Shaping the Integration of Specific Neurons into Damaged Adult Neural Networks for Repair

It was discovered that 30% of all cortical interneurons arise from a relatively novel source within the ventral telencephalon, the caudal ganglionic eminence [36]. Owing to their late birth, these interneurons populate the cortex only after the majority of other interneurons and pyramidal cells are already in place and have started to functionally integrate. In mice it was shown that activity is essential within three days of birth for correct migration, and that after this period, glutamate-mediated activity controls the development of their axons and dendrites [35].

The treatment of patients (Chapter II of [14]) and mice data [35] indicate that during development and repair, selective activity is necessary for activity-dependent neuron migration and neuron axonal arborization and neuron dendrite trees besides weight changes of synapses. Further, functional and structural repair can lead to the repair of physiologic CNS functioning, but it can also lead to pathologic functioning like epilepsy and cancer. It seems therefore that for a repair or building up of physiologic CNS patterns, natural physiologic activity patterns are needed, which are generated by movements, vegetative or physiologic cognitive function patterns. Some natural impulse patterns in afferents, efferents and sub-neural networks (network oscillators) in humans were identified in Chapters III and V of [14]. To reach and train natural organizational patterns in the deepness of phase and frequency coordination, extremely coordinated and complex arm, leg and trunk movements have to be trained like the movement patterns as performed in Figures 31-33. For selective activation and repair of the human CNS, it is necessary first to understand how the human CNS functions and how its patterns are changed by the natural impulse patterns of receptors, informing the CNS about changes in the outside world (Figures 5, 6, 152, 153).

7. Epigenetic Modification for Repair by Movement-Based Learning

7.1. Epigenetic: Adaptation and Repair by Learning

If genomes were the sole determinant of evolutionary development, humans would not have developed any further than mice, and plants would have as much chance of ruling the world as man. Mice and Arabidopsis genomes share the same number of genes, approximately 25,000. In fact, it is the epigenome that has evolved tremendously since the appearance of the first multicellular organisms [21]. The term epigenetic describes the interplay between genes and environment resulting in phenotypes, epigenetic landscape, and adaptation.

Simple organisms, such as bacteria, increase their rate of spontaneous mutations to enable the survival of species in a changing environment. Multicellular organisms use complex mechanisms coordinated by the central nervous system (CNS) to behaviorally adapt to changing environments without having to undergo the difficult process of mutating their genome [22]. Their behavioral adaptation depends on learning, sub-served by epigenetic mechanisms.

Uniquely, humans have the capacity to partially repair their severely injured nervous system by learning. It was shown that in a patient with cerebral palsy, speech could be achieved by learning transfer (Chapter II of [14]). When the patient learned to exercise by herself on the special coordination dynamics therapy (CDT) device, the patient's CNS partly learned from the device's the complex coordination's between pace and trot gait. Her CNS repaired to an extent sufficient to generate the speech pattern, even though it was predicted by different speech therapists that she would never learn to speak. Cognitive function also improved by learning transfer. Continence, which is impaired in very severe brain injury, was also fully repaired in the patient.

Animals have a high capacity for spontaneous repair of their nervous system by the growing of nerve fibers (Chapters I, IV of [14]). Humans, on the other hand, have a high capacity for repair by learning, including the re-learning of the fundamental principles of CNS organization mechanisms such as 'phase and frequency coordination' among nerve cell firings. A limiting factor for movement-based learning in human is structural repair. Although through measuring we know that there is very limited neurogenesis in the human CNS (spinal cord), it is possible to build other neurons (interneurons) through exercising to the limit of endurance. This was indicated by the regeneration of the spinal cord from rostral to caudal following injury; surface electromyography showed with the appearance of first small motor unit potentials that the partial spinal cord repair was probably achieved by neural network repair and not by the regeneration of long tracts such as the pyramidal tract.

To enhance the rate of repair of the human CNS by learning, ways must be found to enhance structural repair, especially neurogenesis. There seem to be two immediate possibilities – the adding of stem/progenitor cells (NPCs) or influencing the epigenetic by learning. In the hippocampus it was shown that neuronal network activity itself increased neurogenesis [24]. If we assume that cells with neurogenic potential exist elsewhere in the brain, albeit to a lesser extent, then the activity itself guides neurogenesis for repair in accordance with the principle that local cellular environments are important in controlling neurogenesis.

7.2. A Gene Expression Pattern Triggered by Excitation in Proliferating Adult Npcs

For weight changes of synapses, growing of axons and dendrites, neurogenesis and building of glia cells during the process of activity induced structural repair, proteins are needed, which have to be generated by appropriate transcription from genes. It seems that natural patterned activity-dependent depolarization of the plasma membrane trigger activity-dependent gene expression programs [24]. Ca^{2+} signaling through $Ca_V1.2/1.3$ channels and NMDA receptors can activate a broad array of rapidly responsive transcription factors. Moreover, excitation could in principle induce release of autocrine factors from the NPCs themselves, leading to recruitment of a host of additional signaling pathways to nuclear transcription factors [24].

7.3. Regulation of Epigenetic Modification for Repair by Movement-Based Learning

To generate repair in the nervous system, it is likely that permanent changes in gene expression patterns are achieved through permanent changes in chromatin remodeling without changes in DNA sequence. The concept of chromatin remodeling addresses a key challenge of how stable changes in gene expression are induced [22] in neural networks to produce long-lasting changes in repair. DNA methylation is one of the many epigenetic modifications that can alter gene expression. Dynamic and reversible DNA methylation may be essential for learning and memory formation and could transmit repair influences onto adult neurogenesis.

Understanding the complex epigenetic regulation of neural activity and adult neurogenesis is integral to designing therapeutic approaches to restore neurogenesis and cognitive functions. It will also give a tremendous insight into understanding how certain environmental or pathological influences, such as stress, physical activity, depression and epilepsy regulate adult neurogenesis [21].

Research in movement-based learning, on the other hand, has to identify how epigenetic mechanisms can be efficiently modified by the performance of specific, corresponding movements or learning processes to improve development (correction en route, which is particularly salient to the treatment of cerebral palsy) and the repair of the human CNS and to avoid pathologic CNS changes like epilepsy and cancer.

7.4. Movement-Based Learning and the Critical Postnatal Period

The establishment of adult patterns of DNA methylation during the postnatal period suggests that critical periods exist after which levels of gene expression are relatively stable [22]. However, movement-based learning in adulthood is capable of targeting the epigenome and altering gene expression and hence repair; but the efficiency of repair may be much lower. CDT should therefore be started as early as possible in the event of CNS injury or malformation. Movement-based learning was administered to a 19-months-old child with a severe CNS injury. The child was nearly blind because of impairments in visual processing, he was afraid of moving and was clearly distressed by doing so. However, exercising on the special CDT device with the mother and physiotherapist was possible. With this administered movement the frequency of crying reduced, especially when being moved for longer than ten minutes continuously. It was as if his CNS was benefiting from this passive exercise and he was being made to feel more comfortable through the improvement of CNS organization in the short-term memory. Even though it is difficult to treat such patients, it is very important to treat them efficiently as early as possible, because the critical period for improving or repairing vision may be missed and the patient may never have a reasonable prospect of repairing his vision.

Young infants can laugh and cry at the same time CDT. They laugh because they like the improvement of CNS organization and they cry because they are getting exhausted. In order to alleviate such exhaustion, interval training is needed.

7.5. Movement-Based Learning in the Prenatal Period

The early environment, consisting of both the prenatal and postnatal periods, has a profound effect on gene expression and adult patterns of behavior [22]. Movement-based learning upon exercising on the special CDT device is normally not possible in the prenatal period. But in premature born babies, the CNS is probably still in a state of prenatal development. A movement-based learning therapy should therefore in principle be beneficial for a repair of the premature CNS, injured for example by hypoxia. Figure 32 shows the special CDT device to administer movement-based learning to a premature born infant. Such a small device may even fit into an incubator.

7.6. Early Treatment in CNS Injury

If there is some similarity between development and repair and the early environment following injury has a profound effect on gene expression and patterns of behavior, then early treatment following CNS injury should be very beneficial for repair. In spinal cord injury, the treatment should be started in the spinal shock phase; the special device in the lying position is as safe as lying in bed if the trunk is not made to rotate. In stroke patients, the therapy should also start as early as possible, especially to improve blood supply to the brain. Putting a patient with severe brain injury into artificial sleep to reduce oxygen consumption is of doubtful therapeutic value.

The improvement of blood supply by early movement based learning therapy may be more successful than trying to reduce oxygen consumption by putting the patient into artificial sleep. Further, through this conservative approach, the therapist partly loses control over the patient. He cannot speak to the patient anymore and cannot monitor the patient's consciousness and brain function. I had a patient with a very severe brain injury. After a bicycle accident, the patient could speak and was able to tell the paramedics his name and address. He was then put into an artificial sleep. When the artificial sleep was terminated, the patient could not speak any more, indicating that the injury had worsened. Below it is reported that children could be taken out of comas sooner following severe brain injury when CDT was administered to them in the coma stage (severe traumatic brain injury; Benjamin, Mario and Andrej).

7.7. Can CDT Influence the Epigenome to Functionally Repair Genetic Defects?

There are two further matters of interest which concern movement-based learning. CDT has never caused long-term pathologic CNS changes and even in patients with Down's syndrome (trisomy 21), physiologic CNS functioning could be improved. It seems as if CDT is even partly able to repair genetic defects by influencing the epigenome of the patient. A possible mechanism of such a repair could be that the epigenome is modified by movement-based learning in the way that the gene with the defect is not recruited by the epigenome. It seems worthwhile to correlate movement-based learning with epigenome changes. Can the modification of the epigenome by movement-based learning block defect genes and recruit different, compensatory genes?

7.8. Dynamic and Reversible Modification of the Epigenetic Landscape in Comparison to Modification of the Landscape of Pattern Formation

The way of dynamic and reversible modification of the epigenetic landscape is probably essential for learning, memory formation and repair of the functional organization of the injured CNS. Movement-based learning is one of the ways of changing epigenetic landscapes. This is because during development movement-based learning is used for developmental "correction en route" and because movements like running, jumping and training balance will have influenced the human genome and epigenome in the last 5 million years, since these movements were essential for getting food and to escape from predators.

It seems likely that CDT has the capacity to change the epigenetic landscape for repair partly even in defects of the DNA sequence. However, not all defects in DNA can be repaired by changing the epigenetic landscape.

Below it is shown that speech could be induced in a patient with severe cerebral palsy by improving the phase and frequency coordination of neuron firing, through exercising on the special CDT device (Figure 112). But the Wernicke's and Broca's areas (speech centers) were not destroyed. These special neural networks for speech were probably only damaged functionally (Chapter II of [14]). If these special neural networks for speech were completely destroyed by injury or malformation, the repair of phase and frequency coordination would not have been sufficient for the induction of speech. An enormous number of neurons and their functional integration into the existing networks would have been necessary for repair. Such a repair is for the time being out-of-scope for this movement-based learning therapy.

If there is similarity between the modification of the landscape of pattern formation of movements (according to the system theory of pattern formation) and the modification of the epigenetic landscape, then changes in the epigenetic landscape can be repaired, but defects of DNA sequences can only functionally repaired by epigenetic modification (recruitment of genes) in some cases, especially when the epigenome can substitute other DNA sequences for repair. There seems to be some similarity between the organization of the genes, accomplished by the epigenome and organization of the CNS according to the system theory of pattern formation. In genetics it is still not entirely clear how the problem of how dynamic organization is accomplished. What controls the organization?

There is a similar problem in human neurophysiology. Although we know a lot about single neurons, little is known about how the tremendously complex CNS organization of the many billions of neurons is accomplished and how it is made to function physiologically.

7.9. Interaction of Neural Activity and Genetic Programs during Development and Repair

The traditional model of brain development broadly consists of two phases during which a rudimentary wiring of the nervous system is laid out, and a later phase during which the rudimentary connections are refined. In this model, the developmental events that underlie the coarse wiring are the result of predetermined genetic programs and occur independent of neural activity, whereas the refinement is a result of interactions between the nervous system and the outside world. In this traditional model, movement-based learning can be seen as a refinement of impaired development. However, the repair of urinary bladder functions and the induction of speech in a five year-old girl with a malformation of the cerebrum and thalamus damage following CDT through learning transfer (Figure 112) cannot just be seen as a refinement of this coarse wiring of the CNS. Therefore, this traditional model has to be re-assessed to accommodate the discovery that neural activity and genetic programs interact to form the composition and organization of neural networks during all stages of development [37] and repair.

Neural networks are quite resistant to perturbation, suggesting that redundancy is built into neural circuits to ensure that the natural activity in the networks is maintained. Pharmacological or genetic disruptions of crucial network components lead to the expression of alternative circuit mechanisms that generate activity similar to the natural endogenous pattern [37]. But in severe brain injury such redundancy is insufficient and movement-based learning by the training of coordinated movements is necessary to support the building and maintenance of natural activity in the circuits. Spasmolytic drugs alter the natural activity in the developing and in networks undergoing processes of repair and may not be helpful for improving development and repair.

7.10. Relative Contribution of Cell Intrinsic Versus Non-Intrinsic Fate Determinants

Even though many of the components of cell identity, such as general intrinsic physiological properties and cell position, are likely determined at the progenitor level, other components such as choice of synaptic partners are likely influenced by local environmental cues.

Therefore, these two mechanisms may contribute to different aspects of cell specification and integration. While earlier decisions, as it were, such as the decision to be a certain neuron are likely established at the progenitor level, later restrictions occur over time as they migrate, integrate into the neural networks, and establish the first contacts with other cells. Genetic profiling of cells at critical developmental stages will bring new insight into interneuron development and subtype specification [39]. A complex set of developmental steps are involved in the integration of interneurons into cortical network. These require the interplay of intrinsic genetic programs and their modulation by cell to cell and matrix to cell interactions.

In Chapter I of [14] it was shown that the target slow muscle fiber was twice chosen to be a synaptic partner. The target muscle cell was first innervated by fast growing thick axons of large motoneurons and then slow growing thin axons of small motoneurons (a second group of motoneurons) migrated to the synapses of the fast conducting axons, pushed their synapses away and established its own synapse of different structure and changed membrane properties and Ca^{2+} secretion properties from the sarcoplasmic reticulum of the target slow muscle fiber. There may be, therefore, several critical periods of innervation and cell intrinsic and extrinsic determinants are intermingled.

Abnormal development of cortical interneuron subtypes during late embryogenesis due to environmental perturbation coupled with genetic abnormalities might represent a primary cause for many neurodevelopment disorders [39]. But the networks can be changed by training certain movements. An autistic child can walk, run and jump, but cannot perform a certain movement for a longer period of time than a few seconds. The property of the CNS to stay in a certain organizational pattern for sustained period of time is absent. By forcing the child to stay in the movement pattern while exercising on the special CDT device with fixed feet and holding the hands in position during turning (so that the patient cannot take the hands from the handles) in some similarity to Figure 31 and giving instructions during coordinated movements, one can often see in the eyes and the face of the child that the therapy is succeeding in reaching deeply into the CNS and improving its organization in the short-term memory. One can observe a marked reduction in anger and frustration in the child. This is because the therapy renders the child more comfortable within its CNS through the improved organization. The child can learn to sustain a movement or cognitive task for longer periods. It is, however, very physically and mentally tiring for the therapist to ensure that the child exercises regularly.

In Figures 47b,c, running stability on a treadmill will be measured in a patient with severe traumatic brain injury. It will be shown that the stability of this movement pattern increased with CDT. The stability of the higher mental functions also improved.

Subtle imbalance in the ratio of excitatory versus inhibitory levels within the cerebral cortex may underlie many neurological and neurodevelopmental disorders such as epilepsy, autism spectrum disorders and schizophrenia [39]. CDT can improve CNS functioning including cases of neurological and neurodevelopmental disorders. It will be shown that a young woman improved her mild epilepsy by exercising on the special CDT device (Figure 44). In a young boy the convulsions reduced, following a temporary increase, after exercising on the special CDT device [40]. Before the discovery of anti-schizophrenic drugs, a working therapy was administered to patients, which only helped a little. CDT is much more effective in repair of schizophrenia than the working therapy of the past.

7.11. Activation of Tumor-Suppressor Gene by Exercise

If movement-based learning can suppress genes by modifying the epigenome, it may also be able to activate genes with anticancer effect.

In the long term, if CDT is performed intensively, the CNS can always be made to function more physiologically and efficiently. Unspecific pathophysiologic CNS functions like spasticity are reduced. In the framework of the system theory of pattern formation, the stability of physiologic functions increased (the potential well became deeper) and the stability of pathologic states like spasticity decreased (the potential well became less deep) (Figure 19). Since the nervous system is involved in nearly all the body's functions, it could well be that CDT enhances specific physiologic functions and inhibits non-specific pathologic functions. Cancer is an uncontrolled unspecific growing of cells. It may therefore be plausible that movement-based learning has an anticarcinogenic effect.

It was reported that exercise might exert an anti-cancer effect by arousing a tumor-suppressor gene [41]. Significant changes in the DNA methylation status of 43 genes occurred in women who engaged, after having completed treatment for breast cancer, in a six month moderate-intensity aerobic exercise program. A correlation between gene expression level and overall breast cancer survival was found in three of these women. One of these three was L3MBTL1, a putative tumor-suppressor gene. In the exercising patients, L3MBTL1 was demethylated, indicating an increase in gene expression. Elevated expression of this gene may be associated with a lower risk of recurrence and improved survival [41].

Having discussed anatomical, electrophysiological, kinesiological, cell physiologic and epigenetic details of the coordination dynamics therapy (CDT) and the assessment of CNS functioning by the coordination dynamic, we shall now proceed to discuss the rate of learning for repair following the application of CDT in healthy humans and in patients with CNS injury. The rate of learning will be measured and compared in order to better understand the repair of the human CNS. A correlation between neuronal network functioning and movements and other macroscopic patterns of human CNS organization is partly achieved. With respect to the healthy CNS, through CDT, we are coming closer to the dream that the spirit may rise above the body by influencing the neuronal networks (the hardware) through movement and higher mental functions (the software).

Chapter III

Development and Motor Learning of the Healthy CNS

1. Normal and Deviant Motor Development and Repair Following Movement-Based Learning

1.1. Outline

Neural development must be considered in any analysis of the repair of the human Central Nervous System (CNS). Firstly, there is some similarity between repair and development and the repair strategies should attempt to recapitulate development and learn from normal motor development. Secondly, because of CNS injury or adverse perinatal events, such as hypoxic ischemic encephalopathy, bronchopulmonary dysplasia, or bacterial meningitis we may find retarded, accelerated or deviant development of motor and other functions.

Some functions may not develop at all, while others may only show a decrease in variability. Combinations of these outcomes may occur within single infants and result in individual motor and other patterns. The deviant patterns are characterized by a lack of variability and assume a thoroughly stereotyped appearance. The learning processes in patients with CNS injury may be hampered by the deficiencies of the neuronal structures on which they are dependent.

One can expect that the rate of learning of injured CNS is smaller than that of a healthy one. The on-going neuronal network development of the CNS in infancy, childhood and adolescence is assessed by the Coordination Dynamics (CD) and the functioning of the CNS improved following Coordination Dynamics Therapy (CDT).

1.2. Longitudinal and Cross-Sectional Study

The first part of development is studied longitudinal in two sibling infants using EMG and motor performance assessment (CD). From 5 years of age on the CD data were collected from pupils from Year 1 until Year 12. Because of inter-individual variability of the cross-sectional study, certain aspects of sequence of development may not come out as pronounced as in a longitudinal study. A general longitudinal study would be better for analyzing development, but it is difficult to organize. Further, the measurement of the CD in the longitudinal study improves CNS functioning and changes the development itself (Figure 35).

For comparing developments in repair over a period of years, patients with cervical spinal cord injury, a cerebral palsy child, and three children who suffered a traumatic brain injury will be the principal subjects considered. It will be apparent that human development cannot be recapitulated by repair. Still there seems to be some CNS organization strategies (to achieve coordinated firing of neurons, change of motor functions) which are common to both development and repair.

1.3. Neural Development

CNS development is mainly measured in two periods. In the developmental period from six months up to seven years of age the healthy infant Jürgen has been followed up longitudinally. From 5 up to 18 years of age, CNS development will be assessed by the coordination dynamics of female and male pupils from one school (preschool plus classes 1 through 12). It turns out that the interpersonal variation of CNS development was so marked that in spite of the large sample size, the true development of function and neuronal network complexity became only partly visible. Even though a longitudinal study of motor development of infants may offer more specific data, the evaluation of a developmental period of twenty years for a study is difficult to organize. Combining data from several schools together (even though further data were collected) was not improving the results, since the CNS development seems to be affected by many environmental factors, including the time of the year, the social situation of the pupils at home and at school, the variation of the rate of development, and physical activities.

Further on, long-term repair of the injured CNS is studied in several case reports. It is then attempted to partly compare CNS development with CNS repair to see whether the repair can recapitulate development and to better understand development and repair of the human CNS.

1.4. Postnatal Development and Repair

With respect to CNS repair, the stepping automatism of the postnatal development is of great importance. The heel strike induced the step automatism of the newborn (Figure 24) partly re-appears following spinal cord injury. The achievement of the partial reappearance of the stepping automatism during supported treadmill walking is the one of the first stages in relearning to walk.

2. Development of the Human CNS in the Age Range from 6 Months to 18 Years

2.1. Intrapersonal Development from Six Months to Seven Years of Age

2.1.1. Special CDT Devices for Babies and Children Up to the Age of Six

It is only relatively recently that special CDT devices became available for babies and children up to the age of six (Figures 31, 32). Coordination dynamics therapy can thus be started shortly after birth in both the healthy and pathologic case. If a baby may have suffered prenatal hypoxia during pregnancy, CDT can be started after birth with a special CDT device (Figure 32). Even premature babies can be treated efficiently by movement-based learning.

The clinical research presented below was undertaken when such small CDT devices were not available.

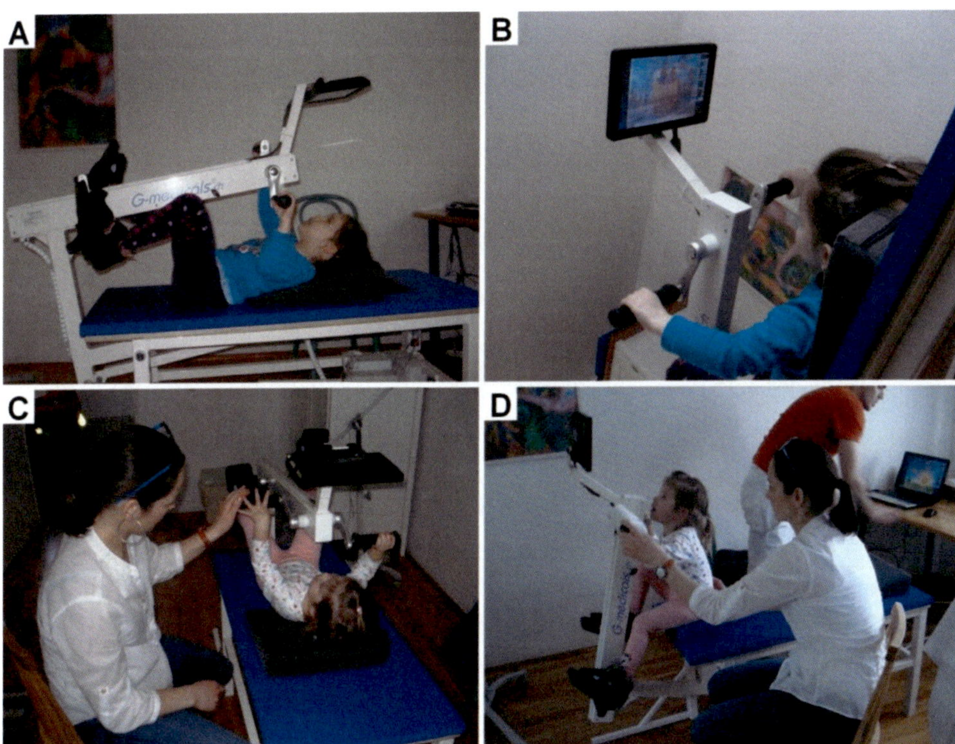

Figure 31. Administration of movement-based learning therapy with coordinated arm, leg and trunk movements to children with cerebral palsy. The coordinated movements on the special CDT device are administered to patients in the age range between 3 and 6 years. Some patients turn volitionally; in others ones the movement has to be supported. To increase the motivation of exercising, the children can watch video films. If they stop turning, the video film stops. The position of exercising can be changed from the lying to the sitting position. If the hand is spastic (C) and cannot hold the leaver, the hand is fixed (D). This special CDT device for measuring and therapy (int.pat.) is produced by the firm: Giger Engineering, Martin Giger dipl.Ing.ETH/SIA, Herrenweg 1, 4500 Solothurn, Switzerland, www.g-medicals.ch.

2.2. EMG Motor Programs

At the age of five months, the infant Jürgen walked more or less in a stepping automatism fashion (Figure 12a). The EMG motor program was not fully developed. The antagonistic action between the right tibialis anterior and right gastrocnemius muscles was not developed (Figure 12b). Three months later, supported walking began to resemble walking somewhat more (Figure 12c), but there was still no antagonicity between the EMG activity of the tibialis anterior and gastrocnemius muscles (Figure 12d). At the age of five, Jürgen could walk nicely by himself and the motor programs likely showed antagonistic action between the tibialis anterior and gastrocnemius muscles. Since the EMG motor program varies among different trials, motor development is better measured by the coordination dynamics (CD).

Figure 32. Special Coordination Dynamics Therapy device for young infants and baby's including premature born ones. Age range of the device 0 to 3 years. Movements can be performed in the sitting (A) and lying position (B). The 18 months old healthy boy is already quite big for the device.

2.3. Neuronal Network Complexity Is Needed For Complicated Coordination Patterns

At six months, the infant Jürgen was too small to be able to turn on the special coordination dynamics therapy (CDT) device. At the age of one the infant was introduced to the device and he liked it. By the age of two, Jürgen was tall enough to exercise on the baby CDT device and he really enjoyed it. However, he was not able to exercise continuously. He repeatedly got stuck after 1 to 3 turns. He solved the problem himself by taking his hands or the feet from the device and avoiding in this way the complicated coordination patterns of arms and legs (with respect to coordination) for which his CNS had not yet developed the necessary complexity (page 92 of [14]). Jürgen's case thus nicely demonstrates that high complexity of neuronal network organization is needed to be able to perform coordinated arm and leg movements with changing coordination between arm and leg movements (pattern change). For turning only with the legs (fitness bicycle) no high network complexity is needed. By the age of three he was able to exercise on the special CDT baby device (Figure 33A); even though the rhythmicity of turning was poor, especially in the backward direction. By five years old, Jürgen was able to turn on the larger children's CDT device in the forward and backward direction (Figure 33B). The exercising in the backward direction was less smooth. The assessed CD will give further insight into the maturing of the CNS during development.

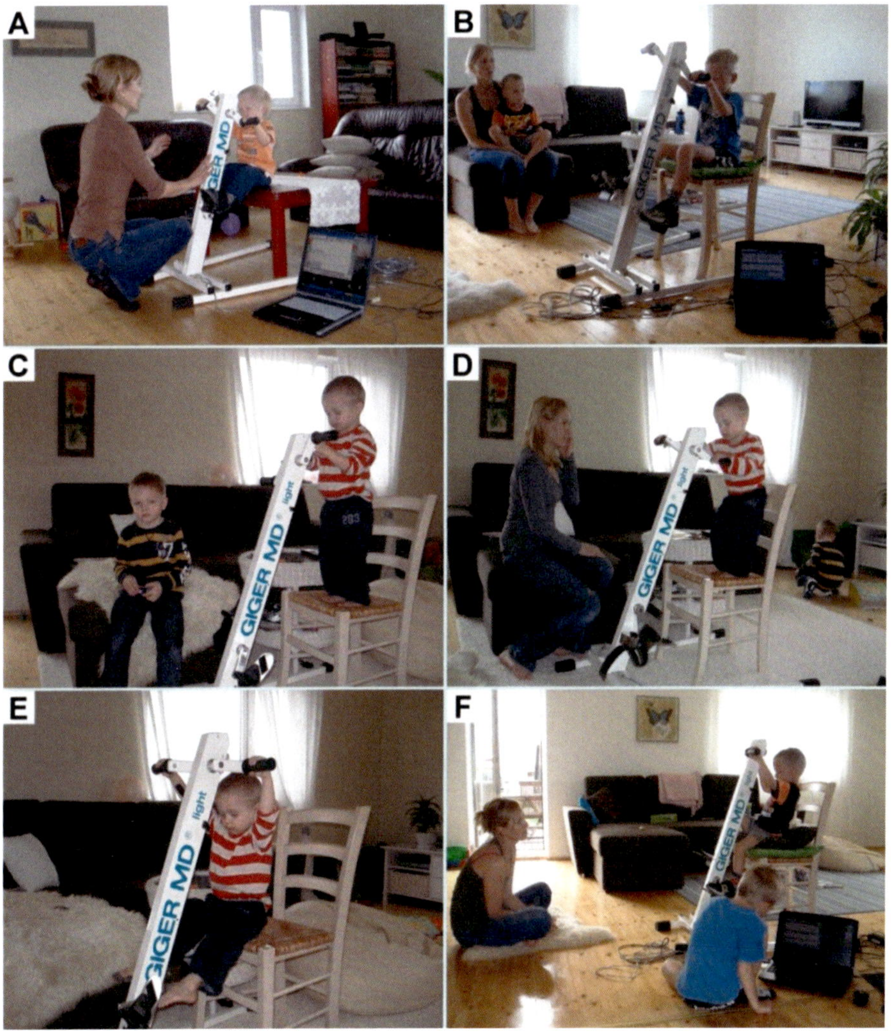

Figure 33. A. Exercising of the 3-year-old Jürgen on a special baby CDT device. B. At an age of 5 Jürgen can turn already on the larger children`s device. B-F. Jonas: Early learning to use a special CDT device. B. From a save position the 2-years-old Jonas watches how the brother Jürgen is turning. C. First he pushes all illuminated buttons. D. Then he turns the leavers for the arm movement. E. Now he is trying to use also the foot pedals. He is not depressed that he cannot exercise because he is still too small for this device (a smaller device was not available). F. At an age of 3 he is from the size able to exercise.

The learning situation for the younger brother Jonas, who is four and a half years younger, was different. He observed his brother Jürgen exercising (Figure 33B). Of course he wanted to emulate what his brother was doing. But the only available children's device was too big for him. So he climbed onto a chair and pushed the entire device's illuminated buttons (Figure 33C). Then he turned the handles of the device (Figure 33D). And then he tried hard to really exercise with arms and legs (Figure 33E). Unfortunately, the instrument was still too large for him. He could not get the feet fully on the pedals. One can nicely see in the face of the two-years-old Jonas how he is experiencing and learning the outside world when playing with the device. At three years old, Jonas became big enough to exercise on the children's device (Figure 33F) and his CNS could be measured (Figure 34A,B). The now 7-years-old Jürgen is looking on the display of the computer to see how the CD traces change when Jonas exercises. This shows how the brothers learn from each other.

With respect to repair of the injured CNS, it is very helpful if the young patient can learn from a brother or a sister and from healthy pupils when going to school. If possible, therefore, cerebral palsy children should attend a normal school and not a school for disabled children. Disabled children have to be motivated and given the opportunity to copy healthy children.

There was actually a problem when assessing the five year-old infants in the pre-school or kindergarten. Some of them were afraid or were too shy and some of them refused to be measured. But in a safe surrounding at home with the mother and motivated by the mother the measurements can more accurately demonstrate normal functioning of the CNS.

2.4. CD Assessment between Three and Seven Years of Age

The generation (increase) of CNS neuronal network complexity during development can be quantified when the child is performing complicated coordinated arm and leg movements and the quality of performance is measured. The child is made to exercise on the special CDT device and the rhythmicity of turning (the quality of performance, the CD) is measured. The best values are taken over a period of 90 seconds of exercising.

Original CD recordings of the three year-old Jonas are shown in Figure 34A,B for forward and backward exercising. For comparison, recordings from Jürgen are shown when he was three years old (Figure 34C,D). It can be seen that Jürgen had slightly better (smaller) values for forward (31.4 against 41.3) and backward exercising (51.2 against 58.4) than Jonas. The probable reason is that Jürgen could exercise more smoothly as he had previously trained a few times on the smaller baby device while Jonas had not. From Figure 34A it can be seen that Jonas could turn better around the easy coordination's pace (P) and trot gait (K). The frequency of turning was higher.

Figure 34E,F shows original recordings of the now seven year-old Jürgen. The performance of these movements improved strongly in the intervening four years. The values for forward exercising improved from 31.4 to 11.8 and for backward exercising from 51.2 to 15.4, that means they improved by approximately a factor of three. As shown in Figure 35, Jürgen could also turn better than the average pupil of the same age, assessed by the cross-sectional study.

This indicates that if healthy pupils would occasionally exercise at school, for example in sport lessons, on the special CDT device, their CNS would improve in its functioning more quickly which would probably also enhance their higher mental functions by learning transfer. The motivation of the leading staff of the gymnasium to strongly support these measurements actually was, to find out how the physical performance can be improved in the pupils and how this physical improvement correlates with higher mental and social functions. A correlation between physical performance and higher mental and social functions could not be found, even though it is believed that physical variability correlates with mental variability: to a healthy mind belongs a healthy body and vice versa. A possible reason for this failure is that CNS functioning depends on very many influences. The young lady with the cervical SCI (Figs.174,186) stated after more than five years of CDT that her intelligence enhanced.

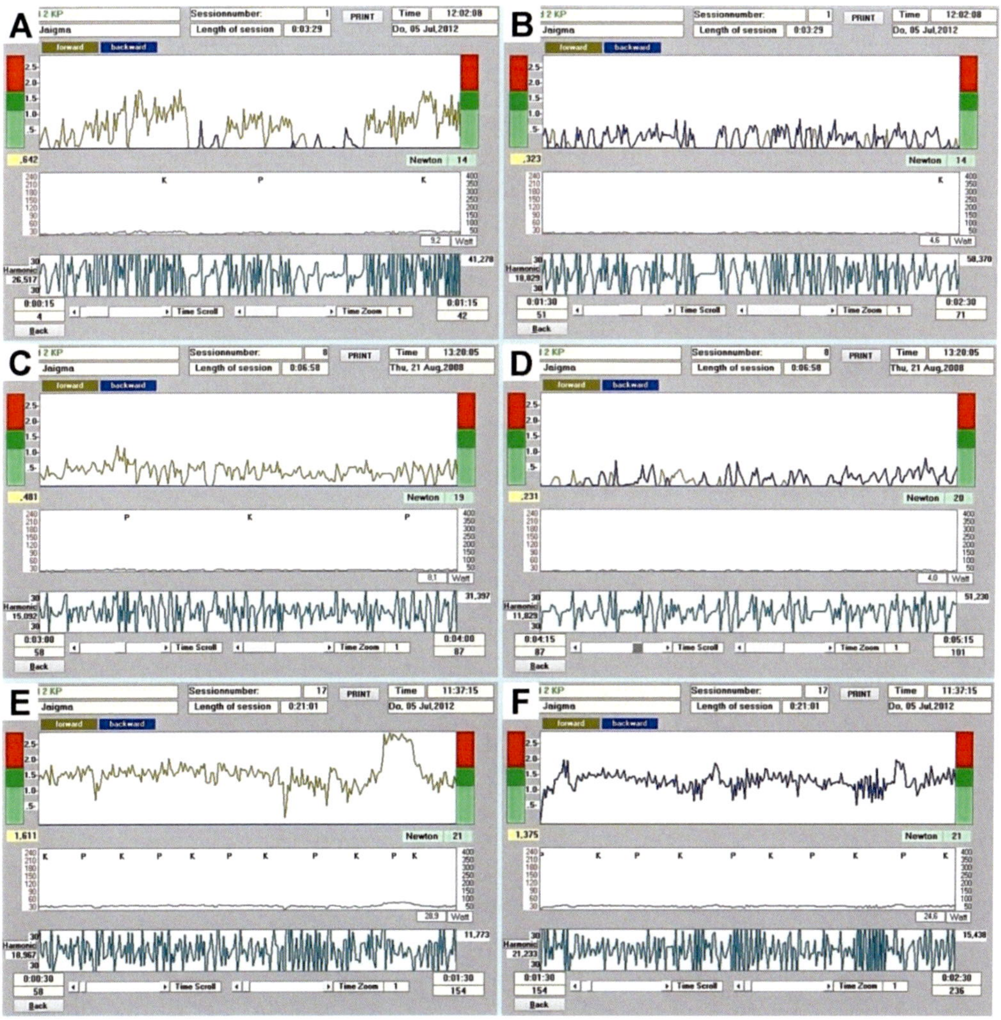

Figure 34. Coordination dynamics recordings of the 3-year-old Jonas (A,B) and the 3-year-old (C,D) and 7-year-old (E,F) Jürgen. Note that the 3-year-old Jonas and Jürgen (A,B,D) make mistakes in the direction of exercising. For example, instead of turning always in the backward direction, they turn a few times also in the forward direction (B,D) as one can see from the different color. Notice further the strong improvement in turning (faster and more smoothly) of Jürgen for the age increase from 3 years (C,D) to 7 years (E,F).

2.5. Transient Rapid Exercise

Before exploring CNS development further through a cross-sectional study, we shall explore an interesting movement performance phenomenon observed in Jürgen. When he became able to exercise sufficiently fast, he also occasionally turned very fast for brief periods (Figure 34E,F). It was as if he was being directed to do so. Such transient fast exercising of children may be a strategy of the CNS to improve its coordinated firing of neurons additionally. During repair, the patient can exercise on a special CDT device to improve the impaired phase and frequency coordination of neuron firing. During normal development, the CNS may use the fast performance of different movements, as for example fast running or crawling, to continuously improve the phase and frequency coordination of neuron firing, which is necessary, because the developing network is in a state of frequent of change and transition so with the increase of connectivity and complexity changing all the time, the coordinated firing of neurons has to be improved continuously. Following severe cervical spinal cord injury, it was shown that with every bit of regeneration of the spinal cord (re-appearance of function), the CD values became transiently worse (Figure 30). During exercise on the special CDT device the need of the CNS for transient fast movement can be clearly seen.

3. CNS Development between 3 and 18 Years of Age, Quantified by Low-Load Coordination Dynamics (Cross-Sectional Study)

3.1. Problems of Assessment

It was shown in Chapter II of [14] that girls and boys have very similar low-load (20N) coordination dynamics (CD) values during development. The values were, therefore, lumped together to get bigger group sizes to better observe the maturing of the CNS with increasing age. The development of the human CNS, quantified by the CD for 20N, is shown in Figure 35. As can be seen from the curves for forward (solid line) and backward exercising (dashed line), the values decreased continuously from 3 to 18 years of age. The decrease is initially dramatic and then only slight.

There may have been a transient increase of CD values at an age of 10 to 11 years, which is marked with an arrow. The size of the age groups is indicated by the numbers at the curves; it varies mainly between 32 and 50 pupils per group. At an age of 14 the group size was only 14, because it was difficult to motivate the pupils. The group sizes for 5 and 6 years (6 and 10) are small because it was difficult to motivate the girls and boys at the pre-school or kindergarten; they were shy and a bit afraid in general. As can be seen from Figure 35, the CD values decrease strongly between 3 years and 9 years of age and then only slowly. The evaluation of CNS functioning in the first 5 years of life is most interesting, because of the fundamental CNS changes (Figure 35), but is also the most difficult to obtain. Further, it was difficult to motivate the infants to exercise on the special CDT device for 10 to 20 min because of the difficulties for them to perform these complicated coordination's between arms and legs. Due to the still under-developed neural network complexity, their CNS had great difficulty in generating these complicated coordination's between pace and trot gait. The control of the supraspinal CNS centers onto the spinal cord neural networks is not sufficiently mature to vary them with enough sophistication to permit them to make these complicated movements. The infants also could not understand why it was so difficult to exercise the instrument; it looked so easy. Other movements, once observed, were quickly learned. Parent is needed to motivate the child.

3.2. Enhancement of Normal CNS Maturation upon Exercising on the Special CDT Device

But as can be seen from Figure 35, the CD values of the infant Jürgen are better than those of the average child and were also better than the values of his brother because he had exercised on the special CDT device several times before. The functioning of his CNS improved faster because of the exercise on the device. Therefore, CNS maturation can be influenced by movement-based learning to improve faster. It is known that CNS functioning and improvement can be modified by cultural influences. It is shown here that Jürgen could turn better than the average child for forward and backward exercising by approximately 30%. The forward values improved from 17.8 to 12.0 and backward values from 22.8 to 15.3. This improvement was partly also possible because Jürgen could start to exercise on the smaller baby device from an age of two onwards. Exercising twelve times made his CNS functioning better than the average child by 30%.

3.3. More 'Correction En Route' of Abnormal CNS Maturation through Exercising on the Special CDT Device

Such an improvement of CNS function is especially important in cerebral palsy children, whose development is anyway retarded. It is actually not only the rate of development which has to be enhanced; retarded, accelerated or deviant development has to be corrected 'en route' by therapy. This is possible also in the very important first five years of life, because special CDT devices are available for all ages (Figures 31, 32, 209, 210). The first five years of life are very important for correcting deviant development. As mentioned earlier, the sooner treatment starts, the higher the rate of repair will be and as stronger the 'correction en route' will be. The performance in Jürgen improved by 30% after a total of just four hours exercising in total (12 sessions of 20 minutes) between the age of two and seven years. It will be shown below that the rate of repair is much slower in the injured CNS. If the infants cannot turn by themselves, the therapist has to help. It is also easier to exercise on the smaller devices, only the integrativity of functions reduces. For older pupils, it will be shown below that the integrativity of CNS organization increases with increasing load. It is the integrated functioning that substantially induces learning transfer.

3.4. Symmetry

As can be seen from Figure 35, the pupils could exercise for all ages better in the forward than in the backward direction. Natural maturation did not correct this missing symmetry. The symmetry can be corrected by movement-based learning, as will be shown below. Symmetry is important especially in repair. This is particularly apparent in stroke patients.

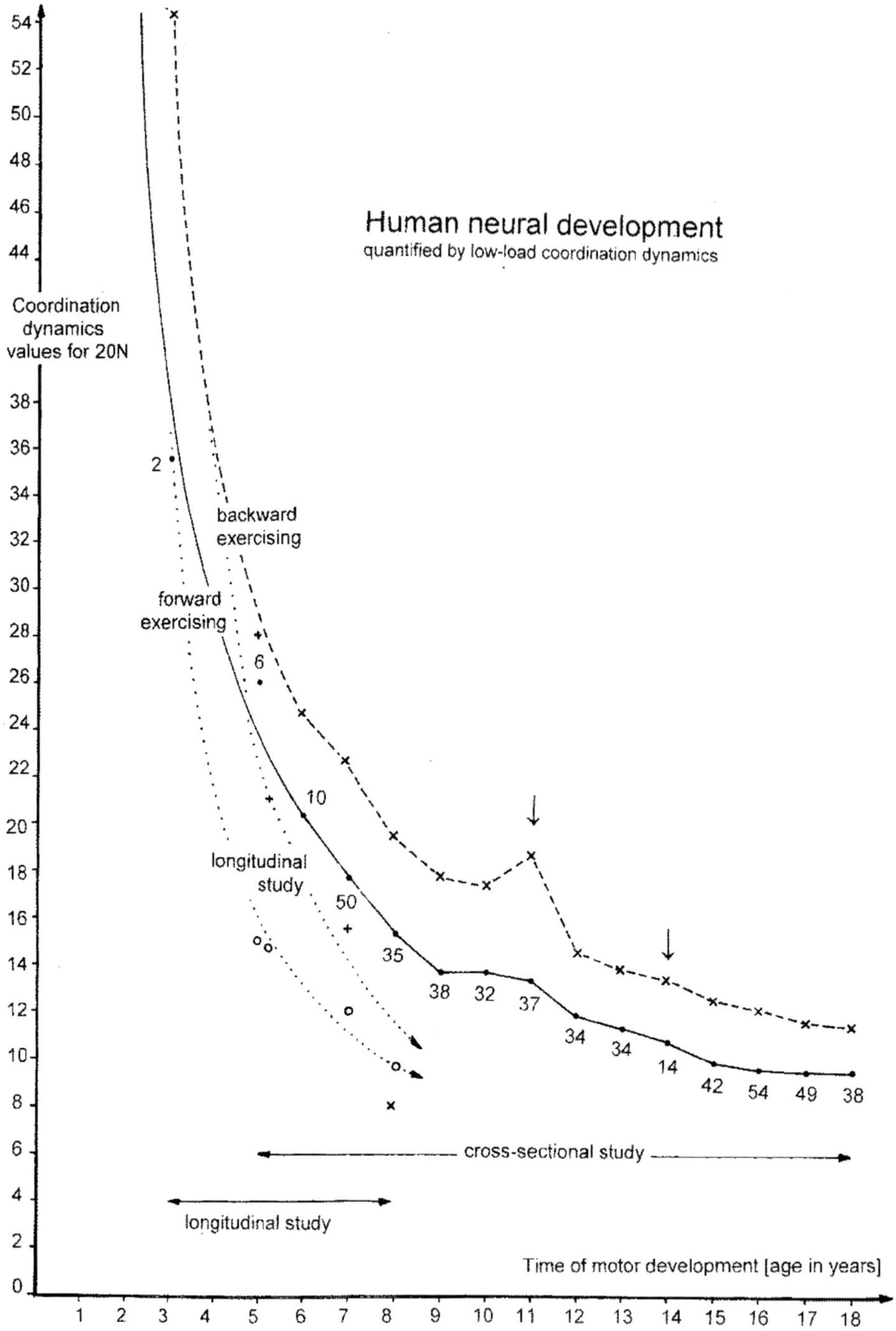

Figure 35. Coordination dynamics values of boys and girls (lumped together) to quantify human neural development. Note that the coordination dynamics values transiently increase at 11 years and that the group size is small at an age of 14 (puberty), as is indicated by arrows. Note further that the coordination dynamics (CD) values for backward exercising during the whole developmental period are worse (higher) and that in the **longitudinal study (dotted lines) Jürgen's CD values become smaller** (better) than the average ones of the cross-sectional study due to the repeated assessment.

3.5. Mistakes Made in the Direction of Exercise during Normal Development: Stability Maturation

As can be seen from Figure 34A-D, at the age of three, Jonas and Jürgen made mistakes as to the direction of turning. The intention was, for example, to turn in the backward direction. But by mistake they turned transiently also in the forward direction. At the beginning they needed instructions from their mother to stay continuously in one direction. With further development, Jürgen noticed the mistakes by himself and corrected the false direction of exercising. By the age of seven, Jürgen did not make these mistakes any more (Figure 34E,F). Also in the cross-sectional study some young infants sometimes made similar mistakes. The ability to consistently turn in a certain direction develops only with time.

The ability to stay in a certain movement pattern has two components: the stability of the movement itself, i.e., the deepness of the potential well and the size of the variability of phase and frequency coordination (Figure 18) and the stability of the pattern intention to convert the intention into the movement pattern. Initially, the child makes mistakes. They are not noticed nor corrected because the intention is forgotten. Instruction is needed. With further neural network maturation, the normal child makes mistakes, but corrects the mistake by itself. The pattern intention is working. Later on, the child no longer makes these basic mistakes and stays stably in the pattern of forward or backward exercising.

3.6. Movement Stability Impairment Following CNS Injury and Abnormal Infantile Development

Following brain injury (not spinal cord injury) and in cerebral palsy the patients sometimes cannot maintain the direction of exercising. With therapy they learn it to stay in a certain direction of exercising. The rate of learning depends on the severity of the injury. Below the stability (Figure 47b,c) of the pattern 'running on treadmill' (Figure 47a) will be measured in a patient with severe brain injury.

Autistic children change their intention every few seconds. They cannot perform a certain movement for a longer time because intention changes all the time. They can keep movement patterns themselves like running, jumping or crawling quite well; it is the pattern intention which is very unstable. It seems that their intention changes are entirely unpredictable. They are not cooperative. But it seems that they can learn, at least to a certain extent, to stay in a certain pattern of movement or higher mental function. However, the therapist has to be very patient.

3.7. High-Load Coordination Dynamics Assessment to Measure Rates of Learning

CNS development between the ages of three to eighteen years was measured by the low-load coordination dynamics value. Exercising against 20N for 21 minutes was easy to do. High-load testing was not popular among the pupils, especially repeated high-load exercising, because it was hard work. The argument that it is healthy did not particularly impress the pupils. But a few pupils did this high-load exercising, which lasted approximately one hour (including the low-load test for warming up). Their data will be compared with those of patients and rates of movement learning, i.e., rates of repair, measured.

4. Motor Learning of the Healthy CNS

4.1. From Development to Movement-Based Learning

It was shown above that CNS function improved in Jürgen against the average infant at the age of seven by 30% by exercising 4 hours on the special CDT device for twenty minutes on twelve occasions over five years. Jürgen could turn more easily (better (lower) CD values) for lower load than 20N or using the smaller baby device. Increasing the load made the turning with arms and legs too difficult. That means for exercising against higher load, it is more difficult for the CNS to learn the different patterns between pace and trot gait. Also, it is more difficult to coordinate finger or limb movements for higher loads.

4.2. Exercising Against Higher Loads Is a Good Measure to Quantify Movement-Based Learning

For older pupils and adult volunteers higher loads can be used to measure better movement-based learning in the normal CNS. The activated functions are more integrated, if the volunteer is exercising against 150 or 200N. For the generation of such power more muscle fibers and upstream more neuronal networks are activated. The whole cardio-vascular performance is activated including respiration. As well as the somatic nervous system, the vegetative nervous system is activated and trained. Because of the activation of the vegetative nervous system, urinary bladder continence functions can be improved and repaired.

4.3. Assessment Is Also a Form of Therapy

The repeated measuring of the low-load and high-load coordination dynamics is analogous to CDT because healthy subjects were move around in their everyday life and also do some sport. It therefore becomes possible to compare this kind of motor learning between healthy subjects and patients with CNS injury. That the measuring of the CD improves CNS functioning (the measurement influences the value to be measured) does not present a problem, because changes occur due to measuring in both the healthy and injured CNS.

The measuring of the CD values in the patient and healthy subject also constitutes therapy, so no time is wasted due to the diagnostic process. It is in fact good training, because the patient is intent on improving their values. The diagnostic is an efficient therapeutic hour for improving CNS functioning, fitness and health.

4.4. Movement-Based Learning for Different Loads in the Healthy CNS

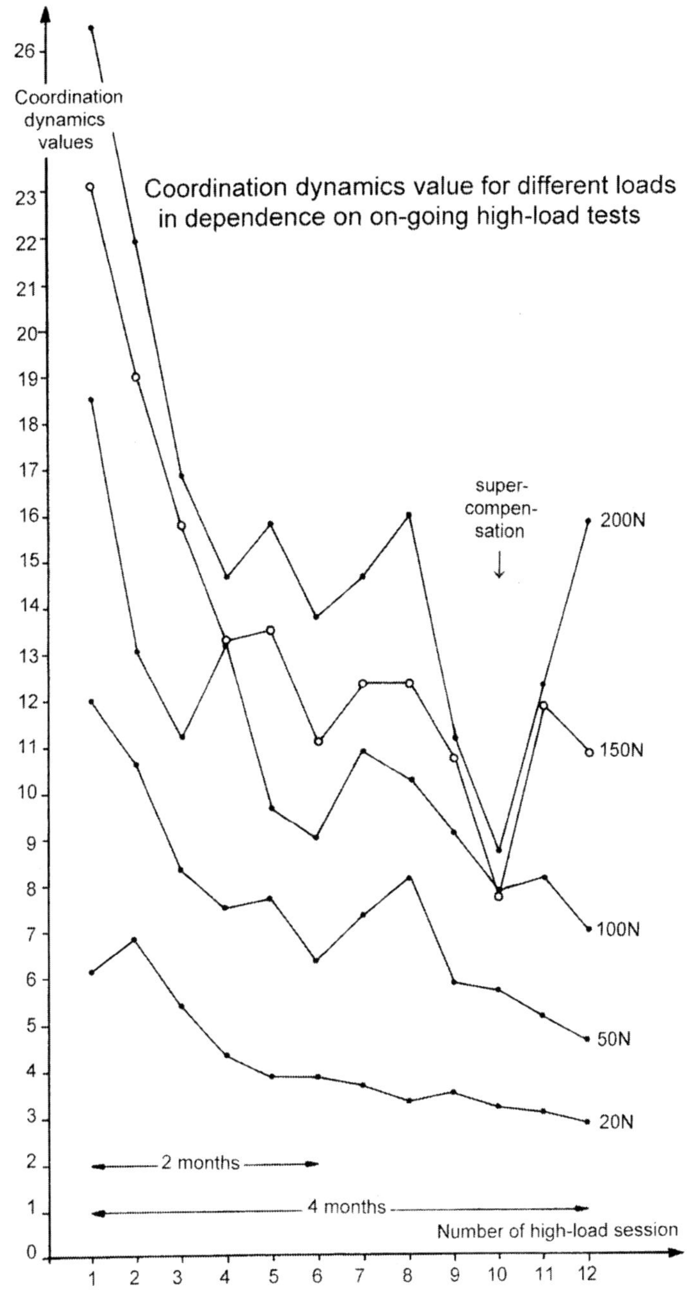

Figure 36. Improvement of coordination dynamics (CD) values against different loads due to 12 times repeated (low- and) high-load coordination dynamics testing and exercising (low-intensity motor learning) in the forward direction in a 15-year-old pupil. The mean time interval between 2 sessions was 11 days and the number of turns per session was approximately 3500. Note the phase of super-compensation around the 10th session; only the CD values for 150 and 200N improved (decreased) strongly transiently.

Figure 36 shows the improvement of the CD values for a 15-year-old pupil for increasing load over the course of repeated sessions. It can be seen that the CD values of the boy reduced (improved) for all loads between 20 and 200N. After the first measuring sessions the CD values reduced (became better) strongly and from then on improved only a little. This strong improvement of the CD values will be used to quantify movement-based learning and is used later for comparisons with patients.

All curves for the loads between 20 and 200N had the same characteristic. With the sixth session, after a period of two months, the strong improvement of CNS functioning for all loads ceased. As will be seen below, this similarity of improvement for all the loads will be different in patients. For the sessions 9th to 11th, the performance of exercising was transiently very good for the high loads of 200N and 150N. Such phases of transiently good performance are called in sport 'super-compensation'.

In a 17-year-old female athlete this fast motor learning was achieved in one month or less (Figure 37). This female athlete learned faster than the normal pupil. The super-compensation was already reached during the 6th to 9th session.

But when this and other pupils did not get better any more, their motivation to improve ended. They also had good reason to stop the repeated measuring, as they had to prepare for examinations. They did not exercise often enough to reach the phase of slow improvement. In Figure 44 (mild epilepsy) the fast and slow phase of learning can be nicely seen and also the occurrence of several phases of super-compensation. The slow phase of learning can nicely be seen when the Author exercised every day for more than one month (Figures 204A,205B).

Between the 6th and 7th session, the female athlete pupil also had to study for examinations and did not sleep enough at night. She suffered from stress caused by the examinations and also had difficulty with pain in the bones from her training of high-jumping. Her performance, therefore, got much worse from the 7th to the 8th session onwards, even though she had no pain while exercising on the special CDT device. This case shows how dramatically performance can worsen if there is stress, not enough sleep and some health problems. In another pupil (Figure 41), who had had only 5 hour sleep the previous night, the performance worsened. In the female athlete this impairment was very strong. During the 11th session, she could exercise for 200N better, than for 150 and 100N which is not normal. Her nervous system was completely mixed up. Her fitness had not declined, because she was continuing with her sports training. For tennis players, transient worsening of performance can be dramatic, especially when also motivation is reduced. As will be shown below, health conditions, motivation and amount of therapy also influences the rate of CNS repair in patients.

4.5. Exercising Coordinated Movements at High Load to Improve Deep Network Complexity by Learning

To improve CNS functioning the developing child or the patient has to exercise on the special CDT device in the forward and backward direction to improve symmetries of network organization. By turning both slowly and rapidly, the movement performance can be improved through the movement-induced afferent input generated by the slow, medium fast (fast fatigue resistant) and fast contracting muscle fibers (fast fatigue) and their corresponding upstream neuronal networks including the premotor spinal oscillators (Figure 8).

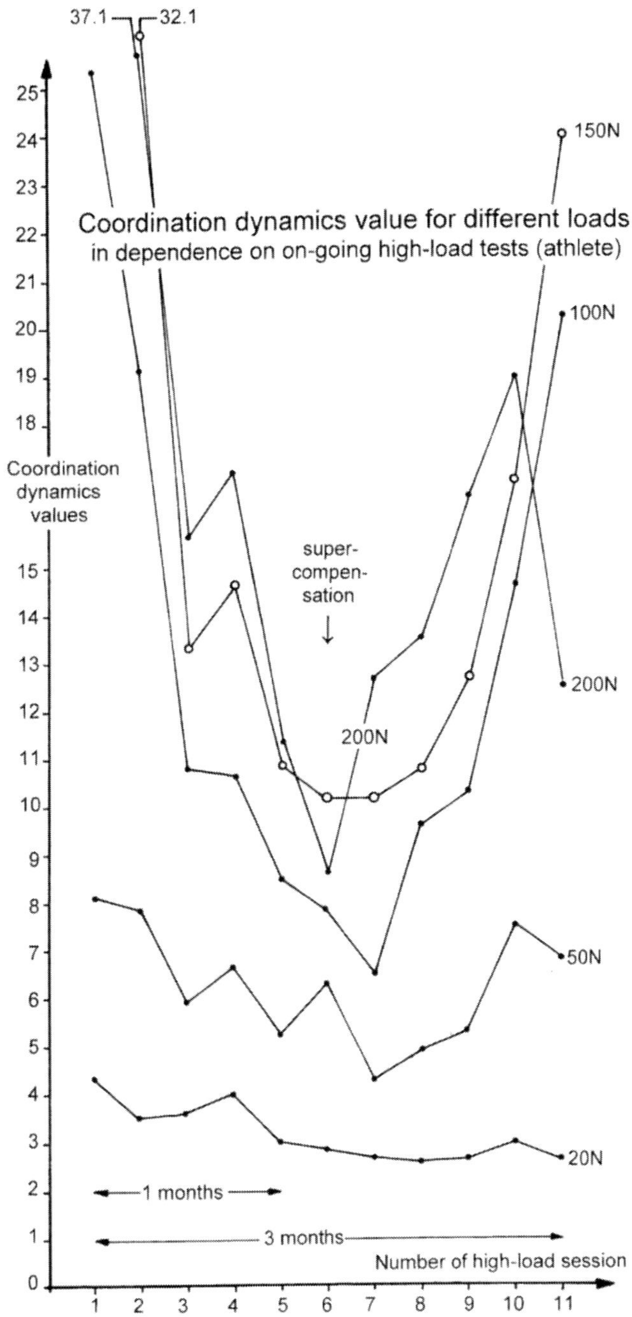

Figure 37. Improvement of coordination dynamics (CD) values against different loads due to 11 times repeated (low- and) high-load coordination dynamics testing and exercising in the forward and backward direction in a 17-year-old female pupil (athlete). The mean time interval between 2 sessions was 9 days and the number of turns per session was approximately 4000. Note the strong increase again of the high-load CD values for 100, 150 and 200N following reduced sleep, examination stress, health problems and lack of motivation. In the first session the CD values for 150 and 200N were with 32.1 and 37.1s^{-1} respectively very high and reduced strongly with the next two assessment sessions to 13.3 and 15.6s^{-1} respectively.

However, exercising against different loads also has to be performed to enhance the integrativity (many neural sub-networks are activated simultaneously and coordinated together) and complexity of neuronal network organization. By increasing the load during longer periods of exercising the slow muscle fibers (and the upstream networks) (Figure 8) are trained which, for example, a marathon runner needs. To get sufficient oxygen, to regulate temperature and cardio-vascular performance, the vegetative nervous system needs to be trained and improved in its functioning in general. This is especially important because of the substantial learning transfer needed to repair trained functions which cannot be deliberately trained in the injured and uninjured CNS.

The high-load assessment and training was performed every one, two, or perhaps four weeks, to see how the CD improved for high-load exercising. The headmistress of the high school also performed this high-load assessment and training and felt very good in her body afterwards. A sport teacher at the school, by contrast, failed the test (he refused to continue at the load of 150N), probably because he did not have sufficient fatigue resistant slow or medium fast muscle fibers (as 100m runners often do not) and insufficient enough motivation to push himself through the high loads. For this high-load assessment and training, one needs the slow muscle fibers, the medium fast and fast (fatigable) muscle fibers (Figure 8). Women can therefore perform this high-load exercising as well as men.

4.6. Improvement of High-Load CD (Movement-Based Learning) and Super-Compensation in Older Healthy Pupils

Figure 38 shows such high-load assessment strategy. The load of exercising was increased from 20 to 200N and then decreased again (Figure 38A). A fit adult or a pupil with a very well- functioning CNS can turn at the same frequency for all loads (no load escape (Figure 38A)) with low CD values (Figure 38B, C). An unfit subject with suboptimal network organization will reduce the frequency of exercising or even stop exercising to escape the load and the CD values will increase (get worse) for higher load exercising (Figure 38D,E,F). In adding up the CD values for the different loads, a value for the quality of high-load performance is obtained. The whole high-load test supplies further information with respect to movement-based learning. Repeated assessments show the improvement of fitness and CNS organization. The assessment of healthy pupils can show improvements in CNS organization by going deeply into the network complexity of CNS organization when exercising at high loads.

Figure 39 shows the high-load CD values with on-going assessment (and no additional training on the instrument) of a 15-year-old male pupil (same as in Figure 36). In Figure 39B the single values of the first and tenth sessions are shown. As can be seen, the hysteresis-like curves improved with on-going sessions. Plotting the high-load CD values (the sum of the single CD values at different loads) against the number of sessions, a curve for forward and backward exercising is obtained (Figure 39A).

First the high-load CD values improve strongly and then reach a plateau with a slight decline (Figure 39A). Before the plateau is reached, a phase is passed through in which the performance is best. This phase of super performance is called 'super-compensation' and occurred in this case during the 10th session. After this super-compensation phase, most pupils stopped. The high-load testing of a 17-year-old female athlete is shown in Figure 40 (this is the same pupil as in Figure 37).

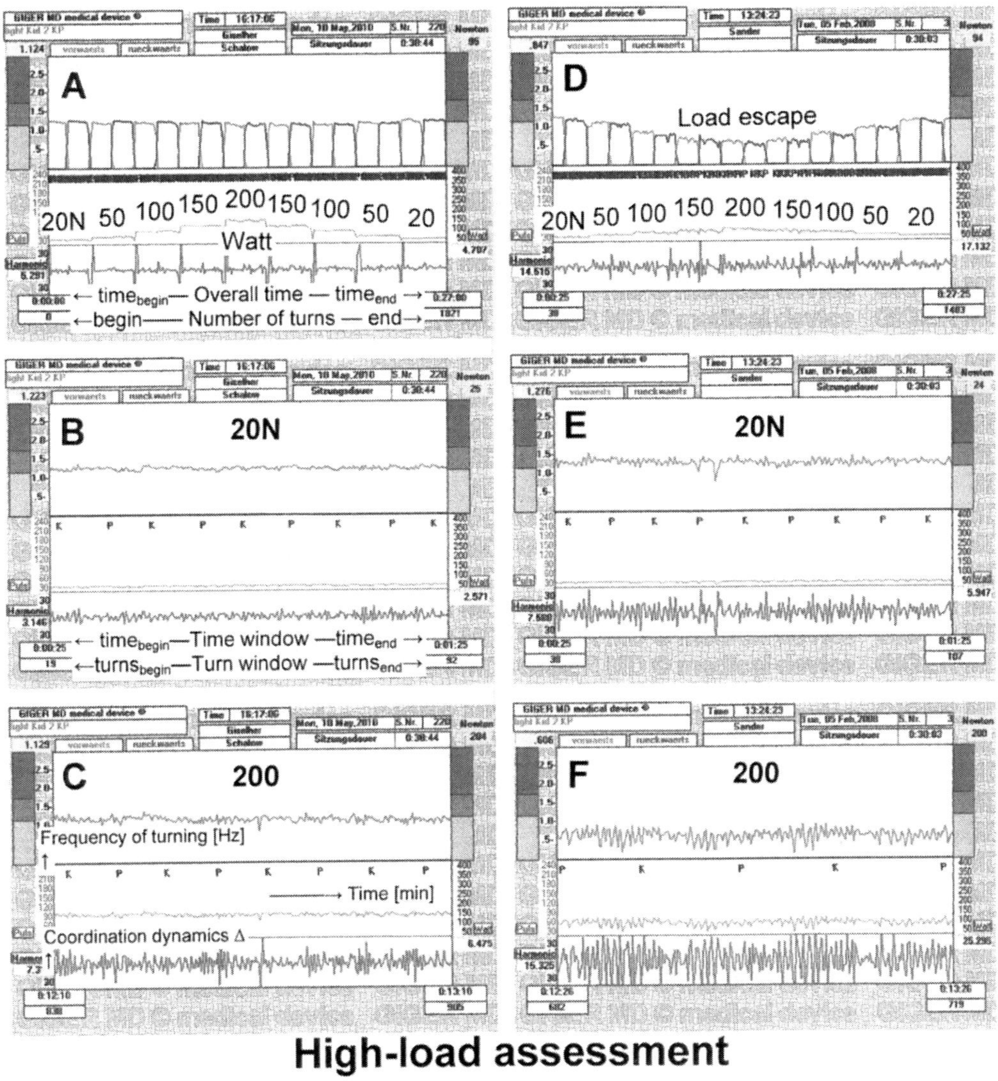

Figure 38. High-load coordination dynamics recordings for a well and balanced functioning CNS of the author (A-C) and a 15-year-old pupil (D-F). In the recordings A through F, the upper traces represent frequency of turning, the middle traces show load changes in Watt, and the lower traces show the coordination dynamics. In A and D, the whole record of 27min duration is shown during force changes in 3min steps (1.5min forward and 1.5min backwards) from 20 to 200N and backwards, and in B,C,E,F, a time window of 1min is shown when the load was approximately 20 and 200N (P and K are the pace and trot gait coordination positions). Note that for force changes between 20 and 200N the power in Watt changed stepwise in the author (A), but not in the pupil who showed a transient frequency reduction (D). Note further that for the load of 200N the pupil could exercise quite well on the special coordination dynamics therapy device for the easy coordination's close to pace and trot gait but not for the difficult intermediate coordination's; the frequency of turning decreased strongly and the coordination dynamics increased strongly (worsened) for the intermediate coordination's, which means that the pupil's CNS had problems to organize the motor control for the difficult intermediate coordination's of arm and leg movements.

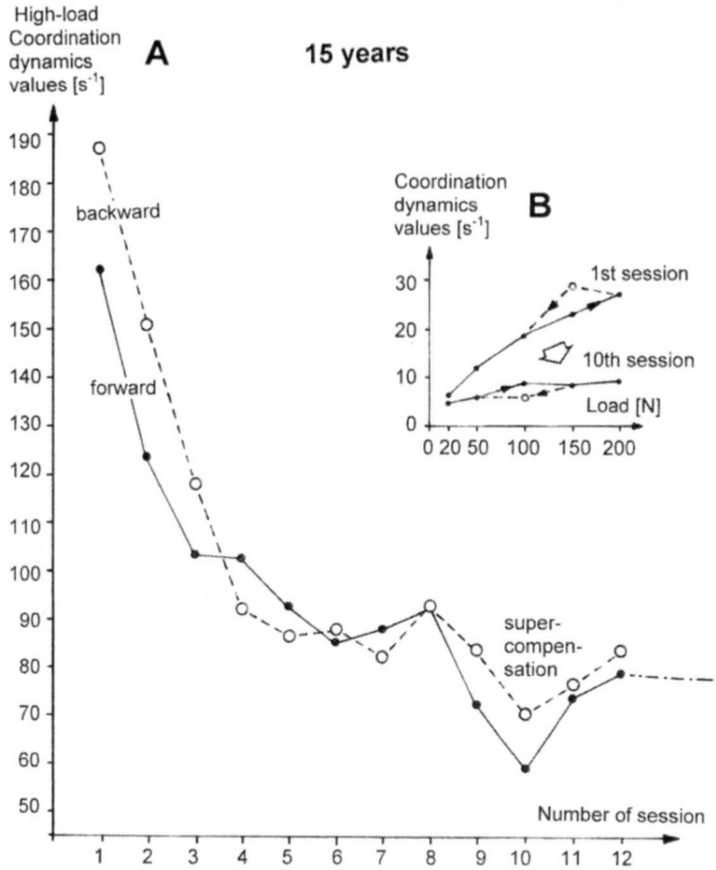

Figure 39. A. Improvement of high-load coordination dynamics (CD) values (for evaluation see B) due to 12 times repeated (low- and) high-load coordination dynamics testing and exercising (low-intensity motor learning) in the forward and backward direction in a 15-year-old pupil. The mean time interval between 2 sessions was 11 days and the number of turns per session was approximately 3500. Note the transient improved exercising (super-compensation) at the 10th session. The dashed-dotted line suggests the further continuation of the high-load CD values. B (inset). Coordination dynamics values (Δ) in dependence on load increase (solid line) and decrease (dashed line) for the 1st and 10th session for exercising in the forward direction. The hysteresis-like curve for the 10th recording shifted to smaller coordination dynamics values. The high-load CD values were obtained by summing up the single CD values, Δ (high-load CD value = Δ20N + Δ50N + 100N + Δ150N + 200N + Δ150N + Δ100N + Δ100N + Δ50N + Δ20N + Δ20N).

Her super-compensation phase was reached during the sixth session. Her plateau values were with approximately 70 a bit lower than that of the 15-year-old pupil ($80s^{-1}$). An 18-year-old pupil had with a plateau value of approximately $60s^{-1}$, the lowest value (Figure 41). It seems that the high-load CD values also slightly decreased with age (CNS maturing). The low-load and high-load CD values depend strongly on the fitness of the person. The 15, 17 and 18 year old pupils (Figures 39-41) were sporty and healthy. The Author, a bit older than the pupils, reached after many years of exercising a high-load value of less than $50s^{-1}$ (Figures 204A,205B).

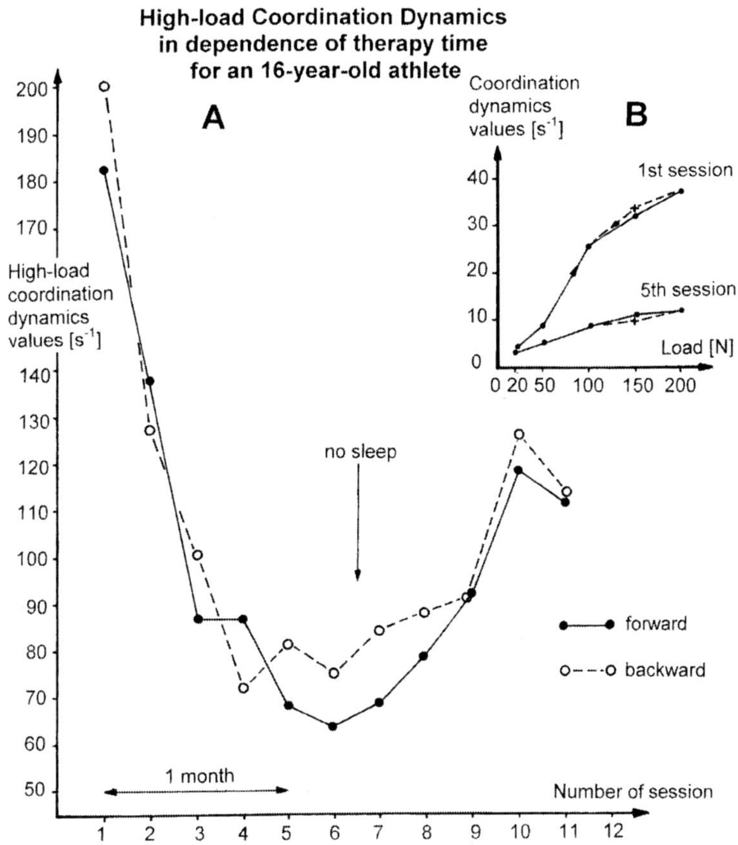

Figure 40. A. Improvement of high-load coordination dynamics (CD) values (for evaluation see B) due to 11 times repeated (low- and) high-load coordination dynamics testing and exercising in the forward and backward direction in a 17-year-old female pupil (athlete). The mean time interval between 2 sessions was 9 days and the number of turns per session was approximately 4200. Note the strong increase again of the high-load CD values following reduced sleep, examination stress, health problems and lack of motivation, indicated by an arrow and 'no sleep'. B (inset). Coordination dynamics values in dependence on load increase (solid line) and decrease (dashed line) for the 1st and 5th session for exercising in the forward direction. The high-load CD values were obtained by summing up the single CD values.

4.7. Movement-Based Learning in the Short-Term Memory Is Fastest in Young Children

To convince somebody that exercising on the special CDT device is really improving CNS self-organization, this person has to be made to exercise on the special CDT device because they will then experience then that this exercising has little in common with a conventional fitness bicycle. This is especially true during backward exercising when the person would feel that his/her CNS has difficulty in generating such movement patterns - and that the movement performance of these patterns improves quickly in the short-term memory.

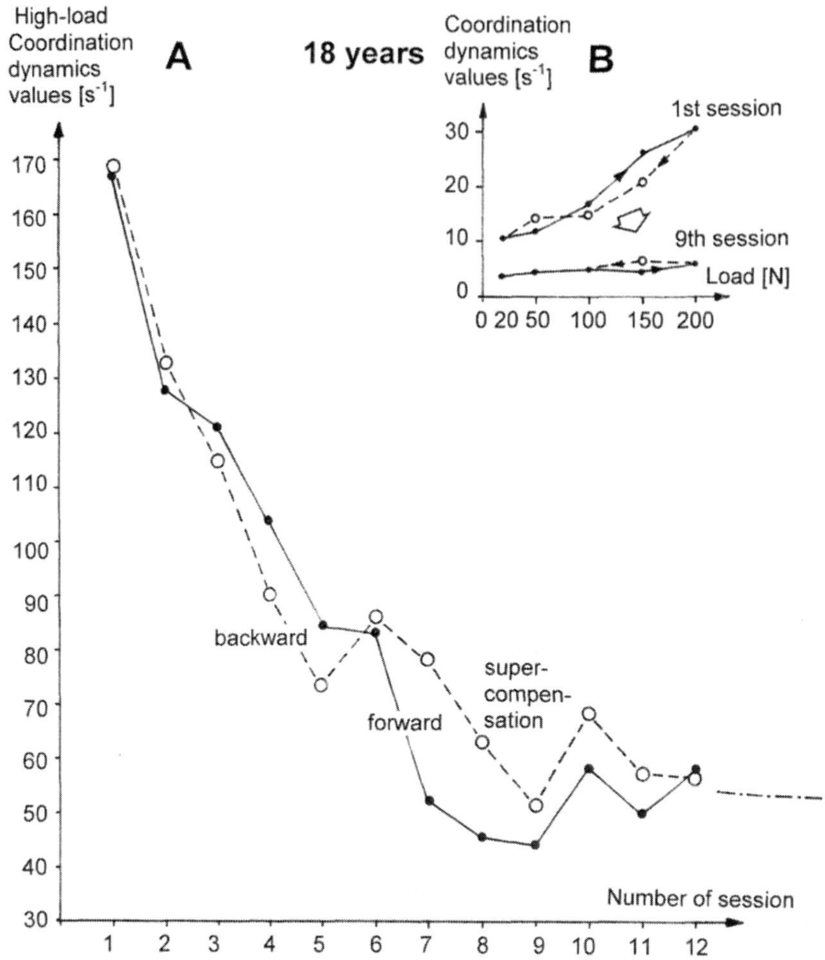

Figure 41. A. Improvement of high-load coordination dynamics (CD) values (for evaluation see B) due to 12 times repeated (low- and) high-load coordination dynamics testing and exercising in the forward and backward direction in an 18-year-old pupil. The mean time interval between 2 sessions was 31 days and the number of turns per session was approximately 4200. Note the transient improved exercising (super compensation) at the 9th session. The dashed-dotted line suggests the further continuation of the high-load CD values. B (inset). Coordination dynamics values in dependence on load increase (solid line) and decrease (dashed line) for the 1st and 9th session for exercising in the forward direction. The hysteresis-like curve for the 10th recording shifted to smaller coordination dynamics values. The high-load CD values were obtained by summing up the single CD values.

It is of interest therefore to measure whether the rate of this movement-based learning in the short-term memory is increasing, decreasing or is the same for all ages. The rate of learning was assessed by the time which was needed to achieve the best CD value within the 21 minutes of exercising against 20N. As can be seen from Figure 42A, the young children learned fastest. For exercising in the forward direction (Figure 42A), the young children were fastest to reach the best CD value. The time to the best CD value increased (with variations) continuously from five to eighteen years of age. This means that more time is needed for motor learning with increasing age; the rate of motor learning decreased with on-going development.

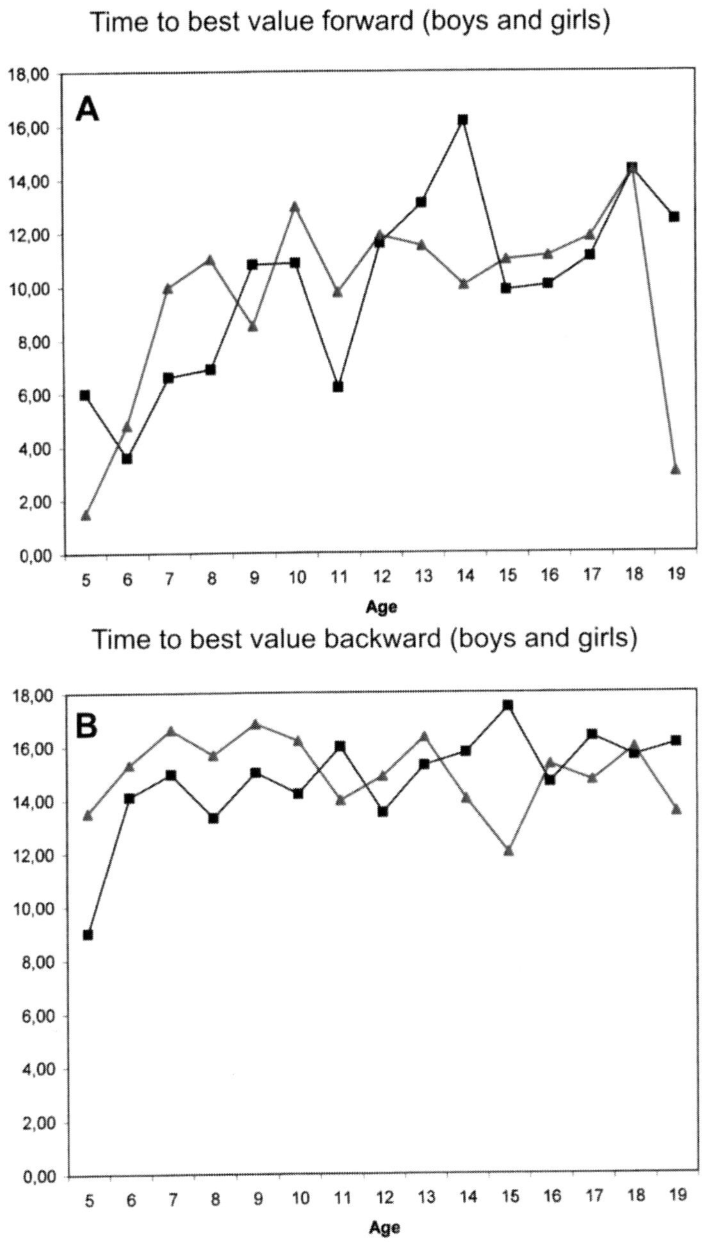

Figure 42. Times till the best CD values are reached during 21min of low-load exercising in the forward (A) and backward direction (B).

For backwards exercising (Figure 42B), which is more difficult than exercising in the forward direction (Figure 35), all ages seem to need the entire 21 minute period to reach the best value. It could be expected that younger children learned faster to get their best value for exercising in the forward direction. However, the young pupils had similar problems to the older ones for exercising in the backward direction was to be expected. During development, the CNS prioritizes the functionally more important act of moving forward rather than establishing symmetries. This prioritization can be observed by the ability of small babies to perform the forwards stepping automatism, but not backwards and the ability in small children to walk forwards, but not backwards.

4.8. Movement-Based Learning in the Short-Term Memory in Spinal Cord and Brain Injury

No real statistics exist for movement-based learning in the short-term memory in the case of an injured CNS. In brain injury, for low-load exercising the best values were mostly achieved after exercising for 15 minutes. The mildly and severely injured neuronal networks need some time for improving its functioning in the short-term memory. In a patient with a complete cervical spinal cord injury, the best values could be achieved at the beginning of the 21 minute period of exercising. Her supraspinal centers were functioning well; she just had to get some information through the injury site to activate the motoneurons physiologically. Also, in incomplete spinal cord injury the learning in the short-term memory was fast, so that the best low-load values were obtained at the beginning.

As will be shown below, the movement-based learning for repair depends strongly on the kind of CNS injury, its severity and what functions have to be relearned. Low-load exercising on the special CDT device, for example, can be learned faster than the high-load exercising, which is going deeper into the complexity of CNS organization (Figure 50).

Now we will turn onto the high-load measurements of patients to meaningfully quantify the rate of movement-based learning in the long-term memory for higher loads, i.e., the rate of repair. It will be shown that there are tremendous differences in the speed of movement-based learning in the healthy CNS and the injured CNS. Repair following therapy of the injured CNS needs very much time depending on the severity of the injury.

Chapter IV

Repair in Mild CNS Injury

1. Repair in Mild Cerebral Palsy

1.1. Improvement of CNS Functioning in Mild Cerebral Palsy upon Continuous High-Load Testing

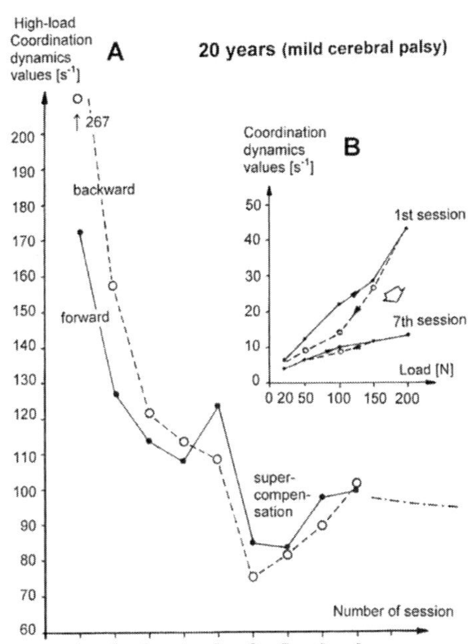

Figure 43. A. Improvement of high-load coordination dynamics (CD) values (for evaluation see B) due to 9 times repeated (low- and) high-load coordination dynamics testing and exercising in the forward and backward direction in a 20-year-old pupil with mild cerebral palsy. The mean time interval between 2 sessions was 5 days and the number of turns per session was approximately 4600. Note the transient improved exercising (super-compensation) at the 7th session. The dashed-dotted line suggests the further continuation of the high-load CD values. B. Coordination dynamics values in dependence on load increase (solid line) and decrease (dashed line) for the 1st and 7th session for exercising in the forward direction.

A 20-year-old pupil with a mild cerebral palsy showed a similar CD value improvement (Figure 43) than the healthy pupils. His super-compensation phase occurred around the 7th session. His plateau value was approximately 100. In comparison to the healthy pupils of the same high school, his values were worse at the beginning and his plateau value was also not as good as in the healthy pupils. All three pupils were not overweight and liked moving. The pupil with mild cerebral palsy (now studying at university) walked an hour to school every day. This fast walking to school improved his CNS functioning, but the high-load exercise on the special CDT device further improved his CNS functioning.

2. Repair in Mild Epilepsy

2.1. Improvement of CNS Functioning in Mild Epilepsy upon Continuous High-Load Testing

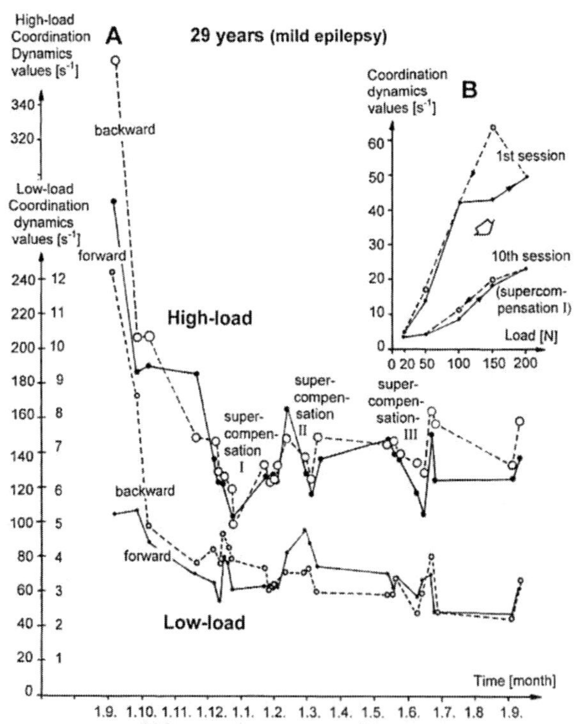

Figure 44. A. Improvement of high-load and low-load coordination dynamics (CD) values upon repeated low- and high-load coordination dynamics testing and exercising in the forward and backward direction in a 29-year-old mother with mild epilepsy. The mean time interval between 2 sessions was 13 days and the number of turns per session was approximately 4300. Note the first transient improved exercising (super-compensation I) at the 10th session; with further exercising sessions further super-compensations (II and III) occurred. Note further that with these 27 exercising sessions a plateau with slight decline was reached. The low-load coordination dynamics values declined in a similar way but showed no super-compensation. B. High-load coordination dynamics values in dependence on load increase (solid line) and decrease (dashed line) for the 1st and 10th session for exercising in the forward direction.

A 29-year-old mother of a pupil with mild epilepsy performed the high-load exercising for a longer period of time (Figure 44). Her super-compensation phase was reached at the 10th session. But because she went on with the high-load exercising to improve her disease further super-compensation periods could be observed (Figure 44A). Her high-load plateau value was around 140; much higher than in the pupils. Also, the single high-load CD values started from much higher values (Figure 44B). Even though her high-load CD values were not as good as the pupils' her disposition to epileptic seizures decreased. Some pupils of the school with mild epilepsy could not be motivated for this demanding training to reduce the frequency of seizures. The organizational effort was too great for them. Movement-based learning was not popular at the high school, even though the head of the school was supporting it. Such CNS and fitness training should be included in school sport lessons.

In Figure 44, the two phases of learning can be seen clearly for high-load testing (the sum of all CD values) and low-load assessment (20N). Initially, the CNS improves rapidly and then slowly reached a plateau with a small improvement thereafter.

2.2. Rate of Learning (Repair) in Mild Cerebral Palsy and Mild Epilepsy

In the female athlete, fast learning lasted approximately one month (Figs.37,40) and in a healthy pupil for two months (Figs.36,39). In the 20-year-old pupil with the mild cerebral palsy the fast learning lasted only a month (Figure 43; the period till up to the 6th session was one month) and in the mother with the mild epilepsy approximately two to three months. The improvement of CNS functioning depends also on the frequency of the measuring sessions. Still, the rate of learning in light disturbances of CNS functioning is not or perhaps only slightly reduced. But, as will be shown below, in severe CNS injuries the situation is completely different. The rate of learning will dramatically reduce.

Chapter V

Repair in Severe CNS Injury

1. Learning for Repair: Outline

1.1. Problems of Learning in the Injured CNS

In severe brain injury (Figure 45) some parts of the brain are damaged more than others and some parts are not damaged at all. Through CDT, the undamaged parts will try to become operational. Depending on the location, extent and type of the damage, it may happen that networks which in themselves are normal cannot bring about physiologic movement because some areas of the brain which are also necessary for the accomplishment of a particular motor function are deficient. For example, certain sub-functions which have been spared from the injury cannot be integrated into larger patterns because other elements of these larger patterns are missing or have not been sufficiently repaired through natural repair mechanisms. In general, both impaired and healthy parts of the brain change over time through therapy and this leads to increased complexity which has direct repercussions on the quality of the learning processes.

The poor quality of movements, stereotypy of posture and motility, the inability to 'discover' new motor possibilities and deficits in higher mental functions in the first years following injury can all be traced back to a lack of learning as a consequence of deficient brain structure. In the case of an inability to 'discover' new motor possibilities, there may be a disturbance in the chain of events because errors in the self-organization of neuronal networks are not recognized (or not repaired), with the result that the processes of repair stop prematurely.

1.2. Problem Solving Therapy by Learning during Repair

It is of importance to recognize the deficiency or absence of particular functions and learning processes. If one recognizes which processes are deficient, one may then be able to offer the patient a cue to start other processes or help him to eliminate errors and to progress in the right direction. It may be sensible to 'teach' the patient specific modes of operation to avoid wrong postures or wrong movement patterns, or to stimulate him to use uncommon modes which may enable him to achieve results which cannot be achieved spontaneously.

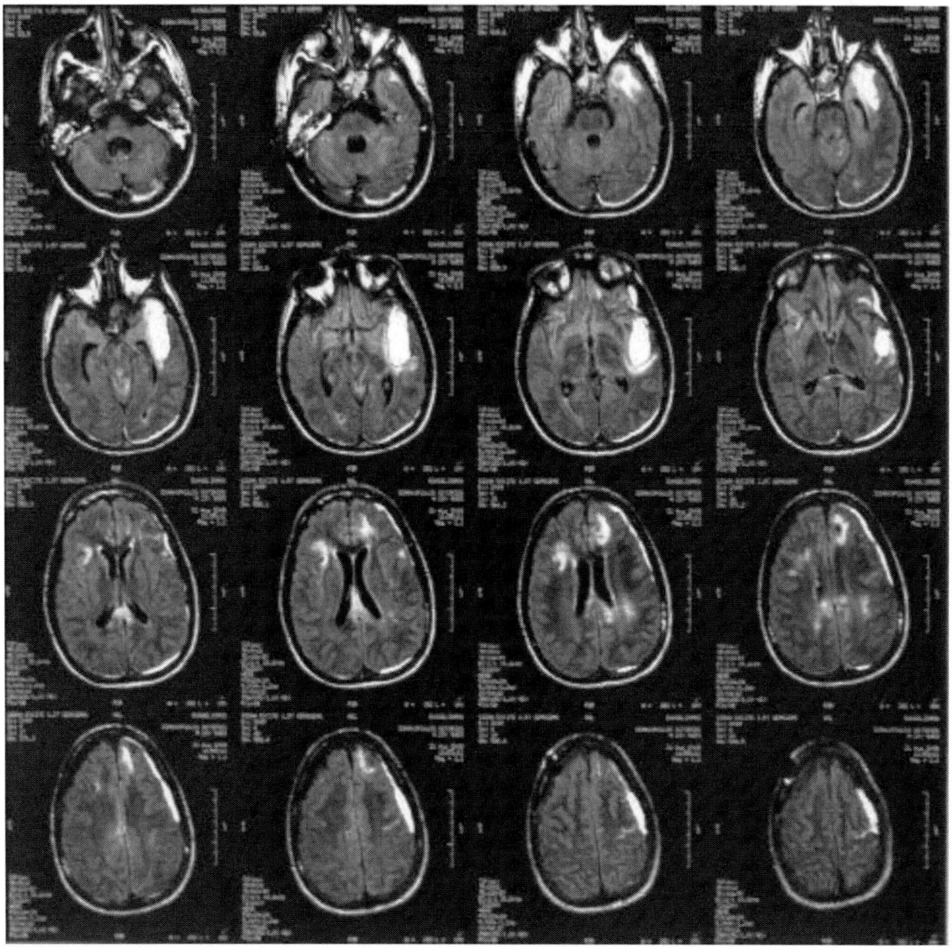

Figure 45. MRI of the patient Sotiris shortly after the severe brain injury caused by a car accident.

1.3. Motor Learning and Problem Solving through Error-Elimination

Motor learning, such as learning to ride a bicycle or play piano, is somewhat different to the learning for CNS repair or optimizing development. The memory of motor skills (nondeclarative memory) such as riding a bicycle is called procedural memory. Nondeclarative memory is formed in brain regions other than the hippocampus, most likely in the sensory-motor cortex and brain stem nuclei. Learning by problem-solving is a means of repairing sub-networks which are necessary for functioning and learning. By inducing trial and error-elimination processes [42, 43] in subunits of the changing nervous system, optimal repair may be achieved. To teach the injured CNS to repair itself by trial and error elimination processes, the CNS has, in similarity to development, to recognize through CDT, which sub-networks, regulation units or sub-loops are not functioning properly or missing and to repair them by error-elimination, including the possibility that other parts of the brain partly take over functions and sub-networks built anew to a limited extent upon structural repair.

1.4. Treatment Principles Embedded in the System Theory of Motor Development

Because of the influence of many subsystems, many possibilities exist to facilitate (motor) behavior. All subsystems contribute to the enhancement or delay of motor behaviors and not only the motor system. In addition to exercising motor functions, instructions or self-instructions may contribute, for example, to the improvement of motor functions. Important questions remain. What degree of deprivation in any subsystem can be compensated by the other systems? Are some subsystems more important than others or is the severity of involvement the restricting factor? It seems that humans are resilient to minor disturbances within any subsystem. The nearly spontaneous repair in mild cerebral palsy and mild epilepsy (Figures 43, 44) provide support to this assumption. Also, the injury following a minor stroke repairs mainly spontaneously. The degree of resiliency is unknown. In severe brain injury there is also some spontaneous recovery in the first and second year following injury. The patient of the following case report (Sotiris) re-learned to walk a little within the first year following injury and no longer needed to be in the wheelchair. But the brain injury was too severe for a substantial spontaneous recovery. An efficient therapy method was needed.

1.5. Retarded, Accelerated or Deviant Motor Functions

Variability in motor performance is a characteristic of normal development. The impaired nervous system is not able to attain such variability in motor performance because the structure of the CNS is inadequate in enabling the patient with brain injury to use various modes of operation for single performance. When there is serious damage to the CNS it often happens that the usual modes cannot be achieved at all; the brain has to resort to unusual modes. Due to the deficiency of the 'hardware', distorted motor patterns arise which fall outside the range of normal variability. These abnormal movements are stereotyped as they are the only movements which the injured hardware is capable of performing.

Due to the reduced variability in motor performance of the impaired nervous system, patients with CNS injury or malformation have great difficulty in exercising on the special CDT devices because the device imposes complex motor pattern variability. But motor pattern variability can partly be re-learned from the device. Due to the deficiency of the neuronal structure, development of motor functions may be retarded, accelerated or deviant. Some functions may not develop at all, while others show only a decrease in variability.

1.6. Learning for Repair by Recapitulating Development

The mechanisms, or strategies, developed by the new-born, infant, or young child are not peculiarities or movement patterns which occur only in the neonatal or later period. They belong to one phase of the developmental course of the nervous system. These patterns are elements which have been tested and trained and which can be modified for use in more differentiated and complex motor behavior during later development. Their integrity must be regarded as a prerequisite for normal development during infancy. Sufficient crawling has to be trained by the infant in order to avoid coordination problems between arms and legs in later life. The stepping movement (automatism) of the new-born (Figure 24) (characterized by the emphasized lifting of the knees) is partly re-appearing in severe cervical spinal cord injury. After partial regeneration [11] (Figure 29), induced following CDT, the stepping automatism partly changes again into volitional walking.

The aim of learning is not only to equip the deviant patterns of the infant or patient with just one efficient mode of motion for each motor performance, but to lay down a foundation in the CNS for adaptive responses to specific circumstances and to induce learning transfer to other movement patterns which cannot be trained. Jumping on springboard and exercising on the special CDT device improves continence in severe cerebral palsy [14] and spinal cord injury [13] by learning transfer [12]. Exercising walking in the backward direction on treadmill improves the forward walking in stroke patients by learning transfer.

The learning process may be hampered by a deficiency of the neuronal structure on which they are dependent. Stimulating the patient to use uncommon modes of operation may enable him to achieve results which he cannot achieve spontaneously. If an adult patient or child cannot jump in anti-phase on a springboard (Figure 2D) continuously, he may learn it by jumping with both feet on the springboard in between anti-phase jumps, for example. Elements of function also have to be learned which are necessary for more complex motor patterns. If one recognizes which learning processes are deficient, one may then be able to offer the infant or patient a cue to start other learning processes or help him to eliminate errors and to progress in the right direction. If treatment strategies become more personalized to the needs of individual infants or patients, it may become more difficult to evaluate the effectiveness of any one treatment program. But if CDT were 10 times more efficient than conventional treatments, differences in effectiveness can be found. Proper treatment of the handicapped infant or patient consists of close co-operation between an active parent and an active therapist.

2. Repair in Traumatic Brain Injury

2.1. Repair in a Group of Patients: Summary

In six patients, aged between 13 to 47 years old, who suffered traumatic brain injury two to seven years ago, central nervous system (CNS) functioning, including motor, vegetative and higher mental functions, could partly be restored by applying coordination dynamic therapy for three months. The repair was quantified by the improvement of walking, jumping and hand function and the direct measurement of the coordination dynamics of coordinated arm, leg and trunk movements. The CNS organization (the low-load coordination dynamics) improved by 69% from 29 to 5.9 within three months of therapy. The mean coordination dynamics achieved (5.9) were slightly higher than those in a population of untrained healthy physiotherapists [50] (coordination dynamics = 5.1 ± 1.2).

Coordination dynamics

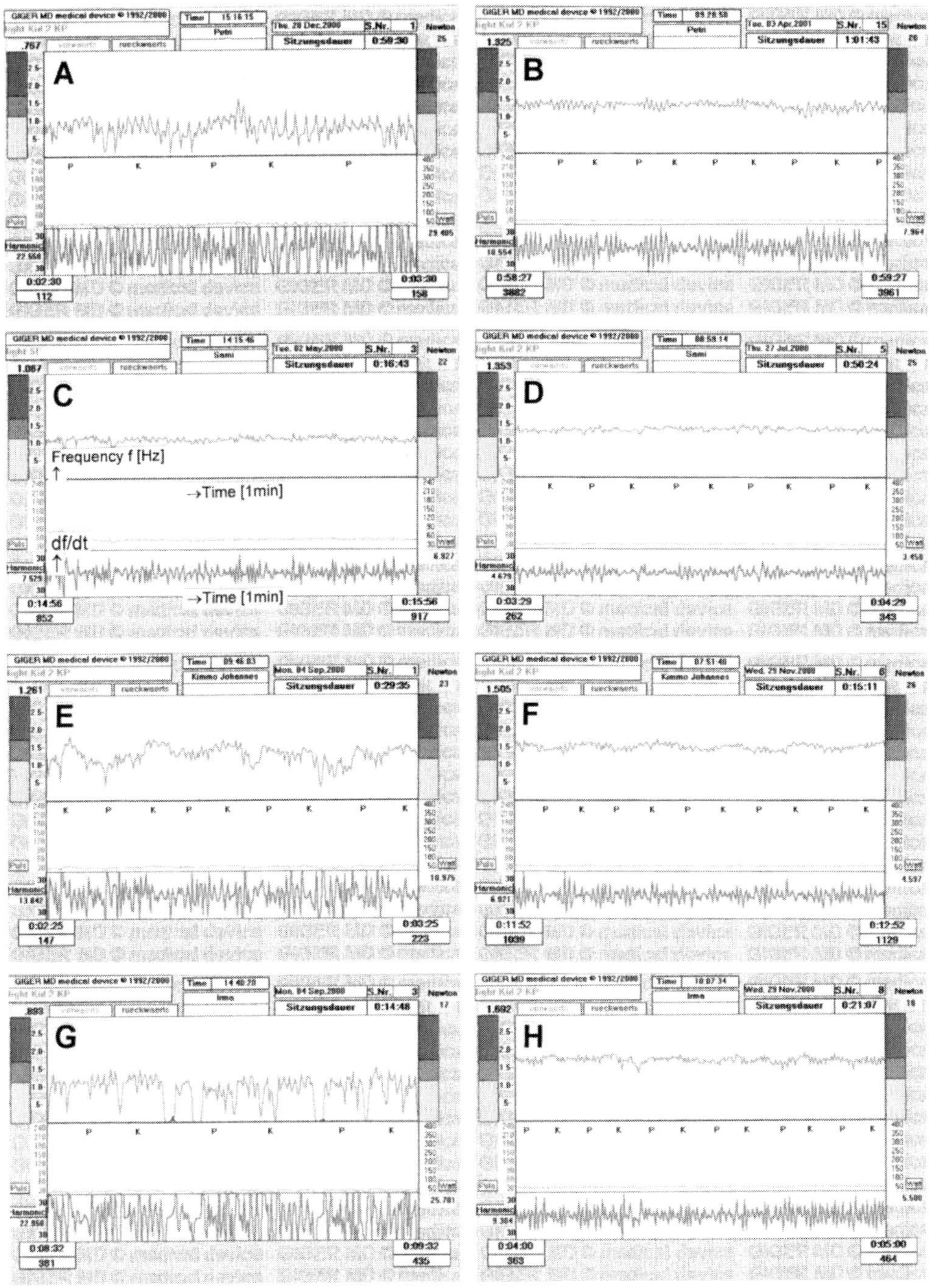

Figure 45a. Coordination dynamic recordings for forward turning of 4 patients with traumatic brain injury at the beginning of therapy (A,C,E,G) and after 3 months of therapy (B,D,F,H). Ordinate, f (upper trace) and df/dt (lower trace). Abscissa, time coordinate (1min from left to right) calibrated by the pace (P) and trot gait (K) coordination's. Note that the amplitude of variation of frequency (Δf) and its time derivative (Δdf/dt) reduced and the frequency of turning increased in all 4 patients.

2.2. Traumatic Brain Injury Recovery in Six Patients

Six patients suffered a traumatic brain injury at ages between 13 and 47 (mean = 30) years, four in a car accident, one in a motorcycle accident, and one as a result of surgery to remove an astrocytoma. They underwent coordination dynamic therapy for 3 months two to seven years after the brain injury with at least 4 hours therapy per day, 5.5 times per week. The relearning in physiologic CNS functioning was quantified by the improvement of movements like walking, jumping and hand grip power (Figure 45c) and by measuring the coordination dynamics at the beginning and at the end of the therapy (Figures 45a,b). It can be seen from Figure 44a that in four of the cases, the frequency of turning increased and the variation of the frequency reduced. The increase of turning frequency and the reduction of the coordination dynamics Δ(df/dt)/f indicate that the injured CNS could generate better coordinated movements at a higher speed.

For the six of the patients, the mean frequency for forward turning increased from f = 0.91 ± 0.21 Hz to f = 1.43 ± 0.15 (mean ± s.d.) and the coordination between arm and leg movements Δ(df/dt)/f improved, i.e., decreased, by 69 ± 12 % (range 50-81%) from 23 to 5.9 within three months of therapy.

Even though the coordination dynamics in one of the six patients (Figure 45aC,D) were unusually good at the beginning of the therapy (Δ(df/dt)/f = 6.9) and nearly as good as in a normal untrained population of physiotherapists (Δ(df/dt)/f = 5.1, see below), the improvement made was still 50 % following therapy.

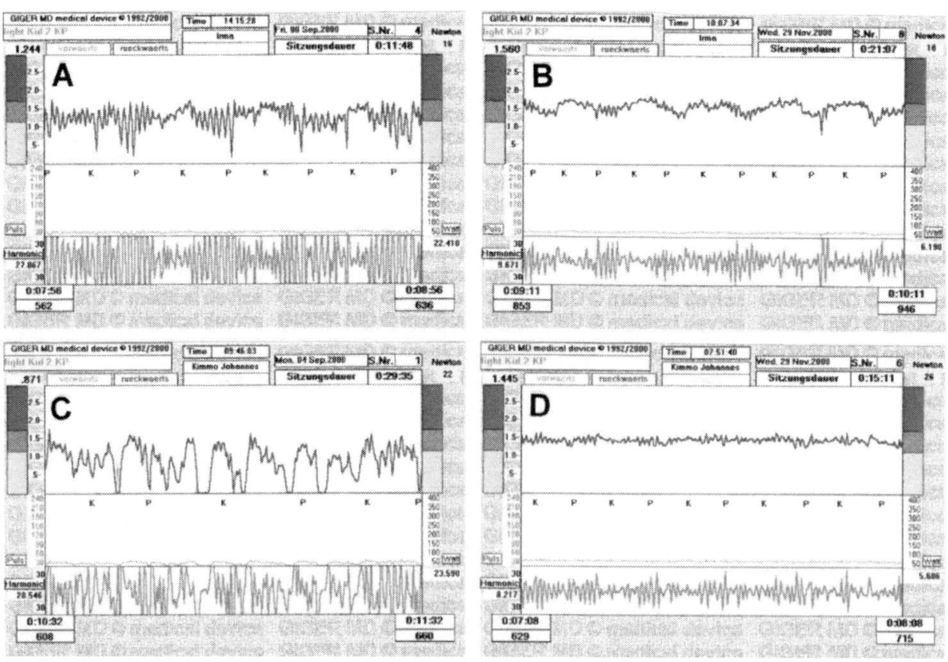

Figure 45b. Coordination dynamic recordings for backward turning of 2 patients with traumatic brain injury at the beginning of therapy (A,C) and after 3 months of therapy (B,D). Ordinate, f (upper trace) and df/dt (lower trace). Abscissa, time coordinate (1min from left to right) calibrated by the pace (P) and trot gait (K) coordination's. Note that the amplitude of variation of frequency (Δf) and its time derivative (Δdf/dt) reduced and the frequency of turning increased in the 2 patients.

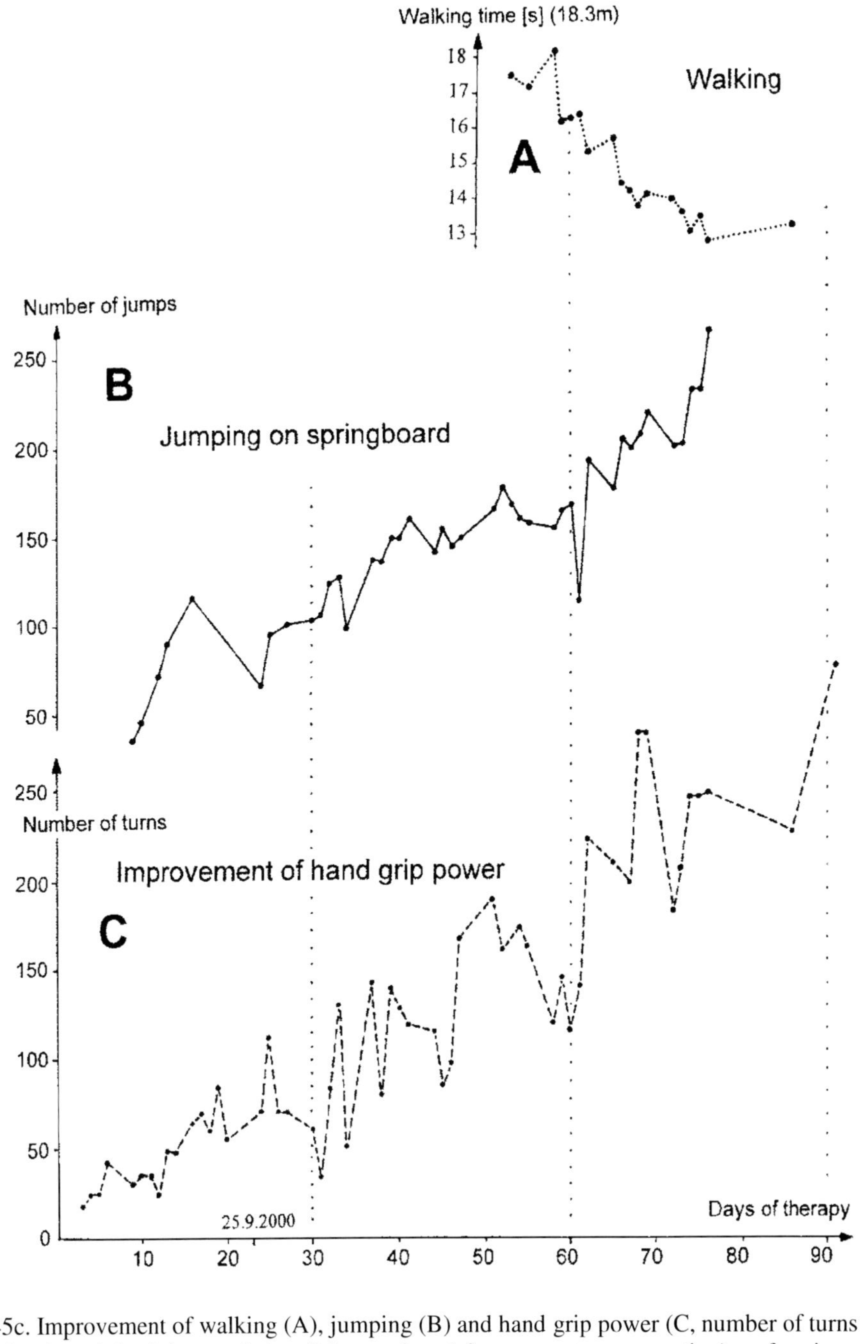

Figure 45c. Improvement of walking (A), jumping (B) and hand grip power (C, number of turns without fixation of the right paretic hand) in a patient (after astrocytoma operation) performing coordination dynamic therapy for 3 months.

The largest value of improvement of the coordination dynamics was with 81%; much larger than the lowest value (50 %). The largest value of the coordination dynamics before treatment ($\Delta(df/dt)/f = 43.1$) was very much larger (approximately 6 times) than the lowest value ($\Delta(df/dt)/f = 6.9$). It is valuable to take the mean of improvement among the group at beginning and end of the therapy period. However, it is better to calculate the improvements in each patient individually and then average them rather than calculating the global mean of the study group because improvement in single patients reflects the history and the individuality of CNS functioning.

In the first patient of Figure 45a Δ(df/dt)/f changed according to pace (P) and trot gait coordination (K). The intermediate coordination's seemed to be more difficult to perform (higher variation or transiently lower frequency of turning). In the other patients, the difference between the easy (pace and trot gait) and difficult (intermediate) coordination's were not so pronounced. When watching the coordination dynamics of a patient in real time on the computer screen, one can see that the organization of the CNS is changing all the time.

For backward turning, the coordination dynamics also improved substantially. In the two cases depicted (Figure 44b), the coordination dynamics improved within three months by 72 and 76 %.

The indirect measurements of CNS organization by observing the improvement of movements also showed substantial progress in all patients. In Figure 45c the improvements of walking, jumping and hand grip power of the fourth patient (Figure 45aG,H) are shown. This patient relearned to walk faster (Figure 45cA) with a better performance, to jump longer on springboard (Figure 45cB) and to turn on the device for a longer time without fixation of the paretic hand (increase of hand grip power, Figure 45cC).

In most patients muscle power, balance, walking speed, function of the weaker hand, general condition and self-confidence became better. Where present, hyper salivation and aggressiveness reduced. In two of the three patients with speech problems, speech improved.

2.3. Coordination Dynamics of Physiotherapists

In six healthy physiotherapists the coordination dynamics were measured [50]. Their age was between 21 and 33 and they were untrained with respect to exercising on the special coordination dynamic therapy device. Their automatic lower frequency of turning was 1.15 ± 0.09Hz (corresponding to a walking speed of approximately 4km/h). Their coordination dynamics for forward turning were Δ(df/dt)/f = 5.1 ± 1.2 (range 3.7 to 6.3) for forward turning. When the author exercised intermittently with a high concentration for a few hours, he could reach a value of Δ(df/dt)/f = 2.5 for forward and backward turning (frequency of turning = 1Hz). His coordination dynamics fluctuated between 3 and 4.

2.4. Coordination Dynamics Improvement in Patients Who Suffered Stroke or Traumatic Brain Injury

In six patients with traumatic brain injury the coordination dynamics could be improved by 69%. In four stroke patients, who obtained the same therapy coordination dynamic therapy for three months, the coordination dynamics improved by 70% [1]. Pooling the original data of the stroke and brain injury patients together, the coordination dynamics in stroke and brain lesion (ten patients) improved by 69 ± 12 % (range 50 to 87) if the patients obtained intensive coordination dynamic therapy for three months.

2.5. More Substantial Relearning and Structural Repair with Longer Therapy Times

It has been measured in stroke patients that when the coordination dynamic therapy (CDT) was applied for longer periods, then the recovery was much greater [1]. A musician even relearned hand and finger functions and could play violin again when the CDT of moderate intensity was applied for just over four years (see below).

In two patients with traumatic brain injury, further substantial recovery was achieved when the therapy lasted longer (at least six months). The movements became better and the coordination dynamics further improved. In two children with cerebral palsy and two patients with spinal cord lesion a substantial rather sudden further improvement occurred after approximately five months of continuous intensive coordination dynamic therapy. It was as if something was growing in the injured CNS which required at least five months maturing.

A now 39-year-old patient suffered severe poliomyelitis sub Th10 at the age of one which resulted in complete paralysis of the lower body. Arm functions were also impaired. After 36 years (at an age of 37), the patient started CDT. After three months of CDT, the patient could move the legs a little when gravity was eliminated. After 7.5 months of therapy, the patient could stand for a short time and walk two steps backwards when supported. After two years of therapy she achieved more leg and trunk functions and the legs started to grow taller. Menisci were built, which she had not before. This therapy-induced growth probably also includes a neurogenesis of motoneurons, since it is believed that poliomyelitis destroys the motoneurons. It seems therefore that neurogenesis in the adult CNS is induced if CDT is applied for periods in excess of five months. Therefore, longer therapy times are needed for patients with severe CNS injuries. Therapy times between six months and a year minimum are needed, depending on the severity of the CNS injury. It would be even better if intensive therapy were administered for more than three years. In the case report below, intensive CDT will be administered to a patient with severe traumatic brain injury for more than three years. It will be shown that substantial repair could be achieved.

Generally, a movement-based learning therapy has to be specific, intensive and sustained. The patient must be pushed to go to his limits and treatment should be started as early as possible. The treatments discussed below will support this. Further indication will be given that CDT is not only capable of repairing the human CNS, but also other body functions are repaired. This is readily understandable because the nervous system is involved in nearly all body functions. CDT even seems to be able to activate tumor suppressor genes to reduce the probability of a reoccurrence of cancer goes beyond repair and towards prevention. The important question is how CDT influences the epigenome to activate a landscape of gene expression for repair and protection.

2.6. Electromyographic Indication for Neurogenesis of Motoneurons

In the above mentioned poliomyelitis patient, electromyography (EMG) was performed on the quadriceps femoris and then compared to the biceps brachii to give an insight by electromyographical means into the improvement of CNS functioning (Figures 4,5 of [2]). After three months of therapy, it is likely that a few α_3-motoneurons (S-type muscle action potentials have very low amplitude) were activated when the patient turned on the special CDT device. After 7.5 months of therapy, the EMG-activity in the quadriceps femoris muscle increased further. The amplitudes and the duration of the action potentials of the motor units point toward the activation of FF muscle fibers, which are activated by α_1-motoneurons. From frog experiments it is known that the thick fast conducting axons regenerate fastest [14(Chapter I)]. The additional rather sudden appearance of action potentials of the FF-type (fast-fatigue, ATPase-type IIB) with ongoing therapy indicates a regeneration process. This argument is supported by the observation that the proximal leg muscles, closer to the CNS, were re-innervated before the more distal muscles. The process of the restoration of spinal cord functions in patients with a spinal cord injury is similar in that it is mostly the proximal muscles that start to work before the distal muscles.

The recording of muscle action potentials of long duration point further towards the building of new motor units. The irregularity of the muscle action potentials in the formally plegic muscles indicates further that new synapses are being built in the spinal cord, since newly built synapses fire first irregularly. The EMG findings support regeneration of the CNS and re-innervation of the leg muscles in the poliomyelitis patient.

2.7. Adult Neurogenesis, Cell Proliferation, Motor Learning and Function

The clinical and the EMG findings in the poliomyelitis patient are only an indirect indication for the building of new motoneurons and possibly other neurons in the CNS. The report of the building of new motoneurons in complete cervical spinal cord injury is more tangible evidence [11]. The biochemical and anatomical evidence concerning neurogenesis strongly suggests that neurogenesis can occur in the adult CNS. However, this does not prove that these new nerve cells are functionally relevant. When working on motor endplates of slow muscle fibers in the frog, in the process of innervating slow muscle fibers I morphologically found healthy looking synapses which were not functioning when measured electro-physiologically (Chapter I of [14]). When applying CDT to patients, it seems that at the beginning of therapy there is some progress taking place which may come from some newly built neurons which were already present in the CNS, but were not being functionally used.

I have discovered that in frog morphogenesis that many motoneuron axons were growing into the pyriformis muscle, but only some of them established stable functioning synapses and survived [14].

There are reports that motor learning enhances adult neurogenesis and cell proliferation in the hippocampal formation of rats and mice [93-95]. Some conclusions on adult neurogenesis are summarized and compared to strategies used in neuro-rehabilitation of patients:

Proliferation and survival of newly formed neurons can be affected by training methods in mice and rats (in the hippocampal formation), especially in cases of suboptimal CNS functioning due to non-use. Movement must be encouraged in patients with CNS injuries.

The survival rate of labeled neurons was more than two times longer in animals which had learned the hippocampus-dependent task than in those learning a similar but hippocampus-independent task. With respect to applicability to humans, this probably means that an injured part of the CNS must be activated during therapy to be repaired.

It is important to distinguish between effects on the formation of new neurons and effects on their subsequent survival. New neurons may be more sensitive than more mature neurons to the effects of activity, and it is possible that the period of maximum sensibility may begin shortly after the neuron is formed. Long continuous motor learning is probably needed for patients to induce neurogenesis *and* to make use of it.

Voluntary exercise of mice in a treadmill increased cell proliferation, cell survival and net neurogenesis. Motor learning may be a specific stimulus for epigenic neurogenesis. This is what the practical experience with patients suggests: for reorganization to be achieved there must be more than just reorganization of the injured CNS. New neurons are needed if some critical parts, for example connecting parts, of the neuronal networks are destroyed. Treadmill running and free running was used in patients, wherever possible, for the reorganization of the injured CNS.

The activation of nerve cells alone is not an adequate stimulus for adult hippocampal neurogenesis. This argument accords with the theory of CDT and is of immense consequence for the methods applied in neuro-rehabilitation. It is of importance which strategies are used during efficient supervised learning to achieve repair. The injured neuronal network parts must be intelligently identified and activated during exercise. The movements and strategies used, such as inter-personal coordination, must be tailored to the patient.

The involvement of the hippocampal formation in learning: direct association between hippocampus-dependent learning and neurons generated in the adult hippocampal formation.

Stressful experiences known to increase levels of adrenal steroids and hippocampal glutamate release diminish the proliferation of granule cell precursors in the dentate gyrus of adult tree shrews and marmorset monkeys. Continually diminished production of new cells resulting from chronic stress or corticosterone treatment may contribute to performance decrements in hippocampus-dependent tasks under some conditions. Performance decrements were also observed in some patients when they experienced stress.

3. Case Report: Severe to Very Severe Traumatic Brain Injury – Patient Sotiris

3.1. Case Report

A 23-year-old male student suffered a severe brain injury in a car accident. He did not lose bladder continence which is an indicator of a very severe brain injury. He needed to use a wheelchair and his higher mental functions including speech were impaired. The patient attended a rehabilitation center for eight months and was able to dispense with the wheelchair. Mainstream medicine had given up on him. The family found Coordination Dynamics Therapy (CDT) on the internet and started it one year after the accident. At the beginning, he trained at least 20 hours per week and he later trained at the limit with 30 hours per week. When the author's involvement was increased and the patient's sister, also an engineering student, started treating her brother, optimal CDT was started. The injury was balanced in that the brain injuries were evenly distributed (Figure 45). He could perform most movements to a small extent. The most significant obstacle to the resumption of a meaningful life seemed to be the impairment of the higher mental functions. The quality of the higher mental functions of this cooperative patient fluctuated. Motor functions and speech improved after three years of therapy. Whether the patient's ultimate goal of continuing his studies in engineering remains to be seen.

3.2. Rates of Repair of Complex Motor Behavior When Exercising Against High-Load

To repair deficient or damaged structures, differentiated and complex motor behaviors have to be activated. Disturbances in the chain of events and missing elements for larger patterns are recognized by the organization of integrative patterns.

Figure 46a shows the improvements of CD value for different loads with on-going therapy. In comparison to the healthy CNS in Figure 36, it can be seen that the curves for different loads show very significant variations (this indicates instability during repair), need much more learning time (low rate of repair) and the corresponding CD values are much higher (lower quality of CNS functioning) than in the healthy CNS.

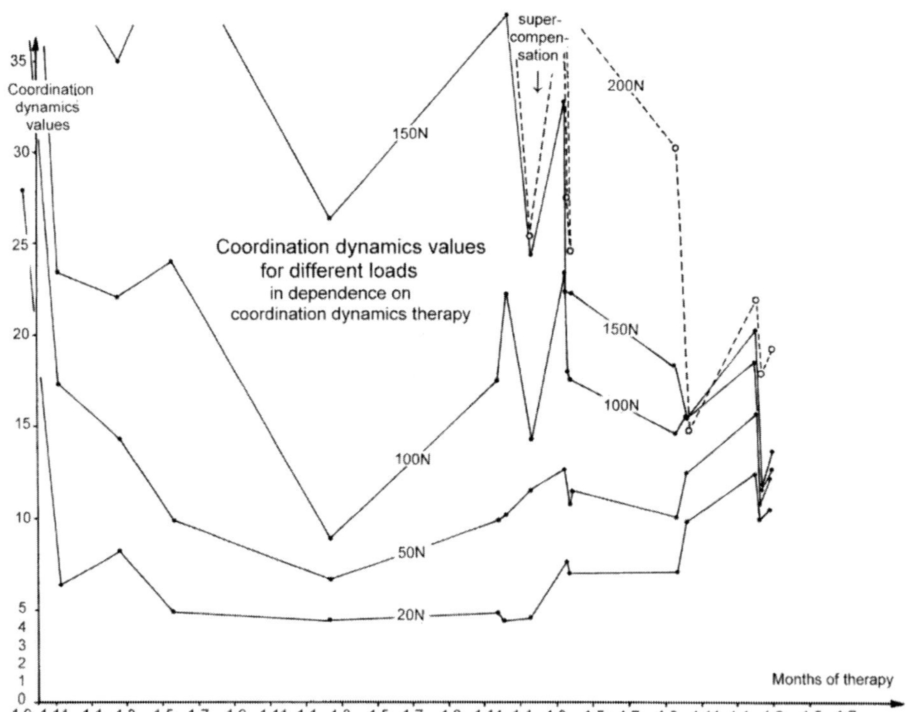

Figure 46A. Improvement of coordination dynamics values of a now 27-year-old patient with severe brain injury following car accident for exercising against different loads during coordination dynamics therapy for several years. Note that the curves for the higher loads vary quite much (instability of network organization), lye at high (worse) CD values and decline much more slowly than those for low loads.

To reach the plateau in learning for 20N, 4 months, for 100N 14 months and for 150N 3 years (36 months) were needed. In comparison to the rate of learning in the healthy CNS in Figure 36, the repair-time needed for this repertoire of motor patterns was approximately 2, 7 and 18 times longer respectively. The rates of repair were approximately 2, 7 and 18 times smaller. In practice, if a healthy subject needs one year to learn a certain complex pattern, the patient would need at least 18 years. The plateau for 200N has not yet been reached. However, if the learning time for 200N were determined, the rate of movement-based learning in this patient would most likely be reduced by at least a factor of 20. This reflects the period of time that may be required to fully repair his nervous system with repertoire of movements that he is currently performing. It is therefore very important to frequently adjust the repertoire of movements to achieve further repair.

The importance of the measurements in Figure 46a is that it can be seen that the repair of differentiated and complex motor patterns, induced by the exercising against higher loads, is much slower. The exercising of such integrative patterns is necessary to activate (reach) and repair all the damaged sub-networks which are needed to set the chain of repair events in train (repair in terms of time) the generation of larger patterns (repair in terms of space) and learning transfer, whereby different patterns share similar neural networks for the performance of distinct functions. This is needed in this patient for repair of the higher mental functions.

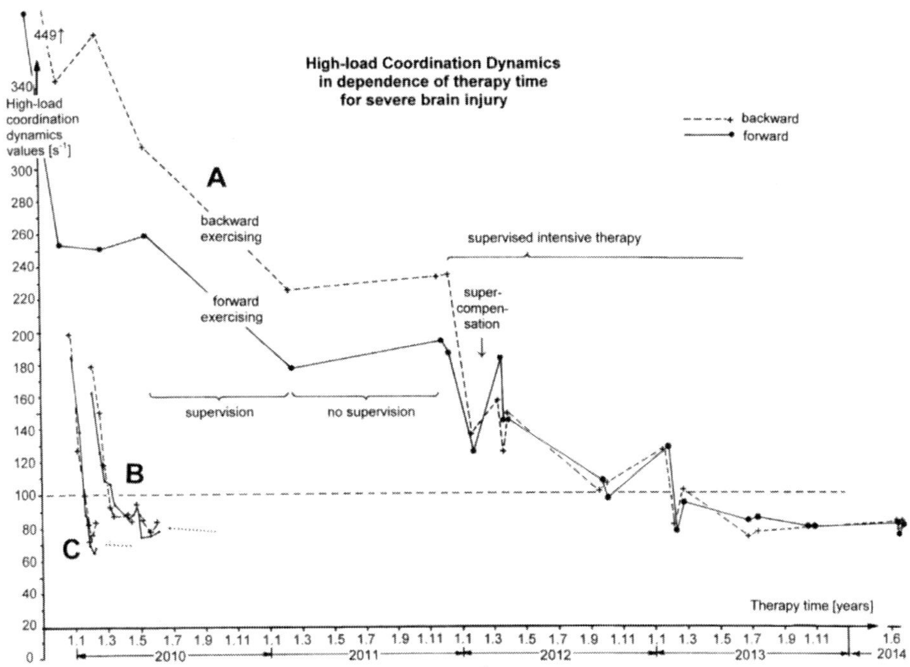

Figure 46b. A. Improvement of high-load coordination dynamics (CD) values in a patient with severe brain injury (same patient as in Figure 46a) upon coordination dynamics therapy for several years. The high-load CD values were obtained by summing up the single CD values for forward and backward exercising, Δ (high-load CD value) = Δ20N + Δ50N + 100N + Δ150N + 200N + Δ150N + Δ100N + Δ100N + Δ50N + Δ20N + Δ20N). B, C. For comparing the rate of repair the improvement curves of the high-load coordination dynamics values of an athlete (C) and a healthy pupil (B) are inserted. Note that the brain-injured patient needed much more time to achieve similar good CD values.

3.3. Repair Rate Quantified by High-Load Testing

More repair aspects can be explored by evaluating all high-load measurements. The overall improvement of CNS functioning while performing CDT can nicely be seen by plotting the high-load values. It can be seen from Figure 46bA that the improvement of the forward-backward symmetry of exercising took 40 months. The healthy pupil learned this forward-backward symmetry in half a month (Figure 39). The rate of learning for this forward-backward symmetry was reduced by a factor of 80 in this patient. In stroke patients, the repair of the right-left symmetry is an even greater problem.

From Figure 46bA, it can further be seen that supervised intensive therapy is more efficient for repair than training without supervision.

It can also be seen from Figure 46b that the overall CNS repair needed longer than three years. The insets B and C of Figure 46b show the repair curves of the normal pupil (B) (Figure 39) and the athlete pupil (C) (Figure 40) adapted to the time scale. It is obvious that the healthy pupils learned faster. The CD value curves reduced quickly to the plateau, indicated by a dotted line. The overall performances of the movements were also better. The CD values for forward and backward exercising started from around 390 to 450 and those from the pupils at 160 to 200.

3.4. Super-Compensation

In the normal pupil the phase of super-compensation occurred approximately after three months (Figure 36) and in the athlete after 1.5 months (Figure 37). The movement-based learning depends in the case of the healthy pupils also on the intensity of treatment, i.e., how frequently the assessment sessions were repeated.

Repeating the high-load test every three days may be optimal for a start of a fit person. With shorter repetition times the CNS gets exhausted but with longer intervals learning is not optimal. In mild cerebral palsy (Figure 43) and in mild epilepsy (Figure 44) the first phase of super-compensation occurred after one month, with something approaching optimal learning, and 3.5 months where learning is not optimal in repeated assessment sessions. In the patient with severe brain injury the probable first phase of super-compensation occurred after 26 months of therapy (Figure 46bA).

In conclusion, in healthy subjects the first phase of super-compensation occurs approximately after two months of movement-based learning. In mild CNS disturbances, the rate of learning is normal or only slightly reduced in accordance with the knowledge that humans are resilient to minor disturbances within any CNS subsystem [43]. In severe brain injury, the rate of movement-based learning is reduced by a factor of 10 to 20 (one to three months as opposed to 26 months).

Obviously, the CNS is not resilient to severe injuries. It will be shown below that in very severe brain injury the rate of movement-based learning is so reduced that certain functions cannot be re-learned at all.

3.5. Repair of the Stability of the Pattern 'Running on Treadmill'

It is not enough to re-learn a pattern. It is important that the patient can maintain a given pattern as long as desired. Therefore, pattern stability also has to be repaired. When the patient with severe brain injury (Figure 45) learned to run on a treadmill (Figure 47a), the stability of the pattern for running was repaired. The stability of the repair was measured. The measure used for stability increase was how long the patient can run without losing the pattern. The longest running period per day was used to quantify repair progress. He could manage his balance problems by briefly touching the rail of the treadmill. However, when he lost the pattern, the treadmill had to be stopped and he had to concentrate for some time before starting again.

Figure 47a. Patient with severe brain injury following a car accident (Sotiris) during running on treadmill.

Jumping on a spring-board (Figure 2D) could also be interrupted by a loss of stability. However, the performance would worsen prior to a loss of stability. Poor performance jumping is not beneficial for learning and increases the risk of damaging the ankle joints. Further, the loss of pattern stability is not exact. While running on the treadmill (Figure 47a), on the other hand, because he could not vary the pattern, he would lose the pattern completely. Therefore, the pattern was either or lost.

Figure 47b shows the improvement of running at 8.5km/h on treadmill. At the beginning he could run only 2min before losing the pattern (trace 03.2012). Eight months later, he could run 30min (trace 01.2013). The longest running time per session (running pattern stability) varied. The absolute variability of pattern stability was increasing, but not relative variability in % (sd values at the traces of Figure 47b).

To see more clearly the running pattern stability, the longest running times per month were plotted. Figure 47c shows the increase of the stability of the running pattern. After two years of therapy, the patient could stay in the running pattern for only two minutes. Ten months later, he could run continuously for half an hour without stopping. The stability of the running pattern had increased by a factor of 15. Now the patient could be motivated to run also in the sports field.

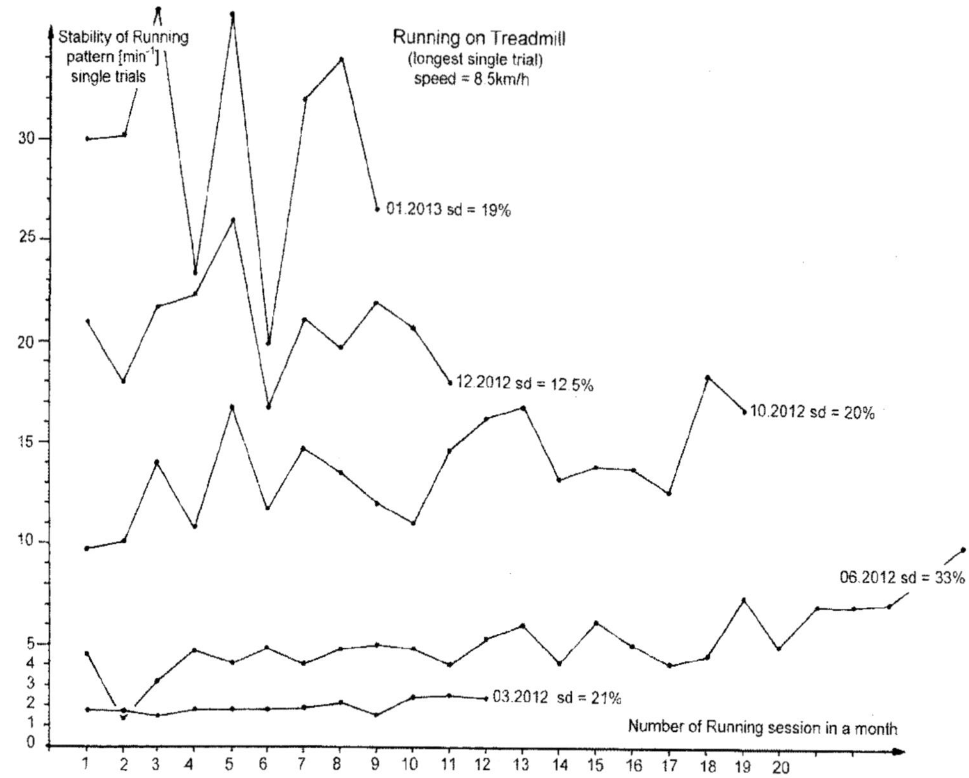

Figure 47b. Improvement of running on treadmill for different months during the years 2012 and 2013. sd = standard deviation in %. The stability of the running pattern in the brain-injured patient is quantified by the longest time in minutes he can run without slipping out of the pattern. Even though the absolute variability of pattern stability is increasing with longer running times (with further treatment), the relative variability in % is varying but not increasing.

As for jumping on springboard (Figure 18), the increase of the running stability can be pictured by a ball moving in a potential well. The stability is higher if the potential well is deeper and the ball (the state of the system) is jumping with smaller amplitude (the variability of the pattern is smaller).

This varying of the network state is given by the variability of phase and frequency coordination among neuron firing. It was shown in Figures 9 and 9e that the neurons of the human CNS organize themselves by phase and frequency coordination. Following CNS injury, this coordination becomes impaired, or more precisely, the variability of phase and frequency coordination sharply increases. The impairment of frequency coordination was shown in Figures 9e and 10 and the impairment of phase coordination is shown in the Figures 48a and b (taken from [12]). In the somewhat healthy case (with respect to the spinal cord) of a brain dead patient, the variability of phase coordination is small (Figure 48a). Following spinal cord injury for example, the variation of phase coordination increases (Figure 48b).

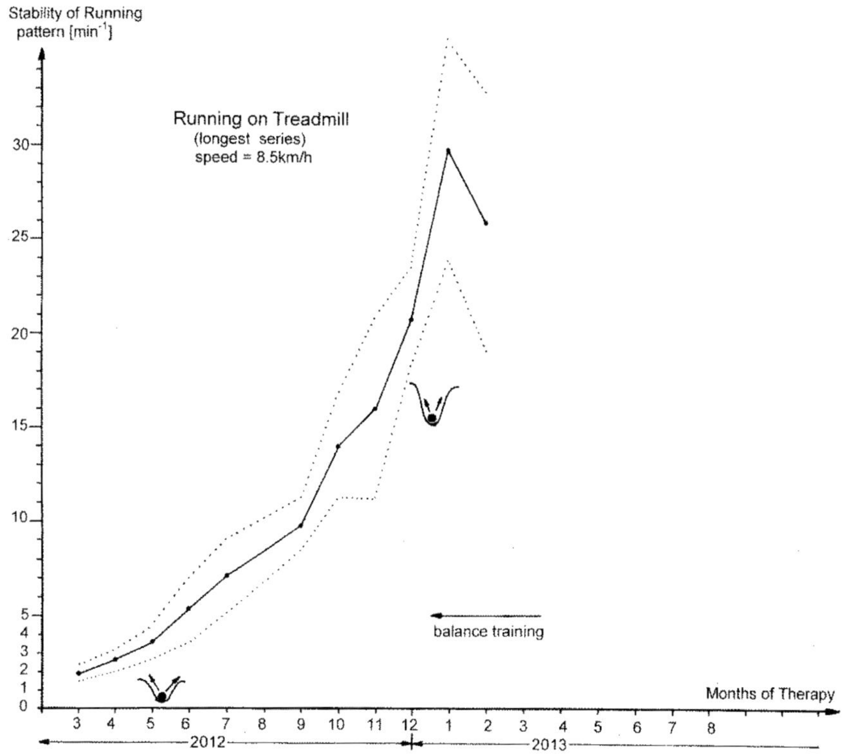

Figure 47c. Stability of the movement pattern 'running on treadmill in the forward direction' in dependence on therapy time in a patient with severe brain injury. The stability of the movement pattern 'running' is quantified by the longest time in minutes the patient could stay in the running pattern with several trials and following a warming-up running. Note that the stability improved strongly. Two insets characterize the increase of the 'pattern stability' with a ball (state of the system) in a potential well. This picturing of stability has its scientific basis in the system theory of pattern formation and human neurophysiology. Within the system theory of pattern formation the stability of a pattern is pictured by a ball (the state of the system) in an attractor layout (here the potential well). The variability of the pattern running (the 'jumping of the ball' in the potential well) is characterized by the lengths of arrows and has its scientific basis in the variability of the phase and frequency coordination of neuron firing (human neurophysiology). Note, with on-going treatment the potential well (the attractor) is getting deeper and the arrows are getting shorter (improvement of phase and frequency coordination).

The increase of running pattern stability of the patient is pictured in Figure 47c by a ball moving in a potential well. For increased running times the potential well is drawn deeper and the arrows characterizing variation of phase and frequency coordination are made smaller. The question is now, what contributed more to the increase of running pattern stability, the deepening of the potential well (according to the system theory of pattern formation) or the reduction of phase and frequency variation? As can be seen from Figure 46b, the increase of the running stability coincides with a decrease (improvement) of the high-load CD values (therapy year 2012). The improvement of phase and frequency coordination will have contributed to the improvement of the running pattern stability, but the main contribution for stability increase will probably have come from the deepening of the potential well, that is to say from a stronger establishment in the networks of the attractor running on treadmill. The increase of the stability of the pattern running was therefore achieved by both a deepening of the potential well for running and the reduction of the variability of phase and frequency coordination as shown in Figure 18 for jumping on springboard.

Healthy humans may tire during running but they do not lose the running pattern, because the stability of the running pattern is very high. It is as if the running pattern is hard-wired, similar to crawling. This patient also occasionally lost the crawling pattern, even though crawling improved strongly during therapy.

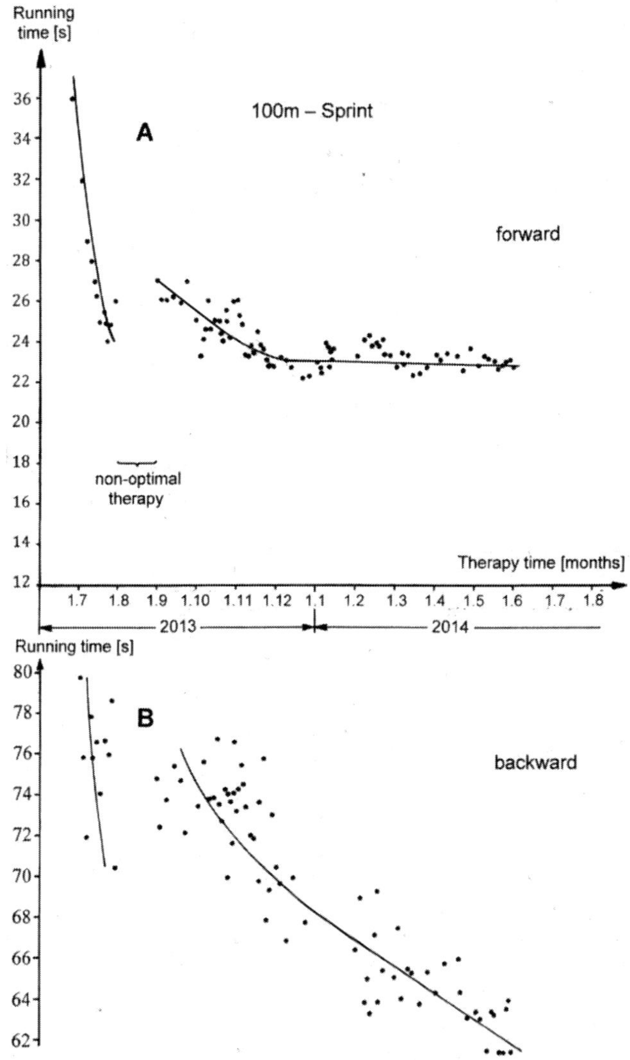

Figure 47d. Times of Sotiris for running in the sports field in the forward and backward direction.

3.6. Change of the Attractor Layout for Movements Following Additional Balance Training

As can be seen from Figure 47c, the stability of the running pattern reduced again when starting special balance training. This finding is in accordance with the system theory of pattern formation. When learning a new movement pattern, then the whole attractor layout of movements is affected. Some movements get better and others may get worse. With this balance training, the stability of the running pattern slightly worsened whereas other movements such as jumping and crawling got better. With further therapy the running pattern should improve further. By the summer of 2013, the patient could run repeatedly in a sports field. Running 100m backwards became possible. Sotiris also started to train with the brain-injured athlete Benjamin (Chapter XI).

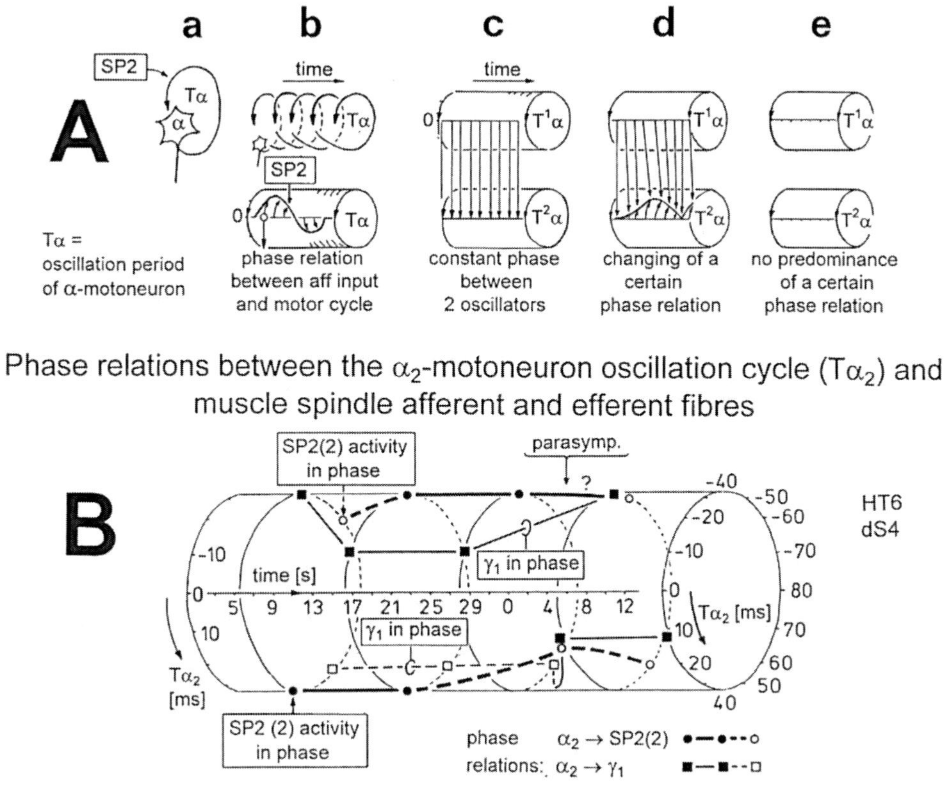

Figure 48a. (A) Derivation of the simultaneous description of interspike intervals and phase relations. (a,b) The oscillation period of an oscillatory firing α-motoneuron is schematically characterized by the length of the loop (perimeter). Successive oscillation periods with ongoing time yield a cylinder. Flashing with a stroboscope on such a cylinder with the same frequency as that of the rotation of the cylinder would make a black spot on the turning cylinder not move up or down. If the frequency of the cylinder or the stroboscope changed slowly, the black spot would move up or down. If the black spot moves from left to right with ongoing time, a curve is obtained. By replacing the flashing of the stroboscope by the occurrence of the APs of the spindle afferent fiber (or another oscillatory firing motoneuron) with respect to the APs of the oscillatory firing motoneuron, phase relation changes are made visible in the lower part of 'b' for a constant oscillation period (cylinder with no diameter changes). (c) A constant phase between two oscillatory firing motoneurons results to a constant line on the cylinder with ongoing time. (d) A changing phase gives a curve on the cylinder circumference. (e) If there is a loss of predominance of a certain phase between two motoneurons (the black spot gets diffused with ongoing time and is then lost) there is no line or curve. (B) Interspike interval and phase data from the brain-dead human HT6 (root dS4) are plotted in the representation of A. Filled dots and squares represent average phases (phase relations); thick and thin lines connect the dots to show trends. Note that the phase relations change only little; the frequency of the sphincter α_2-motoneuron ($1/T\alpha_2$) changes only little - the cylinder does not change its diameter.

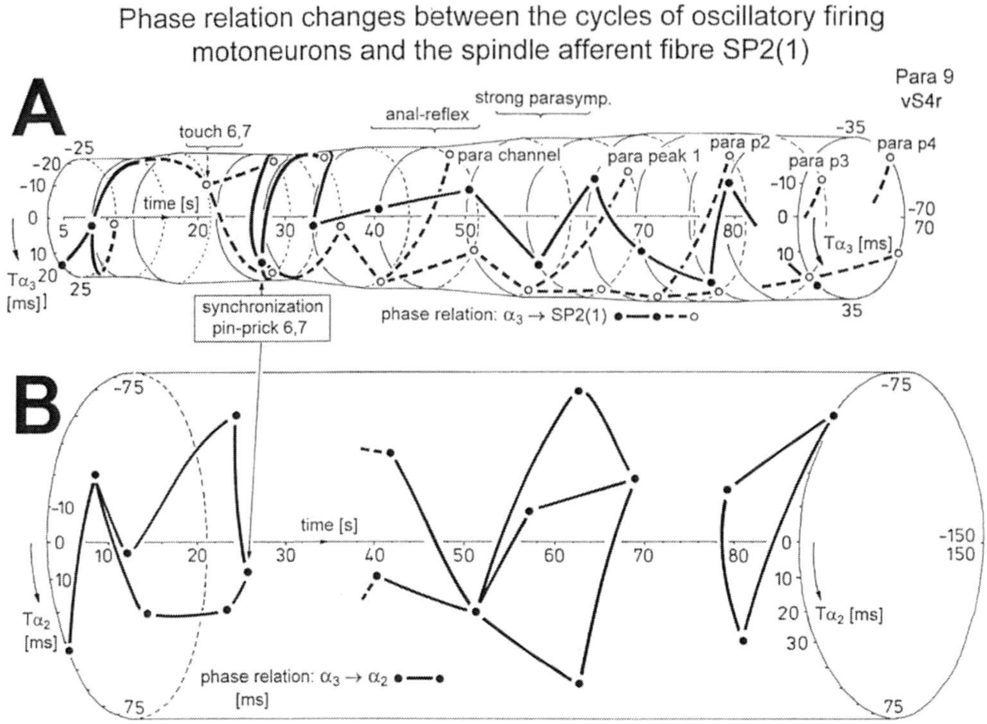

Figure 48b. (A) Phase relations between the secondary muscle spindle afferent fiber SP2(1) and the oscillatory firing α_3-motoneuron, taken from Figs.163-165 (and additional data), are plotted on the oscillation period cylinder $T\alpha_3$ (mean oscillation periods are taken from Figs.164,165) according to Figure 48a. The cylinder is changing its diameter (perimeter) because the oscillation period changes. Phase changes in ms are scaled on the cylinder circumference. The ongoing time (to the right) is scaled on the axis of the cylinder (time intervals are taken from Figure 10A of [12]). Existing phase relations are represented by dots (filled and open (back-side)); lines (filled and dashed (back-side)) only connect the phase relations to show trends. para peak 1, para p2, para p3, para p4 = activity peaks of the SP2(1) fiber due to parasympathetic activation (see Figure 10A right). (B) Phase relations between the α_3 and α_2-motoneurons plotted onto the oscillation period cylinder of the α_2-motoneuron. Dots represent phase relations, taken from Figures 11B,12B of [12]. Note that the phase relations of the paraplegic 9 are much more variable than those of the brain-dead human HT6 (Figure 48aB); also, the number of phase relations changes.

Since the principal goal of the patient at this stage of repair was to improve his higher mental functions, many patterns had to be trained to improve neural network function deep within the complexity of the nervous system's network organization.

3.7. Improvement of Higher Mental Functions

When the patient became able to run in a sports field in the middle of 2013 (Figures 200, 201), running in the sports field was trained every day. The times for natural running improved (Figure 47d). Four months later an improvement of the higher mental functions could be recognized. Before the injury and after the injury during training he sometimes became shortly aggressive.

Typical during high-load testing at 200N he got shortly angry and was shouting at her sister training him. This getting angry did not occur any more from October 2013 on.

This improvement of the higher mental functions has two interesting aspects. First, CDT could improve upon 3.5 years of CDT the higher mental functions in a patient with severe brain injury. Second, this angriness was part of his character and was not caused by the injury. It was found earlier in a patient that the aggressivety caused by a brain injury disappeared within 3 months of CDT. This means that the genetically determined character could be improved upon 3.5 years of CDT. Since the genetic code cannot be changed, it is likely that CDT modulated the epigenetic.

Such finding is important for other diseases of the higher mental functions. If for example a patient is depressive, which is inherited, then this patient has to take anti-depressive drugs for the rest of his life. On the long term the kidneys or the lever will be ruined or other side effects of the drugs will occur. But if the patient performs CDT for 20 hours per week, then probably in one years' time no anti-depressive drugs will be needed any more or its doses can be reduced. And because the not injured CNS learns much faster than the injured one (see below) not a few years are needed for the repair of the higher mental functions if the patient is cooperative.

It is likely that also schizophrenia can be attacked by CDT.

3.8. Impairment of Progress in CNS Repair Due to Interruption of Therapy

Patients are strongly advised not to stop therapy for longer than three days, otherwise the repair will stop and the repair that is stored only in the short and middle term memory may be reduced or lost.

Figure 47d shows the improvement in running times of the patient Sotiris. When the patient stopped running 100m in the sports field and had only performed little therapy in August 2013, his running times were worse when he resumed training (Figure 47d, non-optimal therapy).

The interruption of the therapy for one month worsened his running performance. With the continuation of therapy it took him three months to match his old running times before further improvement was made in his running.

This clearly shows that therapy should not be interrupted for more than a few days. However, this does not necessarily mean that he lost three months of therapy, but he lost some of the repair that he had achieved thus far. When top athletes stop training for one month the sports season is severely compromised. How much the chain of events is disturbed or interrupted at the molecular level is unclear. Interruption should ideally be avoided altogether.

In another brain-injured patient free running was also transiently interrupted, but optimal therapy, i.e., the performance of other movements for sustained periods, was otherwise continued. When the free running outside was re-commenced, the running times had not worsened in the meantime. This means that if a given movement within the whole movement repertoire is transiently left out, for example due to pain or injury, but the therapy is otherwise continued optimally, the patients' ongoing repair is not disrupted.

4. Repair in Incomplete (50% Injury) Cervical Spinal Cord Injury C5/6 – Patient Sten

4.1. Case Report on the Repair of an Incomplete SCI [10,14]

A seventeen year-old male patient suffered a cervical spinal cord injury (SCI) at C5-6 levels in a diving accident. The spinal cord injury was incomplete (ASIA grade D). Some sensory-motor functions were preserved below the level of injury. Initially, the patient received conventional rehabilitation treatment and then started coordination dynamics therapy (CDT) five months after the injury. When CDT was commenced, this patient needed a stick as support for walking, was wearing an orthosis on the left foot and emptied his bladder with intermittent catheterization.

During walking, the left heel was not touching the ground. While jumping on the springboard, his feet had to be supported by the therapist. While turning on the special CDT device, his left hand required fixation.

He received nearly three years of non-optimal CDT. Over the following five years, the patient was intermittently measured up to judge further improvement. One year after the injury the metal support used for spine fixation, was removed.

At the end of the third year of CDT he was able to walk, run, and even jump independently and did not need any internal or external support. He is continent again and has discontinued all medications. In that sense, it can be safely said that the patient is almost cured of his quadruparesis. In the following paragraphs, the clinical aspects and the theoretical concepts, particularly the rate of learning and pattern stability behind this recovery will be explained in detail.

4.2. Quantification of the Anatomical Extent of the Injury with MRI

Magnetic resonance imaging (MRI) of the cervical spine was performed three years after the accident to determine the anatomical extent of the injury. At injury site, the loss of tissue would indicate the severity of the injury, and the presence of residual tissue would indicate the extent of spared tissue bridging the injury site enabling connectivity. The T2-weighted images (Figure 49A) highlighted the pathology to be more extensive than what was seen on T1-weighted images (Figure 49D), and axial images delineated a more extensive injury than the sagittal images (Figure 49A-E).

By comparing these multiple multi-planar images the author estimated the spared tissue to be 50% of the cross sectional area at the site of injury. The grey matter of the central cord seemed to be more damaged than the circumferential white matter. Since in this case, the injury span was less than one vertebral level (relatively modest distance to bridge the injury site), it is possible that a small part of the visible spared tissue was actually from those tract fibers which regenerated during the therapy over the course of three years. It seems that in SCI the grey matter (the neuronal networks) is most vulnerable to pressure and/or lack of sufficient blood supply, whereas the white matter (the tract fibers) survives better.

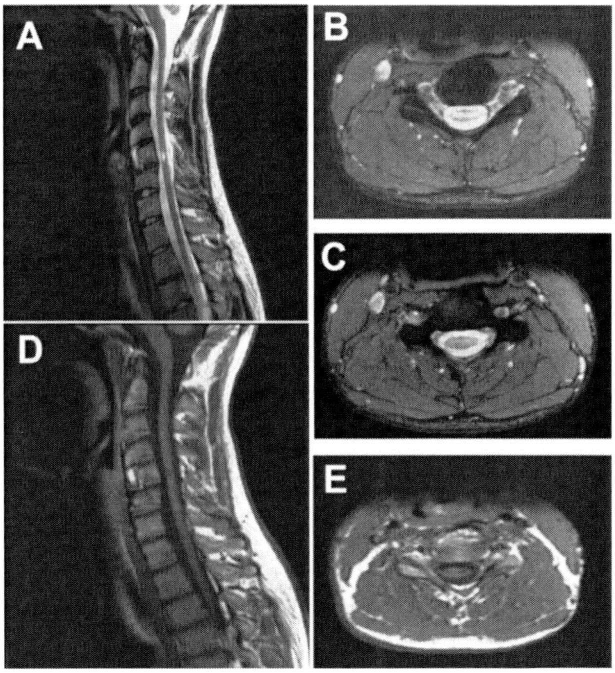

Figure 49. Magnetic resonance images show an incomplete spinal cord injury at C5/6 levels, 3 years after the accident, from a 20-year-old patient who jumped into shallow water. A,D. Sagittal T2 (A) and T1-weighted (D) images. The region of hyper intense signal within the spinal cord of the T2-weighted image (light) indicates no hemorrhagic spinal cord damage. B,C,E. Axial T2 and T1-weighted images of the spinal cord showing a more severe injury than in the sagittal plane. The authors estimated an anatomical injury of approximately 50%.

4.3. Improvement of Coordination Dynamics (CD) Values for Different Loads Following Therapy

The progressive changes in the CD values (mean pattern stability per minute) for loads of between 20N and 150N (and subsequently 200N) over the course of eight years while partly undergoing CDT, reflected gradual improvement in CNS functioning (Figure 50). As explained earlier, improvement in CNS functioning was indicated by the lowering of coordination dynamics values, i.e., decreased arrhythmicity of exercising and increased pattern stability. At the beginning of CDT, as the therapy progressed, there was strong improvement in CD values for loads between 20 and 150N for exercising in the forward and backward directions. This improvement was achieved in spite of non-optimal/unsupervised therapy. Subsequently the patient stopped CDT for more than six months, and then re-started. This break in the therapy resulted in the deterioration in coordination dynamics. Interestingly, when he re-started and continued the therapy, the values improved again especially for exercising at loads of 100 and 150N. Deterioration of CD values after stopping the therapy and improvement after restarting the therapy indicates that even after a few years of therapy, the functional repair of the CNS is incomplete and not permanent. In other words, to maintain an optimal level of improvement of CNS functioning, the continuation of therapy for longer durations is essential.

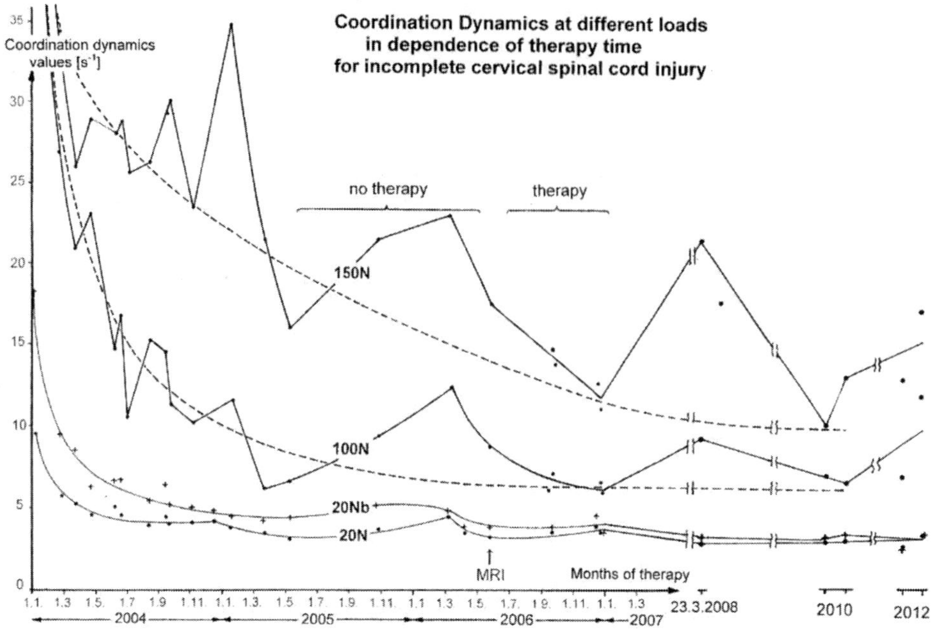

Figure 50. Relation of coordination dynamics values to therapy length for increasing load between 20 and 150N. The loads for forward exercising (dots, 20N, 100N, 150N) are marked at the curves (20Nb = backward exercising (crosses) at 20N). Note that with no therapy the coordination dynamics values got worse and upon administering therapy again the values improved again even 2 years after the accident. After stopping therapy, the coordination dynamics values for 100 and 150N increased again (dotted lines).

One year after finishing the therapy, the high load CD values worsened again in the patient (Figure 50), even though the patient felt more power in arms and legs and the low load values (20N) had not worsen. The re-increase of the high-load values further show that the CNS repair was still incomplete and not fully persistent. The patient would need a maintenance level therapy of about an hour per day. When the patient trained further by himself, CNS functioning varied. With training the patient's CNS functioning further improved, and with little or no training, it got worse. The changes of CNS functioning can be judged by the changes of CD values for 150N between 2007 and 2012 (Figure 50). The patient felt the fluctuations in fitness and movement performance.

4.4. Rates of Repair of Integrated Functions and Learning Transfer

It took the patient approximately four months to reach the slowly declining plateau value curve for a load of 20N while exercising on the special CDT device. Similarly good values were achieved for high loads of 100N after only 24 months (2 years) with on-going therapy and after 36 months (3 years) good values were achieved for 150N (Figure 50). Exercising at significantly higher loads requires recruitment of more muscle groups of arms and legs and also the trunk. Such recruitment of muscle groups of almost the entire body requires the entire central nervous system (CNS) to be activated. Thus, a widespread activation of downstream and upstream motor neurons and their relevant neuronal networks takes place. Such integrated (more systemic, more complex) network activation is important because it not only affects the attractor basin of a particular movement, but also changes the entire attractor layout of the state-space, including the spasticity attractors (Figure 19). As the whole attractor layout changes, also the intrinsic dynamics change, which causes learning transfer from the movement to other CNS functions, like bladder control. The induction of learning transfer by integrated (systemic) CNS activations is very important, since only those networks which can be recruited, substantially activated, trained or reached by learning transfer can be repaired. Therefore to achieve learning transfer, exercising at higher loads is required for longer durations, such as twenty minutes per day.

4.5. Comparison of the Repair Rates between 50% Cervical SCI and Severe Brain Injury

From Figure 50, when exercising against 20, 100 and 150N, the plateau values were reached after 4, 24 and 36 months respectively in 50% SCI. For severe brain injury repair periods of 4, 14 and 36 months were obtained (Figure 46a) for exercising against same loads. The rates of repair for the integrated CNS functions are similar for severe brain injury and 50% cervical spinal cord injury. In severe brain injury, it is largely the supraspinal centers that are repaired. In 50% SCIs it is probably the neural networks of the spinal cord that are repaired and supraspinal centers are functionally reorganized to partly change tract fiber function. Even though in brain and spinal cord injury the whole CNS is reorganized and the repair rates are similar, more stability in CNS functioning can be expected in the case of spinal cord injury because much more CNS matter is left undamaged in spinal cord injury. As Figures 46a and 50 show, the variability and stability of re-learning is better in SCI. The main problem in 50% SCIs is to the extent to which tract fibers can be made to re-wire (see further below). In conclusion, the rates of improving integrated CNS functions in 50% SCI and brain injury are similar, but may be ten to forty times slower than in the healthy CNS (Figure 36).

4.6. Repair Rate Quantified by High-Load Testing

The high-load CD values for forward exercising during therapy are shown in Figure 51. Values for backward exercising were not always available and at the beginning of therapy the SCI patient could not manage the load of 200N. At the beginning of therapy, the high-load CD values improved. The dotted line in the years 2005 and 2006 indicates how further improvement would have continued. It can also be seen that an optimal therapy is needed for a few years. Optimal therapy means the optimal movement-based learning technique has to be performed at the limits with a lot of motivation, no stress and good nutrition. That the patient is continuing by himself is good but it is no substitute for optimal therapy.

According to the high-load curve of Figure 51, the repair time until the plateau is reached is approximately 30 months. In the brain-injured patients it was at least 36 months (Figure 46bA), in the healthy pupil it was two months (Figure 46bB) and in the athlete, only one month (Figure 46bC). The rate of improvement of CNS functioning in 50% incomplete spinal cord injury is approximately 20 times smaller.

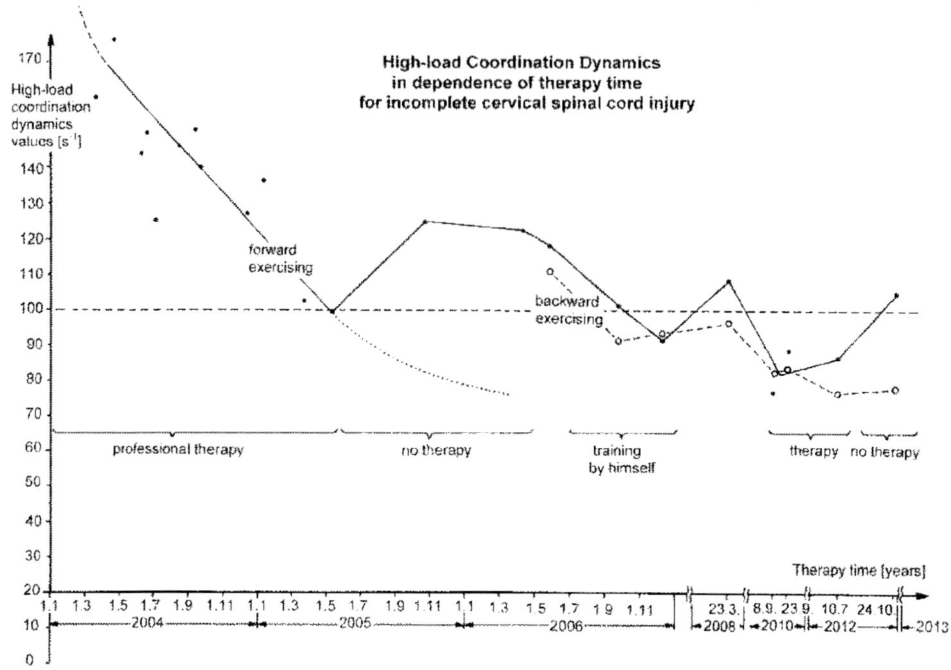

Figure 51. High-load coordination dynamics values in dependence of therapy for exercising in the forward (solid line) and backward direction (dashed line) of a patient with an incomplete cervical spinal cord injury (C5/6) (Sten). Note, with therapy the values decrease (improve) and without therapy increase (get worse). The high-load CD values were obtained by summing up the single CD values, Δ (high-load CD value = Δ20N + Δ50N + 100N + Δ150N + 200N + Δ150N + Δ100N + Δ100N + Δ50N + Δ20N + Δ20N).

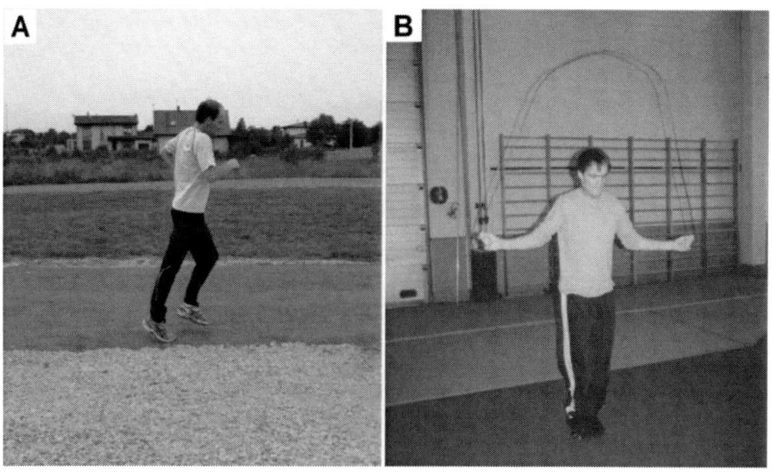

Figure 52. Patient with an incomplete cervical spinal cord injury at C5/6 levels (severity 50%) (Sten) during running in the street and jumping with a skipping-rope.

4.7. Motor Improvement

According to coordination pattern dynamics theory, if CD values improve, there should also be visible improvement in motor performance. The motor training of the patient (which included crawling, walking and running on treadmill in the forward and backward direction, jumping on spring-board,) was guided by the therapist to correct abnormal movement patterns. At the end of the 3 years of therapy, the significant improvement of coordination dynamics values (Figure 50) was accompanied by a substantial improvement in the performance of movements like running and jumping with the skipping rope (Figure 52). Quantitatively, the times required for running 60 and 400 meters decreased and the number of jumps per series increased (Figure 53).

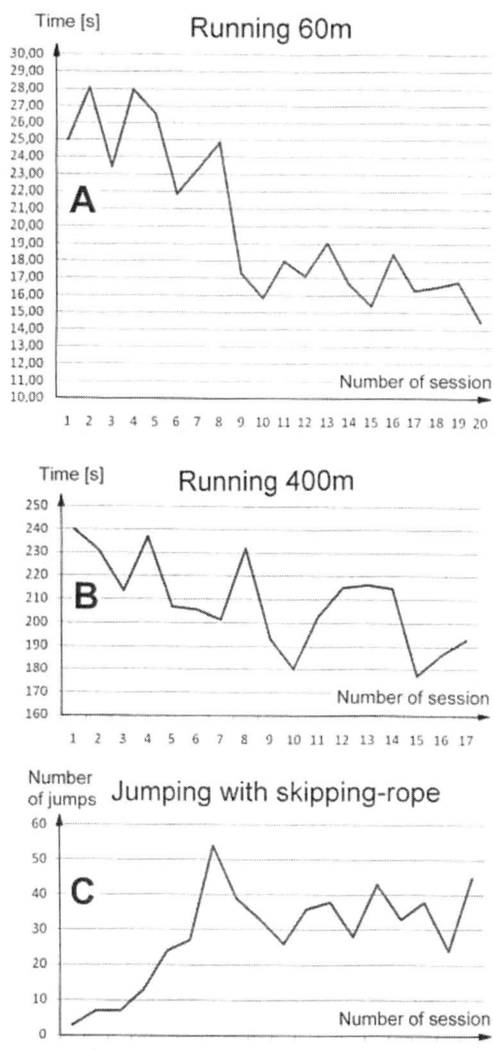

Figure 53. Improvement of times for running 60 and 400m (A,B) and increase of jumps with a skipping-rope per single session (C) in relation to on-going therapy sessions. Therapy period = second half of 2006 (see Figure 3).

4.8. Electrophysiology (sEMG)

With on-going therapy, CD values can be used to measure improvement in CNS functioning to within one percentage point (Figure 50). Clinically, this improvement can be correlated with the corresponding improvement in scores of activities in daily life (Figures 52, 53). However, in order to lay bare the physiological and pathophysiological changes occurring at neuronal network level in the CNS with on-going therapy, surface electromyography (sEMG) is required. Even single neuron activity can sometimes be measured non-invasively by sEMG. The firing pattern of single neurons (motoneurons) can be studied non-invasively in terms of single motor unit potentials.

4.8.1. Pathologic Patterns of Motor Activation

Surface EMG was performed in this patient while he was undergoing CDT. Figure 27 shows sEMG recordings of motor patterns while the patient was exercising on the special CDT device at the beginning of therapy (Figure 27A,B) and again after six months of therapy (Figure 27C,D). The motor patterns were completely abnormal at the beginning of therapy (Figure 27A,B). The amplitude of sEMG activity was high in a muscle proximal to the injury (i.e., biceps brachii), while significantly smaller in the muscles distal to the injury site (i.e., tibialis anterior and gastrocnemius muscles). Note the difference in calibration for displaying the amplitude. The motor patterns in the muscles distal to the injury site C5/6 show obvious improvement after six months of CDT (Figure 27C in comparison to Figure 27A,B).

4.8.2. Motor Pattern Stability

At the beginning of therapy, the motor patterns varied significantly while turning on the special CDT device, as can be seen in Figure 27A and 27B. This variation persisted for as many as six months after starting the therapy. In other words, the stability of the motor pattern of the movement's exercising on the special device by turning with arms and legs was very low at the beginning of therapy. This low pattern stability is also reflected in the high CD values (high arrhythmicity of exercising = low pattern stability) (Figure 50). Clinically this motor pattern variability correlated with the instability of movement performance, observed during walking, running on a treadmill at the beginning of therapy, jumping on springboard and especially jumping with the skipping rope (Figures 52,53C).

4.8.3. Impairment of Reciprocal Relationship of Antagonist Muscles

The natural reciprocal relationship of agonist and antagonist muscles of the leg, namely the tibialis anterior (dorsal flexor) and gastrocnemius muscle (planter flexor), was reasonably well preserved on the left side (Figure 27B,C) but was significantly impaired on the right side. It persisted for as many as six months in spite of therapy (Figure 27D). This impairment of antagonistic muscle action contributed to the difficulties faced by the patient while walking and running.

4.8.4. Measurements of Temporal Stability of Movement Patterns by Pattern Change

As explained above, the sEMG motor patterns were generated and recorded when the patient was exercising on the special CDT and recording device. The mean temporal stability of these motor patterns over one minute (= CD value = mean arrhythmicity value) was used to quantify improvements of CNS functioning with therapy (see above, Figure 50). To measure differences in temporal stability of different movement patterns, the coordinated patterns, involving arm and leg movements, need to be generated with accuracy. This is achieved by the high precision mechanics of the special device for turning movements, when the arms turn at a slightly higher frequency than the legs. The relationship of revolution between arm leavers and legs paddles is 19:18. This difference in revolution produces the changing of coordination patterns of arm and leg movements. Such pattern change must also be seen in the sEMG motor patterns.

The way that the muscle activations changed between the left biceps brachii muscle (arm) and left gastrocnemius muscle (leg) can be seen in Figure 27B. During four gastrocnemius activations (4 turns), the biceps were activated approximately 4.3 times (4.3 turns). This seemingly simple change of arm and leg movements between pace and trot gait is an extremely difficult task for the CNS to generate. All the muscle patterns of arms and legs have to be changed. And if the device is positioned further away from the trunk, then the trunk also has to perform rotating movements in coordination. Since this trunk rotation is changing with the coordination position between arms and legs, all the segmental trunk muscles change their coordination in the rostral and caudal direction. If the CNS is functioning physiologically, then the subject can turn rather easily and smoothly. But if the CNS is injured, then such on-going change of coordination's among arm, leg, and trunk muscles becomes a very difficult task. Patients are often sweating because of the stress the CNS is facing in generating these complicated movements as the pattern keeps changing. As can be seen from Figure 27B, during the time period of the recording, the patient's CNS could manage quite well the change of the coordination between pace and trot gait. Chaotic activation was not observed on either the biceps brachii or gastrocnemius muscle traces during the movement pattern change.

4.8.5. Impairment of Pattern Formation Is Revealed with Integrated CNS Activation

After increasing the load of exercising on the special device from 50N (Figure 27C) to 100N (Figure 27D), the motor patterns of the left tibialis anterior and gastrocnemius muscles partly deteriorated (Figure 27D). The specific patterns and their antagonistic relationship were partly lost. Pathologic synchronization of premotor spinal α_1-oscillators (Figure 8) occurred at 8 Hz, as can be seen from the trace of the left tibialis anterior muscle (Figure 27D). The increasing impairment of CNS self-organization with increasing load of exercising, made visible by the sEMG recordings, is in accordance with the increase of the arrhythmicity of exercising on the special CDT device for increasing load (Figure 50, increase of coordination dynamics values at higher loads).

4.9. Increase in Temporal Instability of Movement Patterns While Exercising against High Load

From the above discussion it is understandable that the increase of the CD values (deterioration of CD) while exercising against high load (Figure 50), is because of the reduction in the temporal stability of movement patterns as shown in sEMG recordings. Kinesiologically, this increased temporal instability of movement pattern formation is apparent when CD traces taken at one minute intervals and compared against increasing load (Figure 28). As can be seen from Figure 28, the arrhythmicity of exercising increased with increasing load. Exercising at 20N (A) and 50N (B), in the recording, the arrhythmicity varied without any specific structure. At 100N (C) and 150N (D), however, structure becomes visible in the CD traces. Note in Figure 28, that the CD values increased with increasing load. While exercising against a load of 200N in the forward direction (E), it can be seen that there is rhythmic increase and decrease of the arrhythmicity of exercising. At times, when the arrhythmicity is small, the arm and leg movement patterns have a high stability. Different patterns of coordinated arm and leg movements (produced while exercising on the special CDT device) have different degrees of stability. Those patterns which are more stable have small amplitudes of arrhythmicity; they are the attractor states among different movement patterns. The unstable movement patterns (with a large arrhythmicity of exercising) are not attractor states (see Figure 20). Sudden instabilities of movement patterns are perceived by the patient and healthy subjects as sudden increases in the resistance to smooth turning. These are sudden perturbations of CNS organization and are felt as a stutter which interrupts smooth exercising.

4.10. Exercising at High Loads Reveals Impairment in the Symmetries of CNS Organization

While the patient was exercising against a high load of 200N, rhythmic changes of the arrhythmicity of exercising were observed during forward (Figure 28E) as well as backward exercising (Figure 28F). Interestingly, the location of attractor states (patterns of high stability) were seen to be situated at different locations from the calibrated states of 'pace' (P) and 'trot gait' (K), as shown in Figure 28E,F. For forward turning, the attractor is located on the right side, while for backward turning, the attractor is located on the left side of the pace and trot gait patterns. This impairment of symmetry for forward-backward movements (different attractor states) became visible only with the very integrated activation of the CNS achieved when exercising against high load.

4.11. Improvement of Symmetries of CNS Organization Increased Pattern Stability

As the impairment of symmetry is diagnosed with the highly integrated (more systemic) activation of the CNS, in order to improve the symmetry, the high load training of those movements is also required. As the pattern stability with respect to symmetry increases, movement performance improves (Figure 18). In the case of an asymmetric injury and subsequent asymmetric CNS organization, the movement pattern 'jumping in anti-phase' can be so unstable that the patient is unable to perform this movement. In terms of system theory, the potential well for 'jumping in anti-phase' in the attractor layout is very shallow or no longer exists (Figure 18B, C). In this patient, initially, only in-phase jumping was possible without support. The anti-phase jumping needed assistance and deep concentration from the patient (Figure 18B, C). With training the patient re-learned to jump in anti-phase; the attractor basin for 'jumping in anti-phase' became deeper and the movement pattern stable (Figure 18A). At the end of therapy, the patient was even able to jump independently with a skipping rope (Figure 52B).

4.12. Repair of Bladder Function - Urinary Tract Infection

Before World War II, most patients with severe spinal cord injury died because of recurrent urinary tract infections. Needless to say that one of the most important functions to be repaired in patients with spinal cord injury is bladder control. Patients are often reluctant to undergo urodynamic evaluation in a rehabilitation center, because of the fear of contracting urinary tract infection with antibiotic resistant bacteria. Development of urinary tract infection is a major set-back to the therapy, as optimal therapy is not possible in when suffering from infections such as these. If urodynamic assessment is not available, some good practical indicators of improvement in bladder function may be seen in the reduction in the incidence of bladder infections and a significant reduction in the volume of post-voidable residual urine in the bladder. Residual urine can be measured by emptying the bladder with a catheter, after the patient has passed urine voluntarily. Residual urine of less than 50 ml indicates good physiological functioning of the bladder.

At the beginning of therapy this patient with an incomplete SCI was using intermittent catheterization for emptying the bladder. A few months after starting the therapy, he started emptying the bladder manually. Initially he had to compress the lower abdomen to empty the bladder fully; but later, more physiological, volitional bladder emptying became possible. Over six months of therapy, the residual urine reduced from 200 ml to 50 ml. Then the patient refused further measurements to avoid unnecessary catheter trauma to the urethral epithelium. The scientific basis of urinary bladder functioning is given in detail in Chapter VII. Bowel emptying was not a major problem in the patient. It is usual to empty the bowel once a day and there is no risk of infection as there would be in the case of the bladder. Since the neuronal network for primary sexual functions are located in the same sacral segment S2-S5 as are the centers of bowel and bladder function, improvement in sexual functions is expected to accompany the improvement of urinary bladder functions.

4.13. Absence of Significant Increase in Muscle Power during Isometric Contraction, in the Face of Improvement of Motor Performance, Indicates That Repair Is Achieved Mainly Through Functional Reorganization

Conventional concepts of neuro-rehabilitation focus on improvement of the strength of the muscles to achieve motor improvement. It was tested whether the improvement of motor performance after CDT was due to an increase in muscle power. The maximal power of isometric contraction of quadriceps femoris muscle was measured. In the spring of 2004, the maximal isometric contraction power was 411N on the right side and 378N on the left side (normal values are around 700N; the muscles had therefore only two-thirds of their normal power). In November 2006, the values were 404N on the right and 391N on the left side. In December 2006 the maximal isometric contraction power was 416 N on the right and 386 N on the left side. In short, at the end of therapy, after 3 years, when the patient had significantly improved and was walking normally, the muscle power of the quadriceps femoris muscles remained more or less the same. It was only the speed of power generation that had increased.

4.14. Improvement in Hand Grip

The power of hand grip improved considerably during CDT. At the beginning of therapy, after only four turns on the special CDT device, the left hand would slip from the handle and needed to be fixed with a bandage. By the end of therapy, the patient could maintain good hand grip even while turning at a high load of 200N. The improvement of hand grip power was also confirmed when measured with a dynamometer. Unfortunately, subsequent readings could not be made due to logistical issues. At the end of therapy, some residual deficits did remain. Considering the time profile, the author feels that the hand grip improvement was mainly achieved through functional re-organization and not by the building of new motoneurons in the grey matter of injured cervical segments C5/7.

4.15. Repairs in the CNS and PNS

From the MR image of the cervical spine in this patient (Figure 49) it is obvious that it is the grey matter that was mainly damaged. It can be seen clearly that a part of the white matter was also damaged, considering the clinically apparent long tract deficits in this patient. As the spinal cord is arranged segmentally, the sensory, motor, and autonomic functions of each segment crucially depend on their connections with the supraspinal centers for all conscious, voluntary or automatic actions. Damage to these supra-spinal connections, as what happens in white matter tract injury, leaves caudal spinal segments partially or totally isolated. Such de-afferentiation of the cord distal to the injury has devastating consequences. This is especially true in the case of the motoneurons of the lumbo-sacral cord. Adequate activation and control of these neurons from supraspinal centers is required to perform walking movements. In the presence of damage to the pre-existent descending pathways, activation of redundant pathways and re-organization in spared descending tracts is required for functional repair. As described above in the patient, improvement in walking cannot be attributed to the increased strength of the quadriceps femoris muscle, which in fact remained unchanged. Therefore, repair is unlikely to be achieved by substantial structural changes like the building of new motoneurons and connections (regeneration). It is most likely that the improved performance was due to functional repair strategies, as for example, sprouting of cortico-spinal axons onto the spared, long descending proprio-spinal tracts that would have re-established the connectivity of supraspinal segments with lumbar motoneurons. In other words, functional repair would encompass plasticity along various levels and subsystems of the CNS such as the cortex, brain stem and spinal cord. For example, cortical sensory-motor areas are known to functionally rearrange [44, 45]. At subcortical level, the rubro-spinal system can reorganize and compensate for the deficits in the cortico-spinal system [45].

The functional improvement in the upper limbs is probably attributable to both structural repair and functional re-organization. The improvement in the upper limb function, seen as lowering of the functional level of injury, within three years of treatment is more related to recruitment of spared neurons and segmental re-organization of connections. A minor re-organization in the peripheral nervous system could have contributed to this functional recovery.

The grey matter injury at segments C5/6 caused death and loss of neurons, glia and precursor cells. Structural repair in this case would involve mobilization of precursor cells to the injury site, building of new neurons and supporting glia. Even though these neurons may get re-innervated by sprouting or regenerating cortico-spinal axons, for these connections to affect functional recovery, the peripheral axons of these motoneurons would have to grow into the arm up to the hand muscles and innervate them (Figure 29). This would take longer than two years, considering the slow speed of axonal growth (1-2mm/day at its best). The strong improvement observed at the end of the third year of therapy could have been partly due to such structural repair. This also stresses the importance of continuing the therapy beyond two years to achieve clinically relevant structural repair.

4.16. Importance of Integrated CNS Activation for Learning Transfer

In the treatment of patients with spinal cord injury, learning transfer is of prime importance especially for the training of those vital CNS functions like bladder control and sexual functions which cannot be trained directly. The repair of bladder function is in fact of highest priority as that can ensure longevity of the patient by reducing life-threatening urinary tract infections which are a common cause of mortality in SCI patients.

Neuronal networks of the somatic and autonomic nervous systems are intertwined upstream and downstream and to different degrees. This sharing of networks is advantageous because only then can the autonomic networks be influenced by the training of the somatic networks. But in order to influence and train the autonomic networks, they need to be recruited and adequately activated during movement therapy. Activation of the CNS must be integrated and systematic. More specifically, activation must be upstream, symmetrical, and across many sub-systems. In this movement therapy such integrated, highly coordinated activation is achieved while the patient is exercising at high loads on the special CDT device. This exercising is similar to the highly integrative activation achieved during running. It takes longer training time for improving the CD of turning against 150 or 200N than is required for turning at 20N (Figure 50), as large parts of CNS need to be coordinated, activated and reorganized.

Within the framework of System Theory of Pattern Formation for Repair, learning transfer can be well understood, by reference to the equation of motion. By influencing the intrinsic dynamics ($\mathbf{F}_{intr}$) with behavioral information like running, jumping, and exercising on special devices ($c_{inf}\mathbf{F}_{inf}$), CNS self-organization can be modified as a whole integrated entity. Unlike as expected in the pattern generator theory [46], while training movements like running or turning on a special device, it is not merely a single circuit that is activated in the neuronal networks, but nearly all networks of the CNS networks. If that were the case, each movement would have to be trained separately.

Only the system theory can easily explain learning transfer. Because of the integrated activation of the entire nervous system, not just a single pattern is formed, but the entire neuronal network is involved, which facilitates learning transfer to other functions; the whole attractor layout is changing.

4.17. The Importance of Symmetries in Impairment and Training

It is well known that asymmetric movements are less stable than symmetric ones [47,48] (Figure 18). Symmetries structure the state-space of movement patterns and influence the stability of movement patterns. The impairment of symmetries in CNS function following injury can be uncovered during the integrated activation of the CNS when the patient is exercising at high loads on the special CDT and recording device (Figure 28E,F). For measuring coordination dynamics, the length of levers and pedals has to be short to reduce kinetic and rotational energies interfering with the true measurement of the arrhythmicity of exercising. Different symmetries can be improved by training symmetric, integrated, and dynamic movements like brisk walking, running, jumping, and exercising on the special device in the forward and backward direction.

The rotational and/or kinetic energies, generated by fast movements, smoothen the movements making them more symmetrical and asymmetries reduce. The coordinated and symmetrical afferent inputs generated by these coordinated movements enter the CNS and entrain the neuronal networks, thus correcting the asymmetries. The therapist has to anticipate, observe, support and correct the asymmetries. Such support, given by the therapist with good interpersonal coordination while the patient is jumping on springboard or when he/she is walking on treadmill, increases the symmetry of the movement and in turn the movement induced afferent input.

4.18. Pathologic Single-Neuron Firing As an Important Cause of Impaired CNS Self-Organization

As shown in the sEMG recordings above (Figure 27D), pathologic synchronization of oscillatory firing motor units occurred intermittently at a frequency of 8Hz. Such pathological synchronization of pre-motor spinal oscillators occurs when the physiological mutual inhibition of motoneurons is not sufficiently strong.

As analyzed and reported earlier in connection with patients with Parkinson's disease [31], such pathological synchronization starts with the oscillatory firing of α_2-motoneurons. Later, the activated α_1-motoneuron oscillators start to synchronize their firing at the same frequency and thus add to the pathological synchronization of the neuronal assemblies. Since α_1-motor units have larger amplitudes than α_2-motor units, the synchronization is largely observed in sEMG recordings only when the α_1-oscillators are activated and their firing becomes synchronized.

This synchronized firing of spinal oscillators indicates that after spinal cord injury, it is not just motor patterns that are impaired, but also the physiological cooperative and competitive interplay of CNS neurons is also impaired, which makes the physiologic self-organization of neuronal networks impossible.

4.19. Repair of Bladder Function

In this patient the repair of urinary bladder functions could partly be due to spontaneous recovery (autonomic functions are known to sometimes recover spontaneously, though much later than the motor functions) and partly due to the learning transfer achieved through the integrated activation of the CNS as described earlier.

Transfer of learning occurred from movement patterns to the patterns of bladder storage and emptying. Rhythmic jumping on springboard (Figure 2) trains the bladder functions most directly. Because of the rhythmic and dynamic load changes, exerted onto the pelvic floor during jumping, the external bladder and anal sphincters (somatic muscles), as a part of the pelvic floor, are rhythmically and dynamically activated and relaxed and passively stretched. The inner sphincter (smooth muscle) is probably trained by the pressure changes in the detrusor muscle. This patient had no dyssynergia (co-contraction of the detrusor muscle and the external bladder sphincter) of the bladder, which made it easier to improve bladder functioning.

As explained earlier in the framework of system theory of pattern formation, the behavioral information, in the form of training coordinated, integrated movements, changed the whole attractor layout of pattern formation, thus inducing learning transfer. However, a far deeper understanding of the human neurophysiology of bladder functioning is essential to characterize optimal behavioral information (required training) needed for the bladder repair. A detailed analysis is given in Chapter VII.

4.20. Exercising on the Special CDT Device Repairs Only Certain Aspects of CNS Organization

Being able to perform well on the special device for high loads does therefore not mean that the CNS is fully repaired. Firstly, good performance at high loads depends also on the fitness of the patient besides going deeply into the complexity of the neuronal network organization, and secondly, turning on the special CDT device quantifies only a certain aspect of CNS organization. This patient with the 50% SCI was, after six years, able to perform on the special device as well as the healthy pupils, but the performance of walking, running and jumping was certainly not as good.

Even following six years supervised and unsupervised treatment, the CNS was not fully repaired. The patient reported that the training of different movements still helped him to make further progress in repair. CDT is in a sense only the beginning of treatment. When basic CNS functions are repaired, many different movements have to be trained. Against the advice of the author, this patient was not training sufficiently on the special CDT device. But there was no need to push the patient, because all important functions were repaired and he was having a normal life. The increase of the 150N CDT value at the end of 2012 (Figure 50) was due to his participation in a triathlon competition a few days before!

4.21. The Problem of Muscle Power Generation

In incomplete cervical SCI (50% injury) sufficient muscle power can be developed by the patient to walk, run and jump. In nearly complete cervical SCI (95% injury, below), the generation of sufficient muscle power is a fundamental problem. Due to limited regeneration through therapy the patient can move different leg muscle and can even walk on a treadmill. But the patient is still a long way from being able to walk without weight-bearing support. The problem of muscle power generation can partly be quantified by the maximum load against which a patient can exercise on the special device. In 95% cervical SCI, after six months of CDT the patient can turn on the special device against a maximum load of 30 Newton and after five years of CDT against 50N. The patient with the 50% cervical SCI could exercise against a load of 200N after two years.

In humans, a motor unit can be increased by approximately 50%. That means a SCI patient needs at least 50% of motoneurons for walking. A rat, on the other hand, can increase the motor unit size up to 10 times, which means a rat can walk with only a few motoneurons, but of course with poor quality and balance. Regeneration experiments in rats therefore, are of little relevance to the treatment of human SCI.

4.22. Tract Fiber Rewiring through Learning at the Limit of Integrated (Systematic) Movements with Complicated Coordination's between Arms and Legs

A detailed analysis of an incomplete SCI is of particular interest because the injury is quite specific and it is possible to understand some aspects of CNS functioning and repair. In severe brain injury, on the other hand, many functions and regulation units are impaired and it is difficult to understand what is really going on during repair. It seems especially profitable to compare the development of the human CNS with the repair processes in different kinds of CNS injury.

In incomplete cervical spinal cord injury descending pathways are also damaged. A movement-based learning therapy will activate redundant pathways for functional repair. But the activation of redundant tract fibers is not sufficient for repair. For further functional repair a re-organization of spared descending tracts is required. It will be shown now that in order to achieve reorganization of spared descending tract fibers, integrated (systematic) movements with complicated coordination's between arms and legs have to be trained at the limits. Such training can be achieved if the patient is exercising against high load on the special CDT device. As can be seen from Figure 51, with varying intensity of training the patient was only slightly better from 2006 onwards. Against the advice of the author, he was not exercising on the special CDT device against high loads. He was walking, jogging, jumping, swimming, cycling and performing everyday movements. From Figure 54, it can be seen that the patient managed to exercise against 200N in the forward (A) and backward (B) direction in May 2006. The frequency of exercising (upper trace) and the coordination dynamics (lower trace, 'Harmonic') varied according to the positions between arms and legs. For some coordination's between pace (P) and trot gait (K) he could turn better (less variation) and not so well for other coordination's.

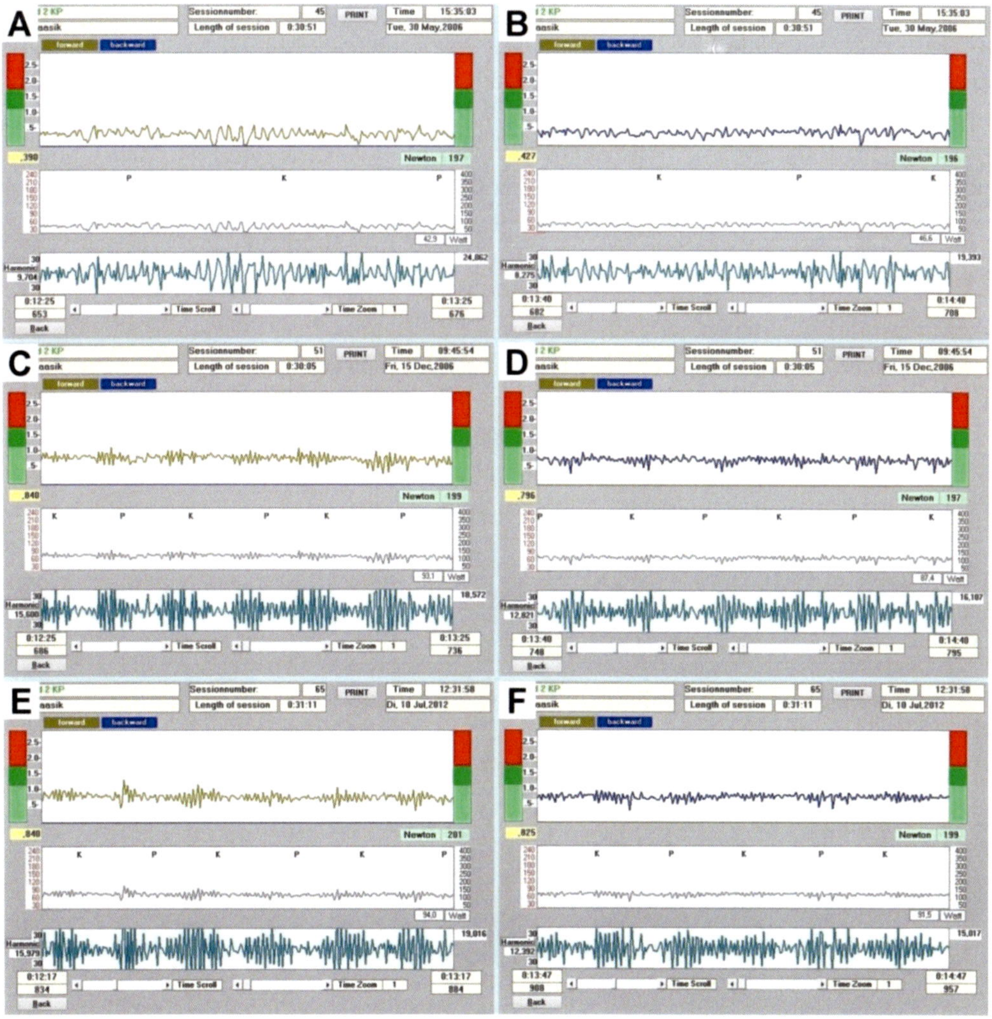

Figure 54. Coordination dynamics (CD) sweep pieces of 1 min when the patient with a cervical SCI C5/6 (at the beginning of therapy 18-years-old) was exercising against a load of 200N in the forward (A,C,E) and backward direction (B,D,F). The frequency of exercising (upper trace) and the CD (lower trace, 'Harmonic') varied rhythmically according to the positions between arms and legs. For some coordination's between pace (P) and trot gait (K) he could turn better (less variation) and for other coordination not so good. Over the years (from A,B to E,F) the CNS of the patient did not learn to manage the difficult coordination's, even though the overall CD values got better (smaller) (Figs.50,51). Further, it can be seen (C,D) that the symmetry between forward and backward exercising was not achieved. The calibrated pace (P) and trot gait (K) coordination's were lying for forward exercising to the right side of the variation of frequency and CD and for backward exercising to the left side. The patient's CNS did not improve substantially with respect to these movements.

The patient had some problems with the difficult coordination's. Through this high-load exercising it became apparent that his supraspinal centers could not sufficiently vary spinal cord neural networks to enable him to perform the difficult coordination's at high loads. He should have trained these integrated difficult movements, but he did not. Six months later in December 2006, it became more obvious that the patient had difficulties with certain coordinated movements between arms and legs in the forward (C) and backward (D) direction. From Figure 54C,D it can further be seen that symmetry between forward and backward exercising had not been achieved. The calibrated pace (P) and trot gait (K) coordination's for forward exercising were erring on the right side of the variation of frequency and CD and for backward exercising to the left side. Six years later during a check-up in July 2012, the CNS of the patient with the incomplete cervical SCI had the same problems with the difficult coordination's between arm and leg movements. Also, the symmetry between forward and backward exercising did not improve substantially; the position of 'P' and 'K' are situated at similar places with respect to the higher variability's (Figure 54E,F). This means that the patient's CNS did not improve substantially. Walking and running were insufficient to re-organize spared descending tract fibers to vary spinal cord networks for generating these complex neural network patterns to make the patient able to exercise these complicated arm and leg movements with performance of a sufficient quality. In spite of his fitness when walking, running and cycling, a further substantial repair of his CNS was not possible. The neural network complexity was not trained sufficiently at the limit for the rewiring of the tract fibers. In short, *treadmill walking alone is not sufficient for repair in patients with incomplete SCI.* And, in any case, treatment must always be performed at the patient's limits of endurance.

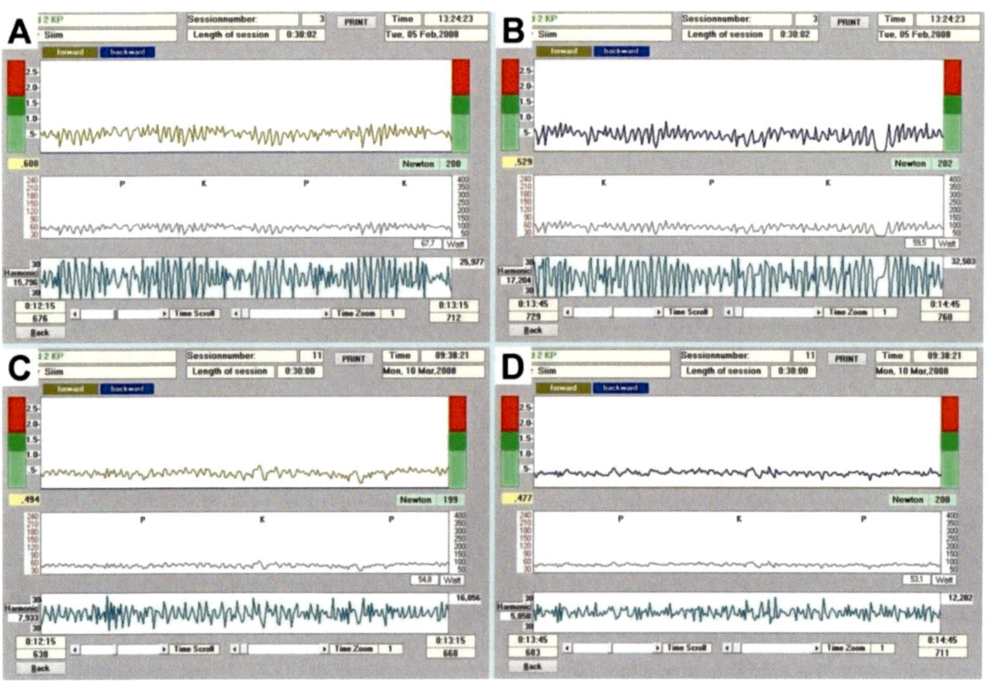

Figure 55. Coordination dynamics (CD) sweep pieces of 1 min of a healthy 15-year-old pupil for the first (A,B) and 5th session (C,D) upon exercising against a load of 200N in the forward (A,C) and backward direction (B,D). The frequency of exercising (upper trace) and the CD (lower trace, 'Harmonic') varied rhythmically according to the positions between arms and legs during the first session (A,B), but not anymore during the 5th session. The CNS of the healthy pupil quickly learned this range of movement patterns within two months (Figs.36,39). The CD values reduced and the pupil could managed with the difficult coordination's easily; no rhythmic changes of frequency and CD can be seen any more in Figure 54C,D.

It is also instructive to look at the uninjured CNS. We take for example again the pupil from Figure 39. When he was exercising against 200N his CNS also had problems with the difficult coordination's between pace and trot gait for forward and backward exercising (Figure 55A,B). But by the fifth session of exercising for one hour against low and high loads, his CNS had learned to manage easily with the difficult coordination's (Figure 55C,D). Within one month his CNS had improved its organization so that it easily generated these difficult coordination's between pace and trot gait. But, of course, his CNS did not have to rewire tract fibers. These measurements show that the rate of learning can differ tremendously between a healthy CNS and an injured CNS. The healthy CNS can learn a task within one month and the injured CNS needs several years to learn; if indeed repair is possible at all as in the case of very severe brain injury or complete SCI. The patient with a severe brain injury (Figures 46a-47d) could repair brain functioning (no substantial tract rewiring is necessary) with respect to the difficult coordination's between arm and leg movements within three years following somewhat optimal CDT with between 20 and 30 hours treatment per week. The rate of learning was reduced by a factor of approximately 36 when exercising against a load of 200N. Initially, the patient could not turn at this load at all.

5. Repair in Complete Thoracic Spinal Cord Injury T11/12 - Patient Avo

5.1. Case Report

A 34-year-old motocross sportsman (Figure 57A) suffered an anatomically complete (according to the MRI, Figure 56) thoracic spinal cord injury (T11-12) during competition. Four weeks after the accident, coordination dynamics therapy (CDT) was started (Figure 57B) in an up-to-date therapy center (Figure 58).

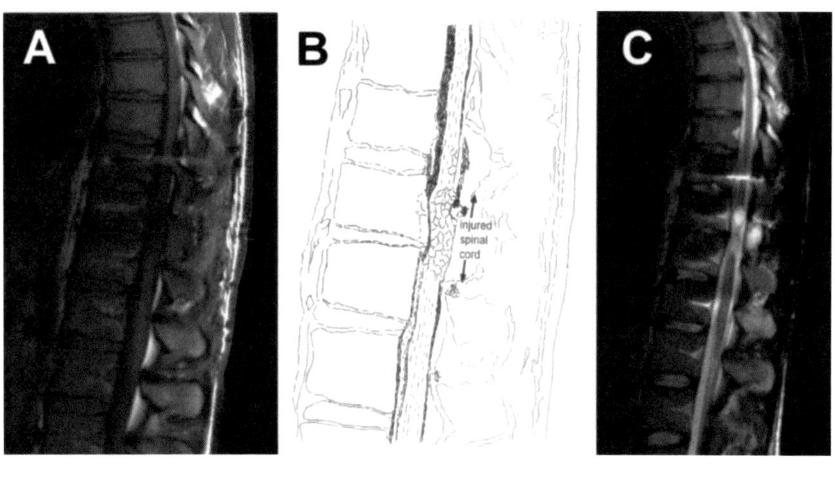

Figure 56. A. MRI of the patient with a spinal cord injury Th11/12 (T1). Over a length of 2.3cm the spinal cord shows no healthy nervous tissue. B. Redrawn injured area (from A) to show more clearly the honeycomb structure. C. MRI in T2. In A,C artifacts present from the 4 titanium screws of the fixation.

The patient was informed that at least three years of CDT is needed for partial repair. From the beginning of the therapy it was obvious that the patient was willing and able to train at his physical limits, as he did as a sportsman (Figure 57A).

Figure 57A. A 34-year-old motocross sportsman during competition before the accident.

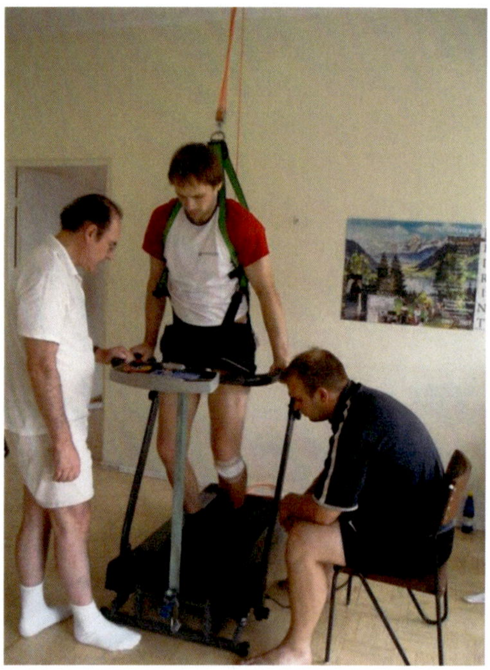

Figure 57B. A 34-year-old motocross after the accident during treadmill walking.. Note that the patient can hold himself on the treadmill. The support is only for safety.

Low-load, high-load, and symmetry coordination dynamics were measured to ensure that the subject trained at his limits without long-term exhaustion. From high-load coordination dynamics hysteresis curves, generated by exercising at increasing and decreasing loads between 20 to 200N, it was determined whether the patient's CNS and body were overloaded.

When the coordination dynamics values for decreasing loads became higher than for increasing loads, then the patient was exhausted.

Since the patient had plenty of upper body strength, the coordination dynamics values for 200N were plotted, in addition to those for 20N (low-load test). 200N loads require that the legs be substantially used to reduce the load-stress. High load 200N coordination dynamics values were therefore the most instructive for showing leg improvements in this patient. Figure 59 shows the coordination dynamics values for 20N (lower trace) and for 200N (upper trace). Schematic hysteresis curves are attached to the 200N curve to indicate whether the patient was exhausted or not. Low-load coordination dynamics values improved quickly and remained constant once the patient reached very good values. The subject's coordination dynamics value of $1.5s^{-1}$ for low-load exercising was much better than that of the average healthy person (around $5s^{-1}$), training for the first time [50]. Therefore, low-load coordination dynamics curve fail to give therapeutically meaningful information.

Figure 58. Some coordinated movements used at the therapy place. Three patients can be seen who had suffered spinal cord injuries and one patient who had suffered a severe brain injury. The exercising is assisted by two physiotherapists. The patient on the left (incomplete cervical spinal cord injury, Sten) is jumping on the springboard to entrain neuronal networks on the assembly level by performing rhythmic, dynamic, stereotyped movements. The patient on the treadmill (Benjamin, sever brain injury) is exercising the automatism walking. Support is provided by the therapist to arm movements to bring them in coordination with leg movements. The patient in the front (formally complete spinal cord injury Th5/6) is exercising the automatism crawling. Support is provided by a therapist to leg movements. The patient on the right (formally complete spinal cord injury sub Th10/12) is exercising highly coordinated arm and leg movements on the special coordination dynamics therapy device in the lying position; the coordination between arm and leg movements is imposed by the device.

5.2. Coordination Dynamics Measurements for Quantifying Improvement of CNS Functioning and Periods of Overreaching and Super-Compensation

The curve obtained from coordination dynamics values for loads of 200N reveal characteristic changes. First, coordination dynamics improved substantially because measured values sharply declined. This improvement may have continued with on-going therapy (indicated by the solid and dashed lines) if overloading or overreaching had not occurred (Figure 59). The hysteresis curve in 'A', attached to the 200N-curve does not indicate overloading. But when pushing the patient to the limits with high-load exercising, the coordination dynamics values increased strongly and the hysteresis curve shows exhaustion (B). Even after two months of therapy the coordination dynamics values for 200N were still very high, but the hysteresis curve (C) no longer shows overloading. At that point, the coordination dynamics values improved significantly and gave rise to a super-compensation period with very good performance. During this super-compensation (period 1), the coordination dynamics values were extremely good. The hysteresis curve (attached to the 200N-curve) does not show overloading (D).

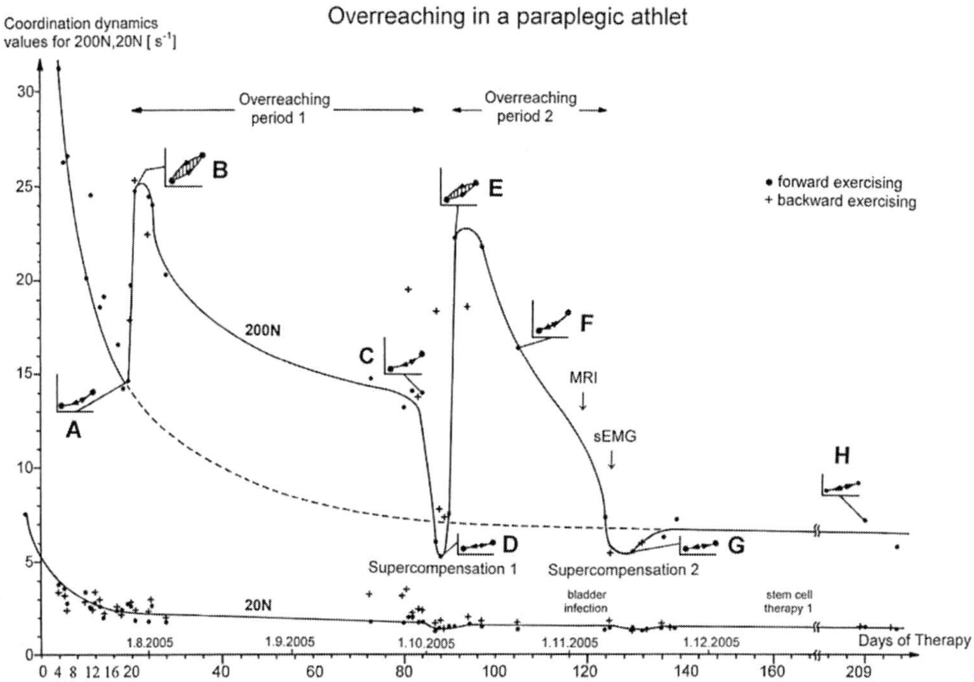

Figure 59. Low-load (20N, lower trace) and high-load (200N) coordination dynamics values in dependence on days of therapy for forward exercising. Dashed line indicates improvement of 200N-coordination dynamics values without overreaching. A-H, schematized hysteresis curves (for increasing and decreasing load) to show exhaustion; details of the hysteresis curves are shown in Figure 60. Periods of overreaching, super-compensation, urinary bladder infection, and first stem cell therapy are indicated.

With a second pushing of the patient to the limits, the coordination dynamics values deteriorated strongly (Figure 59). The hysteresis curve again shows overloading (E). With the reduction of the load, the hysteresis curve shows no overloading (F), even though the coordination dynamic values were still very high. However, a second period of super-compensation occurred, resulting in another strong improvement in of coordination dynamics values. During this second period (Period 2) of super-compensation, the hysteresis curve again indicates no short-term overloading (G).

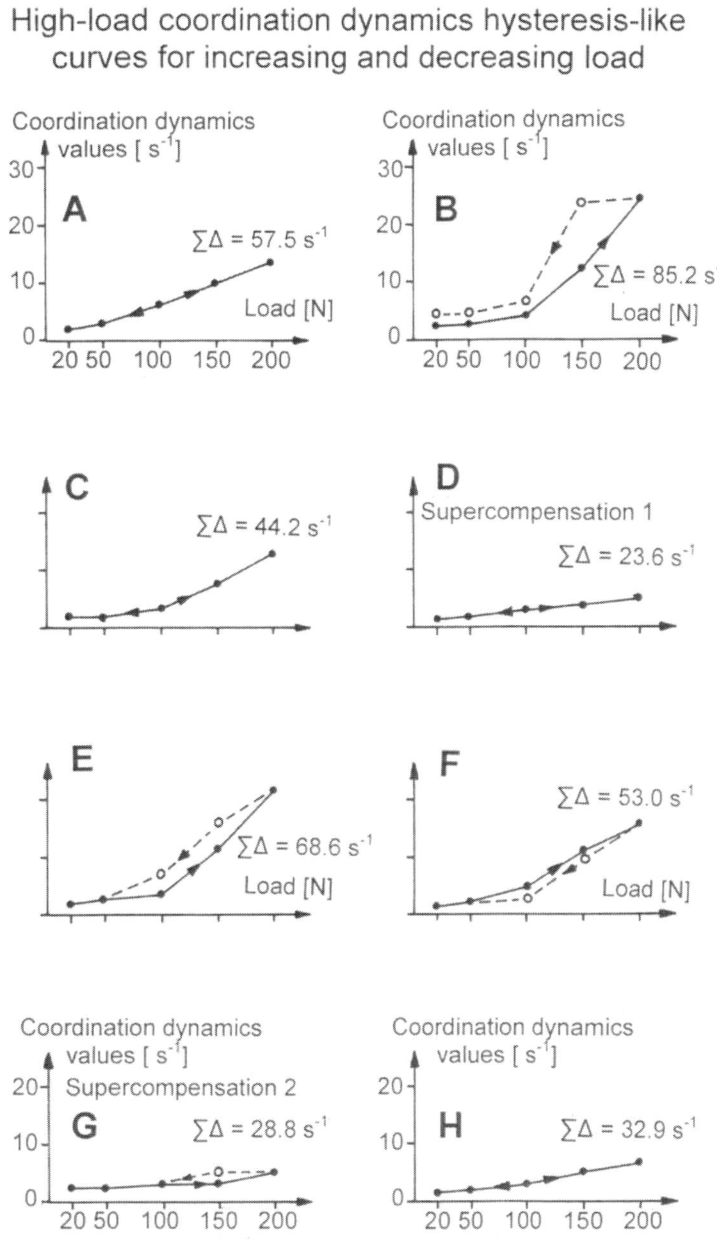

Figure 60. Shape of hysteresis curves in dependence of overreaching and super-compensation. Note that during overreaching (B, E) the coordination dynamics values are higher for decreasing load than for increasing load (indicating exhaustion). Note further that during super-compensation (D, G) no exhaustion occurs and also the sum of the coordination dynamics values ($\sum\Delta = \Delta 20 + \Delta 50 + \Delta 100 + \Delta 150 + \Delta 200 + \Delta 150 + \Delta 100 + \Delta 50 + \Delta 20 + \Delta 20$) for increasing and decreasing load are lowest.

With on-going therapy, the 200N-load coordination dynamics values slowly declined further and the hysteresis curves showed no exhaustion (H). The patient experienced a little fatigue during these overreaching periods. Blood pressure measurements in the morning did not show higher values than in the evening, which would have indicated exhaustion. It was noted that in both periods of super-compensation overreaching (or overloading) ended quantified by the hysteresis curves C and F in Figure 60 prior to sharp declines in coordination dynamics values (Figure 59). The disappearance of overloading in the hysteresis curves before strong improvements in the coordination dynamics values for 200N may indicate that at least two mechanisms were working in the patients' CNS during overreaching. The details of the hysteresis curves of Figure 59 are given in Figure 60 (with corresponding 'A' to 'F').

Periods of increased coordination dynamics values for 200N lasted approximately 72 and 32 days (Figure 59). According to the definitions in sport (see below) the first long period of increased coordination dynamics values would indicate overloading and the second shorter period may indicate overloading or overreaching. But in the case of spinal cord injury, the length of overreaching and overloading periods may be different (longer), since only a part of the body is working properly and at least one part of the CNS, in this case the caudal spinal cord, is functioning pathologically.

Repair mechanisms may also work more slowly, especially shortly after injury. Since the blood pressure in the morning was not higher than in the evening, it is likely that both periods were signs of overreaching and not overloading per se.

5.3. Definition of Overreaching and Overtraining in Sport

Overreaching: An accumulation of training and/or non-training physical stress resulting in a long-term decrement in performance capacity with or without related physiological and psychological signs and symptoms. Restoration of performance capacity may take several days to several weeks.

Overtraining: training at (or exceeding) one's physical limits without adequate rest, inducing general fatigue and muscle breakdown. Restoration of performance capacity may take much longer times (from several weeks to several months).

5.4. Symmetry Improvement of CNS Functioning Achieved by Overreaching

The primary purpose of training the patient to his physical limits was to enhance functional repair and to induce structural repair. It may therefore be possible to find CNS improvements in this patient that were induced by overreaching. Following the periods of overreaching, super-compensation occurred in that period the patient could perform the coordinated movements very well and he felt in good health. An athlete's performance would be optimal during this period.

Since periods of super-compensation were brief, they may not indicate repair. The second period of overreaching was shorter and coordination dynamics values for 200N increased more sharply. This shorter overreaching period may suggest partial repair.

Prior to the first period of overreaching (period 1), coordination dynamics traces for 200N show no sign of symmetrical impairment between forward (Figure 61A) and backward exercising (Figure 61B).

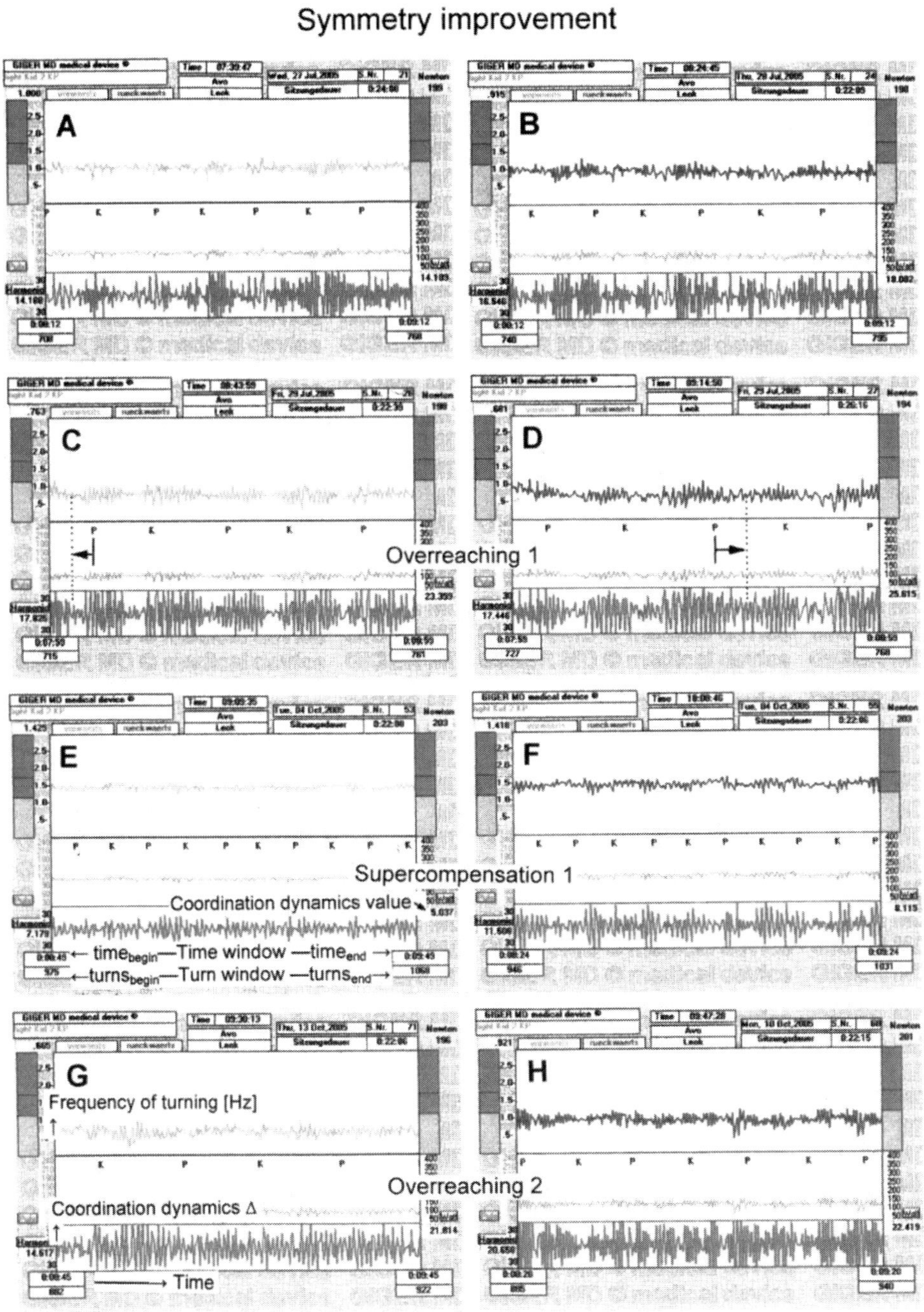

Figure 61. 200N-Coordination dynamics 1min-traces before overreaching period 1 (A,B), during overreaching 1 (C,D), during supercompensation period 1 (E,F), and during overreaching period 2 (G,H). Note the improvement of symmetry impairment of CNS functioning between forward and backward exercising from C,D to E,F. Forward exercising = A,C,E,G; backward exercising = B,D,F,H.

However, when the patient became overloaded a day later, coordination dynamics traces reflect increasing and decreasing arrhythmicity of turning, which differed for exercising in the forward (Figure 61C) and backward direction (Figure 61D). This indicates symmetrical impairment. The smallest arrhythmicity of exercising (highest stability of movement pattern) lies for forward exercising to the left of the trot gait coordination (K) (φ = -60°) and for backward exercising to the right side of the trot gait coordination (φ = +70°) as indicated by arrows in Figure 61C,D. Forwards and backward exercising under overreaching conditions therefore made the symmetry impairment of CNS functioning visible.

During super-compensation period 1 and before the second overreaching period (period 2), the coordination dynamics traces show no rhythmic changes of pattern stability for forward (Figure 61E) and backward exercising (Figure 61F) and thus no more symmetry impairment (as expected). During the overreaching period 2, the coordination dynamics values for 200N are strongly enhanced again, but no rhythmic changes of the arrhythmicity of exercising and no symmetry impairment can be seen (Figure 61G,H). It seems therefore that during the overreaching period 1, an improvement in the symmetry of CNS functioning took place.

In conclusion, training at the limits (quantified by periods of overreaching and super-compensation) led to qualitative improvements in CNS functions in this severely injured patient Avo with a functionally complete spinal cord injury, whose injury also appeared under an MRI scan to be anatomically complete. Whether overreaching induces *substantial* structural repair requires further research.

5.5. Overreaching during Coordination Dynamics Therapy

When administering coordination dynamics therapy to patients, one tries to achieve functional and structural repairs. A basic assumption in this therapy is that all repair mechanisms are activated by training at the physical limes, including the formation of new nerve cells and the building of new neural connections in the adult CNS. While training at his limits a spinal cord-injured athlete displayed the physical characteristics of overreaching, as occurs in athletes while training for sports. In this patient, overreaching while training led to improved CNS function. Symmetries improved during the first period of overreaching. It remains to be determined whether substantial building of new nerve cells can be induced. But it has been shown that some new nerve cells can be built in an SCI patient when training at the limits [11] (Chapter VII). The slight improvement in motor functions may indicate a slight increase of functional connectivity over the injury site.

Since the coordination dynamics hysteresis curves for increasing and decreasing loads (Figure 60) were better than those of uninjured athletes [51], it is proposed that functional repairs achieved these functional improvements. Because if these improvements were due to structural repair, the patient would restored some leg function.

5.6. Stem Cell Therapy

Figure 59 shows how the CNS of the patient improved with on-going therapy. The progress was mainly achieved by an optimization of the functional reorganization of the injured CNS. Structural repair, meaning the building of new nerve cells and/or new long connections, would only have made a small contribution to these improvements What can be achieved by a functional re-organization was brought about by this movement therapy. Strong improvements below the injury site in this patient, who has a rather complete spinal cord injury (Figure 56), could only likely be achieved by building of new nerve cells and new long connections and by their functional integration into the existing injured neuronal networks. This patient would therefore be an ideal patient to demonstrate whether stem cell therapy induces the building of new nerve cells and connections in humans leading to substantial improvements in leg function. Against the advice of the author, this patient went for hematopoietic stem cell therapy after having performed six months of intensive coordination dynamics therapy.

In CDT it is posited that structural repair will only occur after more than six months of intensive therapy, more likely after two years. After the building of new motoneurons following three years of therapy, some finger functions were repaired in the patient Kadri. [11] (Figure 29). However, the more severe the injury, the more difficult it may be to achieve structural repair during CDT, even if the patient is training at capacity with more than 30 hours of training per week, which is what this particular patient was doing.

The changes brought about in the functioning of the CNS after two years of CDT and stem cell therapy are summarized in Figure 62 via plotting the values of coordination dynamics for 20N (low load, lower trace) and 200N (high load, upper trace).

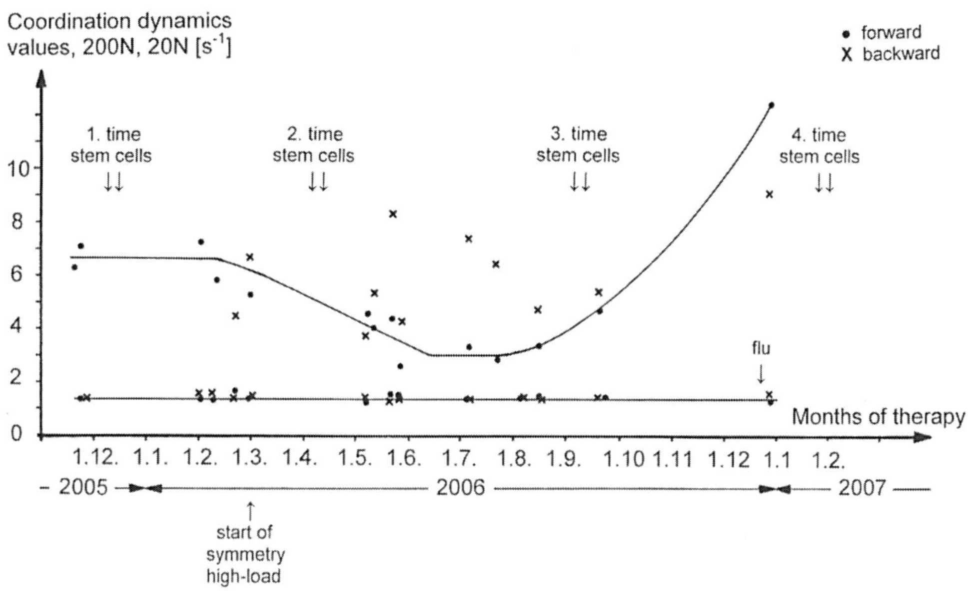

Figure 62. Coordination dynamics values for 20 (lower trace) and 200N (upper trace) in dependence of therapy duration. Notice that the periods of stem cell therapy (injections) did not change the coordination dynamics values significantly.

This Figure serves as the continuation of Figure 59 with the subsequent results of on-going therapy. The values of coordination dynamics values for 20N do not reveal significant changes for the subsequent two years of therapy. At this low load, the patient could easily turn smoothly by mainly relying on his healthy arms. But for a load of 200N, even this fit patient would need to use his legs for optimal exercising in the sitting position on this special CDT device. Therefore, the 200N values of coordination dynamics should show changes due to a structural (and functional) repair of the spinal cord. As can be seen in Figure 62, the 200N values of coordination dynamics did not change markedly with on-going therapy. If there had been structural repair, transient significant increases of the values would have occurred because of the enlargement of the neuronal networks proceeding from the brain to the disconnected caudal spinal cord across the injury site via repaired networks.

After the first session of stem cell therapy, the 200N values of coordination dynamics did not change (Figure 62). When starting symmetry high-load testing and exercising, coordination dynamics for 200N improved (values reduced), which means that the functioning of the CNS improved. The second, third, and fourth stem cell injections had no significant effect on coordination dynamics, i.e., on the functioning of the CNS, either. The slight increase in the values of coordination dynamics at the end of 2006 were due to the exhaustion caused by flu, but not to an enlargement of the neuronal networks in the spinal cord achieved by regeneration (Figure 62), as the high-load (coordination dynamics) hysteresis curve indicated exhaustion (not shown). After the fourth stem cell injection at the end of February 2007, the values of coordination dynamics (until the end of 2007) remained similar to those at the end of 2006.

The patient himself felt no improvement in leg function in response to additional stem cell therapy. The small improvement that occurred during the two years of therapy was achieved with the movement therapy itself.

In conclusion, four sessions of hematopoietic stem cell therapy over a period of two years had no beneficial effect on the recovery of leg function in this patient with a complete SCI at the Th11/12 levels. It even seems that, instead, optimal CDT could have achieved more improvement. For further discussion of stem cell therapy see [52, 53].

5.7. False Hope Generated by Stem Cell Therapy

No real repair was achieved by six months of CDT. Bladder repair would probably have required three years [13] and walking with a little support may have needed five years or more. Even though the sponsors of the former sportsman offered to sponsor a further three years of CDT, the patient stopped optimal CDT because he was mentally broken when the promise of complete repair through stem cell therapy did not materialize. He lost faith in all forms of neurotherapy.

5.8. Rate of Repair

As can be seen from Figure 59, the plateau values for 20N exercising were achieved after approximately one month and the 200N plateau values (somehow similar to high-load values) were reached after two months in some similarity to a healthy pupil (Figures 36,39). But the injury was not repaired. Clinical observations such as the quality of walking or running have to be included in any judgment concerning repair.

It is actually a pity that the patient discontinued therapy. Because of his mental strength, he was able to go over the limits into the range of overreaching to induce structural repair. From Figure 61C, it can be seen by the rhythmic changes of the coordination dynamics (lower trace) that some leg muscles were partly activated during exercising at 200N. Such rhythmic changes only occur for the changing coordination between arm and leg movement, if the leg muscles are being activated. If the legs are only moved passively, there are no coordination changes and no rhythmic coordination dynamics changes. This demonstrated that some small activation of the muscles of the leg was beginning to occur.

6. Repair in Parkinson's Disease

6.1. Neurophysiology: Physiologic Phase and Frequency Coordination among Neuron Firings

To understand the repair strategy in patients with Parkinson's disease, the physiology and pathophysiology of CNS organization must first be considered.

6.1.1. Oscillatory Firing of Motoneurons and Motor Units

Motoneurons can fire occasionally, transiently in an oscillatory manner and continuously in an oscillatory manner [54]. In everyday life, they largely fire occasionally and transiently in an oscillatory manner, depending on the strength and speed of muscle activation. Transient oscillatory firing means that the motoneuron fires for a few oscillations, stops firing and fires for some further oscillations. The oscillatory firing pattern differs according to the motoneuron type. In Figure 17 the different patterns are shown schematically and on an original record taken with the single-nerve fiber action potential recording method and by surface electromyography (sEMG) from single motor units. α_1-Motoneurons innervate FF-type muscle fibers and fire rhythmically with impulse trains consisting of 1 action potential in the range of 10Hz. α_2-Motoneurons innervate FR-type muscle fibers and fire rhythmically with impulse trains consisting of 2 to 5 action potentials in the range of 4 to 7 Hz. The extracellular action potential amplitude of the α_2-motoneurons is smaller than that of the α_1-motoneurons and the FR-type motor units have much smaller amplitudes than the motor unit potentials of FF-type muscle fibers. The α_3-motoneurons innervate S-type muscle fibers and fire in an oscillatory manner at a frequency of around 1 Hz with long impulse trains (up to 50 action potentials per impulse train). The motor unit firing of single S-type muscle fiber motor units could not been safely identified by sEMG because their amplitudes are still smaller than those of FR-type motor units and are thus difficult to identify. The impulse patterns of oscillatory firing motoneurons obtained with sEMG are similar or the same as those obtained with the single nerve-fiber action potential recording method (Figure 17). Since sEMG is a non-invasive recording method oscillatory firing can be recorded easily when using appropriate patients.

It was shown that the oscillatory firing of α_1-motoneurons depends very much on the primary muscle spindle afferent drive and that α_1-motoneuron oscillators are not very stable [31]. The α_2-motoneuronal network oscillators obtain drive from several secondary muscle spindle afferents. The α_2-motoneuron oscillators are more stable since they do not depend so much on the afferent drive. The α_3-motoneuron oscillators receive polymodal input; this means they get input from several kinds of afferent fibers. They seem to be very stable. The problem is that they are difficult to record since the motoneuron axons are thin and the motor unit potentials are small. From the theoretical standpoint, α_2-motoneuron oscillators are most interesting because they are quite stable and they can be measured with the single nerve-fiber action potential recording method and sEMG.

6.1.2. Non-Synchronous Coordinated Oscillatory Firing of α_1-Motor Units

Records obtained with the single-nerve fiber action potential recording method showed that oscillatory firing motoneurons do not synchronize (Figure 16A,B), if not specifically stimulated [55]. The same result was obtained with sEMG from patients with spinal cord injury (Figure 16C-H). Figure 16C shows a recording from a single α_1-motoneuron motor unit of the tibialis anterior muscle of a patient with a complete spinal cord injury sub L3 after trauma which fires in an oscillatory manner at 9.2Hz continuously. The amplitude of the biphasic motor unit was approximately 1mV. The recording quality was so high that even the cross-talk potential in the gastrocnemius muscle can be used to measure the frequency. The same motor unit shown in Figure 16D fired in an oscillatory manner at 19.2Hz. The motor unit action potential could be safely identified from the waveform because of the high quality of the recording. In Figure 16E, the same motor unit is seen firing at 22.4 Hz. Another α_1-motor unit in the tibialis anterior muscle is seen in Figure 16E firing oscillatory at 21.3Hz. This motor unit was also safely identified by its triphasic motor unit potential. The cross-talk potentials in the gastrocnemius muscle showed the same rhythmicity (Figure 16E).

It is concluded that in spinal cord injury α_1-motor units can also fire rhythmically at approximately 20Hz besides firing at 10Hz. This high recording quality could be achieved since there were only a few motor units innervated in the right tibialis anterior muscle, which were probably hypertrophied because this patient underwent coordination dynamics therapy for two years, which included running on the treadmill at between 6.5 and 10km/h for half an hour per day.

Figure 16E shows further that the rhythmic firing of the two motor units at approximately 20Hz was coordinated in the sense that they did not synchronize their firing. Non-synchronous firing is physiologic. In the recordings from a patient with a complete spinal cord injury sub C4/5 (Figure 16F-H) α_1-motor units also oscillated at around 10 and 20Hz and the motor unit firing was coordinated in that it did not fire synchronously. In the muscles of these two patients there were only α_1-motor units that mainly fired since the motoneurons were located at a spinal cord site which was damaged at the accident. These motoneurons were most likely re-innervated as a result of coordination dynamics therapy administered for at least a year. It is proven in frogs [56, 14 (Chapter I)] that the thick fast conduction axons of large motoneurons grow fastest and re-innervate muscles first. Therefore, in regeneration processes, it is the fast systems that start to work first.

In conclusion, motoneurons coordinate their firing in the sense that they do not fire synchronously. This can also be seen in the recordings from the tibialis anterior muscle (Figure 26A-C). The motor units 1, 2, and 3 coordinated their firing so as not to fire synchronously.

6.1.3. Phase and Frequency Coordination between Motor Unit Firing and Building up of A Motor Program

Figure 26A-D shows how motoneurons coordinate their firing and how the CNS generates a motor program during the exercising on the special coordination dynamics therapy device. In the right tibialis anterior muscle of this spinal cord injury patient, approximately five motor units could be activated. The motor units marked 1 and 2 had a low threshold, and motor unit 3 a higher threshold. Their waveforms are shown time-stretched in the inserts of Figure 26A,C. When exercising on the special coordination dynamics therapy device at 200N, it appears that two other motor units could be activated. Here, we concentrate on the firing of these 3 motor units and study their coordination during the exercising on the special coordination dynamics therapy device at increasing loads from 20 to 200N.

The patient whose record is shown in Figure 26A was exercising at 20N. Motor unit 1, with a large motor unit potential, oscillated at around 10Hz and can thus suggested to be innervated by an α_1-motoneuron. Motor unit 2 fired at around 5Hz in between the action potentials of unit 1 to avoid synchronization, but fired in every second interspike interval due to having half the frequency. Motor unit 2 was probably also innervated by an α_1-motoneuron, even though the frequency was low (5Hz) for the oscillatory firing of an α_1-motoneuron. There was a slight inhibition acting on motor unit 1 after the 6th spike since the interspike interval greatly exceeded 100ms. When increasing the load to 50N (Figure 26B), motor unit 1 slightly increased the frequency and increased its activity, and motor unit 2 increased its frequency from 5 to 10Hz and increased its activity. The higher threshold motor unit 3 started to fire occasionally. After the 4th motor unit 1 potential, motor unit 2 stopped oscillating. It is likely that motor program inhibition worked transiently on the oscillatory firing subnetwork in order to reduce the frequency of oscillatory firing or to stop the oscillatory firing altogether. This gave rise to a transient break in the oscillatory firing. When exercising at 100N (Figure 26C), the way in which the motor program was generated can be seen. When an activity burst was activated, motor units were activated and fired in an oscillatory manner for the duration of the burst. Throughout the burst, the frequency increased and decreased. Other motor units were also activated, but the ones with the lowest threshold were activated soonest. The activated motor units avoided synchronization. When the motor program burst was supposed to be terminated, the motor unit firing stopped. They were probably inhibited. When exercising at 200N (Figure 26D), a motor program can be seen in the right tibialis anterior muscle (note the time axis change). Four weeks later, when the patient had a flu (Figure 26E, F), the motor program developed in the right tibialis anterior muscle worsened.

The healthy motor program of the biceps brachii (spinal cord injury sub L3) looks completely different (Figure 26A,B). No coordination can be seen and analyzed. The difference is understandable. In the right tibialis anterior muscle, the CNS could only activate 5 motor units and only α_1-motor units (large motor unit potential and only one action potential per impulse train characterizes α_1-motor unit (FF) oscillatory firing (Figure 8)); no α_2 (FR) and α_3-motor units (S) were innervated. In the biceps brachii muscle, all types of motor units could be recruited in abundance. For 20N exercising (Figure 26A), it was mostly only α_2 (FR) and α_3-motor units (S) (identified by the small motor unit action potential amplitude) and maybe a few smaller α_1-motor units (FF) that were activated in the biceps brachii muscle. For 50N (Figure 26B) motor units of larger amplitude, the α_1-type (FF), were activated. Since recruitment conforms to the size principle in each motoneuron group [57] and the group recruitment depends on the speed of movement, quite a complexity process occurs. Also, the transiently oscillatory firing of motoneurons will be shorter, especially if the patient is exercising very fast. In the partly impaired motor supply of the quadriceps femoris muscle (injury level sub L3) some motor units can be identified (Figure 26B). In the better, left tibialis anterior muscle, no coordination could be extracted from the activity; too many motor units were functioning to allow identification of a single motor unit firing (Figure 26E,F). But many α_1-motor units were activated in the left tibialis anterior muscle up to a size of 2mV (Figure 26F), whereas in the biceps brachii muscle they were still under 1mV (Figure 26D) at a load of 200N.

In Figure 26, relative phase and frequency coordination is shown for motor units in the same muscle, namely the right tibialis anterior muscle of a patient with a spinal cord injury sub C4/5. In Figure 15, phase and frequency coordination among motor units is shown between different muscles. In Figure 15A a single FF-type motor unit of the right flexor carpi radialis muscle is firing approximately at 8Hz, activated on volition. In Figure 15B, the same motor unit is firing at 12.5Hz. This motor unit is firing in relative coordination (marked with the arrows) with another single FF-type motor unit in the left flexor carpi radialis muscle. The phase relations have different values in 'A' and 'B'. Note that the motor unit in the flexor changed its oscillatory firing from 6.7Hz to 10Hz, probably in order to develop more motor unit force. Not every unit potential is correlated. There may be correlation between motor units of other muscles which were not being recorded. In Figure 15C, phase relations are indicated with arrows from the right extensor carpi radialis to the left extensor carpi radialis and in Figure 15D phase relations are indicated by arrows from the left extensor carpi radialis to the left flexor carpi radialis. There is therefore coordinated firing between all motor units activated by volition, when the patient is not exercising on the special coordination dynamics therapy device, but is in position to do so.

During exercise on the special coordination dynamics therapy device at 5N a motor program was partly generated in the right and left extensor carpi radialis muscles (Figure 15E,F). In the right flexor carpi radialis muscle, the CNS attempted with 2 or 3 innervated motor units to build up a motor program in some similarity to that seen in Figure 26E,F (tibialis anterior right). In the left flexor carpi radialis muscle no motor program was generated, even though small motor unit potentials can be seen (marked in Figure 26E by the arrows).

6.1.4. Relative Phase and Frequency Coordination between the Firings of α and γ-Motoneurons and Secondary Muscle Spindle Afferents Recorded with the Single Nerve-Fiber Action Potential Recording Method

It has been shown above that single motor units oscillated or transiently oscillated for high activation and that there was relative phase and frequency coordination between their firings. This means that the innervating motoneurons should have also fired oscillated and that there was phase and frequency coordination between their firings. It is therefore of importance to compare the above data with the measurements, obtained with the single-nerve fiber action potential recording method from cauda equina nerve roots, to see whether the results are compatible. The recordings from the nerve roots offer in addition natural impulse patterns of the afferent fibers, partly driving α and γ-motoneurons.

Figure 25 shows a schematic recording with the single-nerve fiber action potential recording method from a dorsal S4 root of a brain-dead human [58]. From the total afferent and efferent impulse traffic the natural impulse patterns of one α_2-motoneuron, a dynamic (γ_1) and a static γ-motoneuron (γ_{21}) and two to three secondary muscle spindle afferent fibers could be extracted. The natural stimulations performed were pin-pricking (pain) of sacral dermatomes inside the continence automatism zone and urinary bladder catheter pulling. It can be seen that the α_2-motoneuron (α_2(O2)) oscillated with two or three impulses per impulse train and sometimes there was a break in the oscillatory firing as was shown in Figure 26 for the firing of the α_1-motor units (FF). The impulse train for α_1-motor units consisted of only one action potential (Figure 17). Phase coordination's between the α_2-motoneuron, the γ-motoneurons and the secondary muscle spindle afferent fibers are indicated by different arrows and the dotted and dashed lines. It can be seen from Figure 25 that there were many coordination's between the different neurons. The relative phase and frequency coordination therefore not only holds for motoneurons, but for all neurons. It seems therefore that relative phase and frequency coordination is an integrative mechanism for the self-organization of the neuronal networks of the human CNS. Relative does not mean approximate coordination's but very complicated coordination's in the sense that each neuron coordinates its firing with many other neuron firings. Relative frequency coordination includes also the coordination of higher order harmonics.

6.1.5. Coordinated Firing Of Nerve Cells

In conclusion, it was shown above that in humans there are three kinds of motoneurons, namely α_1, α_2 and α_3, which innervate three kinds of muscle fibers, namely fast fatigue (FF), fatigue resistant (FR) and slow muscle fibers (S) respectively (Figure 8). For high activation these motoneurons fire in an oscillatory manner with their own upstream neural networks at frequencies of around 10Hz, 5Hz and 1Hz respectively. This rhythmic firing can be recorded from single motoneurons with the single-nerve fiber action potential recording method and from single-motor units with surface EMG (Figure 17). These three kinds of motoneurons coordinate their firing so as not to synchronize (Figure 16) to generate smooth movements. With nerve cell death, this coordinated firing becomes impaired with dramatic consequences for motor pattern generation.

6.2. Pathophysiology: Impaired Phase and Frequency Coordination in Parkinson

In the CNS we encounter a specific and an unspecific degeneration. This unspecific impairment of phase and frequency coordination can be treated by CDT. The specific impairment of phase and frequency coordination, namely the impaired inhibition among nerve cells can only partly be treated. By performing CDT at the limit, crucial new nerve cells can be built. Unfortunately, however it is likely that more nerve cells are dying than can be build anew. This compounds the fact that older patients with Parkinson's are unable or unwilling to train at limits.

6.2.1. Introduction

In a previous paper [15] and above α_1, α_2, and α_3-motoneurons were shown to fire in an oscillatory manner for high activation and FF, FR, and S-type muscle fiber motor units, they innervate to also fire in an oscillatory manner with the same impulse patterns. It was further shown that there was relative phase and frequency coordination between the oscillatory firing motoneurons and also between the oscillatory firing of the motor units. The coordination was such that the motoneurons or the motor units did not fire synchronously [55] (Figs.16,26) unless exposed to strong and sustained stimulation (Figs.67,68). Such non-synchronous coordination seems to make sense, since synchronously oscillatory firing motor units could give rise to rhythmic muscle movements and tremor. It will be shown that in Parkinson's disease patients the motor units and upstream the motoneurons do synchronize their firing. A mechanism by which muscle movement and tremor is generated by the pathologically functioning CNS will be explained. Since during tremor, α_1-motor units do synchronize their firing with oscillatory firing α_2-motor units to generate tremor, differences in the stability between α_1 and α_2-motoneuron oscillators and the building up of an external loop to the periphery by the premotor spinal oscillators will be clarified (Figure 69), through measurements.

6.2.2. Method to Analyze Parkinsonian Tremor

The methods used to analyze the cause of Parkinsonian tremor and regulatory mechanisms show how spinal oscillators synchronize and communicate with the periphery were the surface electromyography (sEMG, Figure 1C,D) and the single-nerve fiber action potential recording method (Figure 1A). Motor unit firing is recorded from four muscles with two indifferent surface electrodes (plus one earth electrode each) for each muscle using 4-channel sEMG. Single nerve-fiber action potentials were recorded with platinum wire electrodes from thin long lower sacral nerve roots of the cauda equina. With the polarization of the wire electrodes, afferent action potentials show in upward and efferent action potentials in downward direction (Figure 3). From the calculated afferent and efferent impulse traffic in a nerve root, several natural firing patterns, running along single-nerve fibers, were extracted by their waveform and conduction velocity identification. From recording under natural conditions the impulse patterns running into and out of the spinal cord, functions of the human CNS and receptor properties of the periphery were analyzed. The coordinated firing between motoneurons and muscle spindle afferents was studied in detail. For further details of the method, see [14].

Through the single-nerve fiber action potential recording method records can be made from motoneuron axons (and afferent nerve fibers) and with the sEMG from muscle fibers the axons innervate (the motor unit). The single-nerve fiber action potentials have a duration of approximately 0.3ms (Figure 1A), and the single motor unit action potentials have a duration of approximately 10ms (Figure 1D). The nerve fiber action potentials are thus of a much shorter duration than the motor unit potentials. The single-nerve fiber action potential duration varies according to the conduction velocity (Figure 70) and thickness of the fiber, and motor unit action potential duration probably depends on the muscle fiber type and on the usage.

In this neurophysiologic research project measurements are made at the single neuron level (Figures 4, 152, 153) for single nerve-fiber action potentials; at the muscle level (Figures 1C,D,174) for motor programs and single motor unit firing; and at the movement level (Figures 1E,174) when arrhythmicity of turning is concerned (coordination dynamics) during exercise on a special coordination dynamics recording and therapy device. Even though coordination dynamics measurements were not directly used here, they were measured in most cases. To understand how the nervous system is organizing certain functions and dysfunctions, understanding at the neuron and the muscle level is needed. But to measure integrative improvement of CNS functioning, e.g., in Parkinson's disease patients [5], the integrative movement level is necessary, namely measurement of coordination dynamics (coordination between arm, leg and trunk movements over time) (Figs. 78, 80, 83). Also, what side of the patient is more affected by the disease can be measured easily with the coordination dynamics (Figure 80).

6.2.3. The Triggering Mechanisms of Parkinsonian Tremor and Large-Scale Coordination

In Figure 63A,B two mechanisms of tremor induction are shown. The patient was positioned on a chair at the special coordination dynamics therapy device but was not exercising. In Figure 63A one FR-type motor unit in the gastrocnemius muscle (innervated by an α_2-motoneuron) started to oscillate (first triggering mechanism), i.e., fired rhythmically with an impulse train consisting of 3 motor unit potentials. Most likely, the gastrocnemius muscle was not moving. Then more FR-type motor units and maybe some S-type motor units started to fire synchronously (Figure 63C). The gastrocnemius muscle may have moved a bit with a frequency of 4.1Hz. The triceps brachii muscle moved, because a movement artifact was recorded from it (Figure 63C). With the start of FF-type motor unit firing, synchronized with the FR-type motor unit firing (Figure 64A,B and Figure 63E,F, tibialis anterior r.), the muscle moved rhythmically.

A second tremor or muscle movement triggering mechanism concerned FR-type (and may be S-type) motor units synchronizing their oscillatory firing. In the left part of Figure 63B the FR-type motor units oscillated but not in synchronization; no movement artifact at the electrodes of the triceps brachii muscle was recorded. When the FR-type motor units started to oscillate in a synchronized manner at a frequency of 4.2Hz (right part of Figure 63B), then a rhythmic movement artifact (f = 4.8Hz) was generated at the surface electrodes of the triceps brachii muscle. Since the triceps brachii muscle is sited in the arm and the gastrocnemius muscle in the leg, the oscillatory firing networks of the arm (network assemblies in the intumescentia cervicalis) and the leg (assemblies in the intumescentia lumbosacralis) did communicate with each other. There were therefore large-scale synchronization and coordination at least between the intumescentia cervicalis and lumbosacralis, which is a distance of approximately 300mm.

The full muscle movement or tremor was generated when the FF-type motor units (innervated by α_1-motoneurons) synchronized their firing with the firing of the FR-type motor units (Figure 63E,F) (third triggering mechanism). With the synchronization of the FF-type motor units, the muscle movement or tremor frequency changed to match frequency coordination between the driving oscillatory firing of α_1 and α_2-motoneurons. In Figure 63D, the common frequency of the synchronized oscillatory firing α_2-motoneurons (and maybe α_3-motoneurons) was 3.3Hz. With the synchronization of the α_1-motoneurons, the common synchronized frequency increased to 3.6Hz (Figure 63F).

The α_1-motoneurons could have fired in an oscillatory manner at a frequency of 7.2Hz, so that a single FF-type motor unit potential could have fired in coordination with every second muscle activation burst of the tremor. If another α_1-motoneuron oscillated at 10.8Hz, the innervated FF-type motor unit potential could have fired in coordination with every third activity burst.

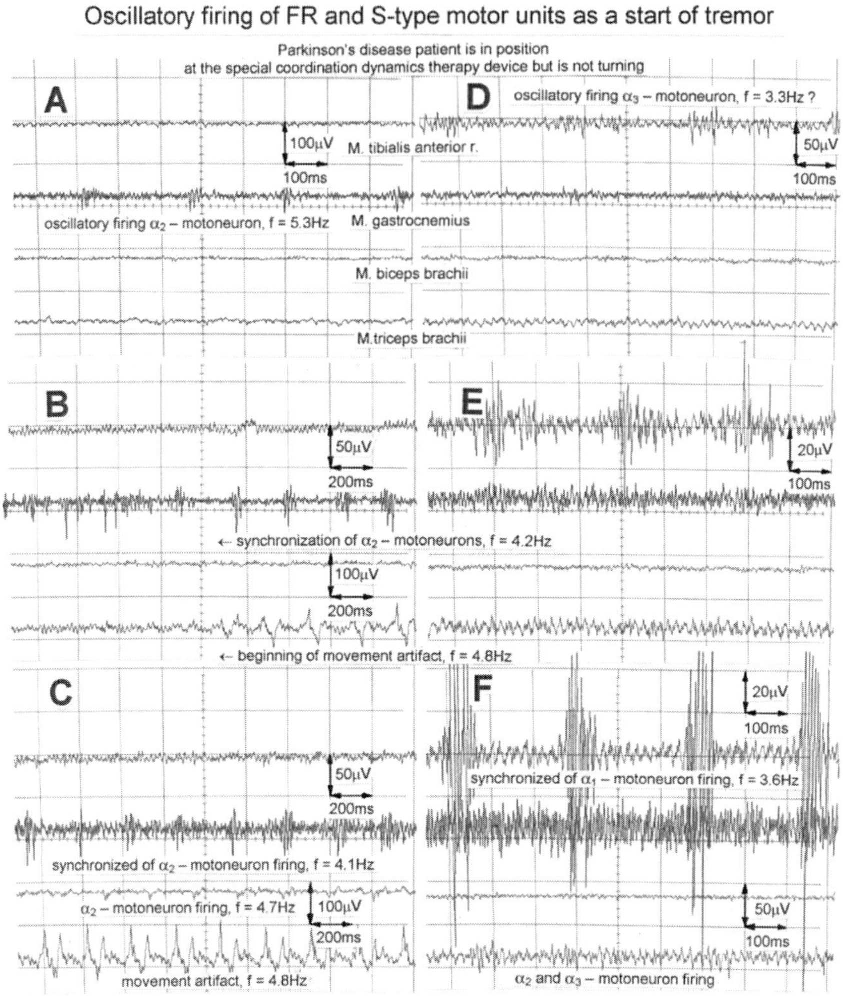

Figure 63. Oscillatory firing of α_2 – motoneurons and the synchronization of oscillatory firing α_2 and α_1 – motoneurons. Recordings were done with sEMG. A. Oscillatory firing of a single α_2 – motoneuron motor unit (f = 5.3Hz; amplitude ≈ 35μV, triplet-firing) in M. gastrocnemius. B. Synchronization of mainly oscillatory firing α_2 – motoneuron motor units (FR, f = 4.2Hz) in gastrocnemius muscle. With the synchronization of the oscillatory firing α_2 – motoneurons a movement artifact was induced in the electrodes on M. triceps brachii. C. Synchronized oscillatory firing of α_2 – motoneuron motor units, mixed with a few α_1 – motoneuron motor units of small amplitude in gastrocnemius muscle. In musculus biceps brachii an α_2 – motoneuron motor unit fires at with 4.7Hz not synchronizedly with the α_2 – motoneuron firings in gastrocnemius muscle. Rhythmic movement artifact in triceps brachii muscle (f = 4.8Hz) indicating an arm tremor of 4.8Hz. D. Possible oscillatory firing of an α_3 – motoneuron motor unit (f = 3.3Hz; amplitude ≈ 10μV; long impulse train) in tibialis anterior muscle; a few α_2 – motoneuron motor unit potentials (larger amplitude) fire synchronizedly. E. Synchronized oscillatory firing of mainly α_2 and α_3 – motoneurons (f ≈ 3.3Hz). F. Synchronized firing of α_1 – motoneuron unit potentials (FF-type) of large amplitude (up to 200μV) with the oscillatory firing α_2 and α_3 – motoneuron motor unit potentials (FR and S-type) in gastrocnemius muscle a few minutes later (resulting frequency = 3.6Hz). Rhythmic firing of α_2 and α_3 – motor unit potentials in the triceps brachii muscle in some coordination to the rhythmic firing in tibialis muscle giving rise to substantial rhythmic movements of tibialis anterior muscle. D, E, and F are consecutive recordings. Recordings from a 70-year-old patient with Parkinson's disease for 6 years (U.H.); tremor on both sides.

In conclusion, in this patient the Parkinsonian tremor started with the synchronized oscillatory firing of FR-type motor units (innervated by α_2-motoneurons). With the synchronized firing of FF-type motor units (innervated by α_1-motoneurons) the full muscle movement or tremor was generated.

The reason that the FF-type motor units synchronized their rhythmic firing and that the FF-type motor units synchronized their firing with those of the FR-type motor units and not vice versa will partly be analyzed below in the course of explaining oscillator stability differences between α_1 and α_2-motoneuron oscillators. α_2-oscillators have a high and α_1-oscillators a low stability.

6.2.4. Synchronization and De-Synchronization of FF-Type Motor Unit Firing with Oscillatory Firing FR-Type Motor Units

In Figure 64, the synchronization and de-synchronization of FF-type motor units (innervated by α_1-motoneurons) with FR-type motor units (innervated by α_2-motoneurons) is shown. In Figure 64A FR and S-type motor units (small motor unit amplitude) fired rhythmically at a common frequency of 3.9Hz in the right tibialis anterior muscle. In Figure 64B,C FF-type motor units (large unit amplitude) fired synchronizedly with the FR-type motor units; large FR-type motor unit potentials appeared on top of the small FR-type motor unit potentials. The common rhythmic frequency is now 3.4 and 3.3Hz. It can be seen that the large amplitude FF-type motor unit potentials slightly changed their synchronization phase with respect to the low amplitude FR-type motor unit potentials; sometimes the low amplitude FR-type motor unit potentials started earlier in the activity burst and sometimes they lasted longer than the large FF-type potentials in the burst (Figure 64B,C). In Figure 64D additionally FR-type motor units started to fire rhythmically in the biceps brachii muscle at a frequency of 4.5Hz. In Figure 64F FF-type motor unit potentials no longer fired in the tibialis anterior muscle, only FR-type potentials were recorded. The rhythmically firing FR-type motor unit potentials in the biceps brachii muscle increased their firing frequency to 5.5Hz. In this patient the tibialis anterior muscle was only moving up and down when the FF-type motor units fired in addition.

In Figure 64E, the rhythmically firing motor units are displayed on a time-stretched scale. It can be seen that the FF-type motor units had a motor unit action potential duration of approximately 8ms (amplitude $\approx$ 150μV) and the FR-type motor unit potentials had a duration of 17ms (amplitude $\approx$ 15μV). The single nerve-fiber action potentials have a much shorter duration (approximately 0.3ms). For details of amplitudes and durations see Figure 70.

6.2.5. Stability of α_1 and α_2-Motoneuron Oscillators

It was mentioned above that rhythmic muscle activity, leading to muscle movement and tremor, started with the oscillatory firing of the too little inhibited motoneurons. The tremor developed then with the synchronized rhythmic firing of the FR-type motor units to which the FF-type motor units, when activated, synchronized. It shall now be analyzed how these synchronizations can be explained by using the data obtained with the single-nerve fiber action potential recording method.

Measurements with the single-nerve fiber action potential recording method (Figs. 1, 3, 4) from an α_1-motoneuron axon and a primary muscle spindle afferent fiber, contained in a sacral S5 nerve root from a spinal cord injury patient, are shown in Figure 65.

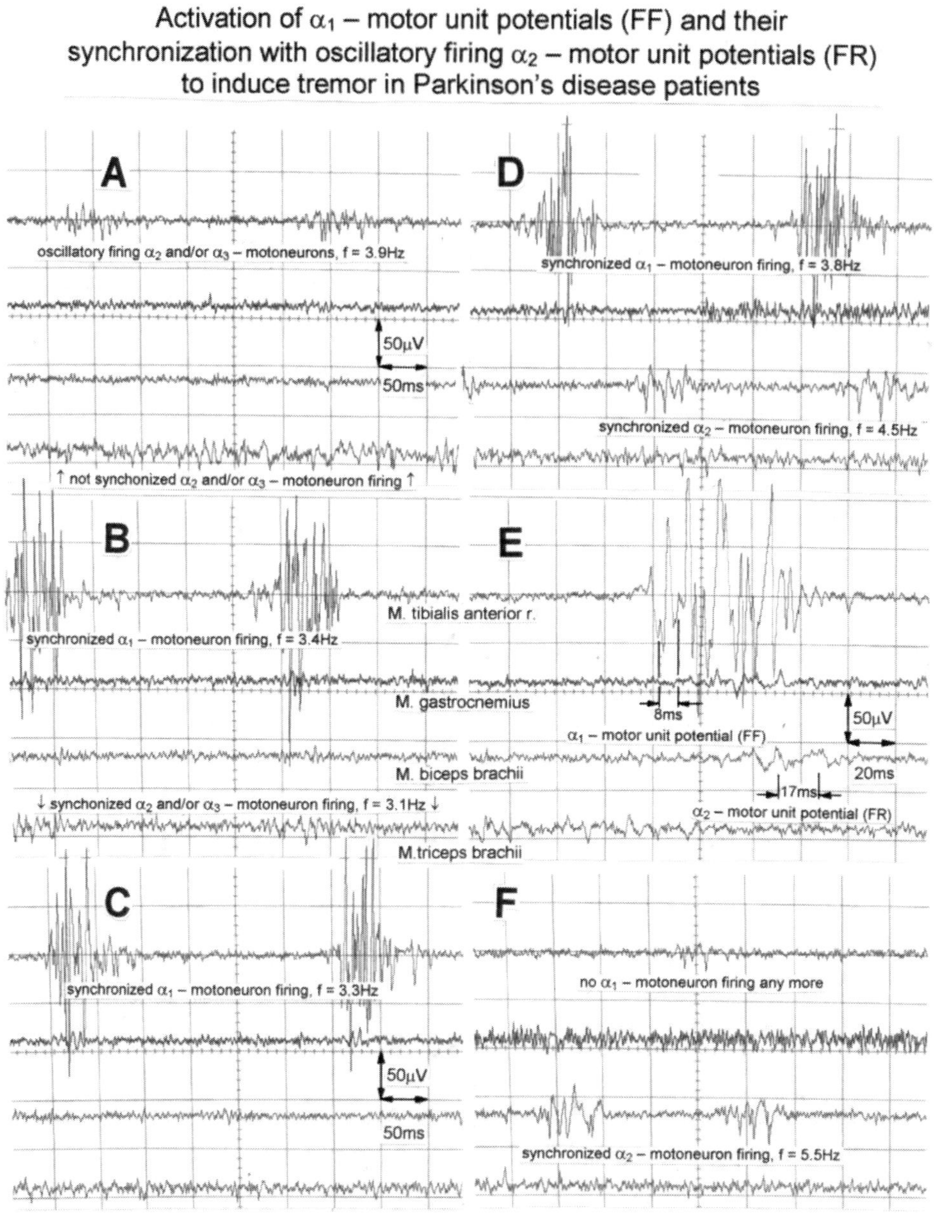

Figure 64. Time course of transient tremor induced by the activation of α_1 – motor unit potentials (FF-type) and their synchronization with oscillatory firing α_2 and α_3 – motoneuron motor unit potentials (FR and S-type). A. Synchronized oscillatory firing of α_2 and α_3 – motor unit potentials in tibialis anterior muscle (f = 3.9Hz). No synchronized firing of α_2 and α_3 – motor unit potentials in triceps brachii muscle. No movement of tibialis anterior muscle. B. α_1 – motor unit potentials are activated and synchronized with the α_2 and α_3 – motor unit potentials (f = 3.4Hz). Tibialis anterior muscle moves visibly. α_2 and α_3 – motor unit potentials in biceps brachii muscle fire partly synchronizedly with the motor unit potentials in tibialis anterior muscle. α_2 and α_3 – motor unit potentials can be seen before and after the α_1 – motor unit potentials in the activity burst. C. Similar firing as in 'B', but nearly no motor unit firing in triceps brachii muscle. D. Rhythmic firing of motor units giving rise to up and down movement (f = 3.8Hz) of tibialis anterior muscle. Synchronized α_2 and α_3 – motor unit firing in biceps brachii muscle (f = 4.5Hz), not coordinated with the rhythmic firing in tibialis anterior muscle.

α_2 and α_3 – motor unit potentials are activated in gastrocnemius muscle. E. Time-stretched motor unit action potentials: duration of α_1 – motor unit potential = 8ms (amplitude ≈ 150μV), α_2 – motor unit potential = 17ms (amplitude ≈ 15μV). F. Cessation of rhythmic firing of α_1 – motor unit potentials in tibialis anterior muscle; no visible contraction of tibialis anterior muscle anymore; some α_2 – motor unit firing still occurring. Synchronized α_2 - oscillatory firing still in biceps brachii muscle. Recordings from a 70-year-old patient with Parkinson's disease for 6 years (U.H.); tremor on both sides; measurements were performed on the more affected right side. During the measurements the hands and feet of the patient were in the position for exercising on the special coordination dynamics therapy device.

An original recording of the α_1-motoneuron nerve-fiber action potential (α_1) and the primary afferent nerve fiber action potential (SP1) is shown in Figure 65A. The schematic pattern of firing and the definitions of oscillation period ($T\alpha_1$) and phase correlation between the extracellular action potentials of the oscillatory firing α_1-motoneuron and primary muscle spindle afferent fiber ($\alpha_1 \rightarrow$ SP1) are shown in Figure 65B. Oscillation period distributions are shown in Figure 65C,E,G and phase correlation distributions between the oscillatory firing α_1-motoneuron and its driving primary muscle spindle afferents are shown in Figure 65D,F,H. It can be seen from Figure 65C,E,G that the oscillation periods varied between 110 and 150ms (frequency variation between 9 and 6.7Hz) depending slightly on different natural stimulation (Figure 68). The phase correlation between the action potentials of the oscillatory firing α_1-motoneuron and of the primary spindle afferent fiber varied between 12.5 and 17ms. The main phase relation variation was between 2 and 3ms.

In Figure 67A another original recording from the same S5 root as in Figure 65A is shown. In addition to the action potentials of the α_1-motoneuron and the primary muscle spindle afferent fiber, action potentials of an oscillatory firing α_2-motoneuron can be seen. Action potentials of secondary muscle spindle afferent fibers do not occur in this recording piece. In Figure 66A,B the different phase relations between the oscillatory firing α_1 and α_2-motoneurons and the primary and secondary muscle spindle afferent fibers are defined. It can be seen from Figure 66C,D that the phases between the α_1-motoneuron and a secondary muscle spindle afferent fiber ($\alpha_1 \rightarrow$ SP2(1)) and between the α_2-motoneuron and another secondary muscle spindle afferent fiber ($\alpha_2 \rightarrow$ SP2(2)) varied much more (phase distribution width ≈ 80 or 120ms) than the phase relation between the α_1-motoneuron and the primary muscle spindle afferent fiber ($\alpha_1 \rightarrow$ SP1; ≈ 4ms). This indicates that the primary spindle afferent fiber drove much more directly the α_1-motoneuron than did the secondary muscle spindle afferent fibers the α_1 and α_2-motoneurons.

The nearly time-locked activation of the α_1-motoneuron by the primary muscle spindle afferent indicates monosynaptic activation, whereas the phase distribution width of ≈ 120ms between the secondary muscle spindle afferent fiber and the α_2-motoneuron indicates more polysynaptic drive and polysynaptic connections indicate that there are more neuronal networks involved. Even though the α_1-motoneuron also got a projection from a secondary muscle spindle afferent fiber, α_2-motoneurons received many projections from secondary muscle spindle afferent fibers with different synaptic strengths, quantified by the phase distribution width and shape [61, 62].

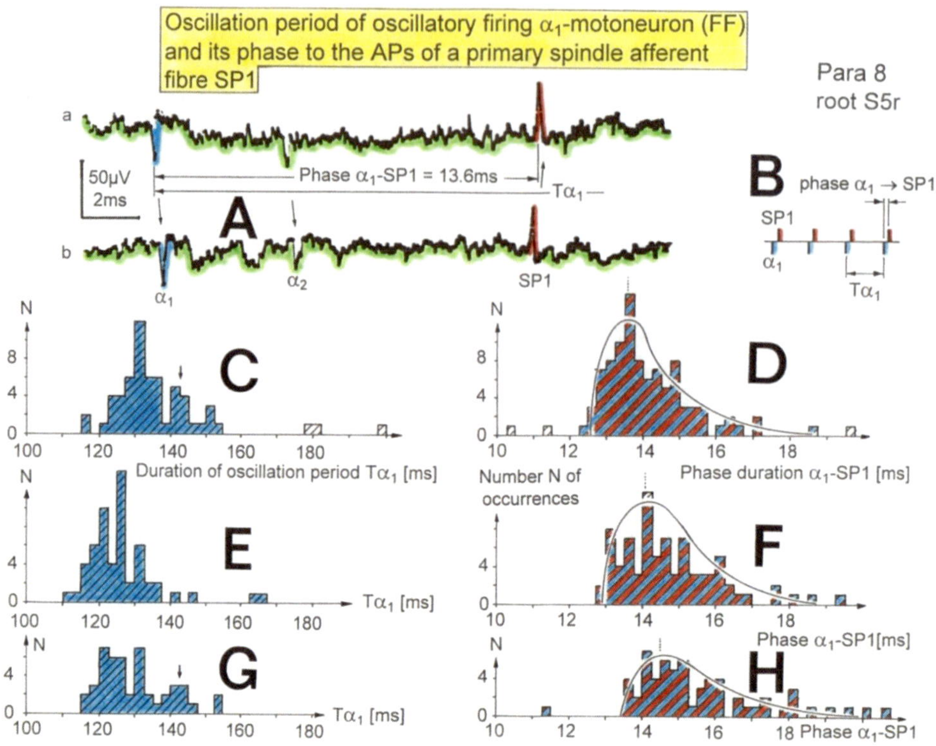

Figure 65. Distribution of the oscillation period of an oscillatory firing α_1 – motoneuron (FF) and the phase to its driving primary muscle spindle afferent fiber SP1. Recordings were taken with the single nerve-fiber action potential recording method. A. Original recording of the action potentials of the α_1 – motoneuron and SP1-fibre. B. Definition of oscillation period $T\alpha_1$ and phase $\alpha_1 \rightarrow$ SP1. C-H. Oscillation periods and the corresponding phases following different stimulation (C;D, oscillation period and phase distributions for the time interval 65-82s; Figure 68A. E;F, distribution for the time interval 1-30s; Figure 7B (strongest afferent drive). G;H, for 31-66s, Figure 68B). In 'C' and 'G', the small arrows mark sub peaks in the oscillation period distributions. Para 8, right root S5.

The differing degrees of the phase correlation indicates that the α_1-motoneuron oscillator was more strongly connected to the peripheral input than the α_2-motoneuron oscillator in accordance with the monosynaptic stretch reflex concept. The stronger dependence on peripheral input of the α_1-motoneuron oscillator is supported by the observation that the α_1-motoneuron oscillator stopped oscillating as soon as the input from the primary spindle afferent fiber was missing for longer than 1 oscillation period (Figure 66E,F).

In conclusion, the α_2-motoneuron oscillators seem to depend less on the peripheral input and are therefore more true oscillators. Rhythmic movements like tremor can therefore be better held continuously by α_2-motoneuron oscillators than by α_1-motoneuron oscillators.

The higher stability of oscillatory firing of α_2-motoneuron oscillators in comparison to α_1-motoneuron oscillators can also be seen when plotting oscillation periods consecutively against time.

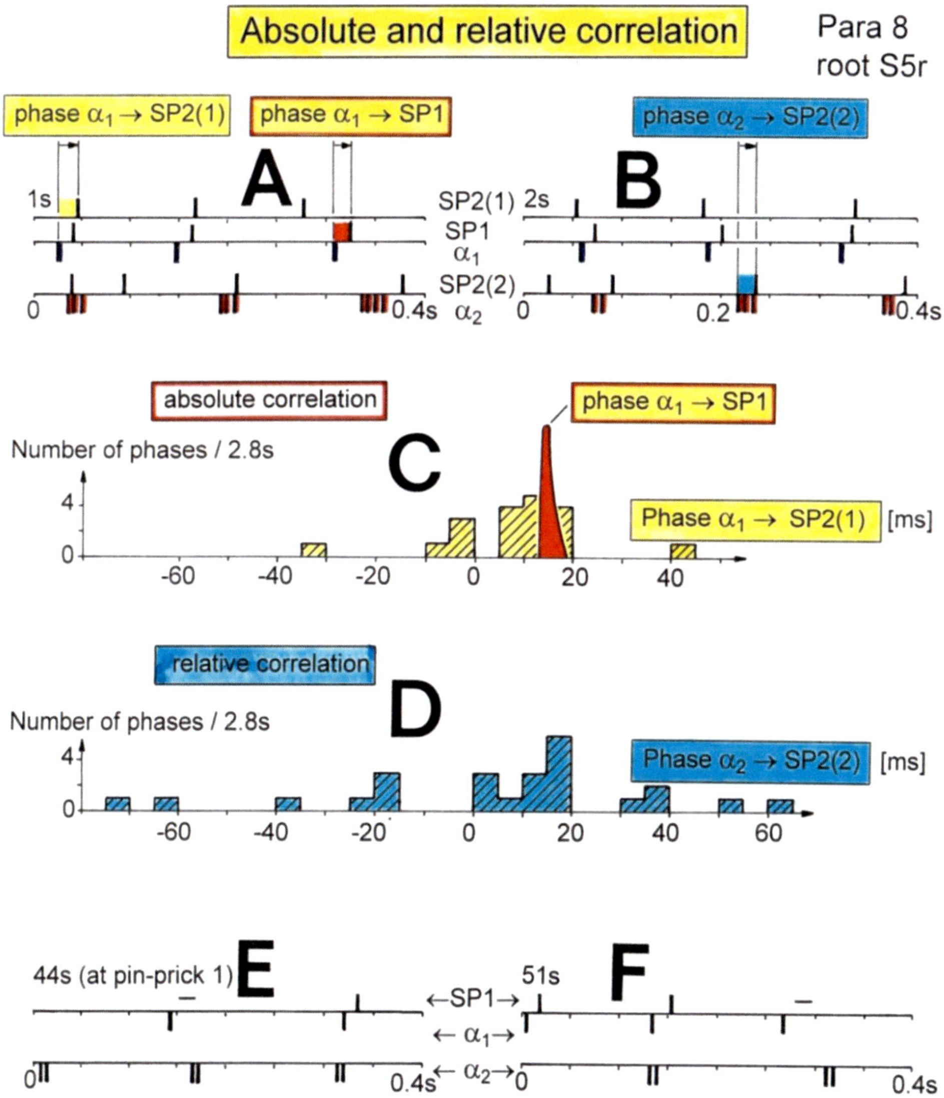

Figure 66. Absolute and relative correlation quantified by phase relations between the α_1 (FF) and α_2-motoneurons (FR) and their driving primary (SP1) and secondary (SP2) muscle spindle afferent fibers. A, B. Definition of the different phases. C. Distribution of the phases between α_1-motoneuron and the secondary muscle spindle afferent fiber SP2(1). Note that the phase distribution $\alpha_1 \rightarrow$ SP2(1) is approx. 40 times wider than that of the $\alpha_1 \rightarrow$ SP1 distribution (phase $\alpha_1 \rightarrow$ SP1 taken from Figure 65). D. Distribution of the phases between α_2-motoneuron and the secondary muscle spindle afferent fiber SP2(2). Note that the phase distribution $\alpha_2 \rightarrow$ SP2(2) is similar to that of $\alpha_1 \rightarrow$ SP2(1) (approx. 4 times wider). E,F. Note that every SP1-action potential is accompanied by a time-locked α_1-motoneuron action potential (AP). Para 8, root S5r [92].

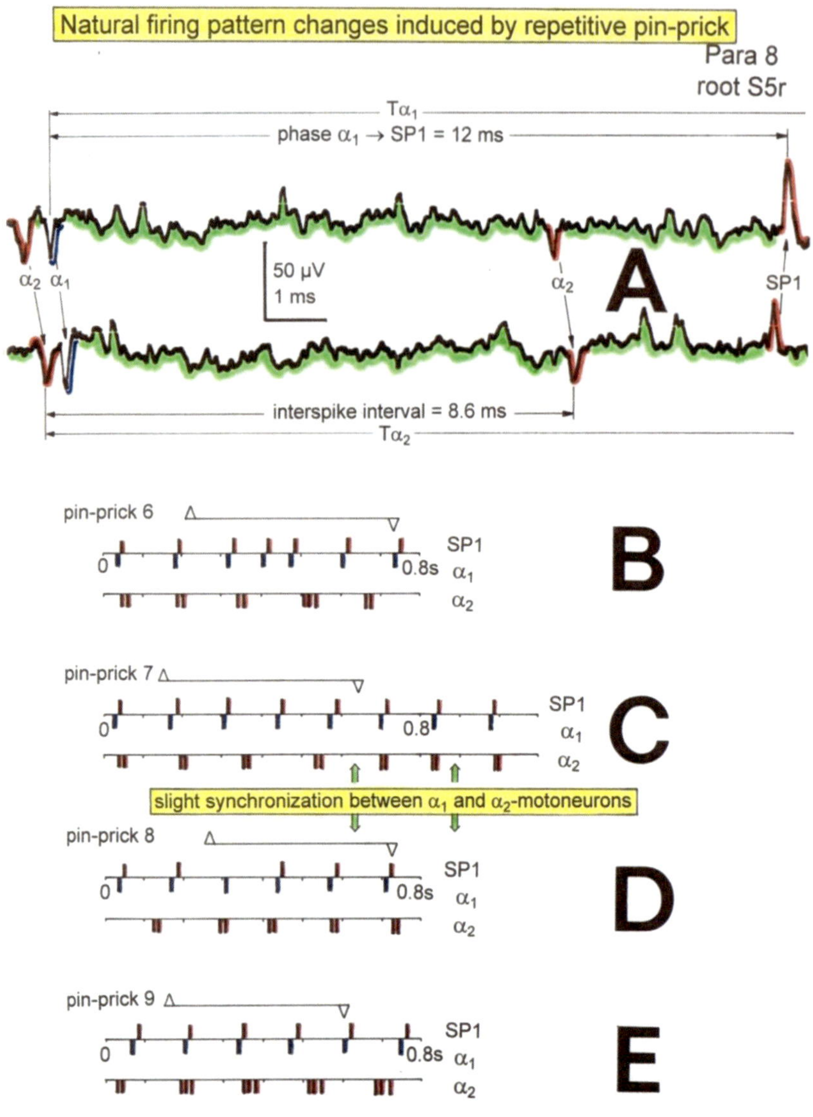

Figure 67. Natural impulse patterns of α_1 (FF) and α_2-motoneurons (FR) and of a primary muscle spindle afferent fiber (SP1) upon repetitive pin-prick. A. Original recording of action potentials (APs) of the α_1 and α_2-motoneurons and the SP1 fiber. B. - E. Firing patterns upon pin-pricking at sites 6,7,8,9 (sacral dermatome, close to the anus, mainly inside the anal continence reflex zone). Note that at pin-prick points 7 and 8 there was partial synchronization between the α_1 and the α_2-motoneurons. Note further the smaller amplitude and the larger conduction time of the α_2-motoneuron APs [92].

In Figure 68 the oscillation periods of the above α_1 and α_2-motoneurons are plotted against time and for different stimulations. It can be seen that the α_1-motoneuron stopped oscillation more often than the α_2-motoneuron. Within the time interval from 50s (Figure 68A) to 66s (Figure 68B) ($\approx$100s time interval) the α_2-motoneuron stopped oscillating 3 times and the α_1-motoneuron 19 times for the same stimulation at the same site. In this recording (with the single-nerve fiber action potential recording method) the α_2-motoneuron oscillator showed more flexibility than the α_1-motoneuron oscillator. When pin-pricking at the anal reflex zone the α_1 and α_2-motoneurons synchronized their oscillatory firing transiently, the α_2-motoneuron adapted the firing to the one of the α_1-motoneuron by reducing the oscillation periods (increasing the frequency) to match the one of the α_1-motoneuron oscillator (time period of 54s to 56s in Figure 68A; Figure 67C,D).

The stability of both oscillators may have been reduced as the recording was taken from a patient with spinal cord injury (Figure 10), where neuronal network functioning is impaired.

6.2.6. Transient Synchronization of Premotor Spinal Oscillator in a Patient with a Spinal Cord Injury

From the schematic impulse patterns shown in Figure 67B-E it can be clearly seen that the α_1 and α_2-motoneuron oscillators transiently synchronized their firing upon pin-pricking the anal reflex zone repeatedly every second. The oscillation period changes can be seen in Figure 68A.

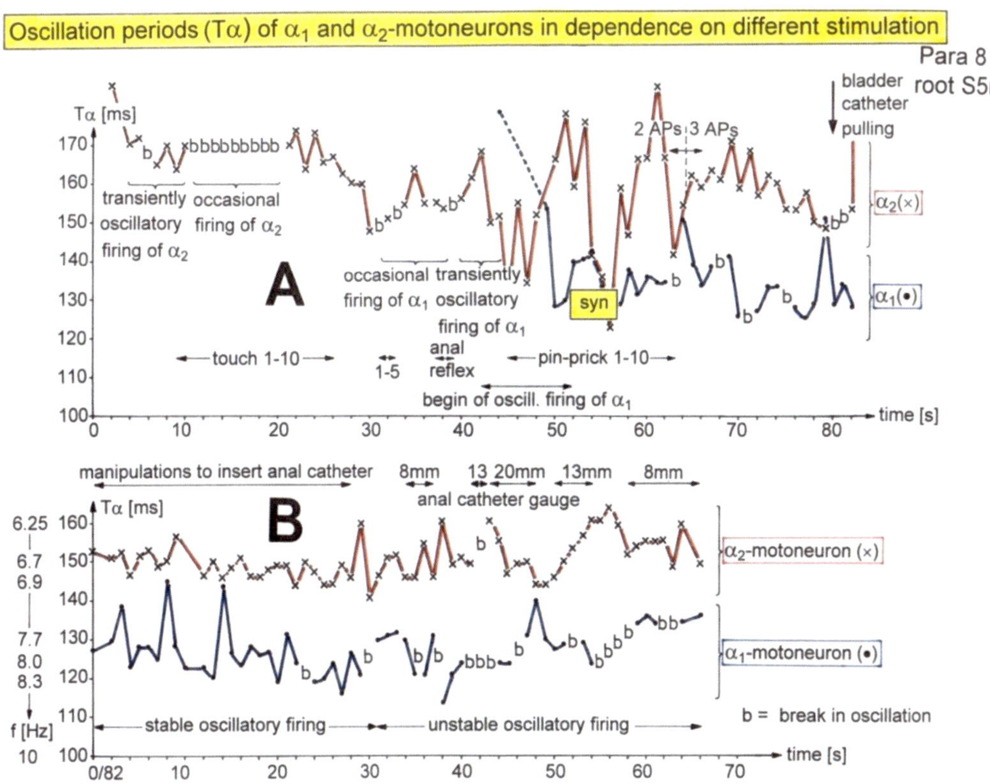

Figure 68. Oscillation periods (Tα) of an α_1 (FF) and an α_2-motoneuron (FR) in dependence on different stimulations; 'B' is a continuation of 'A'. Touch 1 - 10 = touching with a needle at different points in and outside the anal reflex zone. 8,13,20,13,8mm = changing the diameter of the anal catheter from 8 to 13 to 20 to 13 and back to 8 mm (3 gauges of anal catheter were used). syn = partial synchronization of the oscillatory firing patterns of the α_1 and the α_2-motoneurons. 2 APs / 3 APs = impulse train of the α_2-motoneuron mainly consisted of 2 or 3 action potentials respectively. Note that α_1 and α_2-motoneurons first fired transiently oscillatory before firing continuously oscillatory. Note further that the oscillatory firing was most regular upon manipulation at the anus (0-30s in 'B', anal reflex stimulation to secure continence) [92].

Therefore in non-Parkinson patients premotor spinal oscillators can also synchronize. But as soon as the repeated stimulus ceased, the oscillators quickly de-synchronized again. Also, pin-pricking the anal reflex zone (pin-prick 7, 8) induced the continence automatism very strongly. Pin-pricking the peri-anal skin outside the reflex zone (for example pin-pricks 9, 10) did not induce synchronization of the oscillatory firings (Figs.68A,67E). Thus in healthy subjects premotor spinal oscillators seem to synchronize only transiently and only if the repeated stimulus is sufficient to activate the oscillators (to self-organize).

In the patients with Parkinson's disease the rhythmic muscle movement or tremor started in a relaxed position and there was certainly no strong repeated stimulus. It is therefore concluded that two kinds of inhibition were missing in Parkinson's disease patients. First, premotor spinal oscillators did self-organize following very mild stimulus instead of strong, adequate stimulus (missing inhibition to stop high activation without adequate input) and second, the premotor spinal oscillators did synchronize with other premotor spinal oscillators without strong repeated stimulus (missing inhibition not to synchronize).

6.2.7. External Loop of Premotor Spinal Oscillators as a Cause for α_1-Motoneuron Oscillators Synchronizing Their Firing with α_2-Motoneuron Oscillators

It was shown above that several FR-type motor units synchronized their rhythmic firing to trigger muscle movement in a Parkinson's disease patient (Figure 63). But how did the FF-type motor units start to fire and how did they synchronize their firing with the FR-type motor units?

It has been shown in patients with spinal cord injury that oscillatory firing motoneurons can build up an external loop to the periphery via the γ-motoneurons and spindle afferents [63]. Figure 69 shows schematically the communication of premotor spinal oscillators with the periphery.

In Figure 69a sphincteric α-motoneurons oscillate following stimulation from bladder and anal catheters (no rhythmic muscle movement). The γ-motoneurons may fire partly in an oscillatory manner. When touching with a needle repeatedly every second the anal reflex area α and γ-motoneurons partly synchronized their firing. The oscillatory firing α-motoneurons built up an external loop to the periphery via the γ-motoneurons and the secondary muscle spindle afferent fibers (Figure 69b). When stopping the touching, the partial synchronization seized (Figure 69a). With the strong stimulation of pin-pricking at the anal reflex area, α and γ-motoneurons and spindle afferents synchronized their firing in addition to building up an external loop to the periphery (Figure 69e). This means that the periphery was included in the synchronized oscillatory firing including the input from the muscle spindle, joint, and skin afferents (Figure 69f).

If we transpose this external loop measurement to the Parkinson's disease patients, then the reason for muscle movement and tremor can be understood. Because of the lack of sufficient inhibition a number of α2-motoneuron oscillators get activated spontaneously and they synchronize their firing because of the lack of mutual inhibition. Slight contraction of muscles or stretching of muscle spindles, induced by the FR-type motor units, may start to induce the firing also of the primary muscle afferents. They stimulate the α_1-motoneurons to fire.

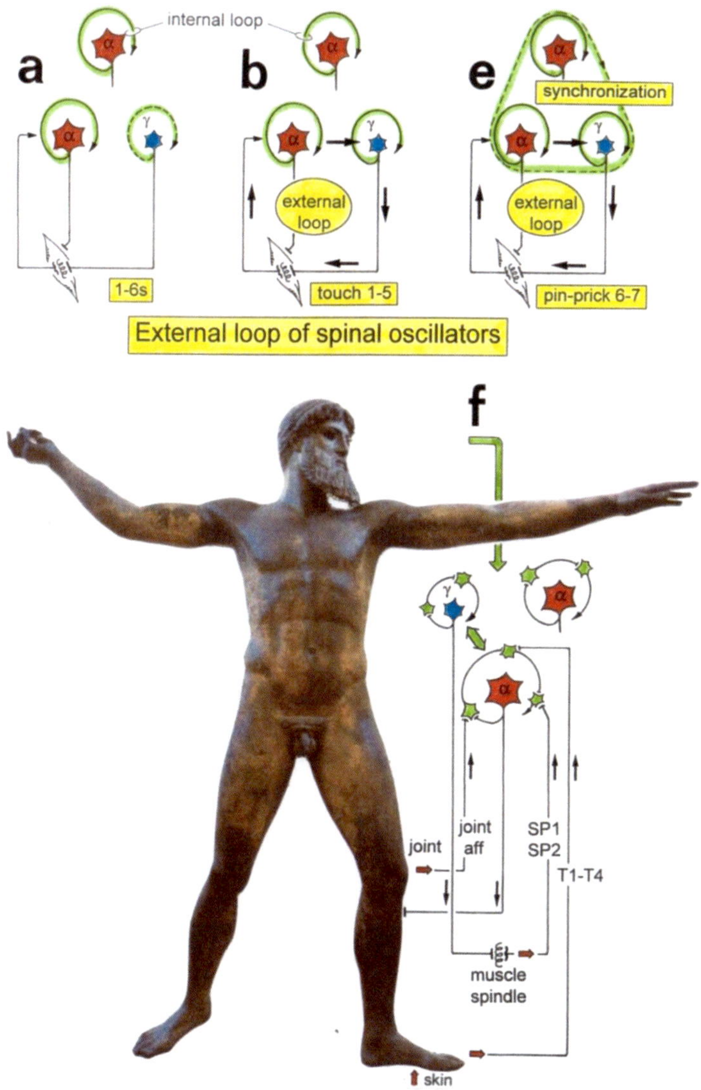

Figure 69. Spreading of oscillatory firing from α-motoneuron neuronal network to include muscle spindles (periphery) and synchronization of different α and γ-motoneuron neuronal networks caused by touch and pin-prick stimulation. (a) α-motoneuron neuronal networks fired oscillatory (solid line loop), γ-motoneuron neuronal network did not or did only partly (dashed line loop), upon no additional stimulation; taken from Figure 3Aa,Ba of (8). (b) Oscillatory firing α and γ-motoneuron neuronal networks built up a phase relation with muscle spindle afferents and efferents (external loop to the periphery, indicated by thick arrows) upon touch 1-5, taken from Figure 1Ab, Bb (8). (e) Oscillatory firing α (internal circuitry loop) and γ-motoneuron neuronal networks (external loop) synchronized (broad peak phase relation) upon pin-pricks 6-7, taken from Figure 1Ae,Be of (8). The dashed line loop represents synchronization. (f) Oscillatory firing α (internal circuitry loop) and γ-motoneuron neuronal networks (external loop) are extended by analogy from the continence muscles to muscles for locomotion. The open arrows indicate that it may be possible to synchronize spinal oscillators by rhythmic afferent input, generated by rhythmic movements (such as jumping on a springboard or running), and to re-preformat the neuronal circuitry by synapse remodeling to fire more physiologically oscillatory to reduce spasticity and improve locomotion. Extensive pathologic movement like tremor may entrain neuronal circuitry to increase tremor movement. The Greek good is a bronze statue of Zeus found close to the cape of Artemision 460 BC.

Via the external loop of the premotor spinal oscillators, where muscle, joint, and skin afferents are included in the rhythmic firing, the α_1-motoneurons fire in synchrony with the oscillatory firing α_2-motoneurons. It is likely that the α_1-motoneurons also oscillated. But it was observed with the single-nerve fiber action potential recording method that also occasionally firing α-motoneurons tend to synchronize their firing with oscillatory firing motoneurons. A macroscopic recording was shown in Figure 1 of [54]. Based on the above measurements of the human nervous system one mechanism is of how tremor is generated in Parkinson's disease patients can be understood.

6.2.8. Tremor Triggering Mechanism

Surface electromyography (sEMG) showed that rhythmic muscle contraction and tremor was generated by rhythmic activation of FF-type and FR-type motor units, innervated by α_1 and α_2-motoneurons respectively (Figs.7,8). Because of the small amplitude of S-type motor unit action potentials, innervated by α_3-motoneurons (for classification scheme, see Figure 7), their role in the induction of tremor is unclear. In Figures 63 and 64 it seems as if the S-type motor units fired in synchrony or coordination with the FR-type motor units.

Figures 63 and 64 show that when tremor started, the FR-type motor units fired before the FF-type motor units. In another patient it seemed as if the tremor started with the FF-type motor unit activity. But since even the motor unit action potential amplitudes of FR-type muscle fiber motor units are five to ten times smaller than the amplitudes of FF-type motor units, it could be that FR-type motor units were not recognized in all cases. Since in the Parkinson disease nerve cells die in the substantia nigra and perhaps additionally in other parts of the CNS, one can expect a large number of different pathologic CNS organizations, depending on the actual degeneration status in the given patient.

The induction of tremor in Parkinson's disease patients cannot be understood unless we differentiate between α_1 and α_2-motoneurons (and α_3-motoneurons), innervating FF-type and FR-type (and S-type) motor units, and without the specific knowledge about oscillatory firing of motoneurons in the premotor network.

6.2.9. Synchronization of FF-Type with Rhythmic FR-Type Motor Unit Firing

It was shown in Figures 63 and 64 that during tremor FF-type motor units were only activated in synchrony with the rhythmically firing FR-type motor units. The induction of FF-type motor unit firing was explained by the external loop of premotor spinal oscillators, that means the inclusion of the γ-loop into the rhythmic firing of the premotor spinal oscillators. But since the FF-type motor units fired only in synchrony with the rhythmically firing FR-type motor units, there was also network synchronization in the premotor network between the α_1 and α_2-motoneuron oscillators, otherwise there would have been also some firing of FF-type units in between the rhythmic FR-type motor unit firing at the beginning of the FF-type motor unit firing, with the building up of the external loop. It seems therefore that the synchronization of the FF-type with the FR-type motor units has a central and a peripheral component. This view is supported by the synchronization measurement between the intumescentia lumbosacralis and cervicalis (Figure 63B). With the synchronization of the FR-type motor units in the gastrocnemius (triceps surae) muscle, a movement artifact occurred in the triceps brachii muscle, indicating that with the synchronization of α_2-oscillators in the intumescentia lumbosacralis, also α_1-motoneurons in the intumescentia cervicalis were activated to fire in synchrony to give rise to movement and movement artifact at recording electrodes on the triceps brachii muscle.

6.2.10. Amplitude and Duration of FF, FR and S-Type Motor Unit Potentials in Comparison to Extracellular Single-Nerve Fiber Action Potentials

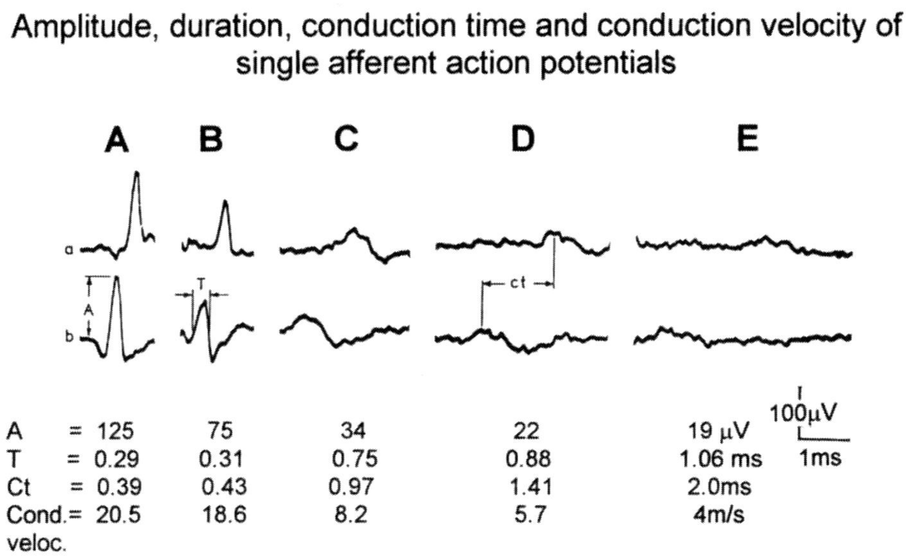

Figure 70. Extracellular single nerve-fiber action potential waveforms of different amplitudes (A), durations (T), conduction times (ct) and conduction velocities (cond. veloc.). The afferent fibers were stimulated by touch, pin-prick, and bladder and anal-catheter pulling and releasing. HT4 (brain-dead human cadaver), 56-year-old at death, 32°C (approximate nerve root temperature). Dorsal S5 root, approximate root diameter = 0.17mm.

The recordings from FF and FR-types motor units in Figure 64E seem to indicate that FR-type motor unit potentials have much smaller amplitudes and are of longer duration than FF-type motor unit potentials. Extracellular single-nerve fiber action potentials have been shown to have different action potential amplitude and duration (Figure 70), depending on the conduction velocity and the nerve fiber diameter [30]. Human nerve fibers could actually be characterized and classified by group conduction velocity and group nerve fiber diameter [49].

It is of importance to know whether in general FF, FR and S-type motor units have approximately similar typical action potential amplitudes and durations, even though motor unit size may increase and decrease during re-innervation and the muscle fiber diameter is dependent on the usage. If one could distinguish better between FF, FR and S-type motor units, innervated by α_1, α_2, and α_3-motoneurons respectively, then one could better analyze natural impulse patterns of the different motor units and their innervating motoneurons. One could better understand the organization of the premotor neuronal networks and further rostral CNS structures like the basal ganglia. Human neurophysiologic research on the CNS functioning has therefore not only consequences for the development of treatments but also for the understanding of the functioning of the human CNS in general.

6.3. Tremor and Clonus in Patients with Parkinson's Disease and Spinal Cord Injury

6.3.1. Patients with Parkinson's Disease and SCI to Analyze The Human Premotor Network

When relative coordination between oscillatory firing motoneurons (and the motor unit muscle fibers they innervate) become impaired, the oscillatory firing motoneurons do synchronize their firing (instead of coordinating their firing to avoid synchronization) and muscles start to move rhythmically and fingers, hands, arms, legs, and face may start to shake. Further, the motor programs and in consequence also the movements are impaired, as can be measured by the coordination dynamics recording method.

Two types of CNS injuries, namely those in patients with Parkinson's disease and those suffering spinal cord injury, will be used to analyze the human premotor network. The patients with Parkinson's disease mainly show an injury to the substantia nigra due to degeneration, whereas their spinal cord is not affected. The patients with a spinal cord injury on the other hand have a healthy CNS apart from the injured spinal cord. Their basal ganglia, for sure, are not damaged by the injury. The differences in motor unit firing between both types of injury will provide us with information on the organization of the neuronal networks of the human spinal cord.

The importance of this research does not only follow from the possible partial repair of the malfunctioning CNS but also because of a better understanding of the functioning of the human brain. In recent publications, the conclusion was reached from the frequencies of tremor in Parkinson's disease patients that neurons in the basal ganglia fire rhythmically [65,66]. But the high activation mode of motoneuron firing has clearly been shown to be firing rhythmically (oscillatory firing). This oscillatory firing of motoneurons originates in the spinal cord, as could be demonstrated by measurements taken from the isolated human spinal cord (in spinal cord injury patients taken from sites caudal to the complete injury) and from measurements of the isolated spinal cord in connection with the lower part of the reticular formation (in brain-dead humans). Therefore spontaneous rhythmic firing of motor units measured by surface electromyography (sEMG) in patients with spinal cord injury or Parkinson's disease corroborates previous measurements with the single-nerve fiber action potential recording method that the rhythmic firing of the motoneurons, and the motor unit muscle fibers they innervate originates in the spinal cord. This does not mean that neurons in the basal ganglia are not firing rhythmically with similar or identical frequencies. Rhythms and rhythm coupling may occur throughout the CNS [196,197]. But it is of importance to know where certain functions and rhythms mainly originate, in spinal cord neuronal networks or supraspinally.

6.3.2. Description of Tremor and Clonus

Physiologic Tremor

The symptom of physiologic tremor can be experienced when the CNS is not functioning optimally as for example when it is overexcited. When one drinks too much coffee and then attempts to dissect a frog under the microscope at a magnification of 20 times, one may see one's hands shaking at a frequency in the range of 5Hz, due to the random synchronization of oscillatory firing motoneurons.

Pathophysiologic Tremor

The symptom of pathophysiologic tremor (strong tremor) can be seen in patients with Parkinson's disease. Muscles may move rhythmically. Arms, hands, fingers, and legs may shake at frequencies in the range of 5Hz. Also face muscles may shake, depending on the degree of the disease.

Physiologic Clonus

The symptom of physiologic clonus can be experienced when leg muscles are exhausted from carrying heavy loads. Then, when you sit down with only the forefoot put on the ground the leg may start to move continuously rhythmically up and down due to consecutive dorsal and plantar flexion's of the feet. The right and left leg mostly do not synchronize their rhythmic movements.

Pathophysiologic Clonus

The symptom of pathophysiologic clonus can very often be observed in patients who suffered a spinal cord injury. Continuous rhythmic plantar and dorsal flexion's of the feet can be strong and continuous.

6.4. Pathologic Motor Programs: Motor Bursts Are Structured with Tremor, Clonus, and Rhythmic Motor Unit Activity

Now EMG recordings are arranged in order to show increasing pathology and complexity of CNS functioning. In slight impairment of CNS functioning, motor programs seemed to be quite normal in accordance with repair in mild CNS injury (see above). But in patients with Parkinson's disease or spinal cord injury some motoneurons fired spontaneously in an oscillatory manner. The more the CNS functioning was impaired, the more pathologic the motor programs. In patients with Parkinson's disease oscillatory firing motoneurons synchronized their firing to give rise to tremor. Tremor and rhythmic firing of motor units occurred in motor burst of motor programs in some muscles. Some muscles were not activated at all or were continuously activated during the motor program. In patients who have suffered a spinal cord injury, muscles were not activated or activated continuously (spasticity) during the motor programs. Motor bursts were partly structured with clonus activity and the oscillatory firing of motoneurons was partly synchronized.

6.4.1. Spontaneous (Uncontrolled) Oscillatory Firing

Figure 71A shows the electromyographically recorded motor program of a patient who suffered an incomplete spinal cord injury while exercising coordinated arm and leg movements on a special coordination dynamics therapy device. The motor program looks quite normal. Still one FF-type motoneuron of the spinal cord fired in an oscillatory manner at 6.25Hz despite receiving no stimulation (Figure 71B). In this case, the motor unit potential shows a complicated structure (Figure 71B, insert), possibly indicating the sprouting of the axon to other motor units. Thus oscillatory firing of motor units can be observed in spinal cord injury patients upon nearly no stimulation.

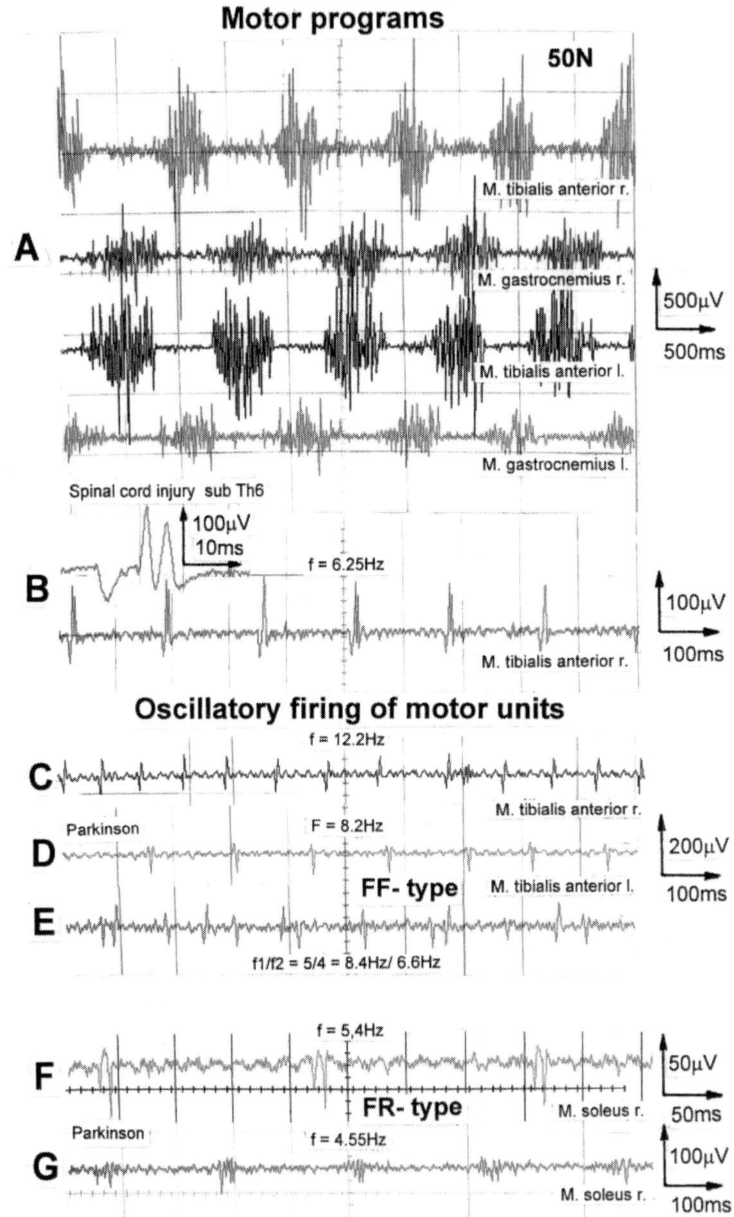

Figure 71. A. Electromyography recordings (EMG) of the motor program of a patient with an incomplete spinal cord injury sub Th6 during exercise on the special coordination dynamics therapy device at a load of 50Newton. The motor program looks normal. B. Oscillatory firing of an FF-type motor unit in the same patient in the right tibialis anterior muscle upon no exercising. Insert; time-stretched motor unit potential (re-innervation potential because of complicated structure). C-G. Oscillatory firing of FF-type (C,D,E) and most likely FR-type motor units (F,G) in 3 patients with Parkinson's disease. In E, two motor units fired simultaneously oscillatory. In F and G, the FR-type motor units fired with 2 and 3 action potentials per impulse train respectively.

Figure 71C,D shows FF-type motor units firing spontaneously in an oscillatory manner at 8.2 and 12.2Hz in Parkinson's disease patients. From the motor unit waveform potential it is clear that it is always the same motor unit that fires. In Figure 71E, we see the same motor unit firing as in Figure 71D, but additionally one further FF-type motor unit oscillating following very little or no volitional activation. The frequency ratio was f_1/f_2 = 8.4Hz/6.6Hz = 5/4. Approximately three minutes earlier the frequency ratio was f_1/f_2 = 7.9Hz/4.8Hz = 5/3. The coordination between these two oscillatory firing motor units is not clear. Figure 71F,G shows two FR-type motor units firing rhythmically in patients with Parkinson's disease with no volitional activation, mostly with 2 and 3 motor unit action potentials per impulse train at a frequency of 5.4Hz and 4.55Hz respectively. Since normally, motoneurons increase their firing rate with the increasing afferent input from occasional firing until they start to fire oscillatory for higher loads, there seems to be a lack of inhibition (as found above), especially in Parkinson's disease patients, to control the onset of oscillatory firing. But uncontrolled oscillatory firing can also be observed in spinal cord injury (Figure 71B).

6.4.2. Uncontrolled Synchronized Firing of Motor Units in Parkinson's Disease Patients

It was shown above that oscillatory firing motoneurons and oscillatory firing motor units coordinated their firing so as not to fire in a synchronized manner (Figure 16B); this may be achieved by some kind of lateral field inhibition. It will be shown below in a Parkinson's disease patient that if this kind of coordinated firing is impaired, some motor units start to fire in a synchronized manner.

In Figure 72B we see mainly FR-type motor units (identified by low motor unit amplitude and FR-type frequency range ($\approx$ 5Hz)) firing in a synchronized manner oscillating in the right soleus muscle at a frequency of 4.55Hz. In the right tibialis anterior muscle, also FR-type motor units oscillated in a synchronized manner antagonistically to the motor units in the soleus muscle. On top of the low-amplitude FR-type motor unit activity, large amplitude FF-type motor unit activity can be seen, which was synchronized. The duration of FR and FF-type activity is marked in one activity burst in the right tibialis anterior muscle. Synchronized oscillatory firing of mainly FR-type motor units can further be seen in the soleus, the biceps brachii, and the triceps brachii muscles in Figure 72D. In Figure 72C,D further synchronized activity of FR and FF-type motor units can be seen in the right tibialis anterior muscle.

It can partly be seen from Figure 72A that this synchronized oscillatory firing of FR and FF-type motor unit's builds up and then ceases. In Figure 72A, the tremor ceases. At first, more and more FF-type units stopped firing and then also the FR-type motor units stopped firing. However, when the tremor starts first the FR-type units fire synchronously in an oscillatory manner to which the FF-type units then synchronize. The process first affects the smaller units with lower amplitudes and then the larger units with larger amplitudes. There is some variation in this rule because of the different distances of the motor units to the recording electrodes.

As can be seen from Figure 72B,C,D, the tremor in antagonistic muscles is mostly antagonistic, probably due to the antagonistic proprioceptive input. Some agonistic tremor coordination seems visible. The seemingly low-amplitude agonistic tremor coordination may also be due to cross-talk of sEMG activity; this means that, for example, some tibialis muscle activity was recorded in the soleus muscle. The coordination of the sEMG-displayed tremor between different muscles may depend on the agonistic and antagonistic afferent input.

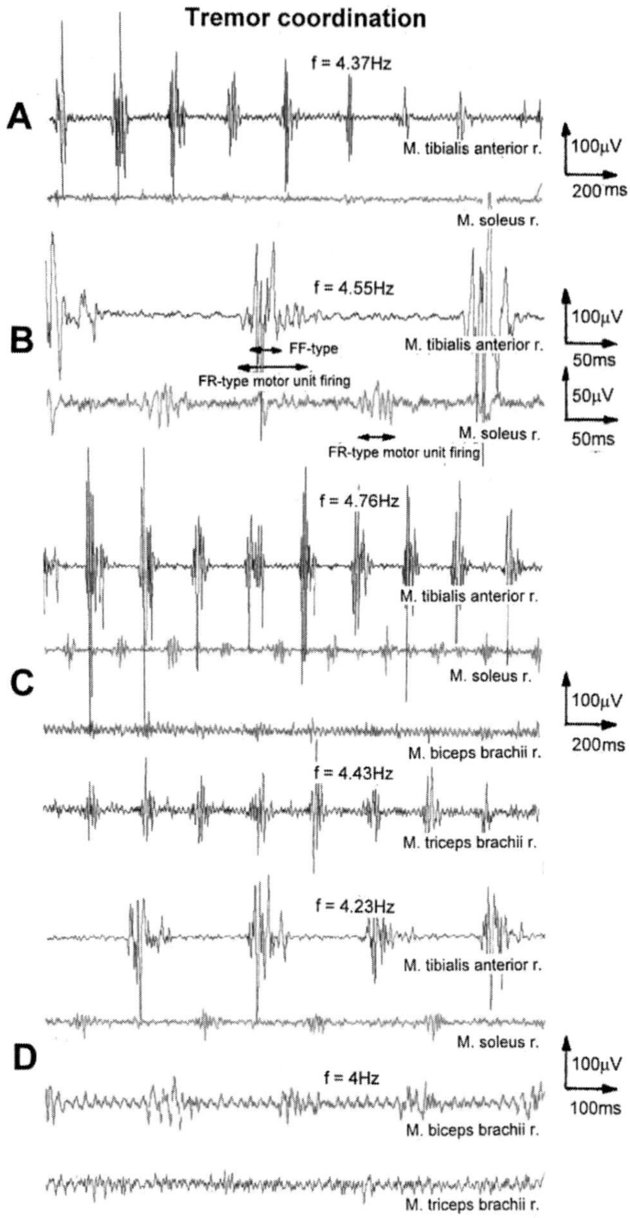

Figure 72. Electromyography (EMG) recordings of tremor coordination and synchronization of FF and FR-type motor unit firing, giving rise to tremor in a patient with Parkinson's disease. A. Ceasing of tremor activity in the right tibialis anterior muscle; no tremor activity in the right soleus, right biceps brachii, and triceps brachii muscles. B. Antagonistic tremor activity in the right tibialis anterior and soleus muscles. The EMG activity in the soleus muscle stems mainly from FR-type motor units and those in the tibialis anterior muscle from FR (small-amplitude) and FF-type (large-amplitude) motor units. No clonus activity in the other two muscles. C. EMG tremor activity in the tibialis anterior, the soleus and the biceps brachii muscles. The clonus activity in the right soleus and the right tibialis anterior muscles shows antagonistic coordination. The activity in the biceps brachii muscle (f = 4.43Hz) was not in coordination with those of the tibialis anterior and soleus muscle (f = 4.76Hz). D. EMG tremor activity in the tibialis anterior, the soleus, the biceps brachii, and the triceps brachii muscles. There was antagonistic tremor coordination between the antagonistic tibialis anterior and soleus muscles and between the antagonistic biceps and the triceps brachii muscles.

6.4.3. Motor Program Bursts in Patients with Parkinson's Disease Structured with Tremor Activity and Motor Unit Oscillatory Activity

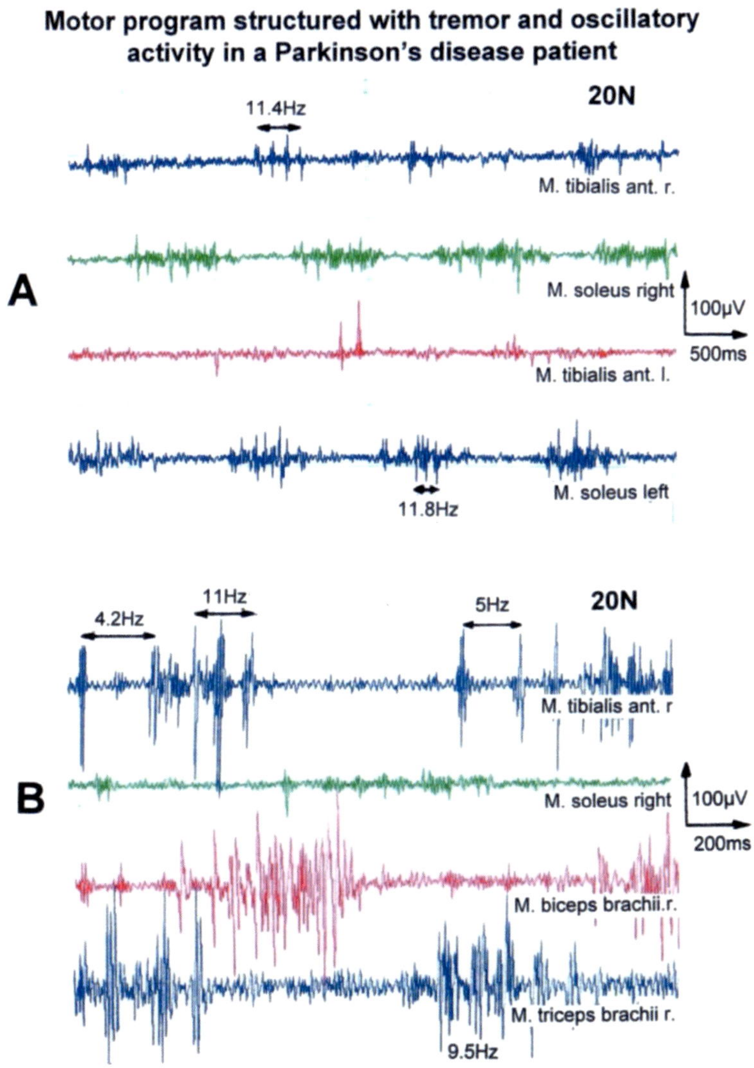

Figure 73. EMG motor programs of a patient with Parkinson's disease during the exercise on the special coordination dynamics therapy device at a low load of 20N. The motor program muscle bursts are more or less structured by rhythmicity (no rhythmic structure in normal motor bursts (Figure 1A)). In A rhythmic activity at a frequency of 11.4 and 11.8Hz is suggested. In B, (faster sweep and partly different muscles) low-frequency rhythmicity of 4.2 and 5Hz (frequency range of α_2-motoneuron oscillatory firing) and higher frequency rhythmicity of 11 and 9.5Hz (frequency range of α_1-motoneuron oscillatory firing) are suggested to be seen.

The impaired coordination leading to spontaneous oscillatory firing and synchronized oscillatory firing of FR and FF-type motor units and tremor should also be seen during movements because also then coordinated organization of the CNS networks is needed.

In the motor program bursts shown in Figure 73A, rhythmic activity can be identified, which cannot be seen in the rather physiologic motor pattern in Figure 71A. The highlighted rhythmic firing at 11.4 and 11.8Hz may indicate oscillatory firing of FF-type motor units innervated by α_1-motoneurons.

In another, more time-stretched motor program recording, such rhythmic firing can be seen more clearly (Figure 73B). Rhythms of 5 and 4.2Hz were most likely induced by oscillatory firings of an α_2-motoneuron, and those of 9.5 and 11Hz were most likely induced by oscillatory firings of an α_1-motoneuron.

Three important conclusions can be drawn from Figure 73. First, the term referring to the symptom 'resting tremor' is not correct. Tremor not only occurs during rest, but also during movement. The reduced intensity of tremor during movements may indicate entraining of neuronal networks resulting in a reduction of tremor in the short-term memory. Second, FF and FR-type motor units, innervated by α_1 and α_2-motoneurons respectively oscillate with their 'Eigen-frequencies' for high activation during motor program burst. This was shown earlier in patients with spinal cord injury for muscles only innervated by a few motoneurons. Third, since synchronized oscillatory firing in tremor can also be seen in motor bursts during movements, two organizations, namely tremor and movement were organized in the neuronal networks at the same time.

6.4.4. Motor Program Bursts in Patients Who Suffered a Spinal Cord Injury, Structured With Clonus Activation and Rhythmic Firing of FF-Type Motor Units

In patients with Parkinson's disease (probably due to the lack of supraspinal inhibition) and in patients with an incomplete spinal cord injury (probably due to the damage of inhibitory tract fibers) it could be demonstrated that few (in spinal cord injury) and many (in Parkinson's disease) motoneurons started to fire spontaneously or nearly spontaneously oscillatory (Figure 71). In addition, in Parkinson's disease patients the oscillatory firing motoneurons partly synchronized their firing (Figure 72), probably due to the missing of some kind of lateral field inhibition. In patients with a spinal cord injury, there was very little or no synchronization in the oscillatory firing motoneurons. Tremor therefore cannot be observed in spinal cord injury patients.

However, marked clonus which is not functionally dissimilar from tremor (Figure 74A) can often be observed in patients with spinal cord injury. Instead of tremor of the hands and feet marked at rest, the feet perform consecutive dorsal and plantar flexion's. The reason for that rhythmic dorsal and plantar flexion is probably rhythmic activation of a very sensitive regulation loop between the feet and the spinal cord rather than synchronization of oscillatory firing motor units.

In Figure 74B a pathologic motor program of a patient with an incomplete thoracic spinal cord injury can be seen. The motor program is poor, because there is no motor program in the right tibialis anterior muscle and the motor bursts in the right and left gastrocnemius muscle are structured with rhythmicity, i.e., with clonus activity. A typical clonus recording not taken during exercise can be seen in Figure 74A.

The uncoordination of motor unit firing can also be seen in the movement patterns in Figure 74. On a stretched time scale the clonus can be seen clearly in the right and left gastrocnemius muscles (Figure 74C). In the left gastrocnemius muscle the motor program burst is structured with three clonus bursts, which are in turn structured by rhythmic activity at 26 and 40Hz which may originate from the oscillatory firing of FF-type motor units innervated by α_1-motoneurons. Since tremor, clonus and rhythmic firing of FF-type and FR-type motor units can be seen in motor programs, they all go through the same final premotor network.

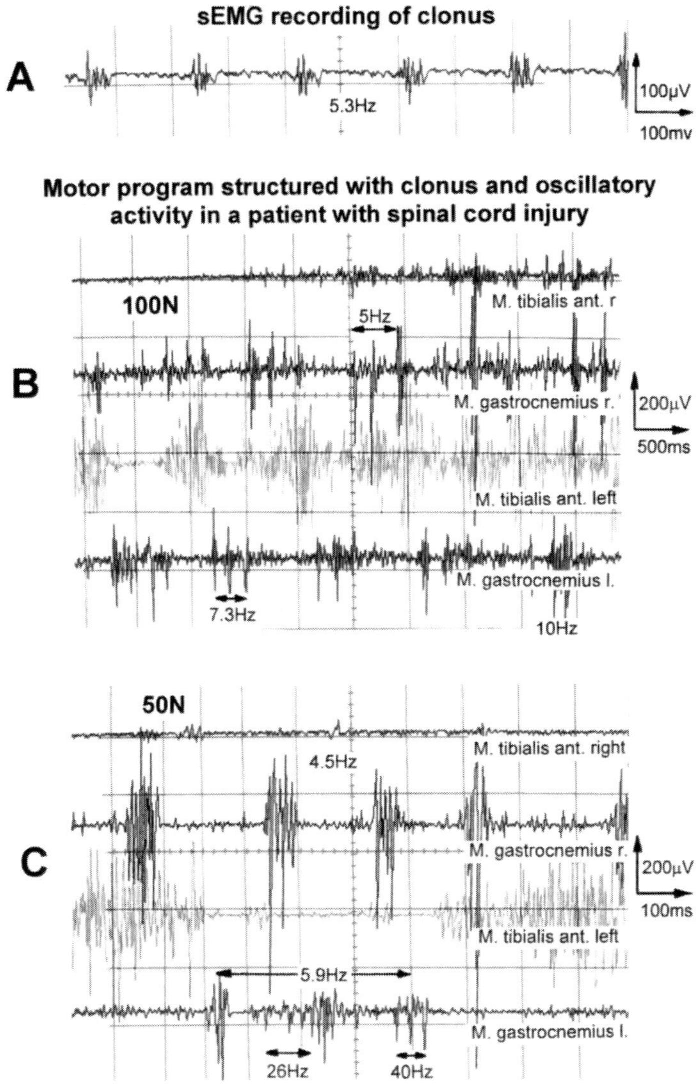

Figure 74. A. EMG recording of a clonus (f = 5.3Hz) in the right tibialis anterior muscle of a patient who suffered a complete spinal cord injury sub Th5/7; the patient was not exercising. B, C. Motor programs of a patient who suffered an incomplete spinal cord injury sub Th4 upon exercising on the special coordination dynamics therapy device at 50 and 100N (medium to high load). In B, motor program bursts are structured by rhythmicity; frequencies of 5 and 7.3Hz are suggested. No motor program in the right tibialis anterior muscle; some motor program structure in the right gastrocnemius muscle. In C (faster sweep), there is no motor program in the right tibialis anterior muscle. Mainly clonus activity at a frequency of 4.5Hz can be seen in the right gastrocnemius muscle. Two physiologic motor program bursts can be seen in the left tibialis anterior muscle (not structured by rhythmicity). In the left gastrocnemius muscle a motor program burst can be seen which is structured by 5Hz rhythmicity (clonus frequency, see clonus in the right gastrocnemius muscle) and higher frequency rhythmicity (26 and 40Hz).

6.4.5. Oscillatory Firing of Motoneurons Originates in the Spinal Cord

Rhythmic firing of FF and FR-type motor units was recorded with no or nearly no volitional activation (Figure 71B-G) in patients with Parkinson's disease and in patients who have suffered a spinal cord injury. These findings support earlier measurements with the single-nerve fiber action potential recording method of α-motoneurons firing oscillatory for high activation, disconnected from supraspinal drive. Recordings from isolated sections of the spinal cord in patients with complete spinal cord injury taken from below the injury level showed less regular rhythmic firing (Figure 10). Recordings from the whole spinal cord connected with the lower part of the reticular formation in brain-dead humans showed more regular oscillatory firing (Figures 10, 25). In the presence of damage to the pre-existent descending pathways, activation of redundant pathways (and may be redundant neurons), reorganization in spared descending tracts and building of new pathways is required for functional repair in spinal cord injury.

6.4.6. Motor Bursts Structured with Rhythmic Activity

When the patients exercised coordinated arm and leg movements on a special coordination dynamics therapy device, rhythmic activity could be observed in the motor bursts for low load in Parkinson's disease patients (Figure 73) and for higher loads in patients who suffered a spinal cord injury (Figure 74). First, this rhythmic activity during motor bursts supports the conclusion that during the motor burst, highly activated motor units fire in an oscillatory manner. Second, this partly synchronized and therefore uncoordinated uncontrolled rhythmic activity was still partly controlled by coordinated afferent input, as rhythmic firing was only present during the motor burst and not continuous. In very severe CNS malfunctioning, such firing can also occur between the motor bursts. The partial control of tremor activity was further indicated by the fact that in different muscles, the tremor was mostly coordinated with respect to agonist and antagonistic muscles, depending on the rhythmic movements of arms, legs and fingers (Figure 72B-D).

6.4.7. Contribution of FF and FR-Type Motor Unit Firing to the Generation of Tremor

There is evidence to suggest that the tremor in Parkinson's disease patients starts with synchronized oscillatory firing of FR-type motor units, innervated by α_2-motoneurons (Figure 63, 64), to which then FF-type motor units (innervated by α_1-motoneurons) synchronize, since the tremor frequencies are in the range of the Eigen-frequencies of oscillatory firing α_2-motoneurons ($\approx$ 5Hz) and the α_2-premoter spinal oscillators have a higher stability (Figures 65, 66, 68). Premotor spinal oscillators probably consist of a neuronal network, with the motoneuron being a part of it [14, 55]. α_2-motoneuron oscillators are less dependent on afferent input, and are therefore more stable (Figures 65, 66, 68, 71, 72, 74). Moreover, the Eigen-frequencies of the muscle-limb mechanics, for example of the arm and hand are also within the range of 5Hz. It then depends on quantitative effects, i.e., which oscillator type contributes to what extent to the rhythmic movement, at which frequencies the arms, hands and fingers shake. Since the contributions of rhythmically firing FF and FR-type motor units changed as did the strength of activation, the tremor frequency varied during the measurements. Since FF-type motor units develop more power, substantial tremor movement was only observed if substantial numbers of rhythmically firing FF-type motor units contributed. After the start of tremor with the synchronization of FR-type motor units, first the small FF-type (and further FR-type) motor units started to fire synchronously, followed by larger FF-type motor units with, on average, larger potential amplitudes. This may indicate recruitment according to the size principle among FF-type motor units. When the tremor ceased the large FF-type motor units stopped firing first followed by the smaller FF-type ones and then also the small-amplitude FR-type motor units (Figure 72A).

Only a little information was obtained concerning the contribution of S-type motor units, innervated by α_3-motoneurons, since their motor unit action potentials seem to be very small to be detected safely by sEMG so far. They even seemed to synchronize their firing with the FR-type motor units. α_3-Motoneuron oscillators fire in the range of 1Hz (Figure 8). Higher-quality recordings are needed to identify S-type motor unit firing safely also by sEMG.

So far, we only have been considering coordination of oscillatory firing motor units. It was observed earlier with the single-nerve fiber action potential recording method that occasionally firing motoneurons had the tendency to coordinate their firing with oscillatory firing motoneurons. The problem arises how to measure coordination among occasionally firing neurons or motor units since they are difficult to identify in the natural impulse traffic of many neurons or motor units. The only feature to identify occasionally firing neurons or motor units is the waveform. The advantage of oscillatory firing neurons is that they can be additionally identified by the rhythmic firing pattern.

6.4.8. Lack of Inhibition As One Reason for Tremor

It was shown using the single nerve-fiber action potential recording method that α_1 and α_2-motoneurons fire oscillatory, that they can synchronize their firing following repetitive stimulation, and that these oscillatory firing motoneurons can build up an external loop to the periphery in the way that γ-motoneurons and muscle spindle afferents get included in the rhythmic coordinated firing (Figure 69) [63]. But the synchronization of oscillatory firing is only transient, and in non-Parkinson patients the building up of an external loop to the periphery could only be observed upon strong repetitive reflex stimulation. It is therefore concluded that patients with Parkinson's disease lack inhibition, so that motoneurons can start to fire in an oscillatory manner upon virtually no stimulation (spontaneously); and, secondly, they lack mutual inhibition between oscillatory firing motoneurons, so that oscillatory firing motoneurons can synchronize their firing to give rise to rhythmic muscle contractions and movements resulting in tremor.

It is likely that the rhythmic firing during tremor starts with α_2-motoneurons because their firing is more stable and their firing frequency range is similar to the Eigen-frequencies of the muscle-limb mechanics. By building up an external loop to the periphery by the synchronously firing α_2-motoneurons primary muscle spindles are probably also activated, which in turn stimulate α_1-motoneurons to fire. Inhibiting input to the premotor network will probably first inhibit less stable oscillatory firing α_1-motoneurons followed by the more stable oscillatory firing α_2-motoneurons.

The results of the measurements performed in this work indicate that in patients with Parkinson's disease and in patients who suffered a spinal cord injury, the inhibition necessary to prevent spontaneous oscillatory firing of motoneurons is missing or impaired. A second type of inhibition is missing in Parkinson's disease patients to avoid synchronization among motoneuron firing.

6.4.9. Possible Repair in Parkinson's Disease Patients

It is difficult to see in detail what integrative repair mechanism could lead to repair of the missing inhibition, which is probably due to cell death, apart from optimizing the efficiency of the remaining inhibiting neurons. On the other hand, it is conceivable that systemic drug therapy to increase inhibition could reduce tremor and improve other pathologic CNS organizations. However, it is difficult to see how a systemic drug therapy can distinguish between the lack of inhibition in some parts of the neuronal networks and the preserved inhibition in others, which is highly likely to be different in every patient.

Below it will be shown how coordination dynamics therapy can improve CNS functioning in tremor and in the CNS generally. The precise capacity of coordination dynamics therapy offers for repair is not entirely known in Parkinson's disease. However, the following shall seek to shed more light on this question. The author could find no electrophysiological measurements of how drugs improve the firing of neurons in the human CNS either. When the author tried to measure improvement of tremor in a patient upon the first application of L-dopa, the side effects were that strong that CD measurements were not possible. It is conceivable that drug therapy and coordination dynamics therapy (movement therapy) in combination could be more successful than each therapy alone. Drug therapy would enhance inhibition systemically and coordination dynamics therapy could work specifically in the detail. This may result in reduction of the necessary drug concentrations and the drug side effects.

The non-invasive measurement of CNS organization by the coordination dynamics [59] to optimize drug application could help to reduce drug concentrations and side effects. The monitoring of CNS organization in Parkinson's disease patients by coordination dynamics measurements could help getting the progressive disease under better control.

A mechanism will be described enabling to reduce tremor in Parkinson's disease patients by coordination dynamics therapy [64].

6.5. Repair-Physiology in Parkinson's Disease Patients

6.5.1. Repair Strategy

It will be shown how highly coordinated arm and leg movements dictated by a special coordination dynamics therapy device (Figure 75) can improve CNS functioning in Parkinson's disease and other patients. The coordination dynamics therapy (CDT) uses the same principle for the re-learning of movements, vegetative and higher mental functions as the CNS uses for its own organization, namely phase and frequency coordination between neuron firing.

When exercising on the special coordination dynamics therapy device, the neurons activated for generating the specific motor program, communicate via phase and frequency coordination with the neurons which simultaneously generate the pathologic network organization, including tremor and entrains them by building up coordinated functional connections. So put simply, the physiologic network organization 'catches' the pathophysiologic network organization and entrains it [64]. Entrainment possibly reduces the efficiency and amount of excitation and increases that of inhibition by changing weight of synapses and building of new connections. The building of new inhibitory nerve cells can probably only achieved following an intensive program of CDT of at least one year in duration.

6.5.2. Reduction of Tremor Muscle Activity in the Short-Term Memory After Exercising on the Special Coordination Dynamics Therapy Device

When the patient was positioned on the special coordination dynamics therapy device (Figure 75), but not exercising on it, rhythmic muscle activity was recorded from the right tibialis anterior muscle. The rhythmic activity occurred at a frequency of 3.4Hz and the amplitude was ≈ 200μV (Figure 76A).

No activity was recorded from the gastrocnemius, biceps brachii and triceps brachii muscles. When the patient exercised on the device, muscle motor programs were recorded from all of the four above mentioned muscles (Figure 76B).

Motor program activity and tremor muscle activity was recorded from the tibialis anterior muscle (Figure 76B). The frequency of the motor program activity during exercise was f = 0.9Hz and the frequency of rhythmic tremor activity was now 1.7Hz.

The tremor activity reduced from 3.4Hz (Figure 76A) to 1.7Hz (Figure 76B) and coordinated its firing with those of the motor program. The tremor amplitude reduced from 200μV to 100μV and the duration of the tremor activity bursts became shorter.

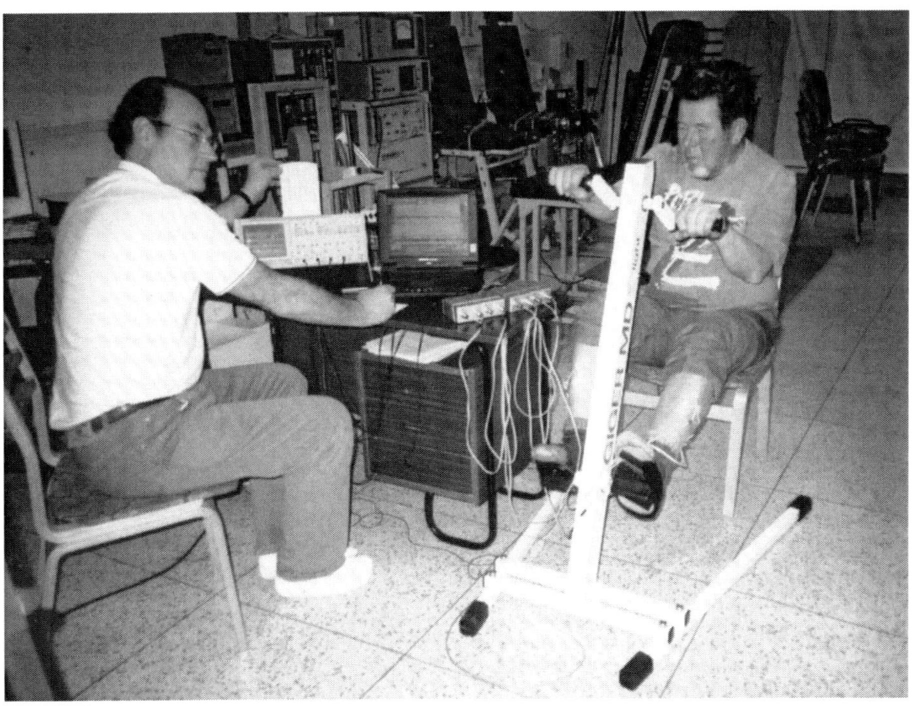

Figure 75. 65-year-old Parkinsonian patient (K.V.) during exercising on the special coordination dynamics therapy device. In this case, sEMG recordings were taken from leg muscles of both sides (tibialis anterior and triceps surae muscles). The recordings were pre-amplified (1000x), displayed on an oscilloscope and printed out.

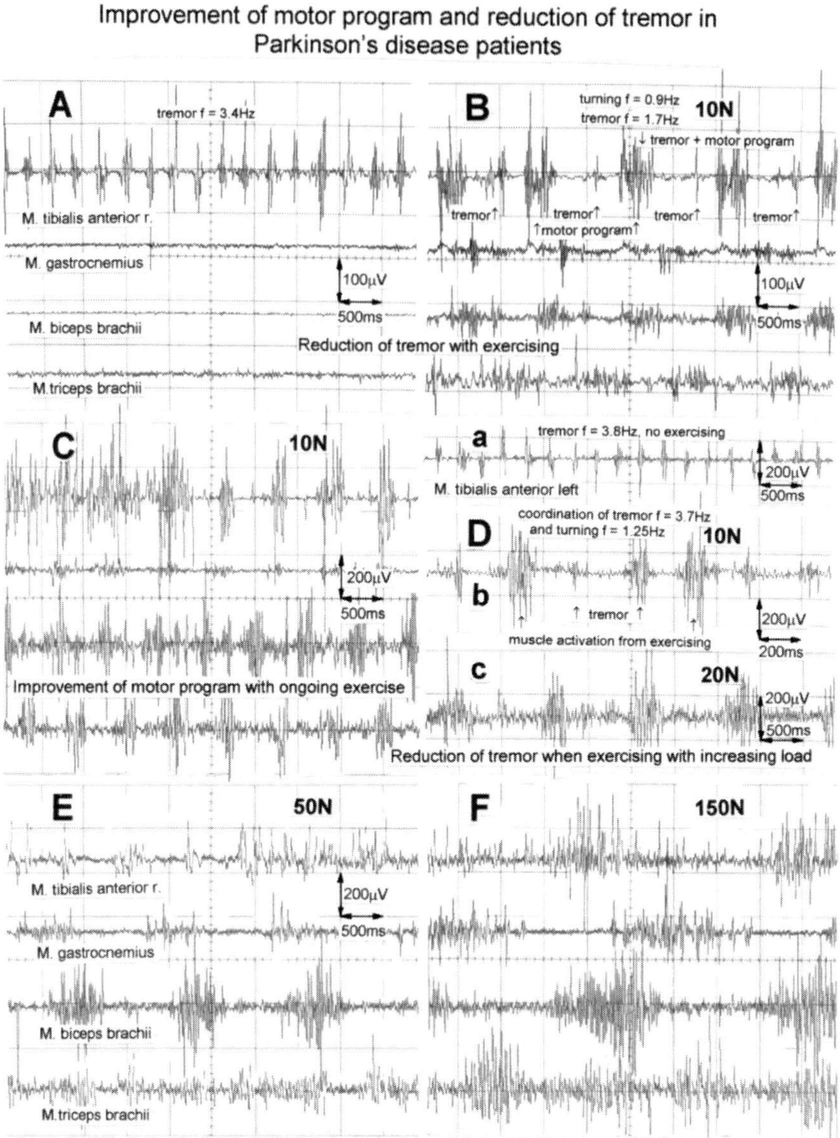

Figure 76. Improvement of the motor program and reduction of tremor in 3 Parkinson's disease patients upon exercising on the special coordination dynamics therapy device. A. Patient is in position at the special coordination dynamics therapy device, but not exercising. Non-voluntary rhythmic muscle activation with f = 3.4Hz is recorded from the tibialis anterior muscle of the right side. B. When exercising on the device at 10N, a motor program appears in all muscles recorded from. The tremor activation in the tibialis anterior muscle reduced in amount and frequency (to 1.7Hz), and the tremor muscle activation synchronized with the motor program. C. Improvement of motor program in the tibialis anterior muscle with ongoing exercising. D. Reduction of tremor in the tibialis anterior muscle with the increasing load when exercising on the special coordination dynamics recording and therapy device: a. no exercising, tremor muscle activation full present. b. Tremor reduced in amount and frequency (from 3.8 to 1.25Hz) when exercising at 10N. c. No strong tremor muscle activation visible at 20N. E,F. Improvement of the motor program in the tibialis anterior and triceps brachii muscles upon exercising on the special coordination dynamics therapy device at increasing load from 50 to 150N. A,B,D, a 70-year-old female Parkinson's disease patient (U.H.), C, a 76-year-old female Parkinson's disease patient (I.K.), and E,F, from a 70-year-old male Parkinson's disease patient (V.H.).

In conclusion, the tremor muscle activity coordinated its firing with the movement-induced motor program and reduced in general during exercise on the special CDT device. The tremor activity was therefore brought under the control of the movement activity and was influenced by it to reduce.

In another recording set from the same patient (Figure 76D) it seemed as if tremor reduction was more pronounced for higher (20N) than for lower load exercise (10N); this is to be expected since exercising at higher load will activate the CNS more integratively, so that the influence of the physiologic CNS organization upon the pathologic organization will be greater. With ongoing exercising on the device tremor reduction could nicely be seen from the reduction of tremor arm movements. At the beginning of exercising, the arm tremor amplitude was large and was afterwards reduced. The patient had tremor of arms, legs and face on both sides of the body. Visible tremor reduction need not mean however that the rhythmic tremor muscle activity had reduced, since the tremor amplitude could have been influenced by movement kinetics. It was therefore important to show through sEMG that the muscle activity leading to tremor really had reduced.

In another measurement (Figure 77E), the tremor muscle activity reduced in the recorded sweep piece; tremor reduced from 5Hz to 1.8Hz.

6.5.3. Improvement of the Motor Program

In two patients with Parkinson's disease the motor program improved after a few minutes of exercising on the device (Figure 76). Figure 76C shows that the motor program improved in the tibialis anterior muscle during that sweep piece when exercising at a load of 10N. From Figure 76E,F it becomes evident that the motor program improved substantially in the tibialis anterior muscle when increasing the load from 50N to 150N. Therefore, the motor program improved with ongoing exercise and with the increasing load provided that the patients could manage the load.

6.5.4. Tremor Changes in Different Muscles

Figure 77A-D shows that the tremor muscle activity changed in different muscles depending on the position, concentration and other influences. The patient was positioned on the special coordination dynamics therapy device but was not exercising on it. Rhythmic tremor activity can be seen in the tibialis anterior, the gastrocnemius, the biceps brachii and the triceps brachii muscles. The biceps and triceps brachii muscles fired in synchronization with a frequency of 4.6Hz (Figure 77A). The tremor muscle activity in the tibialis anterior and gastrocnemius muscles was coordinated in an antagonistic way (f = 4.5Hz), and the activity of the tibialis anterior muscle was synchronized with those of the biceps and triceps brachii muscles. Sometime later there was nearly no tremor activity in the gastrocnemius muscle and the tremor activity in the biceps brachii and triceps brachii muscles were less rhythmic (Figure 77B).

Still at a later point there was no tremor activity in the tibialis anterior muscle, but more activity in the gastrocnemius muscle (Figure 77C). The rhythmic activity in the biceps and triceps brachii muscles increased again. In Figure 77D, similar tremor activity can be seen as in Figure 77A. The tremor frequencies were slightly different. The higher muscle activity amplitudes were mainly due to a different amplitude calibration. Little tremor movements where observed in this patient in comparison to the patient shown in Figure 76.

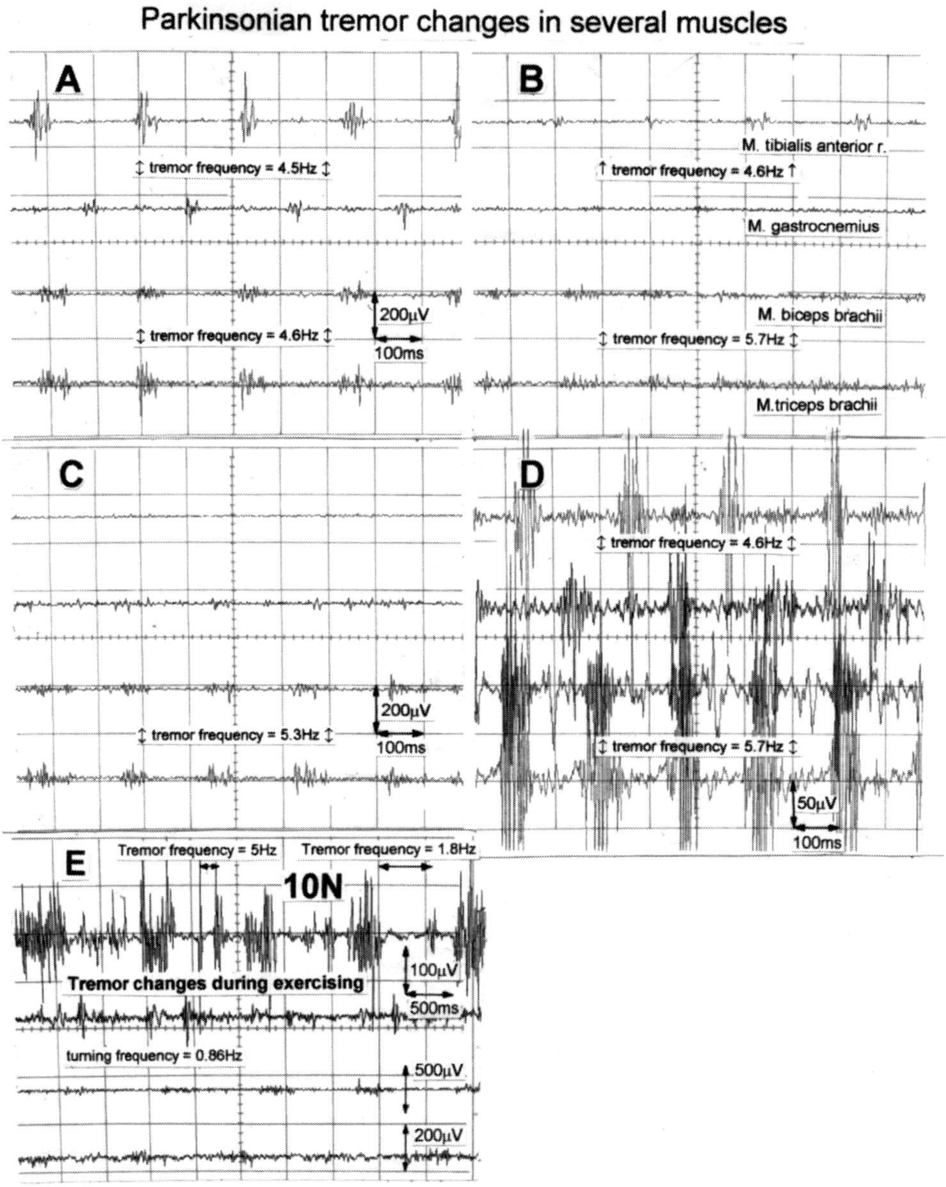

Figure 77. Tremor muscle activation changes in tibialis anterior, gastrocnemius, biceps brachii and triceps brachii muscles in a 65-year-old Parkinson's disease patient (K.V.) positioned at the special coordination dynamics therapy device but not exercising. A. Coordinated antagonistic tremor muscle activation in tibialis anterior and gastrocnemius muscle, frequency = 4.5Hz. Synchronized muscle activation in the biceps and triceps brachii muscles (f = 4.6Hz) synchronized with the tibialis anterior muscle. B. No tremor muscle activation in the gastrocnemius muscle. C. No tremor muscle activation in the tibialis anterior muscle, and activation in the gastrocnemius muscle is approximately synchronized with that in the biceps and triceps brachii muscles, unlike in 'A' and 'D'. D. Tremor muscle activation as in 'A', but different amplitude calibration. E. Tremor changes during exercising (f = 0.86Hz) at a load of 10N. In the upper trace the tremor changes from 5Hz to 1.8Hz during the exercise. In the lower traces only motor program activity can be seen at that amplification. Note that the time calibration in A through D is 100ms, and that in E is 500ms; the tremor activity has a frequency of approximately 5Hz and the motor program, due to exercising, of 1Hz; with a calibration difference of a factor of 5, tremor activity and motor program show some similarity.

In conclusion, it was shown that spontaneous EMG activity was changing in frequency and amplitude with time. Probably, the amplitude change of the rhythmic muscle activity was mainly due to the extent of the contribution from FF-type motor units since FF-type motor unit action potential amplitudes are much higher than those of FR and S-type motor units. The change in the frequency of the muscle tremor activity was due to contributions from all the three muscle fiber types, probably especially from the FR-type motor units since α_2-motor units have the most steady oscillatory pattern.

But changes in tremor frequency may also come from degenerative changes of the ensemble networks generating the oscillatory firing. Premotor spinal oscillators could be shown to partly lose their rhythmic properties as a consequence of CNS injury [55].

6.5.5. Integrative Organization Mechanism to Reduce Parkinsonian Tremor

It is evident from Figure 76AB that during exercise on the special CDT device the tremor muscle activity (Figure 76A) became coordinated with the motor program (Figure 76B). The tremor frequency in the tibialis anterior muscle reduced from 3.4Hz to 1.7Hz at a turning frequency of 0.9Hz, and tremor activity bursts reduced in size and duration (Figure 76B), indicating that less motor unit firings contributed to the tremor activity, which means that Parkinsonian tremor had been reduced in the short-term memory.

The question arises as to how the physiologic motor program can reduce pathologic motoneuron activation, leading to tremor. During the exercise on the special CDT device, CNS neurons organized themselves by relative phase and frequency coordination in cooperation with the movement induced re-afferent input and descending volitional input patterns. The uncontrolled synchronized oscillatory firing of motoneurons and their upstream driving interneurons, giving rise to tremor, were influenced by the physiologic network organization, since the activated neuronal networks for the coordinated arm and leg movements and the tremor activity overlapped. It is likely that some interneurons and motoneurons will have even contributed to both network organizations. The neurons which are directly responsible for the coordinated arm and leg movements and the tremor activity will be directly entrained because they serve both functions at the same time. Other neurons are only entrained by the functional projections they receive.

It has been shown that different activations can cross in neuronal networks via synfire chains [67]. Neurons can therefore serve at a certain moment several functions (organizational states), if the coordinated firing is sufficiently precise with respect to time and space. In reverse logic, if coordinated firing of neurons is substantially impaired, it is difficult for the neuronal networks to organize several functional states at the same time. Injured or pathologically functioning neuronal networks may be able to generate only one network state at a time or there may be a mixture of states taking place or sudden changes from one network state to another one take place. A mixture of movement states, sudden changes between different movement states or patients are only be able to do one thing at a time, for example walking or speaking, can be observed in brain-injured patients.

6.5.6. Neurons Are Coincidence Detectors

A further question which poses itself is why the network state which we may call 'highly coordinated moving on the special device' is more successful in reducing tremor than other movements. The answer is that because neurons are coincidence detectors, which respond to coordinated input more strongly than to uncoordinated input, coordinated input can reach the threshold for action potential generation with less input (fewer postsynaptic potentials) than uncoordinated input (Figure 11).

Also, inhibiting neurons work as coincidence detectors. Therefore, highly coordinated afferent input, generated by highly coordinated arm and leg movements, specifically enhances the efficiency of inhibiting neurons in the short-term memory. Thus one mechanism to specifically enhance inhibition to reduce tremor activity is to use the inhibiting neurons more efficiently through improving phase and frequency coordination in the CNS neuronal networks.

The integrative CNS repair mechanism to improve CNS functioning in Parkinson's disease patients lies in the improvement of phase and frequency coordination of neuron firing, especially of the inhibiting neurons, of the pathologically functioning network parts giving rise to tremor, and its functional integration in the whole CNS neuronal network. By 'linking' the pathophysiologic network organization with the physiologic one, attained during exercise on the special device, the pathologic network organization is entrained and improved.

6.5.7. Neurogenesis of Inhibiting Neurons

Improvement of the CNS functioning in Parkinson's disease patients was achieved by low-intensity CDT [5] (below), in which neurogenesis contributed only a little or not at all to the improvement. It was not possible to motivate the patients to undertake intensive therapy. But if Parkinson's disease patients could embark on an intensive program of CDT, during which new nerve cells are built from stem cells, then it seems most likely that inhibiting neurons would also be built since the inhibiting mechanism in the CNS networks could be activated. The physiologic network organization, generated during exercising on the special device, could inhibit the uninhibited neuron activity, which gives rise to tremor (Figure 76B). Neurogenesis is probably induced in those network parts which are most substantially activated [21, 24, 62, 68, 69].

6.5.8. Integrative Repair Mechanism to Improve CNS Functioning In General

The improvement of CNS functioning through improvement of the coordinated firing of the neurons with respect to time and space does not only hold for Parkinsonian tremor reduction. It is also likely to hold for any malfunctioning CNS, since each injury or degeneration of the neuronal networks is followed by an impairment of phase and frequency coordination of neuron firing. Even the timed firing of neurons in the CNS of healthy humans can be substantially improved including those of athletes (Figs.39-41), since the CNS of healthy humans do not function entirely optimally as can be quantified by coordination dynamics measurements. The connectivity of the CNS neuronal networks is extremely high (each neuron gets on average in the range of 4000 projections from other neurons) that loss of neuron connections or neurons themselves, will change the timed firing of many other neurons directly or indirectly. Large network parts will change their self-organization and physiologic network states will become altered and pathologic network states will occur. Improvement of the coordinated firing of neurons will enhance physiologic network states, weaken pathologic network states such as spasticity and lost physiologic network states such as walking or speaking may re-appear. The integrative mechanism of human CNS organization seems to have a bigger impact on the repair of the injured CNS than molecular mechanisms such as stem cell therapy.

6.5.9. The Necessity of Improving the Coordinated Firing of CNS Neurons in Any Case

It is wrongly believed that the human CNS can be functionally repaired by pure pharmacotherapy by inducing growth of new connections between neurons or building of new nerve cells. Whatever repair strategy is used, the coordinated firing of neurons with respect to time and space must be improved or re-learned. A new connection will only help in the organization of neuronal network states if the information is transmitted in coordination with others. Healthy humans also benefit from improvements in the coordinated firing of neurons with respect to time and space during the development stage of their CNS. Their CNS learns the timed firing of neurons by crawling, commando-crawling, standing up, walking, running, jumping, balance exercises and learning of many specific movements, and probably also by employing the higher mental functions, for example through problem-solving.

6.6. Diagnosis in Patients with Parkinson's Disease

6.6.1. The Rationale of Measuring Coordination Dynamics in Patients with Parkinson's Disease

Above the pathology of the coordination dynamics (CD) in the CNS of Parkinson's disease patients was analyzed in detail with surface electromyography (sEMG). Then a repair strategy was introduced in order to partly repair the aberrant CNS function. The physiologic network organization should 'catch' the pathophysiologic network organization and entrain it. One mechanism of repair is that the entraining reduces the efficiency and amount of excitation and increases that of inhibition (see above). But treatment needs objective assessment to quantify progress. sEMG is suitable to analyze pathologic CNS functioning, but because it is not an integrative CNS assessment, it is not suitable for quantifying treatment progress. As introduced above, CNS functioning can be assessed by the Coordination Dynamics (CD) recording trace and then the CD values give a single value for the quality of CNS functioning. The CD must first be measured and then the treatment is addressed.

6.6.2. Parkinson's Disease

Coordination dynamics therapy is mainly a movement-based learning method in which the CNS re-learns the impaired relative phase and frequency coordination for the self-organization of its neuronal networks and re-learns lost or impaired movements. Even though the learning of the severely injured CNS is approximately 20 to 50 times slower (see above) than that of a healthy CNS [51], it seems that every CNS, including the malfunctioning one, can learn and re-learn. It should therefore also be possible to improve the functioning of the CNS in neurodegenerative diseases by learning.

In the following two sections it will be shown that CNS functioning can be measured and improved in Parkinson's disease. The first part reports on the measurement of the CNS functioning using the coordination dynamics recording method to quantify CNS malfunctioning. In the second part it will be shown that a 35% improvement of CNS functioning can be achieved in Parkinson's disease patients by a low intensity coordination dynamics therapy of two and a half months' duration [5].

Parkinson's disease is a complex disorganization of the human CNS reflecting malfunction of the basal ganglia. There is a degeneration of neurons in the substantia nigra and a decrease in the dopamine content of the striatum and putamen due to a degeneration of the nigrostriatal connections. This view is supported by the effectiveness of dopamine (L-dopa) therapy. The characterizing symptoms are tremor, rigidity and cogwheel phenomenon. Akinesia and postural reflex disorders are also present [70,71,72]. Since rigidity, cogwheel phenomenon, akinesia and postural automatism disorders can often be found in different kinds of brain injuries, the extensive resting tremor seems to be typical for the Parkinson disease. It has been reported [63] that the tremor frequencies in Parkinson's disease (6-8Hz, 8-12Hz) are in the range of the Eigen-frequencies of premotor spinal oscillators (Figure 8). Besides muscle-limb mechanical aspects, uncontrolled premotor spinal oscillators are likely to substantially contribute to Parkinson tremor (see above) even though the cause is widely attributed to the malfunctioning of the basal ganglia.

Since Parkinson's disease patients have substantial coordination problems, as can be observed for example when they change clothes or eat, analysis of CNS dysfunction by CD can be expected to provide further insight into the Parkinson's disorganization. The understanding of Parkinson's disease (for references, see [73]) within the concept of CD will lead to treatment that can result in a substantial improvement of the CNS functioning in Parkinson's disease patients [5].

6.6.3. Coordination Dynamics Measurements

The organization of the human central nervous system (CNS) can be measured in terms of CD, i.e., by the arrhythmicity of turning, when a patient exercises on a special coordination dynamics recording and therapy device (Figure 75). The device dictates the pace and trot gait coordination's between arms and legs *and* the difficult intermediate coordination's between pace and trot gait. The quality of CNS organization is quantified by the ability of the CNS to adapt to the changing coordination's between arms and leg movements. If the patient's CNS can adapt its organization very well, then the patient can turn rhythmically at a steady frequency for all coordination's. If the CNS organization is very poor, then the arrhythmicity of turning is large for all coordination's, which means that the CNS cannot or can hardly generate complicated movements via the motor program. Patients with brain injury can often turn reasonably rhythmically for the easy coordination's pace and trot gait but experience difficulty for the intermediate coordination's, when physiologic functioning of brain neuronal networks is necessary to vary physiologically the neuronal network organizations of the spinal cord (Figs.20,54). The ability to generate easy and complicated coordinated movements is measured integratively (mechanically) by reference to the arrhythmicity of turning (coordination dynamics $\Delta(df/dt)/f$ [s^{-1}], Δ/min [s^{-2}]). Muscles function can also be measured non-integratively by surface electromyography (sEMG). Since during measurement, the patient or volunteer is exercising mostly in the sitting position, muscle activation for carrying body weight is of only little importance. An increase of the load during exercising therefore (by increasing the brake force from 20 to 200 Newton and back again), in combination with the changing coordination's, provides detailed information about the motor program of healthy subjects and patients with pathologic CNS organization. The advantage of the mechanical coordination dynamics is that the CNS organization of these movement states can be easily measured by a single parameter (order parameter = arrhythmicity of turning) and quantifies the integrative action of all activated muscles. With sEMG information is obtained via the motor unit action potentials on the motor program and on the coordinated firing of neurons in the CNS.

6.6.4. Parkinson Group

Coordination dynamics were measured in eight patients in whom Parkinson's disease had set in five to ten years ago and were compared with the coordination dynamics of a control group of similar age to obtain information about the pathologic central nervous system (CNS) organization in order to develop treatment. In a following part it will be shown that the functioning of the CNS in Parkinson's disease patients can be substantially improved by CDT. The average age of the patients was 69 years and the range was 63 to 75.

6.6.5. Low-Load and High-Load Coordination Dynamics in Parkinson Patients and Normal Subjects

Original recordings of low-load and high-load coordination dynamics of a healthy 72-year-old woman are shown in Figure 79A-D. The coordination dynamics look similar to those of subjects in the age range between 18 and 25 years. From Figure 79A it can be clearly seen that the CNS of the elderly woman learned quickly the coordination's for the difficult backward turning for low load. Within five minutes the variation of the frequency (upper trace) reduced strongly, indicating motor learning, namely the ability to turn at a constant frequency. The coordination dynamics in the one minute time window (Figure 79B) shows good coordination for the pace gait coordination (P) and poor coordination (big variation) for the trot gait (K). For higher loads up to 150N (Figure 79C) the frequency reduced. The volunteer tried to cope with the higher loads by frequency reduction; the load in Watt (middle trace of Figure 79C) does not show a nice stair-like shape. In the time window for 100N (Figure 79D) the frequency for forward turning increased for the easy coordination's pace and trot gait and decreased for the difficult intermediate coordination's.

Original recordings of low-load and high-load coordination dynamics of a 73-year-old Parkinson's disease patient are shown in Figure 79E-H. No principal differences to the healthy subject can be seen. At low loads backward turning also improved (Figure 79E, upper trace) but may be not as much. The coordination dynamics in the time window for backward turning was the same for all coordination's (Figure 79F, lower trace). Overall, the high load recording showed (Figure 79G) load escape (decrease of turning frequency with increasing load) in most of the cases measured. For forward turning at 100N (Figure 79H) the frequency increased for the easy pace and trot gait coordination and decreased for the intermediate coordination's between pace and trot gait. Since every CNS is different, the differences in the CD between the patient's CNS and those of the control group woman do not permit of any clear conclusions as to the differences between Parkinson's disease patients and healthy persons.

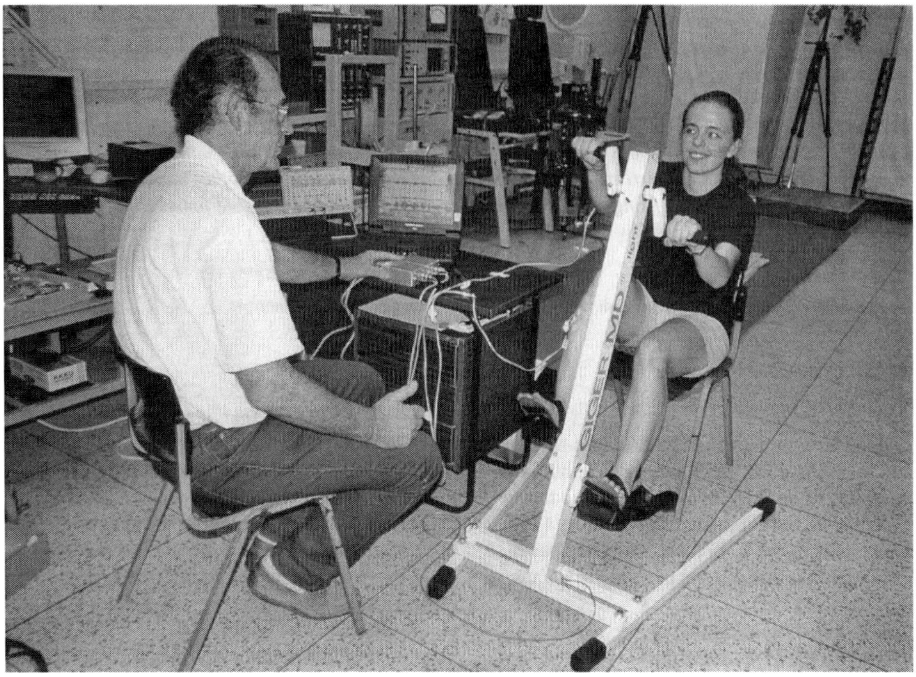

Figure 78. Measuring layout for recording the coordination dynamics by the coordination dynamics recording method and by electromyography: An 18-year-old woman is exercising on the special coordination dynamics therapy device and mechanical coordination dynamics are recorded and displayed on a laptop (the display is showing the frequency and the changing coordination dynamics). Electromyographic activity is recorded from the Mm. tibialis anterior and biceps brachii. Two differential surface electrodes (4cm apart) and one earth electrode (in between) are used for recording from each muscle (they can be seen on the M. tibialis anterior). The electric activity is pre-amplified (1000x; the author GS is touching one preamplifier with the left hand) and displayed on an oscilloscope (Gould, 1504). Printer and plotter cannot be seen. Load changes (increase and decrease) are achieved by the author (G.S.) turning a small breaks knob situated close to the l of 'light' on the device to increase the force from 20 to 200Newton and backwards. The volunteer is only smiling when exercising at lower loads, not at 150 and 200N.

The numerical values of the coordination dynamics however should reflect differences in CNS organization between physiologic and pathophysiologic cases at least because of the resting tremor in Parkinson.

The values of low-load coordination dynamics of the Parkinson patients (5 females, 3 males) for low-load forward and backward turning were: $\Delta_{forward} = \Delta(df/dt)/f = 8.4 \pm 4.2s^{-2}$ (mean ± standard deviation) and $\Delta_{backward} = 16.9 \pm 13.2s^{-2}$. The numerical values of low-load coordination dynamics of the normal subjects of the control group of similar age (4 females, 2 males, mean age = 73) for forward and backward turning were 5.4 ± 2.2 and 11.8 ± 8.1, respectively. As expected with respect to the degeneration of substantia nigra, the Parkinson's disease patients had therefore, poorer (higher) low-load coordination dynamics values. But the intra-group differences were larger than the inter-group differences, making improvement measurements (see the following part) of the coordination dynamics of single cases with ongoing therapy necessary.

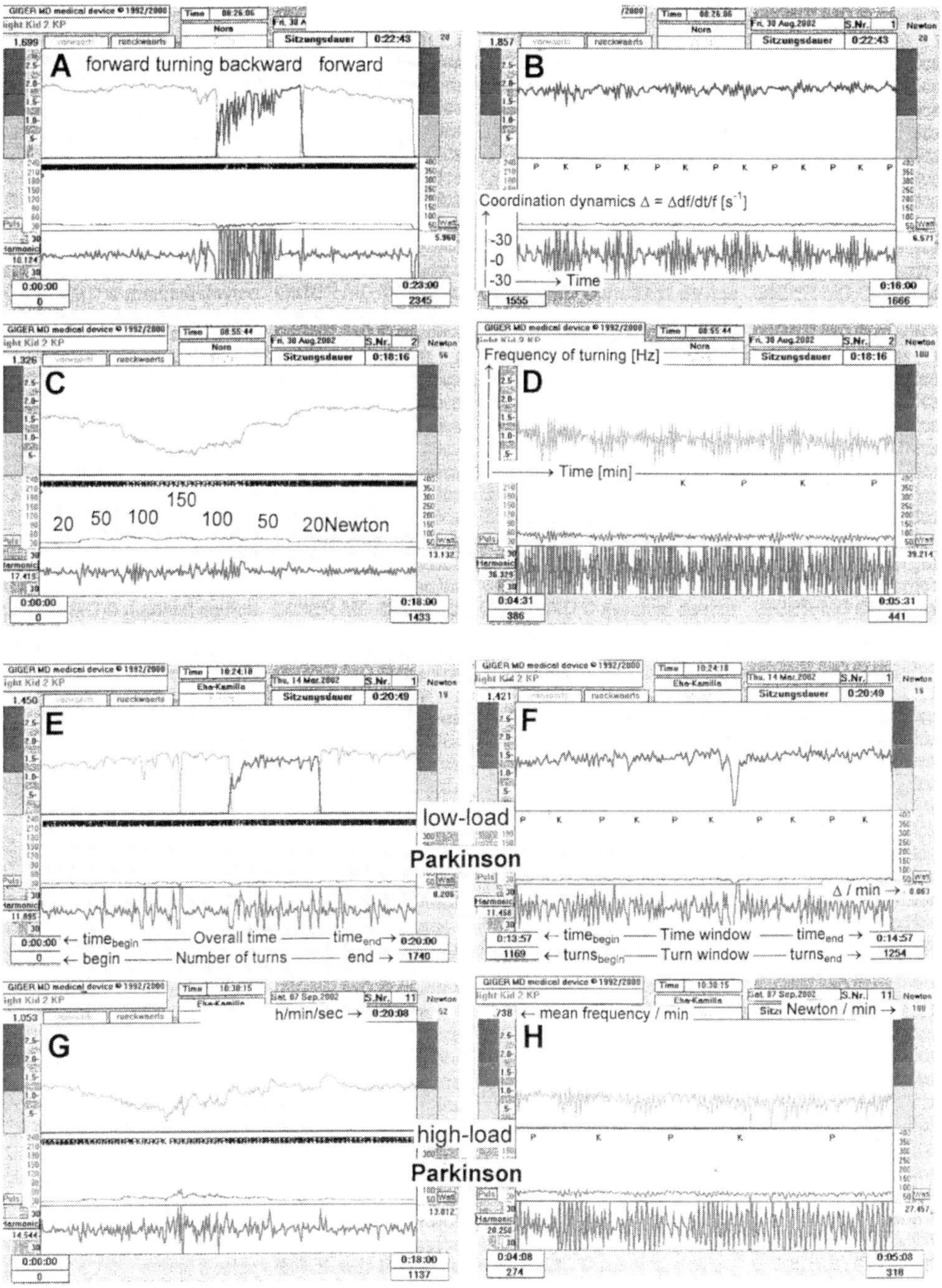

Figure 79. Low-load (A,B,E,F) and high-load mechanical coordination dynamics (C,D,G,H) for a 73-year-old Parkinson's disease patient (E,F,G,H) and a healthy 72-year-old woman (A,B,C,D). The overall recordings are shown on the left panel (A,C,E,G) and time windows of one minute are shown on the right panel (B,D,F,H).

Differences occurred for the high-load exercise on the special CDT device (Figure 78). The volunteers of the control group were all able to turn up to a load of 150N. Even though three subjects had moderate heart problems, no heart problems occurred during 150N high-load exercising! The numerical values of high-load coordination dynamics for increasing loads up to 150N were $\sum\Delta = \Delta(20N) +...+ \Delta(150N) +...+ \Delta(20N) = 114.1$. The Parkinson's disease patients were not able to exercise at loads up to 150N. Only one of them could reach the load of 150 and another one a load of 100N. In general, the Parkinson's disease patients had less strength. This is to be expected: the accuracy of coordination of motor unit firing was more impaired in Parkinson's disease patients, which results in an inefficient CNS organization and reduced muscle strength (see below).

6.6.6. One-Sidedness of Parkinson's Disease

The belief that coordination dynamics may not be diagnostically useful in Parkinson's disease can be contradicted by having patients exercise on the special coordination dynamics therapy and recording device with both legs but only with one hand. When one side of the body was more seriously affected by the disease, the coordination dynamics were more impaired on that side. In seven of the eight patients the one-sidedness was measured by simply turning with both legs but only with the right or left hand. Since Parkinson's disease patients become easily exhausted, the turning with right and left hand was alternated six times to compensate for the reduction or change in the power of turning; mean values of the coordination dynamics for each side were calculated from the three values produced for each side and then compared.

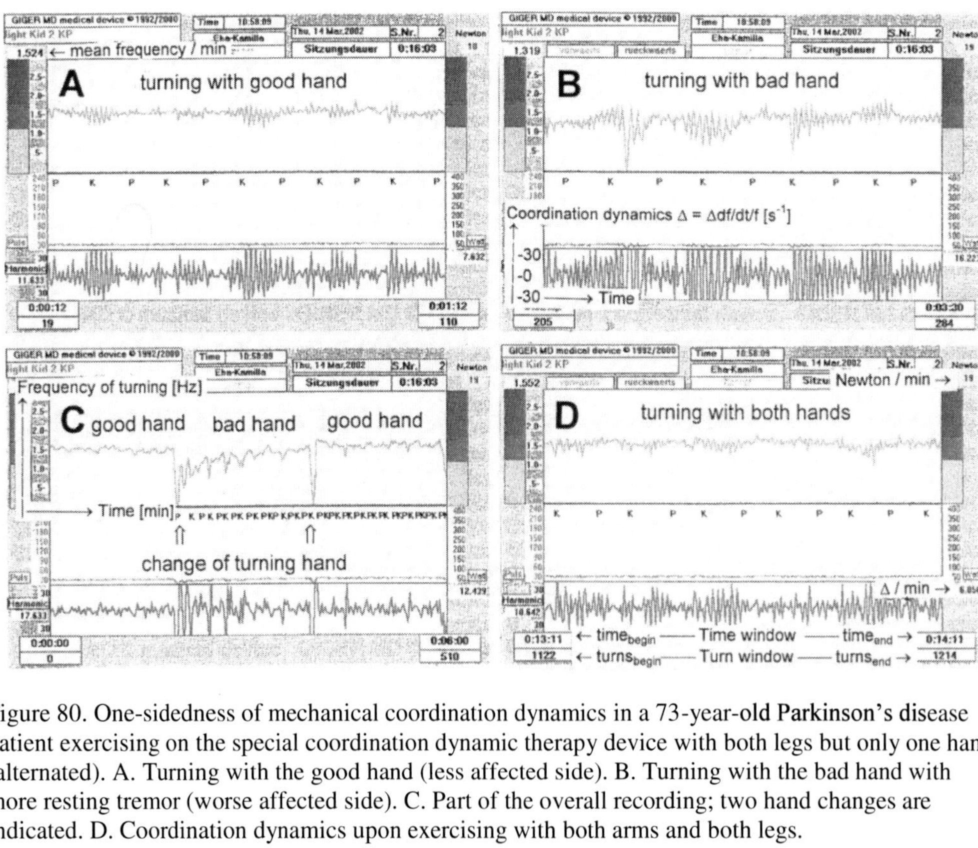

Figure 80. One-sidedness of mechanical coordination dynamics in a 73-year-old Parkinson's disease patient exercising on the special coordination dynamic therapy device with both legs but only one hand (alternated). A. Turning with the good hand (less affected side). B. Turning with the bad hand with more resting tremor (worse affected side). C. Part of the overall recording; two hand changes are indicated. D. Coordination dynamics upon exercising with both arms and both legs.

Original recordings of the one-sidedness of a Parkinson's disease patient are shown in Figure 80. In Figure 80C the coordination dynamics for the hand changing is shown. The patient turned first with the 'good' hand, and the frequency of turning was steady and the arrhythmicity of turning was small and the coordination dynamics value was good. After changing to the 'bad' hand, the frequency (upper trace) got reduced and was changing more, and the arrhythmicity of turning (lower trace) became larger (Figure 80C). After changing back to the 'good' hand again, the frequency stability was becoming better again and the arrhythmicity of turning got smaller. Interestingly, even though the patient was tiring (the coordination dynamics got slowly worse with ongoing turning), turning only with the 'bad' hand was still getting more rhythmical in the short-term memory as can be seen from the increase of the frequency and the reduction of the frequency variation in the upper trace of Figure 80C. As can be seen from Figure 80, the turning with the good hand showed better coordination dynamics ($\Delta = 8.6$; Figure 80A) than with the bad hand ($\Delta = 16,2$; Figure 80B). After turning with the good and the bad hand three times each, the patient turned with both hands (and both legs; Figure 80D) and the coordination dynamics were of course better than when turning only with one hand and both legs ($\Delta = 6.9$ in this case).

The numerical difference between right and left was 27% ± 18% (range 7% to 50%). In two of the seven patients measured the difference was 7% which means that the Parkinson's disease affected both sides similarly, and in five cases the right or the left side was worse affected by the disease (difference 19% to 50%). The higher (worse) values of coordination dynamics coincided with a larger resting tremor in the patients.

6.6.7. Improvement of Hand Coordination in the Short-Term Memory Quantified by a Dexterity Test

The resting tremor is impaired the patients' everyday life, e.g., during putting on clothes or during eating. A dexterity test showed that the hand coordination could be improved by one hour of coordination dynamics therapy. Seven patients underwent dexterity tests (Minnesota Manual Dexterity Tests). The patient had to take out round wooden pieces from one hole and put them to another one. After warming up for ten minutes on the special coordination dynamics therapy device, the first tests were performed. Then the patients were allowed to exercise for one hour on the special coordination dynamics therapy device and the test was repeated. If the CNS function had improved after 1 hour of coordination exercises then the patients should be able to change the pieces faster afterwards. This was indeed the case. There was an improvement in dexterity of 4% ±2%. A coordination dynamics therapy might therefore be an effective method to improve coordination in Parkinson's disease patients.

6.6.8. Impairment of Coordination Dynamics in Parkinson's Disease Patients

The values of coordination dynamics at low loads in Parkinson's disease patients (mean age = 69) for $\Delta_{forward} = \Delta(df/dt)/f = 8.4 \pm 4.2\ s^{-2}$ and $\Delta_{backward} = 16.9 \pm 13.2\ s^{-2}$ were larger by 56% and 44% respectively than for the control group of similar age (5.4 ± 2.2 and 11.8 ± 8.1; mean age = 73). Earlier measurements of coordination dynamics at low loads for young athletes, physiotherapists, gymnasts and musicians were between 5.2 and 6.0 for forward turning and between 6.9 and 10.7 for backward turning [50]. The control group of similar age (mean age = 73) had thus similar values of coordination dynamics at low loads as young healthy subjects (mean age = 21 years). The impaired coordination dynamics for low loads (higher values) are thus a typical neurophysiologic measure for patients with Parkinson's disease.

The low-load coordination dynamics values of the patients were expected to be higher due to the impairment in neuron firing which accompanies neuron cell death due to trauma or degeneration. The coordination dynamics between motoneuron firing, between muscles and between arm and leg movements also becomes impaired [62]. This reasoning is in accordance with the finding that Parkinson's disease patients are more deficient in the regulation of force-time parameters rather than simply in force production of knee extensor muscles [74]. Some identified single motor unit action potentials in the Parkinson patient were of a similar shape to those of a healthy subject (see above) suggesting that most likely the cause of the disease does not originate in the muscles.

6.6.9. Coordination Dynamics and L-Dopa Treatment

If there were instability of dopamine release in Parkinson's disease or if in the treatment of Parkinson's disease carries a risk of dopamine-receptor over-stimulation [75], then CDT could help to stabilize dopamine release and measuring of the CD during L-dopa treatment and dosage could reduce the risk of dopamine-receptor over-stimulation. But it is difficult to understand how a systemic L-dopa treatment can help to avoid the disproportional activation of certain muscle fiber groups in different muscles and right-left differences.

6.6.10. Integrativity of Neuronal Network Activation

It was shown in Figures 76E,F that the amount of muscle fiber activation changed for increasing load when exercising on the special CD recording and therapy device. For a re-learning of physiologic movements a substantial activation is necessary, because only those networks can learn or re-learn which are substantially activated in similarity to neurogenesis. Neurogenesis takes place only in those network parts of the CNS which receive substantial activation.

Different loads and speeds have to be applied to analyze different neuronal network properties of the CNS. Tremor frequency measurements on fingers alone will provide little information about the integrative properties of CNS organization. The amount of coordinated strength that a patient can generate can only be determined by letting the patient exercise against high resistance and measuring the coordination and the motor program of arm and leg movements.

If, on the other hand, one wants to train fine control (such as coordinated hand and finger movements during eating) and high load movement control (such as during walking) coordination for low and high loads has to be trained. Also in sports, like modern pentathlon (shooting, fencing, running, riding, swimming), coordination has to be trained for low and high loads. High load coordination is needed for running and low load coordination for shooting. While comparing the coordination dynamics of athletes and musicians it was found that the musicians were better at low-load exercising and the athletes were better at high-load exercising [50, 51].

The impaired CD found to be present in Parkinson's disease patients, measured with the CD recording method, were in accordance with the impaired motor program measured by sEMG. The data supported each other. The CD (the arrhythmicity of turning) value was the integrative or order parameter for characterizing the neuronal network state 'exercising with arms and legs in a coordinated manner'. The EMG gave detailed information concerning a small part of the motor program and partly on single motoneuron firing in the human CNS. The sEMG did not provide information on the integrative motor program generating the movement; this was provided by the CD during exercise on the special CDT device.

6.7. Repair in a Group Of Parkinson's Disease Patients Achieved by Coordination Dynamics Therapy

6.7.1. A Summary of the Improvement Achieved

Eight patients in whom Parkinson's disease had set on 5 to 10 years earlier underwent low intensity coordination dynamics therapy with on average 4 hours per week for 2.5 months. The ongoing pharmaco-therapy and the conventional fitness training for 1 to 2 hours per week continued. With the coordination dynamics therapy the functioning of the central nervous system (CNS) of the Parkinson's disease patients improved by 35%, as quantified by coordination dynamics measurements. After three months of stopping coordination dynamics therapy, but ongoing pharmaco-therapy and fitness training, CNS functioning worsened again by 21%. It will be concluded that drug-therapy and conventional fitness training alone cannot prevent the worsening of the CNS functioning in Parkinson's disease, but additional coordination dynamics therapy likely can.

6.7.2. Strategy for Repair in Parkinson

Following CNS injury, there is an impairment of the relative coordinated firing between neurons with respect to time and space in addition to the loss of nervous tissue in which specific functions may be generated efficiently. If the relative coordinated firing of neurons is improved by learning, then functional repair will be achieved. It was measured in the human nervous system that α and γ-motoneurons and muscle spindle afferents fired in relative coordination up to 3 to 5 milliseconds (Figure 9). It was further measured that this coordination is partly lost following CNS injury [14,55]. The impaired relative phase and frequency coordination between the firings of neurons can be re-learned when the patient is exercising volitionally on a special CDT device, where the exactly coordinated arm and leg movements are dictated by the device. The activated neuronal networks do re-learn their physiologic self-organization through the training provided by the exactly coordinated movement induced afferent input patterns. Neuronal network parts which are substantially activated learn efficiently.

In the neurodegenerative condition of Parkinson's disease the coordinated firing between CNS neurons is impaired because of degeneration of neurons in the substantia nigra and the degeneration of nigrostriatal connections. As a result of degeneration of neurons and connections, excitation thresholds will change and action potential conduction times along nerve fibers will be prolonged. The timed firing of neurons is impaired. The timed firing of CNS neurons will be varied not only in the basal ganglia and their direct projections but also in the consecutive projections, i.e., in large parts of CNS neuronal networks. Re-learning of the coordinated firing with respect to time and space of the CNS neurons by coordination dynamics therapy may therefore be expected to be successful in improving CNS functioning.

The evidence provided above through measurements indicates that CNS function in Parkinson's disease patients is 56% worse for forward and 44% worse for backwards turning on the special CDT device when compared with healthy individuals. It will be shown here that some of this impaired coordination can be restored.

6.7.3. CDT Plus Drug-Therapy Plus Fitness Training

CDT includes exercising on the special coordination dynamics therapy (CDT) device to improve the relative phase and frequency coordination between neuron firing in the human CNS. The training of automatisms like walking, crawling and running and the exercising of old learned movements like climbing staircases are performed to re-learn movements and to structurally repair CNS networks. Since the Parkinson's disease patients could all walk, climb staircases and undertook conventional exercise for one to two hours per week, CDT only consisted of exercising on the special CDT therapy device (Figure 81).

Figure 81. Measuring layout for recording coordination dynamics with the coordination dynamics recording method and by surface electromyography: A 75-year-old Parkinson's disease patient is exercising on the special coordination dynamics therapy device and mechanical coordination dynamics are recorded and displayed on a laptop. The display is showing the frequency and the changing coordination dynamics. Plots of the display are used in Figs.82,83. Electromyographic activity was recorded simultaneously from up to 4 muscles (4 channels).

The pharmaco-therapy administered consisted of Madopar (250mg L-dopa and 50mg Benserazide, designed to compensate the decreased level of dopamine), Sinemet CR (which includes Carbidopa and Levodopa, to inhibit decarboxylation of aromatic amino acids) and Bromocriptine (includes Mesylate 2.87 mg, to activate postsynaptic dopamine receptors) was not changed.

6.7.4. Low-Load CD Values in Parkinson's Disease Patients in Comparison to Healthy Subjects

Eight patients in whom Parkinson's disease had set on 5 to 10 years earlier underwent low intensity coordination dynamics therapy on average for 4 hours per week for 2.5 months. The average age of the patients was 69 years (range 63 to 75). The progress achieved was measured in terms of the improvement of the organization of the injured CNS, i.e., by measuring the (mechanical) CD and partly by a dexterity test (Minnesota Manual Dexterity Test) to quantify tremor reduction over the long-term.

In the first part of the study (above) the low-load coordination dynamics (CD) of Parkinson patients (five females, three males) were measured at the beginning of the therapy for forward and backward turning; the values obtained were $\Delta_{forward}$ = 8.4 ± 4.2 1/s and $\Delta_{backward}$ = 16.9 ± 13.2 1/s respectively. The values of low-load coordination dynamics for normal subjects of the same age for forward and backward turning were 5.4 ± 2.2 and 11.8 ± 8.1 (mean age = 73) respectively. The Parkinson's disease patients thus showed, as expected in a diseased CNS, higher (poorer) CD values. However, the intra-group differences were larger than the inter-group differences, making improvement measurements of the CD in single patients with ongoing therapy necessary.

6.7.5. Improvement of Parkinson's Disease Patients due to CDT Quantified by Low-Load CD

Original recordings showing the improvement achieved in one Parkinson's disease patient by coordination dynamics therapy are shown in Figure 82A-I. On the left panel, a recording of 20min (A), 1min forward (C) and 1min backward turning (E) was made at the beginning of the coordination dynamics therapy. The frequency (upper trace) and the coordination dynamics (lower trace) show large variations, i.e., the arrhythmicity of turning was large.

The middle panel of Figure 82 (B,D,F) shows coordination dynamics at the end of the 2.5 months. The improvement of the frequency of turning on the special coordination dynamics therapy device and the arrhythmicity of turning (coordination dynamics) can be clearly seen from the twenty minute recording (B), the one minute recording of forward turning (D) and the one minute recording of backward turning (F). The improvement of coordination dynamics for forward turning from $\Delta_{forward}$ = 13.1 (C) to $\Delta_{forward}$ = 7.87 (D) amounts in this case to 40%. The coordination dynamics for backward turning improved from $\Delta_{backward}$ = 31.7 to 10.0 (F), which represents 68%. The improvement of the numerical values of coordination dynamics of all the eight patients varied quite significantly.

For five of the patients, the coordination dynamics $\Delta = \Delta(df/dt)/f$ improved strongly for forward and backward turning ($\Delta_{forward}$ (5) = 43% ± 7%; $\Delta_{backward}$ (5) = 60% ± 7%). In three of the patients, the coordination dynamics improved only little or very little, resulting in an improvement of the values of 8 patients or $\Delta_{forward}$ (8) = 32% ± 18%; $\Delta_{backward}$ (8) = 38% ± 30% (mean for both directions = 35%).

The achieved group coordination dynamics values for forward and backward turning were $\Delta_{forward}$ = 5.5 ± 2.5 and $\Delta_{backward}$ = 9.1 ± 5.2 and thus similar to those of the control group ($\Delta_{forward}$ = 5.4 ± 2.2 and $\Delta_{backward}$ = 11.8 ± 8.1).

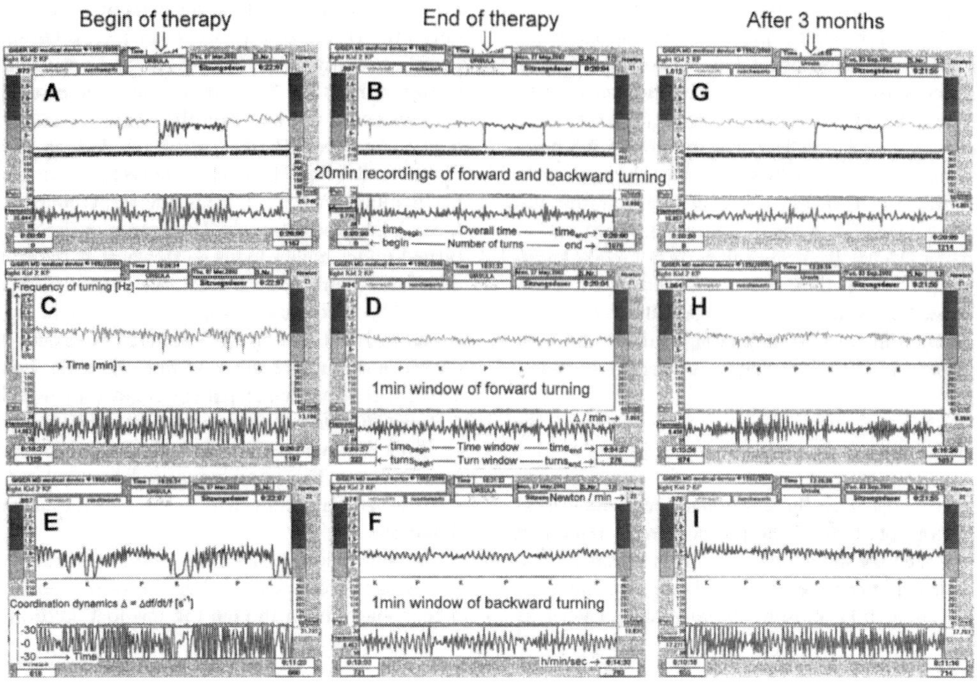

Figure 82. Low-load coordination dynamics recordings from a 70-year-old Parkinson's disease patient at the begin of coordination dynamics therapy (A,C,E) and at the end of the therapy after 2.5 months (B,D,F), and 3 months after the completion of the therapy (G,H,I). A,B,G. Overall recording of 20 min; first 10min forwards, followed by 5min backwards (blue frequency (upper) trace) and 5min forwards turning again. Note the reduction of frequency variation in A for backwards turning, suggesting motor learning in the short-term memory. C,D,H. 1min time windows for best forwards turning. E,F,I. 1min time window for backward turning; Improvement and worsening of coordination dynamics Δ/min can be judged upon easily from the coordination dynamics trace (large arrhythmicity (E), small (F), larger (I)) and from the coordination dynamics values (31.7, 10.8, 17.7).

6.7.6. Worsening of Low-Load Coordination Dynamics Three Months after Completion of the Therapy

Three months after the completion of the coordination dynamics therapy measurements were repeated to see whether the Parkinson's disease patients were able to maintain their improved CD. From the original recordings of Figure 82G,H,I it can be seen that the coordination dynamics worsened again. In the case depicted, the CD for forward turning worsened from 7.9 to 9.1 (Figure 82H) and that for backward turning from 10.0 to 17.7/s (Figure 82I). Overall, the whole group of Parkinson patients achieved worse results: by 22% for forward turning and 20% for backward turning. The mean values of CD for the seven patients who worsened due to the discontinuation of CDT were $\Delta_{forward} = 6.0 \pm 2.4$ and $\Delta_{backward} = 9.3 \pm 6.0$. The improved CNS functioning (by ≈ 35%) achieved previously was partly lost (≈ 21%) (Figure 84). As can be seen from the comparison of the mean worsened CD values, the mean values do not reflect group changes as well as the mean changes for single patients expressed as percentages. A single patient with very poor CD (high values) and little changes can reduce the changes of the whole group.

In conclusion, since drug-therapy and conventional fitness training for 1 to 2 hours per week remained unchanged the important conclusion is that CDT could improve the CNS functioning in Parkinson's disease patients. Drug-therapy and fitness training alone cannot prevent a worsening of the CNS functioning in Parkinson's disease patients. Additional therapy is needed.

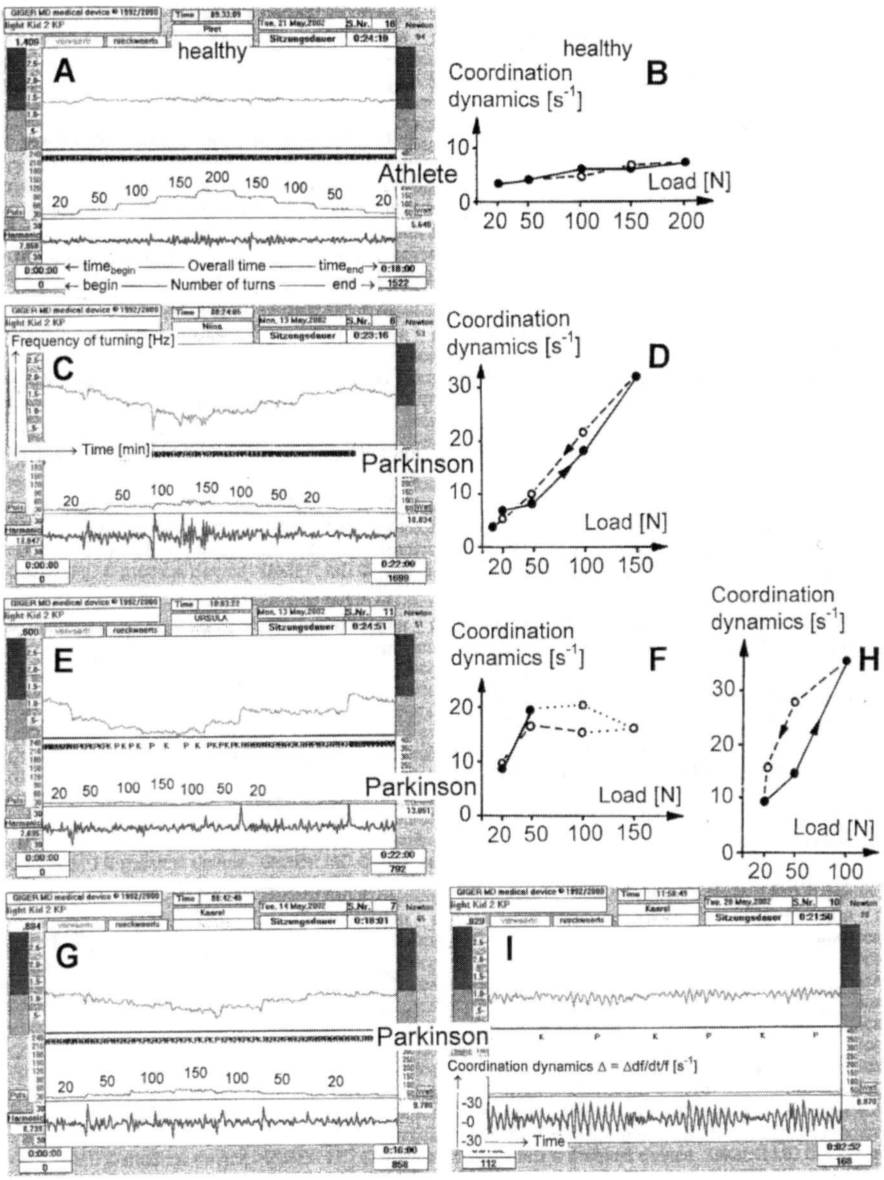

Figure 83. High-load coordination dynamics in a 19-year-old athlete (A,B) and in 3 Parkinson's disease patients (C-I). A,C,E,G. Overall recordings for load increase and decrease. Note that the healthy person can maintain the frequency (upper trace) for increasing load whereas the patients cannot; the load dependence in Watt (middle trace) shows a nice stepwise shape in the recording from the athlete but not in that for the patients. B,D,F,H. Hysteresis-like graphs of coordination dynamics values for load increase and decrease. Note that the curves for the patients (D,F,H) reach higher values of arrhythmicity (coordination dynamics) and that they are unable to exercise at higher loads. I. 1min time window of 'G' showing that the coordination dynamics are good for trot gait coordination (K) and poor for pace gait coordination (P).

6.7.7. High-Load Coordination Dynamics in Parkinson's Disease Patients

At the beginning of the therapy the patients could not cope with loads above 20N. Because of their age they were also not pushed to try. After 2.5 months of the therapy, being fitter than at the beginning of the therapy, the patients were pushed to exercise at higher loads.

Only one patient was able to turn on the special coordination dynamics therapy device at 200N. In Figure 83A,B ideal coordination dynamics are shown, taken from a female athlete. With increasing load up to 200N the coordination dynamics increased only a little (Figure 83B, Figure 83A lower trace). Original coordination dynamics recordings for three patients are shown in Figure 83C,E,G. In all three cases the patients could not maintain the turning speed, resulting in load escape. These three patients could turn under higher loads (Figure 83D,F,H).

The best patient could reach 150N (Figure 83C,D), but the arrhythmicity of turning increased strongly. The patient in Figure 83E,F turned under higher loads so slowly that the coordination dynamics could no longer be accurately measured. The third patient (Figure 83G,I,H) could reach a load of 100N. But his CNS became exhausted (the dashed line is above the solid line, Figure 83H). Interestingly, this particular patient could turn well for trot gait (K) but not for pace gait coordination (P); see Figure 83I - increasing and decreasing arrhythmicity of turning. This phenomenon will be discussed with respect to the lack of dopamine.

In conclusion, apart from one patient all had difficulty in moving arms and legs in coordination for higher loads. Even though the low-load CD improved significantly, the high-load CD remained poor, but were better than at the beginning of the therapy.

6.7.8. No Heart Problems during Exercising at High Loads

Three of the eight patients had slight heart problems. When out of breath, such as when climbing many stairs, they experienced heart problems. The patients were thus instructed to stop exercising as soon as experiencing any heart problems while exercising at high loads. After high-load testing, the patients complained about having not enough stamina when exercising at the highest load but not about heart problems. This may mean that the patients could be fitness trained without endangering their hearts.

6.7.9. Improvement of Tremor As Quantified by a Dexterity Test (Minnesota Manual Dexterity Test)

It could be shown above that hand and finger coordination improved in the short-term memory after exercising on the special coordination dynamics therapy and recording device as quantified by a dexterity test. The question arose whether this improved hand and finger coordination may also be retained in the longer-term memory. When repeating the dexterity tests one month later during the therapy under the same conditions, $3\% \pm 2\%$ better results were obtained for the dexterity parameter.

6.7.10. The Only Therapy Known So Far That Can Improve CNS Functioning in Parkinson's Disease Patients Is Coordination Dynamics Therapy (CDT)

It could be shown that low intensity CDT (for 4 hours per week) for 2.5 months, in combination with pharmaco-therapy and conventional fitness training improved CNS functioning by 35%. When the CDT was stopped after 2.5 months, the CNS worsened 21% even though pharmaco-therapy and fitness training were continued (Figure 84).

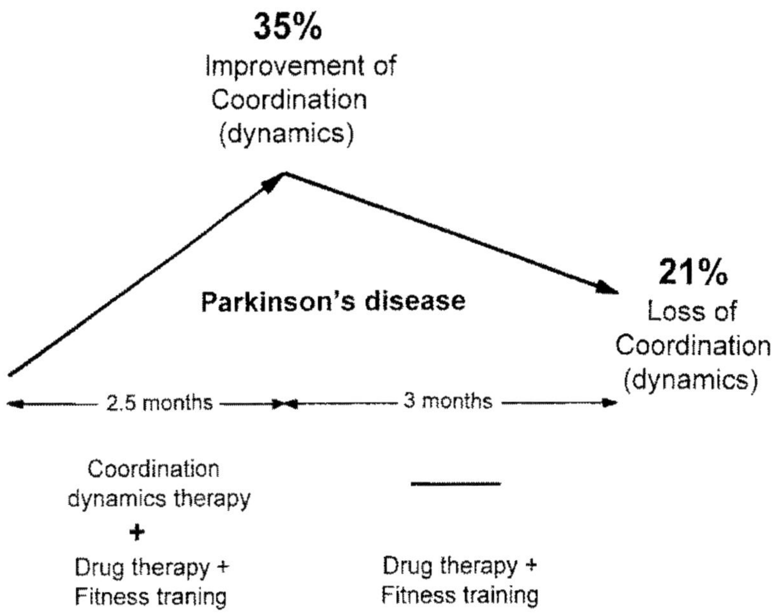

Figure 84. Improvement of coordination dynamics with coordination dynamics therapy and loss of good coordination dynamics after the termination of coordination dynamics therapy.

It is concluded that the only therapy known so far that can improve CNS functioning in Parkinson's disease patients is CDT. The usefulness of a combination of pharmaco-therapy and coordination dynamics therapy has to be investigated. Because of the substantial side effects of anti-Parkinson drugs which subside over time some time, discontinuation of the drugs was not attempted. One patient did not take anti-Parkinsonic drugs at all. He improved by 42% due to the therapy and his condition became worse by 37% after stopping the therapy. It is therefore likely that the improvement of CNS functioning in Parkinson's disease patients was mainly due to a functional reorganization of the CNS achieved by the CDT.

Two other elderly patients with strongly progressive degenerative diseases (an unpublished observation) received intensive CDT (20 hours per week). They lived longer than forecasted by their physicians, but they died soon after the termination of CDT. CDT may optimize the functioning of damaged neuronal networks or optimize them even if the damage to the networks is increasing because of the progressing neurodegenerative disease with loss of neurons and connectivity.

6.7.11. Functional and Structural CNS Repair

Four hours of coordination dynamics therapy (CDT) was applied to the patients because they did not want to exercise more often and longer. Since a patient who suffered severe poliomyelitis, in which the motoneurons are destroyed, substantially improved from the application of an intensive and long-lasting course of CDT of over a year's duration [91]. Long-lasting intensive CDT seems to induce neurogenesis. A very severely affected cerebral palsy patient rather suddenly improved after approximately 5 months of therapy, neurogenesis seems to be induced by intensive CDT lasting longer than five months. New motoneurons were built in a patient with complete cervical spinal cord injury C5/6 [11]. After growing of the motoneuron axons to the arm (Figure 29), some additional finger function re-appeared after three years of intensive CDT.

Neurogenesis therefore, is unlikely to have substantially contributed to the improvements observed in the Parkinson's disease patients as the therapy was not sufficiently intensive and too short. The growth of neurites in the absence of neurogenesis may have slightly contributed to structural repair. But the major part of the improvement was probably due to a functional reorganization, including the activation of non-active neurons.

It would be important to apply intensive CDT (20 hours per week) for longer than 6 months to Parkinson's disease patients and to measure how much CNS functioning can be improved and maintained in the long-term memory.

6.7.12. Pace and Trot Gait and Lack of Inhibition

It can be seen from the frequency and coordination dynamics traces in Figure 83I that this particular Parkinson's disease patient was able to turn quite rhythmically for the trot gait coordination (K) but not for the pace gait coordination (P). This is surprising since with CNS degeneration one would expect that the patient could perform better for the easier pace gait coordination than for the more difficult trot gait coordination. In other patients the easy and difficult coordination's (small and large arrhythmicity) seemed to vary. It is likely that some neuronal network states can be more easily reorganized than others. This will depend on the actual CNS degeneration.

7. Repair Achieved in a Fit 46-Year-Old Patient with Parkinson's Disease Patient Following CDT – Patient Arno

7.1. Case Report

It could be shown above that low intensity CDT (for 4 hours per week) for 2.5 months, in combination with drug-therapy and conventional fitness training improved CNS functioning by 35% in Parkinson's disease patients with an average age of 73. But what rate of repair can be achieved in a young and fit patient who can train more intensively? And what is the rate of repair in this patient through movement-based learning in comparison to healthy subjects? It might be that the degeneration of the CNS is more aggressive in younger patients. It would be of use to produce some rate of repair data to acquire further understanding in general terms as to how much time is required to repair the human CNS and what can be done to live longer with a better quality of life based on this empirical data.

The CNS repair in a 46-year-old patient with Parkinson's disease during intensive CDT (10 to 20 hours per week) for 18 months will be analyzed. As for the group patients, in addition to the drug-therapy and fitness training, CDT was administered and the rate of CNS improvement was measured.

7.2. Improvement of Coordination Dynamics Values for Different Loads upon CDT

Figure 85 shows the CD for different loads with ongoing drug and CD therapy of the fit patient. The 20 to 200N CD values first improved quickly upon therapy and reached the plateau with a small decline as for a healthy pupil (Figure 36), the patient with severe brain injury (Figure 46) and the patient with the incomplete cervical spinal cord injury (Figure 50).

7.3. Variability of the to Be Learned Coordination Pattern Dynamics in Dependence on the Kind of CNS Injury

To reach approximately the plateau, that means the time period of fast movement-based learning, the patient needed 2.5 months (Figure 85). The curves of improvement for the different loads (Figure 85) varied only a little that means they were quite balanced as they were for the healthy pupil (Figure 36) and the patient with the spinal cord injury (Figure 50). This indicates that if the functioning of the brain is not severely affected by the disease; the movement-based learning is rather smooth. In the patient with the severe brain injury (Figure 46a), on the other hand, the curves varied very much, indicating quite an imbalance of CNS functioning during the repair.

Under the system theory of pattern formation, it is the expectation that learning, i.e., the behavioral requirements F_{inf} affect the whole coordination dynamics, including stability, rather than only coordination patterns to which they directly relate. The intrinsic coordination tendencies captured by the intrinsic dynamics influence the performed pattern systematically. The degree to which intrinsic tendencies conflict or agree with the required patterns determines the variability of the performed coordination pattern. In severe brain injury, the intrinsic tendencies conflict more with the learning requirements than in healthy pupils and spinal cord injury patients. The variability of the performed coordination patterns is therefore larger.

7.4. Repair Rate According to the High-Load Testing

It can also be seen from Figure 86A that the overall CNS improvement of this 46-year-old patient with Parkinson's disease needed 2.5 months. In the healthy subjects this rate of learning was measured to be in the order of one to two months. The rate of this motor learning was only slightly slower than in healthy humans.

The problem in patients with Parkinson's disease is not the reduced rate of learning for repair. However, as was discussed above, inhibition is missing resulting in the synchronization of premotor spinal oscillators and tremor. And this missing inhibition only marginally affects the rate of learning. But it may be that something else is pathologic besides inhibition in the CNS functioning of such patients.

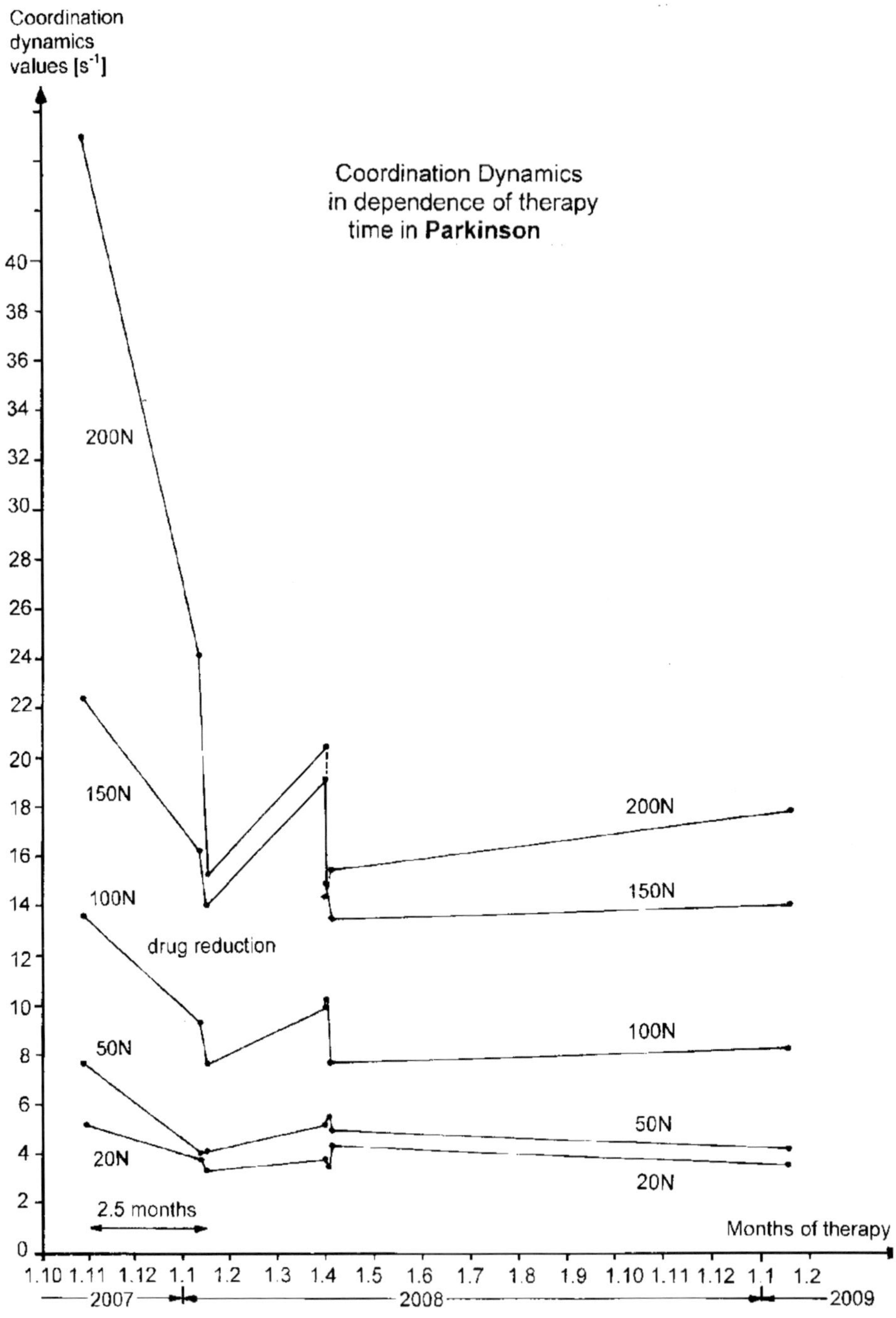

Figure 85. Improvement of coordination dynamics (CD) values for exercising in the forward direction against different loads in a 46-year-old patient with Parkinson's disease upon coordination dynamics therapy.

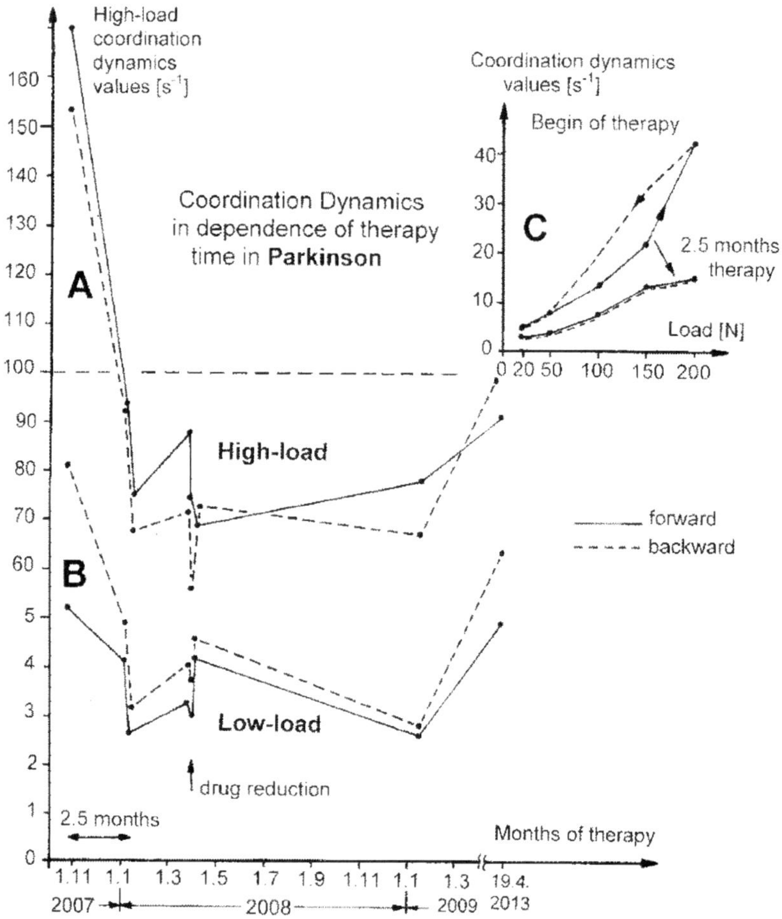

Figure 86. A. Improvement of high-load coordination dynamics (CD) values for forward and backward exercising in a 46-year-old patient with Parkinson's disease upon coordination dynamics therapy (CDT). The high-load CD values were obtained by summing up the single CD values, Δ (high-load CD value = Δ20N + Δ50N + 100N + Δ150N + 200N + Δ150N + Δ100N + Δ100N + Δ50N + Δ20N + Δ20N). B. Improvement of low-load (20N) CD values. Upon transient drug reduction the low-load and high-load CD values reduced transiently. C (inset). Single CD values (Δ) for exercising in the forward direction in dependence of load increase (solid line) and decrease (dashed line) at beginning of therapy and 2.5 months later. The hysteresis-like curve after 2.5 months shifted to smaller coordination dynamics values.

7.5. Systematic (Integrative) Impairment of CNS Functioning Becomes Visible during Coordinated High-Load (200N) Movements

In Figure 85 and 86, it was shown that the CD values for different loads improved quickly (within 2.5 months) and then only slowly in some similarity to healthy cases. But the CD values are only one aspect to describe CNS functioning. In Figure 87 it can be seen that measuring the symmetry of the traces furnishes other information concerning CNS organization and coordination dynamics.

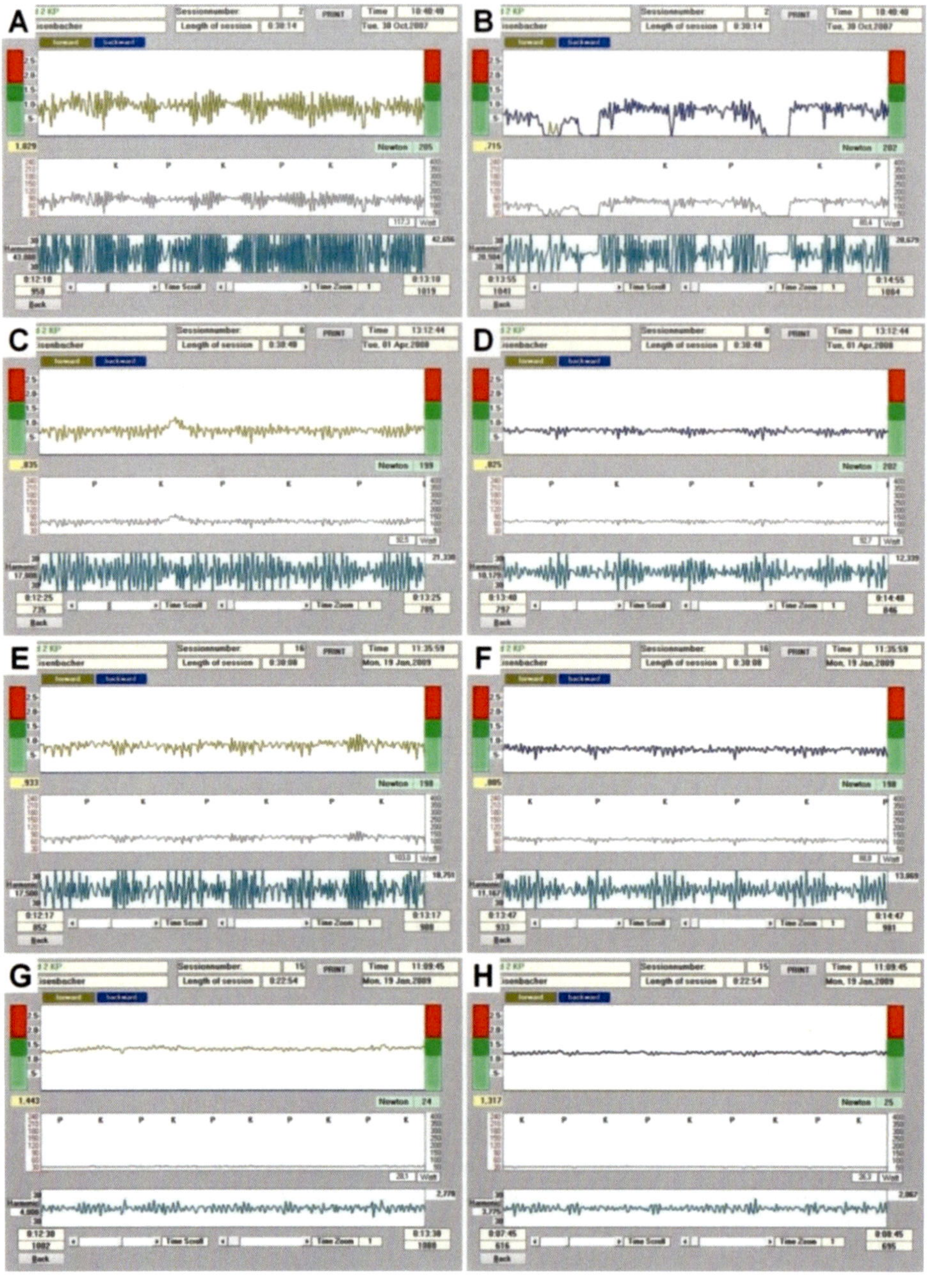

Figure 87. Coordination dynamics (CD) sweep pieces of 1 min when the 46-year-old patient with Parkinson's disease was exercising against a load of 200N in the forward (A,C,E) and backward direction (B,D,F). The frequency of exercising (upper trace) and the CD (lower trace, 'Harmonic') varied rhythmically according to the positions between arms and legs. For some coordination's between pace (P) and trot gait (K) he could turn better (less variation) and for other coordination not so good. Over the years (from A,B to E,F) the CNS of the patient did not learn to manage the difficult coordination's, even though the overall CD values got better (smaller) (Figs.85,86). Further, it can be seen (C,D) that the symmetry between forward and backward exercising was not achieved. The calibrated pace (P) and trot gait (K) coordination's were lying for forward exercising to the right side of the variation of frequency and CD and for backward exercising to the left side. The patient's CNS did not improve substantially with respect to these movements.

At the beginning of CDT it can be seen that the arrhythmicity of exercising in the forward direction against 200N was biggest for the intermediate coordination's between pace gait and trot gait as in the healthy case (Figure 87A). For backward exercising it seems that the highest arrhythmicity occurred just as the pace and trot gait coordination's begin or end (Figure 87B). Five months later with the improvement of the CD values, there was not much difference for exercising in the forward direction between the easy coordination's pace and trot gait and the difficult intermediate coordination's (Figure 87C). For backward exercising the highest arrhythmicity occurred at the pace and trot gait coordination's (Figure 87D) in complete contrast to a healthy case. Further ten months later, the patient with Parkinson's disease had difficulty during forward exercising with the difficult intermediate coordination's (Figure 87E) and for backward exercising still with the easy coordination's pace and trot gait (Figure 87F). Even though the patient was training in the forward and backward direction, the asymmetry between forward and backward exercising could not be improved. The CNS organization seemed to get worse. The high arrhythmicity for the pace and trot gait coordination's while exercising in the backward direction is pathologic. The lack of improvement in the symmetry for exercising in the forward and backward direction cannot be explained by the synchronization of premotor spinal oscillators due to missing of lateral field inhibition.

It can be seen from Figure 87 that the improvement of the forward-backward symmetry could not be achieved within 14 months. The healthy pupil learned this forward-backward symmetry in half a month (Figs.39,40). The rate of learning for this forward-backward symmetry was in the patient reduced by more than a factor of 28. To repair the right-left symmetry in stroke patients is an even greater challenge. From the high-load curve (Figure 86) it can be seen that supervised training is better than training with no supervision. With no supervision from 2009 onwards the high-load value got worse (Figure 86). Supervised intensive therapy is most efficient in repair also in patients with Parkinson's disease.

8. Repair in Stroke

8.1. Summary

In the following, it will be shown that the injured CNS could partly be repaired by coordination dynamic therapy (CDT) in eight stroke patients. The repair is quantified by the improvement of walking, hand function and the direct measurement of CNS organization (the coordination dynamics (CD)). In four patients the improvement of CNS organization (CD) was 70 ± 14% within three months of therapy. In a patient where the intensive coordination dynamic therapy, which consisted of at least four hours therapy per day during which more than 15,000 coordinated movements were performed per day lasted longer than one year, the difficult hand functions could also be substantially improved. It is shown that the CD improved in the short-term memory in a stroke patient by 70% within 30 minutes, when exercising on the special CDT device. As in other learning activities such as schoolwork, which requires the training of higher mental functions in the short-term memory, the repetitively trained CNS functions will also go into the long-term memory, which means the

CNS has been reorganized. Improvements in the short-term memory are used to motivate the patient to perform therapy.

8.2. Recapitulation of General Principles

The purpose of CDT is to reorganize the injured or malfunctioning human central nervous system (CNS). The concept of CDT builds upon the theory of self-organization, pattern formation of neuronal networks, CD and human neurophysiology [14]. Pattern of coordination are generated by the coordinated firing of single neurons and neuron assemblies and are characterized macroscopically by low dimensional collective variables or order parameters whose dynamics are function-specific. Loss of stability leads to switching of patterns. Fluctuation and differential stability govern the switching dynamics among multiple coordinative patterns. If a certain coordinative pattern has a very high stability, it may seem as if this pattern is 'hard wired' and is generated by a pattern generator. The patterns generated in the neuronal networks are a direct consequence of cooperative and competitive interactions between the intrinsic coordination dynamics of the neuronal networks, the intentional impulse patterns and the movement induced afferent input patterns.

The strategy to improve the functioning of the injured or malfunctioning CNS is partly based on measurements on phase and frequency coordination of firing of neurons and neuron assemblies. This phase and frequency coordination on the neural level, the neuron assembly level and macroscopic level, i.e., movement, is partly lost following CNS injury. Coordination dynamic therapy tries therefore to re-establish phase and frequency coordination between the firings of neurons of the neuronal networks by (re-)learning coordinated movements. Coordination dynamics therapy is therefore a learning method, which is based on measurements in the human CNS and has an attendant neurophysiologic theory.

An indispensable part of a therapy is to have tools available to measure the improvement. The improvement of CNS functioning can be indirectly measured by the improvement of movement performances and can be measured directly by the coordination dynamics by using the coordination dynamics recording method.

The CDT is based on five new developments/concepts in human neurophysiology: (1) Self-organization of neuronal networks. The self-organization takes place by the relative coordinated firing of neurons with respect to time and space. The CNS is viewed as a neuronal network which can be changed in its functioning by re-learning. (2) The organization takes place by a specifically changing relative phase and frequency coordination between the firings of neurons. (3) The building of new nerve cells in the adult human CNS and their proliferation with respect to functional aspects, which is led by learning processes including motor learning of coordinated rhythmic movements. (4) The human CNS can learn and store information relating to all manner of things integratively as a second mode of functioning. This means that brain function is not strictly localized as previously thought. This aspect of brain function can be used to reorganize the injured CNS and it is borne out in the clinical work undertaken in CDT. (5) Each CNS can substantially be improved in its functioning through CDT.

In severe CNS injuries, the restoration of physiologic functioning is only possible if the relearning is integrative, coordinated and efficient. The reorganization of the pathologically functioning CNS by relearning can be achieved by: (1) The use of special instruments such as the special CDT device to relearn the lost phase and frequency coordination between the firings of neurons and the lost coordination between the movements of arms, legs and trunk. These movements may relate more to functional reorganization. (2) The training of automatisms, postures and old learned movements, which may remain largely unaffected by the lesion. This may relate more to structural reorganization.

The efficiency of CNS re-organization by relearning is determined by at least four factors: (1) The accuracy of coordination of the exercised movements, which needs to be within approximately 5ms to relearn or improve phase and frequency coordination of neuron firing and to reconnect network parts to recouple arms or legs which may not be moving, for example a plegic arm, and to integrate it into a broader movement like walking. Co-movement (Figure 173) is an example of a reconnection of sub-networks. (2) The increase of the integrativity of coordinated CNS activation by exercising many phase and frequency coordination's as possible simultaneously to entrain more coordination's per time, to reorganize the CNS more globally and to activate and coordinate as many neurons and sub-networks simultaneously to improve also the very integrated functions of the CNS like the higher mental functions. (3) The increase of the physiologic movement induced re-afferent input and intentional impulse patterns to enhance the physiologic self-organization of the injured networks and its communication with the outer world by regulatory processes. (4) The intensity of therapy. At least four hours therapy per day, 5.5 times per week for six months (at least 15,000 coordinated integrative movements per day) must be performed. The 'adaptive machine' CNS has to be forced to adapt.

Instructive training, interpersonal coordination and co-movements are used to enhance physiologic movements in the patient. The movements which are often performed during coordination dynamics therapy are creeping, crawling (commando-crawling), standing from a seated position, jumping on springboard, walking, running and exercising on the special coordination dynamics therapy device in the lying, sitting and standing position.

8.3. Improvement of CNS Organization in the Short-Term Memory

Since substantial reorganization of the human CNS through needs many months to several years it is of importance to have diagnostic means available to give the patients a prognosis concerning the time needed to reorganize their CNS. CDT is a learning method. Therefore, many of the ordinary principles of learning will hold. For example, at school it is believed that if a pupil can retain something in the short-term memory, then it will, with many repetitions at least partly also go into the long-term memory. The same thinking holds here also for motor learning. If a stroke patient can relearn in the short-term memory, then with many repetitions, the things learned are expected to also pass into the long-term memory. Therefore, when a patient asks whether it is worth undergoing CDT, the CD are measured and this can show the patient that the organization of his CNS can be improved in the short-term memory.

8.4. Short-Term Memory in the Stroke Patient

The substantial improvement of the CD of the CNS of a stroke patient is shown in Figure 88. As can be seen from Figure 88A, the frequency variation of turning (upper trace) is very large.

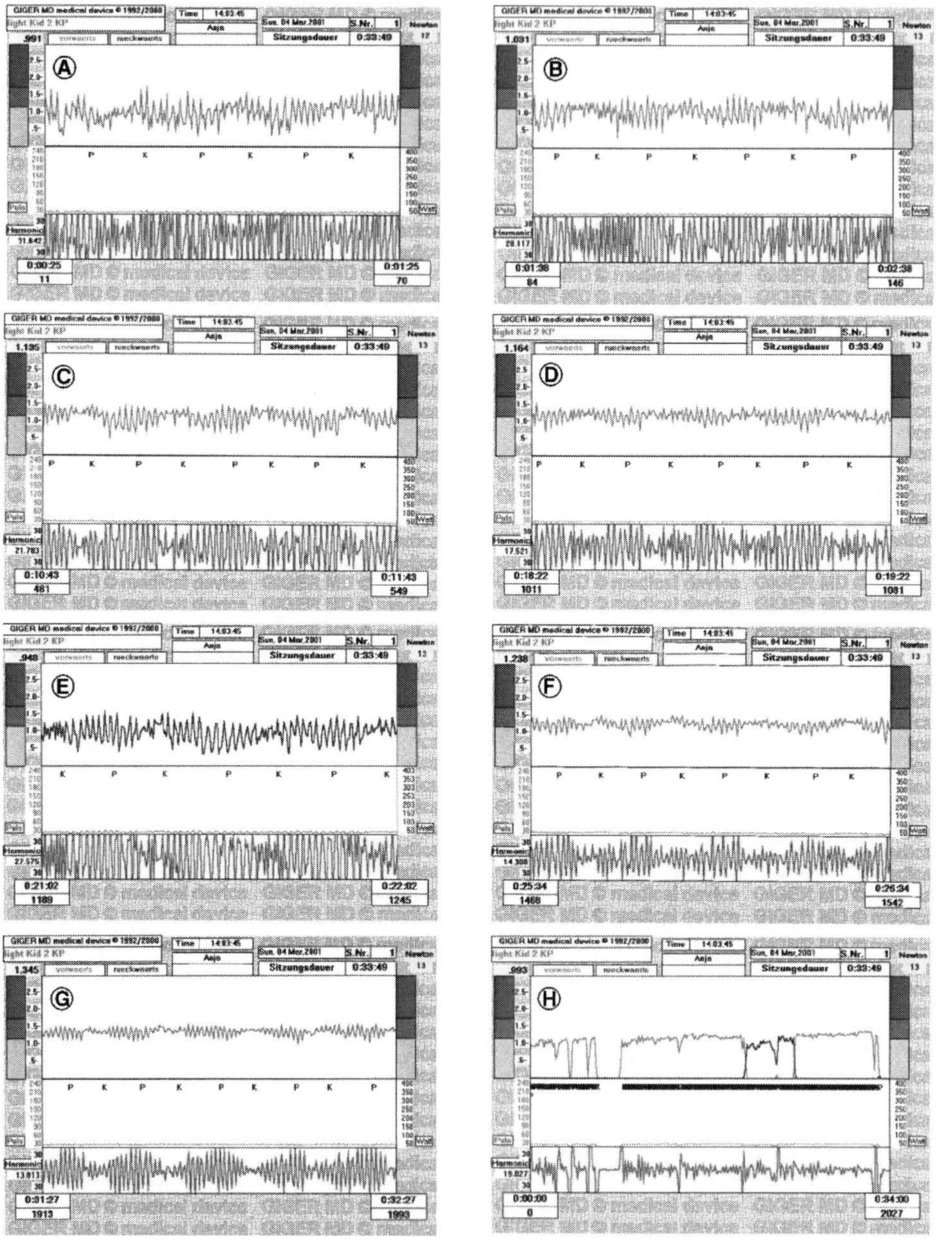

Figure 88. Coordination dynamic recordings from a stroke patient showing CNS organization improvement in the short-term memory when exercising on the special coordination dynamic therapy device. The whole recording of exercising over 34 min is shown in H. Forward (vorwärts) and backward (rückwärts) turning for 1 min interval is shown in A to G. Note the substantial improvement in the short-term memory of frequency f (upper trace) and df/dt (lower trace) between A and G.

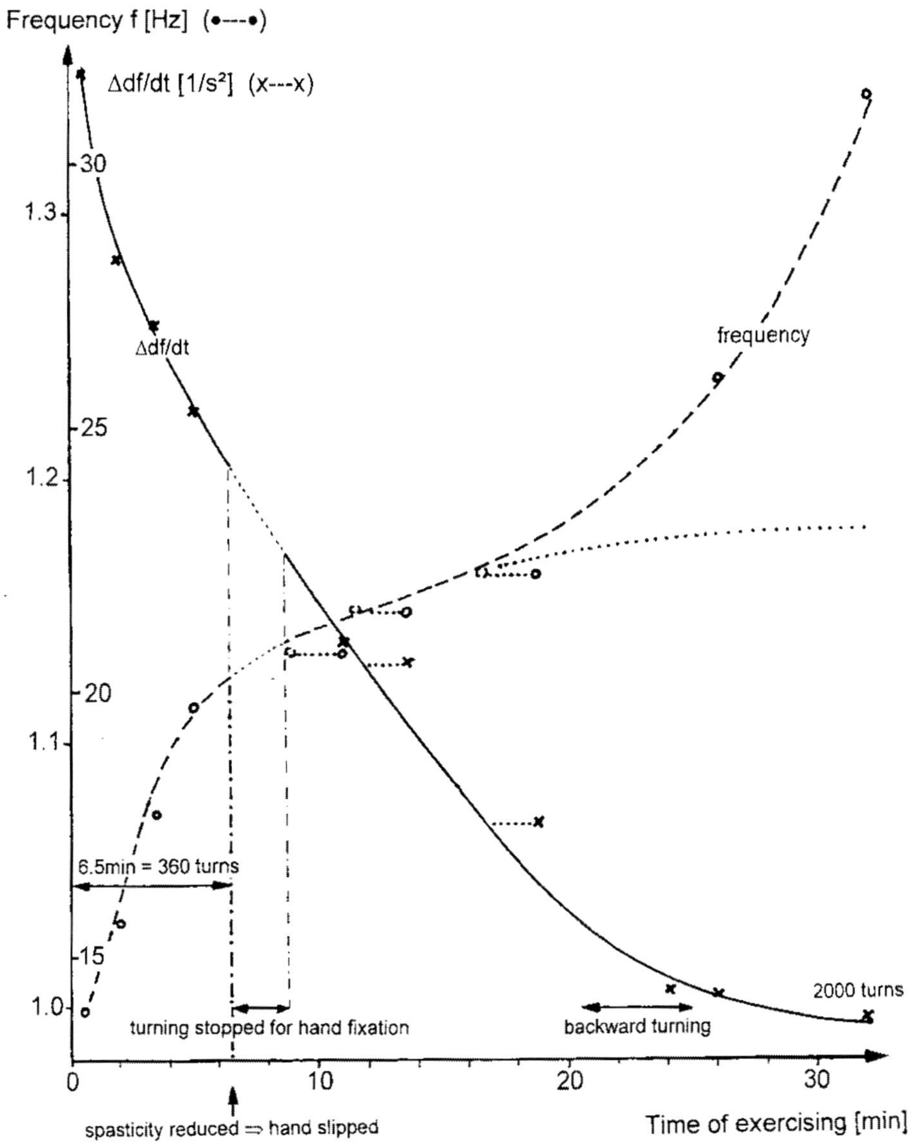

Figure 89. Improvement of CNS organization in a stroke patient (who has never exercised on the special coordination dynamic therapy device before) with ongoing time of exercising. The paretic left hand is first not fixed to the handle till it slips from it after 6.5 min. After fixating the hand, the exercising is continued. The curves for the frequency and Δdf/dt are partly corrected for the time needed for hand fixation. Note that Δdf/dt is reducing strongly (by more than 50%), indicating that the CNS organization of that stroke patient is strongly improving in the short-term memory; the frequency of turning is increasing strongly, which is mostly a good sign for improvements of the functioning of the injured CNS. It seems as if the backward turning is enhancing the frequency strongly. Without backward turning the frequency may have increased less as indicated by the dotted line.

It can be further seen that the stroke patient could perform trot gait coordination (K) better than pace gait coordination (P), since the frequency of turning (upper trace) for the trot gait coordination was higher and the frequency variation smaller (Figure 89B to G). The variation of df/dt is so large, that this trace is not very helpful for analyzing CNS organization at the beginning of exercising (Figure 89A to C, E, lower trace). With ongoing exercising the variation of the frequency reduced and the frequency of turning increased. After 30 min of exercising, the coordination dynamics were much better (Figure 88G). The trot gait coordination (K) became better as did the pace gait coordination (P). The improvement of CNS organization due to exercising for half an hour on the special CDT device can be seen in Figure 88A through G. The frequency variation became smaller as did the variation of its time derivative (lower trace). It can be seen from Δdf/dt (lower trace, 'Harmonic') that the coordinated movements of arms and legs became better because the arrhythmicity of turning reduced. In Figure 88F,G the characteristic structure of coordination becomes visible. The trot gait coordination (K) could be performed very well by the patient. The pace gait coordination (P) was not that good but still much better than at the beginning of the exercise period. The coordination between arms and legs for backward turning in between the forward turning (Figure 88E) was not as good. Larger variations can be seen in both traces.

The kinetic argument that the pace gait coordination must be worse than the trot gait coordination in a hemiparetic patient - because one side is working better than the other will have consequences for the pace gait coordination and not so much for the trot gait coordination, since in trot gait one paretic arm and one healthy leg and one healthy arm and one paretic leg are moving in coordination, whereas in pace gait the paretic side is moving against the healthy side - is not conclusive. Of course, kinetic arguments are coming in, but firstly different kinds of co-movement from right to left are activated by the exactly coordinated afferent input from the moving arms and legs, and secondly the coordination of different stroke patients with hemiparesis can be quite different, as will be seen below (Figure 92).

The plotting of the coordination dynamics of relearning in the short-term memory in a graph (Figure 89) will bring further information. It can be seen from Figure 89 (and Figure 88H) that the frequency of turning increased from 1.0 to over 1.3 Hz, which makes an improvement of CNS organization likely. At the same time, during this half an hour of exercising the arrhythmicity of turning decreased very much, namely from 33 to 15. The arrhythmicity of exercising at 20N in the healthy CNS is between 5 and 6 [50]. Therefore, even though the arrhythmicity of the coordinated movements of the stroke patients reduced significantly, it was still far from normal.

Since, according to the theory of coordination dynamics, with the increasing stability of the attractor 'physiologic movement' the stability of the attractors 'pathophysiologic movements' decreased due to decreased use. It was of interest to see what will happen with spasticity of the left paretic hand if the organization of the CNS improves in the short-term memory due to exercising a very coordinated movement. The author expected a reduction of spasticity in the short-term memory, as it has always been the case till now. In this patient the hand grip spasticity was so strong that when shaking hands it was nearly impossible to release oneself from the paretic spastic hand. Due to this hand spasticity the patient could turn with the spastic hand, because the patient could hold the lever because of spasticity. The interesting question was how long the spastic hand could hold the lever until the spasticity reduces to such an extent that the hand will slip from the lever. After 360 turns on the device in six and a half minutes the left spastic hand slipped from the handle and was then fixed with a bandage. The power required to turn the handle was approximately 13 Newton. Thus, the spasticity became smaller than 13 Newton. The next interesting question was what effect backward turning will have on forward turning. The experience so far in patients is that backward turning enhances the CD for forward turning. Even though backward turning was much worse (Figure 88E) than forward turning (Figure 88D) in the stroke patient, the frequency increased strongly for forward turning (Figure 89). Also, arrhythmicity decreased further.

In conclusion, the improvement of the coordination dynamics (reduction of $\Delta df/dt$) in the short-term memory was 59% (33 to 15). When the patient turned at a frequency of 1Hz, the improvement was 70%.

8.5. Overlap between Coordinated Activation of CNS Neuronal Networks for Movements and Higher Mental Functions Observed in Stroke Patients

Eight stroke patients underwent CDT. The functions on the paretic side improved in the short- and long-term memory and speech, where impaired, was also improved. One patient with impaired speech, who could speak four languages, spoke the first words after stroke in English, which was not his native language, in accordance with reports that later-acquired languages are stored differently in the cortex than the native language [86]. When this patient exercised on the special coordination dynamic therapy device, his leg functions improved in the short-term memory after approximately 30 turns, but his speech only improved substantially after 4000 turns. At the beginning of exercising, the patient was mixing up all four languages and he got angry because he could not say what he wanted and virtually nobody could understand him. But after 4000 turns he was largely able to stick to one language. When he got exhausted after 6000 turns, he began to mix up languages again. Therefore, at the beginning of the CDT an improved communication with this patient was only possible if he had turned a few thousand times on the special CDT device. This clearly indicates that in this patient there was a relationship between the improvements of the coordination dynamics for movements *and* for higher mental functions, indicating that there was interlacing between the neuronal networks generating movement and those generating higher mental functions. In other stroke patients speech improved when the patient had to speak numbers or letters in coordination with the coordinated arm and leg movements. The numbers or letters appeared on a screen changing in coordination with the movement.

8.6. Substantial Improvement of Hand Function in a Stroke Patient After One Year of Coordination Dynamic Therapy – Patient Jan Berky

A 62-year-old well known violinist (Figure 90) suffered a stroke so that his right arm was hanging down and his hand was spastic (Figure 90B). He was no longer able to play violin. The patient trained walking and arm movements a few hours per day to recover from stroke. Due to these efforts, his arm lifted slightly from the hanging position when he was playing violin together with his son in such a way that the musician held the violin and his son was moving the bow in coordination (Figure 90C). By activating the old learned attractor of performing violin music, he activated the right arm a little (Figure 90C).

Figure 90. Coordination dynamic therapy-induced stroke recovery of a now 62-year-old musician made visible by the improvements of the movements. A. Before stroke; the musician is performing with his orchestra. B,C. After the stroke the patient is performing in concerts with his son; he is holding the violin with the left hand and his son is moving in coordination the bow. In B the plegic arm is hanging down and the hand is spastic; in C (later) the now paretic arm is coming a bit upwards (is already a bit flexed), as if the paretic side wants to contribute to the coordinated movements of the old-learned CNS activation 'performing violin music'. D. The patient can just hold the bow by himself 3.5 years after the stroke; the positioning of the fingers is very poor. At this stage coordination dynamic therapy was started supervised by the author. E. With the ongoing therapy the finger positioning is already better for holding the bow; the shoulder is more downwards because the patient has already a bigger amplitude for arm movement when moving the bow. F. Five years after the stroke the movements of the left arm,

hand and fingers are not far from normal. G,H,I. The patient can fully lift the formally plegic arm and can move the fingers (G), can turn the device (hand and arm power came back) (H) and can hold a cup for drinking (I); note in I the picture of former violin playing. K,L. Patient exercising on the special coordination dynamic therapy device; in L he is training the finger positioning for holding the bow when playing violin (in the integrative activation of neuronal networks for relearning also those network parts have to be essentially activated which generate the functions which have to be relearned). M. Patient playing violin again in a concert with his daughter. He is playing the first, she the second violin. The son is on the other side in the background. Note that the positioning of the right hand and fingers is poor because of the stress to organize the concert and performing for the first time again. What cannot be seen in the picture is that the music he played was not ruined by the stress. Two days later, when the stress was over, the positioning of hand and fingers was physiologic again (E,F).

When the patient started coordination dynamics therapy three and a half years after the stroke, including 1500 turns on the special CDT device (Figure 90K), he could just manage to hold the bow in his right hand (Figure 90D). The music he played was poor at the beginning, since firstly he could hardly hold the bow in the hand and secondly the movements of the right arm and hand worked poorly in coordination with the left one. With ongoing CDT, including exercising of the finger positioning on the special CDT device (like in Figure 187, right hand) to hold the bow (Figure 90L), the patient could hold the bow better and the amplitude of arm movements increased (Figure 90D,E,F) and the music he played began to sound better and better.

After eighteen months of coordination dynamic therapy, which now included 10,000 turns on the special device, he could fully lift up the arm and move the fingers separately (Figure 90G) and could hold a cup for drinking almost normally again (Figure 90I). His hand and arm power increased so much that he could turn the device with the formally right plegic hand alone (Figure 90H). When playing violin the positioning of the fingers of the right hand were quite normal again and he was not lifting the right side any more (Figure 90E,F) in contrast to the beginning of therapy (Figure 90D). His granddaughter (Figure 90F,H,L) was telling him that he has to turn more rhythmically when she was looking at the computer display and saw larger frequency variation in his exercise (Figure 90K).

When giving his first concert again following the stroke (Figure 90M, playing by himself), his performance was quite good again, even though the stress before and during the concert was so great that the position of the hand when holding the bow was not ideal. The bow holding was nearly as bad as at the beginning of intensive therapy (Figure 90D) and far from normal in comparison with that of her daughter playing the second violin beside her farther (Figure 90M). Interestingly, the stress transiently ruined the hand and finger movements, but not the music he played. When he was no longer under stress two days later, he could hold the bow properly again (Figure 90F). The stress was therefore able to partly destroy the relearned hand movements, but not the playing of music, indicating that in this case the hand movement and the playing violin was stored differently in the CNS.

The coordination dynamics, which quantify progress in CNS organization directly, are shown in Figure 91. For forward turning, the CD improved by 50% in nine months (Δdf/dt reduced from 16.8 to 8.3). After additional three months of therapy the CD improved further from 8.3 to 7.2. A normal value for the CD (Δdf/dt) is approximately between 5 and 6 (for a turning frequency of 1 Hz). For backward turning the CD improved from 14.6 to 10.6 in three months. Since the normal value lies between 10 and 11 [50], the musician was reasonably good at exercising in the backward direction.

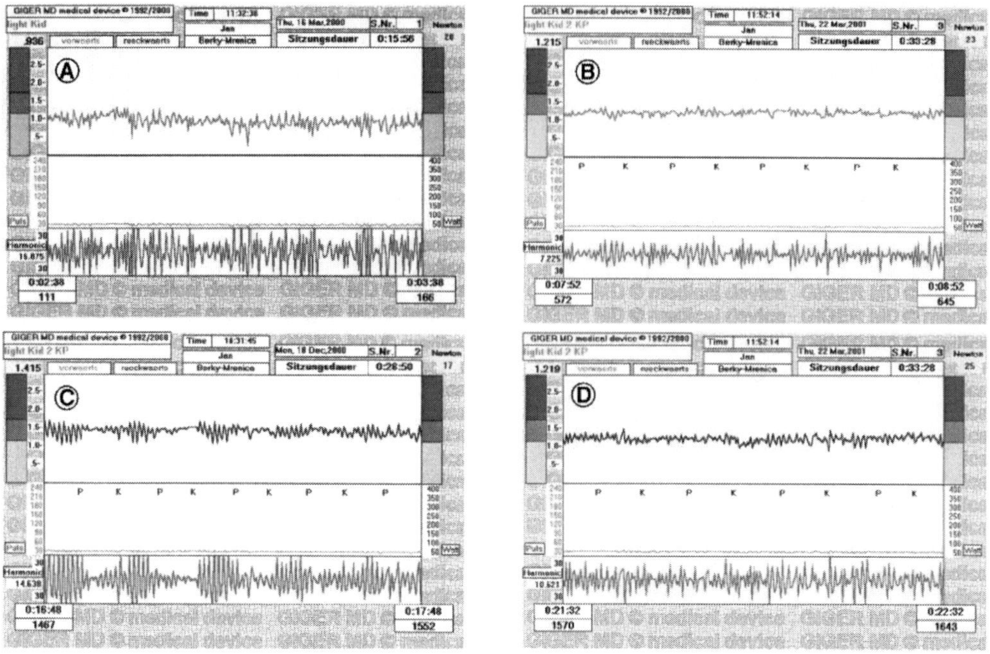

Figure 91. Improvement of coordination dynamics in a musician who suffered a stroke. A,C. Forward turning. D,E. Backward turning.

When standardizing the CD values to the frequency of turning, the CD of the patient improved for forward turning by 60% in one year ($\Delta df/dt$ / f = 18 → 7; = 60%) and for backward turning by 16% in three months.

Such frequency normalization is useful because normally the frequency of turning increases with stroke recovery. Normalization is useful in comparing patients because patients turn at different frequencies. With progressing recovery from stroke, the patients could perform movements faster, due to the more precise phase and frequency coordination of CNS neurons, neuron assemblies and moving arms and legs.

This patient may not be able to relearn the former dynamism of playing the instrument, but he may become able again to play music for children, where not so much dynamism is needed in the hand movement, and may compose music for children. This would give the patient a new perspective for his future life. This case report demonstrates that stroke patients can relearn hand and finger functions.

8.7. Stroke Recovery Quantified by the Coordination Dynamic Recording Method in 4 Patients

Four of the eight patients suffered a stroke at the age between 36 to 57 years and underwent CDT for three months 1.5 to 11.5 years after the stroke with at least 4 hours therapy per day 5.5 times per week. The relearning in physiologic CNS functioning was quantified by the improvement of movements like walking, jumping and hand grip power (Figure 93) and by measuring the CD at the beginning (Figure 92A,C,E,G) and at the end of the therapy (Figure 92B,D,F,H).

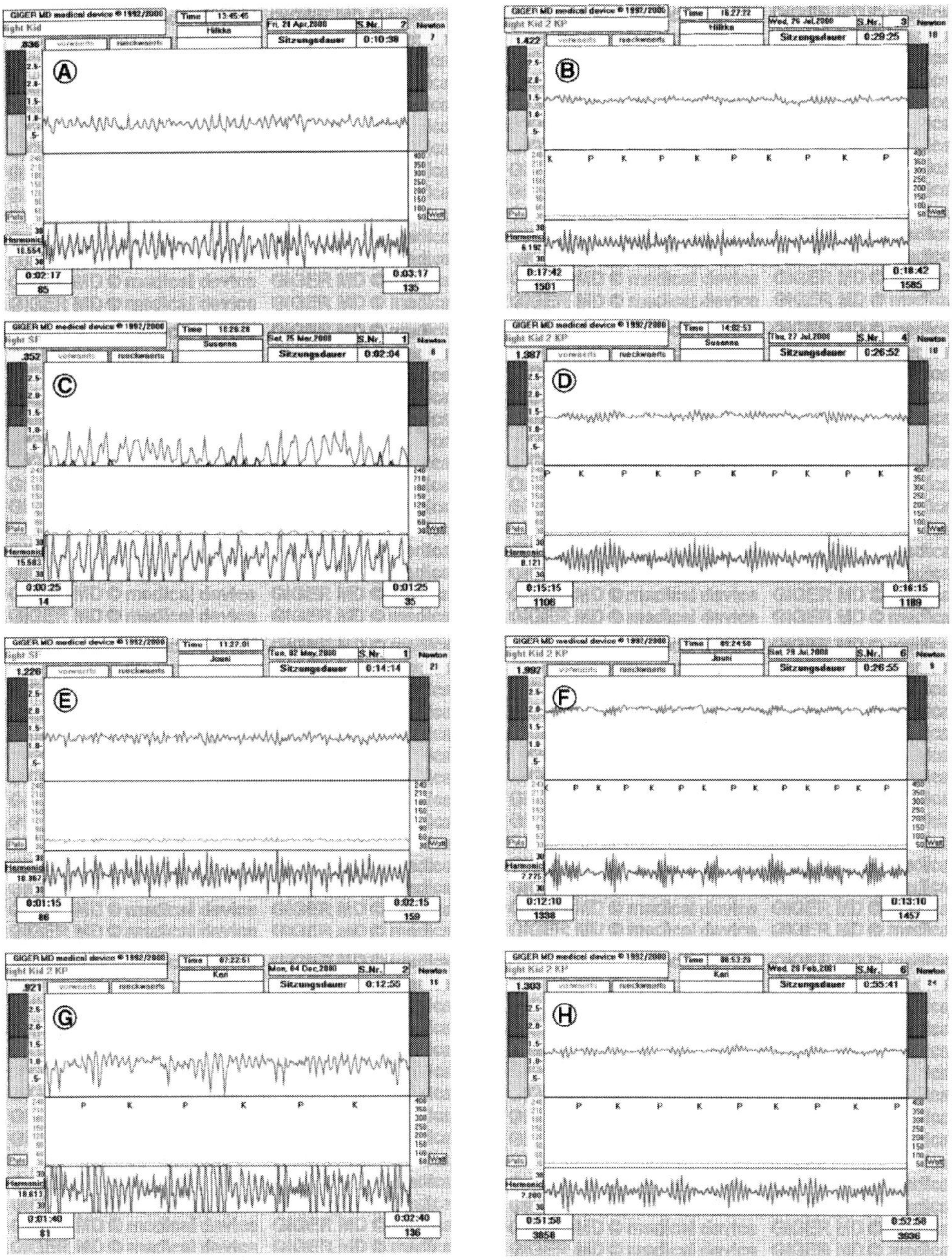

Figure 92. Coordination dynamic recordings of 4 stroke patient at the beginning of therapy (A,C,E,G) and after 3 months of therapy (B,D,F,H). Note that the amplitude of variation of frequency (Δf) and its time derivative (Δdf/dt) reduced in all 4 patients. Ordinate, f (upper trace) and df/dt (lower trace). Abscissa, time coordinate (1min from left to right) calibrated by the pace (P) and trot gait (K) coordination's.

It can be seen from Figure 92 that in all four cases the frequency of turning increased and the variation of the frequency reduced, which means that the patients relearned to turn more rhythmically at a low dynamic level. But also the variation of the time derivative of frequency (Δdf/dt) reduced strongly (lower traces), indicating that also at a higher dynamic level the rhythmicity improved. The increase of turning frequency and the reduction of Δdf/dt indicate that the injured CNS could better generate coordinated movements at a higher speed.

In the first patient, the variation of df/dt reduced more in general (Figure 92B), Δdf/dt had no structure. More specific properties of the CD became visible in the other three patients. Following therapy, the second patient could perform the trot gait coordination between arms and legs well (K) but the pace gait coordination less so (P) (Figure 92D). In the third patient the CNS could generate all coordination's between arms and legs quite well apart from the changing coordination's between trot and pace gait (K→P) (Figure 92F). In the fourth patient easy (Δdf/dt small) and difficult coordination's (Δdf/dt large) can be seen (Figure 92H). However, the easy and difficult coordination's were also changing with time with respect to the pace and trot gait CNS organization calibration (with respect to P and K). Therefore, even though the CD improved very much in all four patients after three months of therapy, the differences between the patients remained significant.

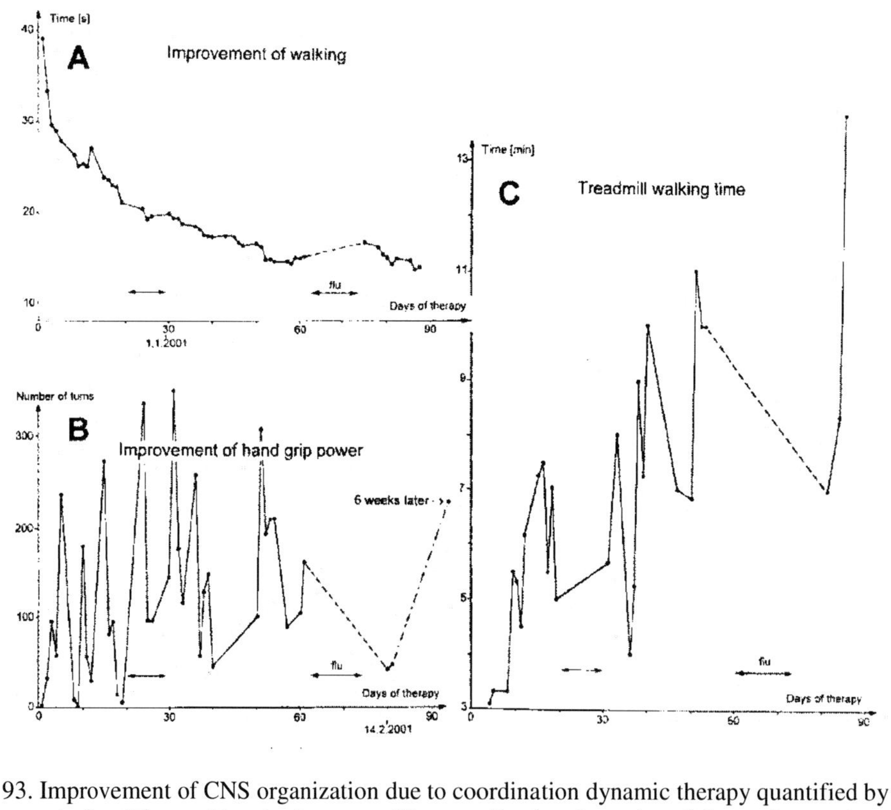

Figure 93. Improvement of CNS organization due to coordination dynamic therapy quantified by the improvement of walking and hand grip power. The coordination dynamics of that patient at the beginning and at the end of therapy are shown in Figure 92G and 92H respectively. A. Walking times for 19.2 m with ongoing days of therapy. Note that a bad flu increased the walking times, whereas a small flu (←→) had no or only little effect. B. Number of turns until the paretic hand slipped from the handle in dependence on the days of therapy. The dashed line indicates that turning without fixation of the paretic hand was not possible due to the bad flu. After the flu, the hand grip power was reduced. Six weeks later (with ongoing therapy) the volitional hand grip power increased again with 218 turnings before slipping from the handle. C. Duration of treadmill walking time (speed = 3 km/h) in dependence on the days of therapy. The dashed line indicates that treadmill walking was not possible during and some time after the flu due to the weakness of the patient.

From the direct measurements of CNS organization (Figure 92), it follows that the normalized arrhythmicity (Δ(df/dt)/f) reduced by 70 ± 14 % (s.d.; range 54-87%). The CNS organization with respect to the CD improved therefore by 70%. Indirect measurements of CNS organization through the relearning of movements also showed substantial progress in all four patients. In Figure 93 the improvements in walking and hand grip power of the fourth patient (Figure 92G,H) are shown. The stroke patient relearned to free walk faster (Figure 93A) with better gait for a longer period (Figure 93C). The hand grip power in the paretic hand became stronger because he could turn longer on the special CDT device without hand fixation of the paretic hand (Figure 93B). A flu infection (marked by flu in Figure 93) contracted after two months of therapy reduced the quality of all relearned movements for longer than the flu symptoms themselves lasted. The hand grip power was transiently completely lost.

8.8. Comparison between Longer and Shorter Therapy Times

Even though the motor functions improved considerably during the three months of CDT in all four patients, there was only little progress in the hand and finger functions on the paretic side. In the musician discussed above, the arm, hand and finger functions improved very much. What then, is the reason that the hand functions improved only a little for the four patients after three months of therapy?

Each injury is different and what the upper (frequency) traces in Figure 92 show is that the repair adapts to these differences between the injuries. It is possible that the musician also had a less severe brain lesion due to stroke. However, the principal reason for the limited recovery of hand function is probably the much shorter therapy time.

8.9. Exhaustion of the Body during the Week When Performing Intensive Coordination Dynamic Therapy

One criterion for the efficiency of the CDT is to push the patient to his limit every weekday, i.e., four to five hours therapy per day, five to five and a half days per week to force the 'adaptive machine' that is the CNS to adapt to fully recruit all possible repair mechanisms including neurogenesis. On the other hand the body, including the CNS, also needs time to recover.

It is generally the case that in the course of therapy, patients get exhausted towards the weekend and have the most energy at the beginning towards the middle of the week. In the fourth stroke patient (Figure 92G,H) the exhaustion could be observed in the recorded walking times for 600m (Figure 94). The patient performed coordination dynamic therapy for four to five hours per day from Monday to Friday. However, he walked 600m seven days a week. It can be seen from Figure 94 that the patient could walk fastest on Tuesday and Wednesday and got exhausted toward the weekend and recovered during the weekend to get better again from Monday to Wednesday. The best movements could be observed in this patient on Tuesday and Wednesday.

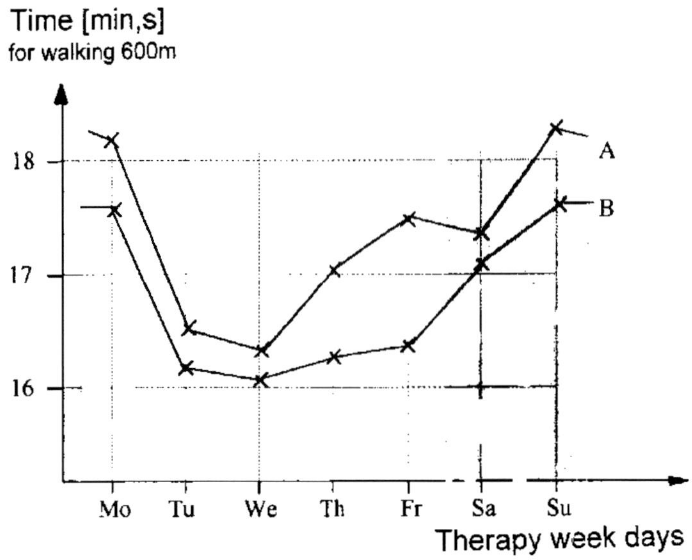

Figure 94. Walking times over 600m from Monday to Sunday (average of 4weeks) of a 49-year-old stroke patient 4 years after the stroke. Walking was performed all 7 days per week, all other exercises only from Monday to Friday. Lower trace (B), when the biggest value is omitted. The best times were achieved on Wednesdays with little more than 16 minutes. Note that the patient is getting exhausted towards the weekend and recovers during the weekend.

8.10. Importance of Measuring Improvement of CNS Organization in the Short-Term Memory

It has been shown that CDT can partly repair the physiologic functioning of the CNS injured by stroke. The improvement in CNS functioning can be quantified by measuring CNS organization with the CD recording method (Figs.91,92) and can be quantified indirectly by the improvement of movements such as walking and hand grip power (Figure 93). Improvement in CNS organization in the short-term memory (Figure 88) in a stroke patient informs the therapist and the patient that the injured CNS of the patient can be improved in its functioning if long and intensive CDT is performed.

The quantification of improvement in the short-term memory is therefore of prognostic value. Patients can also often feel the improvement of CNS organization in the short-term memory. They seem to feel better in their CNS. Spasticity is reduced, walking and hand functions get better and blood circulation, speech and vision become better.

8.11. Necessity for Longer Therapy Times to Improve the Functioning of the Injured CNS

It has been shown that a world famous violinist recovered from stroke through CDT (Figs.90, 91). The critical arm, hand and finger functions also recovered substantially. The improvements in CNS organization were such that it has been demonstrated that 'cure not care' can be achieved.

In general, arm and hand functions recover less well than leg and feet functions. There are three possible reasons for this. (1) During CDT, the patient is forced to use the paretic leg, because otherwise he would not be able to walk. By contrast, the patient does not have to use the paretic arm and hand, because he can do nearly everything with the other arm and his legs. Improvements of arm and hand function can only be achieved through using them. (2) The leg is included more in the walking automatism (stepping automatism), mainly located in the spinal cord, not the arm. Therefore co-movement from the healthy to the paretic side, which improves the paretic side, will be stronger for the leg than for the arm movement. In other words, the cerebral cortex is more involved in arm and hand than in leg and feet movements. An injury of the cortex on one side will therefore affect arm and hand functions more than leg and feet functions. (3) At least in adult human patients the neuronal networks activating finger, hand and arm functions are more complex than the networks controlling foot and leg functions. It is more difficult to repair complex networks than simple networks. This is also a significant difference between the animal and the human CNS. Because the human CNS is more complex in its organization than the CNS of animals, it is much more difficult to repair the human CNS. Therefore, different repair strategies are used in animals and humans: in animals the nerve fiber growing strategy is used for repair and in humans the learning strategy should be used

But why did arm and hand functions improved substantially in the musician (Figure 90) and much less in the other stroke patients? From the recovery of patients suffering stroke, spinal cord injury or brain injury from birth (cerebral palsy, see below), it seems that substantial recovery occurs after many months or several years of continuous intensive CDT. It is as if something needs to be grown in the injured CNS which requires a lot of time (for structural repair, see under stem cells and epigenetic). In the case of injury to the sciatic nerve (peripheral nervous system) we know that it takes two to three years for distal muscles to start to function again, because the motoneuron axons have to grow a long distance down the Bügners bands to the distal muscles, and the growing time required lies in the range of one to two years (growing velocity = 1 to 2 mm/day; growing distance $\approx$ 1m = 1000mm; $\Rightarrow v = s/t$; $t = s/v$ = growing time = 1000mm / 2mm/day = 500days = 1 to 2 years). Further, thick axons regenerate (grow) faster than thin ones [11, 14].

Therefore a possible explanation as to why the patient has to perform intensive continuous CDT for one to three years is that new nerve cells have to proliferate from stem cells and have to be built into the neuronal networks according to functional aspects. Additionally, axons and dendrites have to grow and form functioning synapses. Following CNS injury, many neurons of the neuronal networks will be partially denervated. Partial re-innervation of other old neurons will take place and, through neurogenesis, newly built neurons will be formed. Membrane property changes of synapses and dendrites may take place controlled by the innervating neuron kind. In a frog model it has been shown that in the unexcitable membrane of slow muscle fibers, Ca-channels were incorporated (or unblocked) for excitation in the whole membrane or in localized parts which were controlled by the innervating neuron. A functioning synapse was not necessary for that trophic control [14 (Chapter I)]. Similarly in the CNS of mice, specific neuronal activity is required for the development and migration of specific cortical interneuron subtypes [87]. Therefore many intricate repair mechanisms, including structural and functional ones, will take place at different levels, when the 'adaptive machine' CNS is forced to adapt by pushing the patient to his limits to restore the coordinated firing of CNS neurons with respect to time, space and repair functions. If membranes of dendrites and cell somas of neurons could be partially made excitable in similarity to the slow fiber membrane this would substantially change the timed coordinated firing of CNS neurons.

The most likely explanation for the more substantial recovery of CNS functioning in the case of the musician is that the therapy time was much longer, so that more adaptive growth of nerve cells and neuronal networks and learning was possible.

8.12. Spasticity, Grip Automatism and Hand Grip Power

It has been shown in Figure 89 that hand grip spasticity can be used for holding the handle when turning on the special CDT device. Such spasticity should not be confused with the grip automatisms and volitional hand grip power. The volitional hand grip is fully controlled by the patient. The grip automatism can be activated by the patient, but is otherwise automatic. The grip spasticity is an uncontrolled activation of the hand and is a fully pathologic organization of the human CNS. In principle it is possible to have a mixture of spasticity, grip automatism and volitional hand grip power, depending on the injury.

Spasticity reduces when turning on the special CDT device because the organization of the injured CNS becomes better in the short-term memory. The hand slipped from the handle in the stroke patient (Figure 89) after six and a half minutes of turning because of the reduction in spasticity.

If the patient were able to activate the grip automatism, then the hand would not have not slipped from the handle. If a patient has only little volitional hand power and no grip spasticity and cannot activate the grip automatism, then after some time of turning his hand will slip from the lever because of exhaustion.

Therefore, in this particular case of hand grip spasticity and little volitional hand grip power the hand slipped from the handle after some time of turning. The difference between spasticity and hand grip power is that spasticity cannot be controlled by the patient and volitional power can. Another difference is that with CDT induced stroke recovery, the volitional hand grip power increases (the patient can turn longer on the device (Figure 93B)) and the grip spasticity reduces. Therefore theoretical and practical knowledge are necessary to make judgments concerning hand function in stroke patients.

8.13. Spasticity, Stepping Automatism and Walking

When a patient is walking on a treadmill, different kinds of pathologic and physiologic CNS organizations can take place (in some similarity to hand function) like extensor spasticity, flexor spasticity or rigidity besides the stepping automatism and volitional walking. In stroke patients, most commonly one leg has to be supported by the therapist to enable the patient to walk to improve CNS organization to allow the patient to walk again. Often the stepping automatism can be used to improve supported walking. The stepping automatism can often be activated by bumping with the hand on the forefoot just before the lift-off phase. The overstretching of the knee to use the leg as a crutch has to be avoided, because it hinders the re-learning of physiologic walking.

In some brain injuries the extensor spasticity can so strong that spasmolytic drugs cannot help and it is also impossible for a therapist to flex the leg even if one is trying to induce the flexor reflex.

Then only emotional reassurance may help and also the use of the special CDT device, during which the extremely spastic leg is not involved. This will reduce spasticity in the short-term memory. Extensor spasticity will reduce by turning a few thousand times on the device. Such extreme extensor spasticity has only been observed by the author in hypoxic brain lesion (see below) and not in stroke.

Spasticity can be efficiently reduced in the short-term and long-term memory when exercising on the special CDT device. The quality and strength of exercising will often increase simultaneously. Spasmolytic drugs on the other hand, reduce spasticity and reduce volitional power simultaneously (mostly by blocking the input to the motoneurons). In most cases there is therefore no indication for using spasmolytic drugs. If spasmolytic drugs are administered to patients performing CDT, then the drug concentration can be reduced or the administration stopped following phased reduction.

8.14. The Importance of Avoiding Infections

It has been shown (Figure 93) that flu hindered reorganization of the CNS. Infections, for example of the lung or bladder, generally hinder the improvement of CNS organization for a longer time than the infection itself lasts. Infections therefore have to be avoided as far as possible.

8.15. Quantification Of Coordination Dynamic Therapy (CDT)

The improvement of the organization of the injured CNS in the stroke patients was measured indirectly by the improvement of movements like the increase of walking speed (Figure 93A) or hand grip power (Figure 93B) and directly by the coordination dynamics (CD).

The CDT induced CNS recovery can therefore be quantified, which is important for the patient to motivate him to continue with intensive therapy. The therapist can use the data for adapting the therapy to the progress in CNS recovery and the research worker gets hard data to better understand the functioning of the human CNS and can in this way enhance the efficiency of CNS re-organization.

8.16. Stress

It has been observed earlier and quantified by the CD recording method that stress can reduce or destroy relearned movements in the short-term memory [88]. This stress-induced deterioration of relearned movements was also observed in the musician. When the patient played again in concert for the first time following stroke, he had difficulty holding the bow

(Figure 90M). Two days later, when the stressful period ended, the positioning of the fingers was much better again (Figure 90F).

It is interesting that stress negatively affected the bow hand movement, but did not ruin the music he played. The playing of music was therefore more resistant against stress, which may indicate that the sense for music is stored in a different way in the brain (probably more integratively) than the hand movement. It sometimes happens that a patient with a brain injury, who is no longer able to speak, may be able to sing. Singing may help left-hemispheric stroke patients to recover from non-fluent aphasia through stimulation of the right hemisphere. This finding may challenge the view that singing causes a transfer of language function from the left to the right hemisphere. Instead, both singing and rhythmic therapy (CDT is also rhythm training) patients made good progress in the production of common, formulaic phrases—known to be supported by right corticostriatal brain areas [89].

Why are some neuronal network organizations more resistant against stress than others? In the concept of coordination dynamics all network states are generated by self-organization. Only some attractors are so deep that it seems as if they are hard-wired. If stress worked with noise-like fluctuations on the attractor states, then the deeper attractors would be more resistant against stress (Figs.18,47c). A more conventional concept is that some network states are generated more by a functional organization of the neuronal networks, such as learning data at school, whereas others are more structural, such as the 'escape' automatism running. The finding that in spinal cord injury, when the intumescentia lumbosacralis is injured in which are the networks mainly responsible for generating stereotyped walking and running, the patient can still relearn walking tends to undermine the argument that at least some automatisms are generated by a hard-wired structure. However, this line of argument is not beyond question as it is also possible that structural repair including neurogenesis had occurred in this patient.

These discussions have clinical consequences. It seems so far that the relearning of phase and frequency coordination is achieved more by a functional reorganization whereas working on automatisms largely works towards the structural repair. Exposure to stress may help to distinguish between separate network organizations. Further research has to shed more light on the complexity of CNS repair.

8.17. Relative Phase and Frequency Coordination versus Phase Synchronization

In a recent review article, phase synchronization and large-scale integration has been proposed as a model for the functional organization of the brain [90]. I measured relative phase and frequency coordination between single neurons and neuron assemblies with the single-nerve fiber action potential recording method [14]. Varela et al. argue [90] that there is no direct evidence that synchrony leads to changes in behavior when it is selectively altered. I have measured in the premotor network of the spinal cord that there are selective changes of phase and frequency coordination in the firings of α and γ-motoneurons, secondary muscle spindle afferents and oscillatory firing motoneurons (neuron assemblies) when the continence functions of the urinary bladder and the rectum are altered [14]. Synchronization occurred only transiently. It therefore seems that relative coordination rather than synchronization is the basic mode of organization. Relative phase and frequency coordination also allows a much higher complexity of CNS organization than synchronization alone. When the motoneurons in the CNS oscillated in synchrony, then the whole body would shake. When just a small aspect of CNS organization is disturbed, such as the coordination among the motoneuron firings, tremor enhances strongly as can be observed in Parkinson patients (see above). Besides the use of more clinical methods like positron-emission tomography (PET), functional magnetic resonance imaging (fMRI), transcranial magnetic stimulation (TMS) and measurements of the coordination dynamics between arm and leg movements, basic methods in human neurophysiology, such as the single-nerve fiber action potential recording method also have to be used to get basic data concerning the organization of the human CNS. The organization of the human CNS must be understood at the single neuron level, the neuron assembly level and the macroscopic level in order to properly understand how the CNS generates certain functions and behaviors. Parallel experiments should ideally be performed in both humans and animals to discover whether the wealth of data obtained from animal research is of any value to human research.

9. Repair in Cerebral Palsy

9.1. Summary

Low-intensity coordination dynamics therapy, which included crawling (Figure 95A), treadmill walking, jumping on a spring-board and exercising on a special coordination dynamics therapy device, was applied for three months for four hours per week to eight cerebral palsy patients with an average age of 15 years, ranging from 7-27 years old. All patients improved. The organization of the CNS quantified by the low-load coordination dynamics between arm and leg turning movements, when exercising on the special coordination dynamics therapy device, improved by 46 ± 17% (range 33-60%) for forward and by 48 ± 15% (range 22-66%) for backward turning. The crawling, jumping and walking also improved, though not as much as the CNS organization quantified by coordination dynamics values. The motor programs of the tibialis anterior, gastrocnemius, biceps brachii, and triceps brachii muscles, measured by surface electromyography (sEMG), improved only a little. The evidence suggests that sEMG is a very suitable tool for optimizing movement performance and thus the therapy since sEMG records show under which exercise conditions the recorded motor programs are best. However, coordination dynamics, i.e., the integrative parameter for quantifying CNS organization, is better to show the progress in CNS functioning than either EMG or the simple observation of movement quality. When the patients stopped therapy, the value of their coordination dynamics, worsened 24% after six months. In one patient coordination dynamics therapy was continued intensively for further 3 months, which consisted of 20 hours exercise per week. The value of the coordination dynamics improved altogether by 85% and 82% for forward and backward exercising respectively, and simultaneously movements, vegetative (sleep) and higher mental functions (aggressivity, learning capacity) showed strong improvement. The improvement in coordination dynamics for low-intensity therapy (46%, for forward movements) and additional high-intensity therapy (85%) lie within the recovery range for stroke (70%) and brain injury (69%) after three months of intensive coordination dynamics therapy. There is therefore an indication that the CNS functioning in cerebral palsy patients can be improved by learning as much as in stroke and brain injury. The improvement in the CNS function suggests that cerebral palsy can be partly cured if intensive coordination dynamics therapy is administered for three years. It is further suggested that inabilities, including cognitive impairment, are diseases which can partly be cured rather than inabilities.

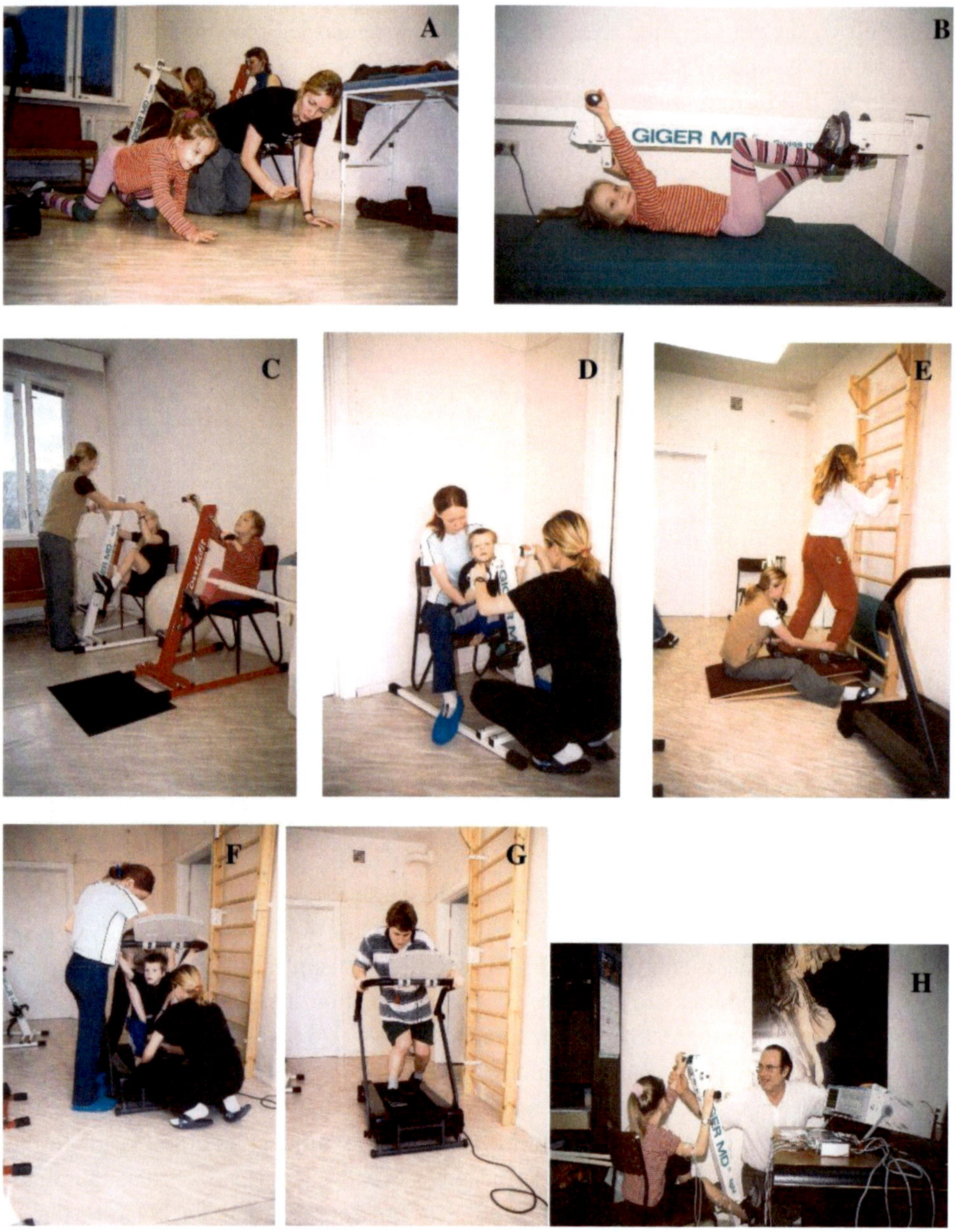

Figure 95. Movements which cerebral palsy patients are expected to perform as part of coordination dynamics therapy. A. A therapist is crawling in interpersonal coordination with a girl with cerebral palsy; her leg and trunk functions are severely impaired. The therapist is concentrating on the hands to keep interpersonal coordination with the patient and to draw her into a better performance. B. Same

patient as in A but exercising now on the special coordination dynamics therapy device in the lying position. In the lying position, the device can be used to stretch the muscle-tendon unit; the left leg of the girl is in the stretching position. C. The therapist supports the patient exercising on the special device in the sitting position to improve the performance (and the motor program of arms and legs) and to motivate the patient to concentrate on the exercise. D. A young patient with severe cerebral palsy while exercising on the special device. The hands are fixed to the leavers and the feet to the pedals. The therapist is mechanically supporting the turning and counts the turns in coordination with the arm movements. The sister of the patient's mother supports the trunk and the legs. With big effort the patient can turn a few times by himself. E. Supported jumping on a spring-board. The patient cannot keep jumping in anti-phase and mixes up in-phase jumping and jumping in rotation. F. Same patient as in D during supported treadmill walking. The mother supports the legs and the trunk, the sister of the mother holds patient's hands to the rails, so that the patient is getting specific afferent input to the hands. The therapist supports the leg movements and pushes the feet (especially the heels for inducing the stepping automatism) down to enhance the afferent input to the soles of the feet; the knees are strongly flexed to prevent extensor spasticity. The therapist motivates the patient by instructions "right – left" or by counting 1 – 2 - 3 – 4 in coordination with the movement. Note that the patient can hold the head better than in D. The patient always enjoyed this treadmill walking very much, even though he was contributing only little to the walking. G. A 19-year-old cerebral palsy patient (student) during exercising on the treadmill with no support. Note that the legs are too much flexed and the left foot rotates inward during lift-off phase. H. sEMG recordings (and coordination dynamics) are taken from a cerebral palsy patient; same patient as in B. The surface electrodes are placed on top of the tibialis anterior, the triceps surae, the biceps brachii and the triceps brachii muscles. The electrical activity is pre-amplified and displayed on the screen of an oscilloscope and printed out. The Author helps when the poorer hand slips from the lever. Patients with cerebral diseases have problems doing two things at the same time (exercising and talking).

9.2. Coordination Dynamics Therapy: Definition

Coordination dynamics therapy (CDT), applied here to cerebral palsy patients, is a movement therapy in which the CNS improves its functioning by learning the improvement of phase and frequency coordination between neuron firing while exercising on a special coordination dynamics therapy device (Figure 95B,C,D) and by learning automatisms and rhythmic dynamic stereotyped movements including, crawling, commando-crawling (Figure 95A), standing from a seated position, walking (Figure 95F,G), running, jumping (Figure 95E) and climbing staircases. Progress is quantified by improvement in the coordination dynamics value, by improvements in relevant movements and by improvement in the motor program measured by surface electromyography (sEMG, Figure 95H)) during the exercise on the special CDT device. The rate of progress may be the most important aspect to the therapy because firstly, it is difficult to motivate the patient to exercise if the improvement is too slow; and secondly, the time for therapy is often scarce, since for example the patients have to attend school, to work or to study. Therefore the CNS of the patients has to be pushed to the limits to adapt because only then can all repair mechanisms available be activated.

9.3. Coordination Dynamics Measurements Used to Quantify Functional CNS Improvement

The coordination dynamics technique measures the arrhythmicity of the turning movements during exercise on the special coordination dynamics therapy device. The device imposes changing phases between arm and leg movements including pace and trot gait coordination's. The variation of the frequency of the turning movements and its time derivative (the coordination dynamics parameter, Figures 99,101,102) quantify the quality of CNS organization. The coordination dynamics parameter is an integrative parameter to quantify CNS organization by mechanical means (kinesiology).

9.4. Movements Which May Be Used to Quantify Therapy Progress

Standard movements performed by the patients are shown in Figure 95. Exercising on the special coordination dynamics therapy device is especially beneficial, since at the early stages of the therapy it is often the only coordinated movement a child with severe cerebral palsy can perform.

If the patient cannot turn independently, because its coordination between arm and leg movements is too poor, then the therapist must support them to perform the movement. (Figure 95C) and has to motivate the patient by giving instructions such as 'right – left – right – left …'. Such supported movements improve the motor program of the patient as can be measured by EMG (Figure 100C) and in this way increases the efficiency of learning, since the CNS learns in a better organizational state. Patients sometimes say: "As soon as I get better rhythm and performance, then running becomes much easier. It is as if I have wings." Additional instructions, in coordination with the movements, make the learning more integrative in that additional auditory input and instructive learning are combined. Patients able to turn by themselves receive such mechanical and instructive support only occasionally. The performed movements are protocolled. Suitable movements are chosen to quantify progress of the therapy (Figure 98). The quality and the improvement of performance of the exercise on the special coordination dynamics therapy device can be made visible by sEMG (Figure 100).

9.5. Surface EMG to Quantify Early Re-Innervation

Four-channel surface electromyography (sEMG) made during the exercise of a healthy person on the special coordination dynamics therapy device gives a motor program like the one shown in Figure 96A. If a patient with CNS injury exercises on the device and sEMG is performed, then the motor program can show different pathologic changes (Figure 96B). The upper trace in Figure 96B shows reduction of spasticity with exercise, but no motor program. A motor program of a muscle means that a muscle shows characteristic increases and decreases of electrical activity with rhythmic movements. The bottom trace in Figure 96B shows muscle activation from clonus and an impaired motor program in the part where the clonus appears. The most interesting feature to be seen in Figure 96B is the sEMG trace from the right gastrocnemius muscle. Muscle activation in the gastrocnemius was only achieved during the exercise on the special device. The patient could not activate the muscle volitionally and no electrical muscle activity could be recorded.

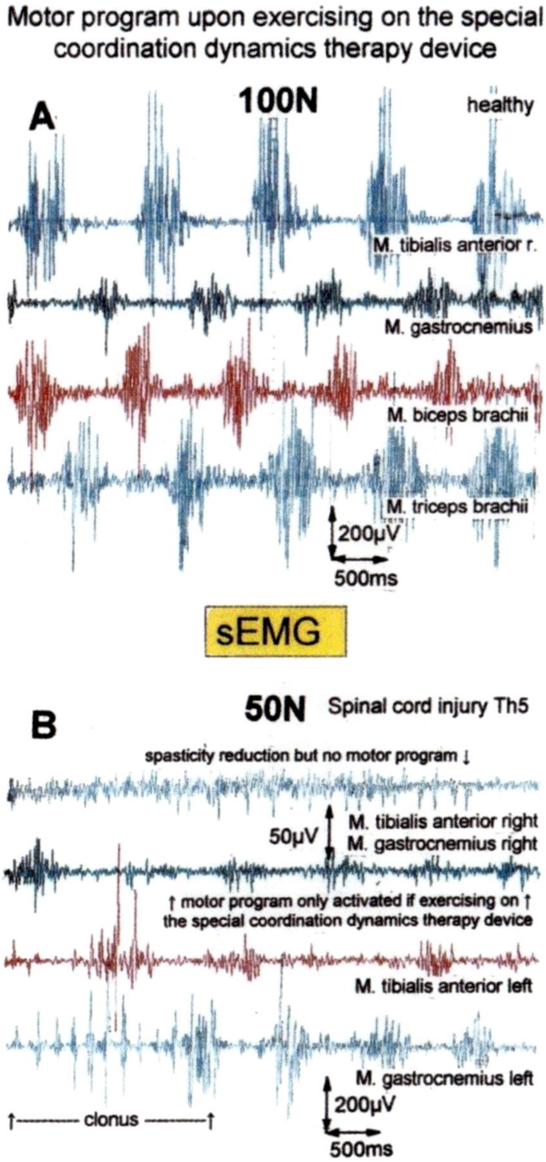

Figure 96. Surface electromyography (sEMG) motor programs of a healthy subject (A) and a patient after a spinal cord injury sub Th5 (B) upon exercising on the special coordination dynamics therapy device. A. Note the phase changes of the motor program bursts between the gastrocnemius and biceps brachii muscles due to the coordination changes between arms and legs when exercising on the device; exercising load = 100N = 100 Newton. B. Pathologic motor program. Upper trace (tibialis anterior right): spasticity reduction, but no motor program development. Second upper trace (gastrocnemius right): motor program only generated when exercising on the device (no muscle activation on volition!). Bottom trace (gastrocnemius left): Clonus muscle activation at 6.1 Hz in the left part of the trace.

This means that early re-innervation of a muscle following injury could not be measured by volitional muscle activation. But when the patient had performed about twenty turns on the device, CNS functioning improved in the short-term memory and a better movement state was built up.

As part of this better movement state, the gastrocnemius muscle was also activated. This could not be achieved by volition alone, despite many attempts. sEMG can therefore be used to optimize the therapy in the sense that movements can be trained under conditions where the muscle motor programs are best. Visual judgements of performance is sometimes not sufficient, especially in cases in which there is nearly no remaining motor program in the muscles due to severe CNS injury.

9.6. Surface EMG As an Aid for Optimizing Motor Performance in Cerebral Palsy

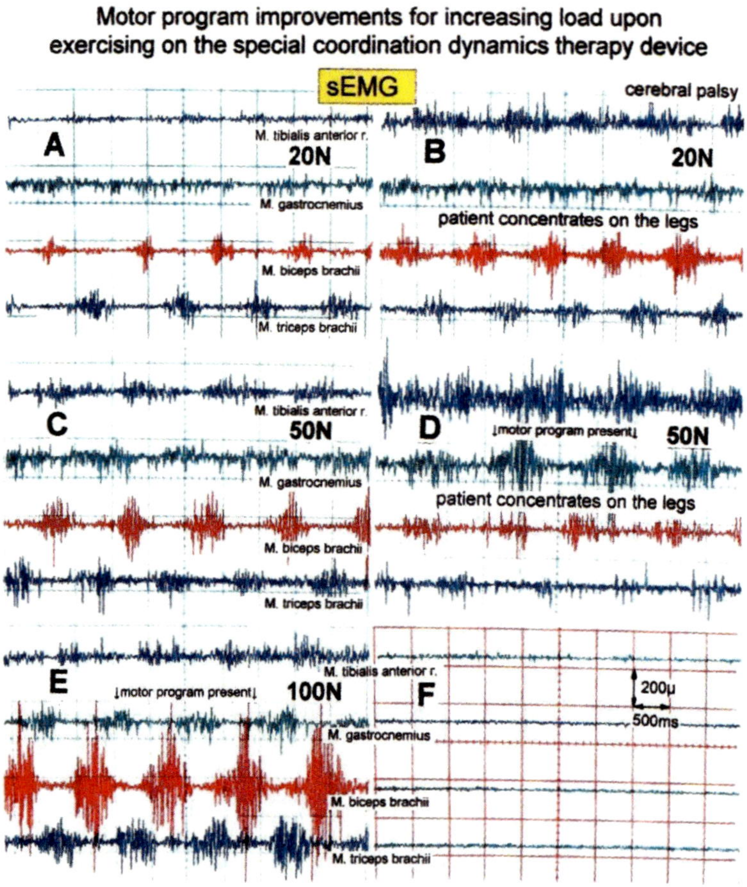

Figure 97. Quality of the motor program in a patient with cerebral palsy (and severe walking problems) in dependence on the load and concentration when exercising on the special coordination dynamics therapy device. Surface electrodes are positioned on the right tibialis anterior, gastrocnemius, biceps brachii and triceps brachii muscles. In F, the patient does not exercise. In B and D the patient concentrated on the improvement of leg movements during the exercising. Note that the motor program is best for the load of 100Newton (E), because a motor program developed in all 4 muscles, including the right gastrocnemius muscle. In D, the motor program in the gastrocnemius improved due to concentration, but the one in the triceps brachii got worse.

Another example of a motor program and movement performance optimization with the help of sEMG is shown in Figure 97. A patient with cerebral palsy exercised on the special coordination dynamics therapy device at the load of 20N. The sEMG recordings of the tibialis anterior and the gastrocnemius muscles show no motor program (Figure 97A). When the patient concentrated on the legs to improve leg movements, a little motor program was achieved in the tibialis anterior muscle but not in the gastrocnemius muscle (Figure 97B). When the patient exercised at the load of 50N, a motor program was developed in the tibialis anterior muscle but none in the gastrocnemius muscle (Figure 97C). But when the patient concentrated at that load onto the legs, a motor program developed also in the gastrocnemius muscle (Figure 97D).

Some additional continuous activity occurred in the tibialis anterior muscle. However, the motor program in the triceps brachii muscle was nearly lost. The best motor program was achieved in all four muscles when the patient exercised at the load of 100N (Figure 97E). When the patient did not exercise and relaxed no electromyographic activity was recorded (Figure 97F). From the point of view of the motor program, it would therefore be best for this patient to exercise on the special coordination dynamics therapy device at the load of between 50N to 100N.

EMG can therefore be used to optimize the performance (the muscle motor programs) during therapy and to increase the efficiency of learning. Quantification of movements with a video camera alone is not sufficient, because movements of body parts can be induced passively. Only EMG recordings show whether a muscle was really activated or not.

9.7. sEMG and Coordination Dynamics for Quantifying Improvement of CNS Functioning by Therapy

Figure 100 shows an example of the quantification of therapy progress using sEMG. However, using sEMG to quantify the progress requires repeated recording to acquire an average picture as there is significant variation. Recording of coordination dynamics avoids this problem. Coordination dynamics are recorded at a low load over twenty minutes and the best 1-min mean value is taken. Variations in motor performance and patient's concentration have only little effect on the best value during the recording. The results indicate that records of coordination dynamics are more sensitive to therapy progress than EMG measurements because they quantify the integrative performance of all activated muscles and the brain stores patterns of muscle activation.

9.8. Treatment and Group of Patients

Eight patients with cerebral palsy underwent low-intensity coordination dynamics therapy for a minimum of three months. They received four hours of therapy per week for twelve weeks. The therapy time was 48 hours in total, including approximately 100,000 coordinated integrative movements. The average age of the patients was 15 years (range 7 to 27). The progress achieved was measured by the improvement of movements, by the improvement of the organization of the malfunctioning CNS, i.e., by measuring coordination dynamics, and was evaluated in terms of the improvement of the motor program as quantified by surface electromyography (sEMG) of the right tibialis anterior, gastrocnemius, biceps brachii, and triceps brachii muscles.

9.9. Improvement of Movements

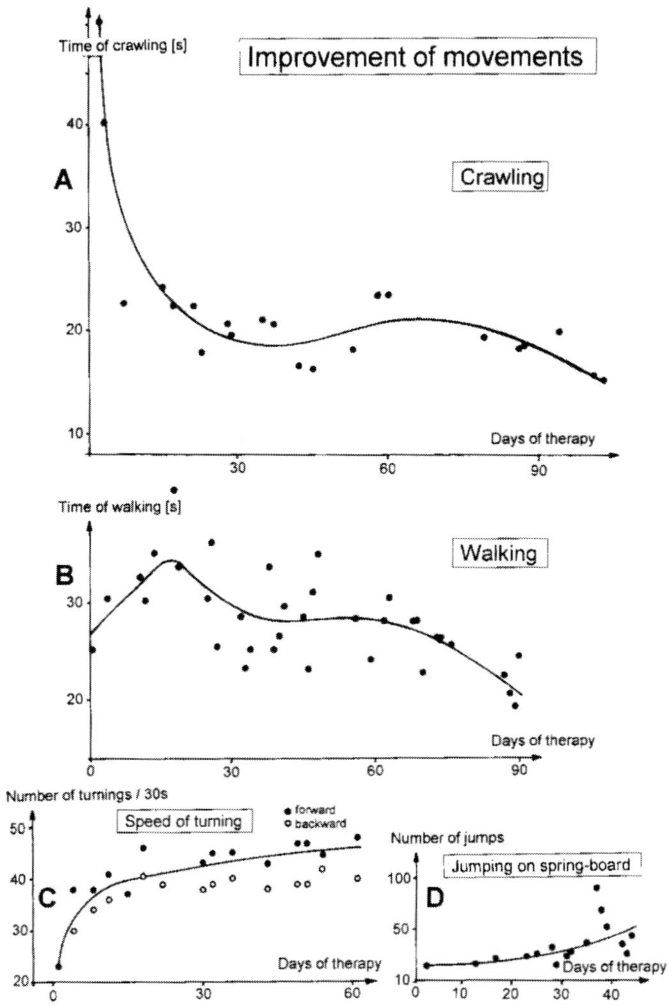

Figure 98. Examples of movement improvements to quantify CNS re-organization. A. Crawling times for 16m with ongoing therapy. B. Walking times for 16m plotted against days of therapy. C. Number of turns on the special coordination dynamics therapy device, when the patient exercised as speedily as possible during 30s. Full dots represent forward movements, open dots backward turning. D. Improvement of jumping on spring-board in anti-phase quantified by the length of the jumping series against therapy days. The dots are mean values of the first 3 series.

Movements performed during therapy sessions were quantified and protocolled. The movements used to quantify the progress were mostly crawling (commando-crawling) (Figs.98A), walking (Figs.98B), exercising on the special CDT device (number of turns per

30s, Figure 98C) and jumping on a springboard in anti-phase (Figure 98D). The crawling, walking and exercising on the device were quantified by the speed. The jumping on the springboard was quantified by the number of jumps per series.

9.10. Improvement of Coordination Dynamics Following Therapy

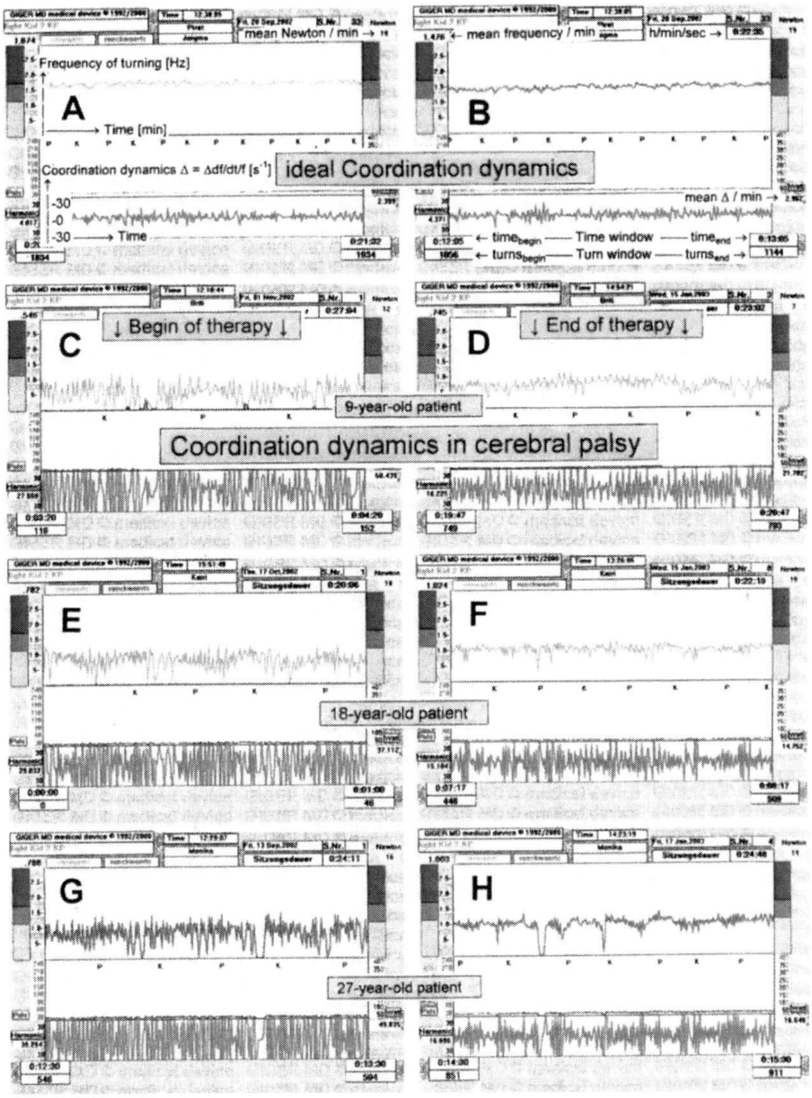

Figure 99. Coordination dynamics (frequency = upper trace, coordination dynamics = lower trace) in a healthy subject (A, B) and in 3 patients with cerebral palsy in the early stages of the treatment (C, E, G) and after 3 months of low-intensity therapy (D, F, H). In A and C-F, the person turned in forward and in B, G, H in backward direction. P = pace gait coordination between arms and legs, K = trot gait coordination. Note the improvement of the steadiness of the frequency and the reduction of arrhythmicity of turning (coordination dynamics) from C to D, from E to F and from G to H.

Coordination dynamics (Δ(df/dt)/f/min [s^{-2}]) were measured for all the eight patients at the beginning and at the end of the coordination dynamics therapy after three months. The improvements achieved in three patients are shown in Figure 99C-H. In the first patient (Figure 99C,D) coordination dynamics value improved for forward turning from 50.4 to 21.6s^{-2} (by 57%). In the second patient (Figure 99E,F) the coordination dynamics value improved for forward exercise movements from 37.1 to 14.5s^{-2} (by 61%). In the third patient (Figure 99G,H) the coordination dynamics value for backward turning improved from 49.8 to 16.6s^{-2} (by 67%). For comparison, nearly ideal coordination dynamics for forward and backward turning are shown in Figure 99A,B. The coordination dynamics values for forward and backward turning are 2.4 and 3.0s^{-2} respectively. In groups of healthy subjects the values are around 5 and 10s^{-2} [50]. The comparison shows that the patients were still far from having reasonably good coordination dynamics values after treatment. The results were in accordance with the improvement of the movements. The movements were far from normal after the treatment. For all cerebral palsy patients the coordination dynamics (Δ(df/dt)/f/min) improved by 46 ± 17% for forward turning and by 48 ± 15% for backward turning within three months (or 48 hours) of therapy.

9.11. Changes of Coordination Dynamics Following Termination or Continuation of the Therapy

The coordination dynamics were measured again on average six months after the therapy had stopped. The coordination dynamics values were worse by 24% in the patients that stopped therapy, whereas a further improvement in coordination dynamics (by 50%) could be observed in the patient who continued with therapy.

9.12. Improvements Quantified by sEMG

The improvements of movements during the exercising on the special CDT device, as quantified by sEMG through the quality of the motor programs, were very small in the seven patients for whom measurements were available. In Figure 100A,B the sEMG of one patient is shown at the beginning and at the end of the therapy. At the end only slight improvement concerned the motor program in the tibialis anterior muscle (Figure 100B). No motor program could be recorded in the gastrocnemius muscle at the beginning and at the end of the therapy, even though the coordination dynamics improved strongly (Figure 99G,H). However, when the exercising was supported, as shown in Figure 95C, some motor program was generated in the CNS for the gastrocnemius muscle (Figure 100C). With support during exercise, the improvement of the motor program in the gastrocnemius muscle was greater, in that some motor program was generated, than with the unassisted treatment (no motor program was generated in three months of unassisted exercise). This shows two things. Firstly, EMG is not very suitable to quantify subtle progress. Secondly, supported exercise is very helpful in improving the motor program in the short-term memory.

9.13. Exercising on a Bicycle Is Only of Little Help in Improving CNS Function

Figure 100D shows an EMG recording from a patient who was only turning with his legs on the special CDT device (load = 20N). This is the movement which bikers perform. Activity can be seen in all four muscles, but no real motor program was generated in the patient under these conditions. Thus, exercising on a fitness bicycle is not a suitable method to induce development of a better motor program in cerebral palsy patients and is therefore only of little help with respect to CNS functioning improvement.

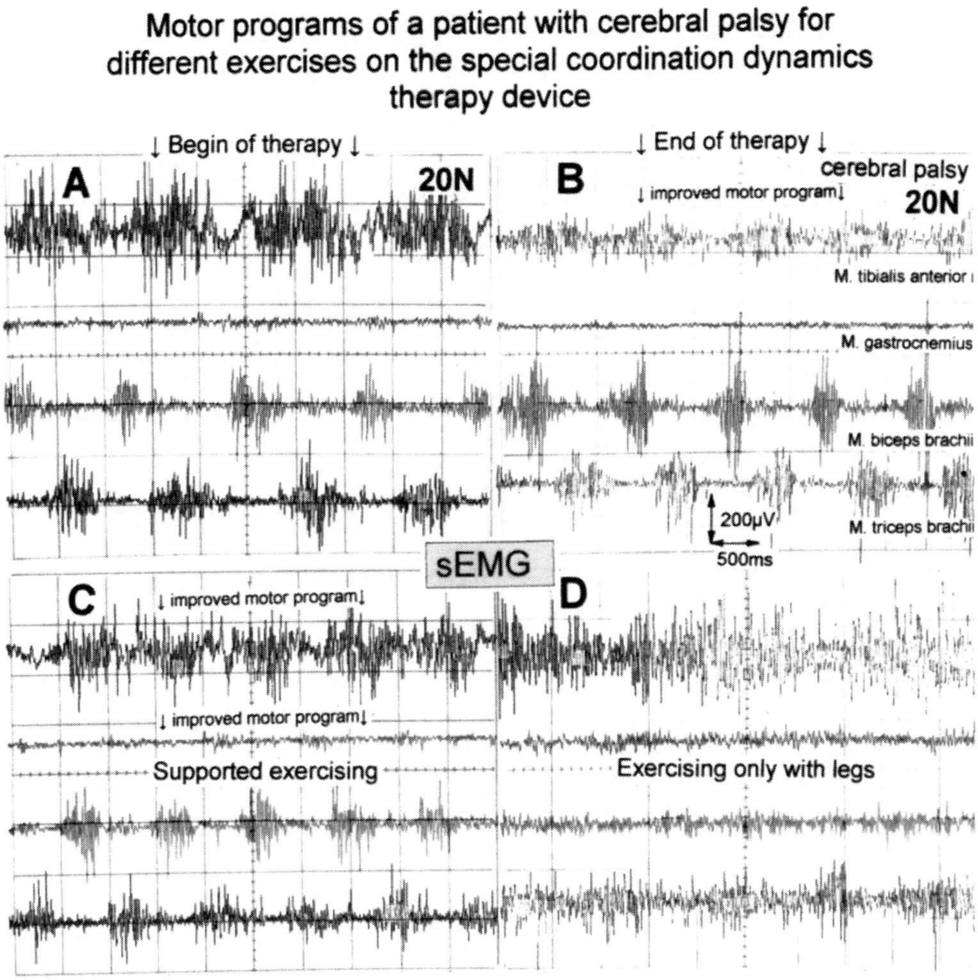

Figure 100. Motor programs (recorded by the surface electromyography technique) of a patient with cerebral palsy upon exercising on the special coordination dynamics therapy device in early stages of the therapy (A), at the end of therapy (B), when supported during turning (C, like in Figure 1H) and when exercising with the legs only (D).

9.14. Improvement in the Higher Mental Functions – Patient Karolin

In one eighteen-year-old patient with severely impaired higher mental functions. Her higher mental functions are comparable to those of a child between 1.5 and 3.5 years of age. Coordination Dynamics Therapy was continued (see Section 9.1) for further three months with high intensity, i.e., 20 hours per week. At the beginning, the patient was so aggressive that the author was afraid to motivate the patient. With ongoing therapy aggressiveness reduced and the motor functions also improved substantially. The mother characterized the improvement of the motor, vegetative and higher mental functions of her daughter as follow:

1. The first response to the therapy was negative. Aggressiveness became stronger. The patient had to be removed from school after two weeks due to aggressivity.
2. The subsequent response after one month of therapy was positive.
    a. Her daughter had suffered from sleep problems throughout her life. She used to wake up ten to twenty times each night. With the therapy she only woke up once to three times per night. This showed a clear improvement in vegetative functions.
    b. Aggressiveness reduced and she was able to bring her aggressive episodes under control more quickly. The incidences of aggression became less frequent, were shorter in duration and stopped abruptly. It seemed as if Karolin wanted to stop them and was able to stop them by herself (improvement of higher mental functions).
    c. She learned to control her body better and achieved better coordination (improvement of motor functions).
    d. The problem of keeping her balance disappeared. She is now able to walk on slippery surfaces. She became able to put her trousers on without sitting down.
    e. Since birth her right side had been weaker. It was not clear whether Karolin was right or left-handed. Now she can control the computer mouse easily with the right hand. It became apparent that Karolin is right-handed. She lost more than 20kg in weight, became fit and strong.
    f. The mental health of Karolin improved due to the treatment or spontaneously. She wants to read now. She can express herself better. She was more willing to attend school than before treatment. She was mostly willing to attend the treatment sessions. After the therapy sessions her mood was improved.

The improvement of the motor, vegetative and higher mental function falls together with an improvement of coordination dynamics. The improvement of the low-load coordination dynamics is shown in Figure 101: the 20min recordings of forward, backward and forward turning movements (from A to D and G) shows an improvement characterized by more stable continuous exercising. In A, the frequency of turning varies very much whereas only relatively little in G. CNS functioning became more stable. The same improvement can be seen in the best 1-min time intervals for forward (B to E and H) and backward turning (C to F and I). Within six months, the values of the coordination dynamics improved from 26.8 (B) to $4.0s^{-2}$ (by 85%) for forward turning and from 21.8 (C) to $4.1s^{-2}$ (I) (by 82%) for backward exercising. The frequency variation (related to frequency coordination) and arrhythmicity of turning (related to phase coordination) improved for low-load (20N) exercising, i.e., low integrative CNS organization.

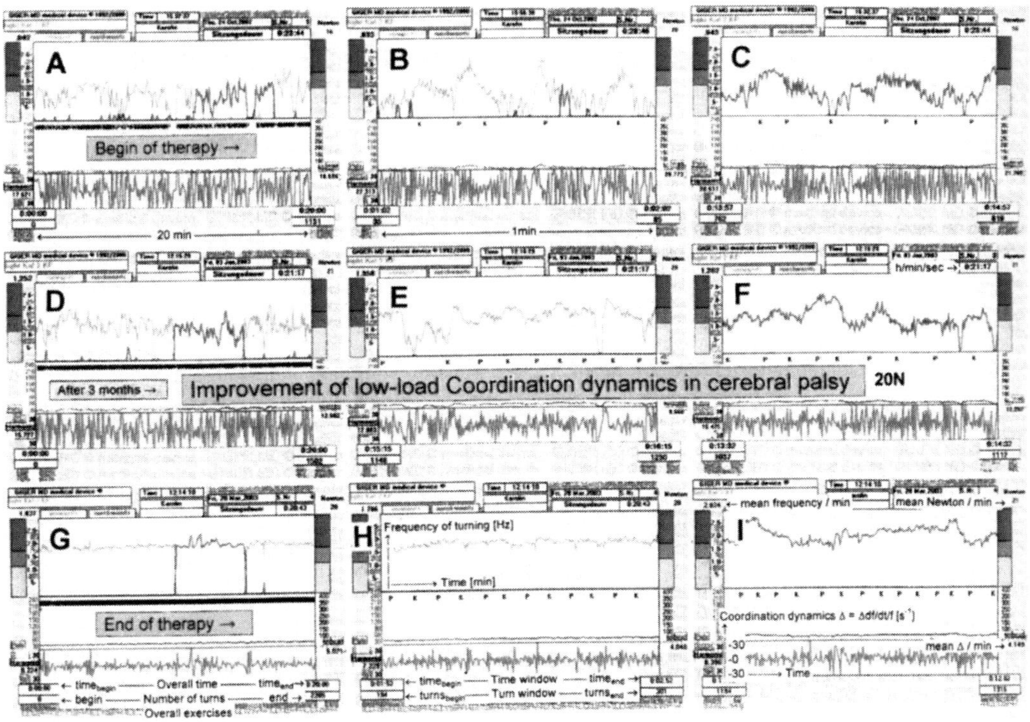

Figure 101. Low-load coordination dynamics in an 18-year-old female patient with cerebral palsy in early stages of the therapy (A, B, C), after 3 months of coordination dynamics therapy (D, E, F; 4 hours therapy per week), and after further 3 months of therapy (G, H, I; 20 hours therapy per week). A,D,G. 20-min recording of forward, backward and forward turning. Note the reduction of frequency variation from A to G. B,E,H. Frequency and coordination dynamics of forward exercising for a time window of 1min. Note the reduction of the frequency variation and the improvement of coordination dynamics from B to H. C,F,I. Improvement of frequency variation and coordination dynamics for backward exercising with ongoing therapy from C to I.

The improvement of high-load coordination dynamics is shown in Figure 102. At the beginning of therapy (A through D), the highest possible load at which the patient was able to exercise was 100N. At the end of the therapy (E through H), the patient could tolerate the load of 200N, and the coordination dynamics values for 50 and 100N were much better. The coordination dynamics for 50N improved from 31.5 (C) to 6.2s^{-2}(G) (by 80%) and for 100N from 65.5 (D) to 25.1s^{-2}(H) (62%). Also for higher loads the turning frequency became more stable and the turning became less arrhythmic.

Substantial improvements of coordination dynamics were mirrored by substantial improvements of the motor, vegetative and higher mental functions. Electromyography was not performed at early stages of the therapy since the author was afraid that the patient may destroy the equipment.

When, after six months, the therapy was reduced in intensity from 20 hours per week to 9 hours per week and was performed very irregularly, sometimes no therapy would be performed for a whole week; the coordination dynamics worsened by 25% and the aggressiveness of the patient increased again. This may indicate that the frequency and regularity of exercising is also important, since nine hours therapy per week should be sufficient to maintain the achieved level of CNS function.

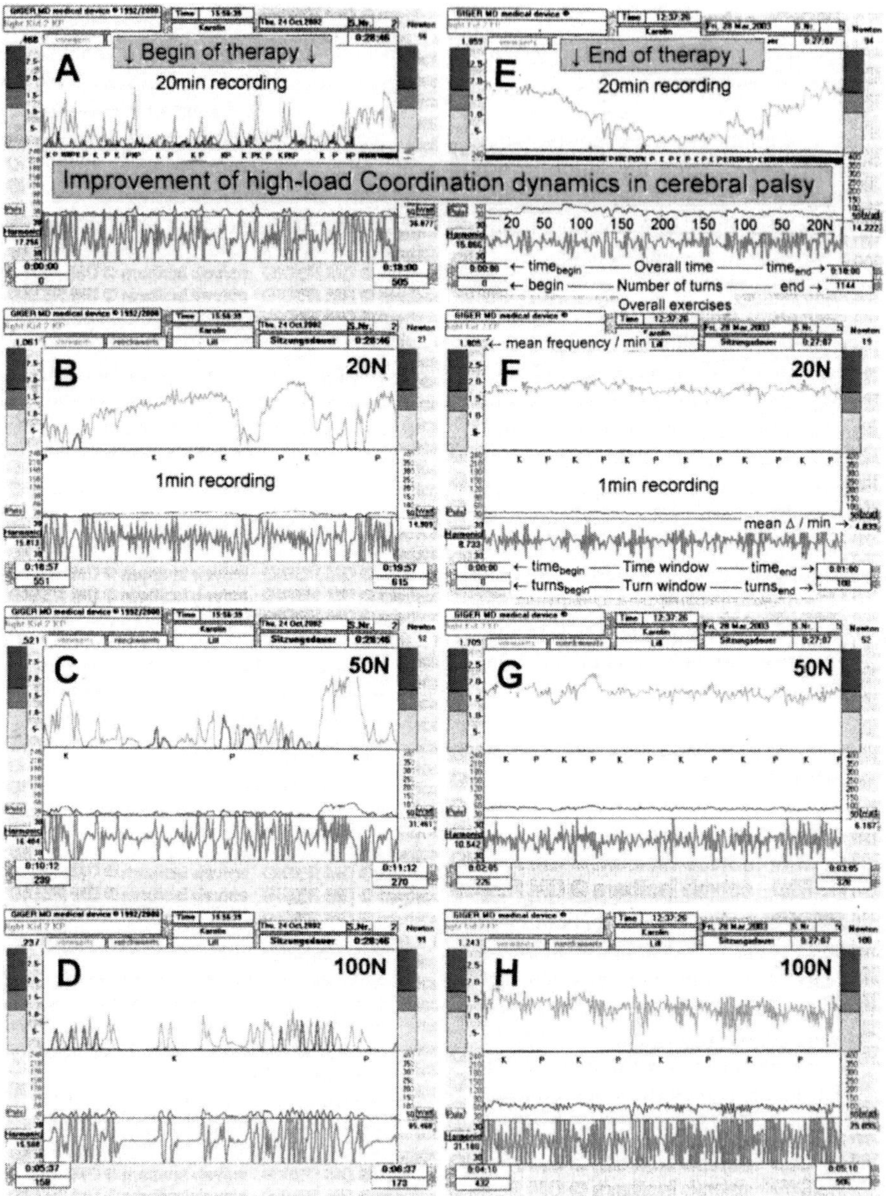

Figure 102. High-load coordination dynamics at increasing load from 20 to 100N of an 18-year-old female patient with cerebral palsy in early stages of the therapy (A through D) and after 6 months of coordination dynamics therapy (E through H). A,E. 18min recording for load increase and decrease. Note that there is no stable frequency in A, but some continuous frequency can be recognized in E. Some stepwise increase in Watt can be seen in E but not in A (Watt trace between frequency and coordination dynamics traces). B-D, F-H. High-load coordination dynamics at the beginning (B-D) and end of therapy (F-H). For the 1min coordination dynamics pairs for 20N (B,F), 50N (C,G), and 100N (D,H) there can be seen a huge improvement in CNS functioning with therapy (more regular frequency; smaller arrhythmicity of exercising (smaller coordination dynamics value)).

9.15. Cerebral Palsy Patients Improve As Much As Stroke or Traumatic Brain Injury Patients

Low-intensity (four hours per week) coordination dynamics therapy administered for three months was shown to improve CNS functioning by 46% as quantified by the values of coordination dynamics during the exercising forward movements on the special CDT device. In one case additional 3-months intensive CDT (20 hours therapy per week) was observed to result in a CNS improvement of 85% [6]. Earlier, it was reported that three months of intensive coordination dynamics therapy improved CNS functioning in stroke patients by 70% [1] and in patients after traumatic brain injury by 69% [2]. There is therefore evidence that CNS functioning in cerebral palsy patients can be improved at least as much as in stroke patients and patients who have suffered traumatic brain injury. Not only can the injured or malfunctioning CNS relearn, but it is also able to learn new movements and to improve vegetative and higher mental functions.

9.16. Adaptive Growth May Change Disabilities into Diseases, Which Can Be Partly Cured

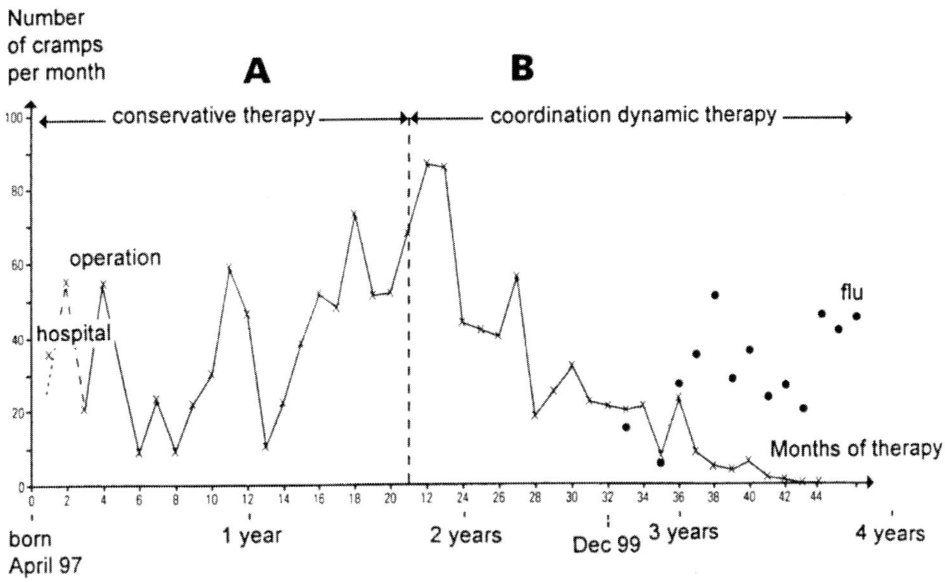

Figure 103. Reduction of affect cramps with loss of consciousness and hypoxia in a patient with myelomeningocele (spina bifida) and brain injury (hydrocephalus) during 2 years of moderately intensive coordination dynamic therapy (no drugs were administered). With the Bobath-therapy the amount of cramps increased (A); during coordination dynamic therapy the amount of severe affect cramps (with loss of consciousness and severe hypoxia) reduced until, at an age of 3.5 years, no severe cramps were occurring any more. Between 3.5 and 4 years of age only light cramps were occurring (no loss of consciousness, only hypoxia, no hand cramps). xxx: = heavy cramps, •••: = light cramps.

This work shows that the CNS of patients with cerebral palsy can improve as much as stroke patients or traumatic brain injury patients. It was shown earlier that the condition of patients with myelomeningocele could be substantially improved [40]. After three years of

CDT, a spina bifida child with a deformed and partly incomplete lower spinal cord, hydrocephalus and a total absence of hip joints, learned to walk with sticks and to speak clearly. The higher mental functions improved. When exercising on the special CDT device in the lying position (like in Figure 95B), the hip joints of the child formed without surgical intervention. Episodes of affective cramps became significant less frequent with the coordination dynamics therapy (Figure 103).

A female patient who suffered severe poliomyelitis at the age of one lost all motor functions below Th10. Thirty-six years later she started CDT by mainly exercising on the special CDT device in the lying position. After three years of intensive therapy her leg functions were re-appearing through re-innervation, her legs started to grow longer and menisci were built (for the first time in her life at the age of 40 years) [91]. Extensive adaptive growth may concern the whole body rather than the CNS alone. However, to induce adaptive growth, CDT must be intensive, which means at least 20 hours therapy per week to push the CNS and the whole body to the limits to induce all repair mechanisms. Adaptive growth consists of more than mere neuroplasticity since new nerve cells and new bone parts (menisci) can be built.

Disabilities, especially those caused by a malfunctioning CNS, do not have to remain disabilities, but can turn into diseases which can be partly cured.

10. Late Repair in Cerebral Palsy – Patient William

10.1. Case Report

Repair in cerebral palsy, as in other injuries or malformations has to start as early as possible. The question is now how much repair can be achieved if early treatment has been missed entirely?

A 24 year-old patient with a pre-natal hypoxic brain injury contacted the Author for treatment. He enquired as to whether there was any point in undertaking CDT given the age of his injury. Since every nervous system can be improved in its functioning, the Author advised the patient to start CDT as soon as possible and postpone his career. One year later, the patient postponed his career and started CDT.

Six months previously, the patient bought a special CDT device for exercising in the lying position. At 24, he consulted the Author and undertook one week of intensive therapy. A year later, the patient committed himself to one year of intensive CDT. The performed movements were mainly free walking and running (Figure 104A,B,C), walking and running on a treadmill (Figure 104D), sky-walking (Figure 104L), jumping on springboard (Figure 104E,F,G), crawling, spider walking (bear walking (Figure 200C) was not possible due to shortened tendons) and exercising on the special CDT devices in the sitting and lying position.

The biggest problems were the pathologic organization of the CNS due to hypoxia, the short tendons and muscles caused by the pathologic movement patterns and the orthopedic operation to adjust tendon and muscle lengths, which had confused the afferent input from the legs to the CNS and the caused scar tissue to form in the muscles. It will take at least a year for the pathologically functioning CNS to correct the false afferent input (Chapter X). The stretching of the short tendons and muscles may need six months if stretching is performed a few hours per day.

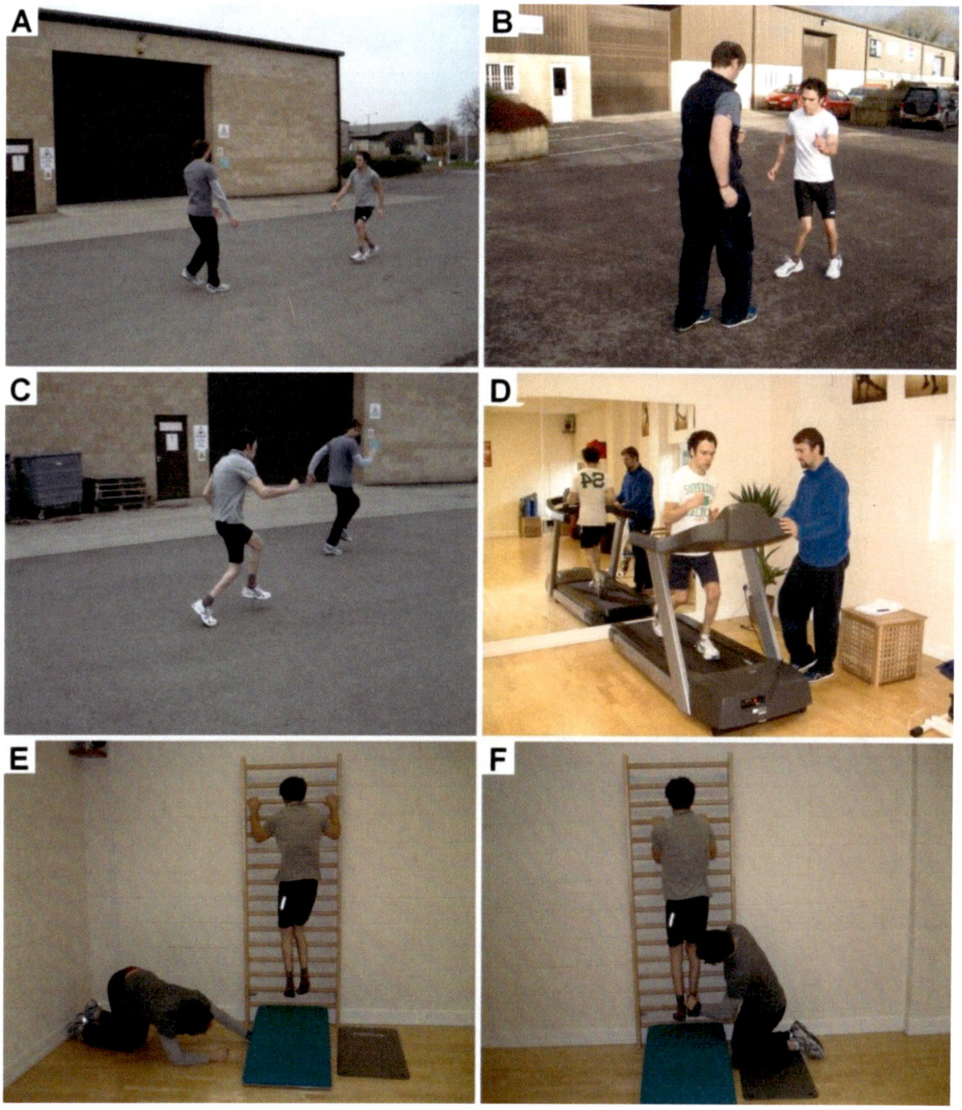

Figure 104A-F. Performed movements of a 25-year-old cerebral palsy patient (William) during CDT. A. Forward walking. B. Backward walking. C. Free running. D. Running on treadmill. E, F. In-phase jumping on springboard. In 'F' the trainer tries to push the right heal down.

After a few months of exercising walking became better (Figure 104A). As the backward walking (Figure 104B) indicates, one of the main problems was the positioning of the right leg, which was mainly caused by the pathologic movement pattern and partly by the short hip flexors, knee flexors and tendons and Achilles tendon. Backward walking is necessary to improve neuronal network symmetry and to improve forward walking. The pathologic forward walking probably became also an old-learned movement which is difficult to change.

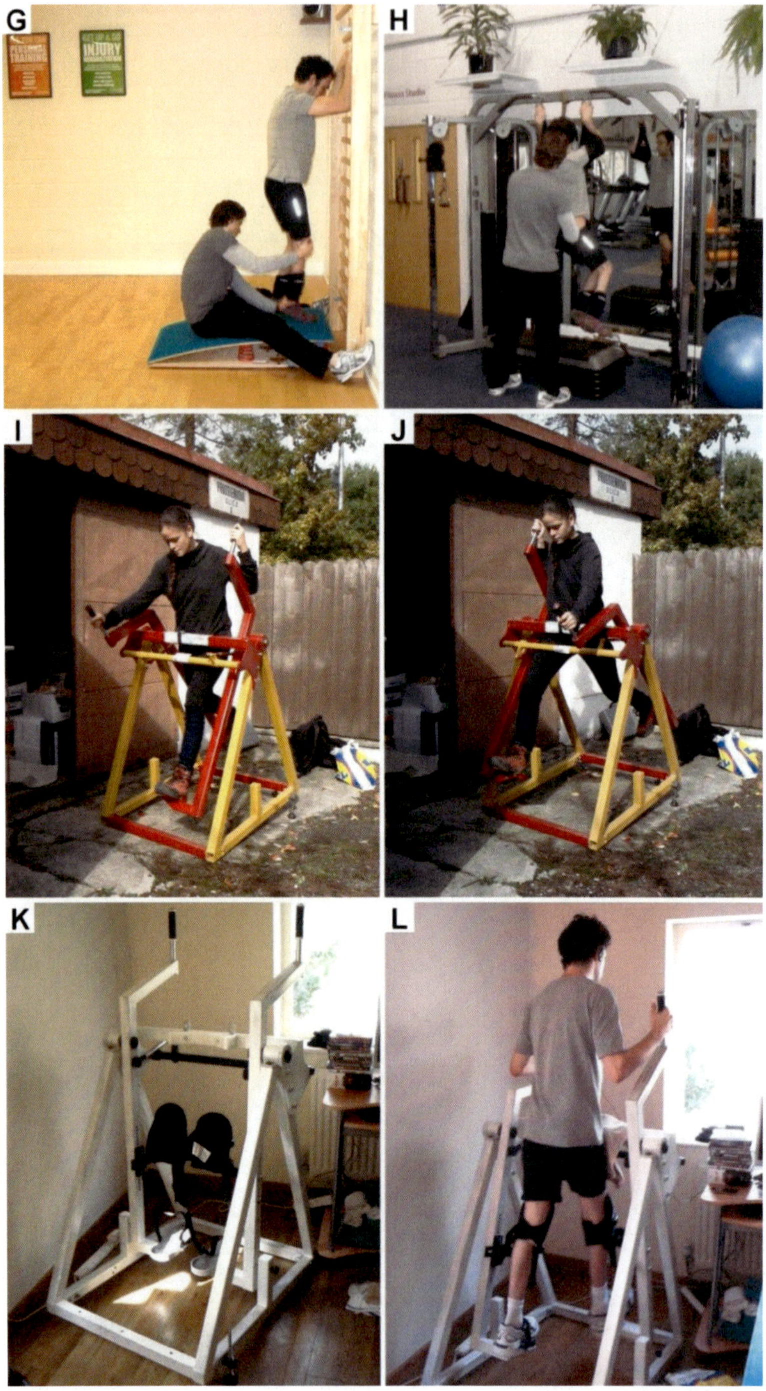

Figure 104G-L. G. Trainer tries to improve the performance of the right leg during jumping on springboard. H. Stretch rotation of the trunk. I, J. Sky-walking of another patient with a small (I) and large stride length (J). The small stride length emphasizes the stretching of tendons and muscles (I) for a more upright walking and the large stride length with a more flexed performance emphasizes the enlargement of the stride length (J) for walking with larger amplitude. K. Sky-walker with the fixations to allow the stretching of the legs and trunk. L. Sky-walking of a 25-year-old cerebral palsy patient in the stretched position. Note the upright position in comparison to Figure 104C.

The forward running in interpersonal coordination was quite good apart from the flexed legs and the flexed back caused by the pathologic running pattern including limited pelvis control. As the stretch rotation (Figure 104H) shows, the flexed back was not caused by joint and tendon problems of the back, but by the nervous system itself. The running on the treadmill (Figure 104D) was a bit different because of the safer running situation and the missing kinetic energy. As can be seen in the mirror, the right leg was dragging on the treadmill. The dragging of the foot could always be heard. Instructing the patient to lift the foot helped a little.

After a few months the patient learned to jump in-phase (Figure 104E), in anti-phase and in rotation. The performance was poor. When jumping in-phase, the body posture was not correct (Figure 104E). But when the therapist stretched the knee (Figure 104F) and dorsal flexed the Achilles tendon (Figure 104G), the jumping performance was much better. The right heel also touched the springboard. The heel strike is important, because the neuronal networks of the stepping automatism are likely activated. The effort of the trainer to support and improve the jumping performance in this manner was great, but becoming less so.

Generally, to implement the full repair possibilities of CDT, tendons and muscles must be stretched sufficiently. The numerous stretch positions for stretching are useful, but they are not sufficient in the case of adult cerebral palsy. An additional long lasting stretch became possible by using the sky-walker.

10.2. Stretching by Use of a Sky-Walker

Figure 104I,J shows the sky-walking of a patient with a mild CNS injury due to a powerful electric shock of 25000V. Additionally to the skin burning, the peripheral and the central nervous systems were injured. For a small sky-walking amplitude the knees are only little flexed (Figure 104I).

For large sky-walking amplitude the knees are more flexed. For our cerebral palsy patient both patterns are of use. The small amplitude sky-walking can be used for the stretching of the hip flexors, the knee flexors and the Achilles tendons. The large sky-walking amplitude can be used for the stretch of those muscles and tendons which limit large amplitude walking and running. For the patient, stretching of the hip and knee flexors, tendons and the Achilles tendons was employed to bring the patient in a genuinely upright position. By introducing knee fixations and Achilles tendon stretching plates by introducing an incline in the foot paddles (Figure 104K) all the relevant tendons were stretched during sky-walking (Figure 104L).

Care was taken not to overstretch and therefore avoid tendon and muscle damage. By sky-walking several times per day altogether two to three hours per day, substantial stretch can be achieved and the trot gait coordination in upright position trained.

His pathologic flexed walking because of shortened tendons and muscles in legs and trunk was counteracted by the sky-walking device which mechanically supported the patient at different points to achieve a good upright position. The performance on the sky-walker looked physiologic. The only problem was how far the muscles and tendons could be stretched without overstretching. Damage delays therapy.

10.3. Progress in Repair of the Brain Injury Quantified by the High-Load Coordination Dynamics Values

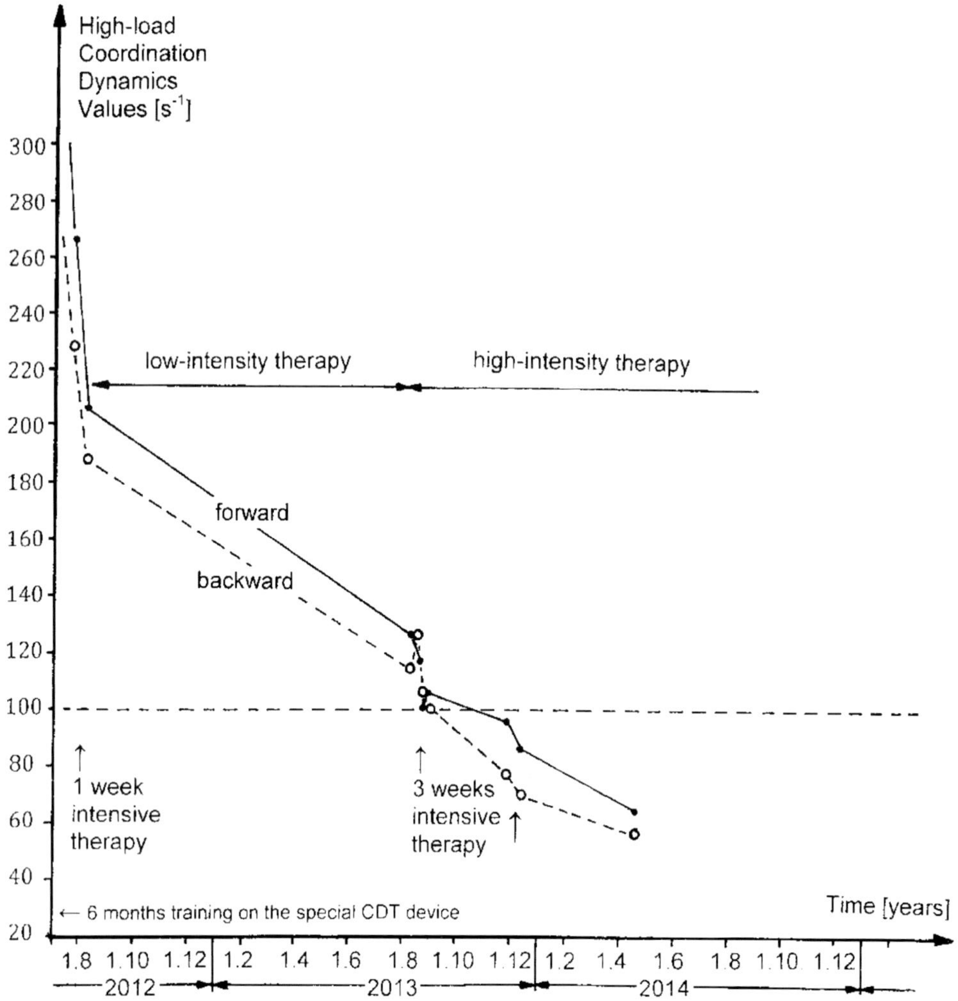

Figure 104M. Improvement of high-load coordination dynamics values in dependence on varying intensity of CDT in a 25-year-old patient with cerebral palsy. Astonishing is the fast improvement of the high-load CD values upon no intensive therapy in between August 2012 and August 2013.

The single week of treatment alone at an age of 24 brought substantial progress in the high-load coordination dynamics values (Figure 104M). With three weeks of training in 2013 at an age of 25, a high-load value of $100s^{-1}$ was reached after very intermittent training on the special CDT device for 18 months (Figure 104M). It seems as if cerebral palsy patients with high intelligence have a better prognosis. This is because they are able to devote themselves to optimal CDT, including high-intensity interval training (HIT). This patient was therefore very different to the non-cooperative cerebral palsy patient Popi (Figs.110-113; see below) with respect to motivation to train at limits.

10.4. Rate of Repair in Comparison to the Rate of Learning of a Former Athlete

A high-load coordination dynamics value was not taken when the patient first started to train on the special CDT device. However, the value was probably around 400, similar to the patient with severe traumatic brain injury (Figure 46b). By training on the device for approximately 15 months he reached a value of 100 (Figure 104M). Taking into account that the patient had already exercised on the special CDT device before being measured, the patient needed approximately 20 months to reduce the high-load value from 400 to 100. One of his therapists, being a formerly active sportsman in cricket, hockey squash and racquets, needed only five days to reduce the high-load value from 130 to under 100. Therefore, the athlete had a much lower starting high-load value and his rate of learning was faster by approximately a factor of 100 (five days in comparison to 20 months). The other therapist had an initial high-load value of 152 at the beginning, which is still a good value.

Interestingly, the low-load values of the trainer with the better values did not reach the low-load values of the patient as quickly as the high-load values. When he was performing the sports, he trained the coordination for higher loads more often than those for lower loads. When comparing the low load and high load values of athletes with those of musicians it was found that the musicians had better low load coordination dynamics values and the athletes had better high load values [50,51].

10.5 How can a Patient Have Better Low-Load and High-Load Values then an Uninjured Person?

Firstly, no CNS functions optimally. With training the person with the healthy CNS would improve his coordination dynamics values quickly. Secondly, even if the patient is exercising integrative movements, not all injured network parts can be reached and repaired because of missing or injured sub-networks. Further, the chain of events for the organization of a larger movement or other pattern is impaired. Thirdly, the movement-induced afferent input to stabilize the pattern induced by intention is pathologic. The movement-induced afferent input can be pathologic because of shortened muscles and tendons, surgical procedures which change the location of receptors and because of a pathologic processing of the movement-induced afferent input. The whole machinery for repair by learning is impaired. Thus, the movement-based learning needs much more time for learning than in a healthy person.

In any case, the therapy starts from the movements the patient can perform and goes in the direction of the movements the patient could not perform at the outset. At the neural network level this means the repair starts from the somewhat functional networks and goes in the direction of the injured or missing sub-networks. The functional building of new network parts is achieved by the generation of new sub-networks that is to say through structural repair or the re-use of existing network parts, i.e., functional repair. The majority of repair comes from functional reorganization which largely consists of changes in the weight changes of synapses.

It is likely that there are also other, as yet unknown repair mechanisms.

10.6 High-Intensity Interval Training

High-intensity interval training (HIT) was performed for example by running repeatedly 100m with five to ten minute intervals. During the intervals the patient was exercising on the special CDT device with low load or walking forwards and backwards. With repetition, the running performance would improve in the short-term memory. But the active relaxation in the intervals by exercising on the special device may optimize the phase and frequency coordination and may help further to improve CNS function in the short-term memory. The active relaxation between running by walking forward and backwards may improve the movement-based learning through giving simultaneous different patterns for learning which increase the integrativity of network learning. Whatever strategy was used, the HIT improved the running performance quantified by the reduction of the running times. Because of the highly and integrative activation of the neuronal networks of the HIT one can expect learning transfer to other patterns including cognitive functions. Jumping on a springboard and running on a treadmill with intervals was also sometimes supported by exercising on the special device during the intervals. HIT will be further discussed when treating a disabled athlete (Figure 201).

Further treatment will show whether the patient can partly compensate for the lack of timely treatment.

Chapter VI

Repair in Very Severe CNS Injury

In very severe brain injury, movement-based learning therapy is confronted with the practical limits of its successful application. The outcome is uncertain. High-load measurements are often not possible and judging progress through the high-load test is impossible. Only improvement judgement and comparison via low-load testing is possible. But conventional physiotherapy certainly has no capacity to improve such severely injured patient.

The severely CNS offers an insight into the way in which the nervous system organizes and repairs itself to an extent that a healthy CNS cannot. Therefore, severely injured patients can be of assistance to the treatment of milder injuries.

1. Hypoxic Brain Injury – Patient Anna

1.1. Case Report Introduction

This case report concerns Anna, a girl who suffered a rather severe hypoxic brain injury [8]. She had again been given up by traditional medicine and was kept inactive and a palliative approach was taken to her care. Through to coordination dynamics therapy (CDT), she could partly recover from the hypoxic CNS injury and return to meaningful life. This patient was previously discussed in an article concerning spasticity reduction achieved without the administration of anti-spasmodic drugs, back in 2002 [77]. The pictures published documented the seemingly hopeless case of a CNS injury. However, substantial recovery was possible, as will now be shown. This case report provides further evidence that the refusal of conventional neuro-rehabilitation to modernise and take account of new research endangers the lives of patients.

1.2. Case Report Summary

Intensive therapy was not offered to the patient by a university clinic on the basis that there were only primitive reflexes left. The patient underwent CDT and could significantly improve; she now can sit, walk, eat, drink, count and speak a few words. During three months of intensive therapy, a 70% improvement in CNS functioning could be achieved, as quantified by the coordination dynamics, i.e., a similar improvement as could be achieved in patients after stroke, traumatic brain injury and cerebral palsy. The improvements in CNS functioning achieved were compared with changes occurring during development. Similarities could be observed insofar as there is an instinct to transiently exercise at a high-frequency to improve CNS function, the inability to concentrate continuously on a certain task and the strong improvement of the coordination dynamics.

1.3 Case Report in Detail

During an accident the 13-year-old Anna suffered an injury of the trachea, and her parents' attempts to ventilate her were unsuccessful. A tracheal tubus was inserted by the emergency staff and proper ventilation was commenced. She had already suffered severe hypoxia. Nine months after the accident, Anna could not walk, stand, sit, speak, eat, and she was not always conscious. Anna was at a care home where she obtained care and one hour of conventional physiotherapy per week, which is not even sufficient to keep the joints mobilized. The parents wanted to take her home, but the extremely strong extensor spasticity of the left leg presented a problem, which they could not handle. Spasmolytic drugs seemed to have no beneficial effect on the extreme extensor spasticity.

It was reported in 2002 that the extensor spasticity could be reduced by approximately 68% in Anna's short-term memory by 35 minutes of assisted exercising (Figure 105) [77]. The extreme extensor spasticity in the left leg became treatable. It was proposed then that it may be possible to return Anna to her parents' home, if CDT were applied to Anna for a longer period of time.

The parents asked for CDT. But the neuro-paediatrician from the University Clinic insisted that there were only simple reflexes left. Later, Anna's parents decided to pursue CDT.

The mother went on to train with Anna, and tried to teach her to swallow food and fluid. She was repeatedly saying to Anna that it was important to learn to swallow so that the tube serving artificial nutrition could be removed. Still later, the mother undertook CDT at her home under the supervision of the Author, which consisted of supported exercising for a few hours per day on the special CDT device.

The extreme extensor spasticity, which was compromising all the other movements, disappeared. The mother could thus start to train more fully with Anna crawling, sitting, and walking.

The mother tried again to get funded CDT for Anna in another country, as this therapy was not available in her country. She showed a physician the report about Anna [77] demonstrating a strong reduction in spasticity in the short-term memory. The authorities responded to Anna's mother's request by offering her daughter a maximum of one month's conventional therapy at a certain rehabilitation centre, which she accepted. After one month of treatment at the rehabilitation centre, Anna's mother continued to train with Anna, especially walking, crawling, eating and drinking. She asked the clinic to remove the tube for artificial nutrition.

Her request was refused because of the risk of aspiration. Then Anna made a decision by her herself. While in the sauna, she pulled the tube for artificial nutrition out herself. Now she eats mainly baby food and food which can be easily swallowed, as she still has some difficulty with swallowing. Nevertheless, Anna had partly re-learned to eat and drink herself (Figure 107A,B). From 2004, she started attending a school for disabled children, where there are six teachers for seven children.

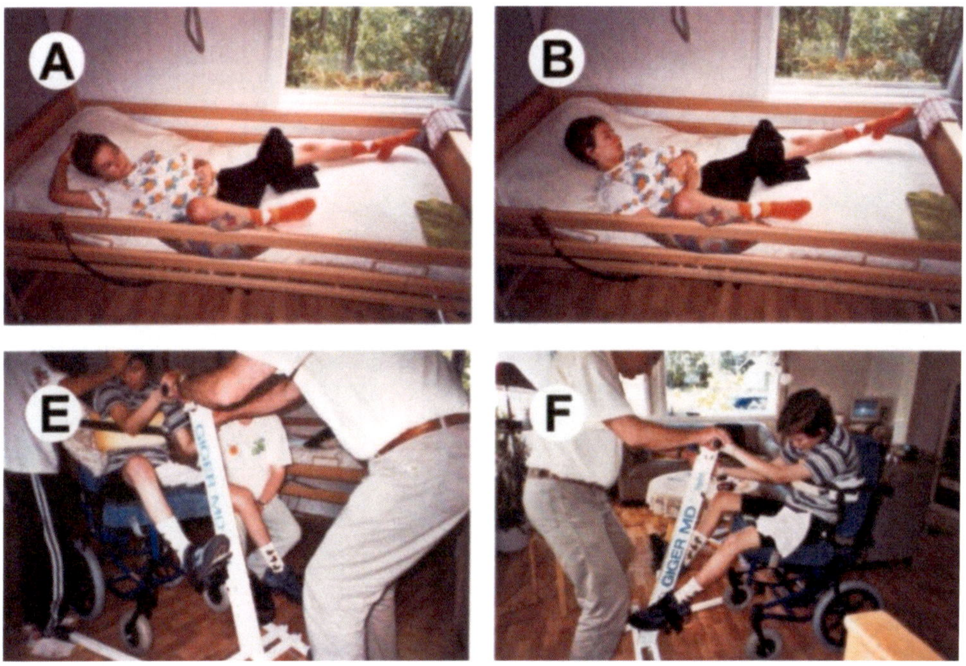

Figure 105. Patient with CNS injury due to hypoxia at the beginning of therapy. A. The typical position when the 14-year-old Anna was lying in bed, probably 20 hours per day: Extensor spasticity in the left leg, flexor spasticity in the right leg, right arm mostly lifted up, hand spasticity. B. Extreme extensor spasticity in the left leg when touching the left leg. E. Exercising with the patient, when only one leg was moving. The hands were mostly fixed to the handles. F. Exercising with the patient, when both legs were moving after non-drug induced spasticity reduction.

When the Author saw Anna in 2004 again to evaluate the functioning of her CNS, the extensor spasticity in the left leg was no longer present. Anna was able to sit (Figure 106B), walk with support (Figure 106C,D) and exercise a little on the special CDT device (Figure 106A) by herself. The functioning of her CNS had strongly improved. Later on, Anna learned to walk better (Figure 107C,D). Still sometimes extensor spasticity occurred in the left hand during walking (Figure 107D).

Anna received an additional three months of CDT at her home. The main intention was to make Anna speak again.

This goal could not be achieved. Anna re-learned to only say two words. But she seemed to be quite close to learning further words with further proper treatment. She re-learned to respond with yes or no by moving the head up and down or from right to left and backwards respectively.

Figure 106. Anna during exercising on the special coordination dynamics therapy device (A), during sitting on a chair (B), during walking, supported by her sister (C) and walking with the Author (D). Note that no extreme extensor spasticity is present any more.

Anna can clearly respond with 'yes' when asked whether she understands English (she is a Finnish girl) and she clearly can tell in this way how many lamps or chairs are in the room. The appearance of Anna's face further improved, even though it was still very different from before.

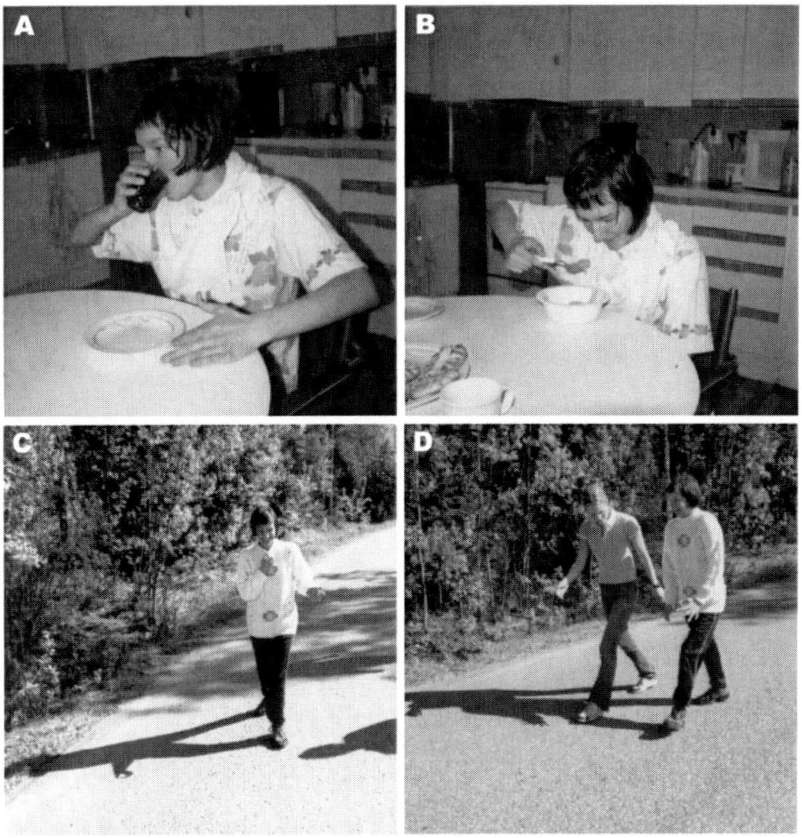

Figure 107. Anna during walking, eating, and drinking at the end of the therapy period.

1.4. CNS Improvements Achieved by a Three Month Course of CDT As Quantified by the Coordination Dynamics

During the three month therapy course, she became able to exercise by herself on the special coordination dynamics recording and therapy device. Her CNS improvements could now be accurately quantified by the coordination dynamics.

In Figure 108, original recordings can be seen of the patient with very severe hypoxic brain injury at the beginning of the three month treatment (A-C), in the middle (D-F) and one month following the course of therapy (G-I). At the beginning of this period, Anna could only exercise the easy coordination's pace and trot gait and got stuck during the difficult intermediate coordination's (Figure 108B,C), for which substantially higher CNS centres are needed to vary spinal cord neuronal networks. In the middle period (Figure 108E,F), the patient could already sometimes manage the difficult intermediate coordination's between pace and trot gait. Following the three months of therapy (Figure 108H,I), Anna could manage quite well with the intermediate difficult coordination's. At the beginning of the therapy, Anna stopped turning approximately every six seconds (Figure 108B,C) which was not only due to the coordination difficulties of arm and leg movements, but also to her difficulty in concentrating on a certain task, in this case, exercising, over longer periods of time. Also during normal communication with Anna, she was only able to concentrate on a certain subject for a few seconds. Repeated strong vocal motivation was needed to maintain her focus.

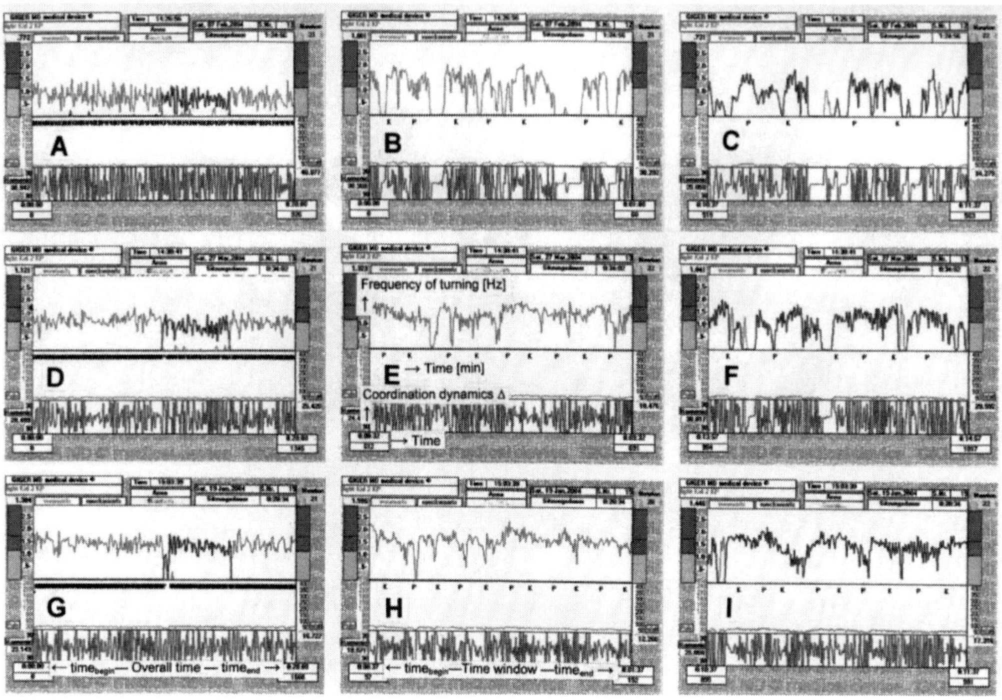

Figure 108. Low-load coordination dynamics recordings at the beginning (A-C), after 1.5 months (D-F), and after 3 months (G-I) of a 3-month coordination dynamics therapy. The recordings were done without touching Anna. A,D,G. 20 min registration: 10min forward, 5min backward, and 5min forward exercising; upper trace = frequency of exercising, lower trace = coordination dynamics (arrhythmicity of exercising). B,E,H. 1min best window of forward exercising. C,F,I. 1min best window of backward exercising. P = pace gate coordination, K = trot gate coordination. Note in 'B' and 'C' that Anna could turn somehow continuously for the pace and trot gate coordination's, but got stuck for the difficult intermediate coordination's.

Figure 108A,D,G presents the overall 20min recordings from the beginning (A), the middle (D) and after the end of the therapy period (G). The frequency trace (upper trace) clearly shows that the mean frequency of 20min forward and backward exercising increased from 0.77Hz to 1.1Hz to 1.4Hz. The variation around the mean frequency was large. The 20min mean coordination dynamics improved from 40.1 to 25.4 to 16.7s^{-1}.

1.5. Comparison of Frequency and Coordination Dynamics Changes to Those Occurring During Normal Development

As can be seen from Figure 109, the frequency (best 1min mean) of exercising increased (A) and the coordination dynamic values (df/dt/f) decreased (improved) (B). The 1min mean frequency of forward exercising (Aa) increased from 0.71 to 1.6Hz (125 %) and the one for backward exercising (Ab) from 0.75 to 1.4Hz (87 %). The coordination dynamics for forward exercising (the best 1min mean value of df/dt/f) (Ba) improved from 40.8 to 12.3s^{-1} (70 %) and the one for backward exercising (Bb) from 42 to 18.3s^{-1} (56 %). The frequency and the coordination dynamics improved more for forward than for backward exercising.

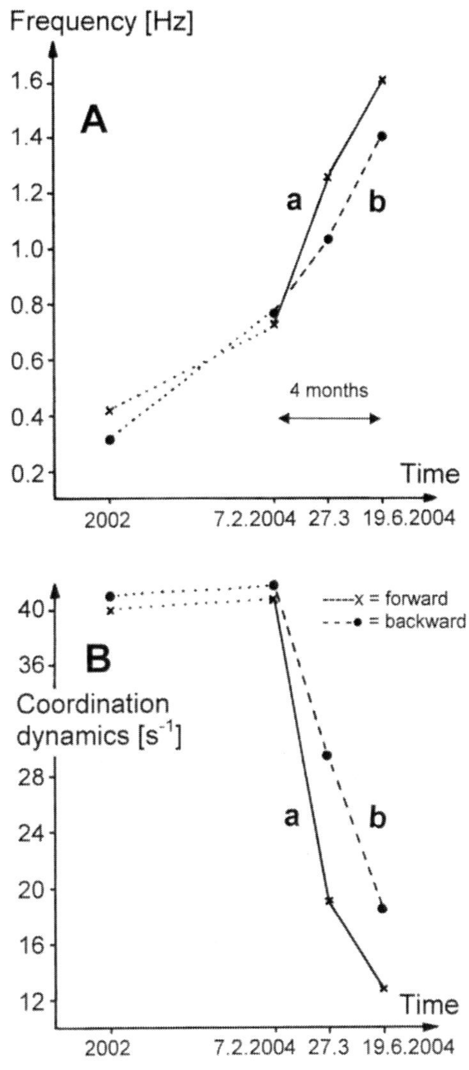

Figure 109. Exercising frequency (A) and low-load coordination dynamics (20N) (B) in dependence on ongoing therapy. The dates on the time base indicate the measurement times.

During physiologic development, the coordination dynamics values decrease (improve) (Figure 35) and the exercising frequencies increase and decrease [14 (page 100)]. The highest exercising frequency observed during normal development was approximately 1.82Hz for forward exercising. This is normally achieved at an age of approximately nine years and is most likely used for improving optimizing CNS function (as quantified by the coordination dynamics). This was not reached by Anna during the therapy period. She could only reach the exercising frequency of 1.6Hz. This frequency speed corresponds to the developmental stage of seven years (exercising frequency ≈ 1.6Hz).

1.6. Improvements after Hypoxic CNS Injury in Comparison with those Achieved after Stroke, Traumatic Brain Injury, and in Cerebral Palsy

During the three months of CDT the CNS functioning, as quantified by the coordination dynamics for forward exercising, improved by 70% in this patient with a very severe hypoxic CNS injury. After stroke, the coordination dynamics for forward exercising could be improved by 70% [1] and after traumatic brain injury by 69% [2] following identical periods of CDT. In cerebral palsy patients, low-intensity (4 hours per week) therapy administered for three months was shown to improve CNS functioning by 46% [6,78]; an additional three-months of intensive coordination dynamics therapy (20 hours per week) in one patient was observed to result in a CNS improvement of 85% [6].

There is therefore evidence that CNS functioning in patients after hypoxic brain injuries can be improved as much as in patients after stroke and traumatic brain injury and with cerebral palsy.

1.7. Improvements after Hypoxic CNS Injury on Comparison with the Changes during Normal CNS Development

Figures 108 and 109 show that CNS function in this patient improved strongly following three months of treatment. This improvement will be quantified in detail below by the increase of the exercising frequency, the reduction of the coordination dynamics values (reduction of best one minute mean arrhythmicity of exercising), and the constancy of exercising.

Improvement of the Frequency

The increase of the frequency up to 1.6Hz (Figure 109A) achieved by Anna and the frequency increase of up to 1.8Hz observed during normal development in children [14, page 100] may indicate a similarity between the improvement of the CNS functioning in Anna and that which occurs during normal development; the CNS uses high-frequency exercising to improve. At the age of nine years, girls are already able to exercise at the frequency of 1.8Hz [14, page 100].

Anna's CNS functions were similarly in certain aspects to girls younger than seven when judged by the coordination dynamics.

The age periods during which exercising at highest frequency occur (at around nine years of age) were not reached by Anna during the three month treatment period.

Improvement of the Coordination Dynamics

For forward exercising, Anna achieved a coordination dynamics value (df/dt/f) of $12.3s^{-1}$. Such value is reached in girls at an age of 12 years (Figure 35). Thus, to some extent, this situation is comparable with that observed in normal girls at an age of 12 years.

Improvement of Concentration

When comparing the arrhythmicity of exercising at the beginning of the three months (Figure 108B,C) in Anna's original recordings with those of a seven year-old girl (Figure 2B,C of [79]), some similarities can be seen with respect to the consistency of intention. The 7-year-old girl could exercise more constantly and could manage the difficult intermediate coordination's between pace and trot gait better than Anna. For forward exercising, the seven year-old child could already turn nearly continuously (Figure 2B of [79]), whereas Anna was not able to. Thus, concentration was also better in the seven year-old child. After the three months of therapy, the consistency with which Anna was able to exercise (Figure 108H,I) was comparable with, or even slightly better than that of the seven year-old child (Figure 2 of [79]).

1.8. Repair of CNS Organization through Learning

Although the constancy at which Anna was able to exercise on the special device after three months of therapy became slightly better than seven year-old children are able to, her ability to concentrate on other tasks was still much worse than that of seven year-old children. For example, she had difficulty in concentrating on conversation with the Author for more than a few seconds. This disparity between the constancy of exercising and concentration on activities and tasks in general can partly be explained by the fact that the diagnostic tool had taught her the movements, but not concentration. Had the children exercised a few times, they most likely would have achieved much better coordination dynamics values. Healthy 20-year-old volunteers were reported to improve their coordination dynamics by approximately 33% after having exercised on three occasions, one hour each [51]. The rate of learning in healthy subjects has been estimated to exceed that in patients with severe brain injury by approximately fifty times [51], which is also an argument in support of this claim. In Figure 46b it can be clearly seen that the rate of repair between healthy subjects and a brain-injured patient differs by a factor of between 20 and 50. It was that a patient with a severe traumatic brain injury could not improve his coordination dynamics any further by exercising solely on the special CDT device. However, when he went through the whole repertoire of CDT movements for three weeks, including all kinds of automatisms, his coordination dynamics improved strongly, even though during that time he exercised much less using the special device [7]. This suggests that exercising on the special CDT device is more than a complex task to be learned. It is a device to partly improve the organization principle of the CNS, namely relative phase and frequency coordination [15].

Improvement of the coordinated firing of neurons with respect to time and space probably requires many different movements with different repair strategies. In Feldenkrais therapy, a certain muscle group or neuronal sub-network organization has to be trained in at least twenty different ways [81] to functionally connect sub-networks integratively.

The integrativity of entrainment of the neuronal networks with respect to time and space seems to be logical and important for the improvement of CNS self-organization. Integrativity with respect to time means that the CNS neurons have to be activated in a certain time period to serve different organizations at the same time. If, for the sake of simplicity, we reduce the functional reorganization to only efficiency changes of synapses according to Hebb learning [80], then only those synapses which are sufficiently activated become stronger and the nonactivated ones become weaker. The, on average, several thousand synapses with their certain strength, through which single CNS neurons communicate with each other, probably have to serve many CNS organizations. The term 'movement therapy' thus does not give much information on the quality of the CNS repair methods. It is important which particular movements are trained *and* the quality with which they are trained when treating the CNS injury. It is also important to train innate movements. It was shown that the efficiency of repair can be increased by a factor of 100 mainly by changing the kind of the exercised movements [7]! To identify the type of movements with the best performance which may contribute most efficiently to CNS repair, motor programs of different movements were evaluated by means of surface electromyography [7].

1.9. Structural CNS Difference between Patients with CNS Injuries and Children during Normal Development

There are fundamental differences when comparing improvements of the CNS functioning between healthy children and patients with CNS injuries. In healthy children, the CNS neuronal networks are structurally connected physiologically corresponding to the developmental stage. It is mainly the functional connectivity that needs to be improved during development. In patients with CNS injuries by contrast, large parts of the neuronal networks have been destroyed or certain neurons more sensitive to hypoxia have been destroyed systematically and may give rise to a strongly pathologic organization, namely spasticity [82-84].

From the aspect of a mainly functional repair of the injured CNS, the CNS functioning will be different from a physiologic case because of the damaged structure. For example, some parts of the CNS may take over lost or impaired functions. If however, neurogenesis is strongly induced during therapy, also structural repair can take place in part. In young patients, in whom the CNS is still developing, a substantial degree of structural repair can be achieved. In young patients we therefore have a further repair tool, namely to influence the development itself to achieve a more physiologic structure again. This argument is supported by the success of CNS repair in children with cerebral palsy [6,78]. Some macroscopic structural repair may have taken place due to the treatment in a patient with a severe traumatic brain injury: after five years of treatment, damage to certain parts of the CNS, as imaged by MRI, disappeared [7].

1.10. Ethical Issues

Anna suffered a very severe hypoxic brain injury at the age of thirteen. Even though CDT was known in Finland at that time, mainstream medicine gave up on her. Her Mother went to the care home several times a week and attempted to help her daughter herself. After she had seen the Author on the TV, she brought Anna to the Author for a consultation. At that time, Anna was suffering from spasticity so strong that it seemed impossible to involve her in any kind of movement exercise. With occasional CDT sessions at home however, Anna's condition improved significantly. Now she attends the school for disabled children. The teachers at the school are prepared to help with the therapy. However, it was shown [7] that the efficiency of home treatment and optimal treatment at a therapy center may differ by a factor of a hundred. Therefore, proper treatment is requires a modern therapy center with up-to-date clinical knowledge. Additional treatment at school will be helpful but it cannot meaningfully substitute proper treatment. When assessed by a university clinic, Anna was considered to have only simple reflexes left; this view has not been revised even though Anna can now sit, speak a few words, walk, crawl, eat and count. The methods that conventional neuro-rehabilitation medicine used to treat Anna have thus proven to be life-threatening for her. It has been previously emphasized with respect to patients who suffered very severe CNS injuries that euthanasia has to be avoided [85].

For the time being, the available diagnostic approaches to severe CNS injuries are insufficient to forecast the outcome in a patient if effective neurotherapy is applied. Therapists and physicians must therefore be extremely careful with their decisions and must be open to learn. Seeing Anna walking in the street (Figure 107C,D), it is obvious that there are more than just simple reflexes remain to her.

A patient (Hofrat Dr. Dieter Cwienk) who also suffered severe brain injury said after partial recovery by CDT:

> "I admire the performance of the medical emergency care. The patient who suffered severe head and brain injury and is subsequently immediately and professionally rehabilitated is fortunate and may find his/her way to life again. A prerequisite for this however is a state-of-the-art and challenging neurotherapy. It exists!
>
> When discharged from the hospital, after having received, during this early stage, sufficient care only to survive, it is easy for the patient to slip into a deep hole. In the majority of cases, this means a life-long burden upon the relatives and society. The patient may not be aware of this at all.
>
> What I mean is that it is not dignified to a human being to merely vegetate somewhere, to slowly decay until he/she dies of a complication, perhaps of pneumonia because of lying in bed for so long a time. If that is what is left to us, why then, don't they leave us to die immediately?"

2. Severe Cerebral Palsy – Patient Popi

2.1. Non-Cooperative Patient

The principal obstacle to brain repair in this patient was and is that she did not and still does not understand the need to work hard to improve her nervous system. The patient learned almost entirely unwillingly. The mother had great difficulty in making her daughter train. Often her hands had to be fixed (Figure 112C,D) to make sure that she was training continuously. It is hoped that her higher mental functions will improve further so that she understands that she has to train by herself and has to become more cooperative for further improvement. EMG recordings would not have been possible in this patient, because she would have damaged the equipment. The recording of the coordination dynamics was possible, when she became able to exercise by herself. The property of being non-cooperative made this patient to be in the class of very severe brain injury. The patient Anna with very severe hypoxic brain injury (see above) could not concentrate to a certain task for more than a few seconds, but she was cooperative. She could be made to stay rather continuously in a certain patter. Autistic children are a big load for the therapist, because it is too difficult to keep them in certain patterns for movements or higher mental functions. Luckily, most patients with CNS injury are cooperative.

2.2. Case Report

Eight months after birth it was recognised that Popi, was disabled because of the occurrence of strong convulsions. The convulsions were treated. Computer tomography and magnetic resonance imaging (MRI) were performed. The given prognosis was that Popi probably would never be able to walk nor speak. One year after birth, the right hand and right leg were not functioning as well as the left. The patient was developmentally delayed and still could not speak. She could only generate vowel sounds.

She did however develop further. At an age of two, she could say a few words. Her level of function then worsened. She stopped speaking. The mother asked for a new MRI. The neuro-paediatrician refused because he thought that nothing would have changed. Following the mother's insistence, an MRI was performed at the age of four. It appeared that the cyst in the cerebrum had enlarged and there had most likely been high brain pressure for one to two years. A shunt was installed. Schizencephaly was diagnosed. The cyst was probably responsible for the high pressure. The mother was advised to have another child. However, the mother started to fight for her daughter. For two to three years, speech therapists from Greece and the USA tried to make the child speak, but without success. They concluded that it was impossible that the patient would learn to speak. At the age of six, CDT was commenced. Speech was induced and motor, higher mental, and vegetative functions improved. When she became able to perform supported jumping on springboard, full continence was achieved.

From birth the positions of the eyes of Popi were not optimal (Figure 111). This minor defect was genetically determined because others in the family also had such a defect. With CDT this pathologic eye positioning disappeared. It seems that some defects caused by genetic factors can be cured by CDT.

2.3. Cerebrum Damage in Relation to the Impairment of Motor and Speech Function

An MRI performed at the age of seven (Figure 110) seems to show a prominent open-lip schizencephaly on the right side (Figure 110A,B) by the connection of an arachnoids cyst to the ventricle system (schizencephalies are clefts of dysplastic grey matter extending from the surface of the brain into a ventricle).

At the point of connection, there is gliosis of the cerebrum from an old ischemic injury. The presentation of mental deficiencies correlates well with the presence of frontal schizencephalies.

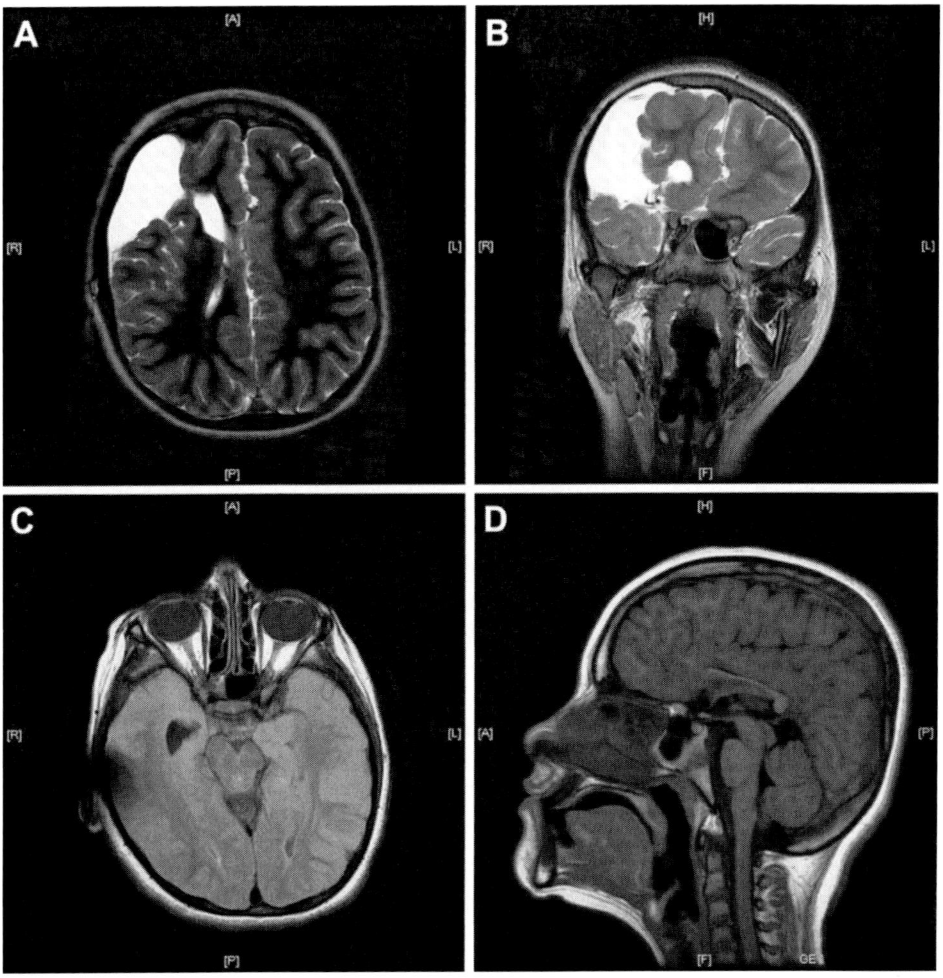

Figure 110. Magnetic resonance images (MRI) of a 7-year-old cerebral palsy child with cerebrum (A,B), thalamus (C), and corpus callosum (D) malformation and/or injury. The cerebrum malformation and/or damage seem to show a schizencephaly. But, as is indicated in the text, the real severe CNS injury with strong deficits in the higher mental functions seems to originate in the damage of the left thalamus (C). The damage of the thalamus may also impair the ascending reticular activating system (ARAS). The third ventricle cannot be seen.

Patients with unilateral clefts, however, as in this case, have a good prognosis for intellectual development, but may have seizures. From the damage to the right hemisphere, one would expect that the functions of the left arm and leg would be more impaired, since each hemisphere is concerned primarily with sensory and motor processes on the contralateral side of the body. However, the right side was more affected than the left side. As can be seen from the image of Figure 110C, the left thalamus (including the ventral part) was also injured, which is likely due to an old ischemic injury, which may have been from an unilateral thalamus infarction. Since also the thalamus projects primarily to the contralateral side, the severe impairment of the movement functions of the right arm and leg were primarily caused by the injured left thalamus and not by the damaged right cerebral hemisphere. Compensation between the right and left hemispheres was limited in this patient because of the dysplastic corpus callosum (Figure 110D). In the right-handed patient, who was only favouring the left hand more (Figure 112B) because of the impaired motor functions on the right side. The undamaged or only little damaged left hemisphere is regarded as being dominant for controlling speech and the Wernicke's and Broca's areas are sited there and should not be severely damaged.

Figure 111. Picture of the 5-year-old patient with very severe cerebral palsy. Note that the positioning of the eyes is not optimal (slight squint).

Wernicke's area, near the primary auditory cortex (Figure 47A of Chapter II of [14]) is important to understanding spoken language. It lies near the angular gyrus, which combines auditory input with input from other senses. The patient seemed to be able to understand the speaking of her Mother, but could not generate the speech pattern to answer a question. The arcuate fasciculus is a fiber tract that connects Wernicke's area to Brocas's area (Figure 47A of [14]). Since the essential step of inducing speech was achieved within three months, too short a time for a regeneration of tract fibers, it seems that the arcuate fasciculus of the left hemisphere was also not severely damaged. Broca's area initiates grammatical speech. It, in turn, lies near the vocalization region of the motor area (Figure 47A of [14]), which issues the specific commands that cause the mouth, tongue, and palate to form words. Since the patient could not initiate speech and, when she learned to speak, could only form words poorly, motor speech coordination was the main impairment. Therefore, Broca's area and the vocalization region of the motor area were not functioning physiologically. Since no clear structural damage could be found, it is likely that the inability to speak was caused by other structures, which communicate with the vocalization area of the motor cortex. Since the injured left thalamus strongly impaired the motor functions of the right arm and right leg, it is likely that the injured left thalamus was also the main cause for the patient's inability to speak.

2.4. Thalamus Injury as the Main Cause for the Inability to Speak

First of all, the thalamus is the largest subcortical collecting point for all exteroceptive and proprioceptive sensory impulses. Since this collecting point is damaged in the patient, at least on one side, we should expect an impairment of several functions. The thalamus is a relay station for all impulses arising in cutaneous and visceral sensory receptors, for visual and auditory impulses and for impulses from the hypothalamus, cerebellum, and brainstem reticular formation, all of which are processed in the thalamus before being transmitted onward to other structures. Through its reciprocal connections (feedback loops) with the motor cortex, some of which pass through the basal ganglia and cerebellum, the thalamus modulates motor functions, including those for speech. If the cerebellum is damaged, dysarthria occurs. If the thalamus is damaged, we expect also dysarthria, at least because the impulses from the cerebellum lead through the thalamus (Figure 47B of [14]). The patient showed dysarthria when she learned to speak. Further, all sensory impulses (other than olfactory impulses) must pass through the thalamus before they can be consciously perceived. Thus, the thalamus was traditionally called "the gateway to consciousness". Lesion of the left thalamus may produce pain syndromes in addition to other sensory deficits. The patient even seemed to sometimes have 'thalamic pain' (pain for which no cause could be found) and most likely had deficits in sensory processing, since even her mother was not allowed to touch her at the beginning of therapy.

The thalamus however, is not merely a relay station, but an important centre for integration and coordination, in which afferent impulses of different modalities, from different regions of the body, are integrated and given an effective coloration. From the damage of the thalamus we expect not only an impairment of the phase and frequency coordination of neuron firing, which occurs in every CNS injury, but also a lack of coordination of the specialist areas including the vocalization region of the motor area, necessary for the generation of speech. This mal-coordination may have contributed substantially to the inability to speak. It is conceivable that the improvement of the neural coordination among neurons, neural ensembles, specialized sub-networks and the improvement in the coordination of the different modalities of sensory inputs lead to the induction of speech in the patient because of her exercise on the special CDT device. Some thalamic nuclei, especially the centromedian nucleus, are also components of the ascending reticular activating system (ARAS), a specific arousal system originating in nuclei that are diffusely located throughout the brainstem reticular formation. Activating impulses from the ARAS are relayed by thalamic nuclei to the entire neocortex. An intact ARAS is essential for normal consciousness.

2.5 Impairment of the Ascending Reticular Activating System (ARAS)

Since the ARAS was impaired at least in the left thalamus, we have to expect impairment of normal consciousness and higher mental functions. The patient showed changing consciousness, dysphoria, and impaired higher mental functions. When the patient learned to speak, she sometimes had a healthy looking face, spoke well and was cooperative. At other times her face appeared unhealthy; she would shout and fail to cooperate. During therapy, it was not possible to keep her continuously in the healthier network state, despite parental intervention (Figure 112C).

2.6. Dysplastic Corpus Callosum and Right Cerebral Hemisphere in Relation to Emotions in Speech

Affective aspects of language are represented in the right hemisphere (this was where the schizencephaly was sited in this patient) and their anatomical organization mirrors the organization of the cognitive content of language in the left hemisphere [96]. Damage to the right temporal area homologous to Wernicke's in the left hemisphere leads to disturbances in comprehending the emotional content of language. Although this pattern of localization appears to be innate, it is by no means completely determined. Young children in whom the left hemisphere is severely damaged in early life can develop an essentially normal range of language function [96]. In spite of the malformation of the right hemisphere (Figure 110A,B) and the corpus callosum (Figure 110D), which connects the two hemispheres, the patient's speech contained emotion. This again supports the view that the severe deficits in her higher mental functions did not originate in the two hemispheres and their connections, but in the thalamus or other brain stem nuclei.

2.7. Coordination Dynamics Therapy (CDT)

At an age of five CDT was commenced (Figure 112). Initially, the therapy was not optimal and not sufficiently intensive. The patient could not turn independently on the special CDT device. Her exercising had to be supported by her mother. The patient was not cooperative. Her Mother, therefore, had to make her cooperative by performing a Punch and Judy show (see Figure 112C). The right leg improved, but the patient did not learn to speak. At the age of six, a three month course of intensive therapy was administered, which amounted to twenty hours per week. When the patient's CNS improved sufficiently to enable her to turn by herself on the special device, she started to speak. Her repertoire consisted of approximately one hundred words. She began to form sentences as taught by the mother. It is concluded that the patient started to speak through learning transfer from coordinated movements, i.e., turning on the special CDT device and supported treadmill walking.

This progress can be expressed in the following way through the System Theory of Pattern Formation for Repair. The attractors for walking and exercising on the special CDT device, and also the attractor layout in general, took a more physiologic form. With the deepening of their attractor basins, improving the stability of these motor patterns, an attractor for speech was built anew or its stability sufficiently enhanced so that the speech pattern could be intentionally recruited by her, enabling her to speak.

When the patient became able to exercise on the special device by herself, coordination dynamics measurements could be taken. The coordination dynamics (CD) values for 20N forward and backward exercising were $\Delta(26N) \approx 40$ poor (Figure 113). A healthy child of her age has values in the range of 12 and 14 (Figure 35). A healthy grown up person has, for forward and backward exercising, approximate CD values of 5 and 10 respectively [50]. A trained person, in this case the Author, has coordination dynamics values of between 2 and 4. Even though the patient had plenty of arm and leg strength, the CD values improved only slowly (Figure 113), indicating that her CNS improved in its functioning only slowly since profound reorganizations had to take place because of the severe damage/malformation and the severe pathologic organization.

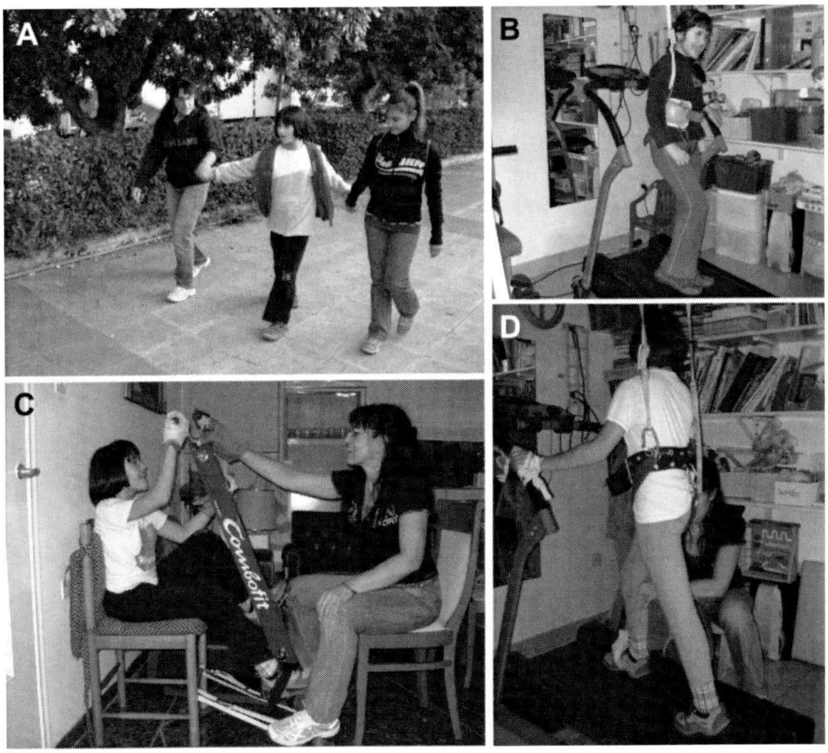

Figure 112. Supported coordinated movements performed by a cerebral palsy child with cerebrum, thalamus, and corpus callosum malformation and/or damage. A. Free coordinated walking with mother and sister. B. Treadmill walking in the backward direction. C. Exercising on the special coordination dynamics therapy device, which imposes exactly coordinated arm, leg, and trunk movements. Hands and legs are fixed, because the patient is not cooperative. When the patient is looking to the doll, her CNS is not only getting very coordinated movement-induced afferent input from arms, legs, and trunk into her CNS, but also coordinated visual input from the movements of the hands. Note, for treadmill walking in the forward direction (D), the mother had to support the right foot and knee for proper performance; for backward treadmill walking (B), no leg support is needed.

2.8. High-Load Exercising on the Special CDT Device

To improve CNS functioning deep in the complexity of CNS organization, the patient had also to turn against higher loads on the special CDT device as much as possible. Over the years of therapy the patient learned to turn also against a load of 100N. But with increasing load the patient became less cooperative and started to complain. The mother was able to make her exercise up to 100N, but the quality was low. Therefore, a real high-load test was not possible. A direct comparison of rates of repair was therefore also impossible. As can be seen from Figure 113 there was only little improvement in the coordination dynamics values for 20 and 50N.

Even though the coordination dynamics values did not improve substantially she became fully continent, speech improved and also her social behaviour improved (see also Chapter II of [14]). It is probable that minor improvements in CNS function in this case of very severe CNS injury could not be measured because of structural neuronal network changes due to repair and in turn the reduction of CD values. Therefore, the improvement in CD values was perhaps obscured by simultaneous structural repair which served to worsen the values. Even if coordination dynamics values do not improve with treatment, some CNS repair may still be taking place.

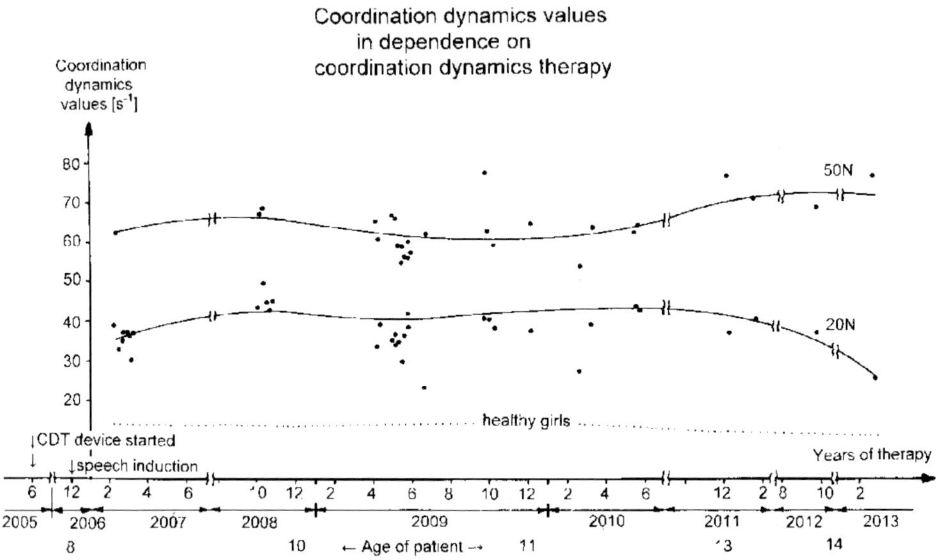

Figure 113. Coordination dynamics values for 20N and 50N of a girl with severe cerebral palsy in dependence on therapy over eight years. The age of the girl is given under the time abscissa. Note that the values for healthy pupils for 20N (dotted line), exercising the first time, are much better (lower).

2.9. Therapy-Induced Structural Repair Disturbs the Stability of Mature Circuitry Transiently

In animal research it was stated that newly build neurons in mature circuits can be integrated into neuronal networks without disturbing stability of mature circuitry: "Adult-born neurons exhibit the same classic critical period plasticity as neurons in the developing nervous system. The transient nature of such enhanced plasticity may provide a fundamental mechanism allowing adult-born neurons within the critical period to serve as major mediators of experience-induced plasticity while maintaining stability of the mature circuitry" [97]. This assertion does not accord with human research. In an adult spinal cord injury patient the coordination dynamics values worsened with each improvement in CNS function and likely with each structural change. Such transient organizational disturbances are quickly functionally repaired with ongoing CDT (Figure 30).

2.10. Improvement on the Cooperation of the Patient

The most significant challenge presented by this patient was that she was not cooperative. However, after six years of treatment there was evidence that the patient had become more cooperative. For short times the patient even exercised a bit with the Author, not just her Mother.

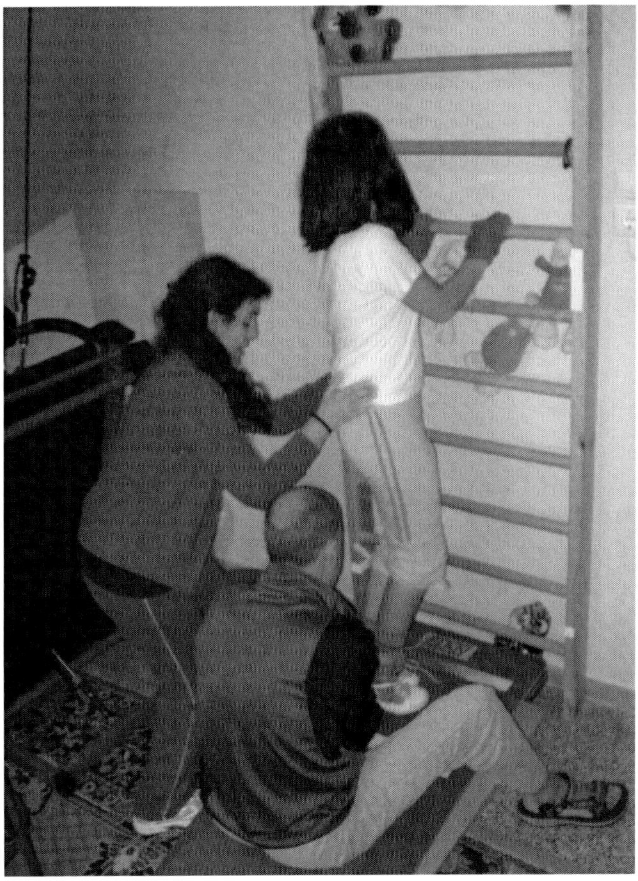

Figure 114. Jumping on treadmill of a 14-year-old patient with cerebral palsy. At the beginning of learning the jumping two persons were needed for support.

The patient began to respond light-heartedly to some instructions. However, the patient still could not tolerate supported exercising, designed to help the patient to turn more smoothly to improve the movement-induced afferent input for improved movement-based learning. Her mood changed from laughter to anger as could be seen from the change in facial expression (Figure 49B of [14]). Cerebral palsy patients are mostly cooperative because they feel the improvement in CNS function in the short-term memory. Babies, once used to the equipment, are often more cooperative than healthy young infants. Conscientious mothers can be extremely helpful in the therapy process and should be included as much as possible.

The ultimate objective was to make Popi cooperative and to allow her to attend mainstream school with healthy pupils, so that she can learn from healthy children, has not so far been achieved despite many years of treatment.

According to the system theory of pattern formation, behavioural information has to be changed, which means additional movements have to be trained to reach the deficient structures. These movements consist of exercising on the sky-walker, jumping on springboard Figure 114), ideally in anti-phase, or even riding a horse. Intensive therapy should certainly be performed during puberty where changes do occur in the CNS. If therapy is stopped during this critical period of development, gained or repaired CNS functions may be lost again.

3. Severe Traumatic Brain Injury on Children – Patients Benjamin, Mario and Andrej

3.1. Rapid Improvement Following Severe Brain Injury on Three Children where Therapy is Commenced in the Vigilant Coma Stage

Three boys, aged nine, twelve and fourteen years old (Benjamin, Mario, and Andrej, respectively), suffered severe brain injuries almost at the same time, two in a car accident and one in a bicycle accident.

In the nine and fourteen year-old patients, intensive coordination dynamics therapy (CDT) was started approximately five to ten weeks after the trauma in vigilant coma stage. The two patients recovered unexpectedly early from the coma and progressed quickly in their recovery of motor functions.

They re-learned running after four months of therapy, even though their locomotor functions were still far from normal. The twelve year-old patient Mario did not obtain intensive CDT, only conservative physiotherapy because of many infections and complicated bone fractures, which were not invasively reconstructed.

The patient recovered much later from the coma (six months as opposed to six weeks) and had severe extensor spasticity (Figure 115E), shortened tendons, problems with several joints, pain, reduced mental functions, grasp reflex in the right hand (Figure 115F, reappearance of infant automatism) and nearly no useful motor functions in his legs. Mario received intensive CDT following a delay of five months. The outcome of each of the three patients will be compared.

It is common knowledge that in brain injury, and most likely in all CNS injuries, including spinal cord injury, rehabilitation should be started as early as possible following the trauma to avoid pathologic reorganizations. If possible therapy should be started in the vigilant coma.

For an overall view of the progress seven to nine months after the injury, all three patients are shown together in Figure 115A. The two patients Benjamin and Andrej who obtained CDT for six months re-learned walking and running (Figure 115A,B) and cognitive functions improved. Mario, who obtained no real CDT for the first seven months following the trauma, could not walk or run and had to be carried from his wheelchair to the training device (Figure 115C). He could hardly stand due to extensor spasticity (Figure 115E) and supination of the feet (Figure 115G). Running improved in the two patients Benjamin and Andrej who received CDT. For the nine year-old Benjamin the running times over 22 m reduced over eight months from 14.4s to 5.2s and for the fourteen year-old Andrej over six months from 18.5s to 6.4s (Figure 116).

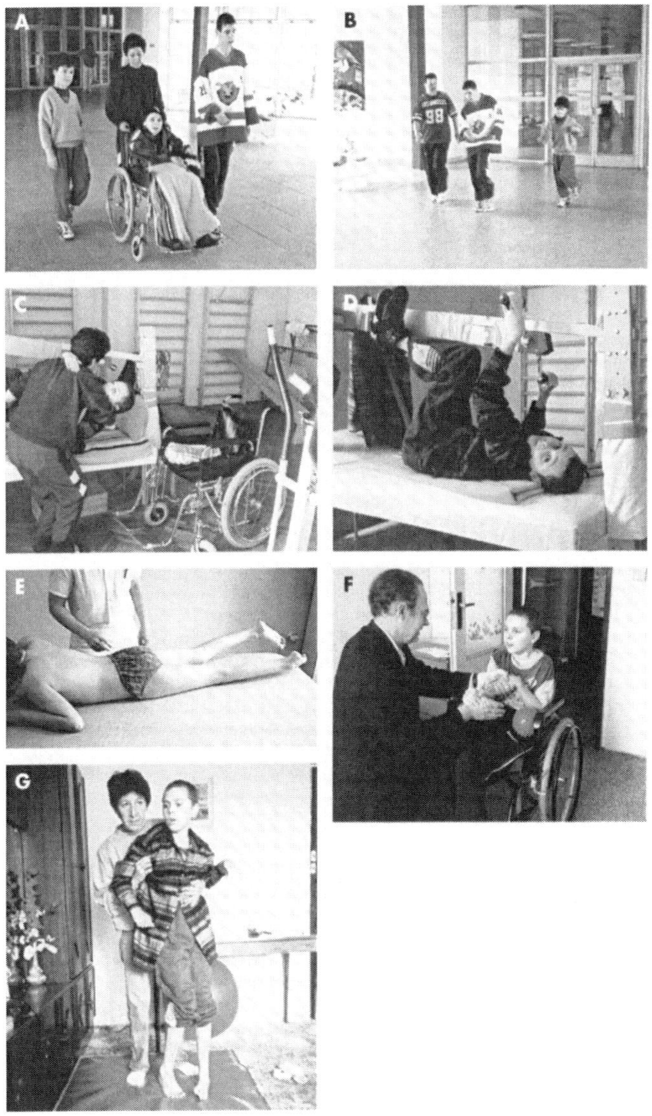

Figure 115. Three patients with severe brain injuries treated at the same time: 9-year-old Benjamin (A, left), 12-year-old Mario (A center, in the wheelchair pushed by the mother), 14-year-old Andrej (A, right)), 7 to 9 months following the accident. The 2 patients who received coordination dynamic therapy for 6 months can walk (A) and run again (B), and the third patient who did not receive coordination dynamic therapy for 7 months following the lesion is still in wheelchair (A) and has to be carried to the coordination dynamic therapy device by the mother (C). In A, the mother is pushing the wheelchair with her child; the mother of the 9-year-old boy and the father of the 14-year-old boy are in the background. In B, the father of the 14-year-old boy is running, for safety reasons, along with his son and is running in interpersonal coordination. D. The 12-year-old Mario (right-sided hemiparesis; quadruparesis) during exercising on the special coordination dynamic therapy device. The right hand is holding the hand-hold quite strongly, probably because of the induction of the grasp reflex (F) in addition to volitional power. E. Extensor spasticity activated in the 12-year-old Mario; the extensor spasticity is only moderately hindering the therapy. F. During shaking both hands with the author, the patient develops relatively strongly the grasp reflex in the right hand (from the hand reaching position the hand slips into the position of holding the thumb of the author; it could be felt how the network organization slipped from the neuronal network state „hand giving„ into the network state „grasp reflex„ in similarity to Figure 107). G. Supination of the right foot during standing (Mario).

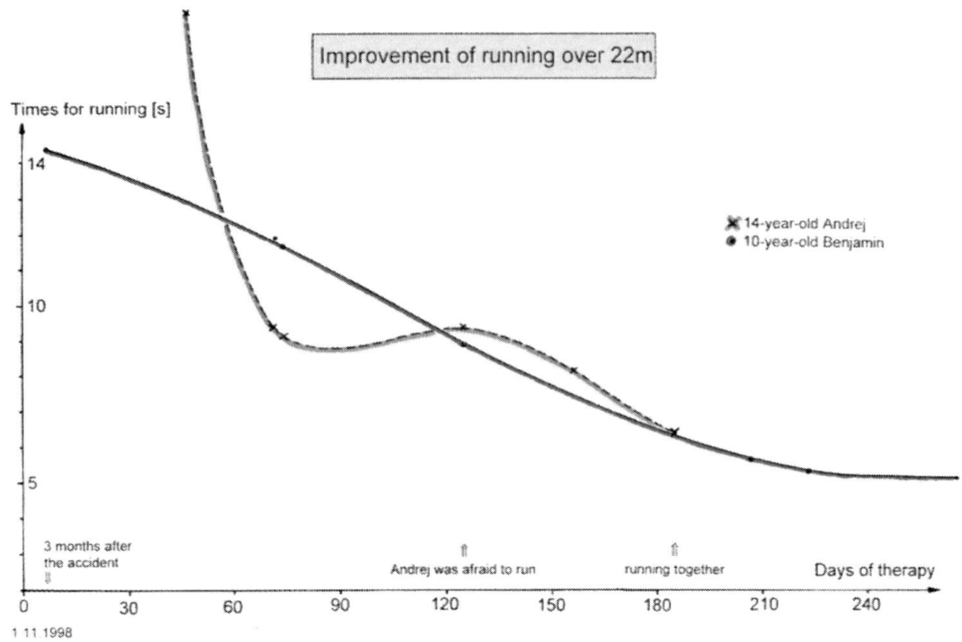

Figure 116. Improvement of 22m running times of the 10-year-old patient Benjamin (severe brain injury caused by a car accident) and the 14-year-old Andrej (severe brain injury caused by a bicycle accident). The transient increase of the running times of Andrej were due to a falling and the fear of the patient afterwards to fall again.

The jumping in anti-phase in Benjamin increased from 25 good quality jumps in succession to over 250 and in Andrej from 7 to 55 (Figure 117). An assessment of the CD was not available at the beginning of therapy.

Benjamin was running whilst playing in the street and was hit by a car driven at 55 km/h. He was thrown several meters through the air. The collided with an iron fence and suffered many injuries, including an open thorax injury on the right side and brain damage of the basal ganglia domain (area = 4 cm x 3 cm) and the frontal lobe. The patient was in a coma for two weeks and subsequently showed some response to stimuli. The intensive care unit medical team forecasted that he would be in vigilant coma for a further one to two months. After ten days of vigilant coma (≈4 weeks after the accident) CDT was started in the recumbent position. Treadmill walking supported by four persons seemed to frighten the patient (Figure 118).

The patient seemed to have left sided hemiparesis; the left arm and the left hand seemed to show signs of spasticity. Due to a humerus fracture, the shoulder joint could only be moved a little. Even though the movements in the left arm and left hand seemed to be painful for the patient, he was cooperative. Because he could not speak, meaningful communication was not possible. When the Author left the patient after three days of therapy, he was stroked by the patient, which may mean that he was thankful for the therapy, even though the therapy caused him much pain.

The CDT was continued by a physiotherapist and the mother, herself a physician. After a further three days, speech returned and the vigilant coma phase ended. The patient said that then he had quite a lot of pain in the left arm and hand during the movements. But because he felt that the therapists and the mother wanted to help him, he helped actively during the arm and hand movements as much as possible, in spite of the pain and being frightened (Figure 118B).

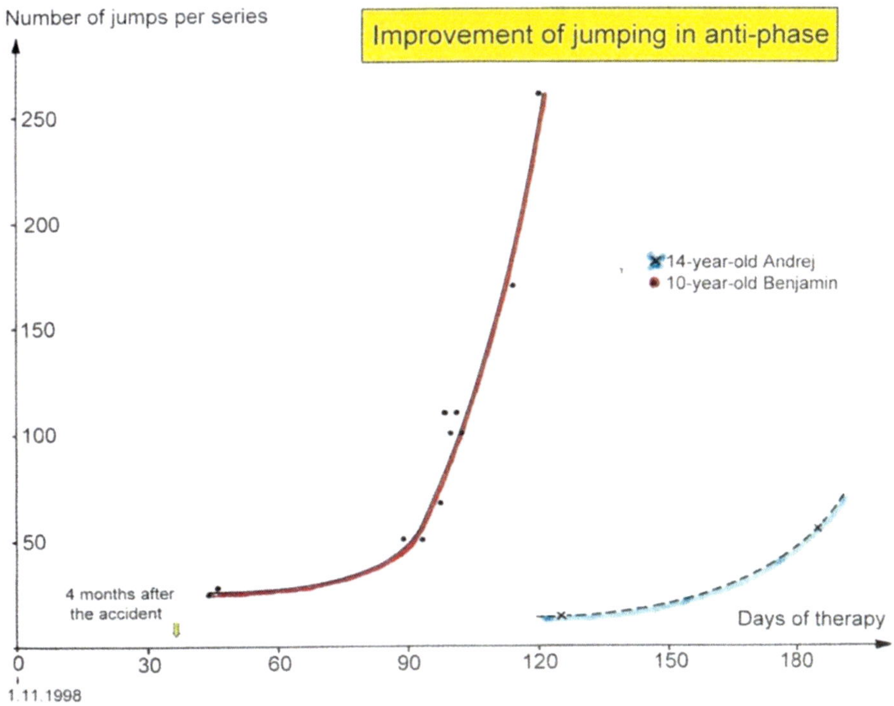

Figure 117. Increase of the anti-phase jumping series of the 10-year-old Benjamin and the 14-year-old Andrej by therapy. The average of the first 3 jumping series is plotted.

The recognition of the reality of the trauma was still difficult for the patient. Often he said to his mother: "Mum, please wake me up, this dream is so terrible" (Figure 119). Depressive thoughts were opposed with an intensive therapy (which was anyway part of the neurotherapy he was receiving), not to give the patient time to think his situation over. Further, training using the special CDT device seemed to have a positive effect on his mood. It became apparent when Benjamin emerged from the vigilant coma, that he was indeed hemiparetic on the left (with a slight quadruparesis), including spasticity in the left arm, hand and fingers and in the left sternocleidomastoids' muscle, which pulled his head to the left side. The organization of the CNS improved continuously with the on-going intensive therapy, which was performed six times per week, several times per day. On the advice of the Author, the parents refrained from botulin therapy to selectively reduce spasticity (by a blockade of the neuromuscular endplates in the domain of the application of the toxin for approximately three months), which was suggested by a neurologist. With the early begin of an efficient intensive neuro-rehabilitation therapy the development of false organization of the CNS (spasticity) can be strongly reduced or abolished; there was therefore no indication for a botulin therapy.

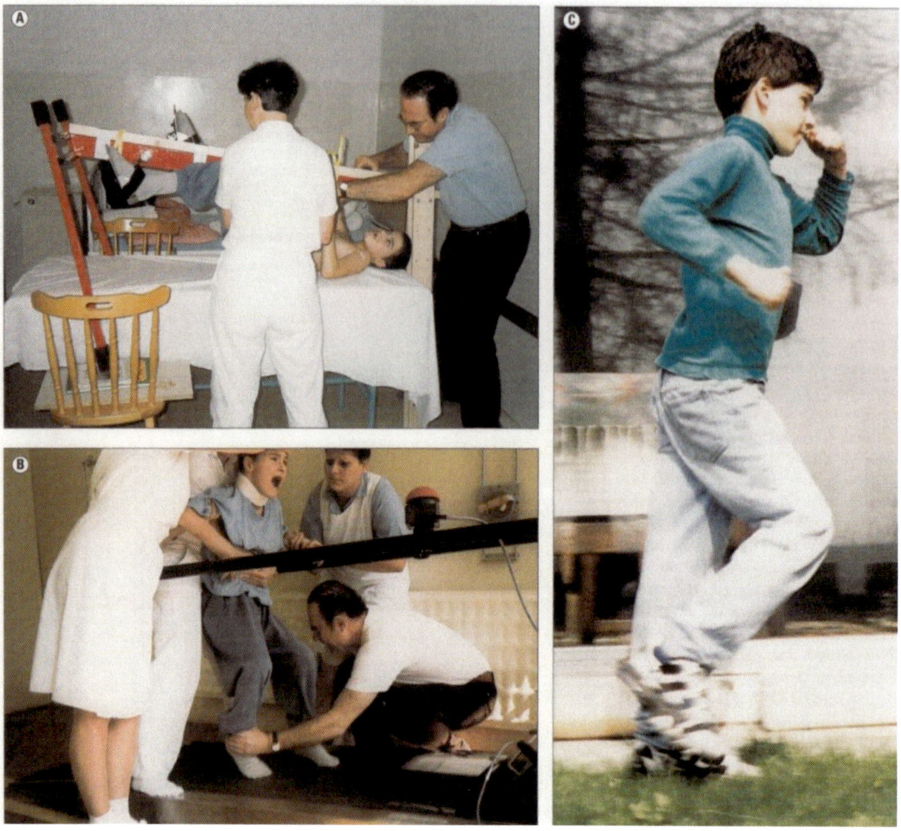

Figure 118. Coordination dynamics therapy administered to a patient with severe brain injury in the vigilant coma state (A, B) and 8 months after the car accident (C). Note that Benjamin was very afraid when treadmill walking was administered to him (B). The Author is supporting the legs.

After three months of CDT, the functions of the paretic arm strongly improved. Thus, the physiologic network states had been strengthened (the corresponding attractor wells had deepened) and the pathologic network states had been weakened (the corresponding attractor wells became more shallow). Only if the left arm can fully be moved in coordination with the other arm and the legs (coupling of the self-organizing network for the right arm and hand movement to the network organization of the other arm and the legs), the spasticity will reduce to an extent that non-rhythmic volitional movements of the left hand and fingers can be fully trained. By oscillator formation therapy, which consists of an entrainment of injured premotor spinal oscillators, achieved for example, by jumping on a springboard, the movements of the paretic arm and leg could be improved.

After six months of CDT (seven months following CNS injury), the paretic arm further improved, the overstretching of the left knee reduced and the patient re-learned running (Figure 115B). It seemed that an essential step forward was achieved in the last three months of therapy, when the patient used the special CDT device in the standing position. At the beginning the patient had great difficulty in keeping his left hand on the lever. Six weeks later, the function of the left hand improved. Several months earlier the patient also had great difficulty in holding the lever of the sky-walker.

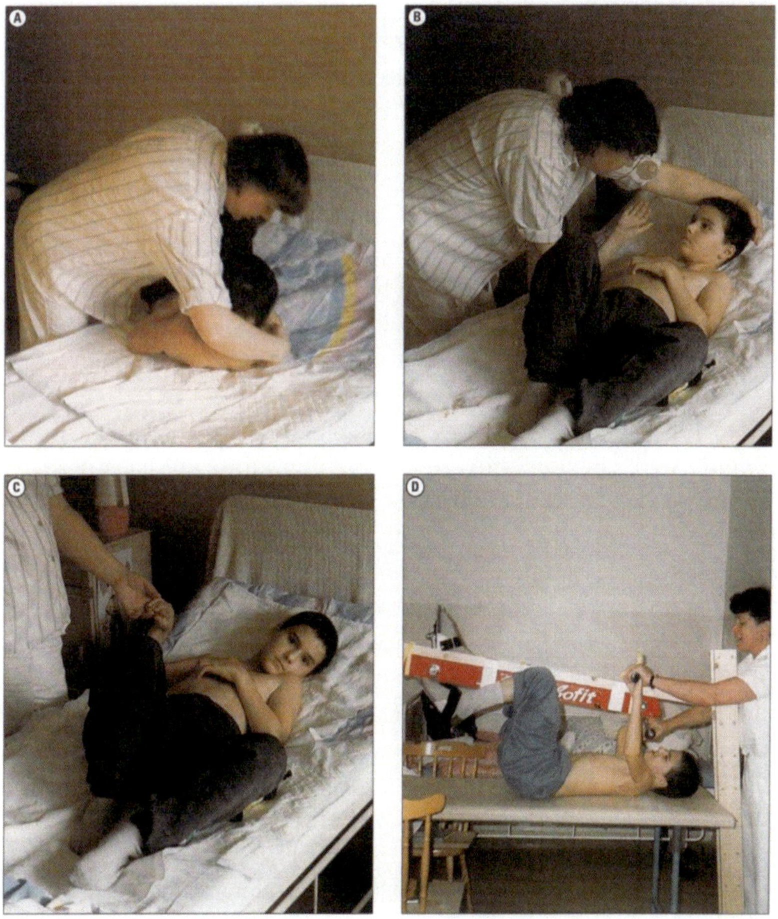

Figure 119. Nine-year-old Benjamin in the vigilant coma state. Patient cannot speak. When he came out of coma he said to the mother: “Mum, please wake me up, the dream is so horrible”.

After six months of CDT the patient no longer needed support when jumping in anti-phase on the springboard, the left knee was no longer overstretched, the positioning of the left arm, hand, head (reduced spasticity of the sternocleidomastoids' muscle) and trunk were better - hemiparesis had partly subsided. With the improvement of the motor functions, cognitive functions also improved. His short-term memory became better. The short-term memory was still not normal because he often forgot what sentences he should translate, when translating two to three sentences from English to Slovak. He enjoyed attending school for a few hours per day. The improvement of the mental functions corresponded with the healthy looking face, as judged by the parents and the Author.

The fourteen year-old patient Andrej was cycling down a steep hill and fell on his head and lost consciousness. He suffered many injuries including contusion of the lungs, obstructions of the airways due to facial and mandibular injuries and brain damage in the region of the right thalamus, the right fronto-parietal region, the frontal periventricular region and left basal ganglia. The coma and vigilant coma stage lasted two and a half months which is six weeks longer than in the nine year-old patient Benjamin, for whom therapy was started earlier. CDT was started two and a half months after the accident in the last phase of the vigilant coma state. When the Author first saw the patient, the patient was aggressive, and the author even thought it was hopeless to attempt to repair the injured CNS because the patient could hardly perform any coordinated movement and his speech was strongly impaired. His parents wanted help their son, so the Author advised them to start CDT using the special CDT device and the treadmill.

At the beginning, the patient had to be carried from the bed and the wheelchair to the training device. Progress was initially slow. However, with the improvement in motor functions and vegetative functions (the intense sweating of the hands reduced) also the mental functions improved. The patient became strongly motivated to train hard to get healthy again. The therapy included crawling and creeping. He started to like the Author and could no longer remember that he had been aggressive at the beginning. It seems therefore that with the more integrated functioning of the neuronal networks, the CNS started to work on a mentally higher level. After four months of therapy, the patient could run (Figure 115B, in the middle) better than he could walk (Figure 115A, right side). After slipping once during running, the patient became nervous and his running was not so good any more. Some viral infections also restricted his therapy.

After six months of therapy, the functioning of the neuronal networks of the patient's CNS was very good in comparison to the following case in which the patient Mario did not obtain any CDT for the first seven months, but was not as good as in Benjamin, because of a later start of the CDT, an illness, a complication with a bone contusion, suboptimal therapy and of the regrettable administration of botulin toxin, which disturbs the movement induced re-afferent input.

The improvement of CNS function in this patient with therapy could also be judged through the reduction of the palmar flexion of the right hand. A farewell photo after six months of therapy demonstrated that the neuronal network organization generating movement, posture and hand functions was influenced by somatic and mental functions. Even though the right hand and right arm functions were still strongly impaired, as could be seen during crawling and shaking hands on the good-bye photo, the right hand was not palmar flexed. In Benjamin, the hand function was better during crawling and was not far from normal when shaking hands, but was a bit spastic on the good-bye photo. In this emotional context, the potential landscape for motivation and emotions after CNS injury [98] was beneficial to the hand function in the case of Andrej but not to Benjamin. Interestingly, one month later it was possible to bring the right hand of Andrej to an aligned position during hand shaking. The right hand could even be slightly flexed dorsally. Thus, the emotion with the present potential landscape for motivation and emotions [98] in this case assisted in the reorganization of his CNS in the short-term memory. Physiologic hand shaking became possible in the absence of an emotional context only one month later. For further details and pictures of patients see [62], case reports 11-13.

Twelve year-old Mario had a car accident when cycling and suffered many bone fractures and brain injuries, when he was run over by a car. Because of many complications of the bone fractures and the brain injuries (additional subarachnoidal bleeding) and several infections, CDT could only be started seven months following the accident, even though the Author was urging all physicians to start the therapy as quickly as possible. The coma and vigilant coma states came to an end six months after the accident, i.e., much later than in the two other cases where CDT was started earlier. Motor and cognitive functions were very much delayed in comparison to those of the other two patients, Benjamin and Andrej, nine months after the

CNS injury. The patient had still to be carried from the wheelchair to the therapy device (Figure 115C,D) because of strong extensor spasticity (Figure 115E) and supination of the feet (Figure 115G), and communication was difficult, even though the patient was friendly and cooperative.

After one month of CDT, motor and mental functions started to improve. It was hoped that the missed early treatment could partly be caught up. It is argued in literature that there is a time window for early treatment [99].

Even if the suggestion of a time window for early treatment were correct, early treatment is important for the following reason: if supervised reorganization is efficient and starts very early following CNS injury, then the reorganization goes straight from the neuronal network organization left by the injury towards a more physiologic state without having to evict entrenched pathological network states. However, if such early therapy is not received, the injured CNS organizes itself in a pathologic manner. Due to the pathologic functioning of the CNS, the mechanical aspects of the body become pathologic; tendons shorten, muscles become spastic or atrophic, joints can no longer be moved and the arms, legs, hands and feet often assume very un-physiologic positions so that it becomes difficult to use them physiologically, and appropriate therapy can only be applied slowly and with many difficulties. Under such conditions, the efficiency of the therapy suffers greatly. With these and other difficulties, the pathologically functioning CNS has to be reorganized in the direction of physiologic functioning.

In conclusion, following CNS injury, the remnants of formerly physiologic functioning network parts are easier to organize for physiologic output than network parts which have started to function pathologically.

Even in Benjamin, it would probably have been better to start CDT ten days rather than four weeks after the injury. Even if tubes, wires and other necessary apparatus hinder intensive CDT, intensive therapy has to be started as early as possible. In severe brain injuries, bone and other reconstructions must be designed in a way to allow an early intensive neurotherapy. Bone fractures, for example, must be reconstructed invasively.

3.2. Stage of Repair in the Three Children One Year After the CNS Injury

The now 10-year-old boy *Benjamin* obtained a further six months of optimal CDT at home. The functions in the left arm and left hand improved further. He could walk and run better after the further treatment. With the left 'bad' hand he could turn over pages of a journal. Benjamin was back to his old school in his old class and was managing well. The strategy to first give priority to the re-learning of the motor functions before going back to school and giving in this way priority to the re-learning of the basic structures of the CNS with the attendant improvement in short-term memory was right in this case. With the improvement of the motor functions also the higher mental functions improved, so that he could better manage at school.

Benjamin was not far from normal again. Even though Benjamin's running was powerful again one year after the accident (Figure 118C), his posture during running was not fully physiologic. It seems therefore that the patient was using the former pathologic posture now at least partly as a habit. The former attractor 'pathologic posture with spasticity' of the injured CNS was still somehow existing in the repaired CNS. The patient had the ability to correct this posture, to switch on volition into the attractor state 'physiologic posture'. The possibility to have a bistability of the two attractors, pathologic posture and physiologic posture can also be found when tackling idiopathic scoliosis in the case of slight disturbances of CNS functioning. When repairing the bladder two attractors, synergy and dyssynergy of the bladder can both appear transiently (see also Chapter V of [14]). A bistability of spasticity and a rather physiologic movement pattern may also exist (Figure 19). Fifteen months after the accident, Benjamin's CNS still had difficulty in organizing the trot gait coordination between arms and legs during walking and during exercising on the special CDT device. When falling, he was still not using protection reaction automatisms. The measured coordination between arms and legs was still not physiologic.

The now 15-year-old *Andrej* only received further conservative treatment for one month. His motor functions did not improve further during this month. The re-appearance of the slight palmar flexion of the right hand may even indicate a loss of re-gained motor functions. A loss of re-learned functions was also found during the therapy of a spinal cord injury (Figs.50,51) and cerebellar injury [9], when therapy was interrupted. With re-starting CDT, the motor functions in SCI and cerebellar injury started to improve again.

Most interesting was the case of the now thirteen-year-old *Mario.* Six to eight months following the brain injury (with no or only little conservative treatment) his motor and higher mental functions were in a seemingly hopeless condition. Because of the pathologic organization of the neuronal networks, including different kinds of spasticity, false posture and positioning of leg and feet (Figure 115E,G), he had to be carried by his mother (Figure 115C). Nevertheless, intensive CDT was commenced.

This included exercising on the special CDT in a recumbent (Figure 115D) and sitting position and treadmill walking. Even though the patient was often crying, especially during treadmill walking, mainly because he was afraid (as he explained later on, when his mental functions had further improved) (Figure 120), the neuro-paediatrician, the physiotherapist, his mother and the Author came to the same conclusion, namely that Mario has to exercise intensively, to have any sensible chance of making further essential improvement in the future.

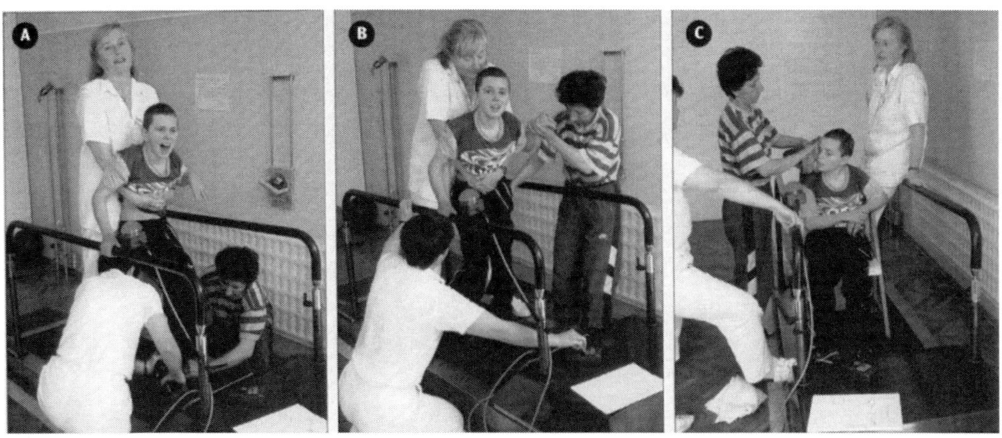

Figure 120. Treadmill training administered to the patient Mario, who suffered a severe brain injury in a car accident. In 'A' the legs are supported, in 'B' not. In 'C' the patient recovers from stress and fear. Mother on the right side, physician behind and physiotherapist on the left side of the photos.

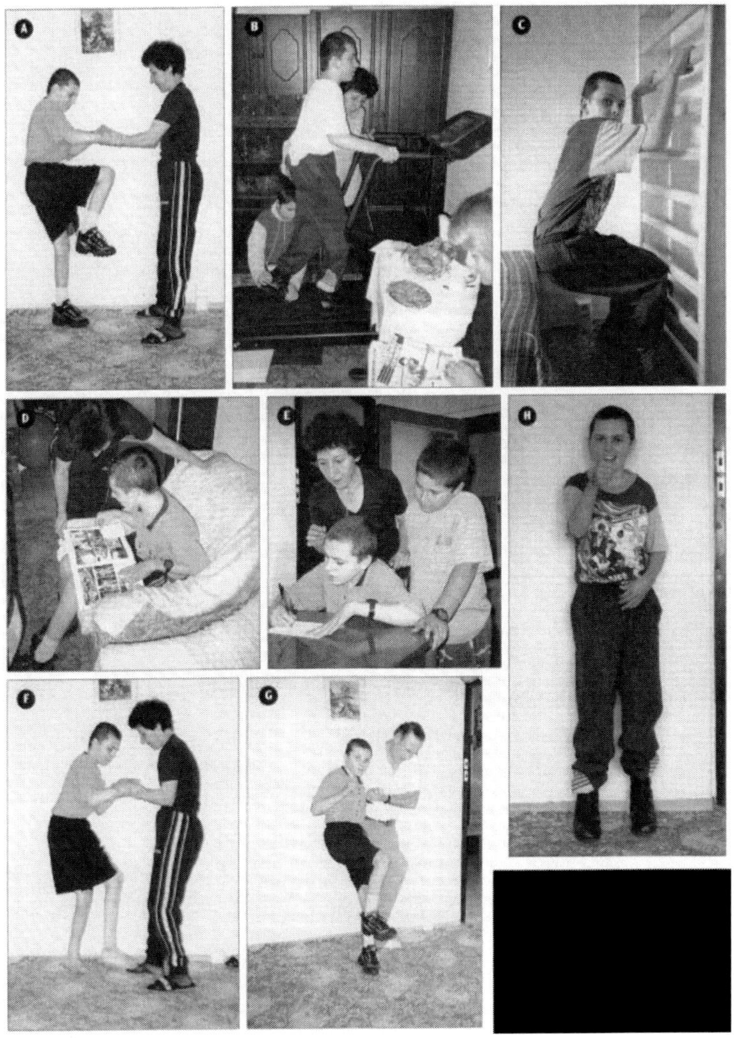

Figure 121. Home treatment of the severely brain-injured patient Mario. A. Patient is walking with the mother. The strong lifting of the knee is due to an activation of the stepping automatism (compare with Figure 24B). B. Patient walking on treadmill, supported by brother and mother. He shows a decisive improvement against the beginning of therapy in Figure 120. C. Training of up-righting movements at the wall bars. In spite of the palmar flexion of the left hand, the left hand can take power over. D, E. The patient can turn over leaves and can write again. F. Patient during bare foot walking; left foot in supination. G. Mario is walking with the Author without the support of the right "good" hand. H. Patient during standing without fear, but leaning against the wall. Optimistic gesture is spontaneously taken.

One year after the brain injury, after four months of rather intensive therapy, essential progress occurred. The patient learned to walk a little when supported. His hand functions improved, so that he could turn over the leaves of a journal and was able to write again. He was now motivated to exercise and he had the feeling that the functional state of his CNS was progressing (Figure 121). His higher mental functions also improved substantially. Eight months after the accident, the patient did not always recognize the Author. Twelve months after the accident, he even spoke some words in German to him (his native language is Hungarian). Mario was back to his old school, but two classes lower. The pupils and the teachers were giving him much help so that he could manage. Further progress depended on the intensity of the continuous CDT. When Mario was separated from his mother for a longer period of time, he was so upset that he had difficulty speaking. If Mario was therefore to be treated further at the clinic, then his mother would have to be with him. 17 months after the brain injury he had difficulty in organizing trot gait coordination between arms and legs.

The case report of the 13-year-old Mario shows that, even when the early CDT is missed, there is still hope that CNS functions can be substantially improved. However, the therapy is more difficult to perform because of the pathologic organizations which have taken place in the CNS in the meantime. The pathologic neuronal network states, which express themselves by spasticity, false positioning of body parts and perhaps by epileptic seizures, make the reorganization of the CNS more difficult and cause partly pathologic movement induced re-afferent input. A further important point is that continuous, intensive and early adequate therapy and not only the nature of the brain injury are crucial for the outcome of similar severe brain injuries. If the therapy had not been continued eight months after the accident, then Mario would not have made such substantial progress.

3.3. Stage of Repair in the Children Six Years Following Injury

Andrej obtained only conventional therapy and did not progress further. *Mario* was trained at home by his Mother and his motor functions improved. When the staff of his school changed, he did not get further help from the school. *Benjamin* went back to his original school class as a disabled pupil. Following improvement, he was able to return to school on the same basis as his classmates. He made a substantial step forward in repair following the administration of one month of optimal CDT in a therapy clinic with thirty to forty hours of treatment per week. Due to the significance of Benjamin's case to CNS repair, he shall be discussed in more detail below.

Benjamin further improved at home, although only a little. He was not able to crawl because of the spasticity of his left hand and arm. When the measuring of low-load coordination dynamics (CD) was introduced, it became possible to follow the evolution of his CNS function. With the introduction of high-load exercising on the special CDT device to activate the CNS more integratively (systemically), Benjamin improved further and became able to crawl. After a certain time, his motor functions did not improve any further. The question arose whether Schalow CDT had come to its limits in respect of the potential improvement of CNS functioning, or whether the therapy had to be updated and performed more intensively. It will be shown below that the CNS repair had by no means reached its limits and those measurements of low-load (Figure 122) and high-load CD (Figure 123) provided further insight into the reorganization changes of the CNS in this patient.

3.3.1. Temporary Impairment of Low-Load Coordination Dynamics and Exercising Rates

When plotting the values of low-load coordination dynamics (CD) measured over five years of home therapy (Figure 122Ba), it can clearly be seen that there was a pronounced worsening of CD values (higher values). Simultaneously, the exercising rate (frequency) showed a transient significant decrease (Figure 122Aa).

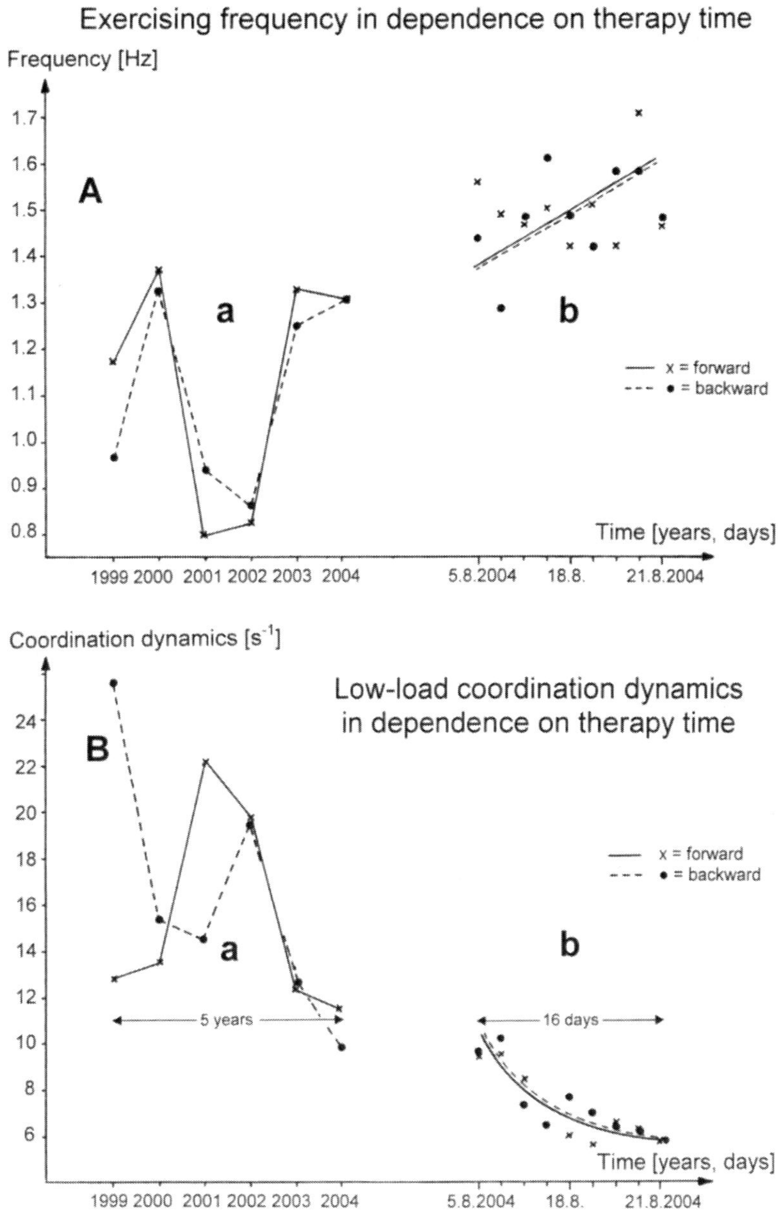

Figure 122. Exercising rate (frequency) and low-load coordination dynamics values with on-going therapy; patient after severe traumatic brain injury. Solid line = exercising in the forward direction; dashed line = exercising in the backward direction. In 'Aa' and 'Ba', the rate and the coordination dynamics changes are shown observed within the years 1999-2004 (training at the patient's home). In 'Ab' and 'Bb', the changes of the rate and coordination dynamics values are shown as observed during 16 days of treatment under professional supervision. Note that the coordination dynamics for exercising in the forward direction (solid line) improved in the 16 days of treatment under professional supervision more than during the 5 years of home therapy.

The pronounced worsening of the coordination of arm and leg movements is explained by substantial changes of CNS self-organization, probably induced by starting high-load exercising. It was predicted, based on the theory of CDT that CNS organization may markedly worsen transiently when the CNS is changing from one organization to another due to the treatment. Such organizational change of the CNS seems to have happened in Benjamin's CNS during the home therapy following the introduction of the high-load exercising. In spite of the principal organizational changes in the CNS, the improvement of low-load coordination dynamics from 12.9 to 11.5 (11%) was comparably small given the duration of the treatment. The low-load CD values (arrhythmicity of exercising) could not be lowered, despite exercising on the special device for at least twelve hours per week. Even the learning effect of these special movements seemed to be very small or non-existent.

3.3.2. Improvement of High-Load CD Following Home Therapy

During home training the high-load CD for increasing and decreasing the load from 20N to 200N improved with twelve hours therapy per week. The hysteresis-like curves for increasing and decreasing load improved from a pathologic (Figure 123Aa) to a physiologic hysteresis curve (Figure 123Ab). The values of the high-load CD improved over the three years of home treatment from 229 to 151 (34%) (Figure 123Ca). This is a small improvement taken to the duration of the therapy (three years).

3.3.3 Improvement of the Low-Load CD Upon Therapy under Professional Supervision

The improvement achieved in low-load CD under optimal conditions is shown in Figure 122. Over the sixteen days of quantified therapy the rate of exercising (frequency) was high and increased slightly (Figure 122Ab). The value of the CD for exercising in the forward direction decreased (improved) from 9.3 to 5.8 (38%) (Figure 122Bb). Initial values taken from uninjured persons are between 5 and 6s^{-1} [50].

3.3.4 Improvement of High-Load CD from Therapy under Professional Supervision

High-load CD improved strongly during this short time interval of therapy received at the therapy clinic, even though during this time Benjamin exercised much less on the special CDT device than at home. The hysteresis-like high-load coordination dynamics curves (Figure 123Ba,b) were comparable to those observed for track and field athletes [51]. The high-load CD improved from 175.3 to 97.3 (44%) (Figure 123Cb).

3.3.5. Comparison of Home Therapy and Therapy under Professional Supervision

The values of the low-load CD improved over the five years of home treatment (1800 days) by 11%, whereas an improvement of 38% was achieved during the sixteen days at the therapy clinic. The values of the high-load coordination dynamics improved over the three years (1100 days) by 34% and by 44% during the sixteen days of institutional therapy. The rate of re-learning for low-load coordination dynamics was 0.61%/100days (11%/1800days = 0.61%/100days) for home therapy and 238% /100days (38%/16days = 237.5%/100days) for institutional therapy place. The efficiency difference in respect of low-load exercising was by a factor of 390 (238%/0.61% = 390). The rate of learning for high-load coordination dynamics was 3.1%/100days (34%/1100days = 3.09%/100days) for home therapy and

275%/100days (44%/16days = 275%/100days) for institutional therapy. The efficiency difference for high-load exercising between home and institutional therapy thus differed by a factor of 89 (275%/3.1%= 89). The efficiency difference in respect of CNS repair, as quantified by low-load (390) and high-load CD (89), between home and institutional therapy differed approximately by a factor of 100.

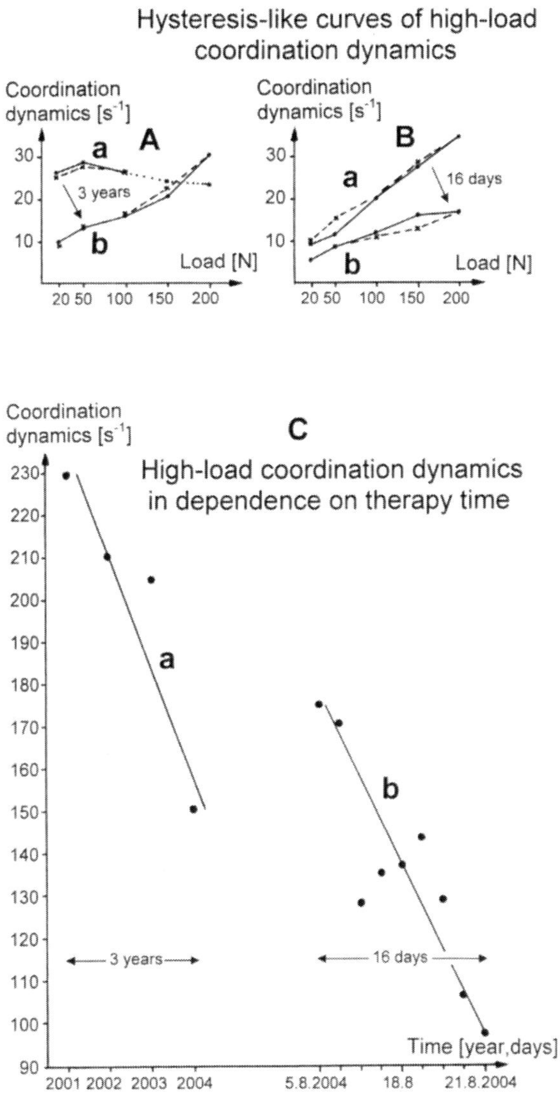

Figure 123. A,B. Improvements (from 'a' to 'b') of hysteresis-like high-load coordination dynamics values during 3 years of home training (A) and during 16 days of optimal therapy at a coordination dynamics therapy place (B); patient after severe brain injury.

C. High-load coordination dynamics values for increasing and decreasing load ($\sum = (\Delta_{20N} + \Delta_{50N} + \Delta_{100N} + \Delta_{150N} + \Delta_{200N} + \Delta_{150N} + \Delta_{100N} + \Delta_{50N} + \Delta_{20N})s^{-1}$) measured during 3 years of home training (a) and during 16 days of institutional therapy (b). Note that the high-load coordination dynamics improved during the 16 days more than during the 3 years before.

3.3.6. Improvement of Movements upon Treatment at a Professional Therapy Place

Not only did CD improve strongly during professional therapy, but walking also got much better. While walking, Benjamin used the left leg much more actively instead of lifting the hip and swinging the leg. The treatment goal could be approached: at first glance you could not see that the patient had suffered a severe brain injury.

3.3.7. Comparison of the Improvement of CNS Functioning after Injury with Physiologic Changes during Individual Development.

It was demonstrated that the CNS functioning, as quantified by the low-load CD values, continuously improves during physiologic development within the age range between 3 and 18 years (Figure 35). During the treatment for repair, the values of the coordination dynamics were shown in this report to temporarily worsen, probably as the CNS was changing its self-organization from pathologic to physiologic (Figure 122Ba). Such temporary worsening of CNS self-organization was predicted by the theory of CD. During the three weeks of intensive therapy, the values of the CD improved in this patient continuously, in similarity with physiologic development.

3.3.8. Oscillatory Firing FF-Type Muscle Fibers

In this patient after severe brain injury, also the basal ganglia on the left side were probably partly injured. Such partial damage of the basal ganglia may be responsible for the difficulty of functional repair. Rhythmic firing of FF-type muscle fibers was sometimes observed in the left tibialis anterior muscle (Figure 124C), suggesting impaired coordination (including synchronization) between oscillatory firing of motoneurons.

3.3.9. Treatment Judgment According to Surface Electromyography (Semg)

In this case, the optimal Schalow CDT included exercising on the special CDT device to repair the impaired phase and frequency coordination by means of imposing very coordinated arm, leg, and trunk movements in the lying, sitting and standing position. The innate automatisms creeping, crawling, up-righting, walking and running were exercised indoor or on beach sand outside for optimal sensory input. Rhythmic dynamic stereotyped movements, like jumping and swinging on the springboard were exercised to entrain premotor spinal oscillators as an entrainment of CNS neuronal networks on the ensemble level. Old learned integrative movements (learned automatisms), like climbing staircases, were also trained. A typical exercising scene at the therapy clinic is shown in Figure 58. All movements were performed with an awareness of the importance of symmetry. To optimally entrain the severely damaged neuronal networks integratively, those integrative movements which were activated most physiologically, quantified by surface electromyography (sEMG), were exercised most frequently. Surface EMG was performed for different integrative movements to ascertain under what conditions critical muscles showed best motor program performance (Figure 124).

Due to an excessive palmar flexion of the left hand, the left brachioradialis and extensor carpi radialis muscles were selected as an indication of the quality of CNS function with respect to the arm and hand movements.

Because of an excessive plantar flexion of the left foot (and an excessive inward rotation), recordings were taken from the left triceps surae and tibialis anterior muscles to judge CNS function with respect to leg movements. The task was now to get the best sEMG motor programs for all four muscles simultaneously to see which integrative rhythmic, dynamic, stereotyped movements entrain the injured CNS optimally.

It is important to improve CNS functioning integratively (systemically). The human CNS is an open system and if only a small sub-network is being entrained, the pathologic organization escapes the entrainment by shifting to another CNS neuronal network site as could be clearly observed in this patient. When the patient was jumping on the springboard and his left hand was in palmar flexion, he was instructed to concentrate on the left hand, to keep the hand straighter. Then, however, the left foot went into plantar flexion and inward rotation. The patient was then instructed to concentrate on getting the heel down during jumping (to perform more dorsal flexion). He concentrated on the left foot and could improve the position of the foot; but then the left hand went again into pathologic palmar flexion. He was then instructed to concentrate simultaneously on the left hand and the left foot; but then he argued that he was unable to concentrate simultaneously on the arm and the leg.

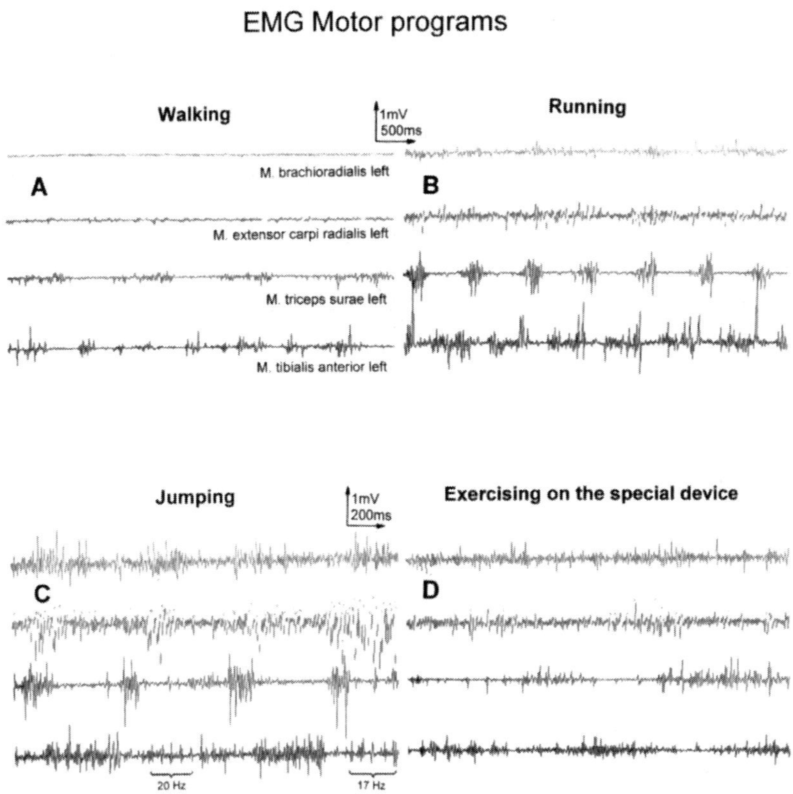

Figure 124. Motor programs registered by surface electromyography (sEMG) during walking (A), running (B), jumping (C), and exercising on the special coordination dynamics therapy and recording device at the load of 100N (D); a patient with formally very severe brain injury. Note that the critical muscles brachioradialis left and extensor carpi radialis left are not activated during walking (A). On the extensor carpi radialis muscle trace, there may be some artifacts on top of the motor program. On the tibialis anterior trace, rhythmic firing of 20 and 17Hz can be seen, which most likely stems from FF-type motor units innervated by α_1-motoneurons.

Patients with severe CNS injuries especially have difficulty in concentrating on two things at the same time. It was therefore important to identify complex movements, which would induce good left arm and left foot positioning, measured visually and by good sEMG motor patterns in all the four muscles. Movements which did not demand very much concentration of the patient were selected because it allowed the patient to perform the movement many times without getting mentally or physically fatigued.

Figure 124 shows the motor programs of the left brachioradialis, extensor carpi radialis, triceps surae and tibialis anterior muscles during walking (A), running on the treadmill (B), jumping on the springboard (C), and exercising on the special coordination dynamics therapy device (D). It can be seen from Figure 124 that the motor programs of the four muscles for running and jumping on the springboard (B,C) were better than those for walking (A) and exercising on the special device (D). In this patient, running and jumping on the springboard thus seemed to be more efficient in respect of CNS repair than walking and exercising on the special CDT device. Improvements of CNS functioning were quantified clinically, largely by improvement in the performance of movements.

Benjamin could walk and run fast after one year of treatment. Only the performance of the left arm and left leg were poor in nearly all movements. The writing performance was poor and his intellectual functions had not fully recovered.

3.3.10. Need for Many Different Movements to Reorganize the Injured CNS

Benjamin did not seem to improve further with home therapy, despite walking and running every day and exercising virtually every day on the special device for one to two hours (approx. 5000 coordinated movements per day). When the patient obtained therapy at the therapy clinic where he exercised many different movements, his coordination dynamics improved strongly, even though he exercised less on the special device than at home. There are two inferences to be drawn from the strong improvement of the CNS functioning during this period. First, after severe CNS injuries, the learning effect of the special movements exercised on the special device is smaller. The special coordination dynamics therapy and measuring device is therefore suitable for the diagnostics of CNS functioning, as improvements of coordinated arm and leg movements are mainly the result of a general improvement in CNS functioning and only little attributable to the learning of the movements themselves. Secondly, single movements, like exercising on the special device or walking on a treadmill will not reorganize the severely injured CNS substantially: Benjamin exercised at home on the special device and was walking and running every day, and he still did not improve further. The therapy has therefore to include many different movements, which have to be adapted to the individual CNS injury of the patient with its unique pathological CNS organization. A specific combination of complex movements and deep knowledge of human CNS organization and repair are needed to achieve CNS repair rather than solely instruments, movements, stem cell therapy or manipulations, broadly targeted electro stimulation, or pure pharmacotherapy.

3.3.11. Need for Optimal Institutional Treatment

The variation in the improvement of CNS organization by a factor of approximately one hundred shows further that an optimal therapy at a therapy clinic is needed for efficient repair of the injured or malfunctioning CNS. Home treatment is helpful but cannot be a substitute for institutional treatment under professional supervision. The difference in the efficiency of the treatment at home and at a clinic, the latter making use of most recent data on CNS repair from human neurophysiologic research is very appreciable indeed. After these three weeks of treatment at the therapy clinic, the Author claimed: “Benjamin, if you come here to be treated for one year, in the end you will not differ very much from the other boys at school any more. You also can exercise at home under my partly supervision, but you may need a hundred years to achieve the same”. Benjamin’s father, a friend of the Author, asked other friends (neurologists and psychiatrists) for their opinion and did not send Benjamin for a year’s treatment.

3.4. Stage of Repair of the Three Patients Eleven Years after the CNS Injury

3.4.1. No Further Repair Achieved by Conventional Therapy, Modest Progress from Supervised Therapy at Home And A Return to Normality from CDT

None of the three patients obtained optimal therapy over the eleven years following the injury.

Against the advice of the Author, *Andrej* obtained conventional therapy and did not progress further. He is now 25 years-old and living at home with his mother and obtains four weeks of psychiatry treatment every year. He was depressed because he was not granted admission to a mainstream school and because of the lack of progress with his motor functions. Conventional neuro-rehabilitation is now twenty years out-of-date and it is more than arguable that his future has been lost due to a refusal to take account of new research.

Mario, now 23 years-old, trained with the mother at home, sporadically advised by the Author. The help from the school stopped, when the staff changed. His motor functions improved further; he relearned walking (Figure 125C). His higher mental functions improved, but are still far from normal. He is living as disabled person with his mother as his carer. The mother reported that he is sometimes aggressive, but not when the Author was present (see below).

Benjamin, now 21 years old, obtained non-optimal CDT. He progressed further in motor and higher mental functions. Six years after the accident he obtained optimal CDT in a therapy clinic for three weeks and made a big step forward in repairing his CNS (see above). He successfully graduated from school (grammar school) and is now studying languages at university. He leads a somewhat normal life and helps to bring better treatment to at least some patients with CNS injury. While studying at the university, he did not undertake the necessary maintenance level of therapy. He felt he was getting worse. He asked the Author to help him. To encourage all three patients to improve their training, the three young men were brought together, to compare them and to motivate them for further training.

3.4.2. Clinical Assessment

Figure 125 shows the clinical state of repair. Benjamin, Mario, and Andrej (from left to right) started to crawl (Figure 125A). Benjamin is crawling quite physiologically and leads (Figure 125B).

Figure 125. Clinical evaluation of three patients, 21- (Benjamin), 23- (Mario), and 25-years-old (Andrej), who suffered 11 years ago a severe brain injury and obtained for 10 years conventional therapy (Andrej), home therapy (Mario), and CDT (Benjamin). A,B. Patient's during crawling. Note the spasticity in both hands of Mario (middle) and the right hand of Andrej (right). C. Patients during walking. Mario (middle) has problems with the spasticity in the hands. D. Mario upon coordination dynamics assessment. Note the strong impression in his face. E. Mario shows proudly the reduction of the hand spasticity, due to the exercising on the special device. F. Eating of the patients after assessment and training. Andrej and Mario are living at home with the mother as unable persons and Benjamin is having a normal life and is studying languages.

Mario is second, because of the problems caused by the spasticity in his fingers and hands on the left side. Andrej can hardly crawl at all because the right hand is in a pathologic position (plantar flexed) (Figure 125B). Afterwards, all three patients walked (Figure 125C).

Benjamin (left) is walking quite physiologically; only the left arm and hand are not fully in the correct position. Mario (center) can walk with some balance problems and is fighting the spasticity of the left hand. Andrej (right) is walking quite physiologically; ten years ago he could even run. Andrej was astonished that Mario had learned to walk over the past ten years.

To compare the CD all three patients were assessed. Note the effort on Mario's face when he was exercising (Figure 125D). When Andrej exercised (Figure 125E) after him, Mario was proudly showing his left hand with no flexion (non-spastic) in the fingers. During crawling

(Figure 125B) and during walking (Figure 125C) the left hand was strongly spastic. The flexor spasticity had strongly reduced in the short-term memory upon exercising on the special CDT device. His own special CDT device worked inaccurately, so that his CNS could not learn anymore from the device the necessary coordinated firing of neurons and reduce the spasticity first in the short- and then also in the long-term memory (Figure 19). His special CDT device had effectively been reduced to a simple muscle training device and could no longer reduce spasticity. After exercising, the patients had lunch together (Figure 125F). Andrej was eating with the left hand, because the right hand was not working properly. To reduce the spasticity of the right hand, he was holding something in his right hand to increase the afferent input from the right hand to reduce spasticity. Mario (middle) could eat quite well with both hands because of the reduced spasticity in the short-term memory from exercising on the special device. For doubtful reasons, a weight was placed around his left wrist to improve hand and arm performance. Benjamin (right) was eating with both hands, but the left hand and arm were a bit spastic. The muscles activated by spasticity can be seen in Figure 125F.

3.4.3. Comparison of the Coordination Dynamics (CD)

The low-load CD (20N) values improved over the period of approximately ten years in Andrej from 21.3 to 13.4 (37% improvement), in Mario from 23.3 to 21.4 (8%), and in Benjamin from 12.9 to 8.2 (36%). Normal values are between 5 and 6 [50]. The improvement (the lowering) of the CD values does not show the whole improvement because at the beginning of therapy, this assessment was not available and the biggest decrease (improvement) of the CD values is normally at the beginning of therapy. Benjamin had the best low-load CD. But as can be seen from Figure 122Ba, they depend also on the repair state of the neuronal networks. During structural repair periods the self-organization of the CNS transiently worsens (see Figure 30). Andrej's CD values improved, but did not reach Benjamin's values. This is to be expected. Andrej obtained poor treatment. With proper treatment Andrej could probably have been clinically as good as Benjamin. Mario's CD values for low-load were quite poor in accordance with his clinical state. However, his CNS was most likely the most severely injured.

Interesting is the 20N-recording from Benjamin (made in 1999) in Figure 126D. It shows two transient high-frequency turning episodes, which can be observed in healthy pupils in the age range between five and eleven years (Figure 34EF). This recording indicates therefore that Benjamin's CNS improved not only from the movement-based learning therapy, but also the developing CNS helped to repair by trial and error eliminating processes, present during normal development. The 50N-recordings of the three young men are compared in Figure 126A-C. Again Benjamin had the best 50N value (8.1), Andrej had the second best (21.7) and Mario the worst (27.8).

By comparing the 100-N CD recordings of Benjamin, Mario and Andrej in Figure 127A-C, it can be seen that those of Benjamin were the best ($14.0s^{-1}$), followed by those of Andrej ($24s^{-1}$) and Mario ($30s^{-1}$); Mario and Andrej were not able to exercise at a load of 200N. When Benjamin exercised again at the limit under supervision, his mean high-load CD value improved from 14.0 (Figure 127C) to 11.2 (Figure 127D); the Watt-curve (middle trace) shows a nice stepwise function, typical in athletes (Figure 72B of Chapter II of [14]) or in persons with a trained and therefore good functioning CNS like that of the Author (Figure 38A).

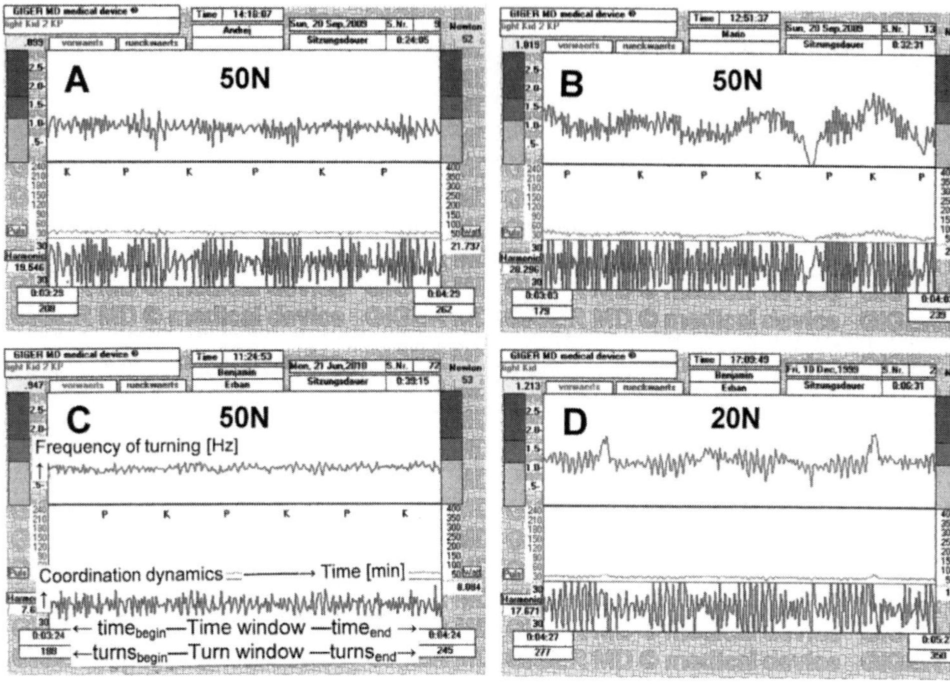

Figure 126. A,B,D. Comparison of low-load coordination dynamics measurements of three patient's to whom different treatments were administered for approximately 10 years. Upper traces = exercising frequency; lower traces = arrhythmicity of exercising (CD); all recordings are for exercising in the forward direction. The CDs in A and B are worse than in C in which the patient was treated by CDT. D. Transiently fast exercising is shown in a 11-year-old patient from the beginning of treatment; same patient as in 'C'.

3.4.4. The Society Has to Look for the Health of Their Children

Only CDT could repair Benjamin's severely injured CNS. The partly supervised home treatment could repair Mario's injured CNS better than conventional neuro-rehabilitation could in Andrej. The Author predicted these outcomes at the beginning of therapy. In these cases the neuro-paediatrician of the conventional rehabilitation clinic wanted to apply CDT. However, she was pressurised by more senior figures to continue with out-of-date treatment. Benjamin was included in an EU program to make videos about children with severe brain injury to show how they can manage in everyday life. The film about him won the competition.

A decision was made not to include treatment, even though proper treatment is essential for managing better later everyday life. Evidently, treatment that can partly repair severe brain injuries was not of interest to the EU administration.

Since the developing (and repairing) CNS has difficulty in generating two patterns simultaneously, such as playing ball in the street and looking out for approaching cars, the necessary complexity is not yet achieved), the society has to take better care for their children in the way that they do not suffer injuries and that the children get adequate therapy if an accident occurs. Some European legal systems still decide that car insurance companies need not pay if a child is hit by a car and the child is at fault. How can a child be at fault when his or her CNS cannot sufficiently organize two patterns at the same time?

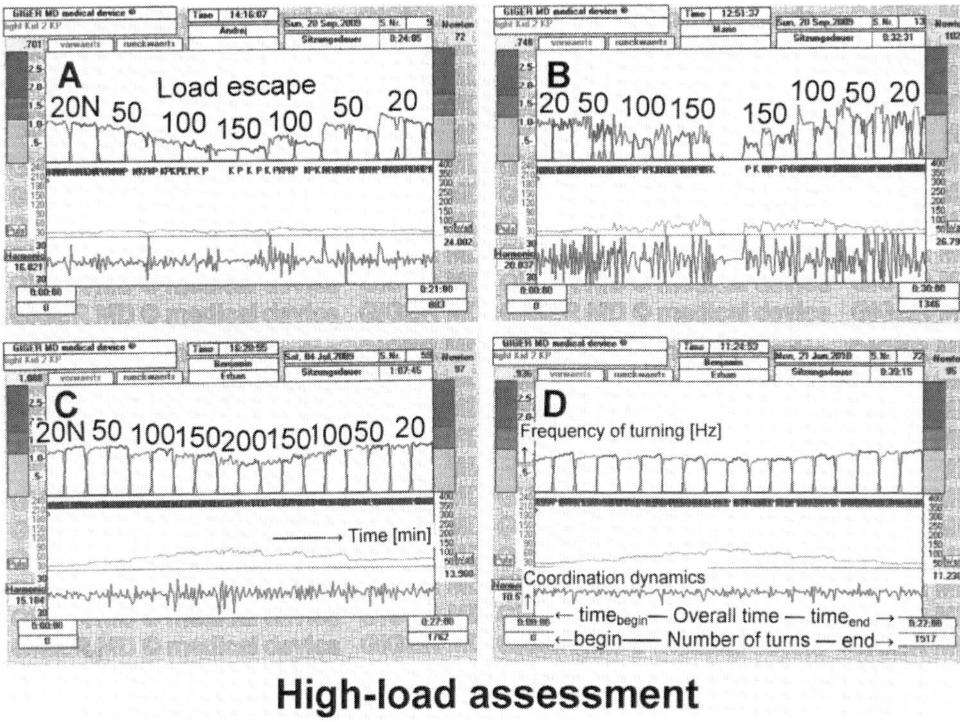

Figure 127. A, B, C. Comparison of high-load coordination dynamics measurements of three patients who suffered 11 years ago severe traumatic brain injuries and to whom different treatments were administered for approximately 10 years. Upper traces = frequency of exercising in the forward and backward direction; time periods between 21 and 30min. Note the reduction of the frequency in 'A' and 'B' (load escape) for loads mainly between 100 and 200N. In 'A' and 'B' the patients could not manage with the 200N load. Lower traces = arrhythmicity of exercising (CD); not the nearly stepwise increase and decrease of the work in 'Watt' in 'C' and 'D', typical for the performance of athletes. The patient in 'C', who obtained CDT, shows the best high-load CD. D. After a few months of intensive training the high-load CD further improved in the patient from 'C' (improved stepwise function).

3.5. Stage of Repair of the Three Young Men Fourteen Years after the CNS Injury

3.5.1. No Further Repair from Conventional Therapy, Some Further Progress from Supervised Therapy at Home, and Normal Life from CDT

None of the three patients obtained optimal therapy during the fourteen years following injury.

Andrej, as Figure 128 shows, his progress, quantified by high-load coordination dynamics (CD) values, improved only a little. Andrej refused be measured again and refused to have any contact with Benjamin in July 2013.

Mario, now 26 years old, lives with his mother and brother as who act as his carers. The overall improvement of CNS functioning can be seen in Figure 128 by means of the improvements of the high-load coordination dynamics values.

Benjamin, now 24 years old, obtained CDT, which was still far away from an optimal therapy. As can be seen from the high-load CD values of Figure 128, he progressed nicely up to 2004. When he obtained in 2004, six years after the accident, three weeks of optimal CDT in a therapy clinic for, he made significant progress in repairing his CNS (Figure 123). After 2004 he stopped intensive training, because his Father stopped sending him for further intensive therapy.

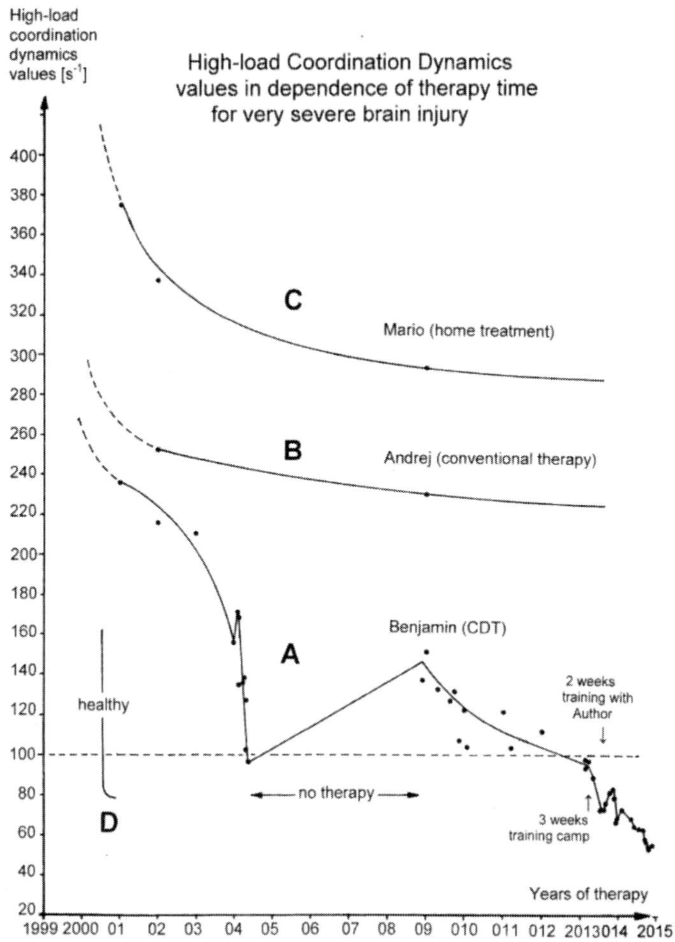

Figure 128. High-load coordination dynamics (CD) values in dependence of therapy in the patients with severe brain injury Benjamin (A) (CDT), Mario (B) (home training) and Andrej (D) (conventional therapy). The high-load CD values of healthy pupils upon repeated exercising on the special CDT device are inserted according to scale (D). Note that the high-load CD values of Mario and Andrej are far away from the healthy case with respect to CD values and time for improvement. The high-load CD values of Benjamin reach the values of healthy pupils but with much more time of exercising.

Sometimes Benjamin would exercise on the special CDT device for one hour but at other times he would do nothing for up to a week, apart from the everyday movements. Six years after the intensive optimal therapy he had the feeling that his condition was worse than it had been in 2004. He also felt that his intellect had declined. He therefore started to exercise up to 6,000 movements per day (2 hours). However, because the special CDT device had lost its precision, its beneficial effect on CNS function was very little. In 2006 he started to train slightly more. In 2009, he bought a new special CDT device and he also started to train more intensively. The functioning of his CNS improved again. As can be seen from Figure 128A, Benjamin's CNS functioning got worse between 2004 and 2009 and thereafter started to improve again. He also started to exercise with the help of coaches from university and improved his CNS functioning further. At the end of 2012 however, he still had not reached the level of the professional treatment in 2004.

Then Benjamin made the fundamental decision that he wants to achieve all that he can with his health and to become better than in 2004. By taking part in a three week training camp to improve running for the competition in the world championship for disabled athletes, his CNS functioning improved and got better than in 2004 with respect to the high-load CD values (Figure 128A). A two week intensive therapy course with the Author improved his high-load CD values substantially. Benjamin achieved high-load CD values normally to be expected of uninjured people (Figure 128D).

3.5.2. Rate of Repair in Benjamin, Mario and Andrej Quantified by High-Load CD

Using the high-load coordination dynamics (CD) values, the rate of repair can be compared with those of the healthy school children. In healthy pupils and athletes the decline of the high-load CD values to reach the point of plateau, one to two months were required (Figs.36, 40,128D). Benjamin needed approximately five years to get his high-load coordination dynamics values under 100 in 2004. If one includes the time during which no therapy was performed, then he reached the plateau in thirteen years. If one takes two months as the average for an uninjured person (Figs.36, 40, 128D) and 60 months (five years) and 156 months (thirteen years) for the repair in Benjamin (Figure 128A), then the repair/improvement in Benjamin needed 30 to 78 times more time. The huge difference of time needed to improve CNS functioning can clearly be seen by comparing the healthy curve (128D, inset) with Benjamin's improvement curve (Figure 128A). The rapid rate of improvement in Benjamin in 2004 with optimal CDT indicates, the rate of repair depends strongly on the efficiency of the treatment (see also below).

Mario and Andrej never reached a high-load value of 100 (dashed line). The efficiency of their treatments was so low that have not reached the repair goal that they will have to live as disabled persons for the rest of their lives unless optimal CDT is administered to them. Even in adult patients with severe brain injury a long time after the injury is sustained, substantial progress can be achieved if efficient treatment is administered.

3.5.3. Comparison of Injury and Repair between Benjamin Mario and Andrej

The curves of the high-load CD values also provide information concerning the severity of the injury and the speed of repair. Mario had at the beginning much higher (worse) high-load values than Benjamin. His injury was probably more severe than that of Benjamin. Andrej, on the other hand, started with values close to that of Benjamin. His CNS was probably similarly damaged to that of Benjamin. But if we compare the curve of Mario with that of Andrej, one can see that Mario started from much higher values than Andrej. His curve declined more than that of Andrej. Even though Mario had a more severe CNS injury than Andrej, the repair of his CNS was more efficient. As stated above, even home treatment was better than conventional therapy. The curves of high-load CD values give something of an overall picture of severity of the injury and the speed of repair (Figure 128).

3.5.4. Comparison between High-Load and Low-Load Coordination Dynamics on Benjamin (CDT), Mario (Home Training) and Andrej (Conventional Therapy)

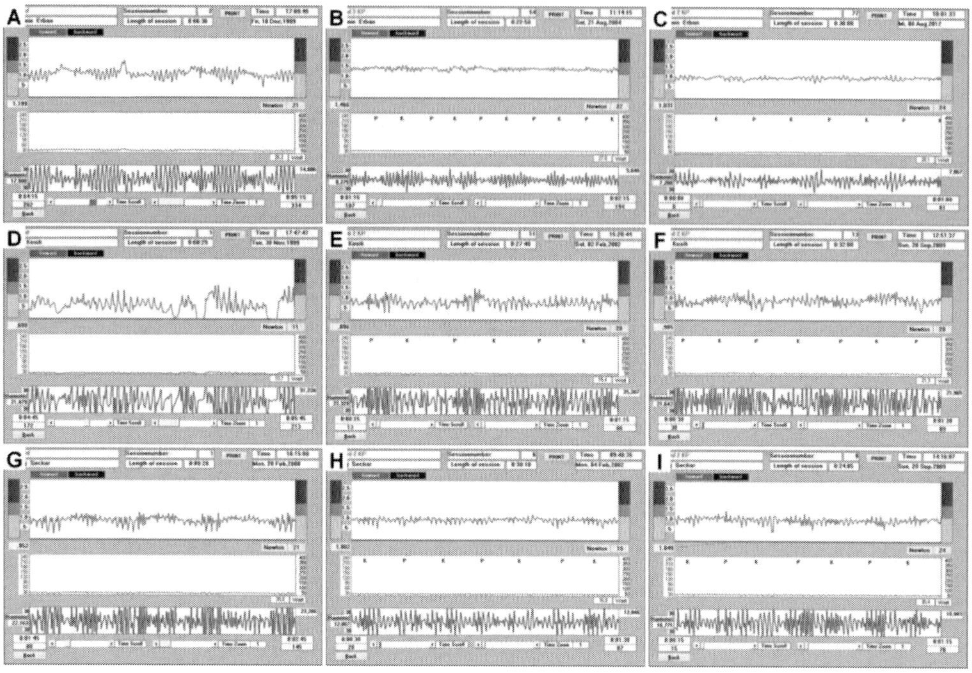

Figure 129. Low-load Coordination dynamics values of the patients Benjamin (A,B,C), Mario (D,E,F) and Andrej (G,H,I) with severe brain injuries in dependence of the years of therapy.

Figure 129 shows the improvements of low-load CD (20N) in the brain-injured patients Benjamin, Mario and Andrej upon CDT, home training and conventional training respectively. The high CD values from the beginning of therapy are not available, because the assessment did not exist at the time. Benjamin (Figure 129A,B,C) improved the values from 14,6 (1999) to 5.6 (2004) and then worsened to 7.1 in 2012. These low-load values show a similar progression to the high-load values. Also the low-load CD values were best in 2004 after optimal CDT treatment. Further, in 1999 (A) transient frequency increase can be seen in Benjamin's recording, similar to that which occurs in healthy children during development between 9 and 11 years. In that respect, Benjamin's development shows normal aspects.

In Mario (D,E,F) the low-load CD values improved continuously at a high (poor) starting point from 31,2 in 1999 to 25.2 in 2002 to 22 in 2009. Andrej improved his values from 23.3 (2000) to 12.6 in 2002 and got then a worse value of 16 in 2009 (G,H,I). Also for the home training and the conventional training the low-load CD values show a similar progression to the high-load values.

3.5.5. Overall Coordination Dynamics (CD) During Follow Up Treatment in Therapy Clinic

If patients come regularly to a physiotherapy clinic and the special device is connected to a computer, CD values and therapy details can be measured, stored and displayed.

Figure 130 shows such overall improvement of a patient (Benjamin) (A) in comparison to an uninjured person (the Author) (B). It can be seen that Benjamin's CD improved, whereas those of the healthy person remained approximately the same value. In Figure 130 the CD values are plotted via the harmonics, which display the rhythmicity of exercising. Higher harmonics values are better. Also the time of exercising, the number of turns (revolutions) and the work (Joule) done are stored. It can be documented over months or years with how much effort was required for the patient to achieve a certain repair state. It is possible to make a record of what the patient has done on the special CDT device.

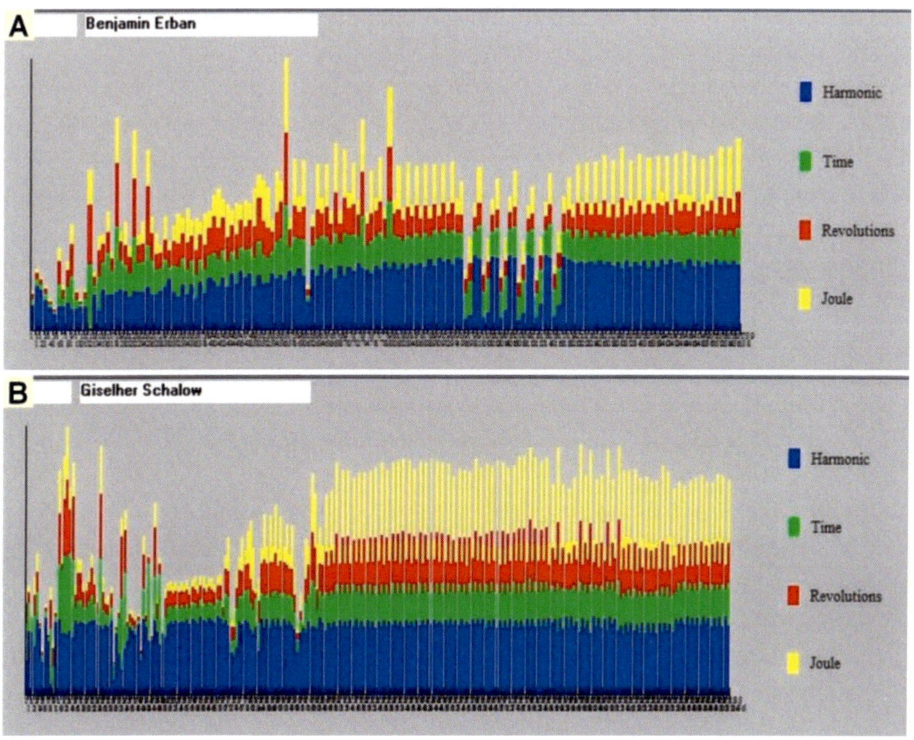

Figure 130. Overall coordination dynamics (mean of CD values for all loads) quantified by the 'Harmonic' (1/CD value = higher values are better), the time of exercising (time), the number of turns (Revolutions) and the done work in Joule are plotted for the brain injured patient Benjamin (A, upper part) and the healthy Author (B, lower part) for comparison. Note that Benjamin's Harmonics are getting better (higher) and those of the Author stay at a certain high (good) value. An overall documentation of therapy improvement with therapy sessions is possible.

Such records can be important in difficult cases. For example, a twenty-five year-old patient with moderately severe brain injury due to hypoxia claimed to be performing CDT largely by herself for at least four hours per day for two years. Her mother discontinued CDT because no improvement could be seen in her daughter. It later transpired that the patient was doing almost nothing.

3.5.6. Comparison between Optimal CDT and Sport Training

For practical reasons the Author was not able to offer the patient Benjamin optimal CDT again as in 2004. Benjamin started to train with university sports coaches in 2009. His high-load CD values improved to a similar extent, if not speed, as during optimal CDT treatment in 2004. With optimal CDT however, he needed eighteen days and with the sport training he needed four years for the same improvement. The repair in 2004 was approximately fifty times faster than that in the period 2009 to the end of 2012 (Figure 128). There are two obvious reasons for this. Firstly, optimal CDT is much more efficient in repair than sports training with respect to the repair of the injured CNS. Optimal CDT therapy includes also sEMG measurements to ascertain under which training conditions the best motor program is achieved (Figure 124). The more physiologic the exercise exercising with most patterns, probably the repair is most efficient. Generally, the coaches had now specific knowledge about human neurophysiology for repair, did not quantify the progress with an assessment like the coordination dynamics measurements and were not using sEMG.

Secondly, Benjamin was approximately six years younger when CDT was administered to him. The rate of repair of the injured CNS depends strongly on the age of the patient. Younger patients have a higher rate of repair. In very young patients the rate of repair is probably tremendously higher. Premature born babies probably have the highest rate of repair, even though it is difficult to administer CDT to them. But special CDT devices are made for very small children (Figs.31,32). In premature born babies the rate of neurogenesis (from endogenous stem cells) is most likely much higher than during later development. Newly born nerve cells, once integrated in the existing networks, will have a greater impact on repair and memory than changing more or less the connectivity only. Substantial 'correction en route' during development is possible and can be used for the repair of the injured CNS, caused for example by the hypoxia due to premature birth.

3.5.7. Treatment Start Already in the Vigilant Coma State

The CNS of the patient Benjamin could be partly repaired by CDT. As stated above, the treatment has to start as early as possible. The possibility cannot be excluded that Benjamin's tremendous improvement is partly due to the early start in the vigilant coma stage (Figure 118B). The same may be true of Mario. Mother, physician, physiotherapist (Figure 120) and the Author all believed in early treatment.

3.5.8. Rate of Repair Depends on Age and Treatment Quality

The progress in repair depends on the severity of the injury, the quality of treatment and the age of the patient. Benjamin offers the possibility to compare the rates of repair for different treatments. Above it was shown that Benjamin needed between five and thirteen years to improve his high-load CD values to what a healthy person learns in two months at the most. The rate of repair in Benjamin was 30 to 78 times slower.

In 2004 it was shown that the repair efficiency in Benjamin could be improved by a factor of 100, when he was 15 years old.

In 2013 a coach in a training camp was able to improve the high-load CD values in Benjamin, below the values of 2004 in three weeks (Figure 128A). But in 2013, the Author was also able to improve the CD values substantially with twelve days of optimal CDT (Figure 128A). The patient reached the values of trained healthy persons (Figure 128D). A comparison of the efficiency of treatment between the coach and the Author seems possible. But the comparison of treatment is complicated by the possibility that the benefit from treatment may be latent to some extent.

Further, by comparing the progress made by the patient in 2004 and in 2013, it seems that the rate of repair was higher in 2004, most likely because the patient was younger then. In 2004 he improved by 80 in 16 days. This means that Benjamin's CD improved by 5 per day. In 2013, he improved by 27 in 90 days, which equates to an improvement of 0.3 per day. In 2004 therefore, the rate of learning was approximately seventeen times faster. However, when the nervous system is already functioning quite well, judged by a high-load CD value of under $100s^{-1}$, then it is much more difficult to improve its functioning further in both the injured and uninjured persons (Figures 128,39).

3.5.9. No Limit of CNS Repair If CDT Is Applied Continuously Over Several Years at the Limit

It will be shown now that there seems to be no limit of CNS repair, if CDT is applied continuously over many years at the limit and is supervised by the Author who can upgrade the therapy if needed.

It is predicted by the System Theory of Pattern Formation that the intrinsic dynamics $\mathbf{F}_{intr}$ can be changed, i.e., repaired by the behavioral information $\mathbf{F}_{inf}(\mathbf{X},t)$ (the training of the different movements) of relative strength (c_{inf}) (volume of a certain movement training) according to the equations of motion of collective variables (the vector $\mathbf{X}$):

$$d\mathbf{X}/dt = \mathbf{F}_{intr}(\mathbf{X}) + \sum c_{inf}\mathbf{F}_{inf}(\mathbf{X},t)$$

It was further predicted that when with a certain training the CNS functions cannot be changed any more, then the treatment has to be changed. However, the System Theory provides no information about the specific changes needed to upgrade the therapy and the intensity to be applied.

It seems that when for example the walking performance is no longer improving, the performed movements cannot reach (activate) damaged networks for repair, which are urgently needed for improving walking. But it could also be that in the chain of events for repair some other networks have to be repaired first before the neural networks for walking can be improved in its functioning. And last but not least, the benefit from newly built nerve cells when exercising at limits seems to take place only after a year of treatment.

With the patient Benjamin it was shown in Figure 123 that through an upgraded therapy in a professional setting, his CNS functioning could be improved further, even though Benjamin thought that he had reached the limit of repair in 2004.

The patient William with cerebral palsy (Figure 104) attempted late repair of CNS functioning. His high-load coordination dynamics improved strongly (Figure 104M). His shortened tendons got a bit longer due to stretching on the sky-walker (Figure 104L). However, his CNS improved only little with respect to better walking. This lack of improvement in walking and running disheartened him so much that he stopped intensive CDT after ten months of therapy, against the advice of the Author.

It is always good practice to repair the CNS because one never knows what complication may come with aging. For example, patients with mild poliomyelitis are often able to walk. In spite of lost motoneurons, the CNS manages to generate the walking pattern. But with aging and degeneration of the CNS, polio patients may lose the ability to walk and have to use a wheelchair (post-polio syndrome).

The patient Benjamin (traumatic brain injury) also had problems with the improvement of walking, running and jumping performance. In 2014, he asked the Author for more supervision. And indeed, during one week of intensive supervision two movements could be found, namely certain jumping (Figure 131B) and walking patterns, which reached those damaged neural networks which contribute to the improvement of the walking and running patterns.

Figure 131. Jumping in anti-phase of a now 25-year-old patient Benjamin who suffered a severe brain injury 15 years ago. A. Left foot is in a pathologic position. B. When jumping against a step, the performance of jumping is much better.

At an age of 24 the main problem in CNS functioning were still the pathologic functioning of the left arm and leg. The free jumping in anti-phase improved the coordinated movement of the left arm. But during this free jumping in anti-phase, the left heel came up (Figure 131A). The physiologic positioning of the left foot is important for all movements and needs to be corrected so that the CNS does not learn a pathologic pattern. Now, by freely jumping against a step, the left foot took a physiologic position and the left arm moved (Figure 131B). Only the left hand and foot were a bit spastic. This kind of upward jumping made the anti-phase jumping somewhat physiologic: The left arm moved and the left foot was in the physiologic position. When jumping a step downwards, the left foot still went into pathologic position.

Interestingly, when climbing staircases upwards, the left foot performed a physiologic movement. When going downstairs the foot positioning was pathologic, in some similarity to the jumping in anti-phase the step up and downwards.

The backward walking performance was quite good in Benjamin. The only problem for covering a distance is that the eyes are on the wrong side of the body. The normal forward walking performance on the other hand was poor and had to be improved. The forward walking could be improved in the short-term memory by simulating and activating the stepping automatism. The stepping automatism (Figs.12a,24) is mainly located in the neuronal networks of the spinal cord and is stimulated for example by the heel strike. The patient had suffered a traumatic brain injury and the stepping automatism should therefore be only little disturbed. When Benjamin was walking in such a way as to simulate the stepping automatism by exaggerating the lifting of the knees during walking and striking the ground with the heel first, the left leg was not swinging out any more and the pathologic inward rotation of the left foot reduced.

It seems likely that Benjamin can further improve his walking and running performance when training these two movements.

The instruction 'put the heel down or do not swing the leg outwards during walking' will not help very much to improve the walking performance. But finding certain movements which functionally reach the damaged networks can.

When Benjamin got further better and his high-load CD values got better, he wanted to compete with the Author and to surpass his good high-load values. The Author cannot compete with the fast running of Benjamin. However, nobody so far can beat the Author in the high-load test where good nervous system functioning, quantified by the coordination of arm and leg movements, is also needed.

3.5.10. Comparison of High-Load CD Values between a Disabled Sportsman and the Author (Aging)

Figure 131c shows the recent high-load (A) and low-load values (B) in sequence of Benjamin (traumatic brain injury 15 years ago) and the Author (over 70). It can be seen that the disabled sportsman Benjamin could reach good (low) high-load CD values, but he is still far away from the Author's values. Recent improvements were obtained in the patient by jumping in anti-phase, including the arms during jumping (Figure 131). With this coordinated arm and leg movements during jumping the performance of the left arm improved. On the other hand, the low-load CD values were comparably poor (high) in Benjamin in comparison to the Author (B). It seems therefore as if Benjamin is succeeding in getting good high-load values partly by using his muscle strength, whereas the Author needs to put the emphasis on good coordination between arms and legs to compensate for less muscle strength. This means that the Author is taking advantage of better phase and frequency coordination between neuron firing in the CNS.

The competition between Benjamin and the Author can also be seen as a competition between a young person fighting for more repair of his injured brain and an older person fighting by use of the nervous system for more rejuvenation of the body. The attempt to reduce the biological age will be picked up below under aging.

4. Complete Cervical Spinal Cord Injury – Patient Kadri

Before turning to even more severe brain injuries, the complete cervical complete spinal cord injury (SCI) will be considered. An incomplete SCI can be largely repaired. As was shown above, a patient with an injury of approximately 50% of the nervous tissue of the spinal cord damaged relearned walking, jumping and running and became continent within six months of therapy (Chapter V.4).

The more severe the injury, the slower the rate of repair. Perhaps it is possible that a SCI patient with 20 to 30% of the cross-sectional area left can re-learn to walk. A patient with a complete SCI, that means 5% of the nervous tissue of the spinal cord remaining following injury, cannot re-learn walking so far within five years of therapy. A patient with a complete SCI will be discussed here. Even in cases in which comparably little nervous tissue is destroyed, as in SCI, the functional implications can be severe. As an introduction to the repair in severe SCI, we shall first turn to groups of patients with largely incomplete injuries.

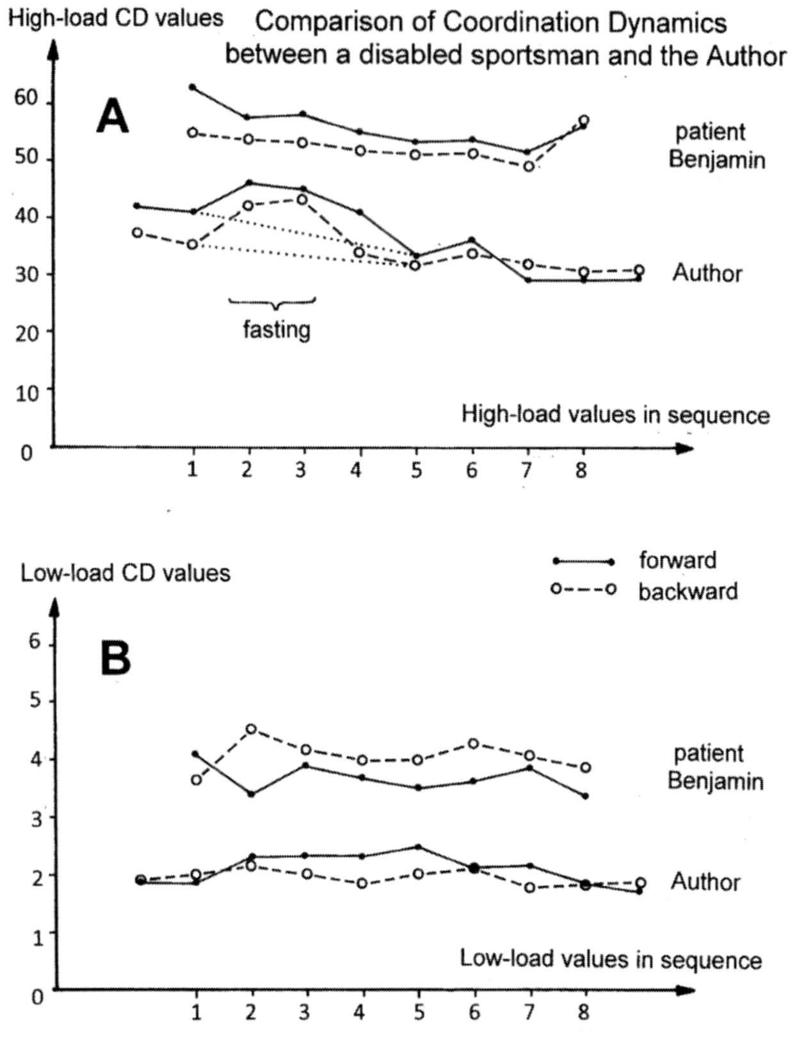

Figure 131c. Comparison of high-load (A) and low-load values in sequence (B) of the patient Benjamin, who suffered a severe brain injury 15 years ago, and the Author.

4.1. Repair in Spinal Cord Injury Achieved by Three Months of Coordination Dynamic Therapy (Eighteen Patients)

4.1.1. Summary

Coordination dynamic therapy (CDT) was applied to eighteen patients (average age 31 years) after a spinal cord injury (SCI) between C4/5 and L4/5; the therapy was administered on average five years after the injury for a minimum of three months. All complete SCIs became functionally incomplete, i.e., motor functions improved below the injury level, including trunk stability and arm, hand and leg functions.

The organization of the CNS, quantified by the coordination dynamics between arm and leg movements, improved by 42% for forward and by 49% for backward moving when exercising on a special CDT device. The improvements of the coordination dynamics were 53%, 32% and 48% for lesions in the cervical, thoracic, and lumbar range, respectively. The plasticity for spinal cord injuries was thus higher when the intumescence alone is injured. Since the coordination dynamics did not change substantially prior to coordination dynamic therapy, further improved with continued therapy and worsened when the therapy was terminated, it is concluded that the improvement of CNS function above and below the spinal cord injury level was due to the therapy. Since in stroke and traumatic brain injury CNS function further improved if the therapy was continued over longer time periods, i.e., longer than three months, it is likely that SCI can be partially repaired if CDT is administered aggressively for three or more years.

4.1.2. Functional Repair in Brain Injury and More of a Structural Repair in SCI Repair

In stroke and traumatic brain injury the injured brain could partly be repaired when administering CDT for three months [1,2].

Further therapy resulted in further repair. Improvements of central nervous system (CNS) functioning could be quantified by improvements of movements and by measuring the improvement (70%) of the coordination dynamics between arm and leg movements. The repair of brain injuries in this study was most likely attributable to functional reorganization.

In SCI on the other hand, significant structural repair is necessary, since connecting tracts in addition to the grey matter are substantially destroyed. Further, in cervical SCI, motoneurons are destroyed which innervate hand and finger functions and in lumbosacral SCI motoneurons are destroyed which innervate leg muscles. A partial cure of the SCI therefore also requires the building of new nerve cells including motoneurons and their functional integration in the neuronal networks of the CNS (Figure 29) [11]. It will be shown here that spinal cord repair can commence after just three months of CDT. For a partial cure of a spinal cord injury however, much longer therapy times are needed. CDT is a learning method, and learning needs time.

4.1.3. Functional Anatomy and Repair Strategy

Following spinal cord injury, neuronal network parts and connections (tracts) become destroyed and the relative phase and frequency coordination between neurons, neuron assemblies and neuronal network parts for the self-organization of the central nervous system (CNS) become impaired [14]. To partly cure the injury, the relative phase and frequency coordination has to be restored by learning. At key functional spinal cord sites new neurons have to be built. In a patient with a cervical SCI, hand and finger functions can only be substantially repaired if new motoneurons are generated or if properties of neurones are changed as there must be motoneurons, which bring the information from the CNS to the muscles (Figure 29).

A partial cure of the pathologically functioning CNS, due to the SCI, by relearning is achieved through: (1) The use of special instruments for relearning of the lost phase and frequency coordination between the firings of neurones, neurone assemblies (for example, spinal oscillators) and neuronal network parts by exercising the impaired or lost coordination

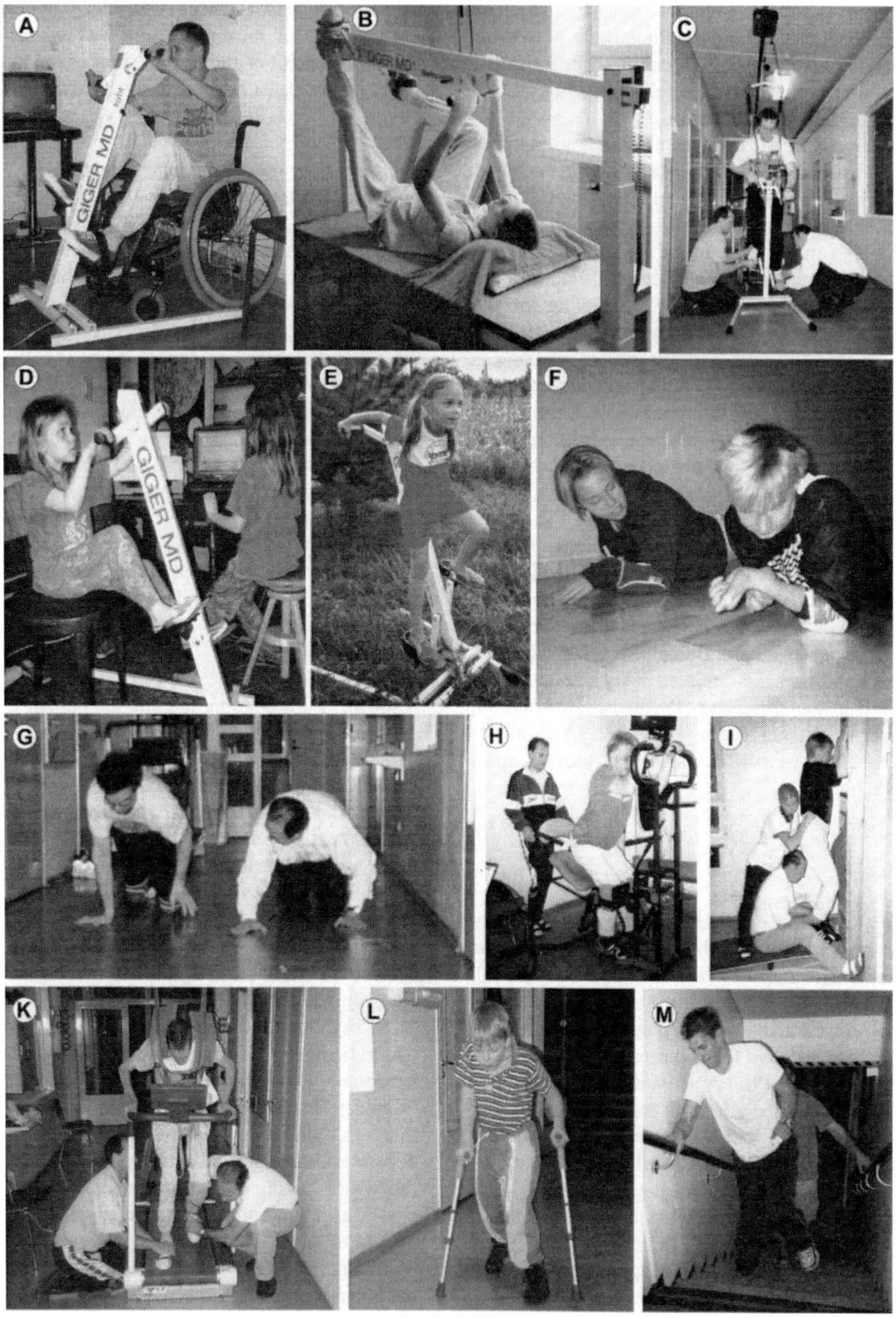

Figure 132. Different movements included in the performance of coordination dynamic therapy. A,B. 21-year-old tetraplegic patient with a complete spinal cord lesion sub C4/C5 exercising in the sitting (A) and lying position (B) on a special coordination dynamic therapy device. In A, the hands are not fixed allowing the exercising and measuring of hand and finger functions (see Figure 133Ab). In B, the hands are fixed. C. 33-year-old paraplegic patient with a spinal cord lesion sub Th6 exercising on a special coordination dynamic therapy device in the standing position. The position of the feet had to be supported (author on the right side). D. Measuring of coordination dynamics. The 'young and dynamic' patients with scoliosis are supervising each other to turn more rhythmically when watching the coordination dynamic traces on the display. The 6-year-old Katharina watches the counter of turnings (notice the concentrated face) and has her moving hands in the field of vision for biological feed-back.

E. Same patient as in D, demonstrating different kinds of coordinated movements on the device. F. 25-year-old tetraplegia patient with a spinal cord lesion sub C5 during creeping exercise for training those neuronal networks which are activating arm and trunk muscles. The therapist is creeping in interpersonal coordination. For quantification of improvement in creeping, see Figure 133Aa. Notice that hand and finger function are not exercised as in A. G. 33-year-old paraplegic patient, lesion sub Th6 crawling in interpersonal coordination with the Author; the movement is quantified in Figure 133C. H. 28-year-old tetraplegia patient with a spinal cord lesion sub C7 during exercising up-righting. I. Same patient as in H during jumping on springboard. One therapist (Author) is supporting the feet and the other one the trunk. Notice that the forefoot of the patient is not touching the springboard, to prevent induction of extensor spasticity. K. Same patient as in A,B during treadmill walking. Therapists are supporting feet and leg movements (Author on the right). By pressing the feet to the ground the movement induced re-afferent input is enhanced. When pressing preferentially the heel when the foot touches the ground and when pushing the forefoot at the lift-off phase, the stepping automatism is stimulated. L. 7-year-old paraplegic patient with a spinal cord lesion sub Th12/L2 during walking with sticks; the walking is quantified in Figure 133E. M. 29-year-old patient during climbing staircases, quantified in Figure 133F. Patient is moving in pace gait coordination (arm and leg on each side are moving in synchrony).

between the movements of arms, legs and trunk. With improvement in the coordination of movements the motor programs are also improved. The motor programs can be measured by surface EMG. Muscle strength also improves because the motor units in a muscle fire in a more coordinated manner. (2) The exercise of automatisms, postures and old learned movements which have been only little changed by the injury. Automatisms are probably most relevant to structural reorganization.

The movements which are most often used for repair in CDT in patients with SCI are illustrated in Figure 132. Figure 132A,B,C shows how phase and frequency coordination is relearned by exercising on a special CDT device in the sitting, lying and standing position. In A, a patient with a C4/C5 injury is shown while training additionally hand and finger functions, and in B, additionally trunk stability is trained by rotational trunk movements. In the standing position (antigravity muscles are substantially activated) movements have to be supported (C). Quite a variety of coordinated arm and leg movements is possible with this device (Figure 132E). The improvement of CNS organization, that is to say, the quality of the coordination dynamics can be easily observed during the measuring process as shown in Figure 132D.

In Figure 132 F to M the exercising of automatisms and old learned movements like commando-crawling, crawling, standing up, jumping, walking and climbing staircases are shown. When walking with support on a treadmill, it is of importance to induce and support the stepping automatism if only rudimentary leg functions remain to the patient. Therapists have to develop a clear view of the level of the patient's nervous system and have to support it so that the best performance of walking is achieved. In accordance with the induction of neurogenesis [93, 94, 100] those neuronal network parts have to be substantially activated where neurogenesis and functional reorganization have to be induced. In other words, for spinal cord injuries the patients have to substantially exercise leg functions (see Figure 132G to M), ideally in coordination with arm movements. Finger and hand functions of patients with cervical spinal cord injuries (Figure 132F) can be trained on a special CDT device (Figs.132A,187F). For stroke patients, Berta Bobath emphasized 35 years ago that the impaired side of the body must be used during therapy [101]. Instructive training, interpersonal (social) coordination and co-movements are used to enhance physiologic movements in the patient.

4.1.4. Injuries of the Patients

Eighteen patients who suffered SCI underwent CDT for at least three months. Their injuries were variously due to car accidents (5), working accidents (4), jumping head first into water (4), motor-cycle accidents (2), playing ice hockey (1), a parachute accident (1) and negligently performed surgery (1). The average age of the patients was 31 years (range 7 to 55), their accident occurred on average five years prior to treatment (range 1 to 23 years), and the injury level was between C4/5 and L4/5. Motor functions improved in all patients. In rostral SCIs (cervical range) mainly trunk stability and arm and hand functions improved. In more caudal injury levels (thoracic and lumbar range) trunk stability, leg and foot functions improved.

The progress achieved was measured by the improvement of movements and was measured by the improvement of the organization of the injured CNS, i.e., by measuring coordination dynamics.

4.1.5. Recovery Ffom Spinal Cord Injury Quantified by Improvement of Movements

Movements performed during four to five hours therapy per day were quantified and protocolled. The movements to quantify progress were mostly the speed of commando-crawling (Figures 132F,133Aa), crawling (Figures 132G,133B,C) and climbing staircases (Figures 132M,133F) and the number of jumps when jumping on springboard in anti-phase (Figures 132I,133D). The improvement of hand and finger functions was quantified by the number of turns without hand fixation when exercising on a special CDT device (Figures 132A,133Ab). Typical movement improvements are shown in Figure 133. The other movements used to quantify improvement were speed of walking with sticks (Figures 132L,133E), duration of supported or unsupported walking on the treadmill and number of times the patient was able to stand from a seated position (Figure 132H).

In Figure 133Aa, a typical improvement in commando-crawling is shown. With ongoing therapy the patient covered in a faster timeframe 7.2m. At the start of therapy, this patient with a complete spinal cord lesion sub C5 needed approximately 65 seconds and at the end 22 seconds. This patient thus relearned to crawl faster, which indicates that arm power and trunk stability improved. Since he also learned to turn longer on the special CDT device without hand fixation (Figure 133Ab), also hand and finger functions improved. Over the three months of therapy the patient could not relearn to crawl. However, the relearning of commando-crawling alone was of practical importance.

In Figure 133B,C, the improvement in crawling in two patients with clinically complete SCIs sub Th5/9 and Th6 is shown. The crawling times for 7 meters shortened from 40 seconds to 8 seconds and from 61 seconds to 9 seconds. In both cases the curves show rather linear progressions. After an initial strong reduction of the crawling times, the times then reduced only slowly. It was as if the two phases of improvement were each underlain by two distinct mechanisms of improvement. Both patients could also commando-crawl.

In Figure 133D, the number of anti-phase jumps of the longest series is plotted against the therapy time of a patient with a complete SCI sub L1. With ongoing therapy this patient could jump longer on springboard, indicating that his leg and foot functions improved. Especially the triceps surae and the gluteus maximus muscles became stronger.

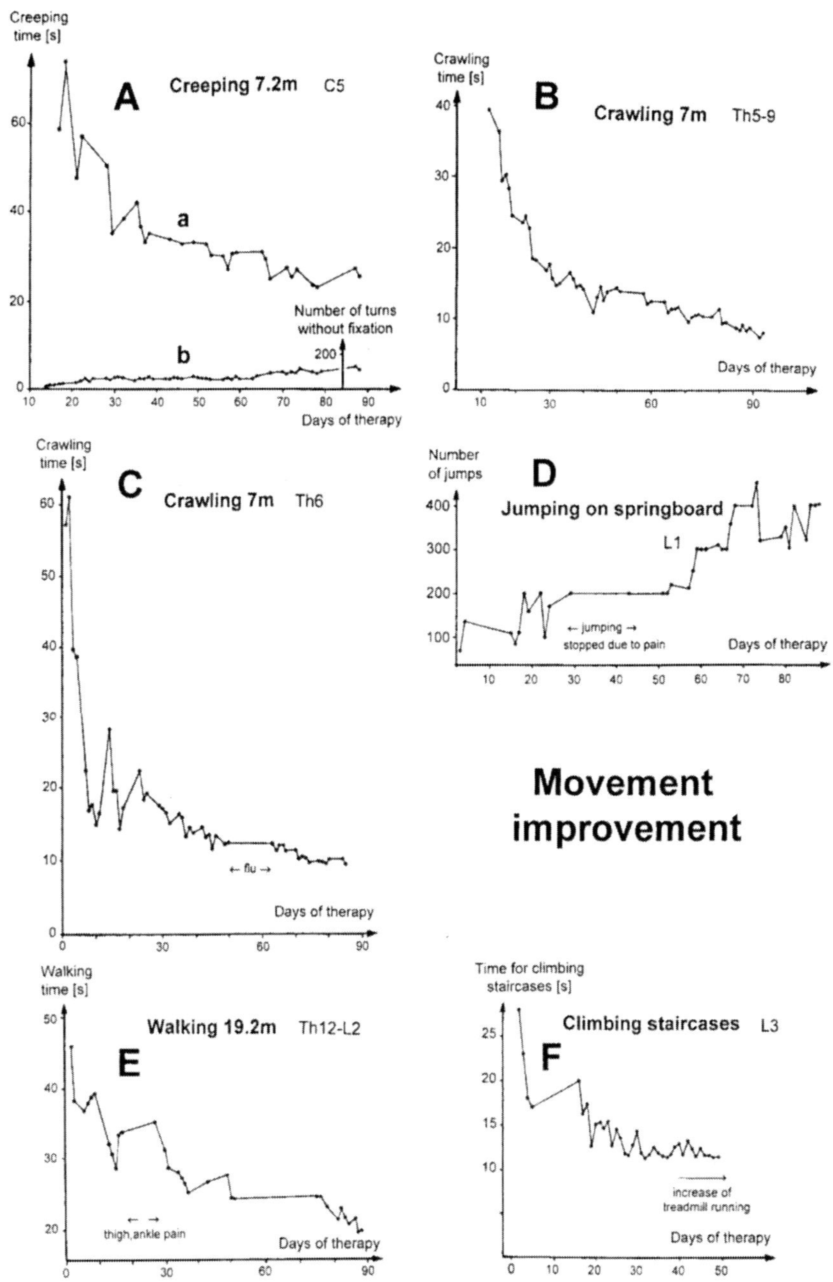

Figure 133. Movement improvements of creeping, crawling, jumping on springboard, walking with sticks and climbing staircases are quantified in terms of time (Aa, B, C, E, F). Improvement of hand and finger functions is quantified by the number of turnings on the special coordination dynamic therapy device (Ab). The jumping is quantified by the number of jumps (D). Notice that the curves for crawling in B,C seem to comprise two components. Notice that pain (D,E) and a flu (C) seem to hinder the progress.

When the patient had to stop jumping for a few days due to pain (Figure 133D) the number of jumps the patient could perform per series stagnated for a short period before improving further. The improvement of walking with sticks is shown in Figure 133E. Within three months of therapy the time needed to cover 19.2m reduced from 46 to 19 seconds, indicating improvement in leg and foot function. Transient pain in ankle and thigh stopped the improvement transiently.

In Figure 133F, the faster climbing of staircases of a patient with a SCI sub L3 is shown. The patient was holding the handrail with his left hand. When holding it with the right hand during climbing (Figure 132M) a very similar curve was obtained. Alternating support with the left and right hand made a continuous rhythmic movement possible. In the latter case, the patient was able to run up the staircase.

4.1.6. Spinal Cord Repair Quantified by Improvement of Coordination Dynamics

Measuring coordination dynamics in patients with a SCI is important for two reasons. Firstly, it can be discerned whether a SCI is complete or incomplete and secondly, changes in CNS function can be quantified.

4.1.7. Diagnosis of Complete and Incomplete Spinal Cord Injury

When a SCI between the cervical intumescence, the site at which rhythmic stereotyped arm movements are located, and the lumbosacral intumescence, the network location of stereotyped leg movements, is complete, the patient can turn on the special CDT device relatively well with the upper body alone with the legs only moving passively. The CD trace does not show rhythmic changes. When the SCI is incomplete or becomes incomplete, then some leg muscles work in coordination with the arm muscles while exercising on the special CDT device. With the change between easy and difficult coordination's (dictated by the device), the quality of rhythmicity (the size of the amplitude of Δ(df/dt)/f) will change. Figure 134A,B and Figure 134D,E show two patients with a spinal cord injury in the cervical and thoracic range, who became incomplete over the three months of coordination dynamic therapy. In Figure 134A,D (lower traces), the coordination dynamics (Δ(df/dt)/f), do not show changing coordination dynamics. Three months later (Figure 134B,E) however, the coordination dynamics show rhythmic changes due to the ongoing changes of the coordination between arm and leg movements when exercising on a special coordination dynamic therapy device. The SCI has thus become incomplete.

4.1.8. Improvement of Coordination Dynamics within Three Months

The coordination dynamics were measured in all patients at the beginning of the coordination dynamics therapy, at the end of the therapy after three months, and again three to six months later, when most of the patients came for a check-up. The coordination dynamics (Δ(df/dt)/f) for forward turning of the three patients with a cervical, a thoracic and a lumbar spinal cord lesion are shown in the lower traces of Figure 134. In the first patient with a cervical spinal cord lesion sub C4/5 (Figure 134A,B) the coordination dynamics improved from 10.3 to 3.8 (by 63%). In the second patient with a thoracic spinal cord lesion sub TH 5/8 (Figure 134D,E) the coordination dynamics improved from 8.1 to 4.8 (by 41%). In the third patient with a spinal cord lesion sub L1 (Figure 134G,H) the coordination dynamics improved from 5.4 to 2.2 (by 59%). For all spinal cord lesions (including cervical, thoracic and lumbar lesions) the coordination dynamics improved by 42 ± 16% (mean ± s.d.) for forward turning and by 49 ± 17% for backward turning within three months of therapy.

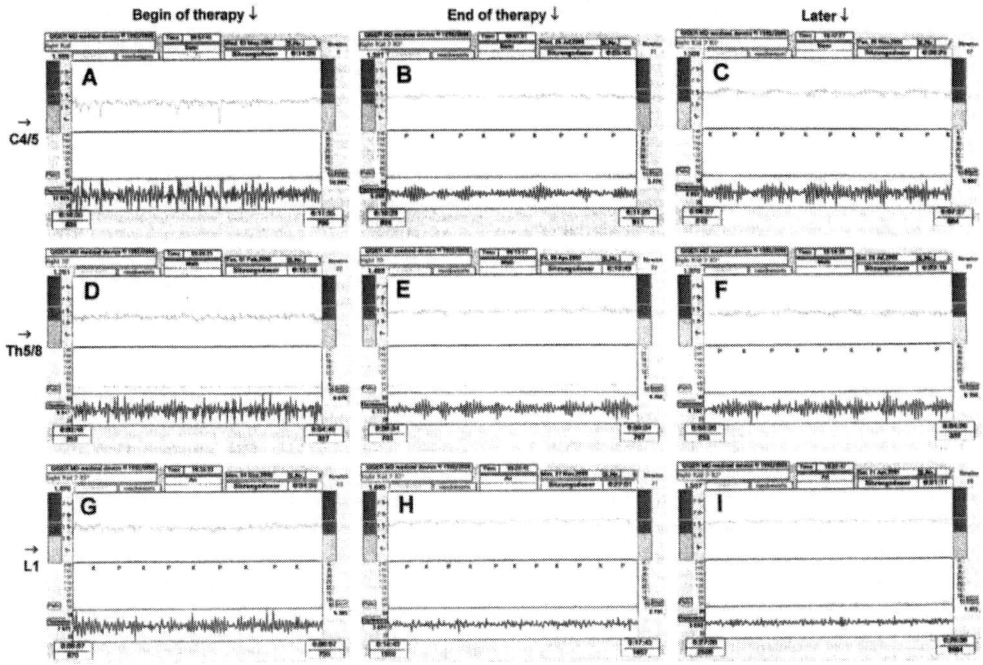

Figure 134. Coordination dynamics of 3 patient with spinal cord lesions sub C4/C5 (A,B,C), sub Th5/8 and L1 (G,H,I) at the begin of therapy (A,D,G), at the end of 3 months of coordination dynamic therapy (B,E,H), and 3 to 5 months later (C,F,I). Notice that coordination dynamics (lower trace = Δ(df/dt)/f) = blue) improve from A,D,G (begin of treatment) to B,E,H (end of treatment); the variation of rhythmicity becomes smaller for each patient. Later after treatment, the coordination dynamics became worse when the therapy had been stopped (C,F), and became further improved when the therapy was continued (I). Some pace (P) and trot gait coordination's (K) are indicated. Upper trace (yellow) = frequency of turning.

4.1.9. Changes of Coordination Dynamics after the Three Month Therapy Course: Improvement with Further Therapy and Worsening with No Therapy

Apart from the five patients who did not come for a follow-up appointment, the coordination dynamics improved in all patients (eight patients) by very approximately 15% over three months if the therapy was continued with varying intensity. The coordination dynamics became worse by very approximately 14% over three months if the therapy in the five patients stopped (Figure 135A). When the patient with the cervical SCI (Figure 134B) stopped therapy, his coordination dynamics became worsened from 3.8 to 5.9 (Figure 134C) (by 20% in four months compared to the baseline value of coordination dynamics (10.3, Figure 134A)). When the patient with the thoracic SCI (Figure 134E) stopped exercising, his coordination dynamics worsened within three months from 4.8 to 6.4 (by 20%) (Figure 134F). When the patient with the lumbar spinal cord lesion (Figure 134H) continued CDT his coordination dynamics further reduced (improved) from 2.2 to 1.9 (by 6% within five months) (Figure 134I).

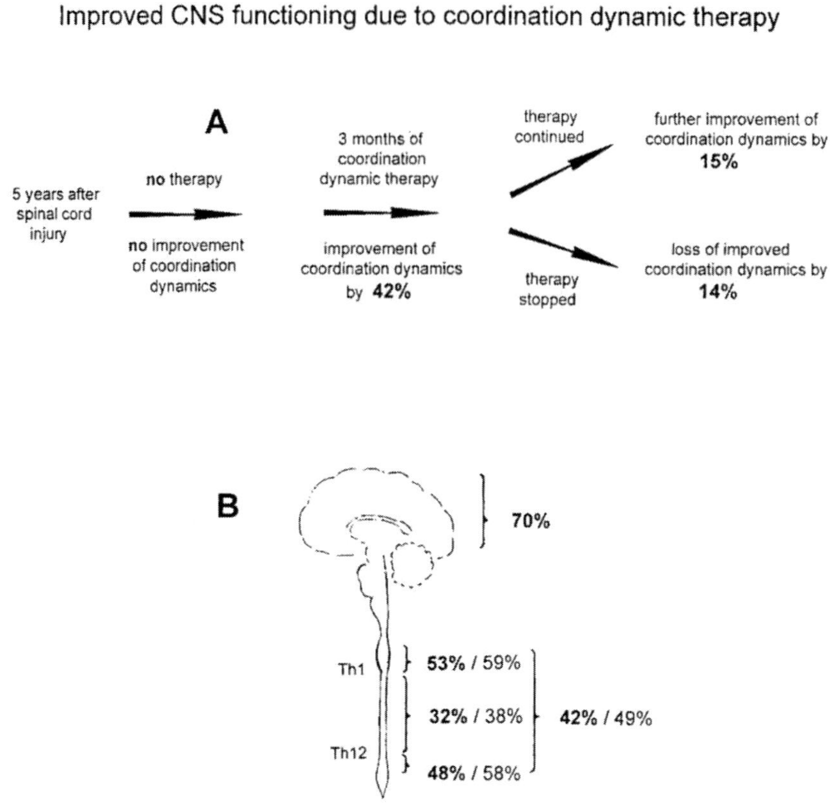

Figure 135. A. Schematic representation of improvements of CNS functioning in spinal cord lesion: with no therapy there is no improvement of the coordination dynamics, after 3 months of therapy there is a 42% improvement, followed by additional 15% (very approximately) further improvement later on, when the therapy was continued, and a 14% loss (very roughly), when the therapy was terminated. B. Improvement of coordination dynamics in CNS lesion due to coordination dynamic therapy for 3 months in dependence on the site of the lesion; first figure for forward and second figure for backward turning. Notice that the improvement is largest for cortex lesions (70%) followed by lesions of cervical (53%) and lumbosacral (48%) intumescence, and for lesions of the thoracic spinal cord (32%) for forward turning on the special coordination dynamic therapy device.

Most patients came a few months before the start of CDT for a consultation to discuss whether it might be of interest for them to undergo CDT to improve their CNS function. During the consultations their coordination dynamics were measured. By comparing coordination dynamics measured at the consultation with those measured as the baseline value at the beginning of therapy it was found that they were rather similar. This means that coordination dynamics did not change very much if no CDT was administered.

Altogether this means that the improvement of coordination dynamics (i.e., the improvement of CNS organization) in patients with a SCI was due to the CDT. Secondly, when the CDT was continued the CNS function further improved. Thirdly, when the therapy was stopped completely the achieved coordination dynamics worsened, indicating that the improved CNS organization in those patients was partly lost again. Therefore, to keep the achieved level of improved CNS organization a certain amount of further therapy seems to be necessary.

Results achieved in some patients in whom the CDT was first continued, then only continued with little effort and subsequently continued intensively again, seemed to suggest that very approximately one hour of intensive coordination dynamic therapy per day six times per week is necessary to keep the achieved level of repair when the patients uses a wheelchair and thus does not move much in daily life. Those patients were turning approximately 4000 times on the special CDT device that means they performed approximately 4000 coordinated arm and leg movements per hour. If patients with caudal SCI were able to walk, i.e., they could use the relearned improved CNS functioning in daily life, the amount of exercise necessary to maintain the reached level of CNS functioning seemed to be smaller.

4.1.10. Improvement of Coordination Dynamics Depends on the Injury Level

When classifying the improvement of coordination dynamics according to the level of the lesion (cervical, thoracic and lumbar) improvements for cervical SCIs were 53 ± 11% for forward and 59 ± 19% for backward turning, for lumbar SCIs they were 48 ± 16% and 58 ± 12% respectively, but those for thoracic SCIs they were only 32 ± 14% for forward and 38 ± 13% for backward turning (Figure 135B).

A better improvement could thus be achieved for cervical and lumbar lesions than for thoracic lesions. This may indicate stronger plasticity potential in the cervical and lumbosacral range than in the thoracic range. The proportion of grey matter (neuronal networks) to white matter (tract fibers) in the spinal cord is larger in the cervical and lumbosacral intumescence in comparison to those in the thoracic portion of the spinal cord.

4.1.1.1 Partial Repair of SCI

It was shown that spinal cord and CNS functioning can be improved by just three months of CDT. The progress in CNS function could be quantified by improvements of movements (Figure 133) and by improvements of coordination dynamics (Figure 134) during exercise on a special CDT device (Figure 132A-D). The coordination dynamics improved by 42% for forward turning and by 49% for backward turning. Since further treatment of patients with SCI resulted in a further improvement of CNS functioning as quantified by coordination dynamics (Figure 135A), and treatment in stroke [1] and traumatic brain injury [2] resulted in a partial repair of CNS injury, it is likely that a SCI may partly be repaired during three to five years of continuous, intensive CDT.

4.1.12. Plasticity Depends on the Site of the Injury

It was further found that improvement of the coordination dynamics was better for the cervical (53%) and lumbar (48%) than the thoracic range (32%) (Figure 135B). In stroke and traumatic brain injured patients [1,2], coordination dynamics improved by 70% within three months of CDT. The speed or the efficiency of CNS repair seems therefore to depend on the site of the injury, and may depend on the plasticity of the damaged network parts. So far it seems that hypoxic CNS lesions and brain stem lesions are very difficult to treat [77]. It is likely that only those network parts which are substantially activated are repaired by reorganization. This is in similarity to neurogenesis, where new nerve cells are only built in those parts of the brain which were substantially activated during training [93,94,100]. Of course, neurogenesis also depends on a variety of factors. Even though it has been shown that the damaged breathing center in the canine reticular formation repaired itself [102], it seems difficult so far to repair brain stem injuries. In one patient with a spinal cord and a brain injury the coordination dynamics improved only by 16% over six months of coordination dynamic therapy [77]. If, for instance the reticular formation is not activating the brain it is not possible to functionally reorganize the cortex because it cannot be activated. Perhaps the reticular formation has to be repaired first. Such a situation may be present in patients in permanent comas following very severe brain injury with attendant high brain pressure. The rate of CNS repair following an injury also seems to depend on the functions of the injured neuronal network parts and their co-operation and competition with other parts of the brain.

4.2. Repair in Spinal Cord Injury (SCI) Achieved through Six to Thirteen Months of Coordination Dynamics Therapy in Four Patients

4.2.1. Longer Treatment to Increase SCI Repair

Above it was shown that five years after SCIs between C4/5 and L4/5 all eighteen patients became incomplete as a result of three months of coordination dynamics therapy (CDT). Trunk stability and arm, hand and leg functions improved below the injury level. The organization of the central nervous system (CNS), quantified by coordination dynamics between arm and leg movements, improved by 42% for forward and by 49% for backward movement when exercising on a special CDT device at load of 20N. Now it will be shown that six to thirteen months of CDT resulted in a partial repair of the SCI in three out of the four patients.

4.1.2. Summary of the Group of Patients

Four patients with a SCI between sub C4/5 and L3/4 underwent CDT for six to thirteen months. One patient with an incomplete spinal cord lesion was cured, two patients with clinically complete injuries were partly repaired, and one patient with a complete spinal injury sub L3/4 became incomplete but showed comparably little progress. Of three patients with urinary bladder problems, only one patient has shown substantial improvement in bladder function. In one patient re-innervation of leg muscles could be documented by continuously measured coordination dynamics.

4.2.3. Strategy of Repair

CDT is based on measurements of the self-organization of the neuronal networks of the human CNS. An increase of the efficiency of this therapy depends therefore on a better understanding of the organization and repair of the human CNS based on further measurements on the single neuron level, the neural assembly level and macroscopic level of the healthy and injured CNS.

Brain imaging research focused on detecting the brain centers involved in various sensorimotor or cognitive tasks, but CNS organization research attempts to understand the self-organization of distributed local networks and their integration and coordination to 'higher' states of processing [90]. A basic mechanism for such self-organization and integration is the measured phase and frequency coordination on the single neuron and neural assembly level in the human CNS (Figures 9,9e) [14-16].

I could provide evidence for such self-organization and integration of the human CNS in the premotor neuronal network of the sacral spinal cord being materialized by relative phase and frequency coordination between the firings of single α and γ-motoneurons, oscillatory firing α-motoneurons (neural assembly) and muscle spindle afferents in the disconnected human spinal cord. These oscillatory firing motoneurons most likely self-organize from distributed local networks, which the motoneuron is part of, to the neural assembly 'network oscillator' (Figure 8) by adequate natural afferent impulse patterns induced by the receptors of the skin (Figure 5), muscles, tendons, joints and urinary bladder (Figure 6) and rectum, among other structures, by movements, continence stimulation of the bladder and rectum or other natural stimulation [14]. In incomplete SCI, these oscillators can also be activated on volition, i.e., by top-down activity as verified by electromyography [103]. These oscillators have a transient, dynamical existence that spans the time at least from one impulse train firing to the next (transient oscillatory mode) in which the neural activity propagates at least once through the oscillator, perhaps by a reverberator network loop [14].

In the process of organization and integration of different parts of the spinal cord, the incoming natural impulse patterns from the receptors of the periphery (bottom-up activity) and the outgoing natural impulse patterns from supra-spinal centers (top-down activity) are integrated with the endogenous activity of the spinal cord by relative phase and frequency coordination.

The interaction for neural organization and integration in the premotor network has been measured so far firstly between the sensory input and the premotor neural network, especially the local networks of the premotor spinal oscillators and γ-motoneurons, and secondly, by the coordinated natural firing patterns of α and γ-motoneurons (Figure 25).

Following spinal cord injury there is a loss of CNS tissue, loss of network connectivity in networks and between networks, and there is impaired phase and frequency coordination for CNS self-organization and integration.

Repair of the injured spinal cord therefore requires the building of new nerve cells (neurogenesis), growing of axons and dendrites to repair network connectivity, and relearning of relative phase and frequency coordination to repair functional connectivity and integrativity.

In the likely event of insufficient new, specific nervous tissue and network connectivity functional connectivity has to be re-learned and functions must be taken over by other neural network parts (plasticity). The effective functional connectivity between sub-networks has to be increased by entraining the neurons with their connections in the network as coincidence and coordination detectors (Figure 11) with extensive repeated coordinated natural input.

CDT primarily assists the repair of functional connectivity by increasing the effectiveness of the functional connectivity when the patient exercises on a special CDT device, and assists the repair of primarily the network connectivity when exercising automatisms. The assumption of responsibility for functions by other network parts (plasticity) is probably trained by re-learning old movements. Since a phase coordination of approximately 3 to 5ms has been measured between secondary muscle spindle afferents and oscillatory firing α-motoneurons in the healthy human premotor network (Figure 9), the special CDT device must dictate exact coordination to the arms and legs so that the induced natural impulse patterns in the receptors in the skin, joints, tendons and muscles are coordinated within 3 to 5ms to entrain the neural networks of the spinal cord so that the relearned phase coordination of the premotor network becomes again within the range of 3 to 5ms. The relearning of frequency coordination (Figure 9e) seems to be achieved if the patient with the SCI is turning the levers of the device as rhythmically as possible. Arrhythmicity of turning (coordination dynamics) has to become as small as possible.

The functions to be relearned, such as walking, have to be exercised. In the concept of macroscopic coordination dynamics, where movements are described by collective variables (in this case by arrhythmicity of turning), this means that in the attractor landscape the attractors for physiologic movements have to be deepened (made stronger and more stable) and attractors for pathophysiologic movements like the different kinds of spasticity have to be made shallower (made less strong and more instable) (Figures 18,19). The stability of a movement pattern is given by the deepness of the potential well for that movement and the variability of phase and frequency coordination, this means how much the movement state (the ball) is moving in the potential well (Figures 18,47c).

The improvement of the mode of CNS organization, namely the phase and frequency coordination, can be measured macroscopically using the coordination dynamics recording method (Figures 136M,137) and improvements of the attractor layout of movements are measured by the improvement in movements like walking, climbing staircases or jumping (Figures 136,138).

The microscopic entrainment of phase and frequency coordination between neurons and neuron assemblies is induced macroscopically by exercising different precise coordination's of arms and legs. This is achieved by exercising the different coordination's between pace and trot gait on the special CDT device. Since the supra-spinal centers are needed for the generation of the intermediate coordination's between pace and trot gait, also the activated supra-spinal neural networks are entrained with respect to phase and frequency coordination.

During isometric contractions the neural assemblies 'premotor spinal oscillators' fire continuously oscillatory in an oscillatory manner (Figure 17). According to the activation strength the multi-frequency α_2-oscillators (f = 1/T, T = 70ms + nAP; nAP = number of action potentials (APs) per impulse train) will change their frequency to adapt their activity to the needs. But during a movement cycle of turning (or walking) the premotor spinal oscillators will be organized for a transient, dynamical existence. Depending on the motor program and the type of oscillator, the existence of an oscillator lasts for a few oscillation periods (Figures 14, 15). For low load turning at 20N, some motoneurons will fire transiently in an oscillatory manner; others will only fire occasionally according to the size principle in each motoneuron group [104].

For high-load turning at 200N, many motoneurons will fire transiently in an oscillatory manner according to the motor program. For low frequency turning (1Hz and lower) more α_2 and α_3-oscillators are activated which are innervating fast fatigue resistant (FR) and slow (S) muscle fibers (Figure 8). For high frequency turning many α_1-oscillators will be activated which innervate fast fatigable (FF) muscle fibers (Figure 8). A sprinter who is turning at 2 Hz on the special CDT device, gets fatigued more easily than a rower turning at a frequency of 1Hz for a high load of 200N, because (1) the load is twice as high for the sprinter, (2) more FF-type muscle fibers are activated for 2 Hz turning than for 1Hz and (3) the sprinter may have a higher percentage of FF-type muscle fibers due to his genetics and training.

4.2.4. Case Reports of the SCI Repair

Four patients who had suffered a SCI underwent CDT for six to thirteen months. In three out of the four patients, movements and CNS functioning improved to such an extent that one can meaningfully speak of a partial repair of the SCI. The specific aspects of repair achieved will be reported for each case separately. Apart from 'Case 3' all the patients were from the group reported earlier that underwent the three month therapy course [3]. When the treatment was discontinued, none of the patients were using spasmolytic drugs.

Case 1: A now seven year-old boy suffered flaccid paralysis due to a SCI sub Th12/L2 at the age of four in a car accident. At the age of six he underwent CDT for six months. The boy arrived at therapy in a wheelchair and, after six months, left the therapy clinic walking with sticks (Figure 136C). The partial repair of the SCI was quantified by coordination dynamics and the improvement of leg and bladder functions. In six months, the low-load coordination dynamics improved by 69 % (from 16.1 to 4.95, Figure 137A,B).

The patient re-learned to walk on the treadmill for prolonged periods of time (time increase from four to twenty-five minutes; Figure 138A) under support and weight reduction like in Figure 136D,E until he could walk without weight reduction (Figure 136B). The recovery of the leg functions was further quantified by the quality of exercising on the special CDT device when he only used his legs to turn the levers (like a fitness bicycle). The number of turns increased from 200 to 1,300 (Figure 138B). Further, the patient learned to walk with sticks faster (Figure 136C). The walking times for 19.2m reduced from 46s to 20s (Figure 138C). The patient learned to turn longer on the special CDT device in the standing position with leg support (Figure 136A). Bladder function improved. Before therapy, the bladder was emptied by intermittent catheterization. Now the patient only needed to empty the bladder every three hours and can empty the bladder volitionally with little rest urine.

The above provides evidence that the urinary bladder can be repaired faster in children than in adult patients with SCI (Figure 175). In the 6-year-old boy the bladder repair needed 6 months and in the 17-year-old lady 30 months. After six months of therapy the boy was able to walk 23 steps when supported by the Author holding his hand. At home the boy mostly walks with sticks. A few months after the treatment the coordination dynamics worsened, which may indicates a further re-innervation of leg muscles (for an explanation see under case report 4).

Case 2: A 25-year-old man had suffered a clinically complete SCI sub C4/5 (medically diagnosed) sustained when diving into water. He could still move three toes 3 to 4mm. He underwent CDT for a year in total, with two breaks of a few months in between. During the breaks he performed 3,000 to 4,000 turns per day on the special CDT device by himself so that he could keep the level of function achieved. With the therapy his cervical SCI became substantially incomplete. He relearned crawling and commando-crawling. After five months he was able to stand with little support (Figure 136F). When walking on the treadmill he needed first support for both legs (Figure 136D) then only for one leg, and eventually he needed no leg support at all (Figure 136E). When re-learning walking on the treadmill, the stepping automatism was induced by pushing the heel onto the ground during heel strike and by pushing the forefoot onto the ground during the lift-off phase (Figure 136D). Hand and finger functions improved slowly with the therapy, which may indicate the building of new motoneurons because in a cervical SCI many motoneurons innervating the muscles for finger, hand and arm functions are destroyed. The poorer right hand improved in the short-term memory following 2500 turns on the special device. The patient learned to turn in a supinated position. His finger, hand and arm functions improved to the point that he can now drink out of a glass and can eat with knife and fork by himself, which is functionally very important. Bladder function did not improve much, but may do so later. Judging from the improvement he made, it is likely that 10 to 20% of the spinal cord was spared.

Case 3: A 67-year-old lady with osteoporosis had suffered an incomplete SCI after a fall. After the first fall the vertebra Th12 broke and she was in very much pain and could no longer walk. She then fell again which caused further injury after which she could no longer move her ankles and the pain further intensified. It is unclear whether the second fall caused any further damage to the spinal cord itself. Because of the complicated spinal situation, the orthopedic surgeon did not want to operate. After ten days in hospital, physiotherapy was suggested to her. Because of the osteoporosis and spine deformation there was not only a SCI in the lumbosacral range but also in the cervical range the spine seemed to touch the spinal cord and may have partially stopped functioning. She turned very gingerly one hour per day on the special CDT device in the lying position. After three weeks she started to turn two hours per day. Later on, she walked additionally with the rolator (Figure 136G). But she was heavily relying on the rolator; the picture shown in Figure 136G was taken later, when she was in a better condition again. Then she re-learned to walk outside with sticks (Figure 136H). Later on, she started to exercise free walking. Her walking times for 25m reduced from 30 to 15 seconds (Figure 138D). After six months of CDT she was cured from the SCI. She can walk quite fast now (Figure 136I), has no further pain and can undertake domestic chores normally again. Her quality of life increased substantially.

Case 4: A 30-year-old patient had suffered a complete SCI sub L3/4 due to a parachute accident. One year after the accident intensive CDT therapy was commenced. In this patient the injury level lowered during one year of therapy, so that the patient learned to run on the treadmill at up to 9.5 km/h (Figure 136L). Some function reappeared in the left triceps surae (S1-S3). However, progress was relatively modest when considering the intensity of the therapy. It seems therefore that large parts of the lumbosacral spinal cord were destroyed and the building of new spinal cord nervous tissue seems to need more time. The bladder function improved only insofar as it went from flaccid to spastic. As will be shown below, in complete SCI substantial benefit from neurogenesis and bladder repair requires in the range of three years of therapy. Of interest in this patient is that the beginning of the re-innervation of leg muscles could be observed in continuous recording of coordination dynamics. After three, five, six and eleven months of therapy the values of coordination dynamics were 4.4 (Figure 137C), 3.4 (Figure 137D, 30% improvement), 6.0 (Figure 137E) and 1.55 (Figure 137F) for 20N, respectively. Thus, the coordination dynamics transiently worsened. A plot of all coordination dynamics values against time of therapy yields a curve which shows three increases. After six months, significant worsening of coordination dynamics for forward and backward movements was followed by a strong improvement of the coordination dynamics for backward movement (Figure 139) and a transient rhythmic change of coordination dynamics for forward turning (Figure 137E), which probably indicates the re-innervation of the leg muscles since newly innervated muscles function poorly at first.

The second slight worsening of the coordination dynamics, which this time was not accompanied by rhythmic changes, coincided with a bladder infection. After twelve months of therapy, a third increase of coordination dynamics appeared connected with rhythmic changes of low-load (20N; Figs.139,137E) and high-load coordination dynamics (146 to 203N; Figure 137G,H) which may indicate a second phase of re-innervation.

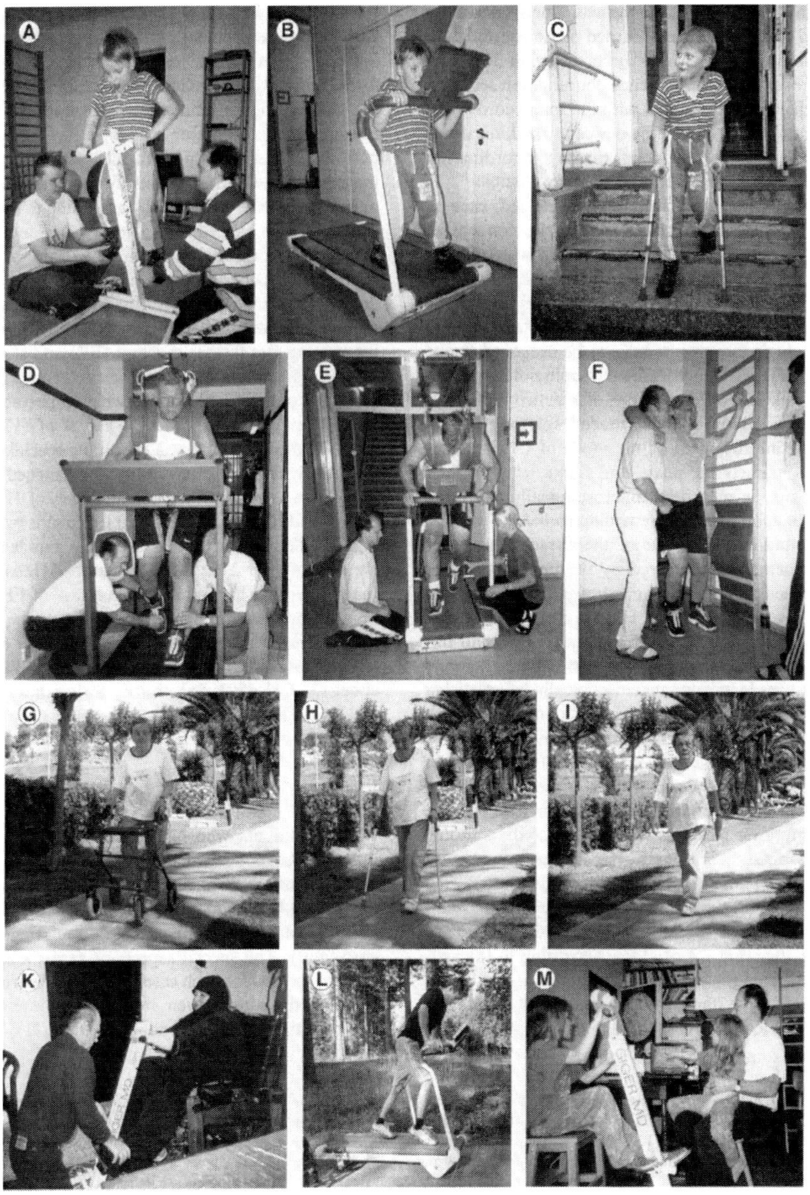

Figure 136. Movements performed during coordination dynamics therapy. A,B,C. 6-year-old patient with a complete spinal cord injury sub Th12/L2 during exercising on the special coordination dynamics therapy device in the standing position (A), during walking on treadmill (B) and during sticks-supported walking down staircases. D,E,F. 25-year-old patient with a clinically complete spinal cord lesion sub C4/5 during walking with both legs supported (D), later on with no leg support (E), and during standing supported by the author (F). In D, the right foot is pushed onto the ground by the Author to enhance the afferent input during the lift-off phase, and the left heel is pushed down to increase the afferent input during heel strike to enhance the stepping automatism. G,H,I. 67-year-old lady with an incomplete spinal cord injury in the lumbosacral range improved her walking with the rolator (G) to walking with sticks (H) to free walking (I) during 6 months of therapy. The pictures were made at the end of therapy. K. First exercise on the special coordination dynamics therapy made by the Mother Superior with a spinal cord injury; assistance provided by the Author. L. 30-year-old patient with a complete spinal cord injury sub L3/4 during running on treadmill. M. Recording of coordination dynamics for children, supervised by the Author.

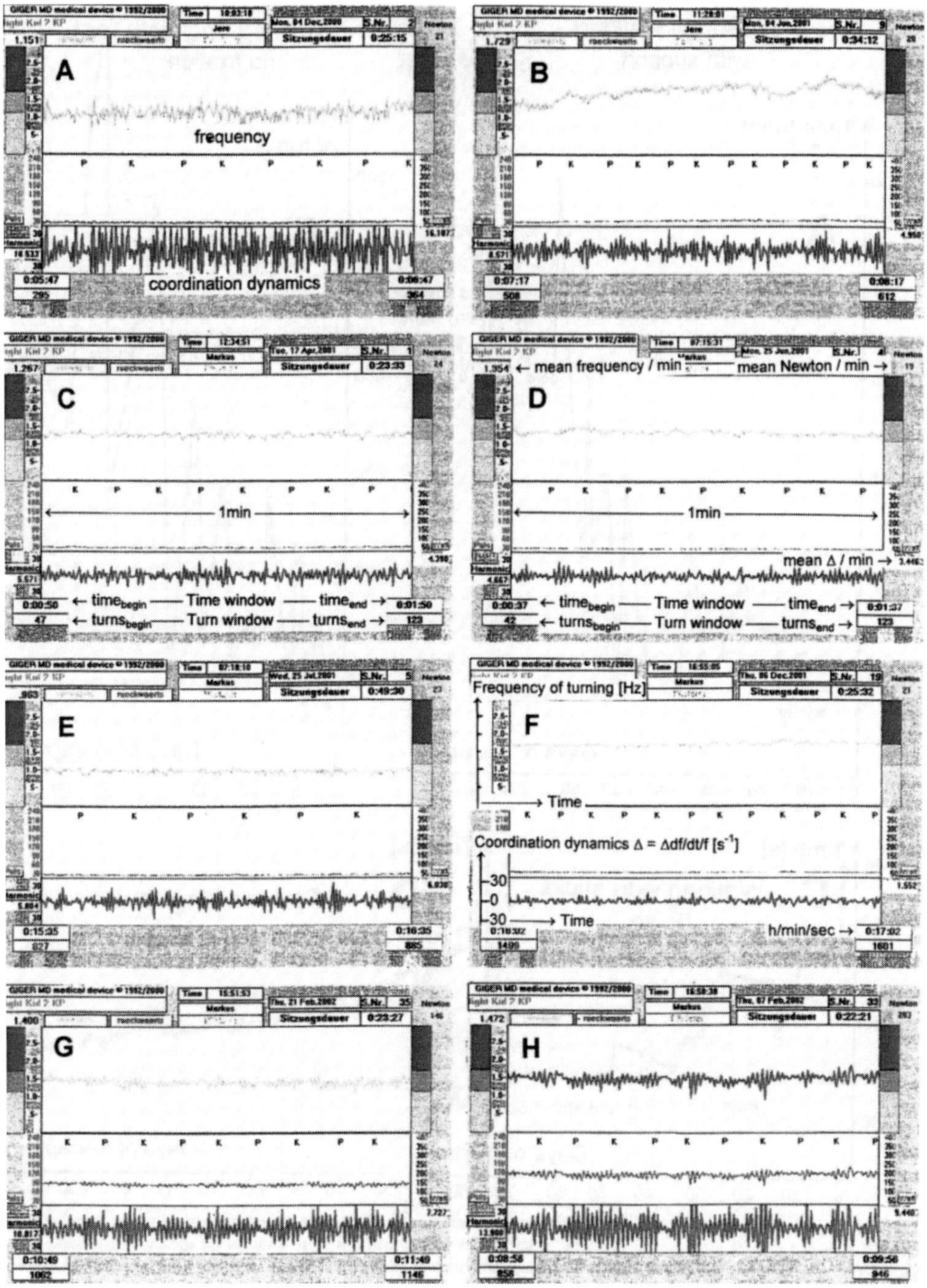

Figure 137. A,B. Low-load coordination dynamics for a 6-year-old patient with a spinal cord injury sub Th12/L2 at the beginning (A) and after 6 months of therapy (B). C-H. Low-load (C-F) and high-load (G,H) coordination dynamics for a 30-year-old patient with a complete spinal cord injury sub L3/4 at different stages of therapy. In E,G,H the coordination dynamics (lower traces) show rhythmic amplitude changes, which may indicate re-innervation of new muscles. P = pace gait; K = trot gait. Coordination dynamics values (mean Δ / min), mean force (mean Newton / min) and mean frequency (mean frequency / min) per minute are indicated.

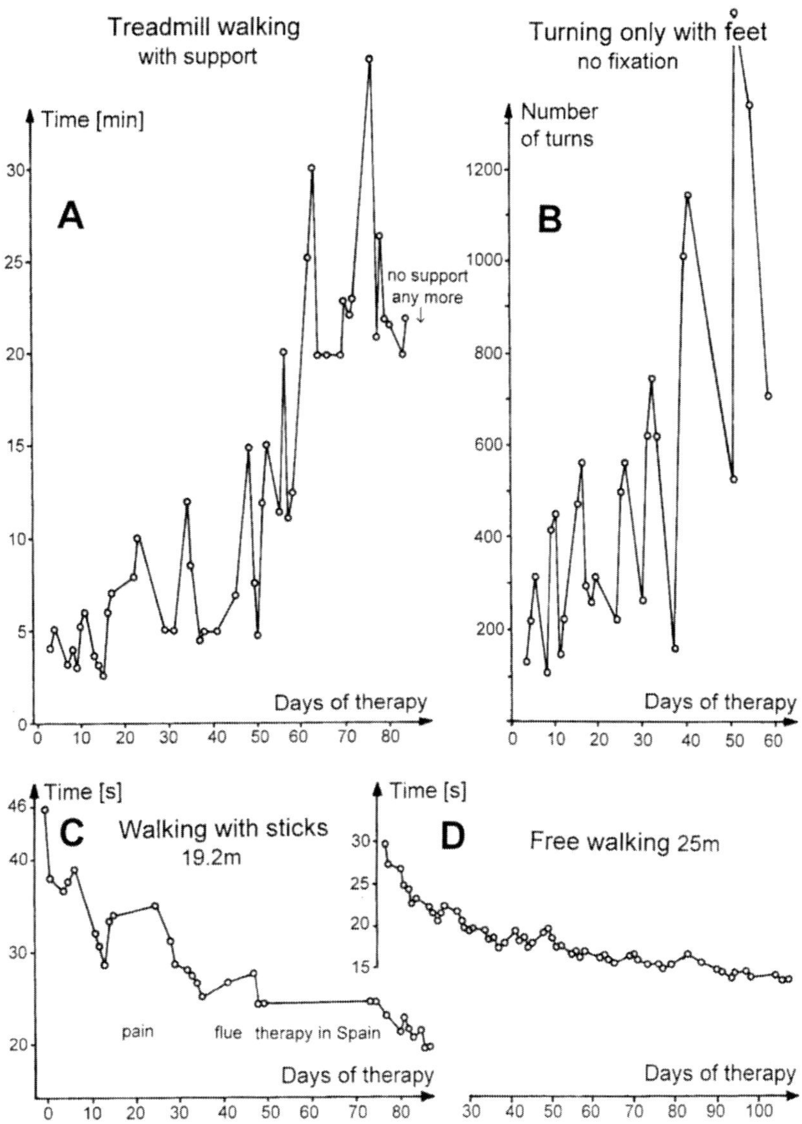

Figure 138. Improvements of movements due to coordination dynamic therapy. A,B,C. A 6-year-old patient with a spinal cord injury sub Th12/L2 learned to walk longer on treadmill (A, from 5 to 25 min) to perform more turns on the special coordination dynamics therapy device only using his legs (B, no fixation of the legs) and to walk faster with sticks over 19.2m (C, 44s to 20s). D. Increase of the speed of free walking (30s to 15s) in a 67-year-old patient with an incomplete spinal cord injury in the lumbosacral range.

Below it will be shown that when CDT is administered to a complete SCI C5/6 over several years, transient increases of the coordination dynamics values can be correlated to changes in the neural networks (Figure 30). The integration of a newborn nerve cell into the adult neural networks, for example will impair the coordinated firing of neurons and increase the coordination dynamics values till the neural networks have compensated for that disturbance by movement-based learning or other repair strategies.

As can be seen further from Figure 139, the coordination between arms and legs was much better in this patient than in healthy humans and, of course, very much better than in untrained patients with a SCI. The values of the low-load coordination dynamics of this patient (Δ = 1.55, Figure 137F) were 74% better than those of athletes (Δ ~ 6.0) whose CNS functioning had not been influenced by CDT. In conclusion, from among the four patients undergoing CDT for six to thirteen months, one with incomplete spinal cord injury could be cured (Case 3), two patients were partly repaired (Cases 1,2) with substantial urinary bladder function improvement in one patient, and one patient has improved only comparably little so far (Case 4).

4.2.5. Building of New Nervous Tissue and New Motoneurons in the Spinal Cord

There was only comparably little progress in the hand and arm functions in the tetraplegic patient (Case 2) and only little progress in the leg functions in the patient with the spinal cord injury sub L3/4 (Case 4). Judging upon the injury level, the cervical intumescence was extensively destroyed in the former case, whereas it was the lumbosacral intumescence that was damaged in the latter, which means that in one case the premotor network for arm and hand function and in the other case the premotor network for the leg and foot functions was substantially destroyed.

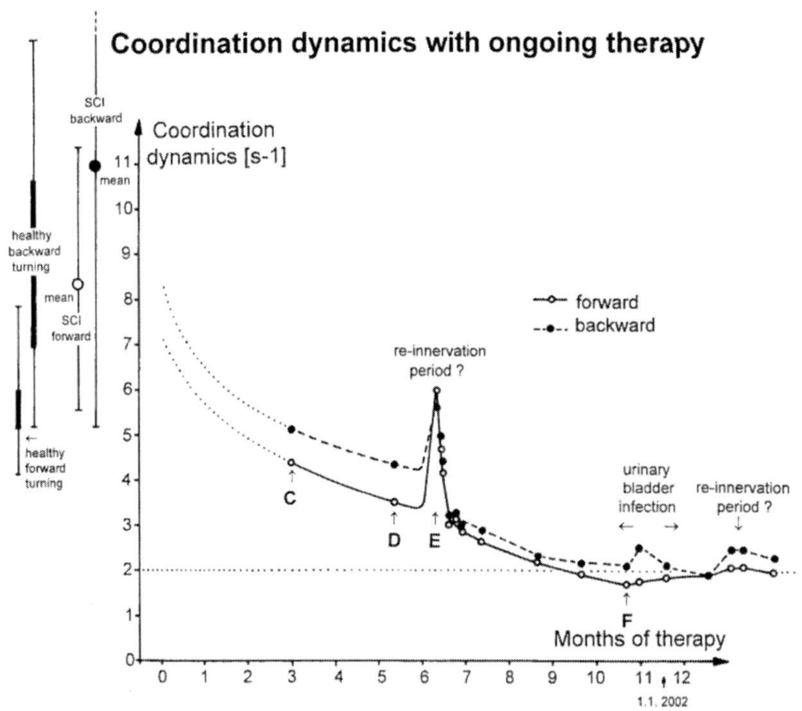

Figure 139. Low-load coordination dynamics in dependence on the therapy duration; 30-year-old patient with a complete spinal cord injury sub L3/4. Low-load coordination dynamics values for healthy subjects and a patient with spinal cord injury (SCI) (22) for forward and backward movements are given in the left part of the Figure. Two increases in the values of coordination dynamics after 6 and 12 months of therapy may indicate re-innervation of new muscles; a urinary bladder infection after 10.5 months also resulted in an increase of coordination dynamics.

More was needed for repair than the effective re-connection of neuronal networks alone. The building of new premotor neuronal networks, including the building of new motoneurons were necessary and the axons of the newly built motoneurons would have to grow down to the muscles (Figure 29B) which would take at least two years. How then was rapid recovery possible in the six year-old boy and not in the twenty-five and thirty year-old patients?

Firstly, every injury is different; the damage to the pia mater and the dura mater may also be different. Secondly, repair in children seems to be better and faster. In children the CNS is still developing and the distances are smaller. In injuries to adults, the spinal cord has to be repaired but in children, both repair and the natural process of 'correction en route' can be used to influence the repair of the spinal cord. This argument is supported by the observation that in a six month-old boy with myelomeningocele (and hydrocephalus) with no leg functions, the legs started to work following the administration of CDT, and after three and a half years of therapy the boy (who is now four) can partly walk, even though he had no hip joints at birth [40].

There is also hope for adults. In an aforementioned case of poliomyelitis [91] a woman suffered severe poliomyelitis at the age of one and at the age of thirty-seven she performed CDT for three years. Her legs started to function and grow again, and her menisci developed for the first time in her life at the age of 42. There seems to be a lot of adaptive growth and CNS plasticity if the CDT is intensive and lasts for a few years. This is also true of the patient who was partly cured of a stroke [1]; the CDT was performed for more than four years.

4.2.6. Muscle Re-Innervation Signs Demonstrated by Coordination Dynamics

In the above cited poliomyelitis patient the functioning of the leg muscles was supported by signs of re-innervation, documented by electromyography. In the patient with the complete spinal cord lesion sub L3/4 (Case 4) re-innervation of the leg muscles is supported by the transient increase of the numerical values of low-load coordination dynamics (Figure 139). However re-innervation of leg muscles is also supported by the specific changes of coordination dynamics, namely by the appearance of rhythmic changes due to the changes of the coordination between arms and legs (Figure 137E,G,H).

Why do coordination dynamics provide evidence of a thoracic SCI becoming incomplete? When the SCI is complete the hands and arms move on volition and the legs move only passively when turning on the special CDT device, and the coordination dynamics show no rhythmic changes because there is no active coordination (changes) between arms and legs. But when leg muscles start to work again, they work first in poor coordination with the muscles of arms and hands. Due to the coordination changes between arms and legs dictated by the device, the arrhythmicity of turning (coordination dynamics) will rhythmically increase and decrease with the changing coordination. In the above patient with complete L3/4 lesion the arm muscles were fully working and some upper leg muscles were working. After several months of coordination dynamics therapy arm and leg muscles were working in full coordination for all coordination's between pace and trot gait (Figure 137C,F). But with the re-innervation of further leg muscles the coordination dynamics started to change rhythmically in amplitude for the changing coordination's (Figure 137E,G,H) because the leg muscles, or parts of them, that started working again were not functioning optimally in coordination during the different coordination's, imposed by the device. The coordination dynamics showed therefore rhythmic amplitude changes for low load (Figure 137E) and for high load (Figure 137G,H). It seems so far that the muscles started to work again first for high load (high threshold for endogenous activity) and then for low load (low threshold for top-down activity). There may be different mechanisms involved in the re-gaining of muscle functions as the repair seems to be both continuous and also occasionally punctuated by major repair events.

The advantage of measuring muscle re-innervation by means of coordination dynamics is that first, they are easy to measure and second, they can be recorded non-invasively and third, re-innervation can be detected in all leg muscle and irrespective of the site. Detecting a re-innervated motor unit by electromyography may be difficult, since needles cannot be inserted in all muscle parts of leg muscles and with surface EMG one gets only the more superficial muscle fibers and malnutrition of the skin due to the SCI is also not beneficial for recording as the skin can be very dry.

4.2.7. The Ideal Patient

In most cases the therapy is not proceeding ideally. Often it is difficult to motivate the patient to do heavy exercise for four hours per day, especially in patients with brain injury of the frontal lobe. Even though children like to move, the progress comes so slowly that their discipline level is low. Often patients feel pain which can originate in the CNS and or in degenerative effects of muscle, joints and tendons, which hinders therapy. Short tendons, stiff joints, false posture including scoliosis, false position of hands and feet are often present and may give rise to pain. Bladder and other infections may slow progress. If everything with the patient is going well, then for organizational reasons, be it money or some other reason, the therapy has to be stopped. Not infrequently this occurs at a moment when substantial progress is beginning to occur and the patient is motivated for further therapy.

An unusual organizational problem is illustrated in Figure 136K. The Mother Superior of a monastery in Crete (Greece) suffered SCI in the monastery following a fall from a tree. It would be unlikely that her fellow nuns would be physically capable of meeting her needs due to their advanced age. Consequently she needed to get into a functional state that will make her more independent. However, how to apply CDT in a monastery? Also, she argued that she had not enough time for therapy because she had to undertake the organizational work. After having been instructed by the Author a few times (Figure 136K) the Mother became able to exercise independently on the special CDT device. By performing 3,000 turns per day for six months, she became able to turn over in bed herself and her diabetes disappeared. For substantial progress however, 15,000 turns per day, five times per week and supported treadmill walking such as can be seen in Figure 136D would have been necessary.

4.2.8. The Importance of an Assessment of the Severity of the SCI by Means of Magnetic Resonance Imaging (MRI)

In the ‘Case 2’ a 25-year-old man had suffered a clinically complete SCI sub C4/5 when jumping into water and made comparable big progress upon CDT. Hand and finger functions, crawling and supported walking improved. It is unlikely that this patient had in fact suffered a complete SCI. The injury was probably incomplete by 10 to 20%. In order to analyze rates of repair more information about the anatomical injury is needed. MRI is needed to estimate the degree of injury, even though not all nervous system damage can be seen in an MRI, particularly in brain injury.

4.3. Repair Achieved in a Patient with Near-Complete Cervical Spinal Cord Injury C5/6 (95% Injury) Achieved through Eight Years of Coordination Dynamics Therapy

4.3.1. Overall Case Report Summary - Kadri

This report describes a case of a twenty-four year-old lady with a severe spinal cord injury (SCI) suffered at the age of seventeen at cervical 5/6 levels (ASIA A), in whom a repair of some spinal cord functions could be achieved by eight years of partly optimal coordination dynamics therapy (CDT). CDT was started two months after the car accident. Magnetic Resonance Imaging (MRI) showed a destruction of almost 95% of the cross-sectional area at the injury levels. The 5% (if at all) spared cord tissue most likely consisted only of sensory tracts, since no motor functions were preserved below the injury level. A near-complete recovery of the important vegetative functions, i.e., urinary bladder control, respiration, and vasomotor tone could be achieved. Her motor and sensory functions also improved to some extent, and she is off all medication. However, her motor recovery was limited and she is still wheelchair-dependent. There is functional and structural (MRI) evidence that the human spinal cord regenerates through CDT.

The movement-based learning therapy included the training of supported crawling, standing up, walking, running, jumping, balance training, and exercising on special CDT devices. The regeneration of the spinal cord started after more than one year of CDT. It was very limited but continuous, and gave rise to substantial functional recovery. The recovery induced by regeneration through CDT was quantified in terms of transient increases of coordination dynamics values, the improvement of motor programs as assessed by surface electromyography (sEMG), the improvement of movement performances and the increase of the spinal cord matter at the injury site seen under MRI. The similarity between the improvement at the cellular and the integrative (network) level during regeneration and development between two patients with different injuries may be achieved by reference to walking. Through comparing the effort required to achieve substantial improvement in this case of severe cervical SCI with the effort required in the case of partial cervical SCI (50% destruction; 50% spared tissue), it is noted that the 95% injury is ten times more intractable. It is inferred that in severe SCI, repair critically depends on the percentage of the spared tissue (tracts fibers and neuronal networks) at the injury site. Improper handling of the patient therefore, such as negligent transport or late relief of spinal cord compression, may give rise to further damage of the cord tissue for which a later administered intensive CDT cannot compensate.

4.3.2. How to Achieve Repair in Near-Complete Cervical SCI

Schalow Coordination Dynamics Therapy (CDT), a movement and learning-based therapy has been shown to be effective in improving the functioning of the damaged central nervous system (CNS) in varieties of conditions like stroke [1], traumatic brain injury [2,7], hypoxic brain injury [8], cerebellar injury [9], spinal cord injury (SCI) [3,4,10], cerebral palsy [6,14] and Parkinson's disease [5]. To achieve even a partial cure in the above intractable conditions, functional and structural repair of the damaged nervous system is required. Such repair can only be brought about by intensive long-term therapy that involves working with the patient at least twenty hours per week for more than three years. To achieve optimal results, the efficiency of this treatment is very important. To enhance the efficiency of the treatment reliable diagnostic tools are needed to tailor the therapy to the type and severity of the CNS injury and to the ongoing CNS changes accompanying the CNS repair during the treatment.

Above it was reported that CDT could achieve repair in groups of patients with SCI [3,4]. Evaluating the effect of stem cell therapy and CDT [53] indicates that to achieve a substantial repair of a severe SCI, functional reorganization alone is inadequate and structural repair is needed for regeneration of the injured tissue to achieve a meaningful behavioral improvement. It is important to distinguish between structural and functional repair. Objective measurements are essential to determine optimal treatment to achieve structural repair. Such measurements are crucial to give an insight into the correlation between the severity of SCI and the extent of possible recovery.

It was shown that if 50% of the cervical spinal cord matter is spared following injury, then the patient can relearn walking, running, and jumping and can become continent within two to three years of even non-optimal CDT [10]. CDT could achieve a near-total repair in this incomplete SCI largely through functional reorganization [10]. The question remains as to how much quality of life a patient can regain if only 5% of the cervical spinal cord matter is spared. If 95% of the spinal cord cross section is destroyed, mere functional reorganization is unlikely to be sufficient for even partial repair. Structural repair, that is to say a regeneration of the spinal cord is also needed. In the adult cat it was shown that as little as 10% of spinal white matter tracts were sufficient to permit spontaneous walking without external support [105]. Since fibers probably regenerate over longer distances in the cat (in the range of 5mm) than in humans (less than 1mm), in humans we can expect that at least 20% of spinal white matter tracts, including motor fibers, are needed to repair walking.

Moreover, since a cat walks on four limbs, continence is not a major issue since its body weight does not rest on the pelvic floor.

Also, in order to gain sufficient muscle power the motor units in the four limbs enlarge far more in rats and cats than in humans. For a discussion of regeneration differences between animals and humans, see [53]. Patients with severe cervical SCI really need repair, since the spinal trauma often leads to devastating medical, psychological, social, and financial consequences. To improve urinary bladder functions, to regain some of hand, finger, and trunk functions, and to become independent in everyday life would be a very significant achievement for a patient with a severe cervical SCI in which only 5% of the fibers are spared. Animal SCI research habitually concentrates on the repair of cases in which approximately 50% of the spinal cord matter is spared (see Figs.2-4 of [106]), even though a near-total repair in such injuries has already been achieved in humans [10]. The real challenge for SCI research is to repair severe cervical SCI.

There is a world of difference in the potential for functional repair if 50% (ASIA D) of a cervical spinal cord matter is spared or only 5% (ASIA between A and B). For a comparison between different treatment methods in SCI, it must therefore be determined how much of the spinal cord matter was really destroyed.

I will show that in this patient with 5% spared spinal cord matter (probably mainly the white matter, because the grey matter is more vulnerable to hypoxia) following severe, motoric complete C5/6 SCI, urinary bladder, breathing, and blood circulation functions could be fully repaired and even motor functions could be partly repaired by three years of optimal CDT. Repair of the autonomic functions is most important.

Before World War II, most SCI patients died of the recurrent urinary bladder infections. If respiratory function repair can be achieved, the incidence of lung infections reduces. If repair of the blood circulation below the injury level can be achieved, pressure ulcers do not occur any more. It should be emphasized that optimal CDT for three years is a tremendous effort to organize, since such treatment needs to be undertaken by the patient themselves as it is not available from state medical care.

Initially, focus will be on the repair of motor functions, breathing and blood circulation. Later, we shall turn to the bladder repair. The necessary human anatomical and human neurophysiological context for repair of the human bladder is fully given in [14].

4.3.3. Details of Coordination Dynamics Therapy in Severe Cervical Spinal Cord Injury

4.3.3.1. Summary of the Therapy

Schalow Coordination Dynamics Therapy (CDT) is a movement-based learning therapy that uses different strategies for repair. The strategies are based on correcting the impaired CNS self-organization (impaired phase and frequency coordination of neuron firing following injury [15]), to compensate for the loss of movements, and to repair the impaired autonomic functions. CDT includes:

1. The training of automatisms such as commando-crawling, crawling (Figure 140C), walking (Figure 140D,E,F,H), and running (Figure 140L) (innate movements which recapitulate development) in the forward and backward direction (for symmetry repair).
2. Training at the limits of physical endurance [52] to stimulate endogenous stem cell reserves [107] to replace lost tissue.
3. Training old-learned movements (i.e., learned automatisms with high stability as climbing staircases).
4. Training of rhythmic, dynamic stereotyped movements such as jumping on a springboard to functionally repair neural assemblies (spinal oscillators) (Figure 144A,B).
5. Training device imposed movements to repair the impaired time and space-coordinated firing of neurons and neural assemblies. The special CDT device for turning (Figure 140A,B,G) is of great importance for repair. It imposes four-limb cyclical rotating movements with changing coordination between arms and legs. Such four-limb movements are the first movements a patient with severe cervical SCI, especially in the lying position (Figure 150) when using help muscles, can perform by himself/herself, whereby neuronal network activation is achieved across the injury site in the rostral and caudal directions.

To understand repair of the integrative functions of the human CNS, the System Theory of Pattern Formation [18, 108] for Repair [10, 48] is used. Strategies used for structural repair are discussed below.

4.3.3.2. Recruitment of Spared Tracts and Neuronal Networks

Training on special CDT devices (Figure 140A,B,D,G,K,150) involves performing integrated movements, in which the arms and legs move in a precisely coordinated manner. As the special devices impose these coordination's, the afferent and efferent inputs have to pass to and from across the injury site to establish coordinated communication between the cervical and lumbar cord segments which are supposed to be the respective centers of the stereotyped arm and leg movements (cervical and lumbosacral intumescences). Such exactness of coordinated movements is important for the activation of denervated neurons below and at the injury site. It is well-known that neurons act as co-incidence and coordination detectors [60]. An action potential is only generated in the axon hillock of the neuron if the cell soma of the neuron receives a sufficient number of depolarization from afferent inputs, which are coordinated up to a few milliseconds (Figure 11).

Non-exact movements will generate asynchronous depolarization from the afferent inputs; the threshold potential for the generation of an action potential will not be reached, and the neuron will fail to fire an action potential along the axon; the neuron fails to get activated. Therefore, on a larger scale in order to activate all spared tracts, neurons, synapses and networks across the injury site, this exactness of coordinated movements of all four limbs is very important. Since every injury is unique and leaves behind different sets of spared fibers and neurons, different kinds of highly coordinated movements need to be trained to activate all spared fibers and neurons distributed in different network activations. Only those spared tracts and neuronal networks can be reorganized which are recruited and therefore only the highly coordinated activity can generate such a maximum of recruitment and functional connectivity over the injury site through the spared tract fibers.

4.3.3.3. Structural Repair

When a patient trains at his/her limit of endurance on the special CDT device (Figure 140G) against very high load, the highly coordinated, integrated movement activates and recruits many of the neuronal networks across the entire nervous system. It also likely stimulates all kinds of repair mechanisms including the recruitment of endogenous stem cells for repair.

With ongoing CDT, the stimulated endogenous stem cells may contribute to repair by various mechanisms such as replacing lost neurons in the injured grey matter, encouraging and supporting the sprouting of severed axons and guiding them across injury site. The stem cells may also support the plasticity mechanisms at cellular and molecular levels, which occur during functional reorganization in spared neuronal networks with ongoing CDT. The commitment of partially denervated neurons may be changed [109] following CDT. Such commitment change could be a powerful mechanism for repair especially in systematic neuron death such as that which occurs in hypoxic brain injury.

Very intensive CDT may also influence selector genes within the stimulated endogenous precursor cells to direct the cells to differentiate along those cell lines [110] which are needed for repair. For example, the formation of inhibitory inter-neurons may be equally important as the formation of motor neurons to reduce the co-contraction in spasticity and improve the antagonistic action of neurons and thereby effectively improving the performance of movements. A similar strategy may also help to reduce tremor in patients in Parkinson's disease [31].

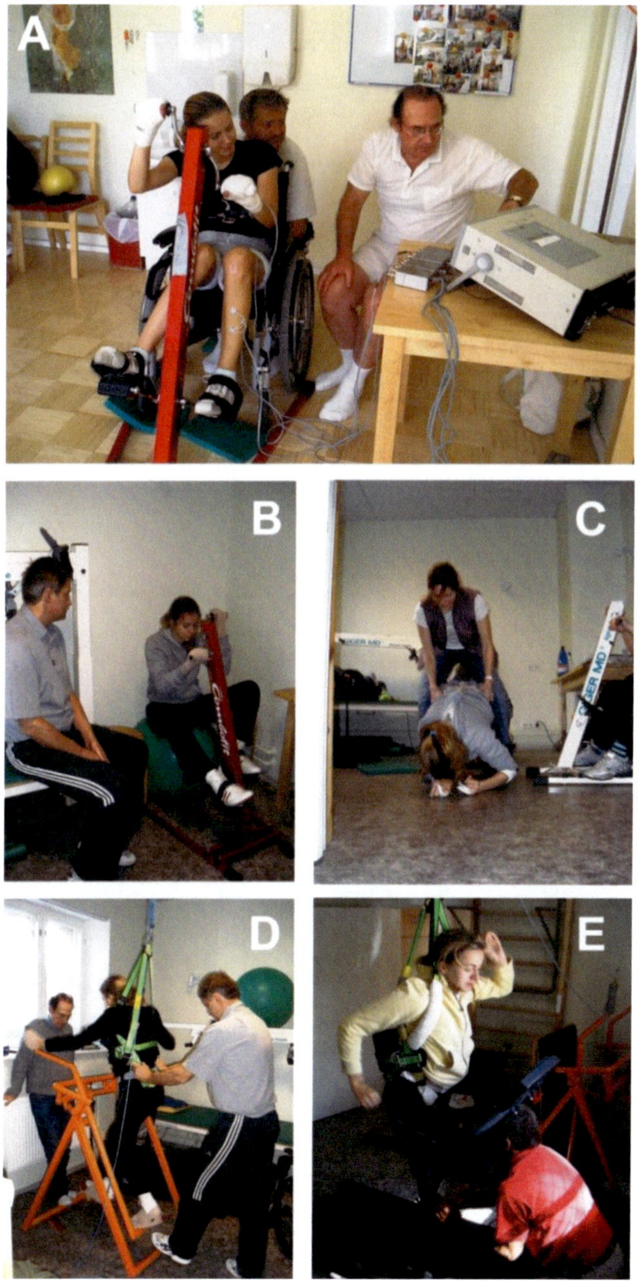

Figure 140A-E. Movements performed during coordination dynamics therapy (CDT). A. Surface EMG recording layout for measuring motor patterns when exercising on the special CDT and recording device for turning. Skin damage healed at the left knee. The author is performing the sEMG. B. After one year of treatment the trunk stability improved that much that the patient could train the balance while exercising on the special device. C. Patient during crawling with support. The legs of the patient are fixed to the legs of the therapist. The skin is still quite vulnerable; injured fingers are covered. D. Weight-supported neuro-walking; device imposed trot gait coordination. The therapist is giving trunk support; at a later therapy stage trunk support was not necessary any more. E. Fast forward walking on treadmill. Leg support is given by one therapist. The patient is inducing the leg movement with the arms to generate very coordinated descending and ascending impulse patterns over the injury site at C5/6 levels.

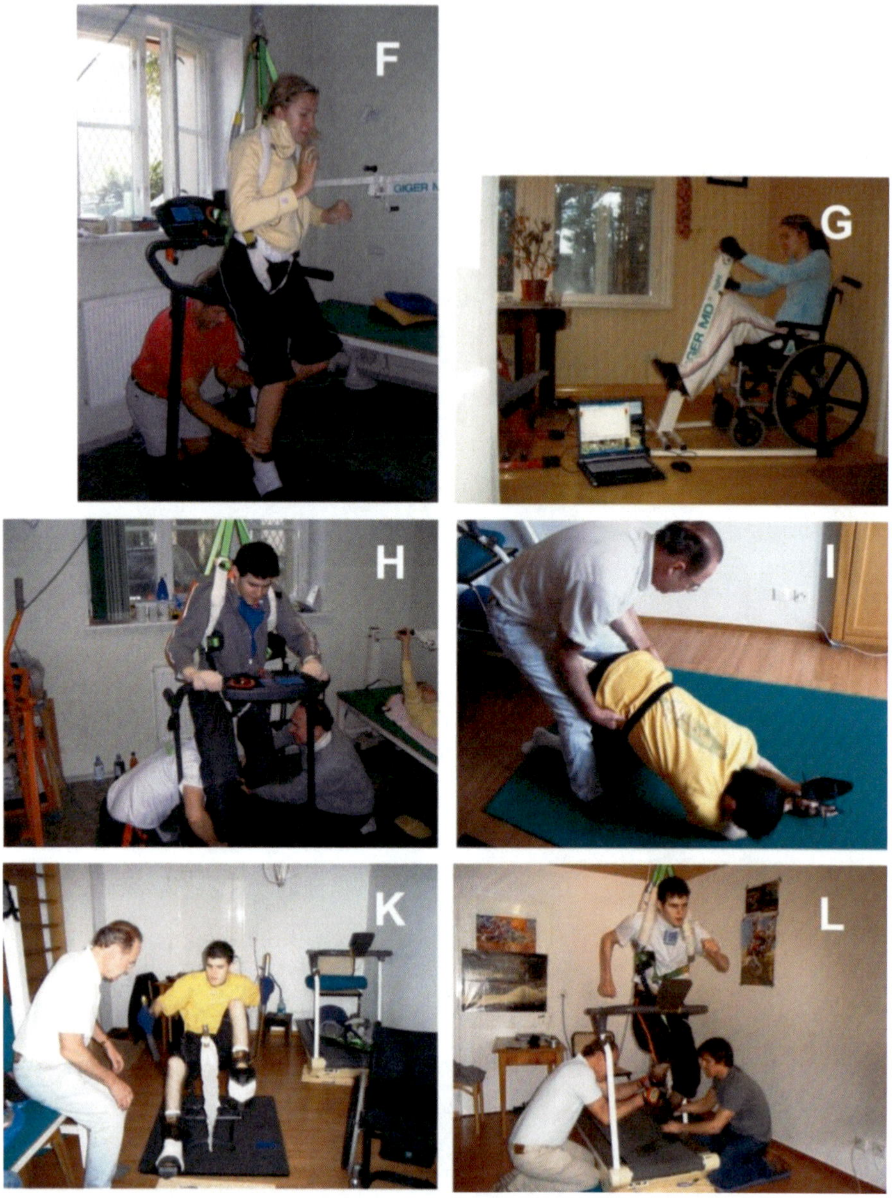

Figure 140F-L. Movements performed during CDT. F. Fast supported backward walking (or running) on treadmill; symmetry counterpart to forward walking. G. Recording layout for measuring coordination dynamics. The patient is sitting actively with trunk stability in the wheelchair and is not slipping out of the wheelchair as during first measurements. H. Supported treadmill walking with 2 therapists and fixed hands (another patient with a severe C5/6 injury, motor bike accident, 5% spared motor tract fibers). I. Supported rhythmic movement from one side to the other; preform of pace gait crawling. An open hand is achieved by support, used in swimming sport. K. Device imposed coordinated arm and leg movements in pace gait coordination. L. Fast forward walking (or running) on treadmill. Leg support is given by 2 therapists. The patient is inducing the leg movement with the arms. Author is on the right side of the patient.

4.3.3.4. Generation of Geographical Landscape of Chemo-Attractive and Chemo-Repulsive Gradients to Induce Homing of Cells and Connections by Performing Innate and Other Movements

One of the very important factors in regeneration within the nervous system is the guidance of nerve fibers and nerve cells to rebuild networks. The proponents of stem cell therapy and pharmacological therapy in SCI often play down this important requirement. It is proposed that CDT - the movement-based learning therapy - will support what may be called natural activity-guided plasticity during repair. CDT achieves this activity-guided plasticity probably by generating a geographical landscape of chemo-attractive and chemo-repulsive gradients for the guidance of growing nerve fibers and homing of cells. The generation of the geographical landscape may have the form of selective expression of endogenous neurotrophins at appropriate sites and with appropriate gradients, which will guide the nerve fibers. Or the geographical landscape is generated by selective homing and differentiation of endogenous stem cells through neurogenesis and gliogenesis. Both processes are orchestrated by the specifically activated spared neurons and tract fibers at the injury site over time and space upon performing supported automatic movements like walking on a treadmill or performing precisely coordinated movements imposed when exercising on special CDT devices (Figs.140A,B,G,K,150). This strategy is very similar to what occurs during nervous system development. The training of innate, automatic movements, like walking, crawling and jumping is likely to produce and re-enforce the development-like chemo-attractive and chemo-repulsive gradients.

Exercising on the special CDT devices primarily entrains the neuronal networks to improve its phase and frequency coordination of neuron firing for improved self-organization by imposing extremely integrative, coordinated movements. Such an improvement of the efficiency of CNS self-organization probably also plays a key role during development for the process of 'correction en route'.

Training on the special CDT device (Figs.140A,B,G,150) is the starting point of CDT, especially for those patients who are either bed-ridden or wheelchair-bound, since every patient can perform these integrative and precise movements largely unassisted and the coordinated firing of neurons must be improved in every patient with an injury of the nervous system. As the geographical landscape of repair processes starts changing and networks reform during CDT, all performable movements should be trained to re-enforce structural and functional repair.

4.3.3.5. Geographical Landscape of Chemo-Attractive and Chemo-Repulsive Gradients, Stimulated by Pattern Dynamics Including the Improved Movement Induced Afferent Input

The geographically precise guidance cues for endogenous neuronal network repair are only sufficiently well activated if the spared neurons and tract fibers are specifically activated in the context of integrative automatisms and other movements. Upon performing integrative automatic movement patterns, generated by the networks including the injury site, the support of the therapist has to bring the SCI patient into a pattern in which all the efferent network parts are activated and the movement induced afferent input is as physiologic as possible so that the geographically precise endogenous signaling cues, present during development or repair, are stimulated. To achieve this optimal support, the interpersonal coordination between the therapist and the patient is extremely important. The therapist has to feel and instantaneously assess the patient's efforts and support what is needed to enable the patient to perform good quality movements. This requirement may keep changing from minute to minute and session to session and over the course of therapy as the patient improves. The therapist has to encourage the patient to put in more effort and focus on the quality of performance and support only there where the efforts are falling short. Therefore, for example, the robot-assisted movements for treadmill walking cannot substitute for the therapist-assisted support. The therapist-assisted treadmill training (Figure 140E,F,H,L) is extremely strenuous for the therapist and is therefore regrettably tempting to avoid. On the other hand, the support required for the exercising on the special CDT device (Figure 140A,B,G) is less demanding for the therapist and is another reason to start optimal treatment with such special devices.

4.3.3.6. Strategy Differences for Structural Repair between Animals and Humans with Respect to Genotype-to-Phenotype Mapping

The nervous system and other phenotypic traits are formed during ontogeny based on the genetic information stored in the cells (genotype). The information specified in the genotype determines those aspects of the nervous system which are expressed as innate behavioral tendencies and predispositions to learn. The inherited genotype can completely specify the phenotypic network; i.e., both the network's architecture and the connection weights (in synapses and other connections) are genetically determined. In this case, the behavior of the network is entirely innate and there is no learning [111]. Or the genotype specifies the network's architecture, but the weights are learned [112]. In still other cases, what has been selected during the evolution are good initial weights for learning or good learning rates and momentums [113].

In humans, the genotype may only partially specify the network's architecture and the connection weights are learned with good weights for learning and good learning rates, apart from the essential life-saving automatisms, which are more or less innate. The price for the higher capacity for learning, higher specific variability and complexity of the networks may be the significantly reduced capacity of innate structural repair of the human CNS in comparison to that of animals. During repair, the changes in gene expression will only partly specify the network's architecture. The main emphasis has therefore to be placed on (movement) learning. The different learning mechanisms may be the main stimuli for repair in humans and not the innate structural repair mechanisms. For the repair in animals, the network's architecture and connection weights are mainly specified by the genotype. The movement learning is mainly only needed to direct the exogenously applied stem/progenitor cells to the injury site [114]. The explicit network's architecture is mainly generated by the genotype-to-phenotype mapping, which includes cell division, cell proliferation from stem cells, and the migration of neurons to reach their terminal position in the CNS.

4.3.3.7. Stepping and Walking Automatisms in Infants

A direct genotype-to-phenotype mapping is biologically implausible. In reality, we cannot predict which phenotype will emerge from a given genotype because of the large nonlinearities present in the mapping process. If the genotype is viewed as a set of instructions, it is not the case that each of these instructions will result in a single network property. Rather, the properties of the network emerge as the result of many interactions among the various instructions and their products [115].

Genotypescan directly incorporate innate behaviors that require no learning, or they can incorporate innate predispositions to learn some behavior. Innate behaviors can be present at birth (congenital) or they can develop during life under the control of genetically specified information.

The stepping automatism in human is an innate behavior present at birth, which does not require learning. After birth this stepping automatism is induced in babies as a control for a healthy CNS at that stage of ontogenesis, and is stimulated, for example, with the heel strike (Figure 141). The stepping automatism is present prior even to birth.

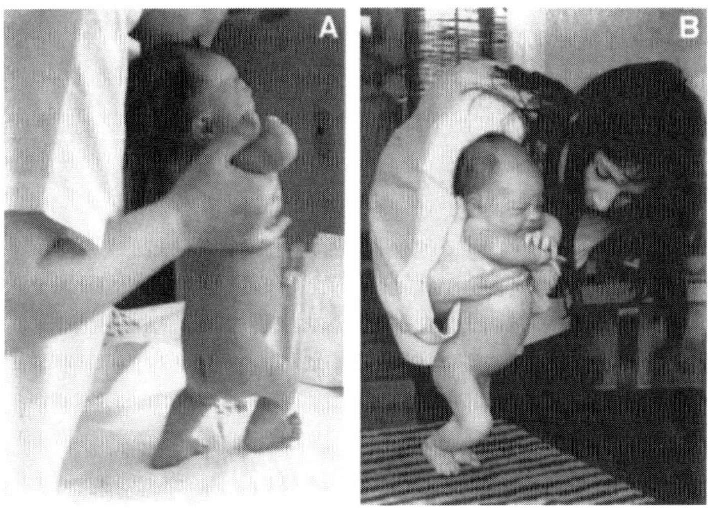

Figure 141. Automatic stepping in a newborn infant. A. The 5-day-old infant, Juliane, performing primary automatic stepping; slight backward posture. The heel of the right foot touched the ground first. B. Infant Juliane, 8-day-old, performing automatic stepping.

The walking automatism develops later under genetic control. It is unclear whether the stepping automatism is replaced by the walking pattern or whether supra-spinal networks use this stepping automatism, modify it and include it in the walking pattern. The walking of the infant in Figure 142a,c seems to show a more gradual change from the stepping to the walking automatism.

At an age of five months (Figure 142a), the walking pattern is characterized by more flexion and high lifting of the knees, more similar to the stepping automatism (Figure 141), than at an age of eight months (Figure 142c). To learn the walking automatism, the proper motor programs have to be learned by the infant in some similarity to patients with spinal cord injury. Necessary connections mainly exist during development but not following spinal cord injury Also, the antagonistic relationship between muscles has to be learned under genetic control. In this case the later developing inhibitory system has not been developed sufficiently so far. The EMGs of the five and eight months-old healthy Jürgen show that the antagonistic action between the tibialis anterior and gastrocnemius muscles has not fully developed at those particular developmental stages (Figure 142b,d); the muscles are still significantly co-activated because of insufficient developed inhibition. This ontogeny context will be employed below for the purposes of comparison with repair.

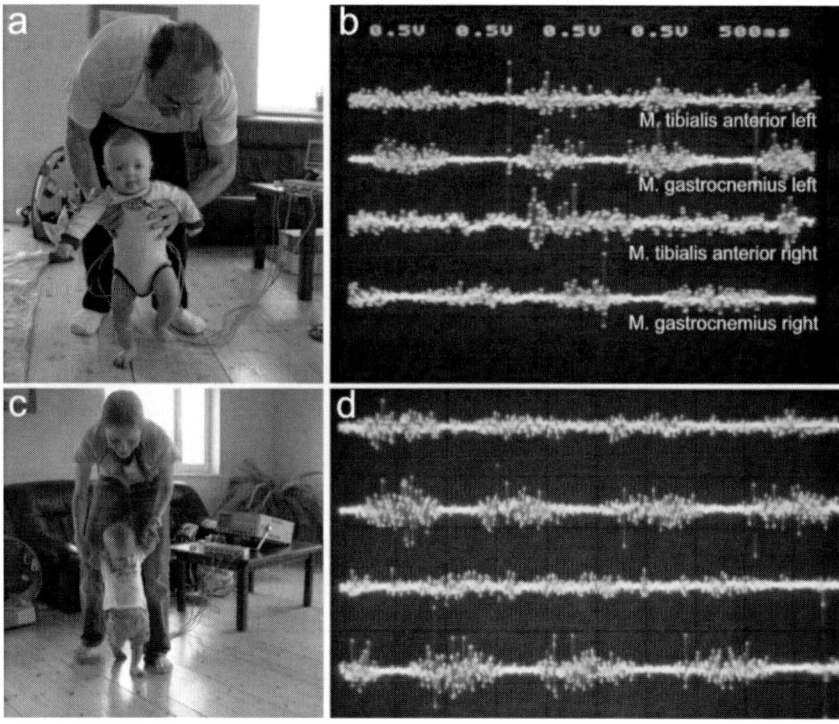

Figure 142. Surface EMG obtained from the healthy 5-months-old (a,b) and 8-months-old old "Jürgen" (c,d) during supported walking. a. Walking resembles automatic stepping, because of the strong lifting of the left knee. The toes of the right foot are plantar flexed, which is not physiologic. b. Surface EMG motor programs of left and right tibialis anterior and gastrocnemius muscles. Note that there is no antagonistic action between the tibialis anterior and gastrocnemius muscles. The right tibialis anterior muscle shows no motor program. c. The walking is more walking like and not so any more much automatic stepping like. d. Better motor programs then 3 months earlier (b). Still there exists no antagonistic action between the tibialis anterior and gastrocnemius muscles. The activation of the right tibialis anterior muscle is a bit better than 3 months ago (b).

4.3.3.8. Re-Appearance of Stepping and Walking Automatisms during Repair

Following severe cervical SCI, parts of the stepping automatism may reappear as the first stage towards repair of walking. Depending on the composition of spared fibers, the stepping automatism is induced either automatically (in some similarity to Figure 142a) or volitionally (in some similarity to Figure 142c) by the patient. To enhance the stepping automatism during training of walking and running, the heel has to strike the ground first (Figure 141A). Because of the absent dorsal flexion of the feet in patients, the heel strike is achieved by positioning the patient at the very front of the treadmill (Figure 140L). By doing that, the forefoot reaches over the edge of the treadmill, and the heel strikes the most anterior surface of the treadmill first. The activation and optimization of the stepping automatism, supported by the volitional emphasis of the patient, is of great importance as it activates as many of the spared tract fibers across the injury site as possible.

With leg support, the therapists can feel the differences in leg muscle activation, depending on whether the patient's arms are fixed (Figure 140H, little regained trunk stability) or move in the naturally coordinated pattern (Figure 140L, trunk stability achieved again). In SCI patients, the stepping automatism can be activated in the forward (Figure 140E) and backward direction (Figure 140F) during supported treadmill walking. It seems that in new born babies the stepping automatism (Figure 141) can only be stimulated in the forward direction. The induction of the stepping and walking automatisms seem to play a crucial role in the repair of the SCI and the therapists have to feel the contribution from the patient during supported treadmill walking for proper support. If the interpersonal coordination between the therapists and the patient is optimal (Figure 140E,F,L), the patients will perceive and report the improvement in his or her performance. It is difficult to see how such motor learning can be replicated in animals.

4.3.3.9. Training of Automatisms to Recapitulate Ontogeny

Another crucial property of the genotype to phenotype mapping is its temporal character. Biological development or maturation does not simply yield an 'individual'; rather, the phenotypic individual is a succession of different phenotypic forms which are sequentially generated by the genotype in interaction with the environment. Interactions with the environment in humans include the early childhood automatisms of commando-crawling, crawling, standing, walking, running, balance training, visual-motor coordination, auditory-visual-motor coordination.

A model in which the genotype/phenotype mapping i.e., ontogeny takes place during the individual's lifetime and is influenced both by the genotype and by the external environment which has been described [116].

Structural repair in human patients induced by CDT attempts to recapitulate development [8, 79, 14] (Figure 35) as in animals research [107].

Therefore, movement-based learning therapy has to include those movements which are performed during development to allow for successful succession of different phenotypic forms which are sequentially generated by the genotype. Movement-based learning is therefore not only needed for functional but also for structural repair and it is of much more importance in humans than in animals.

4.4. Repair of Motor Functions within Three Years of CDT – Patient Kadri

Kadri suffered a severe spinal cord injury (SCI) at cervical 5/6 levels in a car accident. The spine was repositioned and fixed one day after accident and not within eight hours after the accident as it should have been. When the patient woke two days later, she had no sensation and no motor function below the level of C4/5. Some function was preserved in certain arm muscles. Four weeks later, as the spinal shock regressed, some sensation reappeared all over the body. Some more muscles started to work in the upper arm. The injury level lowered to C5/6. Due to the lowering of the injury, it was clear that the spinal shock affected the spinal cord below the injury level and approximately one segment rostral to the injury site. In the following two to three months, some more sensations in the body and some muscle functions in the upper arms returned spontaneously. Dorsal flexion of the hand became possible. No finger functions returned. The bladder did not start working again. Her progress was consistent with usual clinical experience in that the spontaneous recovery is very limited in complete or near-complete SCI (Figure 143). With respect to motor functions, the patient had a typically complete C5/6 injury. Breathing was paradoxical (see below); mainly driven by the diaphragm, innervated by the phrenic nerve. Owing to some spared sensation, the patient was able to feel the support given by the therapist during therapy. A few weeks after the accident, the patient heard from friends about coordination dynamics therapy (CDT) and visited the Author with her parents for the initial assessment of the coordination dynamics. Two months after the accident, CDT was started. It would have been better if she had started CDT a few days after the accident by turning on the special CDT device in the prone position, which is safer than a bed in that the surface does not move as a mattress might. Pathologic CNS reorganization is reduced in early treatment. Also the healing process would have been better with early improved blood supply and micro-circulation.

At the beginning, the therapy was not optimal. There were severe problems with the blood circulation. The patient often got dizzy and the skin below the injury level was very vulnerable. Also, the patient did not believe very much in the therapy at the beginning and discontinued it for a time after two and a half months of therapy (Figure 30, 2005). However, seven months after the accident, the therapy became optimal and intensive with more than twenty hours of therapy per week. This patient started to fight for her future. Blood circulation improved strongly. Skin wounds occurred only occasionally and healed quickly. The risk of pressure ulcers was no longer present. During the second year of treatment the patient trained at her limits with more than thirty hours per week and motor functions below the injury level re-appeared slowly. After two years of optimal and intensive CDT, bladder function returned to normal.

The following description explains in detail the re-appearance of motor functions in relation to the continuously performed coordination dynamics measurements for low load (Figure 30). I will then proceed to discuss bladder repair in the context of the system theory of pattern formation and the neurophysiology of the human urinary bladder (Chapter VII).

4.4.1. Anatomy of the Injury (MRI)

To relate the severity of the anatomical injury of the SCI to the outcome following optimal CDT, the anatomical extent of the injury was evaluated by using magnetic resonance imaging (MRI). The first MRI was performed ten months after the accident (Figure 143A). The MRI seems to show that the screws from the spine fixation touched the spinal cord.

After the removal of the fixation screws, the spinal cord seemed to be free, as can be seen in the sagittal T2 and T1-weighted images taken from a second MRI, even though the spinal cord was quite close to the dorsal vertebra bone (Figure 143B,C). But the patient suffered a partial 'Horner syndrome' during the operation, (sympathetic connections were slightly damaged), which was later cured with ongoing therapy. The author could approximately measure from the MRI that only 5% of the spinal cord matter was spared. In all other sagittal MR images related to Figure 143, no spared spinal cord tissue could be seen at the injury site. Most of the spared tract fibers conducted sensory functions because the patient could not activate motor functions on volition below the injury level. Since the injury extended over the length of only one vertebra, it seems possible that some fibers could regenerate over that distance through an intensive and sustained course of CDT.

Experimental evidence in the rat, which has a much higher capacity for regeneration, shows that the longest single regrown axons had a length of up to 14mm [106]. After three years of CDT, a third scan MRI was performed (not marked in Figure 30). Both sagittal T2-weighted images (Figure 148) show spinal cord tissue at the injury site, which connects the rostral and caudal spinal cord. In earlier MRIs (Figure 143), only one sagittal T2-weighted image showed a connection across the injury site. Therefore, the spinal cord matter had increased at the injury site very approximately from 5 to 8%. Limited regeneration had taken place. Since probably not all regenerating fibers are visible in the MRI, more fibers could have regenerated over the injury site. Further, the spinal cord injury site in Figure 148 looks healthy and shows no cysts, as if a regeneration process was taking place. How the very approximately 3% of regenerated spinal cord matter could give rise to the functional recovery is described below in detail.

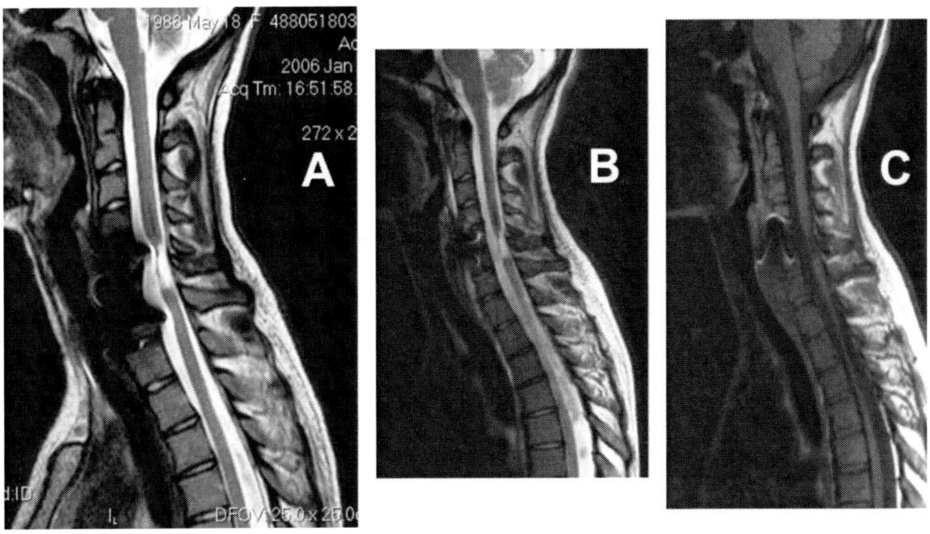

Figure 143. Magnetic resonance images of a severe spinal cord injury at C5/6 levels of an 18-year-old female patient who suffered the injury in a car accident. A. Sagittal T2-weighted image before removal of the spine fixation; note the imaging artifact caused by the metal. B, C. Sagittal T2 and T1-weighted images after removal of the titan-fixation. Note that in A and B there seems to exist only one connection between the rostral and caudal spinal cord, whereas in C two bridges may be seen. On all other sagittal images, no connections could be seen between the disconnected spinal cord parts. The author estimated an anatomical injury of 95% (5% spared spinal cord matter).

4.4.2. Start of Therapy

Coordination dynamics values for 20N were measured (Figure 140G) continuously during the therapy and are plotted in Figure 30. Since in severe cervical SCI the arms and hands have very little strength (the biceps brachii is usually the strongest muscle left after such injury) and the legs cannot contribute due to the complete motor paralysis below the injury site, the patient found it very difficult to turn at a load of even 20N on the special CDT device. However exercising on the special CDT device, at least in the lying position (Figure 150), is the first coordinated integrative movement which such a patient can perform by herself, since no trunk stability is needed and the patient can use other muscles (help muscles) to exercise. While exercising in the sitting position, remaining safely in the wheelchair was a problem. Because of absent trunk stability, the patient was slipping out of the wheelchair. Support for trunk stability had to be given at the beginning.

The absence of trunk control and missing intercostal muscle activation could be seen as the cause of the paradoxical breathing. When the only spared breathing muscle, the diaphragm (innervated by the phrenic nerve from C3/4 segments), contracts during inspiration, the thorax becomes smaller and the abdomen distends. Because of impaired ventilation, there is always a risk of lung infections.

After a few months of therapy, trunk stability improved to the extent that the patient no longer slipped out of her wheelchair while exercising in the sitting position. The intercostal muscles started to work successively in the rostral-caudal direction. Breathing became rather physiologic after two years of therapy. Blood circulation problems (orthostatic hypotension and other dysregulations) during exercising in the sitting position occurred only at the beginning of therapy.

4.4.3. Transient Increase in Coordination Dynamics Values for Assessing Regeneration

At the beginning of therapy the coordination dynamics values for exercising at a load of 20N in the forward and backward direction improved strongly (Figure 30). When the therapy was stopped for some time in summer 2005, the values worsened. After restarting the treatment, the coordination dynamics values improved again. With the removal of the cervical spine fixation, the values became worse, which may indicate improvement of spinal cord functioning. Because of the relief of the slight cord compression, probably exerted by the fixation screws (Figure 143A), the spinal cord functioned more integratively after the operation. The supra-spinal centers now had more access to the neuronal networks of the disconnected caudal spinal cord and the enlarged neuronal networks worked initially with poorer coordination. During therapy, the coordinated firing of neurons improved again as did the coordinated arm and leg movements, quantified by the coordination dynamics value. With ongoing therapy, nine further transient increases in the coordination dynamics value were measured. Each one could be correlated to the onset of a certain new function, indicating that the spinal cord regenerated very slowly in small steps. Each new bit of regeneration increased the access of supra-spinal centers to the caudal spinal cord networks and worsened the coordinated firing of neurons transiently. The newly regained functions were trained strongly in order to substantially activate the newly formed connections across the injury site.

In Figure 30, the instances of transient increase in coordination dynamics values are marked with '1' through '9'. At increase '1' the patient felt more muscles working in the right leg. Also, the sensation in the right leg improved. At peak '2', the left gastrocnemius muscle showed spontaneous contraction. At '3' the antigravity muscles seemed to start to work. The patient was able to rhythmically swing a bit on springboard in the standing position. At '4' partial continence was achieved. The first in-phase jumping on springboard (Figure 144A) became possible at peak '5'. With the large peak '6', the bladder worked better and the lower abdominal muscles started to work. With the transient increase of coordination dynamics values '7', full continence was achieved, even though the feeling of bladder fullness and emptying was still not the same as before the accident. The bladder functioning became more physiological with ongoing therapy. The patient was able to turn in the forward direction for 1.5min at a load of 50N. The leg muscles seemed to help a bit during turning. This increased turning power is attributable to the beginning of the antagonistic action of the tibialis anterior and gastrocnemius muscles (Figure 146C,D). Goose bumps occurred all over the body when walking on treadmill (Figure 140E,F), indicating a further improvement of functioning of the vegetative nervous system. For the first time in two years of treatment the patient was able to achieve at least five turns by using only her legs on the special device in the lying position. Some leg muscles certainly contributed to the coordinated movements. At peak '8', the supported crawling became better and the arms became stronger. At '9', the patient could activate the right tibialis anterior and gastrocnemius muscles a little on volition (Figure 146A,B).

4.4.4. Evidence for A Limited Regeneration of the Human Spinal Cord Through Long-Term CDT

In conclusion, every instance of transient increase in the coordination dynamics value was associated with the reoccurrence of some somatic and/or autonomic functions. In this case of complete motor SCI, the stepwise improvement of motor functions, associated with increase in coordination dynamics values, indicates ongoing regeneration of the human spinal cord induced by CDT. This regeneration seems to take place by the formation of new functional network connections across the injury site in a step wise manner. The amount of reconstructed functional connections at each step is probably very small since only subtle increments in muscle power occurred.

4.4.5. Improvement of Movements

Even though the patient had no motor functions left below the injury level, the motor functions slowly improved continuously in different groups of muscles with ongoing therapy. The trunk muscles started first to work successively in the rostral-caudal direction.

Exercising on the special device for turning in the lying and sitting position was possible from the beginning. The patient mostly exercised when sitting in her wheelchair (Figure 140A,G). When the trunk stability had improved, the patient could also exercise when sitting on a ball (Figure 140B). In that position, the patient trained simultaneously arm and leg movements, trunk stability, and balance.

Supported treadmill walking (Figure 140E,F) started first in the sitting (not shown) and then in the upright position. At first, no muscles of the patient's legs were engaged during the supported leg movements; the movements were passive. Later on, the patient could help during treadmill walking. With ongoing therapy, the patient walked also in the backward direction and could move her arms in coordination with the legs (Figure 140F). The speed for forward walking was often in the range of 5km/h. Sometimes even running on treadmill was trained up to a speed of 11km/h. By moving arms and legs during walking and running, the patient trained automatic coordinated movements, which are generated by neuronal networks including those of the injury site.

The crawling started with the training of the crawling position with support like in Figure 140I. Then the crawling position could be trained without significant support (Figure 144C). Supported crawling became gradually possible (Figure 140C). At a later stage of treatment, the patient trained the exact anti-phase coordination between arms and legs on a sky-walker (Figure 140D). Finger functions were trained, when the patient held the handles as tightly as possible. At the beginning of 'sky-walking', the hands had to be fixed. Later on, fixation was no longer needed. For the patient's nervous system to receive also exactly coordinated movement induced afferent inputs, especially from hands and fingers, the patient's hands and fingers were held tightly to the handles of the sky-walker by the therapist during supported training. When jumping on a springboard (Figure 144A,B) became possible (peak '5' of Figure 30) without weight support (Figure 144A), motor functions improved faster and full continence was achieved. The patient enjoyed this rather free jumping. The feeling to have full weight on the joints was cathartic for her.

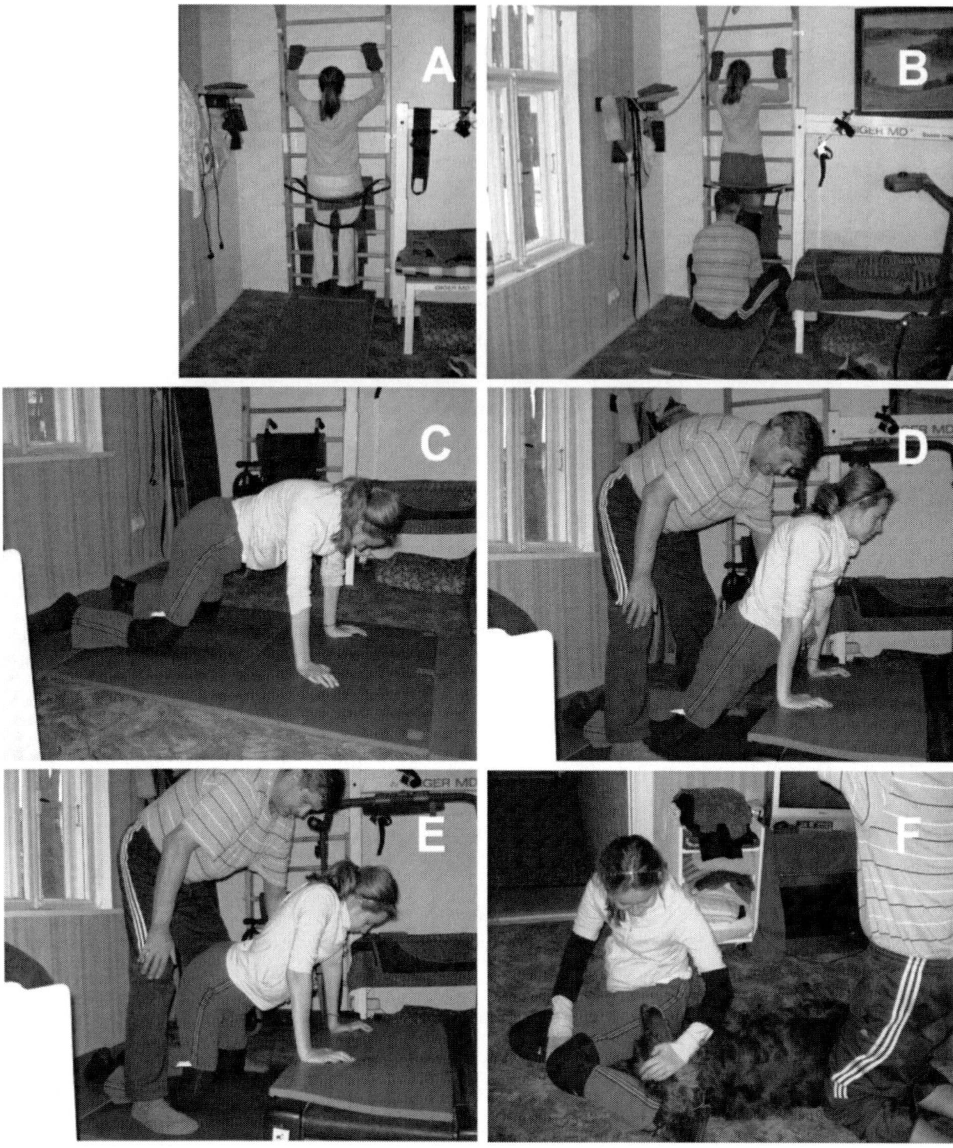

Figure 144. Movements mainly performed by the patient; severe cervical spinal cord injury (SCI); for MRI see Figure 143. A,B. Jumping in-phase on springboard with and without support by a therapist. C. Training of trunk positioning (up-down, right-left, forward-backward). D,E. Up and down movements for training trunk stability; therapist (father) for safety behind. F. Training of trunk stability when playing with the dog; note the natural hand movement and positioning (no functional hand with shortened finger tendons!). Upon 2 years of CDT, the patient had enough arm and hand power for protection, if slipping (not to suffer a second SCI).

In general, the detailed information (anamnesis) of the patient's recurring feelings and movements gave a clinical understanding of the functioning of the patient's nervous system and helped to adapt the therapy to ongoing improvements. Trunk stability was trained with different movements, including movements on the floor (Figure 144C-F). Significant support was needed only at the beginning. The turning of the trunk was also exercised on the floor. Thus the patient became able to turn over in bed at night by herself.

4.4.6. Surface EMG Demonstrating Improvement of Motor Functions

Surface EMG (sEMG) was performed several times when exercising on the special CDT device (Figure 140A) to show re-innervation of leg muscles. After seven months of therapy, sEMG was performed the first time (EMG_A, Figure 30). Following attempted volitional activation of leg muscles, no EMG activity could be recorded. When exercising on the special CDT device for turning movements (Figure 140A), however, gastrocnemius muscles could be activated but not the tibialis anterior muscles (Figure 145A). After nine months of therapy (EMG_B), the tibialis anterior muscles could also be activated when exercising on the special device (Figure 145B). But the EMG activity of the gastrocnemius and tibialis anterior muscles occurred at the same time and showed only minimal signs of a motor program. Therefore there was no antagonistic activity between these two antagonistic muscles. At that time, the patient could nearly stand without support at the wall bars (co-activation of the muscles), but could not turn on the special device with the legs only, for which antagonistic muscle action is required. It seems that the inhibitory loops or networks were not yet established in the neuronal networks at that stage of regeneration to further differentiate the mass contractions into physiological motor programs including antagonistic muscle action. After fourteen months of therapy (EMG_C), the tibialis anterior and gastrocnemius muscles could be activated when exercising on the special device with the hands supporting the knees (pace gait coordination of arm and leg movements; like in Figure 210B). But still no antagonistic EMG activity could be recorded from the gastrocnemius and tibialis anterior muscles (Figure 145C). These muscles could still not be activated volitionally.

After two and a half years of CDT, however, the patient became able to activate the gastrocnemius muscles a little volitionally. The legs also moved a bit upon plantar flexion of the feet and the sEMG (EMG_D in Figure 30) showed activity of some motor units on the gastrocnemius traces (Figure 146A,B). In Figure 146A some FF-type motor units seem to fire transiently oscillatory at a frequency of 20Hz. As expected, the volitional activity exhausted quickly and was irregular, typical of a re-innervation process. Triphasic motor unit action potentials could be recorded (insets in Figure 146A,D), indicating that the re-innervation process occurred in the central nervous system (CNS) and not in the peripheral nervous system (PNS), i.e., in the muscle. In the case of regeneration in the PNS, one would expect more complicated wave forms than the classical triphasic muscle action potential. Clear signs of an enlargement of motor units by sprouting in the muscle were also not found at this stage of re-innervation. Again, the morphology of motor unit action potentials would have been more complex due to the delayed activation of the by sprouting re-innervated muscle fibers.

4.4.7. Regeneration of the Inhibitory System as Quantified by Semg

Varying motor programs could be recorded from the right and left tibialis anterior and gastrocnemius muscles, when the patient exercised somewhat automatically, in that she was not concentrating on the movements on the special CDT device (Figure 146C). The activity in the motor program bursts was relatively low. Upon supporting the automatic exercising volitionally, spasticity occurred first as in Figure 146C (right side of tibialis anterior trace). When, after approximately ten turns, the spasticity had decreased, motor programs occurred with more activity, seemingly similar to those recorded eighteen months earlier (Figure 145B,C). The important difference, however, was that some antagonistic action occurred between the tibialis anterior and gastrocnemius muscles. Before the onset of transient spasticity, the right tibialis anterior muscle was also activated antagonistically relative to the right gastrocnemius muscle (Figure 146C). The left tibialis anterior and gastrocnemius muscles also showed some antagonistic activation (Figure 146D). The tibialis anterior and gastrocnemius muscles did not fully co-contract any more. The inhibition system had started the re-innervate the caudal spinal cord.

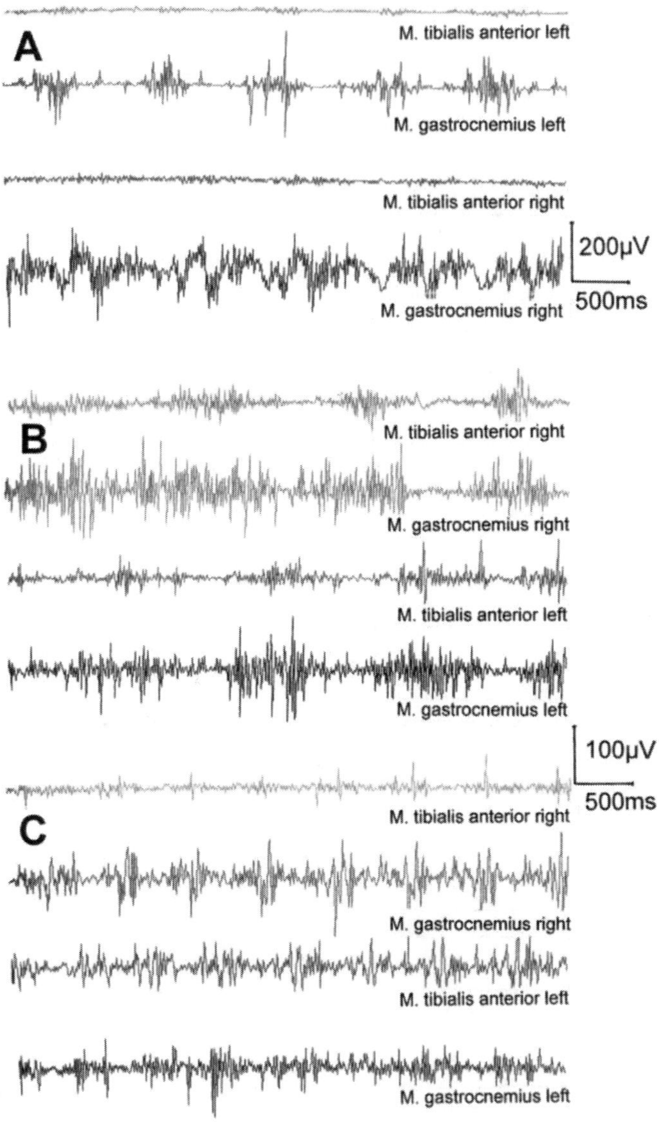

Figure 145. Surface EMG motor patterns of the right and left tibialis anterior and gastrocnemius muscles for exercising on the special coordination dynamics therapy and measuring device with ongoing therapy (for measuring layout see Figure 140A; EMG recording dates are marked in Figure 30). A (EMG_A). The tibialis anterior muscles are not activated. Movement artifacts on the trace of the right gastrocnemius muscle. B (EMG_B). Tibialis anterior muscles can be activated. Tibialis anterior and gastrocnemius muscles of each leg are activated simultaneously = no antagonistic action. Patient obtained hand support during exercising. C (EMG_C). Patient exercises by herself with the hands on the knees (pace gait pattern). Still no antagonistic action between the tibialis anterior and gastrocnemius muscles.

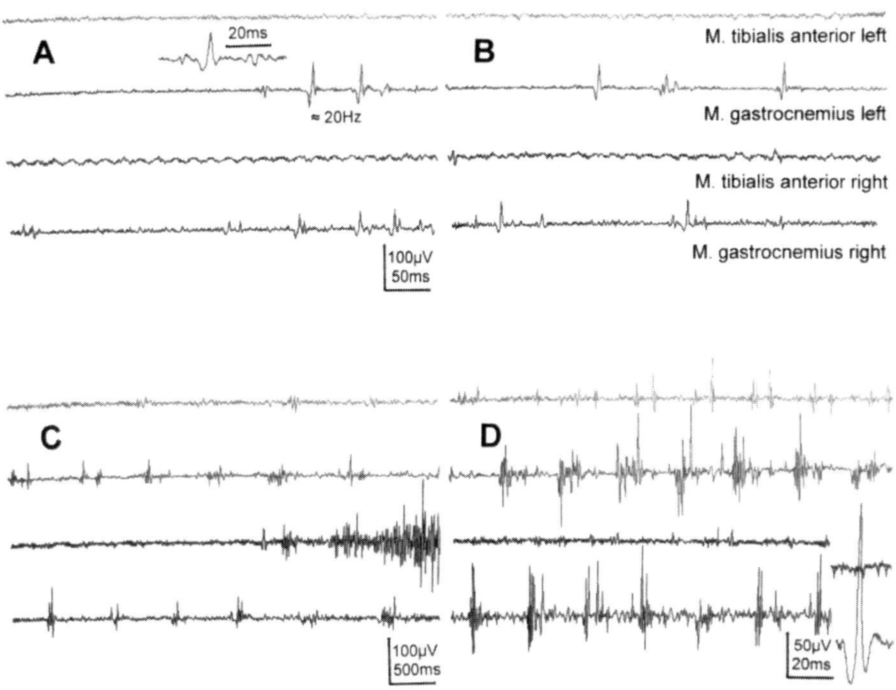

Figure 146. Surface EMG motor patterns of the right and left tibialis anterior and gastrocnemius muscles for volitional plantar and dorsal flexion (A,B) and for exercising on the special coordination dynamics therapy and recording device (C,D) after 2.5 years of coordination dynamics therapy (EMG_D). The inserts in A and D show triphasic motor unit potentials at stretched time scale. In "C" the patient is exercising automatically on the device and in "D" with additional volition and concentration; note the beginning of antagonistic action between the tibialis anterior and gastrocnemius muscles. Note further the beginning of continuous activity, due to starting spasticity on the trace of the right tibialis anterior in "C".

With the onset of volitional control in some leg muscles, the patient was able to vary the motor programs and could reduce spastic motor activity more quickly.

4.4.8. Repair of Autonomic Functions

Even though animal research so far has been paid only little attention to the repair of the autonomic functions in SCI, they are very important for the longevity and the quality of everyday life of the patient.

4.4.9. Repair of Urinary Bladder and Sexual Functions

The most important function for everyday life, the bladder function, could be repaired, including the regeneration of the necessary tract fibers for volitional bladder control. Details will be given below. Since the sexual functions are sited also in the S2 to S5 spinal cord segments, repair of the bladder function includes at least a partial repair the sexual function. The jumping on springboard upon no weight reduction (Figure 144A,B) (which trains the pelvic floor and the external sphincters, which are a part of the pelvic floor), contributed strongly to the repair of the urinary bladder and sexual functions.

4.4.10. Repair of the Blood Supply

Repair of the blood circulation protects the patient from developing pressure ulcers, which can also become life-threatening. Since patients with severe cervical SCI have a very poor blood supply and are often very sedentary, they are very prone to develop pressure ulcers. These ulcers are difficult to heal because of the poor blood supply. The repair of the impaired functioning of the autonomic nervous system in this patient re-established proper blood circulation and heart rate function and decreased the occurrence of pressure sores. For the repair of the blood supply, regeneration of the spinal cord may not be necessary, because the autonomic nervous system has different connections outside the spinal cord via different plexuses including the hypogastric plexus (Figure 153). A network activation (following a network reorganization) of these intermingled plexuses can bypass the damaged spinal cord segments and may contribute to functional repair.

4.4.11. Repair of Breathing

Repair of physiologic breathing was also important. Because of the C5/6 SCI, the intercostal nerves could not be activated anymore which resulted in paradoxical breathing, whereby the respiration could only be achieved by the diaphragm, innervated by the phrenic nerve (from cord segments C3/4). In paradoxical breathing, the ventilation of the lung is poor and there is an ever-present risk of lung infection, which is again life-threatening. With the improvement of the trunk function down to the pelvis, the breathing problem was also solved. The muscles of the thoracic cage started to work again in the rostral-caudal direction. The rostral intercostal muscles contribute most to breathing [117,118]. The repair of breathing indicates some regeneration of the spinal cord since re-innervation of the motoneurons driving the intercostal muscles by supra-spinal centers is required for the coordinated functioning of intercostal muscles.

4.4.12. Partial Repair of Sensations

The different modalities of sensations (Figures 5,6,7) also improved all over the body with ongoing therapy, such as touch, pain and temperature.

4.5. Different Repair Aspects

4.5.1. Regeneration of the Human Spinal Cord

The patient Kadri had suffered a motoric complete cervical spinal cord injury (SCI). No motor functions could be detected below the injury levels C5/6. The approximately 5% spared spinal cord matter was therefore mainly conducting sensations. Within the three years of coordination dynamics therapy (CDT), some motor function of the trunk and legs reappeared. The patient was able to turn using only her legs on the special CDT device. For the in-phase jumping on a springboard with slightly flexed knees (not using the legs as crutches when overstretching the knees) (Figure 144A), trunk stability and some leg muscle functions are needed. These improvements in motor functions after one to three years therefore indicate a limited regeneration of the spinal cord had taken place.

Further evidence for the regeneration of the spinal cord is given in Figure 30 with the transient increments in coordination dynamics value "1" through "9". Each peak correlated to the reappearance of a certain muscle or autonomic function. With the establishment of new connections across the injury site, the CNS sites rostral to the injury gained more access to the neuronal networks of the caudal spinal cord, which increased the accessible networks and increased the coordination dynamics values. The CNS organization was transiently disturbed. With ongoing therapy, the self-organization of the enlarged networks improved and the coordination dynamics values again reduced quickly. With each small step of regeneration, a further transient increase in the coordination dynamics values occurred.

With the continuous recording of the low-load coordination dynamics every few days, a stepwise regeneration of the spinal cord was detected. These frequent measurements did not result in loss of therapy time since for recording and therapy similar instruments are used. Since the patient always wanted to achieve better coordination dynamics values, she exercised with much concentration during the measurements, which means that the rate of motor learning was high during those measurements.

After two and a half years of CDT it became possible for the first time to record volitional EMG activity on plantar flexion of the feet (Figure 146A,B), which provides further evidence for the regeneration of the spinal cord.

The increased spinal cord matter across the injury site after three years of treatment (Figure 148) supports the functional results that the spinal cord was in the process of regeneration. Actually the estimated 3% of regenerated spinal cord matter could account for this significant functional recovery.

4.5.2. Functional and Structural Repair Is Dependent on the Severity of the SCI

It was shown above that CNS functioning could be improved in a group of SCI patients by coordination dynamics therapy (CDT) up to a partial repair [4]. In complete SCI with stem cell therapy and CDT, however, only minor improvement in motor functions below the injury level was achieved [53]. On the other hand, it was shown above that if 50% of the spinal cord matter is spared, a near-total functional recovery was possible following three years of CDT [10]. In this case it is shown that if approximately 5% of the spinal cord cross-sectional area is spared only, still some motor functions below the injury level can be repaired.

Therefore it seems that the extent of functional and structural repair is proportional to the amount of spared spinal cord matter. If 50% of the spinal cord matter is spared, a near-total repair is possible if CDT is administered for a long period of time. If the injury of the spinal cord is ten times more severe, i.e., if only 5% of the spinal cord matter is spared, then only limited repair is possible even if optimal CDT is administered for three years or more. If the spinal cord injury is even more severe, i.e., if only 0.5% of the spinal cord matter is spared, then improvement of motor functions below the injury level is extremely limited [53].

But the extent of the repair probably does not depend only on the percentage of spared spinal cord white and grey matter with respect to the cross-sectional area, but also on the longitudinal extension of the spinal injury. In this case, the longitudinal extension was in the range of one vertebra length. In more extended injury distances, the possible achievable repair will be smaller since fewer fibers will regenerate over longer distances [106].

The repair of the SCI depends therefore largely on the amount of spared spinal cord matter. A patient with an injury with 0.5% spared spinal cord cross sectional area likely has one hundred times less potential of repair than a patient with an injury of 50% spared spinal cord matter after similar treatment. However, this does not mean that nothing can be done in complete SCI (0.5% spared spinal cord matter). If, for example, repair of the bladder, one of the greatest problems in SCI would need only 0.1% of the spinal cord matter to be reconstructed, then the urinary bladder could be repaired. Also, some caudal autonomic networks have connections to rostral ones outside the spinal cord (Figure 153) and can contribute to urinary bladder and microcirculation repair.

4.5.3. Generation of the Geographical Landscape of Chemo-Attractive and Chemo-Repulsive Gradients for Homing of New Cells and Connections

With respect to functional repair, the relationship between the severity of the injury and the outcome following 3 years of CDT is plausible. The more spinal cord matter that is spared, the greater the capacity for functional reorganization. If more connections are spared, other CNS networks can take over functions more easily. The geographical landscape of attractors of pattern formation is less pathological for more spared spinal cord matter and probably improves faster in the direction of an attractor layout with physiologic attractors with optimal treatment. The repair of the attractor 'urinary bladder function' upon CDT will be discussed below.

But what does this relationship between outcome and severity of the injury mean for the mechanisms of structural repair? It could mean that the strength of the general stimulus for the regeneration of the human spinal cord depends on the amount of spared tract fibers and neurons available at the injury site and also on their activation while training at the limit (range of overreaching [52]). For the generation of a physiologic geographical landscape of chemo-attractive and chemo-repulsive gradients to encourage sprouting, guidance of regenerating axons, and neurogenesis, many coordinated integrative movements have to be exercised to recruit all the spared tract fibers through the activation of physiologic movement patterns (Figs.140A-L,144). For specific repair, the microenvironment of single neurons is in the range of 0.1μm (Chapter I of [14]). It may be of importance to have high enough concentrations of attracting and repulsive neurotrophins available in this distance range. The microenvironment of neurons, axons, dendrites, and glia will induce the specific proliferations of stem/progenitor cells and specific homing of these cells and connections to re-establish specific circuitry. This specificity of geographical landscape formation is the basis for the development of a physiologic pattern formation (see above).

If the extent of regeneration is proportional to the number of successfully activated spared and regenerated fibers, then structural repair would become more difficult in more severe injuries and would also significantly depend on the quality of the movements performed.

4.5.4. Repair of Motor Functions

The motor functions improved in the patient, even though the patient was long way from learning to walk again. She is still not independent in life, because the transfer from the wheelchair was not fully achieved. But the independence in everyday life seems possible to achieve if optimal CDT were continued.

Further motor functions are not only needed for independent locomotion, but also for further structural and functional repair of the spinal cord. A significant obstacle to successful treatment of severe cervical SCI is that there are insufficient motor functions spared to train and therefore to repair the spinal cord networks.

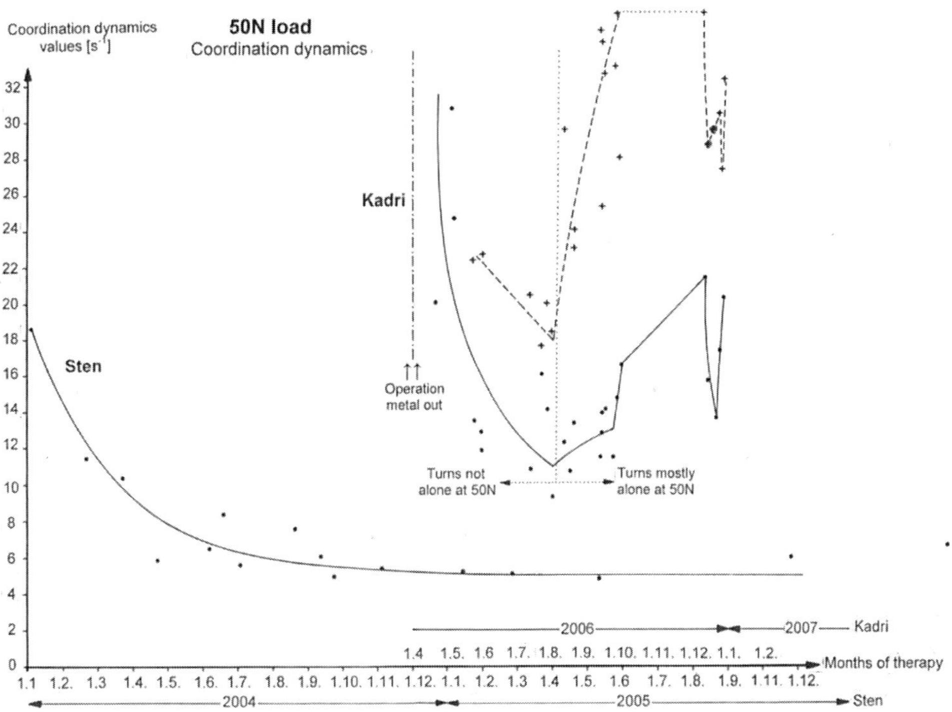

Figure 147. Comparison of the relations 'coordination dynamics values to therapy duration' for a load of 50N between a 50% (Sten) and 95% (Kadri) spinal cord C5/6 injury. Note that the coordination dynamics values for the 50% injury improve smoothly upon therapy whereas the ones for the 95% injury do not improve and are unstable (dashed line = backward exercising).

4.5.5. Lack of Muscle Power

It was shown that with CDT functional and limited structural repair are possible. With ongoing therapy more and more muscle functions reappeared. But for everyday life one needs not only muscle function; one also needs adequate muscle strength.

If 50% of the spinal cord matter was spared, the patient could generate enough muscle power to walk, run, and jump [10]. But if only 5% of the spinal cord matter is spared (Figure 143), then supporting one's own body weight, is a very challenging problem.

The difference in functional muscle power generation between the cases of 50% (Sten) and the 95% (Kadri) injury can be quantified when exercising on the special CDT device. The patient with 50% spared matter (Figure 49) relearned exercising at high loads up to 200N within three years but the patient with 5% (Figure 148) did not.

The patient with 5% spared matter succeeded in exercising at a load of just 50N over three years. The difference in functional muscle power can also be seen in Figure 147. The patient with 50% spared spinal cord matter could easily turn with arms and legs at a load of 50N and the coordination dynamics values continued to improve with therapy. The patient with only 5% spared spinal cord matter however had big problems to manage that load of 50N and the performance was irregular with ongoing therapy (Figure 147). The patient soon became exhausted while exercising at that load.

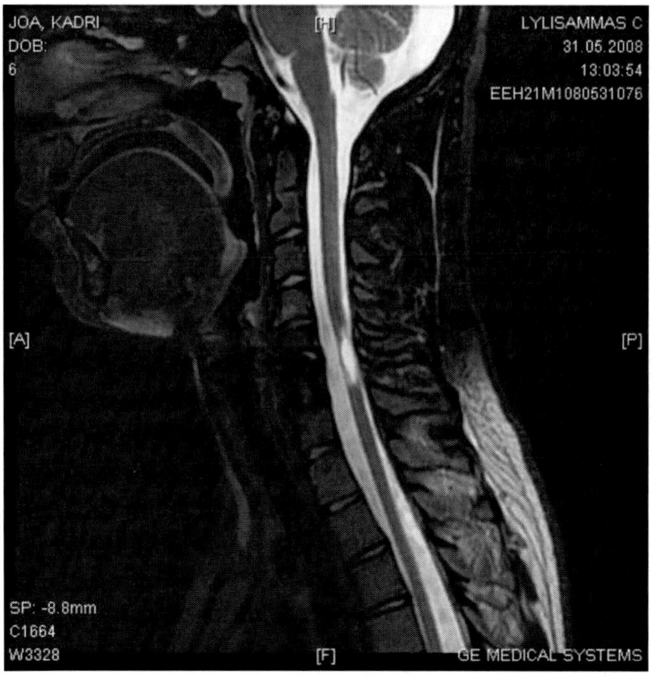

Figure 148. Magnetic resonance image of a severe spinal cord injury at C5/6 levels of a 20-year-old female patient after 3 years of optimal coordination dynamics therapy. The sagittal T2-weighted images seem to show connection between the rostral and caudal spinal cord. In comparison to the images of Figure 143, there seem to be more connections between the disconnected spinal cord parts, because at the beginning of therapy (Figure 143) connections could be seen only on one sagittal T2-weighted image and 3 years later on two images.

4.5.6. Similarity between Repair and Ontogeny with Respect to the Stepping and Walking Automatisms

A newborn baby can step automatically with a heel strike (Figure 141). Later during ontogeny, automatic walking is learned. It seems that the stepping automatism is integrated and modified by supra-spinal centers to generate the walking automatism. In this integration process, at the beginning the walking automatism seems to be more automatic-stepping-like with an emphasized knee lifting (Figure 142a) and becomes more walking-like later on (Figure 142c).

During repair and regeneration following severe SCI, often the stepping automatism re-appears first followed by the volitional walking pattern. However, the muscle power was still significantly weaker than in the less severe injuries. The automatic stepping or automatic walking needed therefore body-weight support in addition to leg support.

Sometimes patients with CNS injury have a pathologic plantar flexion of the toes and have therefore difficulty in getting their feet into shoes. Such kind of spasticity can be reduced by CDT. During development also an infant may develop such a plantar flexion of the toes transiently as seen in the case of the five-month old infant Jürgen in the right foot (Figure 142a). This is in accordance with the as yet absent proper motor program driving the right tibialis anterior muscle (Figure 142b). Three months later (Figure 142c), natural development seemed to have corrected this non-physiologic positioning of the toes. At that time the sEMG motor program of the right tibialis anterior muscle also seemed to be better (Figure 142d).

4.5.7. Impairment of Antagonistic Action between Tibialis Anterior and Gastrocnemius Muscles during Ontogeny And Repair

With the change from the stepping automatism to the walking automatism during ontogeny, strong impairment of the antagonistic action of the tibialis anterior and gastrocnemius muscles could be measured in a healthy five and eight months-old infant (Figure 142b,d).

When the patient with the severe cervical SCI became able to perform supported automatic walking on the treadmill (Figure 140E,F), a synchronized activation of the right and the left tibialis anterior and gastrocnemius muscles was recorded (Figure 145) when the patient exercised on the special CDT device (Figure 140A). But after two and a half years of CDT, the antagonistic action slowly established in the patient's leg muscles (Figure 146C,D). Therefore there seems to be similar problems in the establishment of proper motor patterns during ontogeny and during repair.

4.5.8. Re-Innervation Pattern of Trunk and Leg Muscles Following SCI May be an Example of How Supra-Spinal Centers Establish Control Over Caudal Neuronal Networks During Ontogeny

By measuring continuously the low-load coordination dynamics values with ongoing CDT, it was found that transient increments in the low-load coordination dynamics value occurred continuously (Figure 30), which corresponded to the re-occurrence of motor and autonomic functions.

It is conceivable that when supra-spinal centers take over control of the neuronal networks of the spinal cord during ontogeny, as for example during the process of learning to walk, transient disturbances of the coordinated firing of CNS neuronal network also occur. The disturbances of the coordination dynamics of CNS organization during development are also continuously corrected by learning in the way that the CNS is communicating with the environment via movements and movement-induced afferent inputs. A deprivation of such communication such as too little walking, running, jumping, and training balance during childhood, may lead to deficiencies in the development. In this case, there are not enough opportunities for the CNS to communicate with the environment to enable 'correction en route' to occur.

The movement-based learning during (healthy) ontogeny is especially efficient if it includes highly coordinated movements like fast running and exercising on the special CDT device. The ongoing increases (worsening) in the coordination dynamics value and the subsequent improvements in Figure 30 may therefore give an example how in principle the 'correction en route' during ontogeny takes place. The stronger improvement of the coordination dynamics value of the healthy child Jürgen during ontogeny (Figure 35) (longitudinal study) than in healthy children of the cross-sectional study (Figure 35) supports this view. The health of young children would be improved if they sometimes exercised on the special CDT device during sports, for example.

4.5.9. Endogenous Versus Exogenous Stem Cell Therapy: Distance of Action of Neurotrophins

It has been shown that administered CDT can bring about near-complete repair in SCI if 50% of the spinal cord matter is spared [10]. As shown here, partial repair is also achievable even if only 5% of the cord was spared. But in a complete SCI, CDT has not thus far been very successful, neither has hematopoietic stem cell therapy [53]. The movement-based learning therapy (CDT) is therefore especially successful if a sufficient number of tract fibers are spared to allow substantial learning. It is shown here that, even in a near-complete spinal cord injury, some regeneration of the spinal cord could be achieved, though to a very limited degree. In the future it needs to be estimated how much regeneration is possible and necessary to achieve sufficient repair. This very limited structural repair that could be achieved in the above case also means that the endogenous stem cell recruitment, as a part of the structural repair of CDT, was also very limited. Even the combination of exogenous and endogenous (as a part of CDT) stem cell therapy [53] was unsuccessful in humans.

In a frog model for studying development and repair, it was partly shown (Chapter I of [14]) that the distance of action of attracting and repulsive neurotrophins were probably in the range of 0.1μm. Such a short distance of action is probably difficult to achieve in exogenous stem cell therapy in humans. Further qualified research in animal and human needs to be performed, aimed at estimating the optimal distances of action of neurotrophins. Until such research is undertaken, the Author believes that there is no justification to perform exogenous stem cell therapy in humans. It is unethical to perform speculative experiments on humans under the name of therapy trials.

Chapter VII

Repair of Urinary Bladder and Vegetative Functions

1. Urinary Bladder Repair in General

1.1. Bladder Repair as a Consequence of Human Neurophysiology and Clinical Research

Severe cervical Spinal Cord Injury (SCI) leads to quadriplegia, and autonomic dysfunctions. Continence, cardiovascular performance, and breathing are impaired as well as movement. It will be reported that the aforementioned patient Kadri, who suffered a severe cervical, motoric complete SCI, in whom urinary bladder functions were fully repaired by functional and structural repair (limited regeneration of the cord) following two and a half years of Coordination Dynamics Therapy (CDT). The repair of the blood circulation, demonstrated by no reoccurrence of pressure ulcers, breathing and motor functions was reported above. The mechanism that underlies this important repair of urinary bladder functions is the learning transfer from movements to bladder functions and the direct activation of the bladder and rectum. The human bladder repair is analyzed at the neuron level, the collective variable level (System Theory of Pattern Formation), the movement, and the clinical diagnostic level.

Above it was shown that if 50% (according to MRI) of a cervical spinal cord was spared following injury, the patient could relearn walking, running, and jumping (Figures 52,53) and could become continent. Continence was achieved in the patient within six months. The administered CDT could partly cure this incomplete SCI at C5/6 levels largely through functional reorganization [10]. The question is now, can in a patient with an injury ten times more severe bladder functions still be repaired, if the approximately spared 5% tract fibers of the spinal cord white matter do not lead to the sacral micturition center (no spared sacral range)? A functional reorganization of the neural networks and tracts of the spinal cord is most likely insufficient for bladder repair. A regeneration of the spinal cord would be needed and/or the autonomic nervous system outside the spinal cord would have to take over function.

It will be shown here that in a patient with a severe, motoric complete cervical SCI, the urinary bladder functions could be repaired following two and a half years of optimal and intensive CDT. In the book "Human Neurophysiology" [14], the human neurophysiologic and clinical knowledge was set out to enable understanding of the dysfunction of the bladder by comparing the organization of the neuronal networks of the sacral micturition center between patients with SCI and those of a brain-dead human [12,16,17,119-121]. The data of brain-dead humans are rather physiologic in this respect. It was shown that the natural firing patterns of single somatic and parasympathetic neurons gave rise to the knowledge [119-121] to understand learning transfer from trained movements to bladder functions [12].

It will be shown that the partial solution of the great conundrum of how to repair the human spinal cord, namely the repair of bladder, lies in the undertaking of continuous human neurophysiologic and clinical research. The knowledge necessary to repair bladder functions by CDT cannot only be used in patients with SCI but also in elderly people and young women with stress incontinence.

2. Therapy Method to Repair the Urinary Bladder

2.1. Coordination Dynamics Therapy and Bladder Repair

Schalow Coordination Dynamics Therapy (CDT) is a movement-based learning therapy. Following SCI, the self-organization of the CNS by phase and frequency coordination is impaired and movements, bladder, bowel, and other functions are impaired, become pathologic or are lost. In order to repair the impaired phase and frequency coordination among the firings of neurons and neural assemblies, the patient trains on special CDT devices. For the repair of the impaired or lost movements, the patient trains (if possible) the innate automatisms (which may get genetic support for repair) of commando-crawling, crawling, 'bear-walking', walking, and running in the forward and backward direction (for CNS symmetry [48] repair), old-learned movements like climbing stairs (because they are deeply built into the CNS and have a high stability) and rhythmic, dynamic, stereotyped movements to repair premotor spinal oscillators. Bladder and bowel functions cannot directly be repaired by training and learning. The repeated retrograde filling of the bladder was unsuccessful in the past (private information). However, within the framework of System Theory of Pattern Formation, applied to the injured CNS, bladder functions can be repaired by learning transfer from coordinated movements, especially because the somatic nervous system is substantially interlaced with the parasympathetic division as will be shown below.

2.2. Collective Variable Level of Description of CNS Organization

The System Theory of Pattern Formation for Repair furnishes us with a theory to partly understand the repair of the injured human CNS by learning and learning transfer [122]. Its application to the somatic and autonomic nervous system divisions extends the theory to understand also the repair of bladder functions. By increasing the stabilities of physiologic bladder patterns (bladder synergy) and decreasing the stabilities of pathophysiologic bladder patterns (bladder dyssynergy, spasticity) continence and physiologic micturition can be achieved. The rationale of viewing the deterioration of neuronal network organization from the aspect of pattern stability has its origin in the stability changes of phase and frequency coordination following CNS injury, measured with the single-nerve fiber action potential recording method in humans (Figures 3-6,9).

By cooperative and competitive interplay the many billions of neurons of the human CNS generate dynamics in self-organization in communication with receptors in the periphery which can explain bladder functions. Schematically, collective variables or order parameters capture the collective coordinated activity of the neurons of the CNS. Specific equations of motion (the dynamics) of these collective variables generate the time course of organizational states. If the collective variables are designated by the vector X, the coordination pattern dynamics can abstractly be formulated by the equations of motion [18, 122]:

$$dX/dt = F_{intr}(X) + \sum c_{inf} F_{inf}(X,t) \quad (1)$$

where F_{intr} designates the intrinsic dynamics. The sum is over different types of so-called behavioral information, F_{inf}, as environmental, memorized, or intended behavioral information. The relative strength of the different influences is parameterized by c_{inf}. With the behavioral information $\sum c_{inf} F_{inf}(X,t)$ (intention, instruction, treadmill walking, exercising on special CDT devices, crawling, jumping on springboard, ...) the intrinsic dynamics $F_{intr}(X)$ of neuronal network organization can be changed, including the stability of different patterns, rather than certain coordination patterns itself only [122]. Learning a new pattern involves alterations of the entire layout upon CDT; attractive states are formed anew (Figure 175, continence and micturition states) or previously attractive states vanish (Figure 175, spastic state (spasticity of the external bladder sphincter)). The change of the whole coordination dynamics by the behavioral information is the scientific basis for learning transfer between different coordinated movements and bladder functions for bladder repair.

2.3. Attractor Layouts for Jumping on a Springboard and Switching between Continence and Protection

For jumping on a springboard (Figure 149Aa) the equation of motion of the collective variable φ (angle between the legs) could be solved in the HKB model [123] and an attractor layout constructed for the jumping in in-phase and in anti-phase (Figure 149Ab). By way of analogy an attractor layout for bladder continence and protection reaction (Figure 149Bb) may be drawn. The layout for bladder continence will be used below to show bladder repair by use of the attractor layout during CDT (Figure 175). As can be seen from Figure 149Bb, the enhancement of the continence safety (deepening of the potential well) needed approximately 6ms and the change from the pattern continence to protection reaction needed approximately 100ms. The short time to improve continence pattern stability (6ms) and the much longer time of pattern change from continence to protection pattern (100ms) were obtained from latency differences of the response of premotor spinal continence oscillators following touch and pin-prick (Figure 149Ba). It can be seen from Figure 151A that in nine out of ten cases, the latency of the oscillation period shortening following touch stimulation of the perianal skin was shorter in duration than the oscillation period (≈100ms), represented by the dotted line. The shortest latency was approx. 10ms, when measuring from the afferent volley running in direction of the spinal cord.

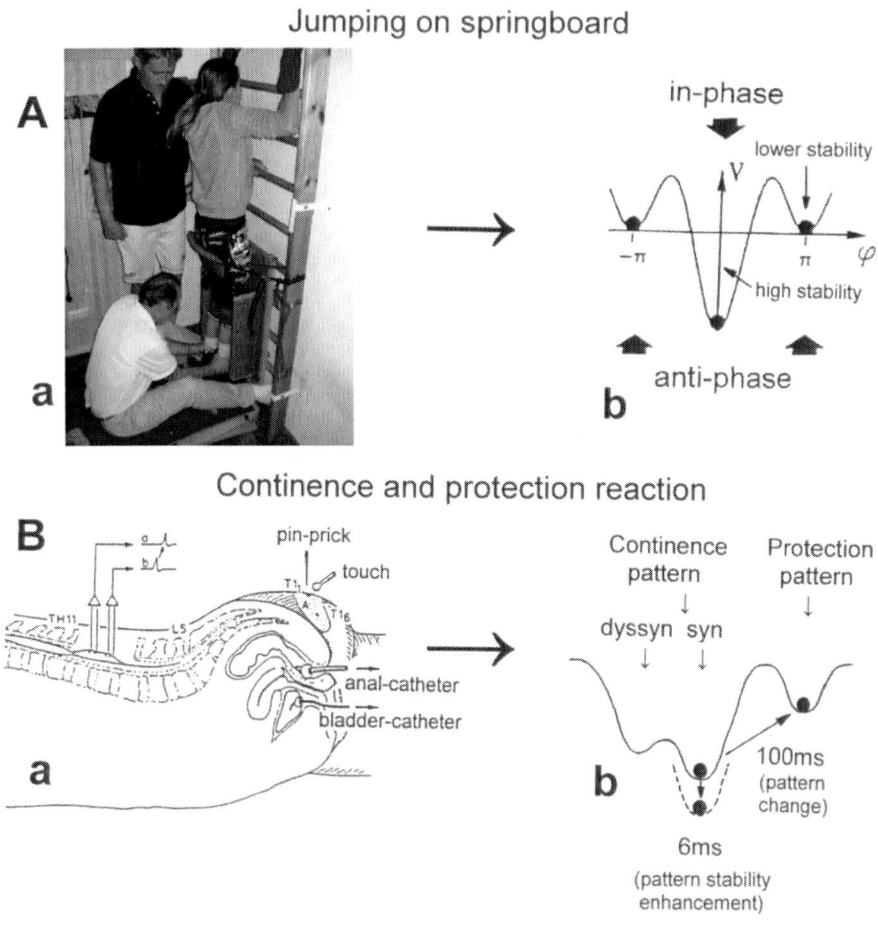

Figure 149. Attractor layout for jumping on springboard in comparison to that of continence and protection.

The latency of the first reduction of the oscillation period, measured from the skin touch stimulation artefact, was longer or shorter than the duration of the oscillation period (Figure 151B). The shortest latency was approx. 30ms. The difference of the latency, i.e., the time from the stimulation artefact (touching of the skin) to the moment when the skin afferent volley passed the recording electrodes, was approx. 20ms. According to the group conduction velocities of skin afferents (T1 skin afferents (PC) ≈ 44 m/s at 36 °C) (Figure 7), most of the time was lost in the touch receptors and the thin nerve fibers connected to them. With a conduction distance of 0.3 m, a conduction time of $t_{cond} = s/v = 300\ mm/\ 44mm(ms)^{-1} = 7ms$ results. The latency to the reduction of the first oscillation period duration following pin-pricking the sites 1 to 10 (see Figure 163B) was mostly longer than the oscillation period itself, irrespective of whether it was measured from the skin afferent volley (Figure 151E) or from the stimulation artefact (Figure 151F). Since not every pin-prick was strong enough to generate pain (stimulation of pain receptors), it is concluded that the latency from the pin-pricking to the first reduction of the oscillation period was longer than the oscillation period. By comparing the latencies to the first reduction of the oscillation period following repetitive touch and pin-prick stimulation of sites 1 to 10, it turns out that following touch stimulation the delay is shorter than the oscillation period, whereas it is longer following pin-prick stimulation. It seems therefore that touching reinforced the sustained stretch reflex of the anal sphincter while pin-prick did not.

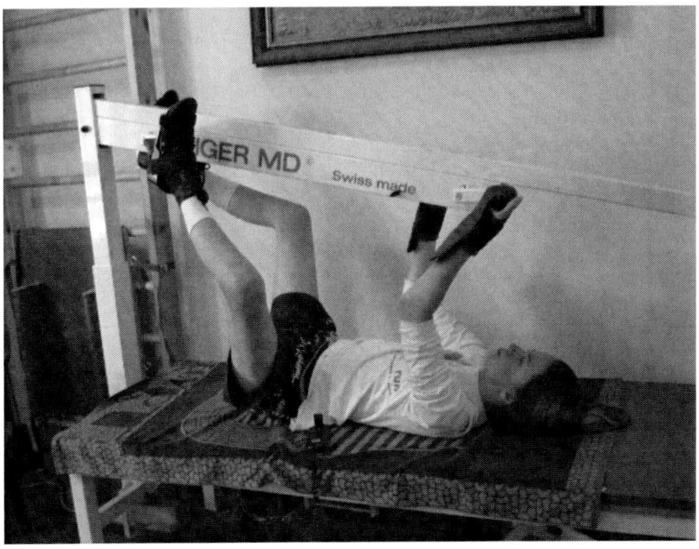

Figure 150. Exercising of a 20-year-old female patient with a severe spinal cord injury at C5/6 levels on the special CDT device in the prone position. Such very coordinated turning with arms and legs is stimulating autonomic functions like bladder, intestine (peristaltic), and blood circulation functions by improving the phase and frequency coordination of the networks driving those autonomic functions.

The first shortening of the oscillation period in comparison to two preceding periods is plotted in Figure 151D,C (touch) and in Figure 151H,G (pin-prick). The reduction of the oscillation period for touch stimulation was between 8 and 28ms and between 5 and 40ms for pin-pricking. The pin-prick stimulation generated a larger transient reduction of the oscillation period. When correlating the latency of shortening to the extent of shortening for pin-pricking (Figure 151H) it seems as if the shorter latency correlated with a smaller reduction of the oscillation period (more touch-like), and the long latencies correlated with the greater reduction of the oscillation period (more pain-like). A similar relation was not found following touch stimulation (Figure 151D). It is most likely that the painful pin-prick stimulation transiently replaced the sustained stretch reflex of the anal sphincter by a protective reaction of the anal sphincter against pain application; the time needed for the reorganization of the oscillatory firing neuronal network was at least one oscillation period long. The relation between the sites of pin-prick (Figure 151G) or touch stimulation (Figure 151C) and the oscillation period shortening shows no clear correlation. The dotted and the dashed lines may show similar latencies. Perhaps the protective reaction against pin-pricking of the anal canal and the pelvic floor is highest at sites 3 and 8 at the border of the anal reflex region.

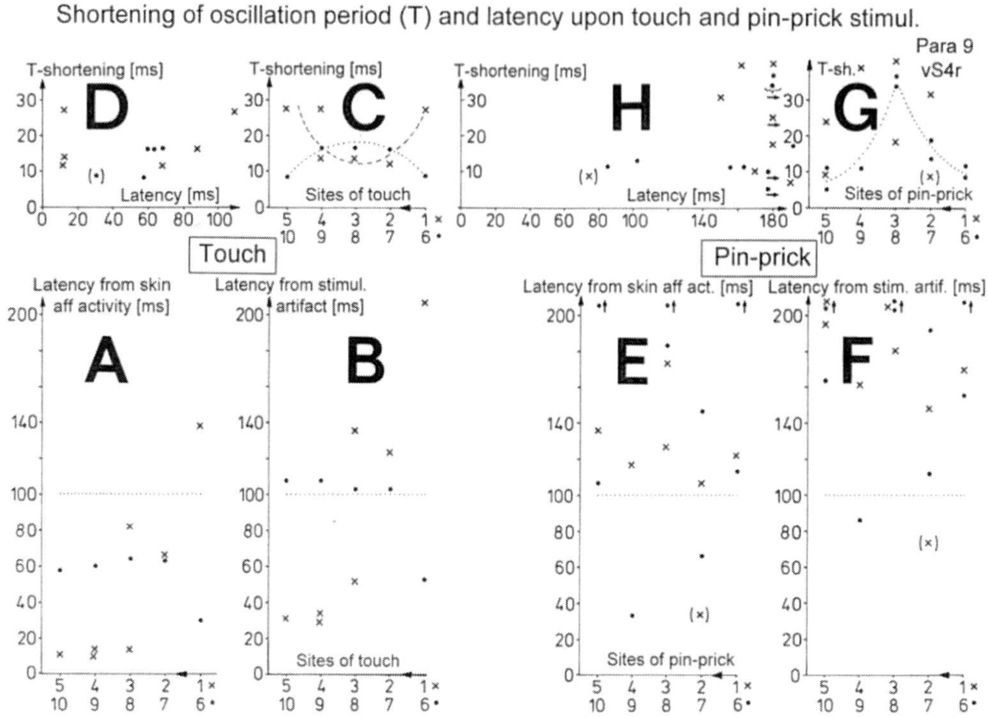

Figure 151. Shortening of the oscillation period T of the α_3-motoneuron (perhaps an α_2-motoneuron upon touch and pin-prick stimulation). A, B. Latency of the shortening of the oscillation period measured from the skin afferent activity (A) and from the stimulation artifact (B) with successive touching of sites 1 to 5 (x) and 6 to 10 (•) as indicated in Figure 163B. Note that the delay is often shorter than the oscillation period. D,C. Shortening of the oscillation period (T-shortening), with respect to two preceding oscillation periods, in relation to the latency of shortening (D) and in relation to successive touching of sites 1 to 10 (C). E.-H. Same description as for A-D, only with respect to pin-pricking. Note that touch stimulation induced a shorter latency for the reduction of the oscillation period than did pin-pricking.

By comparing the latencies from the moment when the skin afferent volley passed the recording electrodes to the first reduction of the oscillation period following repetitive touch and pin-prick stimulation of the sites 1 to 10, it appears that the delay following touch stimulation (≈ 10ms) is much shorter than for pin-prick stimulation (≈ 110ms). It seems therefore that touching reinforced the sustained stretch reflex of the anal sphincter while pin-prick stimulated the protection reaction against pin-prick (pain). It can be evaluated how much time is needed for the reinforcement of the sustained stretch reflex and for the pattern change from the continence state (sustained stretch reflex) to the protection reaction. With a conduction distance of 0.1 m (distance between the recording electrodes and spinal cord), a conduction time for the fastest skin afferents (group conduction velocity of T1 skin afferents (Parcinian Corpuscle) ≈ 44 m/s at 36 °C) of $t_{cond} = s/v = 100$ mm/ 44 mm(ms)$^{-1}$ = 2.3ms is resulting. The conduction time of the action potentials of the motoneurons to cover the distance from the spinal cord to the recording electrodes is resulting also to approximately 2.3ms (the group conduction velocities of the α_2 and α_3-motoneurons are 37 and 50m/s respectively). The processing time to reinforce the continence pattern of the anal sphincter was therefore approximately 5.4ms (10ms-4.6ms = 5.4ms). If we assume a synaptic transmission time of 2.5ms, the processing time in the spinal cord is sufficient for the transmission time of two synapses. The reinforcement of the sustained stretch reflex was may be just a di-synaptic pathway.

The conduction times to and from the spinal cord of the protection reaction are approximately 10ms (7.7 + 2.3), because the fastest pain afferent action potentials needed a bit more time than the touch afferent action potentials (group conduction velocity of the fastest pain afferents = 13m/s; $t_{cond} = s/v = 100$ mm/ 13 mm(ms)$^{-1}$ = 7.7ms). The processing time in the CNS to change from the sustained continence pattern to the protection reaction pattern is therefore approximately 100ms. This time would be enough to involve 40 synapse transmissions. For a reorganization of the neuronal networks for pattern change, conduction times of intra-spinal axons and dendrites (electrotonic conduction, passive conduction) would also have to be taken into account. But it is obvious that the pain application induced a completely different CNS response. That pin-pricking resulted in a longer response time of motoneurons than touching of the perianal skin is not obvious and cannot be explained based on the reflex theory. Being a stronger stimulus, pin pricking (touch plus pain afferent activity, Figure 5) would be expected to have the same or slightly shorter rather than a much longer response time than touching of the perianal skin (touch afferent activity only, Figure 5).

It is concluded that the change from one network state to another one, restricted to the spinal cord (measurement in a paraplegic), needed approximately 100ms.

With respect to movement learning and learning transfer, strong pain has to be avoided during exercise because a certain pattern such as jumping on a springboard has to be trained (and neuronal networks entrained) and not a combination of a movement pattern and a protection reaction. These measurements and estimations show that it is possible to analyze integrative patterns of the human CNS with natural impulse patterns of single-nerve fibers recorded from sacral nerve roots. Natural impulse pattern can explain integrative functions of the human CNS.

2.4. Pattern Fluctuation in the Attractor Due to Variation of Quality of Phase and Frequency Coordination

Since I measured an increased variability of phase and frequency coordination among neuron firing following spinal cord injury in the human spinal cord (Figures 10;48a,b), such fluctuation of coordinated firing has to be included in the equations of motion because it alters the attractor layout and reduces pattern stability. In the Haken-Kelso-Bunz model, the jumping on a springboard can be described in terms of relative phase between the rhythmically moving legs. Without specific behavioral information the dynamical description is defined by a vector field expressing the rate of change in relative phase, $d\varphi/dt$, as a function of the derivative of its potential, $V(\varphi)$:

$$d\varphi/dt = -\,dV(\varphi)/d\varphi + (Q\xi_t)^{1/2} \tag{2}$$

were $V(\varphi) = -\,a\cos(\varphi) - b\cos(2\varphi)$ and $(Q\xi_t)^{1/2}$ is the term representing the phase and frequency variability of strength Q (for further details see page 83 of [14]). To reduce the fluctuation of phase and frequency among neuron firing to increase pattern stability (reduction of pattern fluctuation (ball moves less (Figure 18)), necessary for urinary bladder repair, the patients have to exercise on special CDT devices to re-learn the coordinated firing of neurons. The exercising on special devices has to be included as behavioral information in the equations of motion (formula 2).

2.5. Human Neurophysiology for a Deeper Understanding of Bladder Repair

Since learning and learning transfer during CDT may need a few years, it is of great importance to use such behavioral information which can really repair the urinary bladder. The proper behavioral information for learning transfer was developed by using the single-nerve fiber action potential recording method.

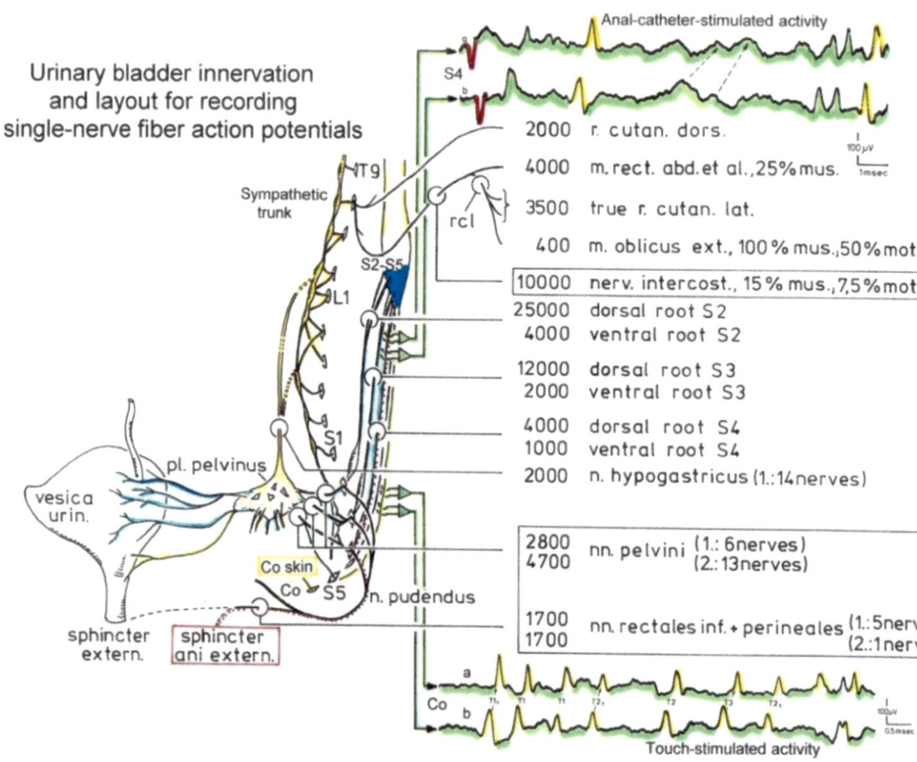

Figure 152. Urinary bladder innervation (anatomy; structure) and recordings of single-nerve fiber action potentials from a S4 and a coccygeal root (electrophysiology; function). Number of myelinated nerve fibers of the nervi rectales inferiores and perinales, the nervi pelvini and the nervus hypogastricus. T = thoracal; L = lumbal; S = Sacral; rcl = ramus cutaneus lateralis, no pure skin nerve; %mus = % of nerve fibers leading to muscles; %mot = % of nerve fibers which are motoric; 1 and 2 = Nerve fiber counts from cadavers 1 and 2. Number of nerves in the bracket gives the number sub-nerves of which the nerve consisted. The innervation pathway of the external bladder sphincter is unclear. The S4 root recording informs about single-nerve fiber activity running into the spinal cord (from the bladder receptors) and running out of the cord to the urinary bladder (bladder efferents) upon natural stimulation. Functions of the urinary bladder functions and the sacral micturition center in the spinal, as a part of the human CNS, can therefore be analyzed. The Co root (no efferents; something like a skin nerve) recording informs about skin receptor activity in the coccygeal dermatome.

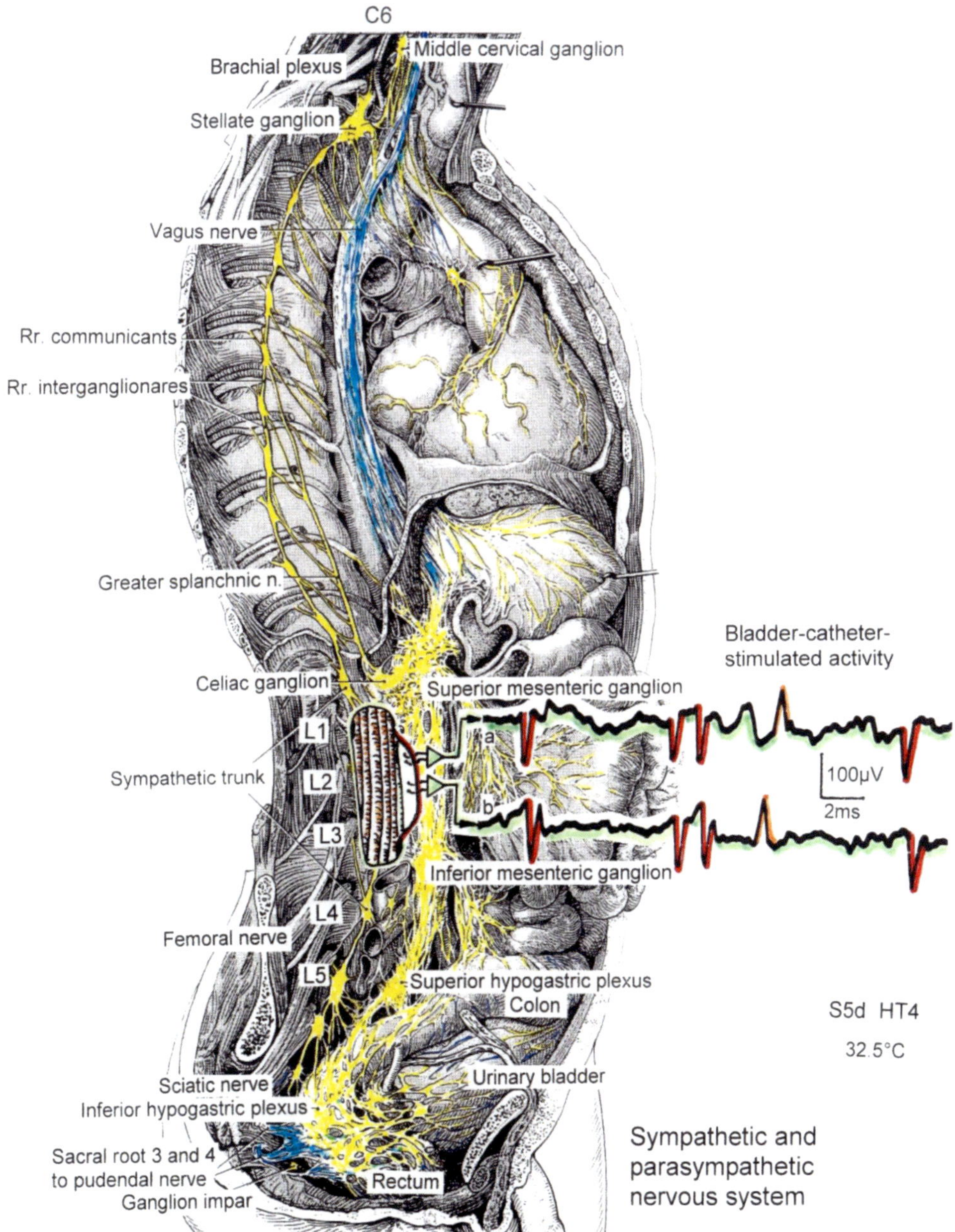

Figure 153. The picture illustrates the complexity of the autonomic nervous system. Connections of the different plexus bypass the spinal cord and offer the structure for a functional repair of vegetative functions (and may be of somatic functions) like urinary bladder control, cardio-vascular performance and breathing. The recording of single-nerve fiber action potentials from human nerve roots on the other hand show that in this complexity of the human nervous system structure it is possible to record activity from several single neurons under rather natural conditions.

By analyzing natural impulse patterns in sacral nerve roots from the somatic and parasympathetic nervous system upon bladder, bowel and skin stimulations, intravesical stimulation, and movements, it was found that jumping on a springboard (Figure 149Aa) and exercising on the special CDT device (Figure 150) are especially efficient to induce learning transfer from movements to bladder functions. The jumping on springboard was specifically designed to repair patterns, including transfer of learning from somatic to autonomic functions, discussed in detail below, and the exercising on the special CDT device was primarily used to reduce the variability of phase and frequency coordination and to increase thus the stability of network states by decreasing pattern fluctuation (Figure 18). Movements are exercised to generate a physiologic attractor layout of patterns again and the coordinated firing of neurons and neural sub-networks is trained to increase pattern stability by reducing the fluctuation of pattern states (Figures 18,170).

The innervation of the bladder is schematically shown in Figure 152. Nerve fiber counts of myelinated nerve fibers indicate the thickness of nerve roots and nerves. Even though single-nerve fiber action potentials can also be recorded from peripheral nerves [14], high-quality recordings can only be obtained from thin and long sacral caudal nerve roots. The very thin coccygeal nerve root contains mainly skin afferents to the coccygeal dermatome and is ideally suitable to analyze skin receptors [14]. A recording is shown in Figure 5. To avoid adding up single-nerve fiber action potentials, the summed activity has to be low to be able to split up the multi-unit recording into several natural single-nerve fiber impulse patterns. The lower sacral nerve roots are suitable for that. An original recording from a S4 root upon anal catheter pulling to stimulate the somatic and parasympathetic nervous systems is also shown in Figure 152. Afferent (main amplitude upwards) and efferent single-nerve fiber action potentials (amplitude downwards) can be seen. The ventral S4 root is an ideal root to record single-nerve fiber action potentials from efferent (running out of the spinal cord) and afferent fibers (leading into the cord) (Figure 3). In caudal sacral nerve roots there are also afferent fibers in ventral roots and also efferent fibers (motor fibers) in dorsal roots. The innervation of the bladder from the sacral micturition center is leading through the nerve roots S2 till S5.

Since the repair of bladder function in patients with SCI probably also requires the reorganization of pathways outside the spinal cord, the connection of the plexus pelvinus to rostral plexuses is shown in Figure 153. The picture allows understanding of just how complex and distributed the autonomic nervous system is. The bladder is mainly innervated by the somatic and parasympathetic nervous system and probably a little by the sympathetic nervous system (internal bladder sphincter). In Figure 154 the innervation of the human organs is shown via the sympathetic and parasympathetic divisions. By recording single-nerve fiber action potentials from caudal sacral nerve roots, only natural impulse patterns in somatic and parasympathetic nerve fibers can be expected to be obtained because it was recorded from the S3 to S5 dorsal and ventral nerve roots.

The complexity of the autonomic nervous system is tremendous (Figure 153). The thin ventral S4 root (motor root) in itself shows quite a complexity (Figure 155). The thick myelinated nerve fibers are prominent. By enlarging the scale, also the non-myelinated nerve fibers become visible (Figure 155B). Their number is in the range of 5-times higher than those of the myelinated fibers [125]. The diameter of the thinnest myelinated fibers from which it is possible so far to record single-nerve action potentials is in the range of 3 to 4μm (Figure 7). To record from these thinly myelinated fibers is difficult but important. To repair of urinary bladder, we must be able to record also from parasympathetic nerve fibers and distinguish their activity from the activity of the different groups of γ-motoneurons to understand bladder function.

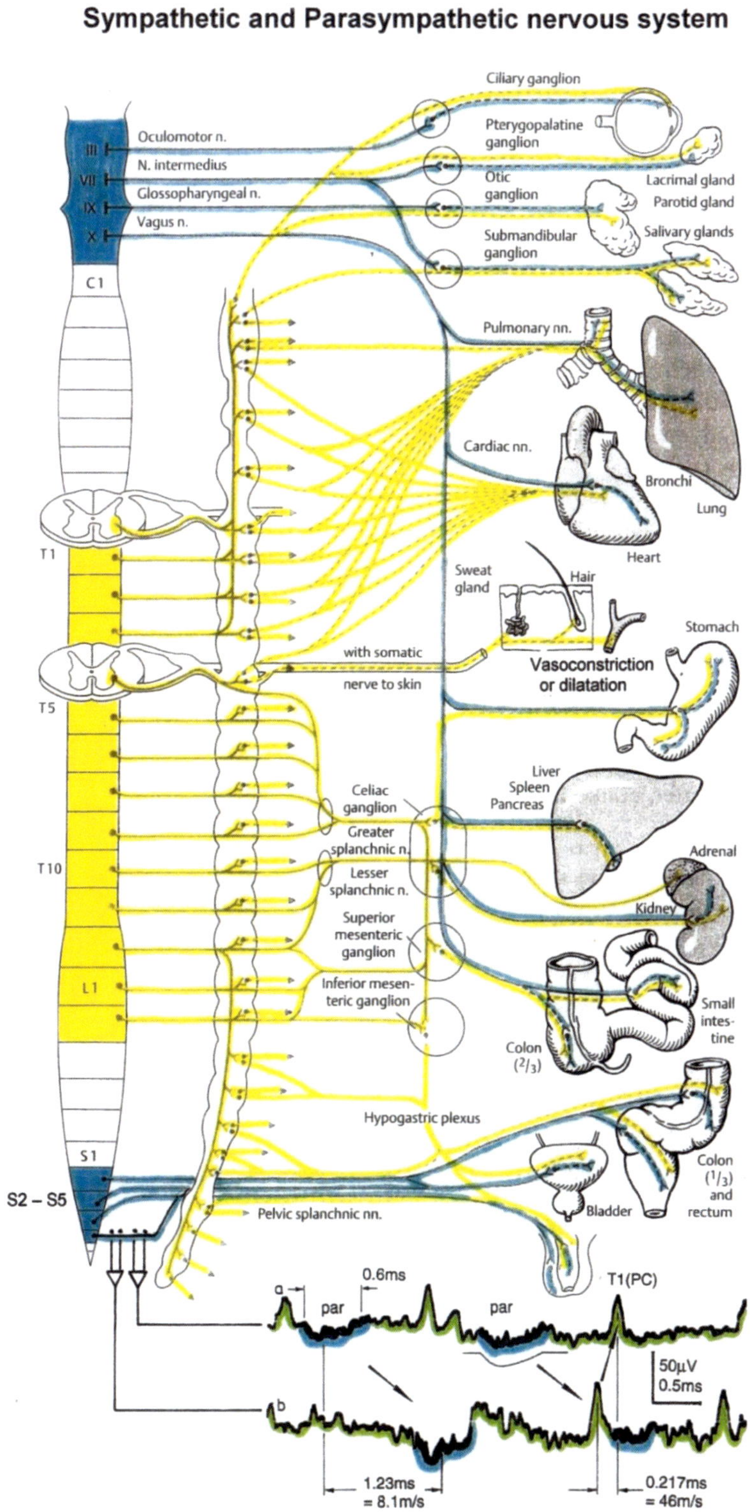

Figure 154. Schematic diagram of the sympathetic and parasympathetic nervous system. Yellow = sympathetic, blue = parasympathetic. The recording from a sacral root shows single action potentials of preganglionic neurons (par) and a skin afferent fiber.

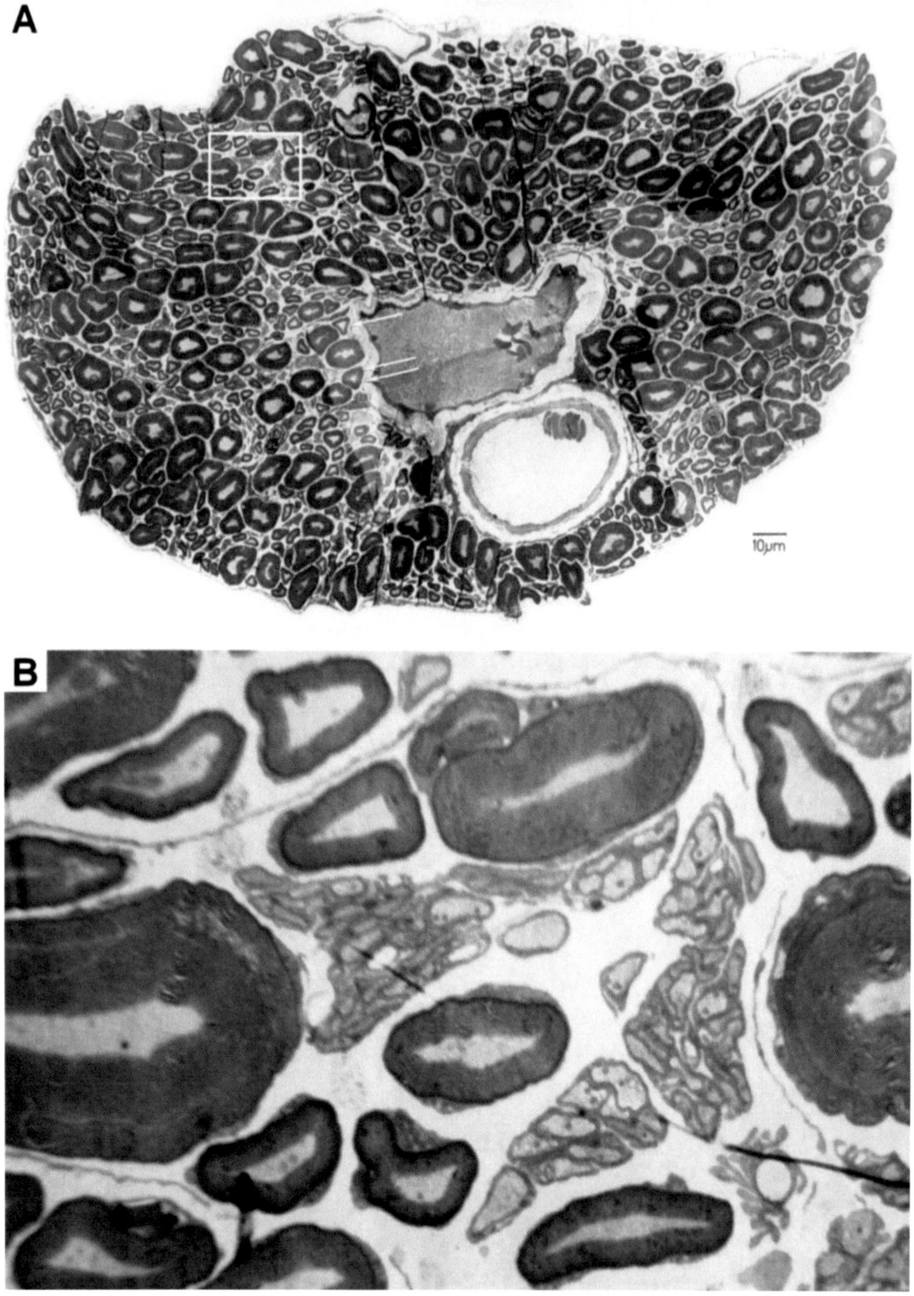

Figure 155. A. Montage of several electron microscope photographs of the S4 ventral root from HT1. Single arrow marks a nerve fiber with a comparably thin myelin sheath; double arrow marks a nerve fiber of same size with a comparable thick myelin sheath. Scale not corrected for shrinkage. B. Enlarged scale to make unmyelinated nerve fibers visible.

Since parasympathetic fibers are thin their single-nerve fiber action potentials are small and the identification of their action potentials in the summed impulse traffic is difficult. Also the γ-motoneurons are thin and its different groups are difficult to separate, especially when the recording temperature is low. In the conduction velocity distribution histogram of Figure

156 (upper right) the parasympathetic fiber groups cannot be distinguished from those of the γ-motoneurons making it difficult to identify impulse patterns of single γ-motoneurons and parasympathetic efferent fibers.

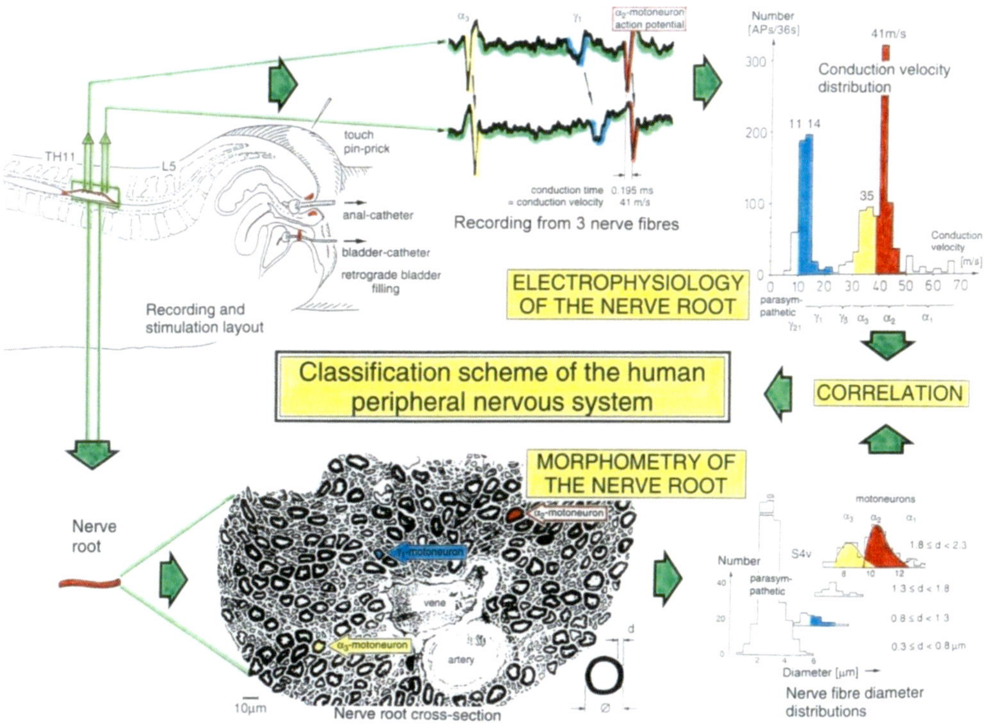

Figure 156. Schematic layout of the classification scheme for the human peripheral nervous system. By recording with two pairs of platinum wire electrodes from a nerve root containing approx. 500 myelinated nerve fibers, a recording is obtained in which 3 action potentials (APs) from 3 motoneurons (main AP phase downwards) can be seen. By measuring the conduction times and with the known electrode pair distance (10 mm) a conduction velocity distribution histogram is constructed in which the nerve fiber groups are characterized by ranges of conduction velocity values and peaks in asymmetrical distributions. After recording, the root was removed, fixated, embedded and stained, light microscope cross-sections were prepared and used to measure the mean diameter and the myelin sheath thickness (d). Distributions of nerve fiber diameters were constructed for four different ranges of myelin sheath thickness. Nerve fiber groups were characterized by the peak values of asymmetrical distributions. By correlating the peak values of the velocity distributions with those of the diameter distributions obtained from the same root, a classification scheme was constructed of the human peripheral nervous system. Brain-dead human HT6.

2.6. Identification of Peaks of γ-Motoneurons and Parasympathetic Fibers in Conduction Velocity Distributions on Log Scale

On a linear scale the conduction velocity peaks may fuse and can be separated by their different functions only partly (Figure 157Da,Ea). By constructing first histogram classes for conduction times however and plotting them on a log scale from the column values, the velocity peaks separated. Figure 157D,E shows that the fused γ-peaks of the linear plots (Da, Ea) split up into different peaks on the log scale (Db, Eabc). Since first histogram classes of conduction times were constructed using a linear scale, it was also possible to study the dynamics of peaks or peak values upon stimulation. The group identification of single-nerve fiber action potentials is necessary to explore human CNS self-organization.

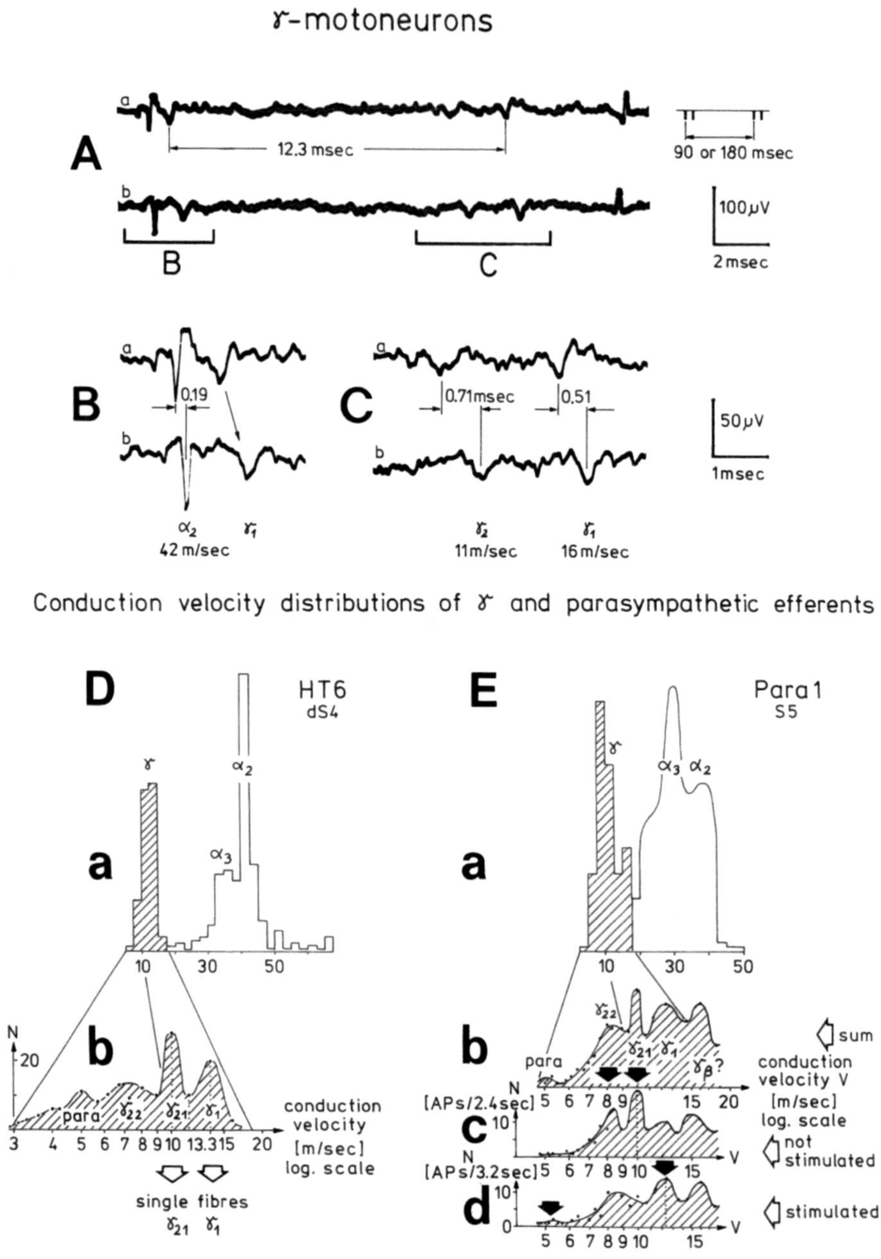

Figure 157. A. Activity from 2 γ-motoneurons and an α_2-motoneuron. The γ_1-motoneuron may have fired in the oscillatory mode for a few cycles with a period duration of 90ms or 180ms. Interspike intervals of the possible impulse train, consisting of 2 APs, are indicated (12.3ms). The insert shows the possible oscillation cycle period. HT6; dS4. B,C. Time and amplitude expansions of A. Conduction velocities are indicated. Note that with the increasing conduction velocity, the AP amplitude increases and the duration decreases. D,E. Conduction velocity distributions of motoneurons for the brain-dead

human HT6 (D) and paraplegic 1 (E). To make intrafusal motoneuron and parasympathetic peaks visible, the main γ-peak of Da and Ea was plotted on a log scale in Db and Eb. The distribution peaks are labeled with the groups, they most likely represent. In E, the distribution Eb is split into the distribution upon no additional stimulation (Ec) and upon additional stimulation (Ed).

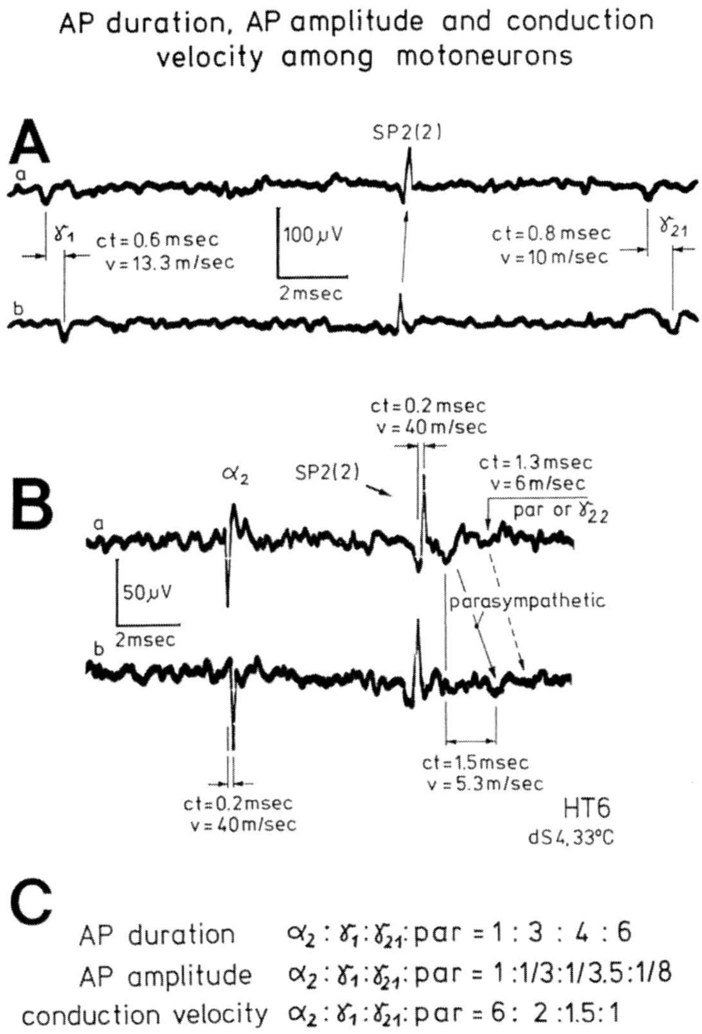

Figure 158. Recordings of single-nerve fiber action potentials (APs) (A,B). C. Relation between the AP durations, AP amplitudes and conduction velocities of single α_2, γ_1, γ_{21} and parasympathetic motoneurons. HT6, dS4-root, 33°C.

Note that in the non-stimulated distribution (Ec) the static γ-motoneuron peaks (γ_{22}, γ_{21}) are highest, whereas under stimulation (Ed) the parasympathetic (Para) and the dynamic γ-motoneuron peaks (γ_1) are highest. When plotting the velocities in Db and Eb logarithmically, the conduction times were first grouped by a conduction time histogram and the column values were then used (conduction distance = 8 mm) to construct conduction velocity distribution curves.

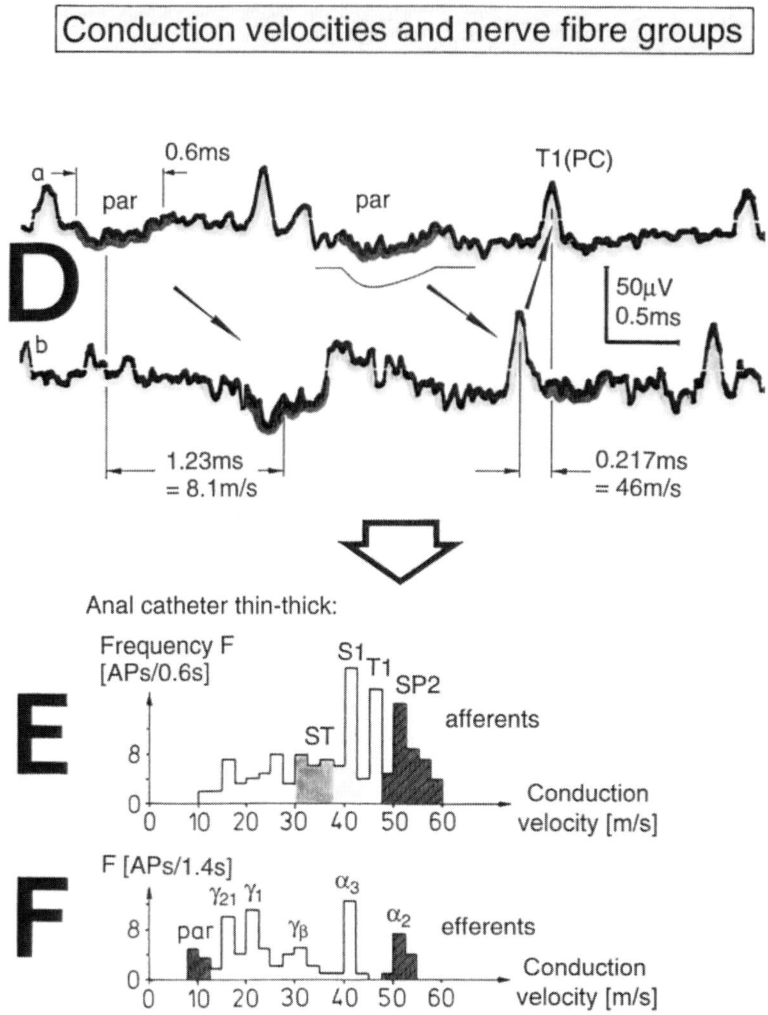

Figure 159. Sweep piece of recording (D) and conduction velocity distributions (E,F) taken from time intervals following a change of a thin anal catheter (∅ = 12 mm) for a thick one (∅ = 20 mm). Note the manifestation of the parasympathetic peak.

Figures 157Db and 157Eb show several distribution peaks in which the γ_1 and γ_{21}-peaks could be identified from earlier functional considerations. By comparing the velocity distributions obtained upon stimulation with those without stimulation (Figure 157E), and reasonably assuming that the dynamic intrafusal motoneuron peaks are higher upon stimulation (Figure 157Ed) and the static ones are higher with no stimulation (Figure 157Ec), a second static γ_2-peak (γ_{22}) can be identified and at least one further rather dynamic motoneuron peak (para) can be seen. This peak distribution observed in the paraplegic patient could also be found in the measurements in HT6 (Figure 157Db). That the "Para" peak contains activity from parasympathetic fibres will be shown later. The identification of parasympathetic efferents contributes to the understanding of urinary bladder functions under physiologic and pathophysiologic conditions and its repair by CDT. The nerve fibre groups obtained from conduction velocity distributions for single fibres are in accordance with the sub-compound action potentials of compound action potentials [124].

The different size of the action potentials of secondary muscle spindle afferents, the different γ-motoneurons and parasympathetic efferent fibres can be seen in the original recordings in Figures 157 through 159. Action potential duration and amplitude vary much and also the conduction velocity (Figure 158).

In the recording from a paraplegic patient in Figure 159 the different groups of parasympathetic fibres and γ-motoneurons do not fuse, probably because of the higher recording temperature during an operation to implant an electrical bladder stimulator. Such a procedure is now obsolete because the bladder can be repaired by CDT.

2.7. Location and Stimulation of Receptors for Continence

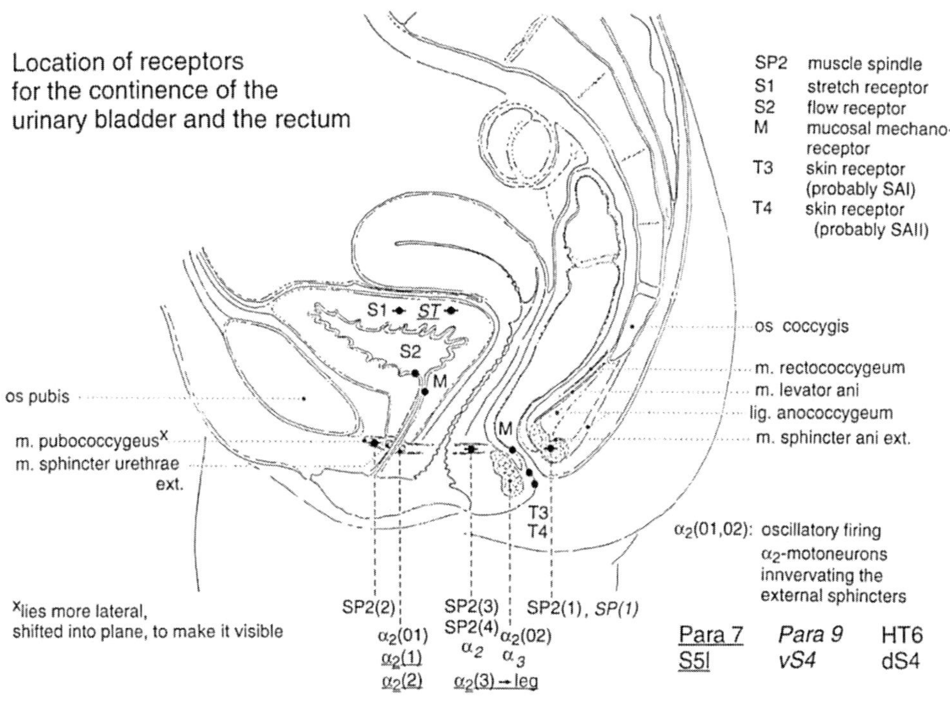

Figure 160. Location of receptors and muscles for the continence of the urinary bladder and the rectum, innervated by motoneurons the activities of which were recorded, in the brain-dead human HT6 (dS4 root), paraplegic 9 (vS4 root) and paraplegic 7 (nerve root S5).

The receptors of skin, urinary bladder and anal mucosa, the muscle spindles which were innervated by secondary spindle afferents, and the muscles which most likely were innervated by α-motoneurons in paraplegics 7 and 9 and in the brain-dead human HT6 are marked in Figure 160 in the lower pelvis. The (rhythmic) activation of these receptors by anal and bladder catheter pulling and skin stimulation of sacral dermatomes, to stimulate simultaneously the somatic and parasympathetic divisions to induce learning transfer, can be simulated and achieved during therapy, if a patient with severe cervical spinal cord injury jumps rhythmically on springboard with no weight support (Figure 149Aa).

Following retrograde bladder filling urinary bladder stretch (S1) and tension receptor afferents (ST) can be stimulated and measured. In Figure 161 the conduction velocity distribution histogram shows the activation of S1, ST and secondary muscle spindle afferents (SP2) for a bladder filling of 750ml. Jumping on springboard with a full bladder or colon and rectum activates the somatic and parasympathetic divisions much more because more receptors are activated. Before exercising on the special device or jumping on the springboard, patients often first empty their bladder and rectum to stay continent. To understand urinary function it is important to study bladder filling at the neuron level.

2.8. Bladder Functioning at the Neuron Level

With the single-nerve fiber action potential recording method it has so far been possible to record single-nerve fiber action potentials from nerve fibers down to a diameter of approximately 3.5 μm in undissected thin long nerve root fascicles. It is therefore possible to record natural impulse patterns from parasympathetic efferents (par), urinary bladder stretch and tension receptor afferents (S1, ST), mucosa afferents from mechanoreceptors of the bladder, the urethra and the anal canal (M), from afferents responding to fluid movement (S2), and from α_2, α_3 and γ-motoneurons and muscle spindle afferents innervating the external striated urinary bladder and anal sphincters (or functionally associated pelvic floor muscles), and to analyze regulatory and organizational mechanisms of parasympathetic neurons and motoneurons in the human CNS.

During a laminectomy, natural impulse traffic to and from the CNS can be recorded from sacral nerve roots. The summed impulse traffic can be split into the natural impulse patterns of single afferent and efferent fibers. Following retrograde bladder filling (as in the clinical diagnostic [14]) and the identification of the neuron type, with the use of the classification scheme for human nerve fibers (Figures 7,156), the natural impulse patterns of identified afferent and efferent fibers can be obtained and analyzed. The obtained natural impulse patterns answered an age-old question: Is it the firing rate of a neuron that codes the information transmitted and processed or does the precise timing of cell discharge codes information? As the different pictures show, the information is coded by specific impulse patterns, including the precise timing and the firing rate.

In response to retrograde bladder filling of a brain-dead human, the self-organization of a premotor spinal α_2-oscillator innervating the external striated bladder sphincter is shown in Figure 161. Because the axon of motoneuron O1 had a recurrent fiber (Figure 161A) at the recording site, each single AP of this motoneuron could be identified safely by the AP of the recurrent fiber. The function of the motoneuron was to secure bladder continence. The activity from urinary bladder receptors, i.e., the activity of bladder stretch (S1), tension (ST) and flow receptor afferents (S2) (Figure 161E), was an adequate afferent input to the motoneuron. Phase relations between the firing patterns of bladder afferent fiber S1(1) and oscillatory firing urinary bladder sphincter motoneuron O1 can be seen in the schematized firing patterns in Figure 161B. They were an indicator that the natural impulse pattern of the S1(1) fiber was an adequate drive of the sphincter motoneuron O1. For retrograde bladder filling up to 550 ml, motoneuron O1 only fired occasionally (Figure 161D,F).

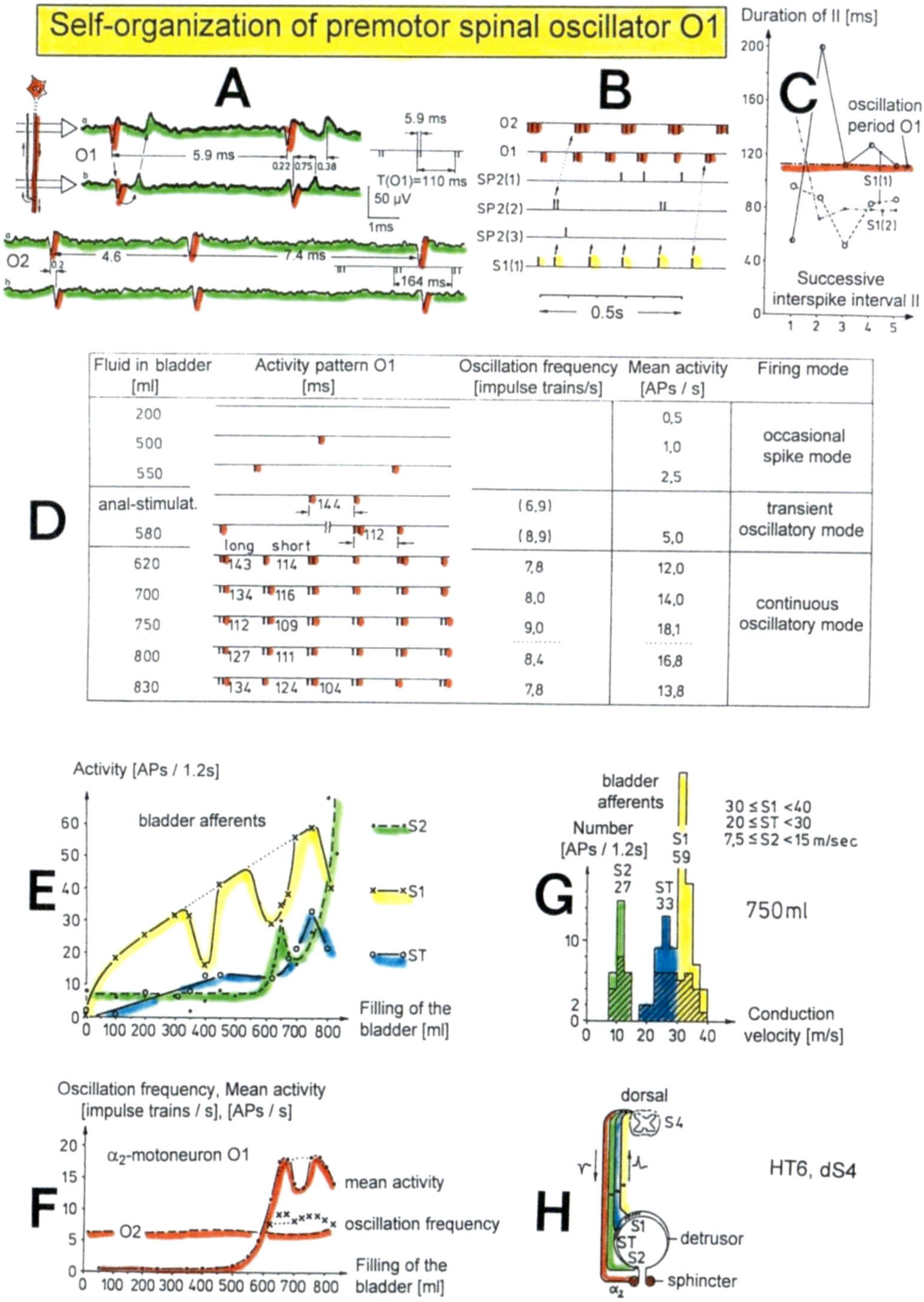

Figure 161. Self-organization of premotor spinal α_2-oscillator O1, which innervates the external urinary bladder sphincter (skeletal muscle). Brain-dead human HT6; recording from a dorsal S4 nerve root.

This was the storage phase of the bladder, during which the intravesical bladder pressure increased only little. For higher bladder filling volumes, the motoneuron switched via the transient oscillatory firing mode to the continuous oscillatory firing mode (Figure 161D) to generate a higher activity (Figure 161F) for a stronger drive of the urethral sphincter to more strongly secure continence when the storage phase was nearly passed and the bladder pressure increased. For bladder filling volumes higher than 800 ml, the activity of the spinal oscillator decreased again; probably, the oscillator became inhibited (Figure 161D). Pain fibers (not shown in Figure 161G) may have inhibited the oscillatory firing to protect the bladder from mechanical damage. The overflow mechanism was initiated. It is likely that fluid entered the trigonum vesicae to activate flow receptors, so that the flow receptor activity (S2) increased strongly (Figure 161E).

The premotor spinal oscillator O1, of which motoneuron O1 is probably a part, was organized by the adequate afferent input (S1, ST, S2, ...) induced by bladder filling. For the oscillator to be formed in the premotor neuronal network in the spinal cord, consisting of motoneurons and interneurons, it needs a certain preformation of the networks (connectivity, synapse efficacies, membrane properties of neurits etc.) and adequate afferent input patterns. In modeling such networks, the organization of the premotor spinal oscillators in the spinal cord neuronal networks cannot be separated from its space-time distributed adequate afferent input patterns, giving rise to self-organization. From Figure 161B it can be seen that the stretch receptor afferent APs (S1(1)) show a relative phase correlation to the impulse trains of the oscillatory firing motoneuron O1. The successive interspike intervals (IIs) of the S1(1) afferents have, on the average, a value similar to that of the oscillation period of oscillator O1 (Figure 161C).

Other premotor spinal oscillators, activating the external striated bladder sphincter not shown here, will also have self-organized themselves by the same or similar afferent input. However, these oscillators sub-serving the same function will be correlated in their firing in the way that they do not fire in synchrony but are distributed in their phases with respect to one another (probably by relative inhibition with respect to each other) to secure that the sphincter muscle does not show rhythmic movements (tremor). Such non-synchrony correlated oscillatory firing of several oscillators has in a small time window been measured (Figure 16B). In the case of a non-injured CNS, this premotor spinal oscillator O1 could also be activated volitionally from supraspinal centers. It would be very interesting to see how the afferent input patterns from supraspinal centers to the spinal premotor network would look.

A. Recordings from α_2-motoneurons O_1 and O_2, firing in the oscillatory mode with impulse trains of 2 (upper recording) and 3 (lower recording) action potentials (APs). The durations of the oscillation periods were 110 (O_1) and 164ms (O_2). The interspike intervals of the impulse trains were 5.9ms (O_1) and 4.6 and 7.4ms (O_2). Motoneuron O_1 conducted at 36 m/s; its recurrent fiber conducted at 21 m/s. The measurement layout is shown schematically. The inserts show the oscillatory firing modes; they have not been drawn to scale.
B. Impulse patterns of oscillatory firing α_2-motoneuron O_2 innervating the external anal sphincter, in relation to the muscle spindle afferent activity SP2(1 to 3), activated by the stretch of the anal sphincter by the anal catheter, and impulse patterns of oscillatory firing α_2-motoneuron O1 innervating the external urethral sphincter, in

relation to the stretch receptor afferent activity (S1(1)) of the urinary bladder, activated by 750 ml bladder filling. Phase relations between APs of SP2(2) and O_2 and between APs of S1(1) and O_1 are indicated by the small arrows.

C. Three series of successive interspike intervals of the 2 stretch receptor afferent fibers S1(1) and S1(2) activated by retrograde bladder filling. The oscillation period of oscillatory firing motoneuron O1, activated only by bladder filling is shown.

D. The firing in the occasional spike mode, the transient and the constant oscillatory firing mode of α_2-motoneuron O_1 in response to filling of the bladder. In the 'activity pattern' column changing durations of oscillation periods are given. The oscillation frequencies in the brackets give the frequencies at the moment of oscillation for the transient oscillatory mode. Downward deflections are schematized APs. Interspike intervals of the close APs $\approx$ 6.0ms (A).

E. Activity levels of stretch (S1) and tension (ST) and flow receptor afferents (S2) (E) and of sphincter α_2-motoneuron O_1 (F) in response to retrograde filling of the bladder. The activity values of the S1, ST and S2 afferents are taken from histograms like the one in G. Filling of the bladder was stopped once between 600 and 650 ml.

F. The small dotted lines represent mean activity (APs/s) and oscillation frequency (impulse trains/s) of α_2-motoneuron O_1 if bladder filling were not stopped in between. Note that the mean activity increases continuously with the filling of the bladder from 550 to 650 ml, even though motoneuron O_1 started to fire in the oscillatory mode from 620 ml on (D). Note further that the oscillatory firing motoneuron O_2 (frequency of firing with impulse trains is shown) is nearly not affected by the filling of the bladder and by the start of the oscillatory firing of motoneuron O_1.

G. Conduction velocity frequency distribution histogram of stretch, tension and flow receptor afferent activity at 750 ml. The activities of afferents S1, ST and S2 are quantified by counting the afferent conduction velocities under the peaks (open plus hatched part), with the conduction velocity limits given in the insert. The counts (27, 33, 59) are given below the peak labeled S1, ST and S2 and plotted into E for the afferent activity at 750 ml.

H. Schematic drawing of the anatomical arrangement of the afferents and the motoneuron O_1.

By recording from the dorsal S4 root of the brain-dead human HT6, impulse trains from another oscillatory firing motoneuron and its driving afferents were measured, which served quite a different function (Figure 161F). The α_2-oscillator O2 (Figure 161A,B) innervating the striated external anal sphincter was activated by secondary muscle spindle afferent activity, induced by the anal catheter-stretched muscle spindles, probably located in the anal sphincter or functionally associated pelvic floor muscles. Also, mucosal and skin receptors within the anal reflex area will induce self-organization of premotor oscillators activating the external anal sphincter to secure anal continence. It is evident from the impulse patterns shown schematically in Figure 161B that the impulse trains of this oscillator O2 show a phase relation and an interspike interval relation to its driving spindle afferent APs. But no synchronized firing can be seen between the sphincter α_2-motoneurons, innervating the external bladder sphincter (O1) and anal sphincter (O2), the CNS networks probably tried to avoid an increased physiologic tremor (macroscopic rhythmic activation of muscles, due to sub-optimal regulation).

Important for the application of human neurophysiology to neurotherapy is the duality of the functions of the sphincter motoneurons and secondary muscle spindle afferents, subserving somatosensory and autonomic (parasympathetic) functions. In animals it was found that also sympathetic fibers innervate muscle spindles [126]. These measurements indicate that also parasympathetic fibers innervate muscle spindles in the parasympathetic innervation area (S2-S5).

Motoneurons innervating the external sphincters of the bladder and the anal canal subserve somatic functions (contraction of the sphincters on volition or for protection reaction) and parasympathetic functions for the coordination of the detrusor function (parasympathetic) and the external sphincter function. This duality of the sphincter motoneurons (and spindle afferents in the parasympathetic domain) makes the pattern change from continence to protection reaction understandable and makes learning transfer between somatic and parasympathetic patterns likely.

The motoneurons build up two phase relations per oscillation cycle with other motoneurons and secondary spindle afferents for somatic activation (Figures 164,167), and buildup 3 phase relations per oscillation cycle when also the parasympathetic division is activated (Figures 165,167, see below). The neuronal networks of the somatic and the parasympathetic nervous systems are interlaced and interact with each other. It should therefore be possible to improve parasympathetic functions when improving somatic functions by coordination dynamics therapy, especially as there is indication that also parasympathetic efferents fire rhythmically.

Repair of bladder function through reorganization of networks seems to be difficult because there is false neuronal network organization in the parasympathetic nervous system (overactive (spastic) detrusor) and the somatic division (spastic external (striated) bladder sphincter), and there is false interaction of the interlacing somatic and parasympathetic networks (detrusor-sphincter dyssynergy: when the detrusor contracts, the external sphincter is also contracting instead of opening (relaxing)). However, repair is possible through CDT (Figure 175). For a clinical bladder function test (Urodynamic) see [12,14].

2.9. Time Course Comparison between the Proposed Parasympathetically Induced Muscle Spindle Afferent Activity and the Detrusor Pressure

It was shown that the activity in parasympathetic efferents can be measured, identified and distinguished from the activity of γ-motoneurons in conduction velocity distribution histograms (Figures 157-159).

Since the action potential amplitudes of parasympathetic efferents is small, it would still be very difficult to analyze the organization of the parasympathetic nervous system and it's coordinated functioning with the somatic nervous system as in the control of the urinary bladder.

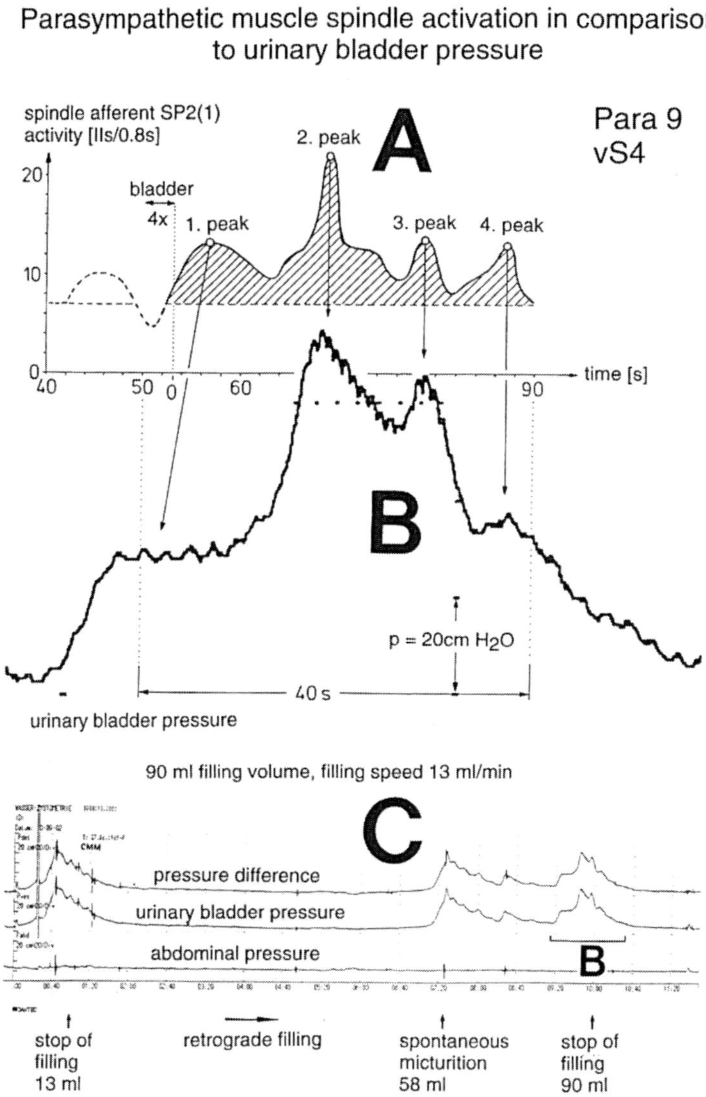

Figure 162. Comparison between changes in secondary muscle spindle afferent activity and detrusor pressure (measured before the surgery; for the implantation of a sacral anterior root stimulator). Paraplegic 9. A. Activity changes of the afferent fiber SP2(1) from Figure 5 following bladder catheter pulling. Approx. mean activity level is represented by a dashed line at 7.5 IIs/0.8s ((APs -1)/0.8s). The activity above the mean is cross-hatched and is proposed to be due to parasympathetic activation. Root vS4. B. Detrusor pressure (pressure difference) changes taken from ‘C’. Note that changes in the detrusor pressure show almost exactly the same time course as do the activated changes of the secondary muscle spindle afferent fiber SP2(1) of ‘A’. Corresponding peaks are correlated by arrows. C. Abdominal pressure (measured as rectal pressure), intravesical pressure (urinary bladder pressure) and detrusor pressure (pressure difference) during retrograde filling before the surgery. One transient detrusor pressure increase, marked ‘B’, is used, after enlargement in ‘B’, to compare with the spindle afferent activity.

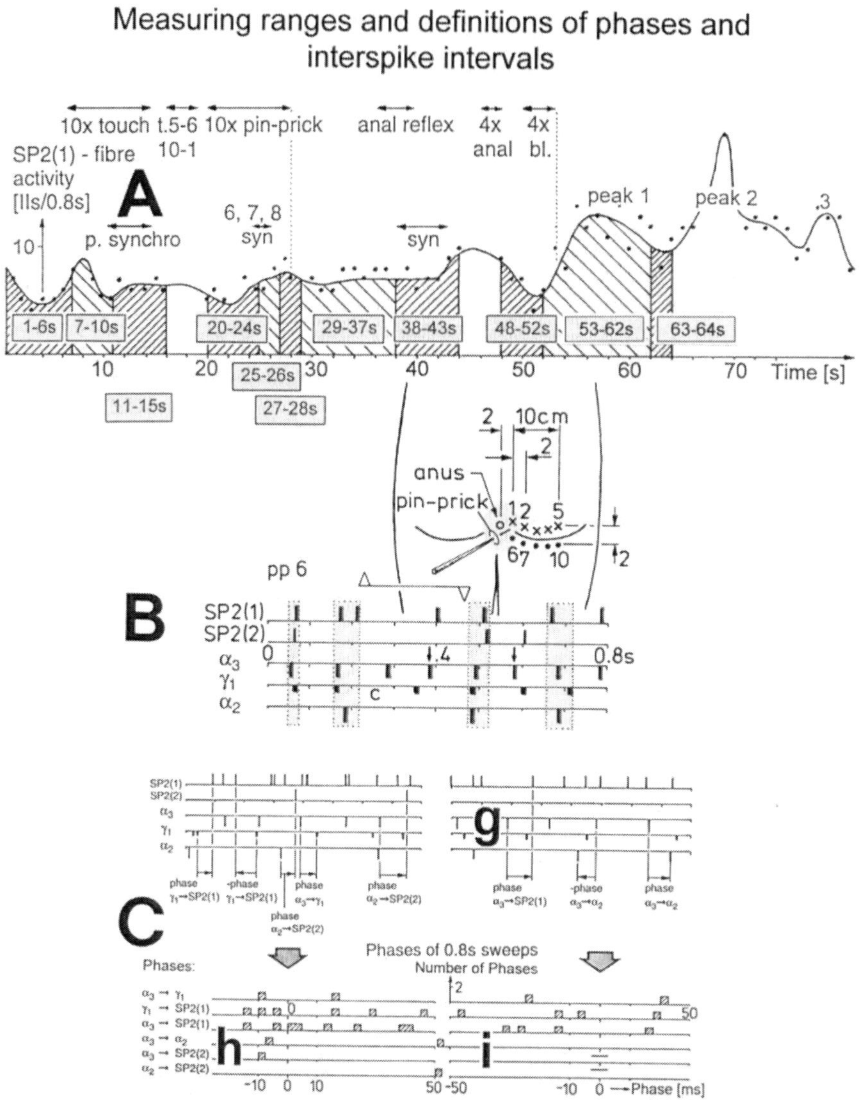

Figure 163. Measurement ranges and definitions of phases for the analysis of phase and frequency coordination's **between motoneurons and spindle afferents. Coordination (synchronization) between** firing patterns can directly be seen in B. (A) Activity level of secondary muscle spindle afferent fiber SP2(1) in dependence on time. 10x touch = touching sites 1 to 10 shown in B; t. 5-6 = touching alongside the skin from site 5 to site 6; 10x pin-prick = pin-pricking sites 1 to 10; anal reflex = anal reflex stimulation; 4x anal = fourfold anal catheter pulling; 4x bladder = fourfold bladder catheter pulling; peak 1- peak 2 - 3 = first, second and third peak of spindle afferent activity due to parasympathetic activation; p. synchro = partial synchronization; syn = synchronization of α and γ-motoneurons and secondary muscle spindle afferents. Note the synchronization of the firing patterns following pin-prick 6 inside the anal reflex area. (B) A set of single impulse patterns of secondary muscle spindle afferents (SP2(1,2)) and α and γ (intrafusal)-motoneurons and sites of stimulation. The small arrows in the impulse pattern of α_3-motoneuron (S) point to a shortening of the oscillation period following pin-prick 6 (pp6). The triangles indicate the beginning and the end of pin-pricking. (C)(g,h,i) Definitions of the phases between the different motoneurons and spindle afferents in 2 sets of impulse patterns (g), and the corresponding sets of phase relation distributions (h,i). Para 9; vS4.

But if some secondary muscle spindles in the parasympathetic range are also innervated by parasympathetic efferents besides somatic efferents (γ-motoneurons), then the activation of the parasympathetic nervous system can also be measured by the activity of secondary muscle spindle afferents. Since the action potentials of secondary spindle afferents are comparably large (thick fibers) the activation of the parasympathetic nervous system can be easily indirectly assessed. It will be shown now that at least some secondary muscle spindles in the parasympathetic range are innervated by parasympathetic efferents (parasympathetic fusimotors). The evidence is obtained by measuring the parasympathetic activation of the detrusor by detrusor pressure changes and by measuring activity changes of secondary muscle spindle afferent fibers and compare the form of the changes of detrusor pressure with the activity changes of a secondary muscle spindle afferent fiber.

Upon retrograde bladder filling in patients with complete spinal cord injuries, the detrusor was reflex activated at rather small storage volumes, as could be measured with urodynamics by the detrusor pressure. In paraplegic 5, the detrusor was automatically activated at a bladder filling volume of approx. 350 ml, and in paraplegic 9 at approx. 10 ml. The reasons for the limited storage volumes are the increased afferent input from the bladder due to a bladder infection, for example, and the strong disorganization of the neuronal networks in the spinal cord. It was found there for the paraplegics 7 and 11 that the first high bladder afferent activity occurred at bladder filling volumes in the operation, at which the detrusor was activated pre-operatively. In Para 7 the first high afferent input occurred at a filling volume of 150ml; the detrusor was activated at 160ml. In Para 11 the first comparatively high afferent input occurred at 50ml filling volume; the detrusor was activated at 80ml.

During surgery under light anaesthesia the detrusor never responded to the filling of the bladder in 120 patients. Also, quick filling with 4°C saline solution did not activate the parasympathetic division. Only in one instance, during surgery on a tetraplegic patient with a lesion sub C3 (artificial ventilation), with no tubes placed in the trachea and extremely light Propofol-anaesthesia, there seemed to be a slight detrusor pressure increase upon retrograde bladder filling. On the other hand, strong (painful) bladder catheter pulling activated the detrusor to contract in paraplegics with no anaesthesia, most likely activated the parasympathetic nervous system in the brain-dead human HT6 (no anaesthesia, possibly partial spinal shock), and seemed to activate the parasympathetic division during very light anaesthesia. It is concluded therefore that painful stimulation of the bladder under light anaesthesia activated the detrusor.

In Figure 162, the increase in detrusor pressure upon retrograde bladder filling before surgery is compared with the activity increase of the secondary muscle spindle afferent fiber SP2(1) following 4 times bladder catheter pulling. Figure 162A shows the undulating activity increase of a SP2(1) fiber. In Figure 162C the cystogram is shown. Upon bladder filling spontaneous micturition occurred several times. If parasympathetic fibers really activated muscle spindles, then the activity increase of the secondary muscle spindle afferent fibers following bladder catheter pulling may have a similar time course as does the bladder pressure increase due to detrusor activation following retrograde bladder filling. To check this similarity of time course, one undulating increase of detrusor pressure (Figure 162C) has been brought to the same time scale as the measured changes in muscle spindle afferent activity (Figure 162A) and transferred into Figure 162B for a direct comparison with Figure 162A. By comparing Figure 162A with Figure 162B, it can be seen that the occurrence of activity peaks of secondary spindle afferents is very similar in its time course to that of the peaks of the detrusor pressure. From the similarity of changes of spindle afferent activity and detrusor pressure (4 peaks) it can be concluded that some muscle spindles in the domain of the sacral parasympathetic nucleus are partly controlled by the parasympathetic division and that the muscle spindle and the detrusor activation have similar time courses. The other retrograde bladder filling induced undulating transient increases in pressure shown in Figure 162C had a similar time course.

Another correlation could be found between detrusor pressure and muscle spindle activation. The bladder pressure in paraplegic 5 increased steadily with no or only little undulations probably because the functions of the parasympathetic nervous system were still relatively preserved. The activity increase of a secondary muscle spindle afferent fibre in paraplegic 5 was also sustained and only little undulating.

Thus there is some indication that some muscle spindles are partly driven by the parasympathetic division. The drive can be by parasympathetic fusimotor activity or more indirectly by somatic fusimotors. Since in HT6, the somatic fusimotors did not change their activity levels strongly with the activation of preganglionic parasympathetic fibres and the activity increase of the secondary muscle spindle afferent fibre SP2(2) followed the transient increase of parasympathetic activity, it is likely that some muscle spindles are directly controlled by parasympathetic fusimotors. The direct control of some muscle spindles by parasympathetic fusimotors is supported by the slow activity decrease following very high activation [120] and the similarity with the time course of the detrusor pressure decrease following electrical stimulation of the preganglionic parasympathetic fibres during the surgery. The slow spindle afferent activity decrease would be difficult to explain by somatic fusimotor activation only. For further details see Chapter V of [14].

2.10. Relative Phase and Frequency Coordination between the APs of α and γ-Motoneurons and Secondary Muscle Spindle Afferents with No Additional Stimulation and Upon Touch, Pin-Prick, and Bladder Catheter Pulling

Above it was shown that natural impulse pattern can explain pattern change that is to say that the generation of integrative patterns can partly be understood at the single neuron level. Natural impulse patterns were related to integrative CNS organizations. How can this aid in understanding why jumping on springboard contributes substantially to the repair of bladder function? A better understanding can be achieved to go deeper into the complexity of the cooperative and competitive interplay among neurons that means going deeper into the complexity of phase and frequency coordination of CNS self-organization. The phase and frequency coordination of coordinated firing will now be analyzed more in detail.

Time correlations of afferent and efferent impulse patterns are easy to detect in the brain-dead individual as the oscillatory firing of α_2-motoneuron (O2) was regularly like an inner clock (Figure 161). The phases of fusimotors and spindle afferent APs could be defined with respect to the impulses of that inner clock. In the paraplegic, the rhythmic firing was rather irregular. The motoneuron firing therefore cannot be used any more as a time reference basis. More phases between the extracellular action potentials of the different fibers are necessary to fully describe the correlation between the simultaneous impulse patterns. In Figure 163Cg, the mutual phases between the APs of the different fibers are defined. Figure 163Ch,i shows the corresponding phase distribution histograms. Since too few phases occurred in a sweep piece of 0.8 s duration, phases occurring in certain time intervals were pooled and plotted in Figures 164 and 165.

In this special pathologic case, the α_2 and α_3-motoneurons fired rhythmically with impulse trains consisting of one action potential (AP), in contrast to the physiologic firing patterns, in which α_2 and α_3-motoneurons fire with impulse trains consisting of more than one AP. The identification of motoneurons by conduction velocity is not absolutely safe since group conduction velocity ranges overlap. It is very unlikely, nevertheless, that one of the motoneurons was an α_1-motoneuron (FF) (missing high conduction velocity and high AP amplitude and no correlation to primary spindle afferent fibers), even though they fire physiologically with impulse trains consisting of one AP. It has been shown that oscillatory firing α_1-motoneurons are mainly driven by time locked primary spindle afferent fibers (Figure 66C). The firing patterns of the α_2 and α_3-motoneurons are strongly pathologic with respect to the length of the oscillation period and the impulse train length so that it is impossible in this paraplegic to identify the kind of motoneurons by their discharge patterns of oscillatory firing; this would be possible were the neuronal network driving the motoneurons to fire in a physiologic manner.

In Figure 164, the interspike intervals (IIs) and the phases are shown for similar time intervals given in Figure 163. Before stimulation, within the time interval 1-6 s, the α_3-motoneuron fired every 100ms, the γ_1-motoneuron every 100 to 130ms, and the SP2(1) fiber every 80 to 150ms (Figure 164Aa). The α_2-motoneuron mostly fired every 300ms and the SP2(2) fiber every 250ms. At that particular time interval, similar phases (phase relation of broad peak type) occurred twice per α_3-oscillation period between the APs of the α_3 and γ_1 axons, between the γ_1 and the SP2(1) fibers, and between the α_3 and the SP2(1) fibers (Figure 164Ba). One phase relation occurred between the impulses of the α_3 and α_2-motoneurons, and two between the α_3 and the SP2(2) fibers. The broad phase relations between discharge patterns are interpreted as interactions between populations of neurons. Following different stimulations interspike interval (II) distributions and phase relations changed with time.

Upon touching sites 1 – 5 (Figure 163B), the IIs of the almost oscillatory firing γ_1-motoneuron reduced in size to be more similar to those of the oscillatory firing α_3-motoneuron (Figure 164Ab). The changing of the different phase relations indicates changes in the interactions between neuronal sub-networks (Figure 164Bb). Upon touching sites 6 to 10 (especially sites 6, and 7 (inside anal reflex area)), the IIs of the almost oscillatory firing γ_1-motoneuron increased again (Figure 164Ac). A transient partial synchronization occurred between the different nerve fibers (Figures 164Bc, 167c).

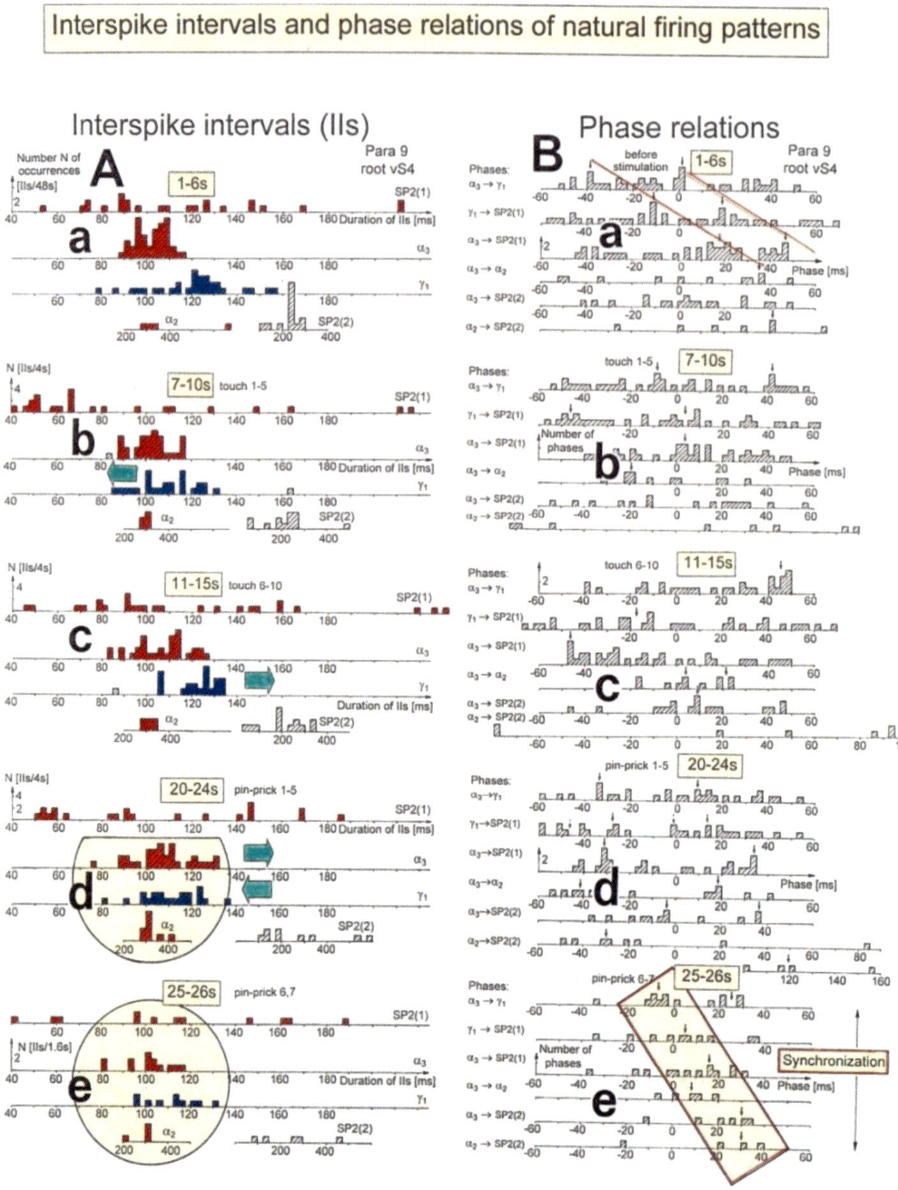

Figure 164. Relative phase and frequency coordination between α and γ-motoneurons and secondary muscle spindle afferents due to touching and pin-pricking sacral dermatomes as in Figure 5G. (A) Interspike interval distribution of spindle afferents SP2(1) and SP2(2), α_2 (FR) and α_3-motoneurons (S) and the dynamic fusimotor γ_1 for different time intervals upon touch, pin-prick and anal catheter pulling. Interspike intervals (IIs) were collected from several sweeps of 0.8 s duration per second. External loop generation and frequency coordination of α and γ-motoneurons and secondary muscle spindle afferents are marked by the semi-circle and the full circle. The large arrows point to the increase and decrease of the mean II of the distribution. Unsafe identification of α_2 and α_3-motoneurons (or vice versa) because of loss of specific oscillator properties. (B) Histograms of the phases between afferent and efferent fibers for the time intervals indicated, upon different stimulation. Phases were collected from several sweeps of 0.8 s duration per second. The small arrows indicate phase relations. Phase coordination is indicated in a,e. Para 9; vS4.

Upon pin-pricking sites 1 - 5, the IIs of the almost oscillatory firing γ_1-motoneuron reduced again to have a similar II distribution as the α_3-motoneuron. The α_3-motoneuron even slightly increased its IIs (decrease of activity), so that the II distribution of the oscillatory firing α_3-motoneuron and the now oscillatory firing γ_1-motoneuron became very similar (Figure 164Ad).

Upon pin-pricking sites 6 and 7 (inside the anal reflex area), α and γ-motoneurons and secondary muscle spindle afferents showed similar II distributions (Figure 164Ae). Only one phase relation was organized per oscillation cycle between the different nerve fibers (Figures 164Be, Figure 167e). A synchronization between the APs of the different nerve fibers occurred as can be seen from the direct impulse patterns (Figure 163B). The occurrence of similar II distributions of, and transient constant phases between, the APs of the α_3, γ_1 and SP2(1) fibers is interpreted in the way that, in its oscillatory firing the oscillatory firing α_3-motoneuron built up an external loop to the muscle spindle innervated by the γ_1 and SP2(1) fibers. The γ-loop became integrated into the oscillatory firing of the α_3-motoneuronal network. Before pin-pricking, the γ-loop, consisting partly of the γ_1 and SP2(1) fibers, also contributed to the oscillatory firing, since on the average there existed phase relations. With the pin-pricking, however, the II distributions also assimilated, so that this γ-loop was directly included into the oscillatory firing of the α_3-network rather than only contributing to the drive of it. The building up of an external loop to the periphery by spinal oscillators is substantially used when a patient with a spinal cord injury is jumping on springboard (see Figure 169, below). Upon pin-pricking sites 8, 9 and 10 (outside of the anal reflex area) and following pin-pricking of site 10, the II distribution of the SP2(1) fiber shifted away from those of the α_3 and γ_1 axons. The oscillatory firing α_3-motoneuronal network had abolished its external loop, even though still getting drive from it. Upon anal reflex stimulation and catheter pulling, the external loop was not built up again (Figure 165).

Following touch, pin-prick and anal reflex stimulation, but not painful catheter pulling, mostly two phase relations existed in paraplegic 9 between the activity of the α_3, γ_1 and SP2(1) fibers per oscillation period (100-140ms) of the α_3-motoneuron (Figures 164,165), but the phase relations changed with ongoing time. With the activation of the parasympathetic division, upon bladder catheter pulling (Figure 163A), three phase relations occurred per α_3-motoneuron oscillation period (Figure 165Bc,e). At the peaks of parasympathetic activation, three phase relations occurred (Figure 165Bc,e), and only two phase relations were present with little parasympathetic activation (times between the peaks 1 and 2) (Figures 163A,165Bd). Even though the functional units, consisting of fusimotor and α-motoneuron neuronal networks and spindle afferents fibers, were rather unstable in paraplegic 9, in comparison to the brain dead human (see below), an important difference between skin (somatic) and bladder (parasympathetic) stimulation occurred. Another (third) phase relation per α_3-motoneuron oscillation period occurred with the activation of the parasympathetic division. The activated parasympathetic neuronal network of the sacral micturition and defecation center seems to have channeled input to the oscillatory firing somatic neuronal network.

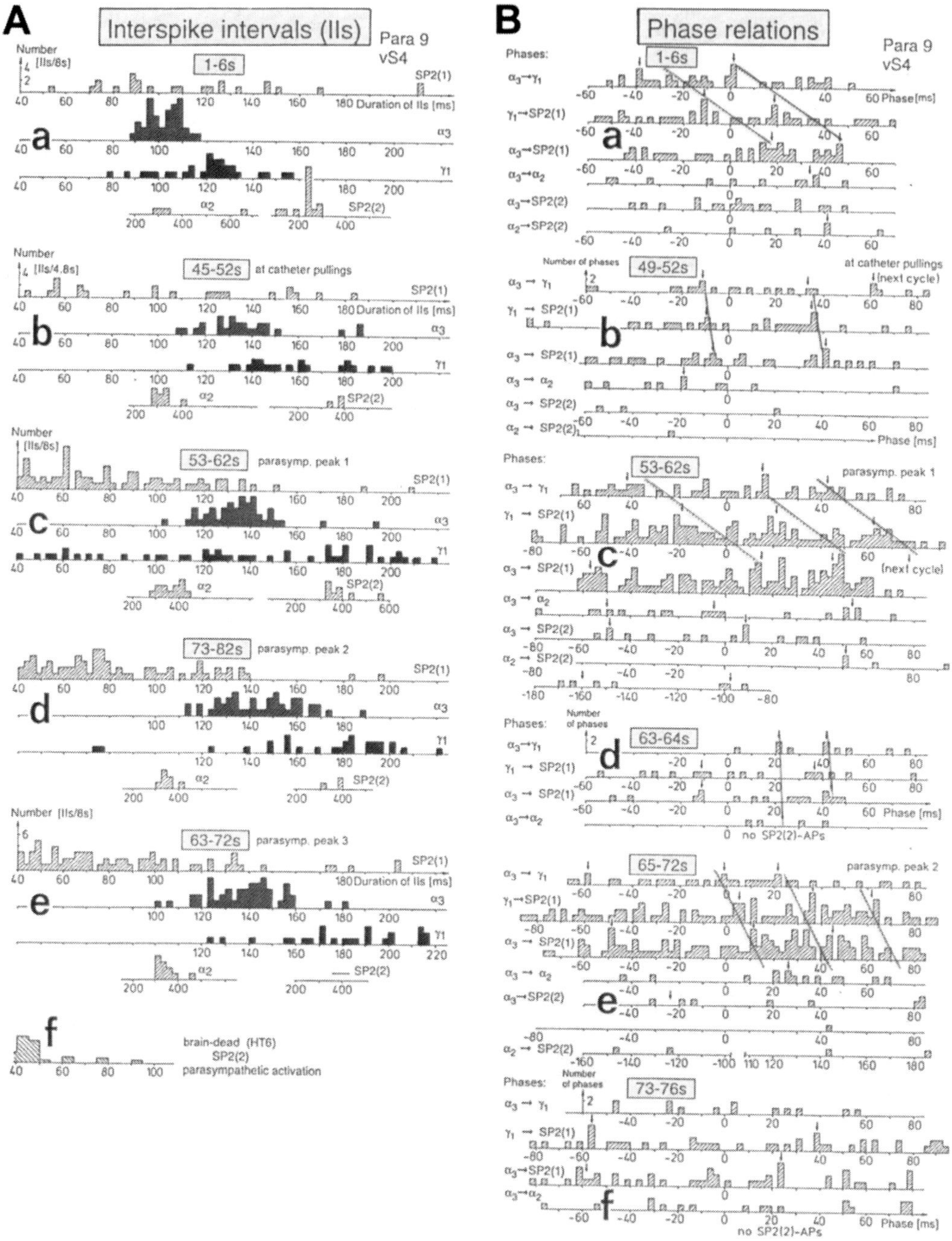

Figure 165. Interspike intervals (IIs) and phase relations for time intervals indicated in Figure 162. For legend, see Figure 164.

2.11. Phase Relation Changes between the Action Potentials of the α and γ-Motoneurons and Secondary Muscle Spindle Afferents in Paraplegic 9 Upon Somatic and Parasympathetic Activation of the Sacral Micturition Center

As shown in Figures 164,165 the number and the values of phase relations changed between the firings of the different nerve fibers upon different stimulations. In the brain-dead human HT6, two phase relations were found between the α_2-motoneuron and the secondary muscle spindle afferent fibre SP2(2) and the α_2 and the γ_1-motoneuron [121]. Also in the paraplegic two phase relations often existed between the firings of the different nerve fibers. It may be that a third phase relation occurs when the activated parasympathetic division channeled an additional input to the oscillatory firing somatic neuronal network. It is therefore worthwhile to further analyze the number of occurring phase relations per oscillation cycle upon different somatic and parasympathetic stimulations.

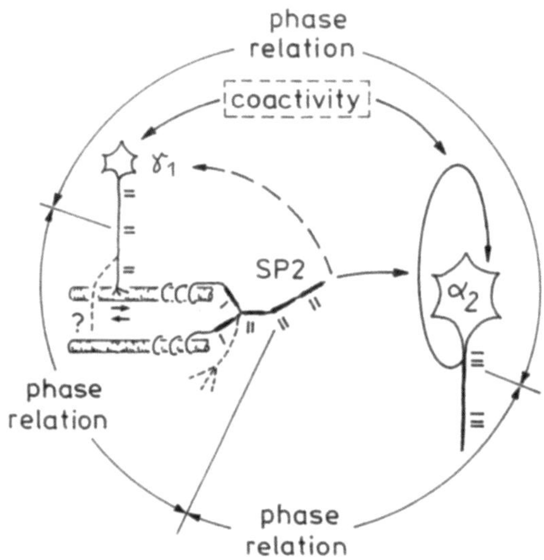

Figure 166. Schematized existing phase relation between α_2 and γ_1-motoneurons and a secondary muscle spindle afferent fiber (SP2). Parallel existing phase relations between other parent afferents and the α_2-motoneuron and between parent secondary spindle afferents are not shown. Phase relation means, the increased occurrence of phases in ms in a certain phase range between the action potentials (APs) of the two compared nerve fibers. The complex afferent and efferent muscle spindle innervation was not tried to show. Small arrows at intrafusal muscle fiber indicate local contraction, which is in nuclear chain fibers readily transmitted to the place of afferent innervation. A possible reason of the doublet firing of the SP2 fiber is pictured to occur from single APs (schematized by bars) of two myelinated endings, not necessarily from pacemaker switching. More endings of the parent SP2 fiber and γ_1-motoneurons are indicated by dashed line branches. "Coactivity" indicates a correlation between γ and α-motoneuron spinal cord circuitries for higher activations.

Since two phase relations occurred per oscillation cycle between the α_3 and γ_1-motoneurons and the SP2(1) fiber (Figure 164Ba) in paraplegic 9, and also their IIs were rather similar, it is concluded that the neuronal networks of the α_3 and γ_1-motoneurons formed together with the spindle afferent fibre SP2(1) a part of a functional unit. The neural ensemble is built by efficiencies of synapses and projections between the convergence of several fusimotors on one muscle spindle and by the divergence of muscle spindle projections onto several rhythmically firing populations of neurons driving α and γ-motoneurons. Such a functional unit is partly pictured in Figure166 and schematized drawn by 3 circles in Figure 167. The α_2-motoneuron and the SP2(2) fiber belonged to another functional unit (another ensemble) (longer IIs and the existence of only 1 phase relation). The two functional units (ensembles) are characterized in Figure 167 by two sets of three circles each. The two functional units interacted with each other as there existed a phase relation between the α_2 and α_3-motoneurons (Figure 167).

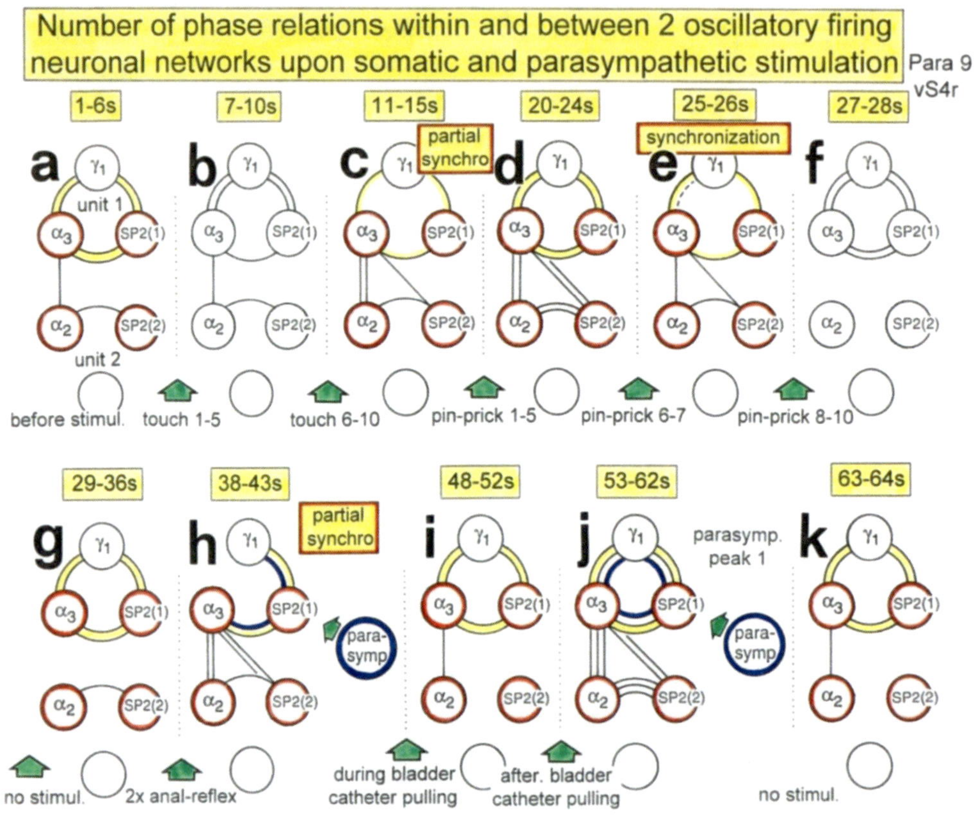

Figure 167. Number of phase relations within and between the two functional units α_3/γ_1/SP2(1) and α_2/-/SP2(2). Time intervals are those of Figure 2A. Note that in 'a' the functional unit 1 is with two phase relations per oscillation period in a stage similar to those seen in the brain-dead individual; with synchronization only 1 phase relation occurred (e) and the parasympathetic division channeled an extra phase relation to interact with the somatic division (j).

Before stimulation (but with the anal and bladder catheters positioned), there were two phase relations in unit 1 (Figure 167a). When touching sites 1 to 5 (Figure 163), only slight changes occurred in the two units with respect to the number of phase relations (Figure 167b). But when touching sites 6 to 10, a partial synchronization occurred (Figure 167c) and functional unit 1 reduced the number of phase relations to one. When pin-pricking sites 1 to 5, two phase relations occurred again in unit 1 (Figure 167d). Upon pin-pricking sites 6 and 7, the number of phase relations between all the components of the two units dropped to one (Figure 167e), and synchronization occurred between the firing patterns (Figure 163B). Since in the brain-dead human two phase relations per oscillation cycle were observed in the functional units [121], it is possible that synchronization and the existence of only one phase relation for 2 to 3 seconds reflected a slight pathologic organization of the networks.

Even though upon touching sites 6 to 10 (Figure 167c) or upon pin-pricking sites 6 to 7 (Figure 167e) only one phase relation existed in unit 1, and synchronization occurred with both stimulations, it was shown above (Figures 149,151) that the touch afferent input organized a different functional state of unit 1 than pin-pricking. The response time till the shortening of the oscillation period was longer than the oscillation period (≈ 100ms) for pin-prick and shorter for touch. It was shown above (Figure 149) that repetitive touch stimulation (most effective inside the anal reflex area) reinforced the sustained stretch reflex of the anal sphincter (continence pattern), and repetitive pin-prick stimulation replaced the continence pattern by the protection reaction of the anal sphincter. The number of phase relations alone therefore only provides limited information on the functional state of the organization of the neuronal networks of the human spinal cord. Measurements of a number of parameters are necessary to yield a rather complete description of the functional state of neuronal networks.

Following pin-prick 8 and 10 and with no stimulation two phase relations existed again in functional unit 1 (Figure 167f,g), in some similarity to pre-stimulation status (Figure 167a). Following two anal reflex stimulations, partial synchronization occurred in the components of the two units, and mainly two phase relations existed (Figure 167h). But the organizational state was still not very similar to the pre- (Figure 167a) or post-stimulation state in unit 1 (Figure 167g), since the parasympathetic division was slightly activated following anal reflex stimulation as was measured by the impulse pattern (increase of doublet activity) of the secondary muscle spindle afferent fibre SP2(1). Therefore, probably one phase relation was due to the somatic activation in similarity to Figure 167c,e and the other phase relation was due to the activation by the parasympathetic division. During bladder catheter pulling (Figure 167i) and with no stimulation (Figure 167k), the number of phase relations and possibly the functional organization, was again similar to the pre-stimulation state (Figure 167a).

Following strong (painful) bladder catheter pulling with a strong activation of the parasympathetic division (time interval 53-62s (Figure 167j)), measured by the increased doublet firing (for doublet firing see page 514 of [14]) of the SP2(1) fibre, the functional organization of the sacral micturition centre of the disconnected spinal cord changed completely. Functional unit 1 was now correlated by three phase relations per α_3-oscillation cycle. The functional unit 2 also showed 3 phase relations per an α_2-oscillation cycle, and interacted with functional unit 1 by 3 phase relations as well (between the α_3 and α_2-motoneurons; Figure 167j (53-62s)).

Between the first and second parasympathetic peak (Figure 163A) at the time interval 63-64s (Figure 167k), the organization form of the two functional units was similar to that before the first parasympathetic activation (49-52s) (Figures 167), only the values of the phase relations changed (Figure 165Bd). With the second strong activation of the parasympathetic division (parasympathetic peak 2, time interval 65-72s) the functional unit 1 was bound together again by 3 phase relations (Figure 163A), in similarity to the first strong activation of the parasympathetic division, measured by the burst firing of the secondary muscle spindle afferent fibre SP2(1) and the increased doublet firing of the SP2(1) fibre [14]. The functional unit 2 was disorganized, but phase relations still occurred between the α_3 and the α_2-motoneurons and the SP2(2) fibre. The α_2-neuronal network and the fusimotor networks, driving the SP2(2) spindle afferent fibre, were integrated differently. After the second strong parasympathetic activation, in the time interval 73-76s (Figure 163A), the functional organization of the two functional units in the spinal cord was similar to that before the activation of the parasympathetic division. Functional unit 2 was slightly disorganized as the SP2(2) fiber strongly reduced its firing.

2.12. The Need to Improving the Stability of Phase and Frequency Coordination to Allow Specific Pattern Formation and Learning Transfer

A young mother, with stress incontinence after giving birth to her first child, could strongly improve her continence status by jumping on springboard in addition to other training. Her CNS is not injured. It was the periphery that required repair by means of changing the CNS.

In severe cervical spinal cord injury (Figure 143), however, solely jumping on springboard (Figure 144A,B) is not sufficient for bladder repair. First, of course, the patient has to regain movement functions back (especially the trunk stability) to be able to perform the jumping on springboard. Further, the self-organization of CNS networks by phase and frequency coordination has to be improved to make learning transfer from movements to bladder functions possible, since in every CNS injury the phase and frequency coordination is impaired. Large instabilities in phase and frequency coordination will not allow specific pattern formation as a basis for learning transfer. However, the stability of phase and frequency coordination can be improved when the patient is exercising on special coordination dynamics therapy devices, especially the one shown in Figure 150. In the following, measurements are presented to understand the necessity of improving the stability of phase and frequency coordination for functional repair.

The importance of stable phase and frequency coordination to allow specific pattern formation and in consequence learning transfer to other patterns can be understood at the collective variable level and the neuron level. The behavioural information F_{inf} of the coordination pattern dynamics, characterized by equations of motion of collective variables, $dX/dt = F_{intr}(X) + \sum c_{inf}F_{inf}(X,t)$, affect the whole coordination pattern dynamics, including stability, rather than only certain coordination patterns. If the behavioural information includes the exercising of extremely coordinated, integrative movements, like exercising on the special CDT device for turning, then the quality of the CNS self-organization can be enhanced by improving the exactness of self-organization, namely the precision of phase and frequency coordination between neuron and neural assembly firing. By improving the precision of organization of the intrinsic dynamics $F_{intr}(X)$, that is, the specific variability of the injured networks, certain patterns do then already re-appear (Figure 18). Neurons often serve more than one network pattern at the same time by time sharing of neuron firing and in this way give rise to learning transfer among the activated patterns. If sub-networks are improved in the organization of one pattern, the organization of the other pattern will also improve. Neurons involved in the organization of breathing and activating intercostal muscles, for example, are also involved in the organization of trunk stability. By reducing spasticity of the trunk (in patients with Parkinson's) breathing will also improve. Similarly, sphincter motoneurons are involved in continence and pelvic floor weight bearing. If the pelvic floor is not trained during pregnancy incontinence sometimes occurs and continues into the antenatal period.

2.13. Phase and Frequency Coordination between Oscillatory Firing α_2-Motoneurons and Their Adequate Afferent Drive in Brain-Dead Human

The relative phase and frequency coordination between the APs of the oscillatory firing α_2-motoneuron O2 and the secondary muscle spindle afferent fibre SP2(1) has partly been shown in Figure 25 and can be clearly seen in the original recordings in Figure 9 taken from a brain-dead human. The firing of the oscillator and the sweep pieces which are shown time-expanded are indicated at the summary trace (Figure 9A). Figure 9B,C shows the action potential (AP) impulse train of oscillator O2 in connection with one of its driving spindle afferent AP. Because of the duration of the phase relation of around zero milliseconds between the firing of the driving SP2(1)-fibre and the impulse train of the oscillatory firing motoneuron O2, the SP2(1)-fibre AP (every second AP of the SP2(1)-fibre) appeared at a similar time as the impulse train. Because the AP of the spindle afferent fibre had a characteristic waveform, it was easy to extract its impulse pattern from the summed impulse traffic of this S4 dorsal root. During touch-induced skin afferent activity (Figure 9A), the activities of the motoneuron and the spindle afferent fibre were covered by the skin afferent activity. After the cessation of the skin afferent activity the afferent and efferent APs were found again at their expected time positions of the regular firings. The phase coordination between the firings of the oscillatory firing motoneuron O2 and the secondary muscle spindle afferent fibre SP2(1) at the time when records B,C were taken, was 1.6ms (3ms - 1.4ms, Figure 9B,C). In Figure 9D, the relative frequency coordination between the firings of the SP2(1)-fibre and the impulse train of the oscillator is indicated. For the time period evaluated, the correlation between the firing of the motoneuron and the spindle afferent fibre was in the range of between 3 and 5ms (Figure 9D).

2.14. Relative Frequency Coordination

In Figure 9e, considerations concerning the relative frequency coordination are extended to the activity of further afferent fibres and γ-motoneurons of the same root. Figure 9eG shows sweep pieces of the original recordings; A through F show the interspike interval distributions of spindle afferents and γ-motoneurons. It can be seen from the overlapping of the oscillator frequency distribution ranges (and the half of it), and from the interspike interval distributions of the afferents that, from the viewpoint of frequency coordination, fiber SP2(1) contributed strongly to the drive of oscillator O2, whereas there was a weaker contribution from other afferents (less overlapping between the distributions of the afferents and the range of the basic frequency or the first harmonic of the oscillator).

Also, γ-motoneurons showed only little frequency correlation at that time period.

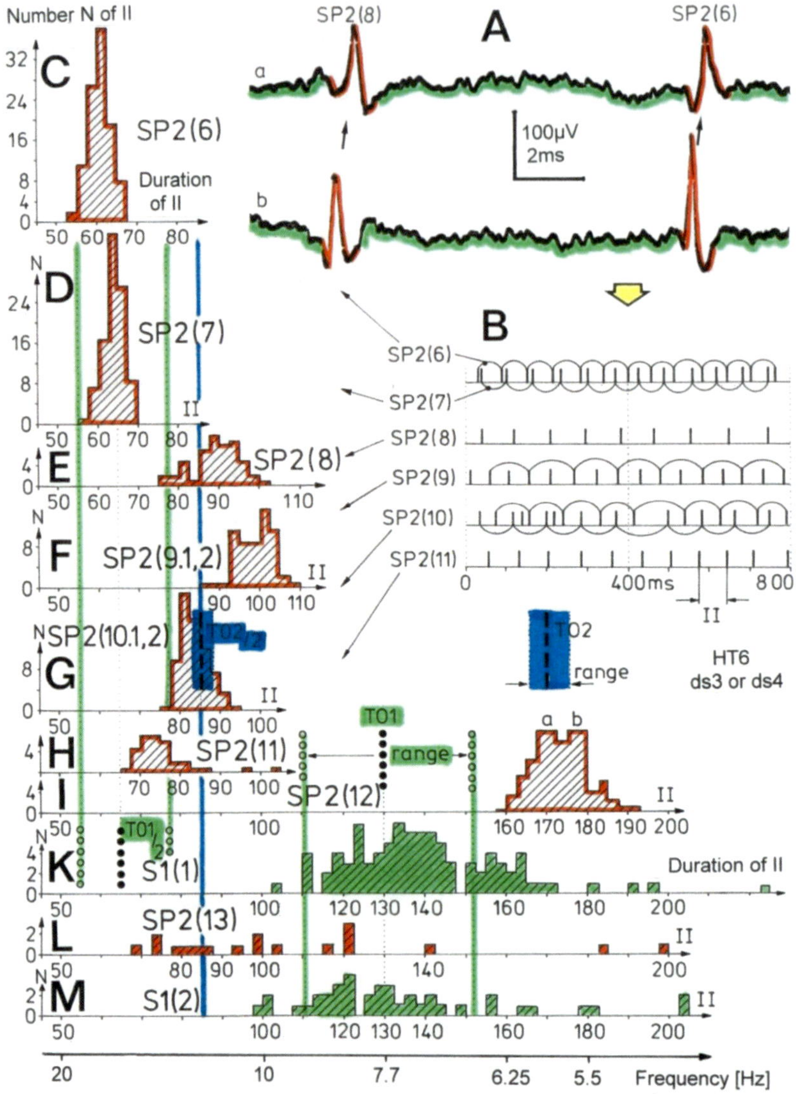

Figure 168. Measurements from brain-dead human HT6 from different spinal cord segments after retrograde bladder filling (700 to 800 ml), with the exception of "I", which was obtained before filling. A. Sweep piece of a recording from a dorsal S3 or S2 root filament. It can be seen that the secondary muscle spindle afferent SP2(6) AP can be distinguished by the wave-form on the two traces from the secondary spindle afferent fiber SP2(8) AP (different amplitude of the three phases of the triphasic APs). B. Simultaneously recorded impulse patterns of the six parent secondary spindle afferents SP2(6) through SP2(11) obtained from dS3 or dS2 root recordings. The impulse patterns of SP2(6) and SP2(7) fibers are not separated to show the similarity of the patterns. The impulse patterns of the parent spindle afferents SP2(9) and SP2(10) are split into patterns of the single endings (single ending activity partly connected by circle lines) with the assumption that single endings of parent secondary muscle spindle afferents should have interspike intervals of duration longer than 50 ms. C to H. Interspike interval distributions of six simultaneously recorded single secondary spindle afferent endings. F, G. Interspike interval distributions of parent fibers, which are the sums of the distributions from the two activated endings. I. Interspike interval distributions of a secondary spindle afferent fiber (SP2(12)) of a coccygeal root. K, L, M. Interspike interval distributions of single-fiber afferent activity from a lower sacral dorsal root. In L, most likely the activity from a secondary spindle afferent fiber is shown. In K

and M most likely the interspike intervals from afferents (S1(1) and S1(2)), innervating stretch receptors of the urinary bladder wall, are shown. In G, H and K, the durations of the oscillation periods (mean and range) of the oscillatory firing α_2-motoneurons are indicated by thick dashed and dotted lines; the motoneurons innervate the external anal sphincter (TO2) and the external bladder sphincter (TO1). The sites of innervation of the oscillatory firing motoneurons are identified (and distinguished from each other) by anal reflex stimulation, bladder filling and catheter pulling. Note that the TO1 and TO2 ranges and their halves overlap with the interspike interval distributions of the secondary spindle and stretch receptor afferents (relative frequency coordination).

Figure 168 shows the interspike interval distributions of more afferents (including the afferents for bladder filling; stretch receptor afferents S1(1), S1(2)) of another root, together with the oscillation period range (and the half of it) of a second α_2-oscillator (O1). By comparing the oscillation periods (and their halves) and their ranges with the interspike interval distributions of the afferents, it can be suggested which afferents made a (frequency coordination) contribution to the drive of what oscillator at that time interval. For example, the S1(1) urinary bladder stretch afferent fibre activity contributed to the drive of oscillator O1 (activating the external bladder sphincter) because its interspike interval distribution overlaps strongly with the range of the oscillation periods of O1. But the S1(1) distribution does not overlap with the range of the oscillation periods of oscillator O2, or with their halves or quarters.

The S1(1) afferent fibre will therefore not have made a substantial contribution to the drive of oscillator O2. On the other hand, the secondary muscle spindle afferent fibre SP2(12) activated oscillator O2 innervating the external anal sphincter, since its interspike interval distribution overlaps with the range of O2 oscillation periods. But the secondary muscle spindle afferent fibre SP2(12) did not activate oscillator O1, as its interspike interval distribution does not overlap with its oscillation period range or half of it (Figure 168).

By comparing interspike interval distributions of afferent fibers with oscillation period distributions it can be estimated what afferents made a (frequency coordination) contribution to the drive of the spinal oscillators. These considerations need no knowledge of the connectivity of the neuronal networks. In the frequency coordination between the firings of afferents and oscillators and among oscillators, entrainment or coordination may occur sub- or super-harmonically (see above). The energy transfer, and therefore the coupling strength will be smaller if the APs coincide in their firing less often. As indicated by my measurements, the coupling and the relative coordination during the self-organization of the neuronal networks of the human spinal cord are of an enormous complexity; this self-organization is induced by sets of mutual impulse patterns from stimulated receptors which are ordered, in time and space (for skin receptors see Figure 5), so as to reflect, in the spinal cord and higher centers, the interplay of the body with the external world.

2.15. Impaired Organization of Premotor Spinal Oscillators Following Spinal Cord Injury as an Indicator for Pathologic Network Organization

Following a spinal cord injury, the oscillatory firing networks lose specific properties. The Eigenfrequencies of the premotor spinal oscillators change from a narrow to a broad frequency band (Figure 10). Self-organized α_2-oscillators fire physiologically at an Eigenfrequency (varying within a small frequency band as probably indicated with the hatched distributions in Figure 10) with impulse trains consisting of 2 to 3 action potentials. Following brain death, this Eigenfrequency band enlarges (black area in Figure 10). Following a spinal cord injury, the Eigenfrequency band enlarges strongly and includes in this case the frequencies between 4 and 14Hz for firing with 2 or 3 action potentials per impulse train (Figure 10). The premotor spinal oscillators have lost their specific properties and could now be excited at frequencies at which they physiologically would not be excited.

2.16. Explanation for a Spastic External Bladder Sphincter

In the brain-dead human HT6 of Figure 168, the premotor spinal oscillator TO1 could according to the overlap of the narrow frequency band harmonically (for simplicity only this case of activation is considered) only be driven by the activity of the bladder stretch receptor afferents S1(1) and S1(2). This oscillatory firing α_2-motoneuron, innervating the FR-type muscle fibres of the external bladder sphincter, is getting activated with the activity of the stretch receptor afferents, when the bladder is filled to secure continence. All the secondary muscle spindle afferents cannot activate the external bladder sphincter harmonically, because their frequency bands did not overlap (frequency band overlap in the frequency range 6.5 till 9Hz).

Only the stretch receptor bladder afferents can activate the external bladder sphincter to secure continence. Following a spinal cord injury, the Eigenfrequency band broadens from 6.5-9Hz to 4-14Hz. According to Figure 168, also the secondary muscle spindle afferents SP2(8), SP2(9.1,2), SP2(10.1,2), and SP2(12) could activate the motoneuron innervating the external bladder sphincter. The striated external bladder sphincter can be activated by secondary muscle (and probably other) afferents, not involved in continence. This broadening of the frequency band could be one reason for spasticity (continuous contraction) of the external bladder sphincter. Our patient of Figure 144, with a severe cervical spinal cord injury (Figure 143), had a spastic external bladder sphincter when therapy was commenced. She was continent, but could not empty the bladder – the emptying of the bladder was achieved by a suprapubic catheter.

2.17. Reduction of Spasticity of the External Bladder Sphincter

If one could improve the functioning of the sacral micturition centre and reduce the frequency band of the oscillatory firing premotor spinal oscillators, innervating the external bladder sphincter, then this kind of spasticity of the external bladder sphincter could be reduced. I named this training ‘oscillator formation training’. It will be shown (Figure 175) that this spasticity reduction could be achieved by learning and learning transfer. But, as will be shown (Figure 175), the reduction of the spasticity of the external bladder sphincter is only the first step to repairing bladder functions. Following CNS injury (including spinal cord injury) not only the frequency coordination, but also the phase coordination among neuron firing becomes impaired.

2.18. Stable Phase Coordination in the Brain-Dead Individual

Existing relative phase relations among α and γ-motoneurons and secondary muscle spindle afferents in a paraplegic are indicated in Figures 164 and 165 by small arrows. In Figure 167 the natural impulse patterns were used to show the interaction between the somatic and parasympathetic nervous system divisions in the sacral micturition center. Next, the representation of the phase relations (shown in Figures 164,165) is changed to reveal phase stability which makes it possible to find reasons for the pathologic organization of neuronal networks in paraplegics. We shall start with a physiologic case, specifically a brain dead individual. To make the phase relation changes better recognizable with time, a representation of phase relations is used which comes from the measuring of the speed (frequency) of rotation.

The speed of rotation of a turning cylinder with a spot on its surface can be measured with a stroboscope. If the stroboscope flashes light with the same frequency as the cylinder is turning, the spot on the circumference seems to stand still. There is a constant phase between the two frequencies (frequencies are same or multiples of each other). If the phase relation changes, the spot will move.

If no phase relation exists between the turning of the cylinder and the flashing of the light, no spot will be seen. In similarity to stroboscopic measurement of frequencies of turning cylinders, the phase relation between two oscillatory firing spinal oscillators is pictured in Figure 48aA. A time axis is introduced on the horizontal line, to make phase relation changes visible in dependence on time.

In Figure 48aAa, the loop excitation is pictured for this oscillator model. In Figure 48aAb, the phase relation between the SP2 fibre activity incidence and the oscillatory firing is pictured on the circumference of the oscillation period cylinder of the oscillator. Figure 48aAc,d,e shows different phase relations, namely a constant phase relation (c), a changing phase relation (d), and no phase relation (e). In Figure 48aB, phase relation changes are plotted between an α_2-motoneuron and the activity of a secondary muscle spindle afferent fibre and between an α_2 and a γ_1-motoneuron. The data were taken from Figures 4 and 5 of [121] of a brain-dead individual (probably normal with respect to the number of phases per oscillation cycle and with respect to phase changes). It can be seen that there were two phase relations per α_2-oscillation cycle and that the phase relation changed only little with time. The phase coordination between the firings of the α_2 and a γ_1-motoneuron and the secondary muscle spindle afferent fibre was stable.

2.19. Unstable Phase Coordination in the Patient with a Spinal Cord Injury

In Figure 48bA,B different phase relation changes are plotted from Figures 164 and 165 with respect to the α_3-oscillation cycle (A) and the α_2-oscillation cycle (B). It can be seen that the different phase relations changed strongly in value over time (upon different stimulation), and that also the number of phase relations per oscillation cycle changed. The phase stability of the cooperative and competitive interplay among neurons became impaired. Whether the change of the number of phase relations from 2 to 3 following the activation of the parasympathetic nervous system in the sacral micturition centre (Figures 165,167) is physiologic is not clear.

2.20. Difference of Phase Stability between a Brain-Dead Human and a Paraplegic

The most obvious difference of the phase relation changes between the above mentioned brain dead human and the paraplegic was that in the paraplegic the phase relations varied very much, whereas they changed only a little in the brain dead human. The strong phase relation changes in the paraplegic can be interpreted as instability in the organization of neuronal networks. The correlation of neuronal sub-networks was unstable in relation to those of the brain-dead human. Assuming that the neuronal network organization and functioning was rather physiologic in the brain dead human with respect to the firing patterns of the premotor spinal oscillators, the functioning of the networks became unstable following spinal cord injury. The more frequent occurrences of changes of phase relations between the different nerve fibres in combination with the changing number of phase relations per oscillation may mean that sub-networks reacted and interacted more quickly and easily with others according to the afferent input. Especially because the oscillatory firing networks lost specific properties, their resonance frequencies changed from a narrow to a broad oscillator frequency band (Figure 10), which means that the oscillators were not excited at a certain frequency any more, but by a broad frequency band. They were now being excited at frequencies at which they physiologically functioning oscillator would not. Over-activation and mass effects could be the result. On the other hand, certain networks could escape from driving afferent influence by changing their phase by phase escape to avoid interaction. Functionally distant networks are no longer reached, which also would result in a loss of specific properties. Therefore, because of the loss of specific properties, some interactions could have occurred more easily and others not at all.

W.R. Hess tried in 1944 to compare biological order and human society [128]. The behaviour of spinal oscillators described above may be compared to a mob following a populist ideology. There could be similarities between the organizations of the human nervous system and the organizations between very many individual nervous systems.

2.21. Change of the Neuronal Network Organization Following Spinal Cord Injury - Pathologic Network Organization

Following spinal cord injury the spinal cord neuronal networks have been observed to change their organization, and this can be quantified by six changes of organization.

1. Following spinal cord injury, the spinal oscillators strongly widen their oscillation frequency band (Figure 10), which means that their firing lacks rhythm. An increase of the more irregular oscillator firing can also be seen in the broadened distributions of the interspike intervals of the impulse trains of oscillatory firing α_2 and α_3-motoneurons [55]. The loss of specificity of oscillatory properties will at least partly

be due to the loss of supraspinal inhibition because muscles below the paretic spinal cord injury are over-activated.

2. Following the injury, the phase relations among the oscillators and between oscillators and their driving afferents (for example, secondary muscle spindle afferents) become very instable (Figure 48b).
3. Because of the widening of the frequency bands (Figure 10) and the instability in phase coupling (Figure 48b), the oscillators partly lose their rhythmicity and their coordination and cooperation properties. The rhythmicity of movements and their coordination and coupling of arms and legs is reduced. The loss of rhythmicity and the loss of coordination can be observed easily when measuring the coordination dynamics.
4. The spinal oscillators are not under full volitional control by the patient any more (Figure 48 of [62]). Sometimes, paraparetic or tetraparetic patients can switch on motoneurons to fire in an oscillatory manner, but cannot switch them off again.
5. Following natural stimulation, the recruitment of motoneurons in the occasional firing mode (low activity mode of motoneuron firing; see Figure 161D for bladder filling up to 550ml) according to the size principle in each nerve fibre group changes [57,129]. The level of motoneuron activation increases following spinal cord injury (probably due to loss of inhibition) and the slowly conducting α_3-motoneurons (S) are recruited before the faster conducting α_2-motoneurons (FR), which is pathologic.
6. The coupling of α and γ-motoneurons changes in strength following natural stimulation [57].

The loss of rhythmic oscillator properties and also the change of motoneuron firing in the occasional firing mode (when the oscillators have not been self-organized because of low activation) indicate deterioration in the self-organization of the spinal cord neuronal networks below the injury and can at least partly explain the pathologic CNS network organizations of spasticity and increased clonus.

The impaired phase and frequency coordination's, measured among α and γ-motoneurons and secondary muscle spindle afferents by an increased variability of coordination, was introduced in a first approximation in the equation of motion at the collective variable level of description for jumping on springboard ($d\varphi/dt = - dV(\varphi)/d\varphi + (Q\xi_t)^{1/2}$) by an variability increase of strength Q (see above). This increased variability was pictured by the increased fluctuation of pattern states (pictured by a ball) in an attractor basin (Figure 18B).

2.22. Re-Learning of Phase and Frequency Coordination

This partly impaired phase and frequency coordination between the firings of single neurons and neuron assemblies (Figures 10,48b) can be expected to have consequences in the coordination between arm and leg movements, because motoneurons innervate muscle fibres, and rhythmic coordinated firing of single motor units has been measured electromyographically with surface electrodes [15-17] (Figures 13-15). Indeed, the coordination between arm and leg movements is partly or fully lost following CNS (brain and spinal cord) injury and is often not taking place at all in the malfunctioning CNS.

This partly impaired phase and frequency coordination at the single neuron level, the assembly level and the macroscopic level can be measured macroscopically when the patient is exercising on a special coordination dynamic therapy device (Figure 174) on which arms and legs turn with a slightly different frequency (transmission 19 (arms) : 18 (legs)). The phase coordination between arms and legs is imposed by the device. The loss of phase and frequency coordination between arm and leg movements becomes visible and measurable by the arrhythmicity of turning when the patient is exercising on the device. During a turning cycle and during the change of coordination between arms and legs the turning frequency increases and decreases. This frequency variation (df/dt:f; f = frequency) can be recorded, quantified and displayed on a computer screen (Figure 174).

During the functional reorganization of the injured CNS of patients, the relative phase and frequency coordination of neuron firing has to be entrained as exactly as possible by the movement induced afferent impulse patterns from the receptors (learning through feedback information) to restore coordination in the range between 3 and 5 milliseconds (Figure 9). The device has therefore to impose the exercising patient coordination in the millisecond range for the different coordination's of arms and leg movements between pace gait and trot gait. The simple pace and trot gait coordination's can often be performed easily by the patient, but not the difficult intermediate coordination's (pattern change). Therefore, the continuous change from the easy to the difficult coordination's and backwards diagnoses the extent to which the patient's nervous system can produce both the easy and the difficult organizational states. If the movement state can be easily generated by the neuronal networks of the CNS, then the frequency variation of turning is small during the turning cycle, and if the movement state is difficult to organize by the CNS, then the frequency variation is large.

Because it is sometimes the case that in the pathologically functioning CNS, intermediate coordination's are often easier than pace and trot gait are, a coordination coordinate is needed to judge the coordination dynamics. The pace and trot gait coordination's are used for the calibration, since both coordination's between arm and leg movements occur naturally during rhythmic coordinated (automatic) movements like commando-crawling, crawling, walking and running. The quality of CNS functioning can be measured by pattern change, namely the coordination dynamics.

The imposed coordination of arm, leg and trunk movements performed while exercising on the special coordination dynamic therapy and measuring device, is in accordance with the coupling possibilities of premotor α_1 (8-12Hz), α_2 (6-9Hz) and α_3-oscillators (0.4 (may be down to 0.1) -4Hz), even though the frequencies are only a relative coordination parameter, whereas phase is an absolute coordination parameter.

When the hand levers are turned at between $\approx$ 0.4-1.5 Hz, the resulting frequency difference in turning between arms and legs is approx. 8.5 Hz (low α_1-oscillator frequency or high α_2-frequency) for low hand turning frequency of 0.5Hz (low α_3-frequency). A slower turning of the hand levers would train directly more the premotor α_2-oscillators ($f < 8.5$Hz). Faster turning of the hand levers (higher α_3-frequency) would train directly the α_1-oscillators in the higher frequency range ($f > 8.5$Hz). Therefore, similar frequencies appear with respect to the frequency of turning on the device for measuring CNS organization and reorganization as have been measured for premotor spinal oscillators.

2.23. Improvement of the Organization of Motoneurons in the Occasional and Oscillatory Firing Mode

When turning the levers steadily with medium or high strength the premotor spinal oscillators in the patient's CNS are entrained because premotor spinal oscillators self-organize themselves for high activation (Figure 161D). The turning of the levers is therefore an oscillator (or assembly) formation therapy. The members of premotor oscillator assemblies (most likely the motoneuron and interneurons) are entrained to improve activation and inhibition by adjusting, for example, the efficacies of the corresponding activated synapses. By turning the levers with little strength at approx. 0.4Hz (releasing the power (load)-setting knob), the motoneurons get only partly organized into premotor spinal oscillators. The motoneurons are firing mainly in the occasional firing mode and are trained for a better recruitment according to the size principle (rhythmicity of repeated recruitment $\approx$ 0.4Hz (Figure 15 of [62])). Between the motoneuron firings in the occasional and oscillatory firing modes, a better coordination of both firing modes is entrained (Figures 15,16 of [62]).

2.24. Building Up of External Loops to the Periphery by Premotor Spinal Oscillators

It will be shown now that with the building up of simultaneous phase relations between α, γ and SP2 fibres and the assimilation of II distributions (coordination's of rhythms), an external loop of premotor spinal oscillators is built up to the periphery, which makes it possible to directly influence the firing of spinal oscillators by a rhythm training. The somatic and parasympathetic pattern organizations in the sacral micturition centre can simultaneously be entrained by jumping on springboard (Figure 169g) (including the stimulation of movement (Figure 169f) and bladder receptors (Figure 169h)) to allow movement-based learning in the continence and movement patterns and to induce learning transfer from movements to urinary bladder functions. A repair of neuronal network patterns of the functionally disconnected sacral micturition centre in spinal cord injury seems possible.

In Figure 165Ba it can be seen that there existed two phase relations between the firings of the α_3, γ_1 and SP2(1) fibres, which means that the γ-loop, including the γ_1 and SP2(1) fibres, contributed to the drive of the α_3-oscillator (probably of α_2-type). However, since the II distributions are different (Figure 164Aa), the γ-loop was not a part of this spinal oscillator; it was only contributing to the drive of it, as pictured in Figure 169a. Since the II distribution of the fusimotor γ_1 was often rather broad (for example Figures 164Aa,165Ac), its driving network was not or only almost oscillatory firing (the dashed line loop in Figure 169a). Upon touching sites 1 to 5 approximately every second, phase relations occurred between the α_3, γ_1 and SP2(1) fibres, even though reaching a different value (Figure 164Bb,c) and the II distributions of the α_3 and γ fibres (Figure 164Ab) and the SP2(1) fibre (Figure 164Ac) became assimilated. The γ-loop became directly connected to the oscillatory firing network: the premotor spinal oscillator (the network driving the α_3-motoneuron; the α_3-motoneuron is a part of the network) built up an external loop to the periphery (Figure 169b). Upon pin-pricking sites 6 and 7, one phase relations occurred between the firings of the α_3, γ_1 and SP2(1) fibres (Figure 164Be) and also the II distributions assimilated (Figure 164Ae).

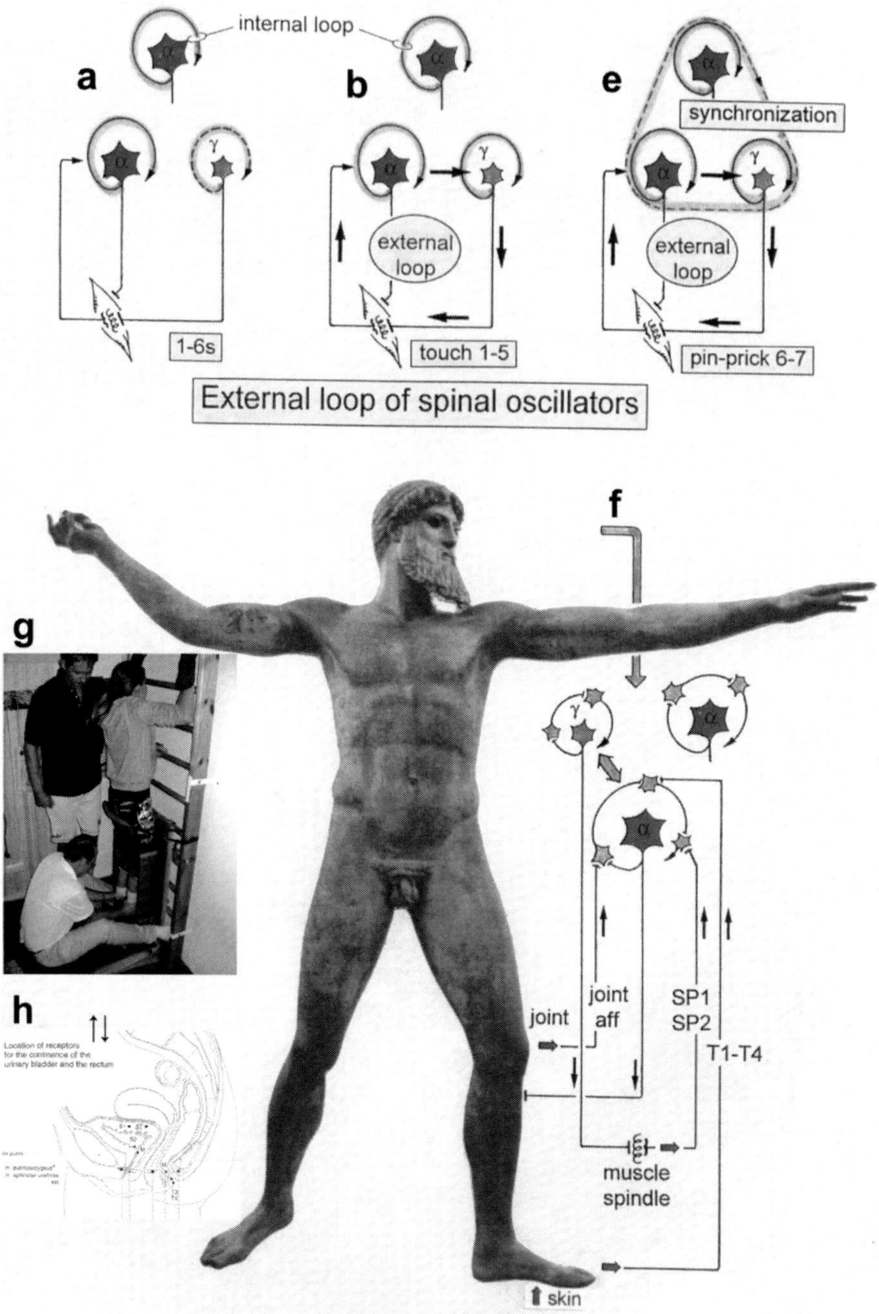

Figure 169. Spreading of oscillatory firing from α-motoneuron neuronal network to include muscle spindles (periphery) and synchronization of different α and γ-motoneuron neuronal networks caused by touch and pin-prick stimulation. (a) α-motoneuron neuronal networks fired oscillatory (solid line loop), γ-motoneuron neuronal network did not or did only partly (dashed line loop), upon no additional stimulation; taken from Figure 164. (b) Oscillatory firing α and γ-motoneuron neuronal networks built up a phase relation with muscle spindle afferents and efferents (external loop to the periphery, indicated by thick arrows) upon touch 1-5. (e) Oscillatory firing α (internal circuitry loop) and γ-motoneuron neuronal networks (external loop) synchronized (broad peak phase relation) upon pin-pricks 6-7 (Figure

163B). The dashed line loop represents synchronization. (f) Oscillatory firing α (internal circuitry loop) and γ-motoneuron neuronal networks (external loop) are extended by analogy from the continence muscles to muscles for locomotion. The open arrows indicate that it may be possible to synchronize spinal oscillators by rhythmic afferent input, generated by rhythmic movements (such as jumping on a springboard or running), and to re-preformat the neuronal circuitry by synapse remodeling to fire more physiologically oscillatory to reduce spasticity and improve locomotion. Extensive pathologic movement like tremor may entrain neuronal circuitry to increase tremor movement. The Greek good is a bronze statue of Zeus found close to the cape of Artemision 460 BC. (g) Supported jumping on the springboard in anti-phase. The patient with the severe cervical spinal cord injury is supported by the author. (h) Location of receptors for the continence of urinary bladder and rectum stimulated upon jumping on springboard (Figure 160).

The spinal α_3-oscillator had built up a full external loop to include the γ-loop in its oscillatory firing (Figure 169e). Since there was transient synchronous firing of unit 1 with unit 2 (including the α_2 and SP2(2) fibres (Figures 164Be,167e)), probably also the α_2-oscillator built up an external loop to the periphery.

2.25. Extension of the External Loop Generation of Spinal Oscillators to Non-Continence Muscles

If one extends the integration of the γ-loop in the oscillatory firing of spinal oscillators innervating striated continence muscles to muscles activated for locomotion (Figure 169f), then for example during jumping without weight support (Figure 169g), sky-walking without weight support (Figures 2B,194C,D,E), running (Figure 47C) or other coordinated, rhythmic, stereotyped, dynamic movements, at least oscillatory firing α_2 or α_3-oscillators build up external loops to the periphery. It seems partly possible during jumping on springboard (Figure 144A,B) to synchronize spinal oscillators with the jumping rhythm, mainly given by the properties (Eigenfrequency) of the springboard. Especially the α_3-motoneuron networks can be entrained efficiently by jumping on springboard, since the 'Eigenfrequency' of the springboard and that of the α_3-oscillator network are both in the range of 1Hz (for entrainment, see below). The extension of premotor network organizations serving bladder functions to premotor network organizations serving movement functions is most likely justified, since rhythmic firing of single FF and FR-type motor units, which are innervated by α_1 and α_2-motoneurons, has been recorded by surface electromyography (sEMG) during rhythmic, coordinated movements (Eigenfrequency around 1 Hz) in patients with spinal cord injury (Figures 14-16). Upon exercising on the special CDT device (Figure 150), single FF-type motor units in arm and leg muscles coordinated their rhythmic firing with respect to phase and frequency. These single-motor unit recordings performed with sEMG support the recordings from single-motoneuron axons (Figure 17).

2.26. External Loop of Premotor Spinal Oscillators and Rhythmic, Dynamic Stimulation of Motor and Bladder Functions

While jumping on a springboard (Figure 169g) (and other rhythmic movements like sky-walking or running) premotor spinal oscillators organize themselves to fire transiently in an oscillatory manner according to the motor pattern and build up an external loop to the periphery (Figure 169). If the frequency of the rhythmic movement has an integer relationship to the Eigen-frequencies of the premotor networks and more rostral networks, these premotor networks get entrained for more specific self-organization (see below). When jumping on a springboard (Figure 169g) not only the motor networks get activated; also the external sphincteric motoneurons, innervating the external bladder and anal sphincter, as a part of the pelvic floor, are rhythmically activated to counteract the rhythmic weight changes of the intestine. Further, the rhythmic, dynamic, stereotyped up and down movements stimulate stretch, tension, flow, and mechanical receptors of the bladder (detrusor and proximal urethra) (Figure 169h). This rhythmic movement-related sensory input with ≈1Hz bears similarity to the sensory input stimulated by bladder and anal catheter pulling with ≈1Hz during the measurements (Figures 163,165). Repetitive phase relation changes in and among neural ensembles will occur in some similarity to the changes following catheter pulling (Figure 165B). Since the neurons involved in the generation of movement and continence (and micturition) patterns (especially if the neurons serve both functions at the same time) are synchronously, rhythmically activated, the pathologic bladder patterns get entrained from the rather physiologic jumping movement, in some similarity to co-movement. The synchronized activation of the somatic and parasympathetic networks allow efficient learning transfer, since the neurons work as coincidence or more general as coordination detector (Figure 11). If there is fluid in the bladder and material in the bowel and rectum, the continence stimulus is stronger. Also walking and running will stimulate and change the intrinsic dynamics of the micturition and defecation centres, but not as strong as the jumping on springboard. The walking and running on a treadmill in severe cervical spinal cord injury is performed under weight support, whereas during jumping on springboard no weight support is needed (Figure 169g). The sympathetic nervous system division, probably innervating the internal urinary bladder sphincter (smooth muscle), will also be activated synchronously. Since the frequency of jumping is around 1 Hz similar to the 'Eigenfrequency' of the α_3-motoneuron oscillators, these oscillator networks should be entrained most efficiently (see below).

2.27. Entrainment of Premotor Spinal Oscillator Networks by Rhythmic Movement-Induced Afferent Input and Inputs from Supraspinal Centers

If one approximates for high activation spinal neuronal networks into premotor spinal oscillators (driving the motoneuron) and propriospinal oscillators (generating by coupling with one another movement patterns) then premotor spinal oscillators can be handled in a first approximation as single linear oscillators. The premotor spinal oscillators and the spinal pattern generating networks are self-organized and driven by afferent and supraspinal inputs. When training dynamic, rhythmic, stereotyped movements, the premotor spinal oscillators approximated as linear oscillators are driven by movement-induced afferent input from the periphery (mainly the legs) and surrounding pattern generating networks, and possibly supraspinal inputs. These spinal oscillators and most likely their neuronal network can be entrained at least by use of the external loop for a better self-organization. But the functional repair becomes only physiologic if volitional control (by ascending and descending tracts) is intertwined.

If one assumes that loop circuits do not only exist between the premotor spinal oscillators and the periphery, but are a general structure within the CNS, then motor learning involves the formation of loop circuits (or better loop network circuits) between the cortex and the periphery involving the sensory cortex and the thalamus. When a linear oscillatory system is driven by an external periodic input its response contains both frequency components. This is also, in general, true with nonlinear oscillators. However, in this case, if the external frequency is close to the Eigenfrequency of the oscillator itself, then it is possible to have a response at the external frequency only. This phenomenon is known as entrainment or synchronization. It is of paramount importance with respect to biological oscillators because it allows them to 'latch on' to the environment. Thus a rhythm with a free-running period of 24.7 hours may be synchronized to 24 hours when exposed to the natural sequence of day and night.

An oscillator with one degree of freedom can be described by the equation:

$$d^2x/dt^2 + f(x, dx/dt) + \omega_0 x = E \cos\omega_1 t$$

(x = variable, t = time, f = nonlinear term, $\omega_0 = 2\pi f_0$ = frequency of the oscillator, ω_1 = entrainment frequency, E = amplitude)

There exists a set of paired values of amplitude E and the absolute difference in frequencies $\Delta\omega = |\omega_0 - \omega_1|$ such that the output of the system only contains the frequency ω_1. Figure 170 shows a typical example. Entrainment occurs in the shaded part of the plane. If E and $\Delta\omega$ are below the curve (outside the shaded area) usually the frequencies ω_0 and ω_1 are present. In the case that $\Delta\omega = 0$ but E is very small the phase of oscillations is not influenced by the input. For further details, including van der Pol oscillator see [127].

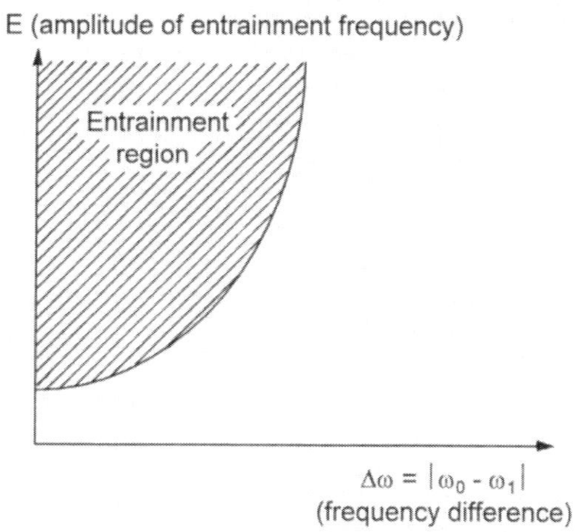

Figure 170. Illustration of the relation between the amplitude E of the external input and the difference in frequencies $\Delta\omega$ which produces entrainment of a nonlinear oscillator.

Upon jumping on springboard, the entrainment frequency ω_1 (the jumping frequency) is 0.9 to 1Hz, close to the Eigenfrequency of the premotor spinal α_3-oscillators ($\approx$ 1Hz). The α_3-

oscillators and the networks they are integrated in are entrained directly. For the entrainment of the α_2-oscillators (Eigenfrequency ≈ 6Hz) and α_1-oscillators (Eigenfrequency ≈ 10Hz) sub-harmonic and super-harmonic entrainment has to be considered. When a nonlinear oscillatory system is driven by an external periodic input z_k, the entrainment can be harmonic (z_k itself has the oscillation period T of the oscillator; case of the α_3-oscillators), sub-harmonic (z_k has a period which is an integer multiple of T, mT; case of the α_2 and α_1-oscillators) or super-harmonic (z_k has a period which is an integer fraction of T, T/m) [127]. With the increasing order of sub-harmonic entrainment, the entrainment strength reduces for the same coupling strength or entrainment amplitude E. For the α_2 and α_1-oscillators the entrainment is sub-harmonic and the entrainment strength therefore reduced. In α_2-oscillators, two entrainment phases per oscillation period were mostly observed, which correspond to 'in-phase' and 'anti-phase' coordination of arms and legs, which enhances entrainment. Also, the change of the number of phase relations between the neural assemblies in the sacral micturition centre may indicate changing entrainment or coupling communication between the somatic and parasympathetic nervous system divisions. However, oscillator models are still far away from human network properties. The jumping on springboard is very rhythmic and stereotyped (frequency of jumping ≈1Hz). But firstly, the movement-induced afferent input enters the network at different levels (premotor neuronal network, propriospinal oscillatory network, brainstem network and higher centres); secondly oscillators can be driven from different sources; and thirdly often the rhythmic input patterns consist of impulse trains with increasing interspike intervals and with delays between the responses from different receptors and receptor types. For natural sets of simultaneous impulse patterns of numerous skin receptors induced by touch or pin-prick see Figure 5, and for natural impulse patterns of secondary muscle spindle afferents see Figure 25. In simulations of networks consisting of populations of interacting oscillators the known natural afferent input patterns have to be used, and oscillator network structures have to be used which give the measured output patterns under both physiologic and pathophysiologic conditions. Even though the oscillator models used are far away from human self-organization network forms, their results are still very useful for the interpretation, under certain conditions, of measured data and to understand better the jumping on springboard training (as a part of the coordination dynamics therapy) which contributes substantially to the repair of urinary bladder function. For jumping on the springboard in in-phase or anti-phase and for sky-walking the amplitude/stride length has to be large to be efficient for repair by movement-based learning.

2.28. Stimulation of the Parasympathetic and Somatic Division via Their Receptors of the Pelvic Floor and Intestine to Induce Learning Transfer from movements to Urinary Bladder Functions for Cure

It was shown that the parasympathetic division can be activated besides the somatic one by anal and bladder catheter pulling (Figures 163,165). The activated parasympathetic neuronal network channeled input to the rhythmically firing somatic neuronal network via an additional phase relation (Figure 167j). This joint stimulation of the somatic and parasympathetic division is used in coordination dynamics therapy to use movements which substantially activate the somatic and the vegetative nervous system to induce learning transfer from movement to bladder patterns to cure urinary bladder function. Upon jumping on springboard, without weight support in severe cervical spinal cord injury (Figure 171), the pelvic floor (including the sphincters) is mechanically stimulated (stretch of muscle spindles) under the weight of the intestine. Also the stretch and tension receptors of bladder and rectum are activated upon the rhythmic up and down movements. The desire or urge to void is also strongly stimulated when the patient is exercising on the special coordination dynamics therapy device, especially in the lying position (Figure 150). But this stimulation of the parasympathetic and somatic divisions is related to the improvements in the short-term memory of the phase and frequency coordination of the neuronal networks of the sacral micturition centre including the functional units pictured in Figure 167. Both, the rhythmic mechanical stimulation of the continence organs (Figure 171) and the improved CNS organization in the short-term memory upon exercising on the special device (Figure 150) lead to a strong activation of the sacral micturition centre which may lead in turn to strong increases of the heart rate and blood pressure, which are vegetative symptoms to inform supra-spinal centres about bladder fullness in complete spinal cord injury, if the bladder is filled and if there is plenty of material in colon and rectum.

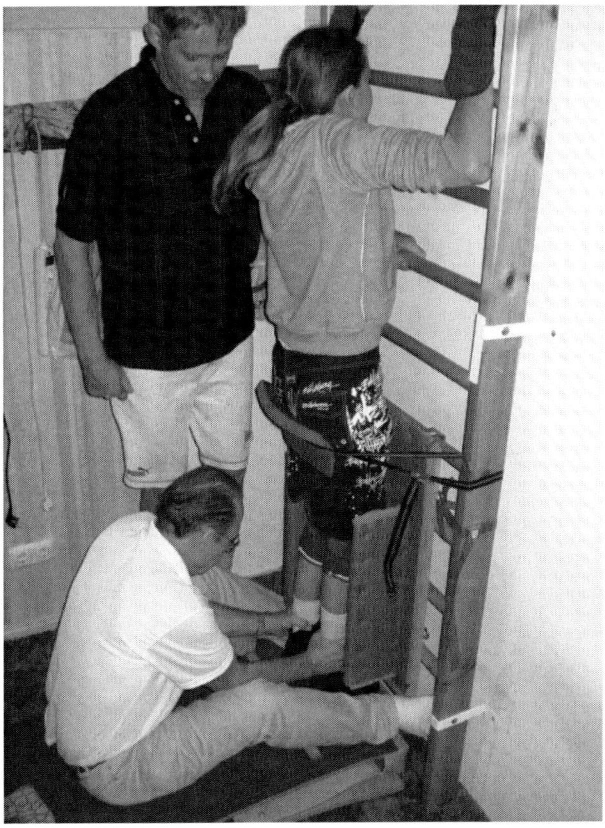

Figure 171. A. Patient with a severe cervical spinal cord injury (C5/6) during rhythmic jumping on springboard. The Author is supporting the legs. Upon jumping, the different receptors of the intestine and the pelvic floor, including the external urinary bladder and anal sphincters, are rhythmically activated. Such repetitive mechanical stimulation has some similarity with the repetitive bladder and anal catheter pulling and activates simultaneously the somatic and parasympathetic division and increases learning transfer.

2.29. Structural and Functional Repair

Functional and structural repair are intertwined. To induce and enhance the structural repair in the human CNS (necessary in severe SCI), a different strategy is used than in animals. In rats for example, one is trying to enhance the growth of nerve fibers over the injury site. In humans it is necessary to induce structural repair by learning in the long-term memory and learning transfer, for which a high complexity and variability of CNS neuronal networks is needed. To induce structural repair (including network rewiring and neurogenesis) in the spinal cord and supra-spinal centers by learning in the long-term memory is in accordance with current thinking about long-term memory in the cortex [130] and a relation between neurogenesis and learning [131].

2.30. The Lack of Evidence for Spontaneous Regeneration in the Human Spinal Cord

In lower vertebrates the CNS fully regenerates following injury [132]. It hardly needs that injury to the spinal cord is a devastating condition and that the human spinal cord does not regenerate spontaneously, even after twenty to thirty years, only the complications increase. An evolutionary explanation for the absence of the human nervous system's regenerative capacity when compared to animals, especially lower vertebrates, is that this is an unselected by-product of gaining the increasingly complex nervous systems that selection pressures have favored over time [107]. CDT uses this complexity (and variability) of the human nervous system (namely the learning) to partly overcome the 'loss' of CNS regenerative capacity and does not pursue the nerve fiber growing strategy.

2.31. Structural Repair in Animals: The Nerve Fiber Growing Strategy

If the spinal cord of a rat is injured, nerve fibers try to grow over the injury site. Since the distance of growing is too short and the number of growing fibers is too small, experiments have been undertaken to attempt to enhance the power of regeneration to get more fibers to grow in greater numbers and over longer distances. Nerve growth factors are used. Applied stem cells may proliferate to neurons and their neurites may cover the injured site. If one believes that regeneration over the injury site is inhibited, then one tries to inhibit this inhibition. If scar tissue formation inhibits the regeneration, then the scar tissue is removed. This nerve fiber growing strategy is logical with respect to animals, because in less complex nervous systems it is often simply a question of growth, since nerve fibers partly regenerate to specific sites they innervated before the injury. The learning capacity on the other hand is limited. A rat cannot relearn if the nerves to extensors and flexor are transposed [133]. The monkey can relearn [134] and the human can relearn quickly [135]. Further, it took more than two years to repair bladder function through learning in severe cervical SCI (see below). The entire lifespan of a rat is not much longer than the time needed for mere bladder repair in humans. A learning strategy would not be very helpful for the rodentine CNS unless it were the case that the rate of learning in rats is higher than in humans.

2.32. Structural Repair in Human: The Neuronal Network Learning Strategy

The learning capacity of the human CNS is unparalleled. Further, the human life span affords more time for learning and re-learning than for most other species. To simplify the explanation, take for example an incomplete thoracic SCI. For repair, the treatment strategy must be focused on the intrinsic spinal cord networks consisting of the intumescentia cervicalis, the intumescentia lumbosacralis and the connecting spinal cord parts. For repair the whole subnetwork over the injury site is substantially activated, including the injured networks and the functionally surrounding uninjured networks by exercising rhythmic, dynamic, stereotyped, coordinated arm (mainly located in the networks of the intumescentia cervicalis) and leg movements (mainly located in the intumescentia lumbosacralis). To activate many, ideally all, of the spared fibers over the injury site, different coordinated arm and leg movements have to be exercised such as walking, running (Figure 172), crawling (Figure 140C) and exercising on the special CDT device (Figure 150). The crawling, walking, and running are innate automatisms, which may strongly induce genetic support for repair (by changing the expression of regeneration-associated genes) and the exercising on the special device (Figure 150) is primarily improving the stability of the impaired phase and frequency coordination. At the beginning of therapy, the repair is largely functional. However when activating the spared fibers at the limits (range of overreaching [52]), structural repair seems to be induced. The stimulus for improving the network connectivity over the injury site (not necessarily in the spinal cord) is the specific motor learning process. Those connections should be enhanced or newly built through sprouting building of new nerve cells from adult-born stem cells and changes in the efficiencies of synapses, needed during that specific motor learning process. Ongoing pre- and postsynaptic activity in the injured area during the learning induced loop circuit formation likely has an important role in guiding the axons that extend from adult-born neurons. Local target-derived growth factors could attract regenerating fibers, or facilitate growth of regenerating fibers along highly active spared fibers, stimulated upon movement-based learning (range of overreaching [52]). Just placing stem or progenitor cells to the injury site seems not to be helpful in enhancing repair [53]. The more coordinated the movements, the more the coincidence principle of neurons (Figure 11) is engaged and the more of the relevant fibers and synapses can be activated at their limits at the injury site or those functionally connected to it. To enhance specifically the communication over the injury site, the patient has to concentrate (for example) on the automatic walking or running (Figure 172). He/she substitutes partly missing connections in the network activation of the automatic walking by using volitional connections in the learning process (volitional support in the cranial → caudal direction). The therapist, supporting the legs during treadmill walking, brings the patient into the intended automatic walking or running pattern (Figure 172) by contributing his own power at moments when additional power is needed during the walking cycle. In this way, strong and rather natural afferent impulse patterns (Figures 5,169) enter the spinal cord for processing in the caudal → cranial direction to keep the walking or running pattern activated and to give information to higher centers. In this way the functional and structural repair of the injured circuits may be initiated without disrupting the consolidation of the spared intact neuronal networks. The functionally recruited spared and overloaded intact fibers and the synapses activated in (network) loop circuits most likely give the stimulus for specific structural repair. The automatic walking pattern, the patient intends to perform activates preferentially intrinsic spinal neuronal networks. Patients with incomplete SCI, with some remaining sensory functions, sometimes feel the activation of the intraspinal circuitry.

Figure 172. Running on treadmill of a 20-year-old patient with a severe cervical spinal cord injury at C5/6 levels (Figure 148) after 3 years of optimal coordination dynamics therapy (CDT). Leg support is given by 2 therapists. The patient is inducing the leg movement with the arms; in this moment only with the left arm. Author is on the right side of the patient.

According to this reasoning, the fewer the remaining fibers, the fewer fibers are available for activation for the purposes of functional and structural repair. In other words, the more severe the injury, the more difficult the structural repair.

By administering the neuronal network learning strategy, an attempt is made 'teach' the injured nervous system of the patient to repair itself by learning.

In the process of organization and integration of different parts of the spinal cord during the re-learning of movement and continence patterns, the incoming natural impulse patterns from the receptors of the periphery (bottom-up activity) and supraspinal centers (top-down activity) are integrated with the endogenous activity of the spinal cord by relative phase and frequency coordination. For the integration of newly built circuits, the coordinated firing of neurons has to be continuously re-learned by exercising on special CDT devices.

Since neurogenesis in the human CNS seems to be extremely limited [136], the movement-based learning therapy (CDT) has to be optimal and continuous to induce the building of new cells from stem cells and to induce the self-renewal of the few available stem cells. A break in CDT of a few months will stop the structural repair and a re-initiation of the neurogenesis may need at least a few months.

2.34. Co-Movement (German: Mitbewegung)

The co-movement of limbs or other body parts of the body can be induced by the simultaneous input from receptors in the range of a few milliseconds. The neuron is a coincidence or coordination detector (Figure 11). With the simultaneous input from afferent fibers to a neuron the threshold to generate an action potential at the axon hillock is reached earlier than with uncoordinated input. Communication between right and left or rostral and caudal body parts can be achieved with less active connections.

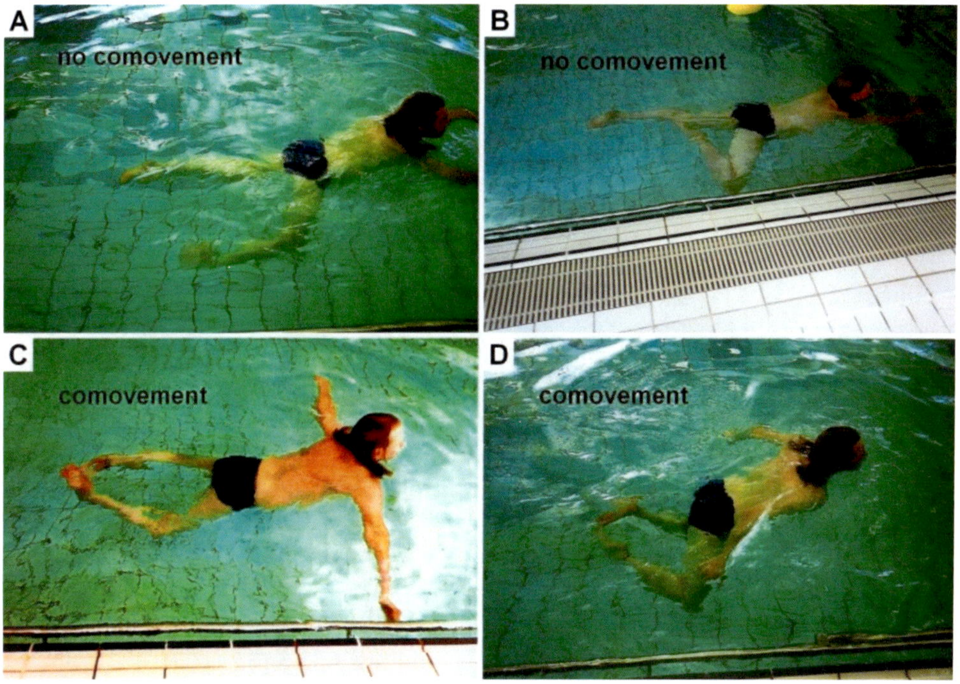

Figure 173. Co-movement between legs in a patient with an incomplete cervical spinal cord injury during swimming. When the patient was not closing the legs before flexion only one leg moved (A,B). When he closed the legs and the legs touched each other, both legs moved (flexed) (C,D). The simultaneous touch input achieved a coupling from right to left and made the not moving leg moving.

The co-movement will be demonstrated here between legs. A patient with an incomplete SCI was able to swim (Figure 173). However, because he was not closing the legs before flexion, only one leg moved (A,B). But when he closed the legs and the legs touched each other, both legs moved (flexed) (C,D). The simultaneous touch input achieved a coupling from right to left and made the not moving leg moving. This shows that simultaneous input will help to improve the performance of a movement.

When exercising on the special CDT device (Figure 150), co-movement will also be induced from muscle spindles and other receptors to make the 'poor' leg move better. In SCI or stroke the knees will come better into a central position during the repeated coordinated arm and leg movements performed on the special CDT device. This co-movement works for all exact coordinated movement patterns, especially in in-phase and anti-phase movements. Exact coordinated instructions and coordinated visual input (as for example due to interpersonal coordination [18]) also enhances the earlier generation of action potentials and thus the improvement of the performance of coordinated movements.

2.35. Judging the Functional Stage of CNS Repair for Therapy Updating

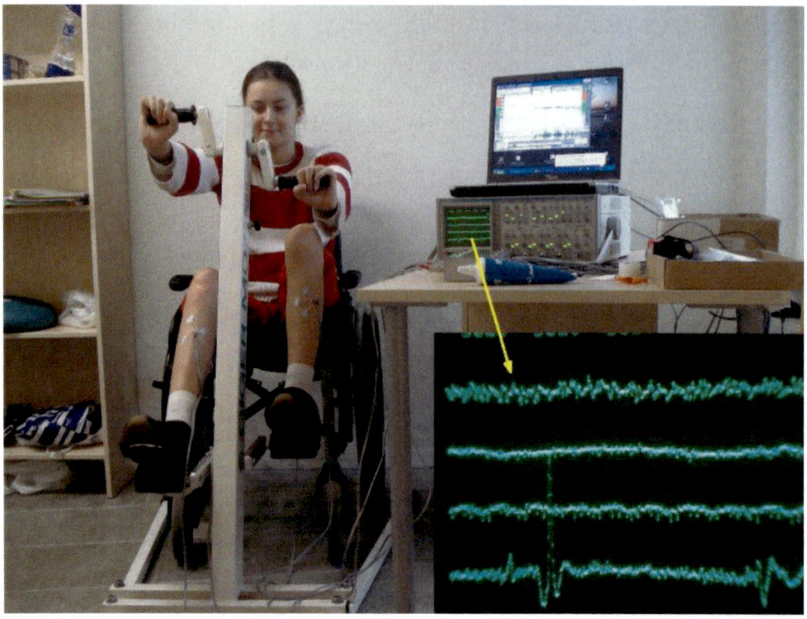

Figure 174. Layout for measuring coordination dynamics (arrhythmicity of exercising) between arm and leg movements, displayed on the laptop; for the intermediate coordination's between pace and trot gait, the fluctuation of the network states is larger. The recording of sEMG activity (displayed on the oscilloscope) from the tibialis anterior and other muscles is also shown. The inset shows single motor unit action potentials on the lowest trace. The recordings are taken from a patient with a motoric complete cervical spinal cord injury C5/6.

The stage of repair in patients is quantified by different means in order to understand how the patient's CNS is functioning and how to adapt the therapy for optimal repair. Objectively the quality of CNS self-organization is measured by recording the coordination dynamics when the patient is exercising on a special device (Figure 174). The stability of coordinated arm and leg movements is measured by the arrhythmicity of exercising on the device while performing the continuously imposed changes of the pattern of coordination between arm and leg movements between pace (in-phase) and trot gait (anti-phase). Surface EMG is occasionally performed (Figure 174). The performance of other movements, as for example walking, is judged visually for optimal performance. Further, the supervising therapist also performs supported treadmill walking with the patient to feel the volitional and automatic contributions of the patient's CNS to the walking performance and to give instructions to the therapists how to assist the walking with respect to achieve optimal performance (Figure 140L). Detailed anamneses of the patient's feelings with respect to the working of new muscles and further re-appearance of sensitivity, gives a clinical understanding of what is changing in the patient's CNS during the therapy-induced repair. By comparing the actual knowledge obtained from the patient with the continuously measured and graphically pictured parameters (as for example shown in Figure 140M) and the research knowledge and practical knowledge of similar patients in mind, the movements performed during therapy can be updated.

2.36. Practical Aspects

In order to undertake effective and continuous treatment in severe cervical spinal cord injury, it is important to reduce the occurrence of bladder infections as quickly as possible. They obviously weaken the patient and obstruct treatment. The blood supply of the lower body has to be improved as quickly as possible to avoid pressure ulcers and make the skin resistant to allow movements, otherwise simply firm touching of the patient's skin will lead to bleeding. The special coordination dynamics therapy devices in the sitting and lying position (Figure 150; functioning muscles (help muscles) can be used to enable the patient to perform the movement) are especially useful for starting the treatment and to improve the autonomic functions to improve the blood supply of the lower body (especially the skin) and to get sufficient blood supply to the brain to keep the patient conscious. The start of the treatment is especially difficult in severe cervical spinal cord injury, since nearly no motor functions are spared for training. The earlier the treatment is started, the less muscle atrophy occurs in both muscles that are still innervated and those that are not. There may even be a critical period for regeneration in the spinal shock phase. For practical reasons it is almost impossible to commence treatment at the ideal time.

3. Urinary Bladder Repair (Kadri)

3.1. Case Report with Respect to Bladder Repair

The seventeen year-old female patient Kadri suffered a severe cervical spinal cord injury in a car accident. No motor functions remained below the injury level of C5/6 and the patient had impaired sensation. From the MR images the author estimated that approximately 5% of the spinal cord matter was spared (Figure 143). When the spinal shock faded away, it became obvious that no motor functions remained below the injury level but uneven sensitivity remained more or less all over the lower body. Two months after the accident CDT was started.

Following two and a half years of CDT, sensitivity improved and some motor functions returned below the injury level, indicating that some regeneration of the spinal cord had occurred. The connectivity over the injury site, according to the magnetic resonance imaging (MRI), may have increased to 8% (Figure 148). Above I reported about the improvement of the motor functions of this patient; the improvement was compared to a patient in whom 50% of the spinal cord matter was spared (Figures 49,52). How bladder repair was achieved shall now be discussed.

3.2. Transient Increases of the Coordination Dynamics Values Coincide with the Recovery of Motor and Urinary Bladder Functions

The changes of low-load coordination dynamics values are shown in Figure 30 over a three year therapy period. Transient worsening of the coordination dynamics (higher values) coincide with the improvement of motor functions and probably indicate successful regeneration of a small number of nerve fibers. Following the largest peak of transient increase of the coordination dynamics values (indicating the strongest transient regeneration), the bladder started to function rather physiologically again after three years of CDT.

3.3. Time Course of the Improvement in Bladder Functions from CDT

In Figure 175 I give the stages of bladder repair over two years of therapy. Changes of functions of the detrusor (bladder) and the external and internal bladder sphincters are extracted from a detailed anamnesis and are pictured by an evolution of the attractor layout with the re-learning of bladder functions.

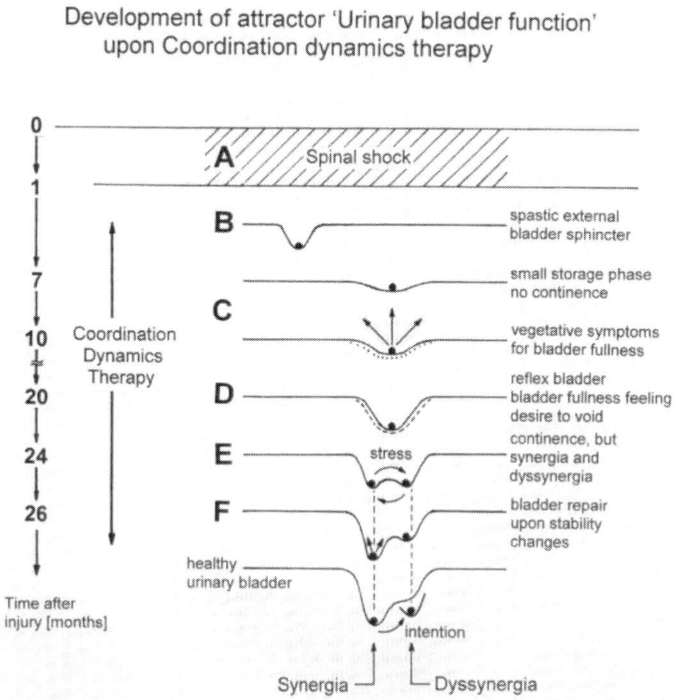

Figure 175. Evolution of the attractor layout of bladder functioning induced by learning transfer from movements to bladder functions upon CDT. The region around each local minimum of the potential landscape acts like a well that weekly traps the system into a coordinated state. Black balls correspond to stable minima of the potential. With learning the pattern 'spasticity of the external bladder sphincter' vanishes and the patterns for bladder functioning ('synergia' and dyssynergia') appear anew and gain their physiologic stability (physiologic deepness of each basin of attraction). The corresponding attractor layout for physiologic bladder functioning is given. Fluctuation of pattern state (the black ball) (C), and their decrease (F), due to the impairment of phase and frequency coordination of neuron firing, is pictured in 'C' and 'F' by long and short arrows. Dotted and dashed lines indicate the re-occurrence of bladder sensation. Note that more than two years of optimal CDT were needed for bladder repair.

1. After the operation (fixation and alignment of the broken cervical spine) a lying catheter was installed in the patient. The patient was suffering continuous infections and fever.

It is understood that the bacteria 'creep up' the lying catheter into the bladder, especially in females due to the anatomy of the urethra, to give rise to ongoing infections in spite of antibiotic therapy. Prior to the advent of antibiotics, patients would often die of such infections. It is thanks to Sir Guttmann from Breslau [137] who stopped or reduced these infections by introducing intermittent sterile catheterization. The introduction of antibiotic therapy helped further.

2. One month later, when the spinal shock had relented, a suprapubic catheter was installed and no more infections occurred. But the bladder did not show any physiologic functions. The patient had no feeling of bladder fullness, no desire to void and did not feel when the fluid was leaving the bladder. The catheter was used for emptying (when opened) and storing (when closed) the urine. Since no fluid was leaving the bladder through the sphincters, the external striated sphincter was spastic (continuous contraction) and the internal sphincter (smooth muscle), as a part of the detrusor, was probably not working.

When functioning physiologically, the internal sphincter (smooth muscle, slow acting, probably a part of the detrusor) maintains continence for low and medium bladder pressure. For high bladder pressure and sudden bladder pressure increases, which occur for example during coughing, the fast external sphincter (innervated by α_2-motoneurons and consisting of fast fatigue resistant muscle fibers (FR)) contracts to secure continence. If the striated external sphincter is not working properly, patients often suffer from so-called stress incontinence.

3. Seven to eight months after the accident (end of 2005), the fluid was leaving the bladder by itself after a small storage phase. This means that therapy had reduced the spasticity of the external sphincter. The patient was now incontinent. So far, the spastic external sphincter had mainly stopped the fluid from leaving the bladder by its spasticity. The internal sphincter started to work a bit to allow a small storage phase. When the fluid was leaving the bladder there was at first no feeling of fluid movement. Later on the patient had some feeling of fluid movement. Flow receptor afferents (S2) most likely started to work. Three months later the suprapubic catheter was removed. The patient started to use diapers.
4. Twenty months after the accident (beginning of 2007, after 18 months of therapy) the patient felt bladder fullness and the desire to void. The stretch (S1) and tension receptor (ST) connections had probably started to work again. The vegetative symptoms of bladder fullness information (sweating and sudden heard rate increases, probably transmitted by plexus connections; Figure 153) were replaced by bladder fullness feeling and the desire to void.
5. 5. The patient became able to press the fluid out of the bladder. To get all fluid out, the reflex bladder had to be activated a bit, by tapping, touching or massaging the skin above the bladder, which is the reflex skin area for the bladder reflex. Sometimes body positioning was used to influence the bladder reflex. With these maneuvers, the desire to void reappeared and the patient could empty the bladder further.

Often patients to whom no CDT is administered train the bladder-emptying reflex. The reflex is stimulated by tapping the skin above the bladder. But if, for example, the external sphincter is spastic as it was in Kadri, it may not be possible to generate a well-functioning bladder reflex.

6. After the appearance of the desire to void, the patient became able to hold the fluid for 30s to 1min. The quality of continence was subject to variation. This means that the external bladder sphincter, which can be controlled volitionally, slowly started to work, albeit irregularly. The feeling of bladder emptying became similar to that of before the spinal cord injury.
7. Twenty four months after the accident (spring 2007), the bladder started to function rather physiologically again. After a storage phase, the fluid came out by volition. The detrusor started to work fully. But if the patient was pressing too much at the end of bladder emptying to get all fluid out (to reduce residual urine), the external sphincter contracted. The external sphincter co-contracted with the detrusor. Detrusor-sphincteric dyssynergia of the bladder occurred. But when she then activated the reflex bladder by tapping or touching the reflex bladder area, the desire to void reappeared and she could empty the bladder fully. The rest urine was not measured.

At this stage of bladder repair two patterns existed: the synergia of the bladder, in which the detrusor and external sphincter contracted antagonistically, and the dyssynergia of the bladder, in which the detrusor and the external sphincter co-contracted. The synergia pattern was for emptying the bladder and the dyssynergia pattern was the pathologic pattern. The pathologic co-contraction of the external sphincter with the detrusor occurred more easily when there was less fluid in the bladder and the patient had to press more, inducing stress to the CNS.

Both bladder emptying patterns exist in the healthy bladder. The synergia pattern is for emptying the bladder and the dyssynergia function is for stopping micturition. However, both patterns are normally under volitional control.

8. After two years of CDT, the patient was fully continent again and could empty the bladder volitionally. The time interval from the first feeling of a desire to void to elimination was one hour. The patient no longer needed to use diapers. The patient had never used pharmaceuticals which are supposed to improve bladder function. The repair of bladder function was achieved by the re-learning of bladder functions including transfer of learning from the movements jumping on springboard (Figure 171), treadmill walking and running (Figure 172) and exercising on the special CDT device (Figure 150). The strong improvement of urinary bladder functions occurred when the coordination dynamics values strongly increased (got worse) (Figure 30), indicating regeneration.

For patients with incomplete spinal cord injuries, the ability to maintain continence until toilet facilities can be accessed. Improvement in bladder function can also be judged by how long the patient can hold the fluid following the first desire to void. The patient was able to withstand the desire to void for up to one hour after two years of CDT.

The feeling of the lower abdomen, which was poor after the accident also improved strongly at the time of full bladder repair. The patient could experience sensation in the lower abdomen very strongly, both inside and out and could also feel the abdominal muscles working. At that time, also the finger functions became very subtly better and her supported treadmill walking improved (see below). During walking on treadmill, and during other movements, goose-pimples appeared all over the patient's body. It seems that an overall improvement of vegetative and somatic functions occurred at the time of full bladder repair.

3.4. Attractor Layout Changes during Urinary Bladder Repair

Within the framework of System Theory of Pattern Formation, the repair of the urinary bladder functions can be understood and visualized by changes in the attractor layout.

One month after the injury, once the spinal shock faded away (Figure 175A, only the pathologic bladder pattern 'spasticity of the external sphincter' was present (Figure 175B). Six to ten months later, the spasticity of the external sphincter reduced and a small storage phase of the bladder re-appeared as a first sign of bladder repair (Figure 175C). Twenty months after the accident, the reflex bladder pattern organized itself with bladder fullness feeling and the desire to void (Figure 175D). Twenty-four months after the accident, the attractor layout showed two attractors, the bladder synergia (the detrusor action inhibits the external sphincter) and the dyssynergia (co-contraction of detrusor and external sphincter) (Figure 175E). The state of the system, indicated by the ball) switched easily from the attractor synergia to the attractor dyssynergia. Twenty-six months after the accident, the stability of the attractor synergia had increased and the stability of the attractor dyssynergia had decreased (Figure 175F). Micturition could be induced and stopped intentionally as in healthy individuals.

3.5. Conclusion

Bladder functions could be cured in severe cervical SCI after two and a half years of CDT. Since the injury was motoric complete and the cord was 'free' in the spinal canal (Figures 143,148), some regeneration in the human spinal cord had likely taken place.

4. Different Learning Aspects of Urinary Bladder Repair

4.1. Integrative Functions and Central Pattern Generators for Bladder Repair

It was shown in this patient that urinary bladder functions could be cured after two and a half years of coordination dynamics therapy (CDT). Three important steps were achieved. First, the patient regained volitional bladder control. Second, a physiologic attractor layout for bladder patterns could be generated; and third, the dyssynergia of the bladder could be cured by increasing the stability of the synergia pattern and decreasing the stability of the dyssynergia pattern. The stability changes of the two bladder functioning patterns can be understood within the framework of system theory of pattern formation and human neurophysiology but not in the framework of central pattern generators (CPGs).

4.2. Similarity of Development between the Ontogenetic Landscape for Locomotion and the Evolution of the Attractor Layout for Bladder Functions Upon Learning

The development of the potential landscape of bladder functions following repair (Figure 175) shows some similarity to the development of the ontogenetic landscape for locomotion (Figure 176). Also the attractor layout for locomotor functions develops from a monostable to a bistable or multistable structure. With learning, it is not only attractors which appear anew but some attractors also vanish. During the re-learning of bladder functions, the attractor 'spasticity of the external bladder sphincter' disappeared. There is some similarity between repair by re-learning and the development of sensory-motor functions during ontogeny. Following spinal cord injury, the stepping automatism and the bladder reflex re-appear through learning. During ontogeny and repair, the automatic stepping is changed by supraspinal control into walking and running patterns and the bladder reflex (or automatism) develops into proper micturition and continence functions.

4.3. Vegetative Symptoms Are No Autonomic Dysreflexia

Exercise does not normally pose any special risks to people with SCI, aside from the initial period. Fractures or muscular injuries occur more often in sport than in patients with SCI. It is argued that people with SCI have atypical physiological responses to exercise, for example abnormal heart rates and sudden increased sweating, which can limit their ability to sustain intensive exercise [138]. When a patient cannot feel the fullness of the bladder and has no desire to void, then the bladder can only inform the patient through vegetative symptoms (through plexus connections), namely sudden heart rate increases and sweating.

Most patients have sufficient awareness of their vegetative symptoms and can empty their bladder before intense exercise. However, if patients are unaware of their vegetative symptoms or ignore them, small blood vessels can burst in the brain during sudden blood pressure increases.

Patients with SCI have to re-learn their body functions. One of the most significant signs of progress for the patient with whom we are concerned was that, following twenty months of CDT, when the vegetative symptoms were no longer required to judge bladder fullness as she could feel the need to void. To treat a patient with severe cervical SCI, obviously theoretical and practical knowledge is needed for the repair of motor and autonomic functions.

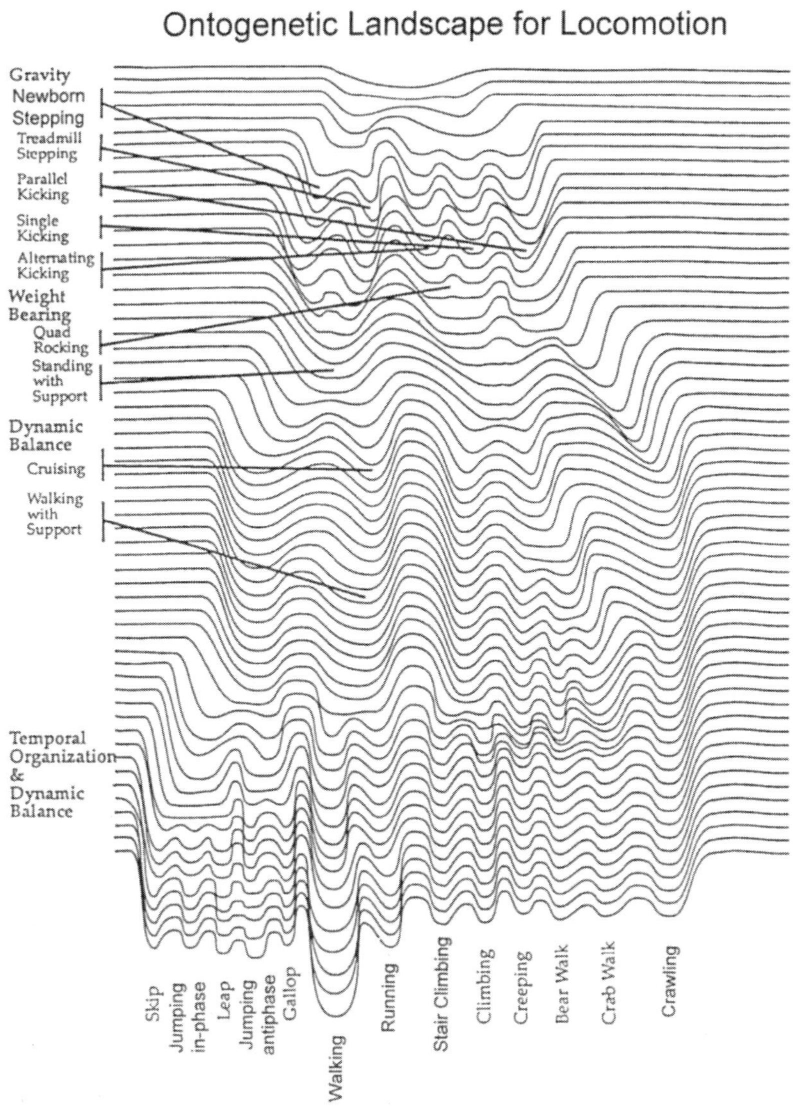

Figure 176. Ontogenetic landscape for locomotion. The evolution of the attractor layout for the different movements develops similarly as the re-learning of urinary bladder functioning upon CDT. With permission of Esther Thelen [198].

4.4. Bladder Repair in Children with Spinal Cord Injury

Nervous system complexity increases not just across phylogeny but also across ontogeny. The loss of regenerative capability with age demonstrates the cost of the complexity [107]. In children with spinal cord injury it should therefore be easier to repair bladder functions, since the nervous system is less complex and the rate of learning higher. A four year-old boy suffered flaccid paralysis due to a SCI sub Th12/L2 in a car accident. At the age of six years he undertook CDT for six months. The boy came to the therapy center in a wheelchair and left it walking with sticks (Figure 136C). Bladder function improved. Before the therapy the bladder was emptied by intermittent catheterization. After six months of therapy the patient was continent for at least three hours at a time and could empty the bladder volitionally with little rest urine. A more likely explanation for why children are able to re-learn lost functions faster may not lie solely in the fact that the nervous system has not reached its full complexity, but that the repair during development is partly due to a correction of the development 'en route'. The increase of the nervous system complexity, including neurogenesis and growing of axons and dendrites, can be used by movement-based learning for functional and structural repair. Young cerebral palsy patients therefore have a good prognosis for repair through CDT [6].

4.5. Continence in Aging

Incontinence problems in elderly people can be caused by a variety of factors. A generally lesser known reason is that the sacral micturition center is not receiving sufficient blood supply through the radicular-medullary arteries. The thickest feeder artery to the sacral micturition center, the A. spinalis magna (great radicular artery of Adamkiewicz), supplies 68% of the blood [139] and is often very atherosclerotic. A combination of CDT and a reduction of arteriosclerosis may improve continence in elderly people.

4.6. Therapy Induced Regeneration

It has been reported that SCI is a devastating condition for which there is, as yet, no cure [106,114], even though it has been documented that SCI can partly be cured by CDT [3,4,10]. In most patients, the repair was largely functional as opposed to structural. The important implication from this case reports is that even in a motoric complete cervical SCI, bladder function can be cured, through which some regeneration in the human spinal cord takes place in addition to mere functional repair.

In experimental animal-based SCI research the spinal cord is only partly severed (spinal hemi-section; see discussion to the pictures in [106,114]. However, partial SCI should no longer form the focus of such research as it has been shown that a partial SCI can be largely cured [4,10]. The real challenge is the severe and apparently complete cervical SCI. Researchers in the field of animal research seem loath to handle bladder incontinence, which would result if the whole spinal cord were transected in rats, cats or dogs. Emptying the bladder of animals several times per day is not an appealing part of research work. Added to this, regeneration will self-evidently be much more difficult to achieve, and therefore to demonstrate results, in complete SCI.

4.7. Extent of Regeneration of the Human Spinal Cord

It is important to establish how many regenerating axons support the functional improvement obtained in the patient with the motoric complete cervical SCI. When in adult rats as little as 1-2% of the corticospinal tracts was spared following electrolytic lesion, performance of forepaw reaching recovered [140,141]). These observations imply that regeneration of a small percentage of axons and functional reorganization of the CNS, including the intrinsic spinal cord networks, could lead to a significant recovery of function. In adult cats as little as 10% of spinal white matter tracts were sufficient to permit spontaneous walking without external support [142]. When large proportions of spinal tracts were transacted in human patients owing to intractable pain, disturbances in motility were small [143].

Since during CDT, those fibers needed for function were stimulated to regenerate by coordinated sensory-motor learning, around 1 to 3% of the white matter regenerated. Since the trunk muscles and also the intercostal muscles started to function again, as shown by the absence of paradoxical breathing, as well as some upper leg muscles, probably less than 1% of the white matter regenerated for the activation of the neuronal networks for driving the pelvic floor (of which the external bladder and anal sphincters form part), the detrusor and the bowel. It seems that only a very small percentage of fibers needs to be successfully reconnected with their targets to mediate bladder recovery. Since the spinal cord was severely injured over only one vertebra length (Figure 148), axonal growth and elongation will have been shorter than one vertebra length, depending whether atypical synaptic relay circuits were formed at the injury site. This additional small amount of regenerating ascending and descending fibers may not be visible in the MRIs of patients with approximately 5% (sensory) spared white matter.

That these established connections are mainly activated ectopic pathways through root connections, sympathetic chain or plexus connections (misdirected haphazardly during development) is unlikely, since the functions of trunk and leg muscles reappeared from the rostral (below injury level) to caudal direction, indicating a regeneration in general.

4.8. Endogenous Adult Neural Stem/Precursor Cells for Spinal Cord Repair

The partial structural repair in this patient was probably achieved by activating intrinsic self-repair processes. The efficient and intensive CDT may have generated a local microenvironment in which endogenous adult stem/precursor cells self-renewed, proliferated, differentiated and migrated to contribute to the CNS repair. But what are the explanations as to why this intrinsic repair mechanism is so slow and so limited?

Firstly, if the sciatic nerve in humans is injured, repair also requires more than two years. Secondly, the therapy should have started in the spinal shock phase to avoid complication such as the growth of cysts and pathologic reorganization of the CNS. If the patient could have started to exercise one week after the accident or indeed earlier on the special CDT device in the lying position, which provides a safe position for the reconstructed spine during arm and leg movement (Figure 150)), the blood supply and the circulation of the cerebrospinal fluid would have recommenced and improved much faster following the injury. Damage through active secondary processes would have been reduced and adult neural stem/precursor cells could have reached the injury site earlier. Thirdly, if there is similarity between development and repair (see below), it may be that a critical window for more efficient repair does exist. A ‘plasticity window’ could have already been closed before optimal CDT was started. Fourthly, the complexity of the human CNS may not allow a faster regeneration to avoid instability in CNS self-organization (see below). Fifthly, it has been reported that, without stimulation, none or only undetectable levels of neurogenesis take place in the adult neocortex [136].

4.9. Combination Therapies to Enhance Regeneration

It is widely accepted that no single approach will prove sufficiently powerful to achieve successful regeneration [106,107]. A methodology combining the most effective individual therapies is required. However, it has been shown in this that CDT alone could cure bladder functions. This single human therapy was able to induce successful regeneration. I also believe that a combination of therapies may act synergistically to enhance regeneration. But first, it is not enough to ask for combinational treatments, one also has to support them and second, each single approach must have a solid scientific basis and must be a safe treatment without unacceptable side effects. A combination of stem cell and CDT could enhance regeneration, since the neuronal network repair strategy, used in CDT can generate a microenvironment at the injury site for the self-renewal, proliferation, differentiation and migration of neural stem/precursor cells, even though a first therapy trial was not successful [53]. This movement-based learning therapy (CDT) probably stimulates the endogenous stem cell reserves for structural improvement. Additional administration of embryonic or adult neural stem/precursor cells may then enhance the rate of structural repair.

4.10. Scientific Basis for Therapies to Repair the Human Spinal Cord

Basic scientific research should provide a rational basis for tailoring specific combinations of therapies. It has been reported that neuro-rehabilitation should have a scientific basis [114]. CDT has a theoretical basis [14]. Reviewers often cite human research with questionable scientific merit while at the same time arguing that neuro-rehabilitation methods must have a scientific basis.

It has been measured with the single-nerve fiber action potential recording method that the phase and frequency coordination among neuron firings becomes impaired following SCI (Figures 10,48a,48b). One important consequence of this impairment is that the variability of phase and frequency coordination increases and CNS organization becomes unstable. A macroscopic consequence of this instability of cell communication for self-organization is that movements and other patterns are changed and become unstable. This instability of movement patterns can be measured macroscopically and improved when the patient is exercising on the special CDT recording device (Figure 174).

It is not only CNS injury that can cause instability in CNS organization. The uncontrolled building of new nerve cells and growing of neurites will also induce instability of phase and frequency coordination of CNS self-organization if the newly formed connections are integrated in the existing injured networks. This instability in CNS functioning will lead to an increase of the instability of physiologic patterns and an increase of the stability of pathologic patterns like the different kinds of spasticity in SCI and epileptic seizures in brain injury. Structural repair of the spinal cord (or CNS) in humans must therefore be a carefully controlled specific mechanism to maintain the delicate balance between plasticity and stability. In Figure 30 it can be seen that with every bit of reoccurrence of muscle function, i.e., the successful regeneration of a few fibers, the coordination dynamics values got transiently worse (the movement patterns were transiently more unstable), until the movement-based learning therapy re-established stability again. These transient instabilities are marked in Figure 30 with numbers. Following the largest period of transient instability at 'Peak 6', bladder repair was achieved.

Methods sensitive enough to detect potentially small increments in function have been long sought after [114]. However, Coordination dynamics measurements can quantify small increments in function and regeneration (Figure 30).

Structural repair therefore has to be highly specific and controlled at the injury site to avoid instability, which means the growing of nerve fibers has to be enhanced and inhibited according to the functional needs of the network repair. If neurobiologists really succeed in inducing or inhibiting the massive grows of nerve fibers and if these drugs were applied to humans, uncontrolled nerve fiber growth may occur and the human CNS may become unstable in its self-organization. Following the successful administration of stem cells and nerve growth factors to patients, there is therefore not only the risk of teratocarcinoma and the toxicity of side effects but also the risk of inducing extensive instabilities in the self-organization of the human CNS. To quote the poet Goethe: "Die Geister die ich rief (neurogenesis, gliogenesis and neurit growth), werd ich nun nicht los" (uncontrolled growing and generation of CNS instabilities) (Zauberlehrling).

The risk of causing instability in CNS organization is much greater in the human as opposed to the animal CNS due to the much higher complexity and variability of the human CNS organization, necessary for learning.

Brain imaging research focused on detecting the brain centers involved in various sensomotor or cognitive tasks, but CNS organization research attempts to understand the self-organization of distributed local networks and their integration and coordination to 'higher' states of processing [90].

Extensive learning potential needs specific variability in the system. In order to avoid instability in changing self-organization, there must be stability control. Partly disrupting this stability control may undermine CNS function. In young patients with severe brain injuries strong instabilities can be observed and measured [8].

If in lower animals the circuitry is rather fixed and the nerve cells are rather specifically connected, then the repair needs only the appropriate cells substituted and the connections to be repaired. But if in the human CNS the specific, fixed wiring is rather limited, then the repair has to be controlled by network functions; uncontrolled neurogenesis and growth of nerve fibers may be dangerous.

4.11. Similarity between Development and CNS Repair with Respect to Activity Dependent Migration of Neurons and Network Maturation/Repair

It has been shown in frogs that there is close similarity between the innervation of muscles during development [144] (Chapter I of [14]) and re-innervation following denervation [145,146]. Two types of motoneurons with thin and thick axons communicate with each other and with the two target muscle fibers to achieve the proper innervation pattern, namely that the small neurons with thin axons innervate the slow muscle fibers and the large neurons with thick fast conducting axons innervate the twitch muscle fibers. This four-cell communication is of interest, because when the thin axons reach their target (the slow muscle fiber), the thick axons of the large motoneurons have already placed their motor endplates on their target muscle fiber (the twitch muscle fiber) but also on the slow muscle fibers, which are the target muscle fiber of the thin axons of the small motoneurons. The large motoneurons activate both types of muscle fibers in a physiologic way according to the developmental or repair stage. The thin axons, reaching the target later, communicate with their target muscle cell (the slow muscle fiber), change their (slow) muscle fiber properties and push the synapses from the large motoneurons away. During repair the same target finding process takes place. The re-innervation recapitulates development. Regenerating axons recapitulate developmental guidance during recovery from the denervation injury.

It has been shown that electrical activity regulates development in a variety of species and in varies structures [37], including the retina [147-149], spinal cord [151,152] and cortex [151]. Within the mammalian cortex specifically, the development of dendrites and commissural axons in pyramidal cells is activity-dependent [152,153]. It was found that some cortical interneurons, owing to their late birth, in some similarity to peripheral innervation in frogs, populate the cortex only after the majority of other interneurons and pyramidal cells are already in place and have started to functionally integrate. Such developmental period shows some similarity to a repair scenario. In mice it was demonstrated that for the (for CGE-derived reelin (Re)-positive and calretinin (Cr)-positive) interneurons, activity is essential before postnatal day 3 for correct migration, and that after postnatal day 3, glutamate-mediated activity controls the development of their axons and dendrites [35]. It was further shown that the engulfment and cell motility 1 gene (Elmo1), a target of the transcription factor distal-less homeobox 1 (Dlx1), is selectively expressed in Re^+ and Cr^+ interneurons and is both necessary and sufficient for activity-dependent interneuron migration. These findings reveal a selective requirement for activity in shaping the cortical integration of specific neuronal subtypes [35].

Experimental evidence indicates that interneurons are electrically active shortly after their birth and participate in the early network activity that may contribute to circuit maturation in the neonatal cortex. It was demonstrated that altering the level of neuronal excitability in vivo within genetically targeted CGE-derived interneurons has profound consequences on multiple aspects of the development of select subtypes within this population, as well as their associated gene expression [35].

If we compare neonatal development in mice with CNS repair in humans, then natural activity is needed in the neuronal networks of the injured CNS for repair to achieve physiologic functioning. In the absence of proper movement-based learning therapy, neurogenesis (as for example of pyramidal cells and interneurons) will probably be very limited, the migration of neurons will be incorrect and the development of axonal arborizations and of the complexity of dendritic trees will be pathologic. The injured CNS will reorganize itself into a pathologic organization. The therapy, which offers physiologic activity for repair, has to reorganize the pathologically functioning CNS into a physiologically functioning one. If coordination dynamics therapy were commenced in the intensive care unit for example on the special CDT device in the lying position (which is as safe as the bed), then the functional and structural repair would move the nervous system from its injured state towards a physiologic state, partly omitting the 'detour' to pathologic states.

Which activities are needed to induce neurogenesis, the necessary cell migration, the building of the necessary dendritic trees and axonal arborizations at the places where repair is most needed? In the context the building of new motoneurons, it is shown in Figure 29 how the activity is brought to the repair sites. Is it possible to generate the required activity to achieve the building of specific neurons at distinct stages of repair and is it possible to link the modulation of gene expression to neuronal activity during repair? During mice development there is indication that genetic programs initiated at the progenitor stage are modulated by activity [35].

Two approaches have been emphasized in recapitulating development in the injured CNS: re-establishing crucial developmental cues in the correct pattern to guide regenerating axons, and maximizing the sprouting and plasticity of intact fibers through sensory feedback rehabilitation techniques [107]). A faithful recapitulation of developmental path finding and circuit-refining mechanism is likely to be beneficial to repair in animal SCI, but may only partially hold in humans (see below). By measuring the self-organization of neuronal networks upon pattern change (by using the coordination dynamics assessment) during human development (Figure 35) and human CNS repair (for example Figure 46b), similarities were found between the attractor layouts. More human research has to be performed to learn how the human CNS optimizes and corrects its functioning 'en route' during development.

In the patient Kadri I have shown that by learning, learning transfer, and high motivation, bladder functions could be repaired including some structural repair. More generally, I report how the mind is taking part in the functional and structural repair of its neuronal networks by which it is generated.

Almost as soon as the first computers were made, the analogy with the brain was acknowledged. The model of the brain as a computer, with hard-wired connections that are adaptive and flexible mostly with respect to the software and the information that is distributed over the hardwired network still dominates. The genius property of the human CNS seems to be that by sustained learning in the long-term memory the network circuitry can be changed and is not fixed as in computers. "The spirit can rise above physis".

I believe that the property of improving and repairing the CNS structure by the mind (and movement therapy) is unique to humans. Therefore, not all repair mechanisms of the human CNS can be explored by animal research. Even monkey experiments cannot substitute for human CNS research. If nerves, tendons or muscles are transposed, the rat cannot relearn [133], the monkey can learn after several attempts [134] while the human learns immediately [135]. The relearning capacity in the human CNS with respect to functional reorganization to compensate for changes in the periphery is used in neurosurgery and microsurgery. But as is shown in this book, the rate of learning varies hugely between the healthy CNS and the injured one. The rate of learning may differ by a factor of 20 to 100, depending on the severity of the CNS injury. Orthopedic surgeons operate on patients to achieve a better positioning of muscles and limbs. Patients with a healthy CNS learn quickly to use the altered limbs. Patients with cerebral palsy cannot relearn in a reasonable time, because their rate of learning is slower by a factor of 20 to 50. Before an elongation of the tendons they may be able walk poorly. However, following an operation, the positioning of the legs may be better, but they can no longer walk.

Some patients understandably try to recover from SCI through sheer force of will. They concentrate on the movement and performing the movement in their minds again and again and the natural impulse patterns, generated in the patient's brain, run against the injury site in the rostral-caudal direction. But I have not heard of any success with respect to repair in complete SCI by using only the mind.

Firstly, at least some fibers probably have to be spared to secrete target-derived growth factors and secondly, very precise movements have to be performed repeatedly for a long time. Further, to substantially activate the injured neuronal sub-networks for repair, coordinated movement induced afferent input is also needed. For network repair, the networks have to be intrinsically activated integratively and at physical limits. One kind of input, as for example concentration on a certain task, is insufficient for repair.

4.12. CNS Development and Repair in Animal and Human: Growth, Guidance and Learning / Why Animal Research Alone Cannot Cure SCI in Human

Two approaches have been proposed to recapitulate development in the injured CNS: Re-establishing crucial developmental cues in the correct pattern to guide regenerating axons, and maximizing the sprouting and plasticity of intact fibers through sensory feedback rehabilitation techniques [107]. Re-establishing at least part of the developmental pattern of guidance molecule expression would contribute to any regenerative approach for treating CNS injury in adult animals [107]. Secreted and membrane-bound isoforms of guidance molecules interact with receptor complexes. A recurring theme shared by many of the ligand-receptor combinations directing tract guidance is that interactions can result in growth cone attraction or repulsion, depending on the receptor subtype and intrinsic state of the neuron. This intrinsic state differs among not only neuronal subtypes, but also developing neurons, adult neurons and regenerating neurons. Generally, developing neurons possess the intrinsic state most suited for rapid axonal elongation and target finding.

Circuit Connections and Plasticity

In many developing neurons, a surplus of axons initially reach their targets. Many preliminary synapses form only to retract soon after. The winners of this game are determined by competition for limiting target-derived growth factors. Eventually, excess neuritic branches are pruned, successful neuritic branches stabilize, and many cell bodies that lack victorious nerve terminals undergo apoptosis.

The Flexibility of Immature Circuits

Once axon path-finding and the pruning process have selected for the formation of appropriate synaptic connections, further plasticity occurs in an activity-dependent manner. This results in the potentiation of some connections and inhibition of others. Through learning, neuronal networks become progressively more organized.

Disruption of sensory experiences (behavioral information) within certain age windows results in circuit reorganization, allowing the organism to adapt within the new sensory information including the information generated by sensory-motor coordination. In barn owls, over several weeks they adapt not only regaining auditory-visual coordination but also visual-motor coordination [154]. The ability to adapt correlates negatively with age – owls over 200 days old are unable to regain auditory-visual coordination [155]. Critical periods define the age windows during which various circuits retain the plasticity to sensory deprivation.

Generating Spinal Cord Plasticity

Plasticity applies also to the intrinsic spinal cord circuits that mediate locomotion. Spinal cord neuronal networks generate coordinated activity between groups of agonist and antagonist limb muscles on opposite sides. The networks allow ambulation to become a nearly automatic neural program. Newborn babies (Figure 24) and anencephalic human infants can step automatically. After birth, Jürgen (Figure 12a) could step automatically forward but not backwards. This automatic stepping automatism and walking automatism (including arm and leg movements) in the forward and backward direction is used extensively in CDT. The role of axon guidance molecules in coordinating proper neuronal network connections in animals has been emphasized [107]. However, coordination in human infants is very poor. Its improvement depends crucially on the training of coordinated movements. Eight-month-old Jürgen (Figure 12c) has very poor coordination between the motor programs of tibialis anterior and gastrocnemius muscles (Figure 12d, absent antagonicity), even though he is healthy and quite developed for his age. At the age of three, the coordination between arm and leg movements is, as in other healthy children, still poor as is measured by the coordination dynamics in Figure 34CD. Temporary deprivation of sensory feedback to the rat neuronal networks in the postnatal period results in permanent walking and swimming deficits, defining a critical period for neuronal network plasticity [156]. Humans can re-learn walking and other movements throughout life following incomplete SCI.

Inhibitory neurotransmitters have a crucial role in plasticity. The transition from synchronous (in-phase, pace gait) to alternating bilateral rhythmic limb movement (anti-phase, trot gait) late in gestation depends on glycinergic signaling [157]. Serotonin, acting through 5-hydroxy-tryptamine (5-HT) receptors, delays the maturation of GABA-mediated inputs into lumbar versus brachial spinal circuits [158]. Human babies first learn pace-gait and then the trot-gait crawling. Human infants who missed the crawling period may have coordination problems throughout life unless CDT is administered. The crawling automatism is used in CDT.

Closing the Plasticity Window and Learning Plasticity

Understanding the mechanism underlying critical closure period will form the basis for approaches to re-establishing plasticity in the regenerating adult nervous system. Not only neurotransmitters contribute to the definition of plasticity windows. Extracellular chondroitin sulphate proteglycans (CSPGs), produced mainly by astrocytes, for perineural nets around inhibitory interneurons in the visual cortex, coincident with the closure of the critical period window [159, 160, 161]. CSPGs inhibit neurite outgrowth, probably blocking the synaptic structural dynamics that partly underlie plasticity [161]. Enzymatic CSPG removal re-opens the critical period [161]. Therefore, maturing astrocytes and oligodendrocytes have a role in consolidating neuronal plasticity.

The mechanisms that evolved to consolidate plastic circuitry directly contribute to the inability to regenerate after CNS injury. Hence, plasticity and regeneration are intertwined. The most effective therapeutic approach in SCI repair in animals will untangle these pathways to allow the regeneration of injured circuits without disrupting the consolidation of existing intact circuits. In humans, the mechanisms that consolidate the plasticity in the circuitry may contribute to the control mechanism to avoid instability of physiologic patterns and to avoid the development of pathologic self-organizations, resulting in epileptic seizures for example. But the learning process offers a well-balanced mechanism for functional and structural repair without inducing instabilities in the CNS self-organization. Small instabilities do occur during the regeneration of the human spinal cord, as can be seen from Figure 30. These small re-occurring instabilities are resolved through the afferent input provided by CDT.

Neural Response to Injury and Differences in Mind-Structure Communication for Repair in Animals and Humans

CNS injury, disease, and malformation wreak havoc on the intricately choreographed neural circuitry. Unlike the many similarities across species in patterns of neural development, responses to injury and repair differ greatly. Responses to injury also vary in individual organisms, depending on time, the anatomical location and type of injury. The most effective spontaneous injury responses occur in simpler vertebrate species, at younger developmental ages, and in the peripheral nervous system (PNS) rather than the CNS. A detailed understanding of the factors responsible for these differences provides insight into the challenge of improving treatments for adult mammalian and human CNS injury. The main difference in repair between animals and humans is that repair in animals mainly occurs spontaneously and in humans mainly volitionally. The capacity in animal mind for inducing non-spontaneous repair is very limited. Training devices rather obviously do not exist. Rats, frogs, and lampreys cannot turn on the special CDT device to improve the phase and frequency coordination for the improvement of CNS self-organization (Figure 150), cannot jump on a springboard (Figure 171) for rhythmic natural stimulation of the sacral micturition center to improve continence, cannot perform supported treadmill walking such as that in Figure 172, and cannot be instructed and motivated during therapy.

Translating Animal Data to Human Data

To make use of animal research for the improvement of human diseases, a correlation has to be established between animal and human data (Chapter IV of [14]). For correlating nerve fiber groups for example between rats and humans, there must be a detailed classification schemes for rat nerve fiber groups and human nerve fiber groups (Figure 7). By comparison we can see then, what nerve fiber groups in the human PNS can also be found in the rat PNS. Such comparison is not yet possible, because an up-to-date classification scheme for a rat or a cat PNS does not exist.

The existing classification schemes [162,163,164,165] are too inaccurate for meaningful comparison. Animal research needs to improve their basic data. To make use of the regeneration data obtained from animals, also regeneration data of humans are needed for comparison. Such data do not exist, except in a few isolated instances.

To make use of animal data to cure diseases, parallel studies have to be performed in animals and humans and similar treatments and measurements for quantifying progress have to be used. The switch from animal experiments to human trials is to fall into an oversimplification and is too dangerous for human patients.

4.13. Measuring Functions and the Differences between Those of Animals and Humans

As emphasized above, in many developing neurons, a surplus of axons initially reach their targets. Winners of this game are determined by competition for limiting target-derived growth factors. Once axon path-finding and the pruning process have selected for the formation of appropriate synaptic connections, further plasticity occurs in a pattern-activity-dependent manner. This results in the potentiation of some connections and inhibition of others. Through learning, neuronal networks become progressively more organized.

These steps also hold during the development of the frog (tadpole stage), when the faster developing large neurons innervate slow and twitch muscle fibers first. The slow muscle fibers also have fast contracting properties and the membranes are generating action potentials, as the twitch muscle fibers do, even though the morphology is different (no M-line). When the slower developing small motoneurons reach the slow muscle fibers, they 'push' the synapses of the large motoneurons away and change membrane and contracture properties, but not the morphologic properties. The membrane cannot generate an action potential anymore and the muscle holds a contracture upon depolarization. The formation of appropriate synaptic connections therefore occurs twice (first with the arrival of the fast and then with the arrival of the slow conducting axons) and the specificity is only relative in that frog model. In humans, there are three kinds of muscle fibers, innervated by three kinds of motoneurons (Figure 8). The specificity of correlation between muscle fibers and motoneurons is unclear.

The intricate changes during development can only be explored using at least morphology and electrophysiology. The regeneration following a section of the sciatic nerve in the adult frog follows the same scheme, as can be measured with intracellular recording techniques and Ca transient measurements (Chapter I of [14]). To follow up the morphological changes during re-innervation would be difficult, because the denervated synapses are resolved at different speed, which would make it difficult to distinguish between newly build synapses and old denervated synapses. Only single-cell electrophysiology can follow up the re-innervation processes safely.

In humans, the measuring methods are quite different for comparing development and repair of SCI. As non-invasive methods, electromyography and coordination dynamics measurements can be used, beside movement performance measurements. Also the times for development and regeneration in humans are quite different to those of the frog, lamprey and rat.

Figure 12d shows the EMG motor programs of the right and left tibialis anterior and gastrocnemius muscles in the eight month-old healthy Jürgen (Figure 12c). The single muscles can generate motor programs. But the important antagonistic action between the tibialis anterior and gastrocnemius muscles is still absent. There is still much more co-contraction than antagonistic action between the muscle activations. The antagonistic action between the tibialis anterior and gastrocnemius muscle was also impaired in a patient with a severe cerebellum injury (Figure 21). In the patient Kadri (SCI), the generation of motor programs and muscle power was the greatest challenge.

At an age of two years, Jürgen was able to exercise on the special coordination dynamics therapy device. He could only perform approximately five turns continuously. The frequency of turning and the arrhythmicity of exercising changed very much. At an age of three he could turn better (Figure 34CD). By the age of five, he could turn much more smoothly (Figure 34EF). The coordination dynamics values of Jürgen with development are given in Figure 35 (dotted lines). The mother could of course exercise smoothly. The patient Kadri with the severe cervical spinal cord injury could turn quite smoothly at a low load of 20N (Figure 30). But she managed to turn for higher load (50N) only with high arrhythmicity (Figure 147) (see below).

The research examples of development and regeneration in animals and humans indicate that they almost belong to different worlds with respect to CNS organization, measurements of the organization, and the repair following injury. The development of the organization of the human CNS needs different movements to be trained for proper development and is not simply a case of 'correction en route' but rather a more fundamental conditioning of the direction of CNS development. Two-year-old Jürgen can crawl, walk and climb, but his CNS is unable to generate the complicated coordinated movements between arms and legs, when exercising on the special device upon changing the patterns continuously between pace and trot gait. The CNS of his mother can easily generate these intermediate coordination's between pace and trot gait. Therefore there is a world of difference between the complexity of CNS organization between Jürgen and his mother with respect to coordinated movements. In the patient with the cervical spinal cord injury (SCI), the supraspinal CNS is working well; the main problem after some regeneration of the cord is the insufficient muscle strength.

4.14. Understanding Bladder Repair Based on Human Neurophysiology

At a time when the spinal shock fades following injury, most patients with a severe spinal cord injury (SCI) present with a spastic bladder (Figure 175). Some patients with a less severe injury have a small storage volume in the bladder. Many also have to rely vegetative symptoms to detect bladder fullness. Some patients learn a reflex bladder to empty the bladder. But only patients with incomplete SCI can a feel when the bladder is full and the desire to void. Simultaneous bladder synergy and dyssynergy is a repair stage or a sign of incomplete SCI with tract fibers left to the urinary bladder. Kadri's bladder functioning changed upon repair from the state of a complete SCI to an incomplete SCI. The repair change from dyssynergy to synergy can be understood by comparing the interplay between the somatic and the parasympathetic nervous systems. In brain death, the sacral micturition center is not disconnected from pontine micturition center and synergy should therefore be present, which was actually the case. In patients with severe SCI on the other hand, a dyssynergy is present. By measuring the different organizations of neuronal networks of the parasympathetic and somatic divisions in both cases, we can learn what has changed into pathologic functioning following the SCI. Then we will look for possible pathways for repair.

4.15. Activity of Bladder Afferents to Retrograde Bladder Filling

First we shall discuss the afferent input to the bladder upon bladder filling in the physiologic (brain-dead human) and pathologic case (SCI). Can some pathology in the bladder input be seen in the patient with SCI?

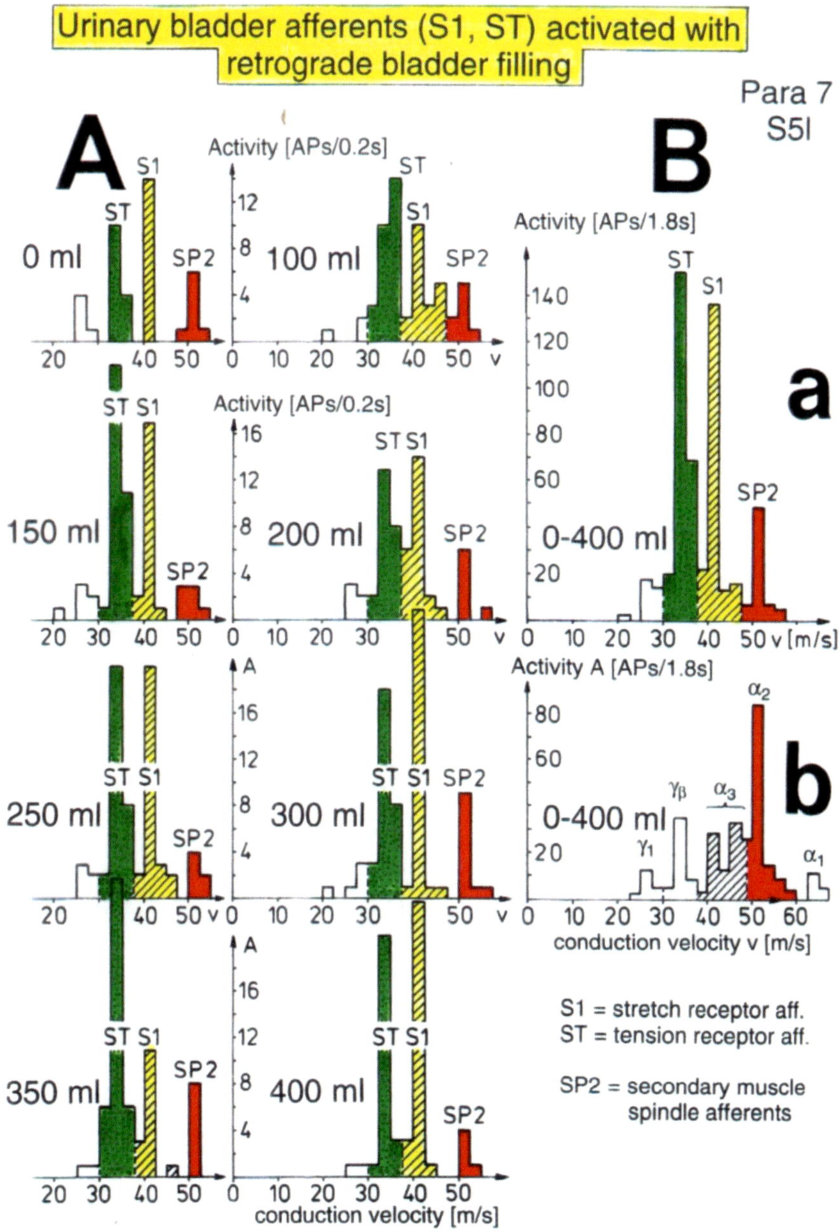

Figure 177. Conduction velocity distribution histograms for stretch (S1) and tension receptor afferents (ST) and secondary spindle afferents (SP2) for different retrograde filling stages of the urinary bladder (A). Summed histograms for afferents (a) and efferents (b) are plotted in B. α_1, α_2, γ_β, γ_1 represent velocity distributions of α_1, α_2, γ_β and γ_1-motoneurons. Para 7, left nerve root S5.

For measuring bladder afferent activity increase upon retrograde bladder filling, the bladder was emptied and then filled while recording. With the groups of nerve fibers identified from conduction velocity distribution histograms (Figure 177B), the increase in activity of the urinary bladder stretch (S1) and tension receptor afferents (ST) to retrograde bladder filling can be measured. Figure 177 shows the activity increases following filling of the bladder of Para 7, via a catheter, with up to 400 ml fluid. The bladder was filled to the maximal acceptable value as determined pre-operatively by compliance (bladder filling volume / bladder pressure) measurements. Figure 177B shows the sum of the individual velocity distributions for different filling stages for afferents and efferents. The peaks are identified by two calibration relations; α_2-motoneurons conduct with the same velocity as the secondary muscle spindle afferents (~50 m/s); and α_3-motoneurons conduct approximately with the same group velocity as the stretch receptor afferents S1 (41 m/s) at about 35.5°C. Figure 177A gives velocity distributions for certain bladder filling stages. The most important obvious feature is that the bladder stretch and tension receptor afferents fire considerably even with no bladder filling.

By setting up conduction velocity borders of the S1 and ST groups according to the distribution curve, the activity of each group could be measured for different filling volumes and plotted (Figure 178A). As can be seen in Figure 178A, the stretch (S1) and tension receptor afferents (ST) in the left root S5 fired with 15 APs during 0.2s at no bladder filling. The bladder afferents increased their firing with the increasing bladder storage volume. At 160 ml filling, a transient very strong firing of ST afferents occurred. This was the filling volume at which before the surgery the detrusor was activated in an urodynamic measurement. It was not possible to activate the detrusor with retrograde bladder filling during surgery, even though the light anesthesia as paraplegics with a complete spinal cord lesion feel no pain.

From a ventral S3 root in paraplegic 11, a similar undulating activity increase of S1 and ST bladder afferents was obtained with retrograde bladder filling. When the bladder was empty, the afferents (S1 and ST) fired already with 10 APs/0.2s. At 50 ml filling, they fired with 23 APs/0.2s, at 200ml with 6 at 250ml with 25 and at 500ml with 12 Aps/0.2s. The detrusor was activated pre-operatively at a filling volume of 80ml (continuously undulating contractions (0.08Hz) started). As in the para 7, the first high afferent input falls approximately together with the onset of the pre-operative detrusor contraction, when filling the bladder. The occurrence of bladder afferent APs in the ventral S3 root was higher (166 APs/2.2s) during the bladder filling than those of α-motoneuron APs (70 APs/2.2s); the secondary muscle spindle afferents fired with 64 APs/2.2s. The speed of filling in Para 11 was 300ml/min. In this ventral S3 root there were at least a few percent of the fibers afferent.

4.16. Comparison of Bladder Afferent Activity Increase to Retrograde Bladder Filling between a Paraplegic and a Brain-Dead Human

The pathology of the bladder afferent activity in the paraplegic 7 analyzed in Figure 178A, with dyssynergy of the bladder, is fully revealed by a direct comparison with the activity increase of the bladder afferents following retrograde bladder filling in brain-dead human HT6 (Figure 178B). The dependence of the bladder afferent activity on the bladder filling volume in paraplegic 7 shows four main differences in comparison to those in HT6. (1) The bladder afferents already fired with no bladder filling which was not the case in HT6. (2) The activity increased more smoothly in HT than in the paraplegic, even though bladder filling was stopped twice in HT6 but was not stopped in Para 7. (3) In the Para 7 the tension receptor afferent activity was higher than the stretch receptor afferent activity which was not the case in HT6. (4) The bladder afferent activity was higher in the paraplegic than in the

HT6. By taking into account the different measuring times of the activity, the ST-activity in Para 7 was approx. 40 times higher than those in HT6 (210APs/1.2s in Para 7 against 5APs/1.2 in HT6). The recording conditions are not directly comparable. However, there are on average more bladder afferents in a root dS4 (HT6) than in a root S5. Still, the 40 times higher tension receptor afferent activity is probably overestimated. It is nevertheless safe to conclude that in some paraplegics the bladder afferent activity, especially from tension receptor afferents, was several times higher than that in the brain-dead human, the latter probably representing the physiologic case in this respect. The finding of increased afferent activity in two paraplegics is in accordance with the experience that in brain-dead humans activity from bladder afferents was recorded only occasionally, whereas in paraplegics the activity from S1 and ST-afferents was always obvious and prominent in the velocity distributions, rather independent of the extent the bladder was filled. Paraplegics with dyssynergy of the bladder suffer from high bladder pressure. This increased afferent activity will have consequences on the excitation of the CNS activating the striated sphincters of the bladder and the rectum.

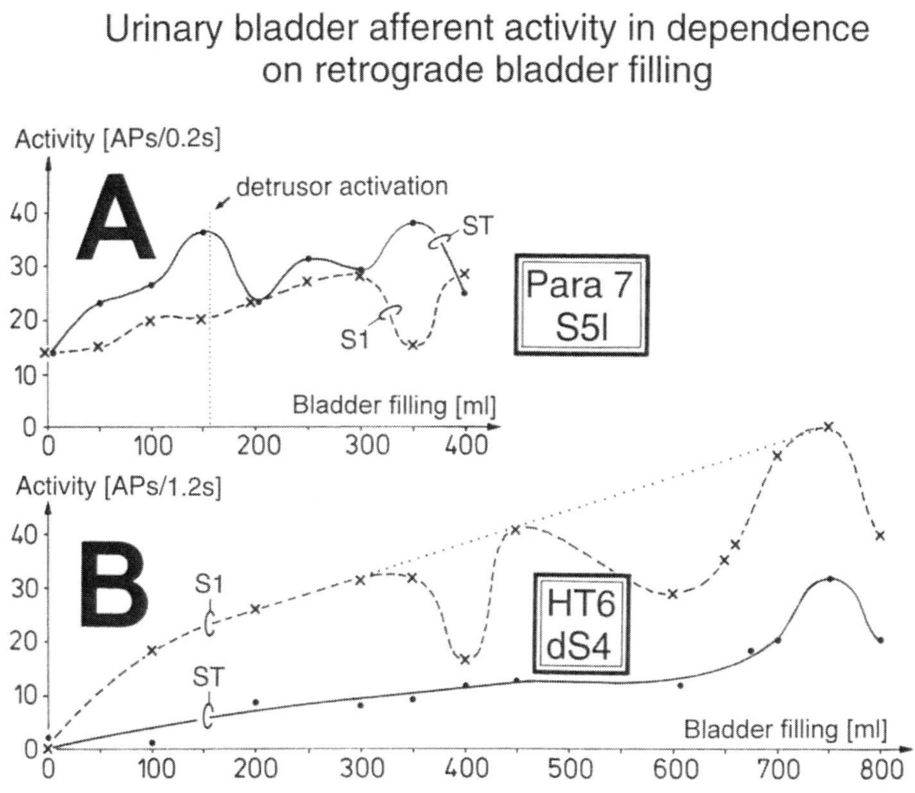

Figure 178. Afferent activity increase of stretch (S1) and tension receptor afferents (ST) to retrograde bladder filling in paraplegic 7 (left root S5) in relation to those of the brain-dead human HT6 (dorsal root S4). APs = action potentials. The speed of filling was 130 ml/min in para 7 and 100 ml/min in HT6.

4.17. Increased Bladder Afferent Activity a Reason for Urinary Bladder Dysfunction

A dramatic increase in the activity of the stretch (S1) and tension receptor afferents (ST) was shown with certainty in two paraplegics (Figure 178). Also, in several other patients' bladder afferent activity seemed increased, as the bladder afferent activity was prominent for even small bladder filling volumes. Nearly all patients who underwent surgery had a much smaller storage volume of the urinary bladder, but not necessarily a reduced compliance (filling volume / pressure ratio). Following urinary bladder de-afferentiation the bladder storage volume increased as did the compliance. However, there are also paraplegics with a storage volume of 50 ml and no increased compliance. They are said to have hyper-reflexia of the bladder.

Thus, there are patients with a stiff and infected bladder in whom the afferent activity of the bladder is much higher. The afferent activity is even often present with an empty bladder. The increased bladder afferent activity causes the detrusor being activated too early. On the other hand, there are patients with no increased compliance and no infection, and a detrusor which already contracts at filling volumes of 50 ml.

It will be shown that there are important changes in the somatic and parasympathetic neuronal network of the sacral micturition centre and in the coordination between them; this can account for an early activation of the detrusor and the external sphincter. As was shown above most of the complications of urinary bladder functions can be avoided and urinary bladder functions be cured if CDT is started early after the spinal cord injury.

5. Coordination Impairment between the Somatic and Parasympathetic Nervous System Divisions in the Human Sacral Micturition Center Following Spinal Cord Injury

5.1. Physiology and Pathology of Bladder Function

After having shown that the identification of somatic and parasympathetic nerve fibers by the group conduction velocity and the group nerve fiber diameter is possible (Figures 157-159) and had not changed following spinal cord injury (SCI) [14], it will be analyzed here, on the basis of natural impulse patterns, travelling into and out of the sacral micturition center, what became pathologic in the organization of the central nervous system (CNS) following spinal cord injury.

Normal voiding is a voluntary act which results in sustained contraction of the bladder (detrusor) and relaxation of the urethra until the bladder is empty. To enable fluid flow along the urethra it is necessary that the pressure in the urinary bladder exceeds that within the urethra lumen. Under normal circumstances, in order to initiate micturition, a fall in urethral pressure immediately precedes a rise in pressure within the lumen of the bladder. Usually this pressure rise is produced by active contraction of detrusor smooth muscle at the onset of micturition. The extensive vesical part of the pelvic plexus (Figure 153) and the profuse distribution of autonomic motor nerves emphasize the importance of the autonomic nervous system in initiating and sustaining bladder contraction during micturition. Immediately prior to the onset of micturition, the tonus of the striated external bladder sphincter is reduced by central inhibition of its somatic motor neurons located in the third, fourth and fifth sacral spinal segments. This inhibition is mediated by descending spinal pathways originating in higher centers of the CNS. The central integration of the nervous control of the bladder and urethra is essential for normal micturition.

Immediately following complete SCI there is a period of spinal shock that lasts about three weeks or longer. This period is characterized by muscular flaccidity and a loss of spinal reflexes. The recovery of reflex activities below the level of the injury occurs at different times. If bladder reflexes reappear they differ in some important respects from those in a normal individual. The 'autonomic bladder' contracts in response to distension, but the power is rarely that of the normal bladder, and the residual volume increases. Lesions above the level of the brain stem centers lead to involuntary voiding that is coordinated with sphincter relaxation (detrusor-sphincter synergy). Lesions below the level of the brain stem centers but above the lumbosacral spinal cord lead, after a period of bladder paralysis associated with spinal shock, to involuntary reflex voiding that is not coordinated with sphincter relaxation (detrusor-sphincter dyssynergy).

The coordination between the detrusor and the external bladder sphincter will now be compared between a paraplegic patient with an injury below the brain stem and detrusor-sphincter dyssynergy (pathologic) and a brain dead human with partly destroyed brain stem and an at least partial detrusor-sphincter synergia (partial physiologic).

If the SCI is complete it is believed that paraplegics develop vesico-sphincteric dyssynergy because of the disconnection from the pontine micturition center which is responsible for the coordination of the detrusor and the external bladder sphincter. In cases in which no detrusor-sphincter dyssynergy has developed, the spinal cord injury is believed to be incomplete, so that supraspinal centers can still coordinate bladder functions. The external bladder sphincter will not be 'co-activated' with the detrusor in response to distension. The view that the important function of coordination is performed by spino-bulbo-spinal pathways has incorporated anatomical ideas of a hierarchy of functions of increasing complexity, the further rostral one goes in the CNS. Still, the coordination for the mutual inhibition of detrusor and external bladder sphincter may be located in the sacral micturition center. This dys-coordination problem shall now be analyzed using the single-nerve fiber action potential (AP) recording method. As was shown in the patient Kadri (Figure 175), such a dys-coordination can be repaired by movement-based learning. The human nervous system has sufficient complexity for re-learning. Rats and cats may not have a sufficient network complexity for such repair [133,135].

Learning transfer is successful in repairing urinary bladder function, since the somatic and the autonomic divisions dovetail with one another with respect to jumping on springboard and exercising on the special CDT device. The necessary connection to the pontine micturition center for the coordination control of the detrusor and the external bladder sphincter is achieved upon CDT by limited regeneration of the human spinal cord, recruiting ectopic pathways through root connections, sympathetic chain or plexus connections (misdirected haphazardly during development) and re-organization of plexus outside the spinal cord. It is believed that the autonomic division of the nervous system is an elaborately organized division of the CNS with representations at all levels. At each level the somatic and the autonomic systems dovetail with one another. For animals, reviews by de Groat [166] provide a basis for understanding how the filling and voiding functions of the bladder work. He states that the primary stimulus for micturition is bladder distension, which induces reflex activation of the parasympathetic excitatory outflow to the bladder, depression of the sympathetic inhibitory outflow, and depression of the somatic efferent output to the external sphincter; secondary reflexes elicited by the passage of urine through the urethra may reinforce these primary reflexes and facilitate the complex emptying of the bladder. When the micturition reflex is exceeded, the parasympathetic reflex pathway through the brain stem is active. For an introduction to the physiology of the lower urinary tract, see [167].

5.2. Pathologic Organization of Neuronal Networks in the Sacral Micturition Center Following SCI

In detrusor-sphincter dyssynergy of the urinary bladder the somatic external bladder sphincter is activated at the same time as the detrusor (smooth muscle) with the consequence that the bladder cannot be emptied. This dys-coordination between the somatic and parasympathetic nervous system divisions in the human sacral micturition center is reflected urodynamically in the simultaneous increase of the detrusor pressure and the electromyographic activity of the pelvic floor. The detrusor-sphincter dys-synergia is analyzed by comparing the natural impulse patterns of secondary muscle spindle afferents (SP2) contributing to continence (SP2 fiber activity changes are similar to detrusor pressure changes (Figure 162)) and sphincter motoneurons in a brain dead human with those in a patients with SCI. In the brain dead human the sphincter motoneurons, sub-serving continence, were inhibited at a time, when preganglionic parasympathetic efferents and a SP2 fiber increased their activity (physiologic). In the paraplegic the sphincter motoneurons were not inhibited (pathophysiologic). In the brain dead human, an SP2 fiber showed doublet firing (interspike interval (II) 10 to 14ms) for low level parasympathetic activation and multi-ending regular firing for high parasympathetic activation. In one paraplegic with strong bladder dysfunction, the multi-ending regular firing was replaced by a repeated burst firing with a shortest II of 0.2ms (transmission frequency = 5000Hz) [14]. The pathologic firing patterns of the SP2 fibers, the detrusor-sphincter dys-coordination, and hyper-reflexia in paraplegics are most likely a result of neuronal network changes in the parasympathetic and somatic nervous system divisions of the sacral micturition center after SCI. As in every CNS injury the phase and frequency coordination is impaired in the paraplegic (Figures 10,48b) in comparison to those in the brain dead human (Figures 9,9e,10,48a). We shall now focus on the detrusor-sphincter dys-synergy.

5.3. Differences in the Impulse Patterns of Secondary Muscle Spindle Afferent Fibres between Paraplegics and the Brain Dead Human HT6

The time course of the increase in the secondary muscle spindle afferent activity, induced by the parasympathetic nervous system in muscle spindles contributing to continence, is very similar to that of detrusor pressure (Figure 162). The detrusor-sphincter dyssynergy is therefore analyzed by comparing the natural impulse patterns of single secondary muscle spindle afferents (SP2) and sphincter motoneurons in a brain-dead human (physiologic) with those in patients with spinal cord injury (pathologic). The parasympathetic nervous system was activated by painful bladder catheter pulling.

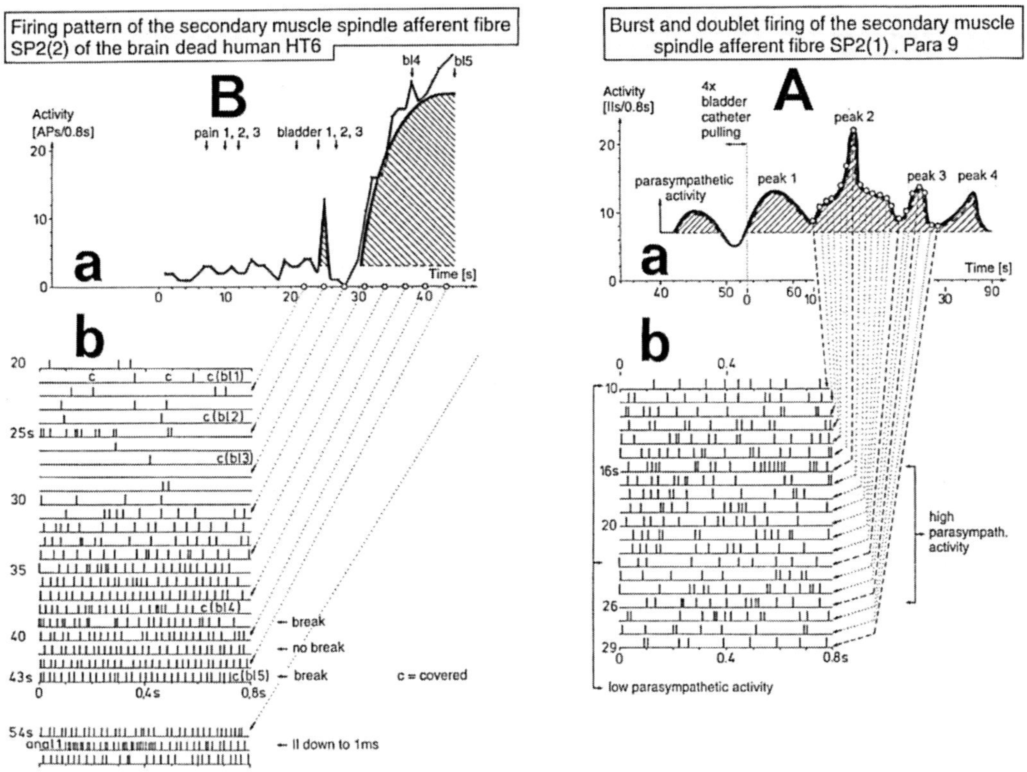

Figure 179. Comparison of the natural impulse patterns of single afferent fibers between paraplegic 9 with dyssynergia of the bladder (A) and the brain-dead human HT6 (B) with synergia of the bladder. Activity increase of the spindle afferent fibers following bladder catheter pulling (a) and the natural impulse patterns at different times following catheter pulling (b). A. Note that at activity peaks (a) there was burst firing in paraplegic 9, (marked by the arrows, 16s, 26s), and at low activity (a) there was no burst firing (10s, 23s). B. Note that even though burst firing appeared also in the brain-dead (25s), the activity increase sustained with similar interspike intervals (IIs). break = break of oscillation of sphincter motoneuron. c = covered = skin afferent action potentials covered SP2 fiber impulses due to bladder catheter pulling (bl 1 - bl 5).

Figure 179 shows the impulse patterns of parasympathetically driven secondary muscle spindle afferents in paraplegic 9 and the brain-dead human HT6. In the brain dead individual (Figure 179B), the activity was low before strong bladder catheter pulling and increased strongly continuously and regularly following stimulation. In paraplegic 9, the activity level was higher before bladder catheter pulling and increased more slowly, in an undulating and irregular manner following catheter pulling. The burst-like firing seen in paraplegic 9 did not occur in HT6. The doublet firing was less regular in Para 9. Specific properties in the drive of the spindle seem to be lost in the paraplegic. Even though a burst-like firing also occurred once in HT6 (Figure 179Ba), the afferent fibre in HT6 could generate a sustained high regular activity level, whereas the fibre in paraplegic 9 could not.

The increase in the detrusor pressure in paraplegic 9 during retrograde bladder filling was very pathologic, since there was nearly no storage phase and the pressure increased transiently and in an undulating way (Figure 162C).

It is satisfying that the results obtained with the single fiber AP recording method from impulse patterns of afferent and efferent fibers support the results obtained urodynamically. The impulse patterns however will provide additional insight into the organization of the human nervous system under physiologic and pathophysiologic conditions.

5.4. Detrusor-Sphincter Synergy of the Bladder in the Brain-Dead Human HT6, and Dys-Synergy in Paraplegic 9

The measurement of parasympathetic activation of the detrusor by activity changes of secondary muscle spindle afferent fibers (the spindle is innervated by parasympathetic fusimotors) allows an analysis of detrusor-sphincter dys-synergy using the natural simultaneous impulse patterns of secondary muscle spindle afferents and sphincter α_2-motoneurons (and γ-motoneurons).

Figure 180A shows that in the brain-dead human HT6, whose parasympathetic preganglionic neurons increased activity (Figure 180A,b) upon bladder catheter pulling, the SP2(2) fiber activity increased strongly (Figure 180a), whereas the α_2-motoneuron innervating the external anal sphincter discontinued its oscillatory firing (Figure 180Ac), which is a measure for a strong activity decrease. A α_2-motoneuron, innervating the external (striated) bladder sphincter, was not activated. This means that with the activation of the detrusor the sphincter motoneurons were relaxed by inhibition. Thus, the brain-dead human HT6 had a detrusor-sphincter synergy of the bladder.

In paraplegic 9 who showed strong activity increase of the SP2(1) fiber, there was no sphincter relaxation following bladder catheter pulling (Figure 180B). The secondary muscle spindle afferent fiber SP2(1) increased its activity in an undulating manner (Figure 180Ba). The parasympathetic fusimotors, driving the muscle spindle innervated by the SP2(1) fiber, probably were not continuously active as suggested by Figure 180Bb, in contrast to the parasympathetic activity observed in HT6 (Figure 180Ab). The other secondary muscle spindle afferent fiber in paraplegic 9 (SP2(2), Figure 180Ba) slowly reduced its activity upon bladder catheter pulling.

This spindle afferent fiber was not connected to the continence of the bladder. It is likely that its spindle was not parasympathetically innervated and was sited in leg muscles or parts of the pelvic floor muscles not contributing to continence. The α_2 and α_3-motoneurons (Figure 180Bc) showed a high activation, which is expressed in their oscillatory firing, and probably contributed to the continence of the bladder and the rectum. These likely sphincter motoneurons did not reduce their activity following parasympathetic activation, as can be seen from the SP2(1) fiber activity (monitoring parasympathetic activity).

These motoneurons were not inhibited and the external sphincters were probably not relaxed. The somatic fusimotor γ_1 (Figure 180Bc) increased its activity transiently and slightly upon painful bladder catheter pulling, in similarity to a γ_1 fiber in HT6 [58]. The measurements in paraplegic 9 indicate a loss of the inhibitory action of the detrusor onto the sphincter motoneurons.

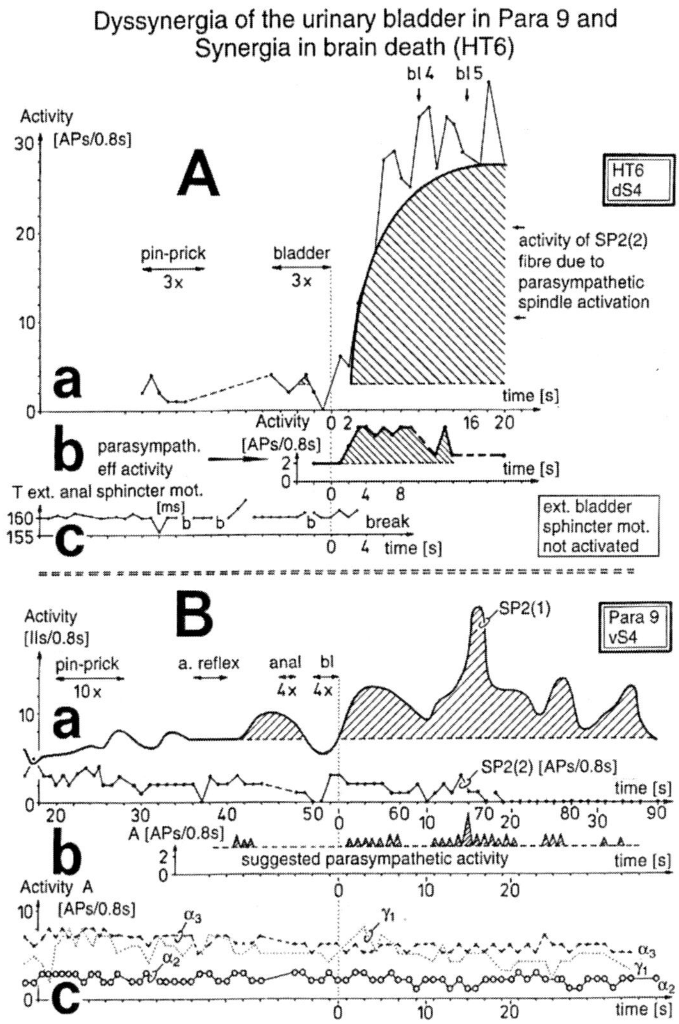

Figure 180. Direct comparison of secondary muscle spindle afferent and motoneuron activities between brain-dead human HT6 with no dyssynergia of the bladder (A) and paraplegic 9 with dyssynergia of the bladder (B). A. Simultaneous measurements of activities of secondary muscle spindle afferents (a), parasympathetic preganglionic motoneurons (b) and oscillatory firing (high activity mode) of a sphincter motoneuron innervating the striated anal sphincter (c). Note that with the transient activity increase of the parasympathetic fibers (b) the secondary muscle spindle afferent fiber increased strongly its activity (a) for minutes, and the oscillatory firing sphincter motoneuron discontinued its oscillation (c) to reduce strongly its activity. bladder 3x = 3 times bladder catheter pulling. T ext. anal sphincter mot. = oscillation period of the sphincter α_2-motoneuron innervating the anal sphincter. For further details, see Chapter V of [14]. B. Simultaneous measurements of activity of secondary muscle spindle afferent fibers SP2(1) and SP2(2) following anal (anal 4x) and bladder catheter pulling (bl 4x) (a), and the activity changes of an α_2-motoneuron (FR) and α_3-motoneuron (S) and a dynamic fusimotor fiber (γ_1) (c). Note that following bladder catheter pulling (and probably parasympathetic activity increase), the spindle afferent fiber SP2(1) (most likely contributing to continence) increased its activity in an undulating manner (a), whereas the SP2(2) fiber did not (probably not connected to continence) (a), and the α-motoneurons did not reduce their activity (c). The dynamic fusimotor γ_1 transiently increased its activity similarly as in HT6 measurements. In similarity to 'Ab', the suggested parasympathetic activity increase is pictured (b). a.reflex = anal reflex stimulation. IIs = interspike intervals; IIs/0.8s = (APS – 1)/0.8s (the activity measures IIs/0.8s and APs/0.8s differ by '1').

The time constant for the activity decrease of a spindle afferent fiber following parasympathetic activation was 31s in a paraplegic and approx. 40s in a brain-dead human (Chapter V of [14]). It is concluded that the muscle spindles are unchanged following spinal cord injury. The pathologic firing patterns of the SP2 fibers are thus probably a result of neuronal network changes in the parasympathetic and somatic nervous system divisions of the sacral micturition center.

In conclusion, in a brain-dead human the sphincter motoneurons sub-serving continence were inhibited at a time, when preganglionic parasympathetic efferents increased their activity (physiologic) for 10s and an SP2 fiber increased its activity for several minutes. In a paraplegic with a strong bladder dysfunction, the SP2 fiber activity increased, due to parasympathetic activation, lasted for approx. one minute, showed undulations, and its amplitude was smaller than that measured in a brain-dead human. The sphincter motoneurons were not inhibited (pathologic).

6. Functional Morphometry to Find Possible Pathways for CNS Repair in the Peripheral Sympathetic and Parasympathetic Nervous Systems

6.1. Pathways and Networks for Functional and Structural Repair

By what pathways and network reorganizations could bladder repair (Figure 175) be achieved in the patient Kadri with motoric complete SCI C5/6? As can be seen from Figures 180Ab and 180Bb, the parasympathetic activation of the bladder had changed following injury. The vagus nerve originates above the injury level and is innervating the intestine down to the colon (Figure 154). Only the caudal 1/3 of the colon is innervated through sacral parasympathetic division according to anatomy books (Figure 154).

If the peripheral rostral and caudal parasympathetic nerves systems are intertwined with networks, then due to reorganization, the detrusor could be activated via the vagus nerve. The sympathetic chain with its different plexus (Figure 153) is unlikely to be a pathway because the highest contribution to the sympathetic chain originated from the T1 level which is clearly below the injury level.

The blood supply to the skin was impaired following the SCI, causing the skin to be very delicate. At the beginning of therapy, the skin damaged very easily (Figure 140C). With six months of CDT, the skin was not vulnerable any more, which indicated that the blood supply of the skin had normalized.

No pressure ulcers occurred. If the blood supply through the arterioles is controlled via the sympathetic nervous system, then it is difficult to understand why it was damaged at all, unless the sympathetic chain was also injured. The repair of the blood supply and bladder function can only be understood if the autonomic nervous system is much more complex (Figure 153) than anatomy books indicate.

6.2. Relation between Nerve Fiber Diameter and Myelin Sheath Thickness

In order to gain further information about nerve fiber populations innervating the bladder, and the anal sphincters and possible connections to supra-injury levels (functional anatomy), nerve fiber diameter frequency distribution histograms were determined from a cadaver. To identify possible nerve fiber classes in the nerve fiber diameter distribution histograms, the diameter spectra were split according to different myelin sheath thickness ranges.

To choose feasible ranges the relation between nerve fiber diameter and myelin sheath thickness of the pelvic splanchnic nerves (nn. pelvini) was constructed and shown in Figure 181. The myelin sheath thickness roughly increased with the diameter. But the correlation points indicate that there may be different populations of nerve fibers with their own correlation between myelin sheath thickness and diameter.

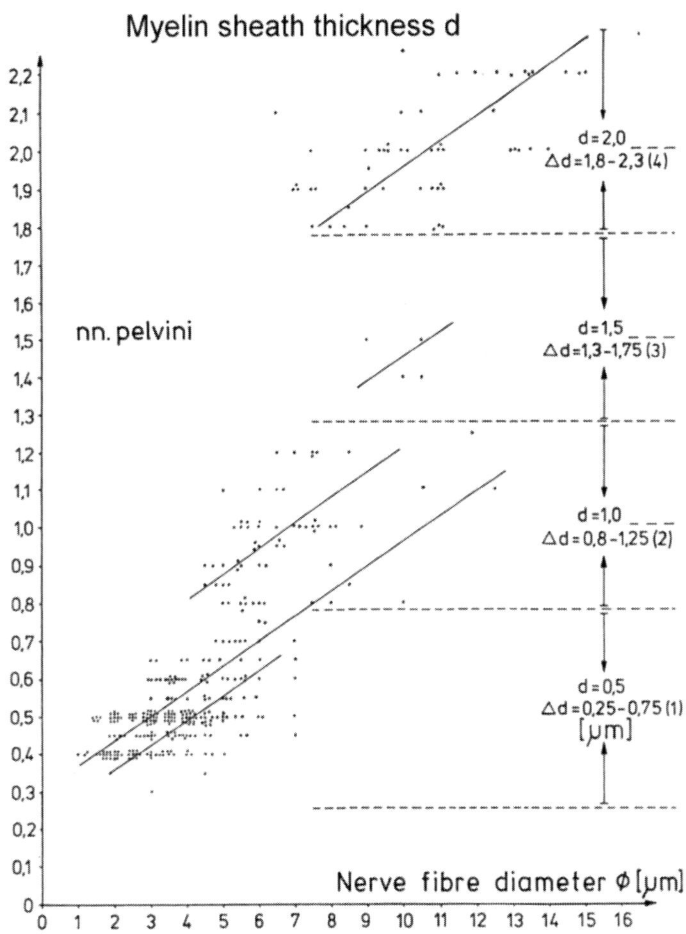

Figure 181. Myelin sheath thickness in relation to the nerve fiber diameter Ø of the Nervi pelvini. For definitions of d and Ø see Figure 182Aa. Lines indicate suggested correlation between d and Ø for different nerve fiber groups. Note, there may be several nerve fiber groups in one class ($0.3 \leq d < 0.8\mu m$) and a nerve fiber group may extend into another class. Correlation dots in class $0.3 \leq d < 0.8\mu m$ are shifted slightly because of closeness.

Correlation lines for possible nerve fiber populations are drawn in into Figure 181. It can be seen from the large scatter of the correlation points that it is difficult or impossible to identify different nerve fiber populations on these morphological data alone. The myelin sheath thicknesses were therefore placed in four classes characterized by the myelin sheath thickness ranges $0.3 \leq d < 0.8$, $0.8 \leq d < 1.3$, $1.3 \leq d < 1.8$ and $1.8\mu m \leq d < 2.3\mu m$. Nerve fiber diameter distribution histograms were then calculated for these myelin sheath thickness ranges. With the chosen classification ranges it is still possible that a certain population of nerve fibers covered more than one myelin sheath thickness range, as is indicated in Figure 181.

In other nerves it also seemed as if nerve fiber groups have their own correlation between fiber diameter and myelin sheath thickness. As Figure 181 indicates, the choice of the myelin sheath thickness ranges was not arbitrary. The split histograms separate different nerve fiber groups better than the whole diameter histogram. Perhaps the best method of identifying nerve fiber groups from morphometric data is to correlate the nerve fiber diameter, the myelin sheath thickness (or better the number of myelin sheath lamellae) and the number of occurrences in a three-dimensional space. The method of plotting data in Figure 181 is a first step towards a three-dimensional representation.

6.3. Nerve Fiber Diameter Distribution Histograms of Urinary Bladder Innervating Nerves

6.3.1. Nervi Rectales Inferiores and Perineales

Figure 182C shows the nerve fiber diameter distribution histograms of the nervi rectales inferiores and perineales, for the 4 myelin sheath thickness ranges according to Figure 181. In Figure 182Cb the characteristic cross-section area and in Figure 182Ca the corresponding spectra are shown.

To analyze the diameter spectra of the nervi rectales inferiores and perineales, the spectra of a L4 ventral nerve root are shown for comparison in Figure 182Ba, with the identified three α-motoneuron peaks for myelin sheath thicknesses between 1.8 and 2.3μm. The nerve fibers of a L4 dorsal root are distributed differently over the four myelin sheath thickness ranges (Figure 182Aa). The majority of fibers is thin and has a thin myelin sheath and lay therefore in the histogram with sheath thicknesses 0.3 and 0.8μm. Since this spectrum shows two peaks, at least two populations of very thin afferent nerve fibers can be expected to exist in this range. For the mixed (afferent and efferent) nerve rectales inferiores and perineales (Figure 182Ca) the majority of fibers lie in the class with myelin sheath thicknesses between 0.3 and 0.8μm. Several fiber groups can be expected in this histogram. From the nerve fibers of the histogram with $1.8 \leq d < 2.3\mu m$ an estimation of the number of nerve fibers leading to the sphincter muscle fibers can be obtained. Sphincter muscles have muscle fibers mainly of the slow twitch type (FR) (Figure 8). These muscle fibers are innervated by α_2-motoneurons (Figure 8) with a nerve fiber diameter range of 9.5 to 12μm (hatched in Figure 182Ca).

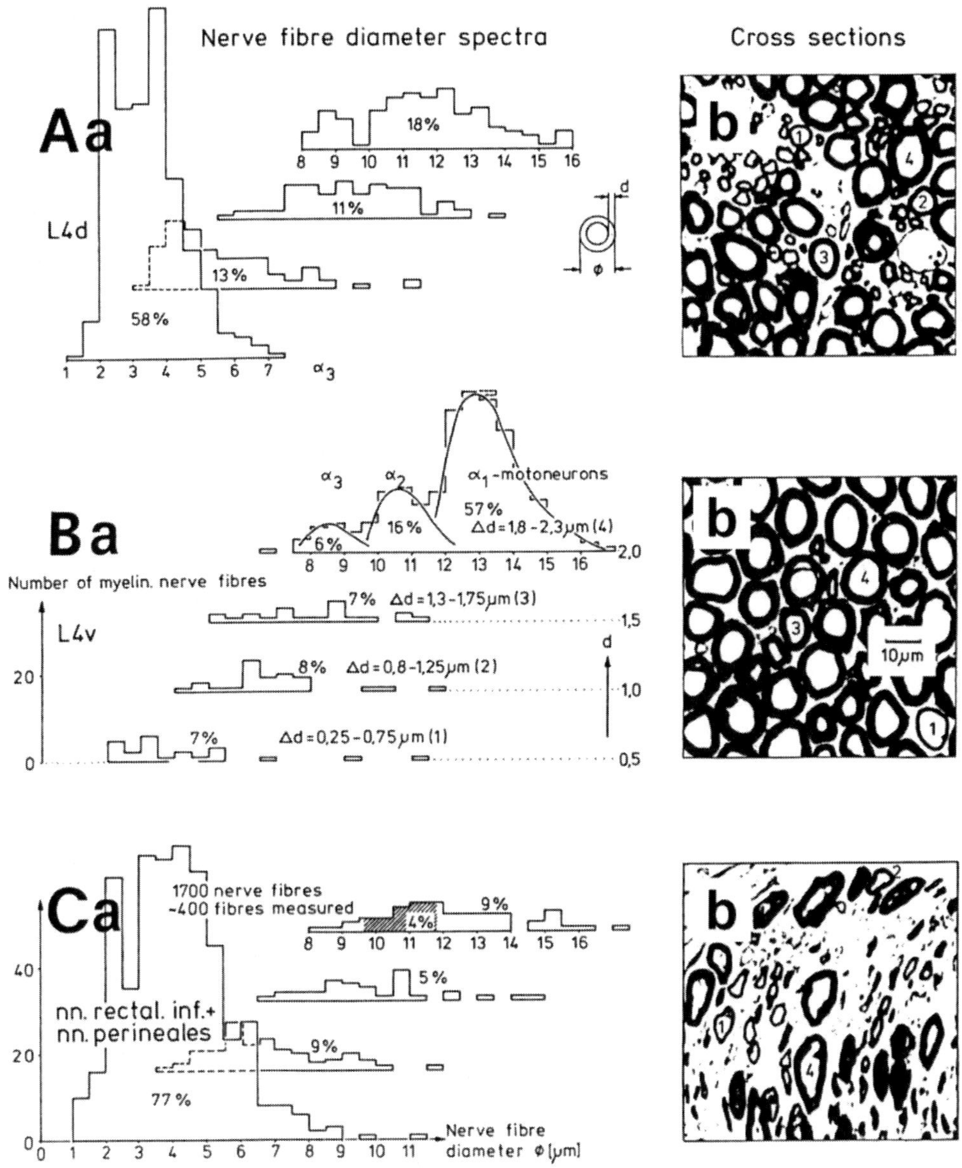

Figure 182. a. Nerve fiber diameter distribution histograms classified by 4 classes of myelin sheath thicknesses as indicated in Ba. % indicates percentage of fibers in classes or subgroups. b. Corresponding characteristic cross sections. A few fibers are numbered by their myelin sheath thickness range they belong to. Dimension scale for A, B, and C is drawn in Bb. 8% shrinkage correction. For the definition of fiber diameter Ø and myelin sheath thickness d see insertion in Aa. – A. Nerve fiber diameter spectrum of a 4th dorsal lumbar root of a 47-year-old female human cadaver removed 2 to 5 hours after death, 660 fibers were measured. B. Spectra of a 4th ventral lumbar root (same case as in A). 320 fibers were measured. In the myelin sheath thickness range $1.8 \leq d < 2.3\mu m$ the distribution curves of the 3 α-motoneuron classes are drawn into the histogram. C. Spectra of the Nervi rectales inferiors and perineales. Note, the majority of fibers (77%) has a very thin myelin sheath ($0.3 \leq d < 0.8\mu m$) with a relatively large amount of thick fibers. In the histogram of very thick myelin sheaths ($1.8 \leq d < 2.3\mu m$) the diameter range of α_2-motoneurons, to which sphincter motoneurons belong, is crosshatched (4% of the fiber).

In these nerve fiber diameter and myelin sheath thickness ranges are about 4% of the nerve fibers of the nervi rectales inferiores and perinales, namely 68 fibers (0.04 x 1700). This number of nerve fibers contains α_2-motoneurons and afferent fibers with the same nerve fiber diameter and myelin sheath thickness ranges. If one assumes a similar distribution of afferent and efferent fibers in this myelin sheath thickness range ($1.8\mu m \leq d < 2.3\mu m$), then one can expect to have an upper limit of sphincter α_2-motoneurons leading to the anal sphincter and possibly to the external bladder sphincter through the S3 to S5 ventral and dorsal roots of 30 fibers on one side. The number of additional necessary γ-motoneurons will not exceed 30 since there are no or only few spindles in the sphincter muscles [168]. The α-motoneurons to the pelvic floor muscles normally do not lead through the pudendal nerves [169,170].

6.3.2. Nervus Hypogastricus

Figure 183Aa shows the nerve fiber diameter spectra of the hypogastric nerve (plexus). The majority of myelinated fibers are contained in the spectrum with myelin sheath thicknesses between 0.3μm and 0.8μm. Most sympathetic efferent fibers will be unmyelinated since the hypogastric nerve lies caudal to the sympathetic chain (Figure 152). The two nerve fiber peaks at about 2.3 and 4.3μm lie at similar diameters than the two peaks of afferent fibers in the L4 dorsal root of Figure 182Aa and contain therefore most likely afferent fibers. It is likely that a few preganglionic parasympathetic fibers will be intermingled. Their diameters lie also in that range (Figure 183Ba). The hypogastric nerve contains more thinly myelinated thick afferents in comparison with the afferents of the L4 dorsal root (Figure 182Aa; $0.3\mu m \leq d < 0.8\mu m$).

6.3.3. Pelvic Splanchnic Nerves (nn. Pelvini)

The pelvic splanchnic nerves of Figure 183Ca have nearly no fibers in the spectrum for myelin sheath thicknesses between 1.3 and 1.8μm. The fibers in the myelin sheath thickness ranges $0.3\mu m \leq d < 0.8\mu m$ and $0.8\mu m \leq d < 1.3\mu m$ mainly represent parasympathetic afferent and efferent fibers. For the efferent parasympathetic fibers this becomes clear by comparing Figure 183Ca with Figure 183Ba. For the afferent parasympathetic fibers such direct comparison is not possible. The large numbers of thick fibers in the myelin sheath range $1.8 \leq d < 2.3\mu m$ (hatched in Figure 183Ca) is unexpected. It is likely that these fibers are all of parasympathetic afferent type. One would expect the thickest parasympathetic afferents innervate mechanoreceptors, flow receptors, Pacinian corpuscles (distributed over large areas in the tissue [171,172]) and tension receptors in the detrusor wall. Their diameters probably range up to values of about 12μm (Figure 7). The largest at 11.3μm and the existence of nerve fibers with diameters larger than 12μm suggest that there are also motoneurons and may be a few spindle afferents in the spectrum with myelin sheath thicknesses between 1.8 and 2.3μm. These thick nerve fibers could innervate the external urethral sphincter.

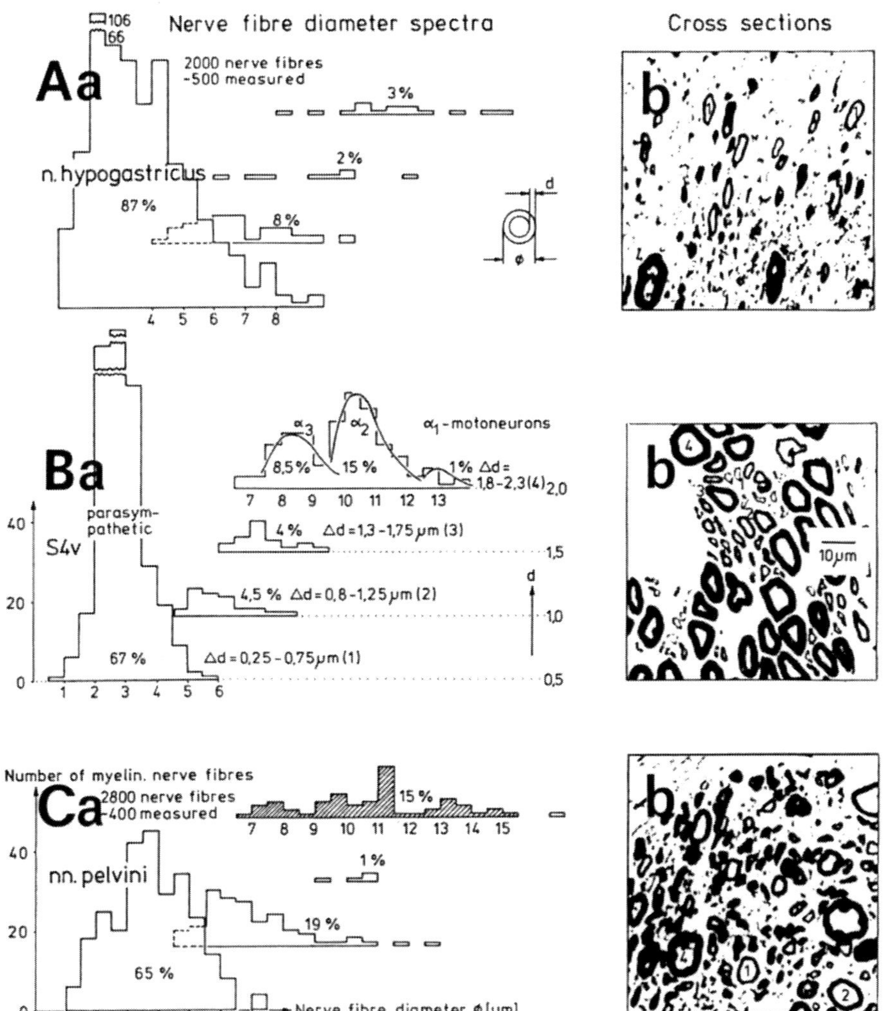

Figure 183. a. Nerve fiber diameter distribution histograms of the Nervus hypogastricus, a S4 ventral root and the Nervi pelvini classified by 4 ranges of myelin sheath thicknesses, indicated in Ba. % indicates percentages of fibers in classes or subgroups; b. Corresponding cross-sections. A few fibers are numbered by the class of myelin sheath thickness they belong to. Dimension scale is drawn in Bb. 8% shrinkage correction. – A. Nerve fiber diameter spectra of the Nervus hypogastricus. Note, the nerve fiber diameter histogram class $0.3 \leq d < 0.8\mu m$ (87% of the fibers) shows a large proportion of thick fibers. B. Nerve fiber diameter spectra of a 4th ventral sacral root. In the distribution histogram of class $1.8 \leq d < 2.3\mu m$ the distribution curves of the 3 α-motoneuron classes are drawn. C. Diameter spectra of the Nervi pelvini. Note, 15% are thick and have a very thick myelin sheath (class $1.8 \leq d < 2.3\mu m$, cross-hatched). α_1, α_2, α_3, SP1, T1 = peak group diameters of α-motoneurons, primary spindle afferents and T1 skin afferents.

6.4. Number of Efferent Parasympathetic Fibers in the Pelvic Splanchnic Nerves

The parasympathetic peak of S4 ventral root fibers of Figure 183Ba contains about 380 fibers. The number of parasympathetic efferent fibers in the pelvic nerves on one side is therefore about 800, if one assumes that parasympathetic fibers are contained in three roots

(S3, S4, S5 or S2, S3, S4) and that 20 to 30% of the fibers are ventral root afferents ((380 x 3) x 0.7 ≈ 800). Taking the value for myelinated fibers in the pelvic nerves (nn. pelvini) (Figure 152) and disregarding the existence of possibly a few thick somatic motoneurons, then one finds that among myelinated fibers in the pelvic nerves about 20% are efferents. In the monkey it was found that the parasympathetic nucleus on one side contains about 410 motoneurons [173].

6.5. Innervation of the External Bladder Sphincter

In the pelvic splanchnic nerves and in the hypogastric nerve (hypogastric plexus) are thick fibers with a thick myelin sheath ($1.8 \leq d < 2.3\mu m$) (Figure 183). At least for the pelvic nerves it is unlikely that all these fibers are afferent fibers. In the pelvic nerves these fibers make up 15% (hatched in Figure 183Ca). The possibility that a thin direct muscle branch to the pelvic floor [174,175] was removed with the pelvic nerves is unlikely, since the direct branch to the pelvic floor tends to lie more laterally and the pelvic nerves more medially [176] and such thick fibers were found in all sub-nerves. It is possible that these thick fibers of the pelvic nerves innervate the external bladder sphincter. It has been reported that the external bladder sphincter could also be innervated through a direct branch from the S3 and S4 roots [175] or via the pelvic nerves [177]. In five male cadaver dissections, the Author could not find the branch from the pudendal nerve or nervus dorsalis penis leading to the external bladder sphincter, situated caudally to the prostate gland. There is uncertainty of the innervation pathway to the external bladder sphincter. Caution is advised in any attempt to learn more about the innervation of the human urinary bladder through comparison with animal data, since it was reported that similarly named muscles (for our purpose the external sphincter muscle) may not have homologous functions in humans and cats [178]. The representation of the bladder in the sacral nerve roots is different in dogs and humans [179-181].

6.6. Thickness of Nerve Fibers at Different Distances from the Soma (Tapering of Nerve Fibers)

Nerve fiber diameter distributions of nerve roots and peripheral nerves (outside the spinal canal) were compared here, assuming that the nerve fibers have the same diameter along their length. If the peaks in Figure 183Ca at diameters of 7.8 and 9.8μm ($1.8 \leq d < 2.3\mu m$) are really the distribution peaks of α_2 and α_3-motoneurons, then these values would indicate that the motoneuron axons are distal slightly thinner.

In the comparisons made here it was assumed that the reduction in nerve fiber diameters at distal parts is small in comparison to other errors or variations. The reduction of nerve fiber with increasing distance from the soma is small (0.2%/per 13cm) [125,14].

6.7. Tapering of Afferents, γ-Motoneurons and Preganglionic Parasympathetic and Sympathetic Fibers

The rostral and caudal nerve-fiber diameter distributions of the same root (Figure 184) and other cases seem to indicate, that the γ-motoneurons and the preganglionic parasympathetic fibers taper only little or not at all. The tapering of the preganglionic sympathetic fibers leaving the spinal cord via the TH10 to L2 roots is more difficult to measure since these roots are much thicker than the lower sacral roots in which the parasympathetic fibers are contained. Since however, the preganglionic sympathetic and parasympathetic root fibers had very similar nerve-fiber diameter distributions rostral and caudal (Figure 185), it seems that also the preganglionic sympathetic fibers (contained even more caudally in the ramus albeus), taper only very little or not at all. The much thinner and unmyelinated postganglionic sympathetic fibers contained in the ramus griseus have not been measured here.

6.8. Unmyelinated Nerve Fibers in a Ventral Sacral S5 Root

In the cross-sections of the ventral S5 root of the paraplegic patient 5 (24 years of age) unmyelinated nerve fibers were observed (Figure 186A).

Because of the existence of ventral root afferents in the lower sacral nerve roots, the finding of unmyelinated fibers was to be expected. Since in different cross-sections there seemed to be sometimes mainly thick or mainly thin unmyelinated fibers, mean diameter of unmyelinated fibers were measured, and diameter distributions were constructed (Figure 186B). The distribution in one area of unmyelinated fibers showed three peaks (solid line distribution) at 0.25, 0.95 and 1.2 μm. In another part of the cross-section with mainly thick unmyelinated fibers (dotted line) the distribution showed only peaks at 0.95 and 1.2 μm. Because of the 3 distribution peaks at least 3 groups of unmyelinated nerve fibers seem to exist in the lower sacral nerve roots, which may belong to the somatic and parasympathetic nervous system divisions.

6.9. Numbers of Unmyelinated Nerve Fibers in Relation to Myelinated Fibers

In six areas of the ventral S5 root of the paraplegic patient 5, the number of unmyelinated fibers was measured. The fiber number varied quite much according to whether thinner fibers with a group diameter around 0.25 μm were contained or more thickly unmyelinated fibers were contained with probable group diameters of 0.95 or 1.2 μm. In the chosen areas (approx. 23 μm x 17 μm), the number of unmyelinated fibers ranged from 42 to 173, and the number of myelinated fibers from 10 to 16. The relation between unmyelinated and myelinated fibers ranged from 3.5 to 14. The mean ratio was 6.6. The ratio between unmyelinated and myelinated fibers depended therefore strongly on the number of thinly myelinated fibers contributing to it.

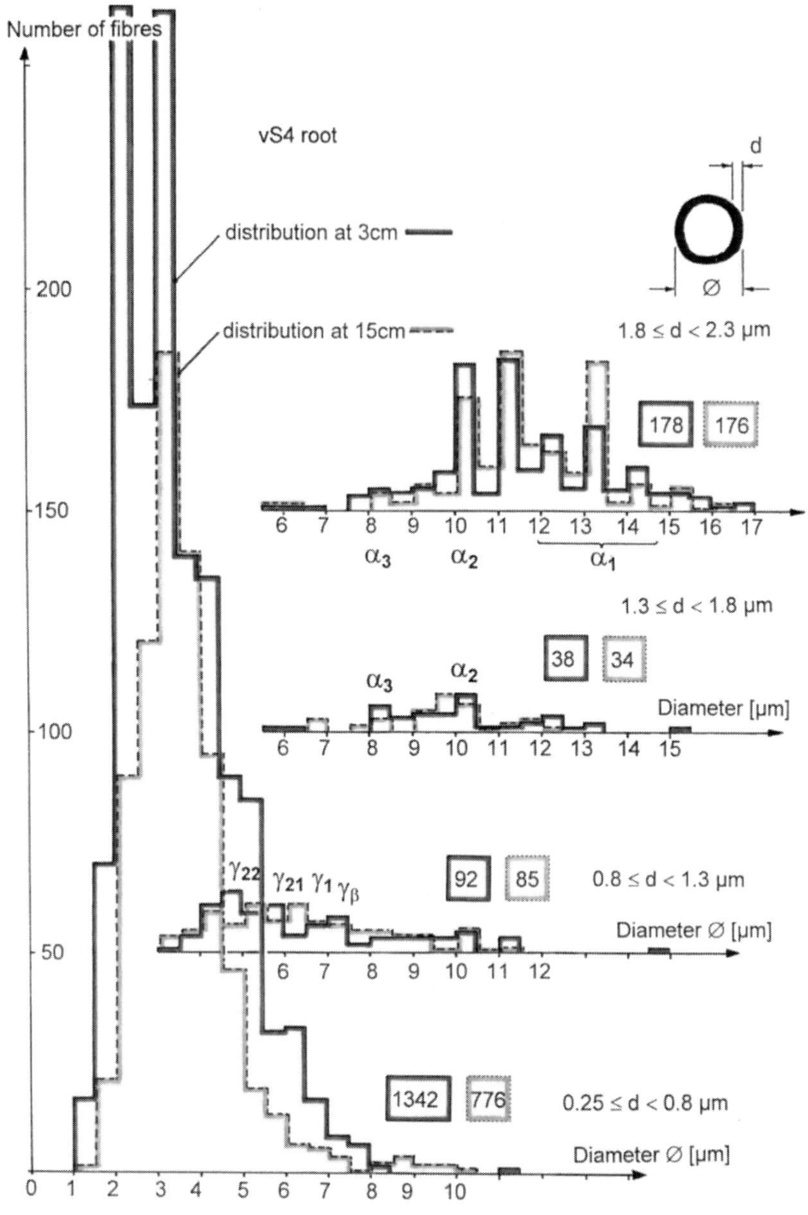

Figure 184. Nerve fiber diameter distributions for 4 myelin sheath thickness ranges 'd' of the rostral (solid line) and the caudal (dashed line) cross-section of a ventral S4 root. Nerve fiber groups are partly marked at the corresponding distribution ranges or peaks. Figures in the solid and dashed line rectangles give fiber numbers from the fiber distributions of certain myelin sheath thickness ranges rostral and caudal. α_1, α_2 and α_3 = α_1 (FF), α_2 (FR) and α_3-motoneurons (S). γ_1 = dynamic γ-motoneuron. γ_{21}, γ_{22} = static γ-motoneurons.

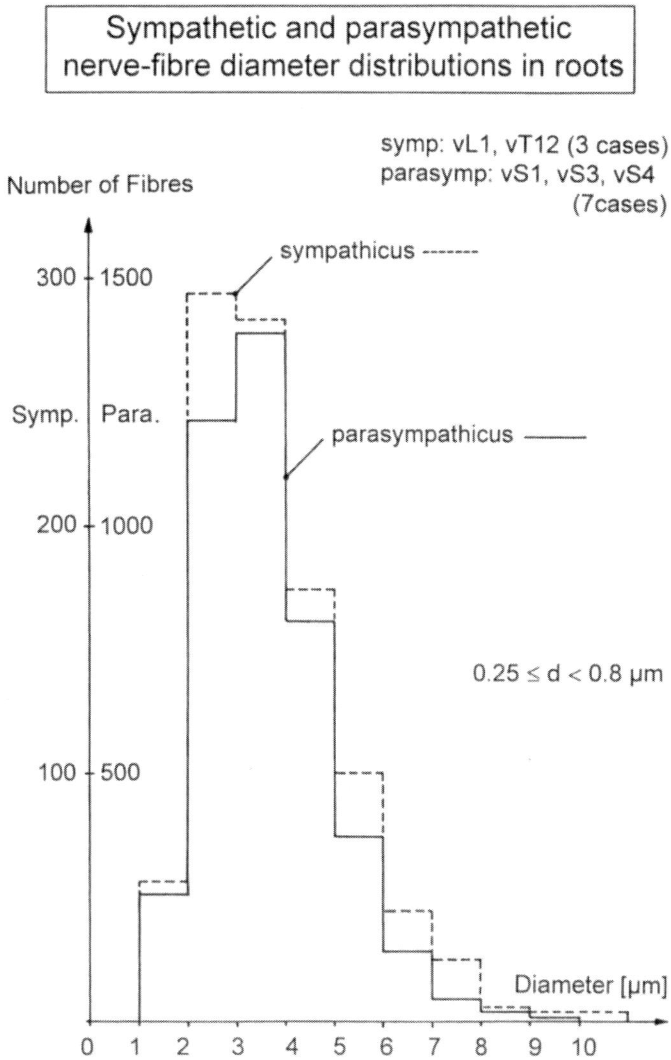

Figure 185. The sum of sympathetic (dashed line) and parasympathetic (solid line) efferent nerve-fiber diameter distributions of several roots indicated in the upper right Figure. Myelin sheath thickness between 0.25 μm and 0.8 μm. Note that the sympathetic and parasympathetic distributions are very similar. Sympathicus (symp): vL1, vL2 (3 cases); Parasympathicus (para): vS1, vS3, vS4 (7cases).

6.10. The Need for More Human Functional Anatomy

As was shown above, the number of unmyelinated nerve fibers exceeds the myelinated by a factor of 6.6. Nothing was discovered concerning the self-organization of the human nervous system with respect to unmyelinated nerve fibers which most likely have small cell bodies. It can be expected that the autonomic nervous system significantly contributes to the functional repair of the SCI and brain. More detailed anatomy data are needed of the human nervous system especially with respect to unmyelinated nerve fibers.

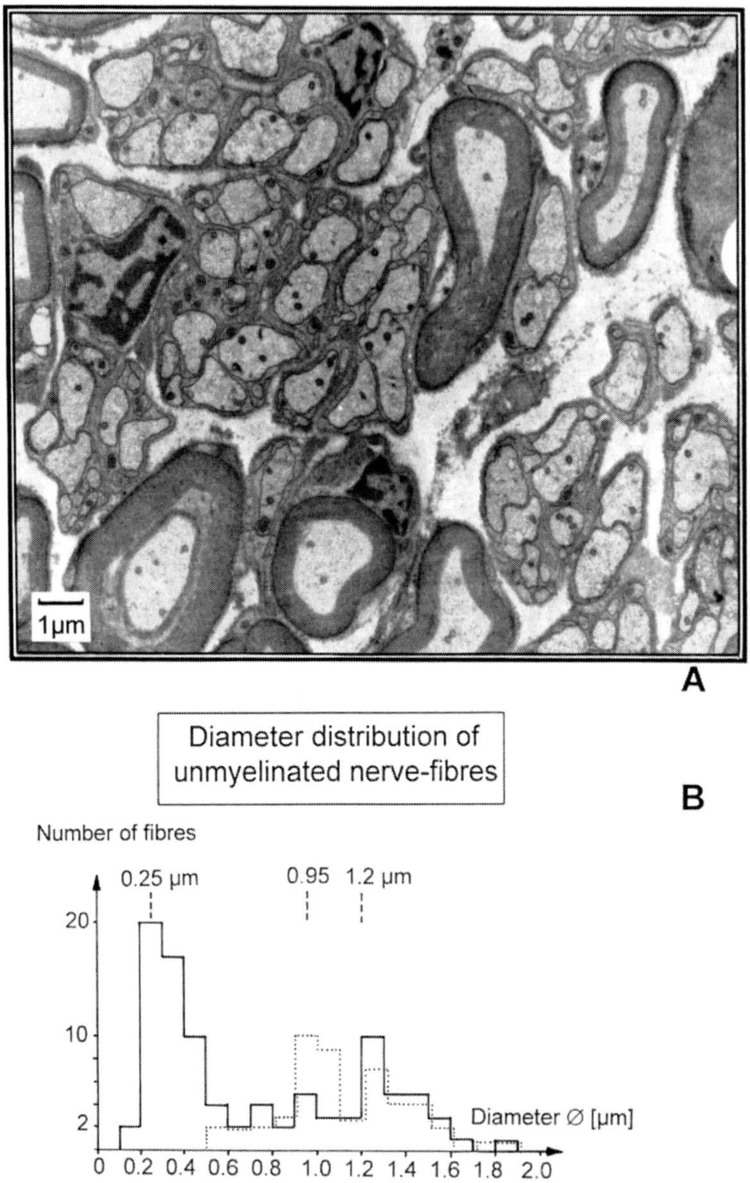

Figure 186. A. Section of a ventral sacral S5 root of a 24 years old female paraplegic patient (para 5). Thin and thick unmyelinated nerve fibers can be seen besides myelinated nerve fibers. B. Nerve fiber diameter distributions of unmyelinated nerve fibers. The solid line distribution was taken from an area which the section in A was a part of. The dotted line distribution stems from an area, in which mainly only thick myelinated fibers were present. Note that the peak at 0.25 µm is missing in the distribution marked by the dotted line.

Chapter VIII

Building of New Neurons in the Human CNS upon Long-Term Therapy

1. Building of New Motoneuron in the Human Spinal Cord to Improve Finger Function

1.1. Structural and Functional Repair to Improve Finger Functioning

Cervical spinal cord injury (SCI) can be partly cured (including movement, bladder, sexual, and blood circulation functions) via the administration of sustained 'Coordination Dynamics Therapy' (CDT). In incomplete cervical SCI, repair is mainly achieved by a functional reorganization including an improvement of the impaired phase and frequency coordination for the self-organization of the neuronal networks of the CNS. However, in motoric complete SCI at C5/6 levels, repair has to include structural repair. Limited regeneration of the cord, which started approximately a year after the accident, could be achieved in the patient Kadri following three years of optimal CDT. But a further problem in SCI at C5/6 levels is that also the motoneurons in the grey matter are destroyed at the injured site, with the consequence that the fingers cannot move any more. There is evidence that after three and a half years of optimal CDT, the fingers started to function again. From brachial plexus and cauda equina injuries it is known that hand and leg functions, respectively, start to re-appear two years after the accident. The growing of motoneuron axons, after injury, along the Bügners bands in peripheral nerves to re-innervate hand and foot muscles therefore needs at least two years. The reappearance of finger functions in the patient with the cervical SCI after three and a half years of optimal CDT indicates that the building of new motoneurons in the human spinal cord needed a year and a half (endogenous stem cell therapy) and the growth of the axons down to the muscles and the building of new synapses needed at least two years. The belief that motor functions can reappear in a few days or a few months in patients with SCI following exogenous stem cell therapy cannot be reconciled to the actual times needed for repair.

Above it was shown that if 50% of the cervical spinal cord was spared following injury then the patient could relearn walking, running and jumping (Figure 52) and could become continent within one year. The administered CDT could almost cure this incomplete SCI at C5/6 levels mainly by functional reorganization. The question is now, how many functions can be repaired, if the SCI is 10 times more severe, that means if approximately only 5% of the tract fibres of the spinal cord are spared? It was shown that in the patient Kadri with a motoric complete SCI at C5/6 levels breathing, cardiovascular performance and urinary bladder functions could fully be cured after three years of CDT. Also some motor functions the patient could regain such as trunk stability, breathing muscles, and leg functions.

Limited regeneration of the human spinal cord in motoric complete cervical SCI, which requires the growth of nerve fibres over the injury site, is not sufficient for repair. Due to the destroyed motoneurons and propriospinal neuronal networks at the injury site (Figures 148, 29), hand and finger functions are lost. The question is, can CDT induce the endogenous stem cell reserves to generate new motoneurons to improve finger functions? Since it has been reported that neocortical neurogenesis in humans is restricted to development [183,184], it can be expected that there is little or no neurogenesis in the injured human spinal cord.

It will be shown that in this patient with a severe cervical SCI some motor functions in both index fingers re-occurred after three and a half years of optimal CDT. This regaining of finger functions is important for three reasons. First, every little bit of repair of finger functions is improving the quality of life of patients with a complete cervical SCI. Secondly, scientifically such repair is of great significance, since this means that new motoneurons can be built in the human spinal cord. Thirdly, by what means can neurogenesis in humans be induced and enhanced and what is the timescale for, and the extent of, repair?

1.2. Repair Aims to Be Achieved in General

With CDT, various forms of cortical, brainstem, spinal cord and peripheral autonomic nervous system plasticity occur that contribute to compensatory recovery. The application of CDT to SCI has several aims: to bridge any cysts or cavities (see Figure 148; Figure 29D); to replace dead cells (for example, to provide new motoneurons and myelinating cells); to create a favourable environment for axon regeneration; and to functionally reorganize the structural repaired spinal cord and its controlling supraspinal centres for physiologic functioning. Major challenges for endogenous stem cell-mediated repair after cervical SCI, stimulated by CDT, are controlling the migration, self-renewal, survival, integration, and differentiation of endogenous cells, and the path finding of motoneuron axons from the cervical cord to the denervated hand and arm muscles (Figure 29). These cells should promote functional recovery by reconstituting damaged circuits, re-myelinating axons, and increasing plasticity and/or axon regeneration. CDT probably also induces axonal growth of ascending and descending spinal tracts.

Improvement following CDT occurs because after SCI, the spinal circuitry below the injury site does not become silent but maintains active and functional neuronal properties, and can respond to peripheral input from below the level of injury. Actually all the circuitries in and circumferential to the injury site are functionally integrated in CNS functioning (Figure 29D). The repair is most efficient if the activation is induced by physiologic patterns; and these physiologic patterns have to be generated in the patient's CNS and have to become stable and stored in the long-term memory. The stability of the patterns is achieved by training physiologic movement patterns and by training the coordinated firing of neurons, which became impaired following injury. One of the principal aims of human repair neurophysiology is to identify those movements, which have the highest capacity for repair with on-going CDT. Finger functions in motoric complete cervical SCI are trained in connection with integrated coordinated arm and leg movements (Figure 187F).

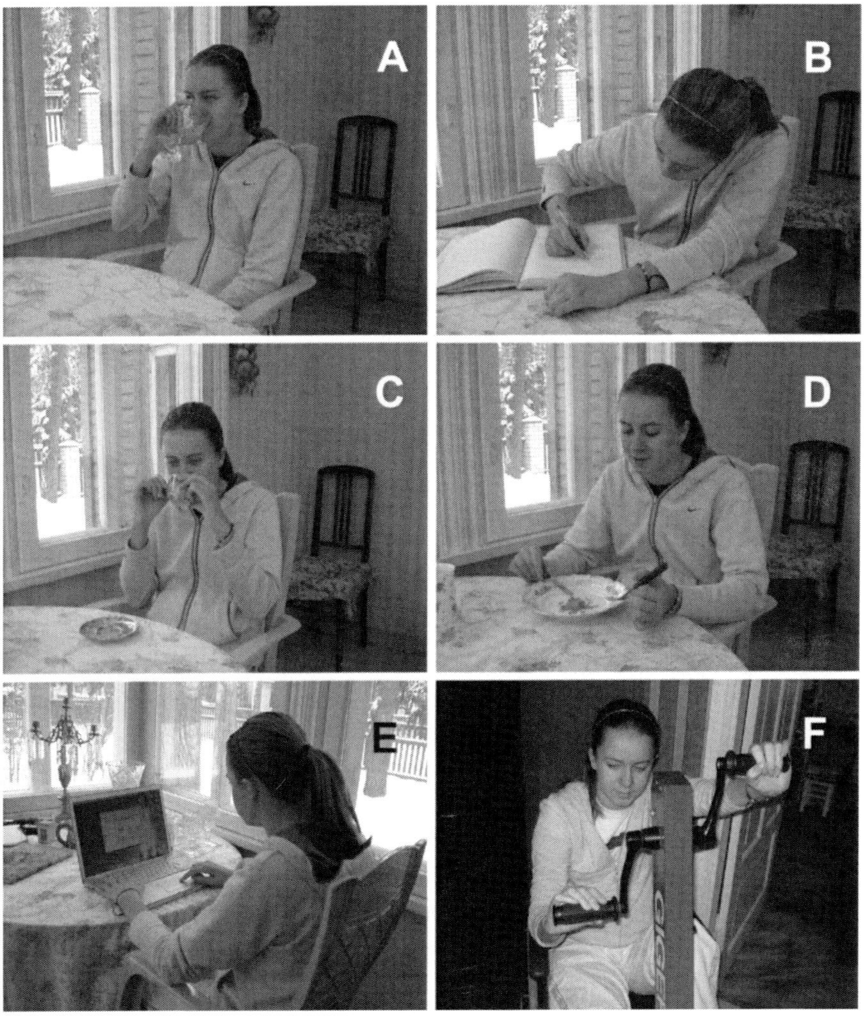

Figure 187. Finger functions of a patient with motoric complete C5/6 SCI achieved upon CDT. Even though the fingers have practical no power, they are useful in everyday life. Destructive manipulations (so called functional hand = shortening of finger tendons) are not needed for getting function (A-E). In 'F' the finger functions are trained in connection with coordinated arm and leg movements to activate the neuronal networks across the injury site (Figure 29D).

1.3. No Evidence for Spontaneous Regeneration in the Human Spinal Cord

In lower vertebrates the CNS fully regenerates following injury [185]. An evolutionary explanation for the 'loss' of CNS regenerative capacity in humans is that this is an undesirable by-product of gaining an complex nervous systems that selection pressures have favoured over time [107]. However, animals, unprotected by a caring society, may seldom have the time to recover from brain or spinal cord injury therefore obscuring the extent to which their nervous system can recover. CDT utilizes the complexity and variability of the human nervous system (namely the learning) to partly compensate for the 'loss' of CNS regenerative capacity.

1.4. Repair of Propriospinal Circuitry

In Kadri, I will concentrate here on the repair of propriospinal circuitry at the injury site. The repair of the propriospinal circuitry is of special interest in cervical C5/6 levels, since those motoneurons are destroyed which importantly innervate lower arm, hand, and finger functions. Further, to identify the repair mechanisms in the human spinal cord by the novel CDT is of high importance to further improve the efficiency of the treatment. The repair of finger and bladder functions has not been tackled by animal and biological research so far as can be seen from recent review articles [106,107,114,186].

1.5. Recovery and Repair of Sensory Function

Following three years of CDT, motor functions re-appeared in the rostral-caudal direction, indicating a limited regeneration of the human spinal cord. The recovery of sensory functions started in the caudal-rostral direction in the spinal shock phase, which lasted approximately three to four weeks. Two weeks after the accident, the patient felt first the Achilles tendons followed by feeling in the toes. After the termination of the spinal shock phase, the re-occurrence of the sensitivity was rather complicated with respect to topography because of the uneven distribution. The last location in which the patient got sensitivity back was the abdomen, at a time when bladder function was cured, two years after the accident (Figure 30).

The first sensation the patient experienced after the spinal shock was pressure and then touch. Later on, pain and temperature sensation reoccurred. The later re-appearance of pain and temperature feeling can be understood on the basis of functional anatomy. In Figure 29B the spared white matter is cross-hatched pictured. The distribution of spared white matter is taken from the MRI (Figure 143,148). The later reappearance of pain and temperature may mean that new sensory fibers were crossing the injury site adjacent to the spared sensory fibers, indicated in Figure 29B by open arrows. Looking at the cross-section of the human spinal cord (Figure 29A), it seems that the tractus spinothalamicus anterior, carrying pressure and touch information, was mainly spared by the injury. With the regeneration, parts of the tractus spinothalamicus lateralis, carrying pain and temperature information, were repaired. Growing of nerve fibers and perhaps homing of cells are attracted by the functioning fibers crossing the injury site. The funiculus posterior also seemed to have improved in its functioning. The spared white matter at the dorsal part of the cord cannot be seen well in the T2-weighted MR images in Figure 148. But on the T1-weighted images some spared connections can be seen between the proximal and distal spinal cord parts (Figure 143C).

The growing of nerve fibers across the injury site adjacent to the by the movement patterns activated fibers is pictured in Figure 29D by short arrows pointing into the white cavity adjacent to the longitudinal long arrow in and beside the spared tract fibers. It seems that training at the limit activates spared and newly regenerated fibers and stimulates new fiber growth. Target-derived growth factors seemed to have been released from the over-activated fibers in the rostral and caudal direction upon exercising for repair adapted movement patterns at the limit. Based on the reoccurrence of the reported functions, the mechanism of sensory and motor repair can be partly understood.

1.6. Partial Repair of Index Finger Function

After three and a half years of optimal CDT a complete different mechanism of repair became visible: the index fingers started to work again and this was probably achieved by the building of new motoneurons. Improvement of finger functions is important for everyday life (Figure 187).

The only hand functions the patient had regained within the first year following the accident (mainly spontaneous recovery) were the dorsal flexion and a weak plantar flexion. The dorsal flexion was achieved by the extensor digitorum muscle (Figure 188A,B,C), innervated by the nervus radialis (ramus profundus) and having its motoneuron cell bodies in the spinal segments C6-C8, which fits with the spinal cord injury level at C5/6 levels. The pripriospinal premotor network for activating the extensor digitorum is sited just above the injury cavity (Figure 148, Figure 29D) in the grey matter (butterfly). The weak plantar flexion is probably caused by the flexor digitorum superficialis, innervated by the median nerve, having its motoneuron cell bodies in the spinal segments C7-Th1. The very weak muscle power is to be expected, since most motoneurons will have been destroyed by the injury.

After three and a half years of therapy, the patient could rather suddenly repeatedly lift the index finger first on the right and shortly after also on the left side. No other fingers could be moved separately. The dorsal flexion of the index finger is activated by the musculus extensor indicis (Figure 188D,E), innervated by the radial nerve, ramus profundus, from the spinal segments C8-Th1. But these are the segments, which were destroyed and are forming now the cord cavity (Figure 148, Figure 29D).

Why did another extensor start to work and not another hand flexor (or the flexor digitorum superficialis got much stronger), as for example the musculus digitorum profundus, innervated by the median nerve and ulna nerve from the cord segments C6-Th1? If we look at the somatotopic ordering of motoneurons in the ventral horn of the grey matter, we can see that the extensors are sited most ventrally, whereas the flexors are sited more in the direction of the dorsal horn (Figure 29C). When the patient is performing coordinated leg and arm movements on a special device, including the fingers (Figure 187F), then the tractus spinothalamicus anterior and posterior are activated alongside the cavity (Figure 29D) and the working parts of the funiculus posterior (ascending activation) and some regenerated axons, activating trunk and some leg muscles (descending direction) across the injury site. Because the patient is holding the handle with the fingers during the coordinated arm and leg movements, the extensor digitorum is also activated and sensory input is reaching the cord from the fingers and hand. Grows factors could have been secreted from the sensory fibres in the tractus spinothalamicus anterior/posterior and the premotor network of the extensor digitorum to attract stem/progenitor cells for homing. The attraction for homing is indicated by open (Figure 29B,C) and small (Figure 29D) arrows. The flexors, on the other hand, are far away from supposed target-derived growth factor concentrations (Figure 29C), and also, in the middle of the spinal cord cavity there are no mechanical structures, as for example basal membranes where cells could attach to. According to this reasoning, the flexors can only come much later with further building of new propriospinal circuitry. It is therefore important that the patient understands the therapy and is activating all by the injury spared and newly formed fibers and premotor networks circumferential to the injury site (the cavity) (Figure 29D) to prepare the way for homing of cells and growing of axonal and dendritic arbors by performing all possible movements, especially the coordinated and integrative ones (according to the System Theory of Pattern Formation and the coincidence principle (Figure 11)). The guidance mechanisms at the injury site to support 'selection-at-a-distance' will be further analyzed further below.

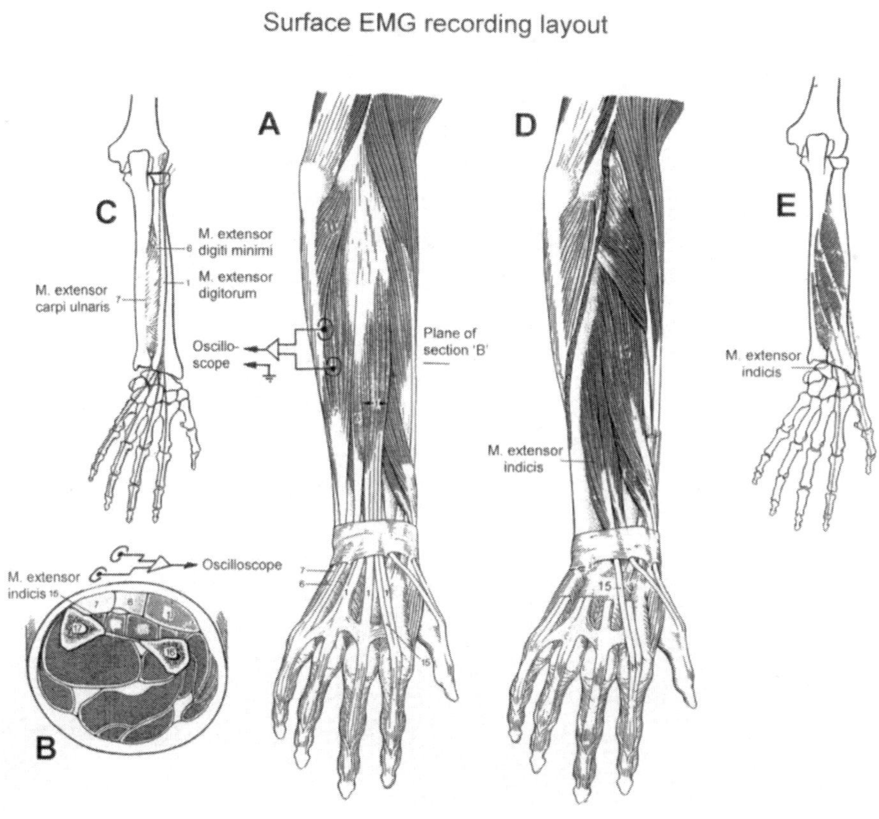

Figure 188. Drawings of lower arm muscles with positioned surface electrodes for differential EMG recording.

1.7. Is the Index Finger Function a Separated Function of the Extensor Digitorum Muscle?

It is possible that the index finger function is due to a separation of function from the digitorum muscle? However, the index finger started to function suddenly. Secondly, no other finger could perform an independent dorsal flexion. Thirdly, since the dorsal flexion of the hand is generated by the last working motoneurons above the injury site in the caudalion direction, the premotor network of the extensor digitorum will also have been damaged, and with a rather simple connectivity it may not be possible to separate single finger functions from this muscle, which is probably possible in healthy humans.

1.8. Three and a Half Years to Regain Index Finger Function

Why did it take three and a half years of optimal CDT to regain some finger functions? From brachial plexus injuries it is known that functions start to re-appear after two years. This results to a regeneration speed in the PNS of less than 1mm/day ($v = s/t = 500mm/700days \approx 0.7$ mm/day) for the thickest, fastest regenerating axons. To get index finger function back, new motoneurons must be built and their axons must find their way to the rootlet, may be guided by activated axons leading to the extensor digitorum, and grow along the Bügners bands to the flexor indices and form there functional motor endplates (Figure 29B). The muscle fibers of the extensor indices were most likely atrophied and needed another three months to regain full power.

The building of new motoneurons from endogenous stem cell reserves, therefore, needed approximately one year to eighteen months. A further two and a half years were needed to get function back because of the slow growing speed and the long growing distance required of 0.5m. Some physicians make patients believe, that following the exogenous stem cell therapy, they can get functions back after a few days or months. Such unscrupulous promises are not supported by the data of human physiology.

1.9. Surface Electromyography (sEMG) of the Extensor Indices and Tibialis Anterior Muscles

To further understand the repair pattern of the spinal cord in this patient, electromyographic activity was recorded from the extensor indices (above the injury) and the tibialis anterior muscle (below the injury) (Figure 189). The electrode arrangements for indifferent recording can be seen in the Figures 140A, Figure 174 (tibialis anterior) and Figure 188A,B (extensor indices).

As can be seen, single motor unit potentials could be recorded from the extensor indices (Figure 189A,C) and the tibialis anterior (Figure 189D) muscles. Even though the extensor indices is lying below the extensor carpi ulnaris (Figure 188A,B,D) and the tibialis anterior is sited directly under the skin, the largest motor unit action potentials of the extensor indices have 400μV (Figure 189B,C) double as high amplitude than those of the tibialis anterior (200μV; Figure 189D; note different calibrations). Physiologically, a motoneuron of the extensor indices innervates approximately twenty muscle fibers and in the tibialis anterior approximately two hundred. The single motor unit action potentials recorded from the tibialis anterior muscle should therefore be much larger than those from the extensor indices, namely over 1000μV (Figs.13,14); but they were not. The unphysiologic small amplitude of the motor unit action potentials of the tibialis anterior and the different onset times of function are giving further insight into the regeneration process taking place in the spinal cord.

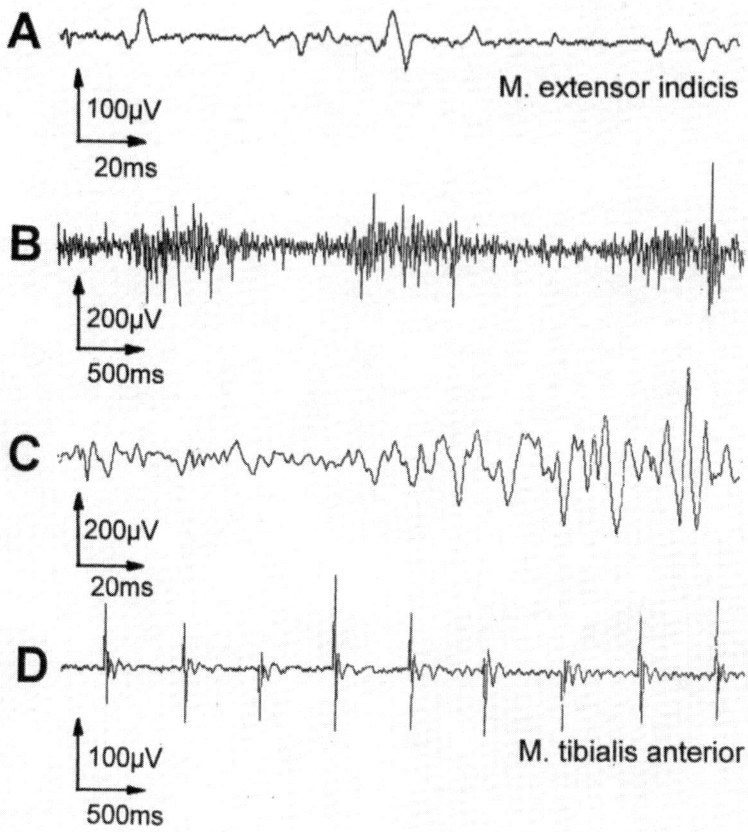

Figure 189. Surface EMG recordings from the extensor indices (A,B,C) and tibialis anterior (D). Note the different calibrations and that the activity of the extensor indices (B,C) has a higher amplitude than that of the tibialis anterior (D). The motor unit in the tibialis anterior fires at a frequency of 1.8 Hz; this frequency falls into the frequency range of oscillatory firing α_3-motoneurons innervating under physiologic conditions slow (red) muscle fibers.

1.10. Evidence for Network But Not Tract Fiber Repair

For the volitional activation of the tibialis anterior, some regeneration of the spinal cord was necessary. Some axons had regenerated over the injury site to reconnect the distal spinal cord part. This regeneration was faster than the building of new motoneurons and the growing of their axons along the Bügners bands in the median nerve to re-innervate the musculus indices to induce finger function. But why did the tibialis anterior show only small motor unit action potentials, which cannot be explained only by atrophied muscle fibers?

Human muscles have FF, FR, and S-type muscle fibers, which are innervated by α_1, α_2, and α_3-motoneurons respectively (Figure 8). The large, fast conducting α_1-motoneurons innervate the FF-type muscle fibers, show large motor unit action potentials and are integrated only in small networks. The thick tract axons from the cortical pyramid cells will preferentially innervate these α_1-motoneurons to generate high muscle power fast. The FR-type motor units generate smaller motor unit potentials than the FF-type motor units and the innervating α_2-motoneurons are integrated in larger networks and have specific premotor networks, generating for high activation stable oscillations. The S-type motor unit action potentials are most likely small; the innervating α_3-motoneurons can be activated polymodal and are integrated in large neuronal networks.

The small motor unit action potentials recorded from the tibialis anterior indicates therefore, that the recorded activity was mainly generated by the activated smaller α_3, and α_2-motoneurons, integrated in larger and more complex networks. An oscillation frequency of 1.8 Hz of a motor unit action potential in the tibialis anterior muscle (Figure 189D) supports that view. α_3-motoneurons fire oscillatory in the range of 1.8Hz (Figure 8). A change of the commitment of motoneurons, due to a partial denervation is also possible. This means that the regenerating axons over the injury site were not giving rise to a partial regeneration of the pyramidal tract, but contributed to a re-innervation of the neuronal networks below the injury site. After three and half years of optimal CDT, the repair was probably mainly achieved by network repair, including a substantial functional reorganization for generating physiologic functioning, and not by the repair of tracts. Further, regeneration of the axons of the pyramid cells over the injury site down to the motoneurons, sited in the lumbo-sacral cord segments, would probably take much more regeneration time, if indeed it is possible at all. The regeneration of the motoric complete cervical SCI had been so far achieved by the growing of some axons over the injury site, including the building of new neurons and formation and activation of new pathways, combined with a substantial functional reorganization to regain physiologic functions.

At this stage of spinal cord repair it is unclear whether the regeneration-associated genes have the capacity to partly restore original spinal cord structure. However, in order to regain finger functions, new motoneurons are needed; network repair through the growth of axon and dendrite arbors is not sufficient.

1.11. Spinal Cord Repair through CDT and Increasing Complications in the Absence of CDT in C5/6 SCI

It is generally known that in severe cervical SCI, aside from some spontaneous recovery in the first year following the injury, there is no recovery of function in the following years of injury without proper therapy. There is therefore no need for controls. However, there will always be controls as many patients do not receive CDT.

Another male individual suffered a very similar cervical SCI, except that a few more fibres were spared. One year post-injury, the patient heard about CDT and his coordination dynamics were measured. However, for organizational reasons the patient could not undertake supervised CDT. He was advised as to what he could do independently at home with help from his family. He received funding to buy the special CDT device on which he regularly exercised.

He was also received funding to perform upright movements on a device called 'Locomat'. He was able to train crawling by himself. The patient reported that his motor functions improved a little.

Two and a half years after injury the patient contacted the Author for a consultation. The patient Kadri was also present; they know each other. Kadri was optimistic and feels that she is healthy and does not need a physician and only needs further motor functions to become independent. The control patient, sitting opposite to her had a bladder infection due to the lack of bladder repair. The measured coordination dynamics clearly indicated that he lost functions since his last consultation the injury level had become a little more severe. He had lost some hand functions. The MR images showed new cavities in the cervical cord above the injury and the enlarged spinal cord pressed against the bones of the spinal canal, which impairs cord functions. Surgeons put forward the possibility of placing a tube into the cord to reduce the pressure of the cerebrospinal fluid, probably occurring because of impaired movement of the cerebro-spinal fluid. However, placing a bypass into the cord (in similarity to bypasses in increased brain pressure) is risky, because the spinal cord at the injury site may be further damaged by this procedure, including newly regenerating fibres. The patient also was not keen to opt to have the metal fixation removed to increase neck mobility and perhaps improve the circulation of the cerebrospinal fluid to stop further cyst formation because his friend had lost his speech in such a fixation removal. Kadri suffered partial Horner syndrome following the fixation removal.

2. Re-Learning and Long-Term Memory

2.1. Re-Learning of (Movement) Patterns and Re-Learning of Disturbed Pattern Stability

The above case report shows that there is regeneration of the injured spinal cord, including growing of axonal and dendritic arbors and building of new motoneurons. A better understanding of the underlying repair mechanisms may accelerate the learning process of this movement-based learning therapy.

The new motoneurons and their premotor neuronal network can be generated in the spinal cord most likely from stem/progenitor cells, perhaps in similarity to the increase of neural network complexity during development. But how would the newly created cells with their new dendrites, axons, and synapses, functionally integrate into the spinal cord and supraspinal CNS networks without disturbing the existing networks. And indeed, changes in the circuitry disturb the self-organization of the neuronal networks, quantified by the coordination dynamics measurements, shown in Figure 30. Each enlargement of the network, related to the repair of a certain function, increased the instability of network organization, quantified by the increase of the coordination dynamics values. By the movement-based learning, however, the stability was re-established following each instability increase. The movement-based learning therefore serves two functions: it repairs the network patterns by learning and it repairs the instability of network organization following each network repair (due to network enlargement). Both repair and learning strategies may involve different remodeling strategies.

These two repair functions, namely functional/structural rewiring and pattern stability improvement, are pictured in the pattern attractor layout for jumping on a springboard (Figure 18). Due to spinal cord or CNS injury, the geographical landscape of pattern formation is deformed (pathologic, Figure 18B,C) and the pattern states (the balls) show large fluctuation (long arrows indicate large network fluctuation). Following repair, the attractor layout becomes physiologic again, the attractor basins (potential wells) get deeper, and the network state fluctuations become smaller (short arrows) (Figure 18A).

2.2. Neurogenesis and Structural Repair

From the beginnings of modern neuroscience in the late 19th century, it was assumed that the human CNS became structurally stable after development and remained that way throughout life. A fundamental aspect of this stability was that no new neurons were added to the CNS in adulthood.

One reason why neuronal replication in the adult was considered unlikely was the complexity of most neurons; with their highly branched dendrites and polysynaptic axonal combinations, they were considered to be terminally differentiated and unable to re-enter the cell cycle and divide, strictly from a mechanistic view. This mechanical problem of accommodating the dividing neurons has been overcome by the discovery (in animals) that neural stem cells exist throughout life in the adult brain and can renew and can give rise to new neurons, astrocytes, and oligodendrocytes, just as in the developing brain [187].

A second problem was conceptual. If neurons were able to divide, how would the newly created cells with their new dendrites, axons and synapses, functionally integrate into the brain without disrupting existing circuits? As shown in this human research project, newly created cells with their neuritis can be integrated in existing circuits after injury by movement-based learning.

Above it was measured that new neurons could be built in the human CNS. The building of new motoneurons, innervating the index finger and their full integration into the adult CNS took three and a half years. There is therefore strong evidence of postnatal neurogenesis in man, albeit to a very limited extent. The plasticity and adaptability for the integration of the new motoneurons cannot only be accomplished by "strengthening synapses" without requiring substantial reorganization. The structural repair plasticity in the adult human CNS to make the motoneurons functioning most likely includes synapse formation and elimination, outgrowth and retraction of dendrites and axons, and the generation of the appropriate interneurons, including Renshew cells. At least a simple premotor network has to be built to make the new motoneurons function physiologically.

Was this structural repair induced in response to learning, or only with injury? The movement-based learning therapy probably includes both, a response to injury and a response to learning. Since long time periods are needed for the repair, presumably longer-range connections are also built.

The appropriate behavioural and environmental stimulation to achieve this structural repair was the movement-based learning therapy, named Coordination Dynamics Therapy (CDT). With non-optimal movement-based learning, received by the control patient, there was only slow and inefficient structural repair, including a likely low birth rate of new cells with a low rate of survival. The site of origin of these adult-generated motoneurons (and other neurons) is unclear. If not generated in the spinal cord, there is no information concerning how these new neurons migrate from their birth site to their ultimate functional destination.

The integrative functions of the human CNS can be put into the form equations of motion of the collective variables $dX/dt = F_{intr}(X) + \sum c_{inf} F_{inf}(X,t)$. With the behavioural and environmental information, F_{inf}, the intrinsic dynamics, F_{intr}, can be changed and repaired.

Within the System Theory of Pattern Formation for Repair, the change of the intrinsic dynamics includes also structural changes. This system approach for understanding CNS self-organization offers some understanding how the human CNS organizes itself and an understanding of the stability of network states can be obtained. However, this theory does not help to find the environmental and behavioural information with which it is possible to repair the human CNS. For repair the whole knowledge of human, animal, and biological neurosciences seem to be needed.

The rate of neurogenesis in the adult human is probably low when compared with the rate during development. If the adult-generated cells were to have the same relative influence as those generated in earlier life, then it is difficult to imagine how they would be important. If on the other hand, these newly build cells have the unique property to be 'placed' into sites where they are urgently needed for the most important repairable functions, then their impact on the structural repair could be substantial.

Therefore structural repair guidance mechanisms are needed in place to support 'selection-at-a-distance'.

2.3. Guidance Mechanisms

If there were no guidance mechanisms during the structural learning process in place to support 'selection-at-a-distance', an exhaustive search through the space of all accessible wiring diagrams would be required; such exhaustive physical research is intractable [130]. The guidance of grows of axons and dendrites is achieved through activity-dependent release of diffusible factors from axons (and dendrites) and possibly through changes of electromagnetic fields, generated by the propagation of action potentials in spared axons of tracts (white matter) and neuronal networks (grey matter) and possibly electrotonically conducted post-synaptic potentials in dendrites.

For physiologic (re-)wiring physiologic activity-dependent stimulations are most likely needed to generate physiologic concentration distributions of target-derived grows factors (with concentrations and gradients according to geographical landscapes). Similar arguments may hold for electromagnetic field densities. Only physiologic movement patterns were administered to Kadri in order to generate rather physiologic network organizations around the injury site (Figure 29D); an attempt was made to include the injury site into the network organizations. The structural repair seemed to build up from the spared tract fibres and neuronal networks around the injury site (Figure 29B,C,D). Untreated injuries may be associated with pathological network changes, which might release mechanisms of CNS rewiring that lead to pathologic functioning (pathologic network states, spasticity).

2.4. Structural Repair by Learning in the Long-Term Memory

Coordination Dynamics Therapy (CDT) is a movement-based learning therapy. The functional and structural repair is probably achieved in response to prolonged injury and prolonged learning. The repair for activating the musculus indices needed three and a half years of learning. The experience is that re-learned functions are not lost again. The re-learned finger function became an acquisition of a long-term memory and is an irreversible step in cell differentiation. The building of new motoneurons (and probably other neurons) must be addressed in the context of what dictates the limits of the brain's capacity to learn.

It was proposed [131] that "permanent changes in gene expression may be better suited to determine in a permanent manner, the number and properties of the synapses of a neuron and thereby the way in which that neuron interacts with others. After such change in gene expression, much of the enormous plasticity normally afforded by synapses may come to an end. The cell, now a reliable repository of information, may have lost the ability to acquire new long-term memories. According to this outlook, acquisition of a long-term memory is akin to an irreversible step in cell differentiation, leading to a lasting change in circuit properties. The acquisition of a long-term memory may be one thing, previous learning standing in the way of new learning. Those neurons, and not their individual synapses, would be the unit of long-term memory, and the number of neurons that can be modified in this manner would determine how much can be learned and remembered. The replacement of older neurons by new ones that are, in turn, transient and replaceable, offers a vehicle for the constant and spontaneous rejuvenation of key CNS circuits".

The repair of damaged CNS circuits by learning for long-term memory uses the replacement of older neurons (the damaged ones) with new neurons. In this view the structural repair, including the building of new neurons, used the replacement theory of learning in the long-term memory. The re-learning of automatisms, like crawling and walking, may give rise to the generation of new neurons that hold key positions in the acquisition of long-term memories for movements. If some of the neurons, serving movement functions, have more than one function, they assist in the re-learning of automatisms and may support the re-learning of other trained movements during CDT.

The rate of neurogenesis in the adult human spinal cord is very low, probably because the learning capacity of the spinal cord is low.

Adult neurogenesis is important because it results in a continual influx of neurons that are, at least temporarily, immature. Because these immature cells are structurally plastic, they are highly susceptible to changes in the environment [188], including movement-based learning. Exercise increases hippocampal neurogenesis and improves learning in mice [189]. Whether newly born cells participate in learning and memory has been discussed for newly born cells in the hippocampus and various types of learning and memory [190].

3. Repair within 8 Years of CDT with No Intensive Therapy Any More: No Substantial Progress Any More

3.1. Cessation of Aggressive Therapy

Within a few months of regaining some finger function in the right and left index fingers (Figure 190) Kadri made the decision to finish high school and then to study further. The author agreed with that decision, because it could not be expected that she could re-learn walking within a few years' time, which would be the next goal in repair. With further intensive therapy more new motoneurons would be built and more functions would have been repaired in the legs. But walking would probably not be achieved, because for walking much more muscle power is needed. Since the motor units can only be increased by approximately 50%, something like 50% of the motoneurons for walking would have to be brought under supra-spinal (supra-injury) control. This achievement seems to be impossible for the time being.

Kadri finished high school and started to study and did not train any more at her limits. While at school and university, she trained with 50% of the necessary effort and trained irregularly. Sometimes she did not train at all for short periods.

Figure 190 shows the low-load coordination dynamics (CD) values for 20N for forward and backward exercising over the following years. Figure 190 is the continuation of Figure 30. The forward and backward CD got worse when she could not train her limits anymore in 2009. When she trained more, the CD values improved and when she did less she got worse. With flu her CD values got worse (Figure 190). When she did not exercise at all for two weeks her CD values got much worse. Interestingly, when she had good CD values, the CD values for forward and backward exercising were rather similar and when the CD values were bad, the CD values for backward exercising were worse than those for forward exercising. It was as if the repair had only partly reached the structural repair.

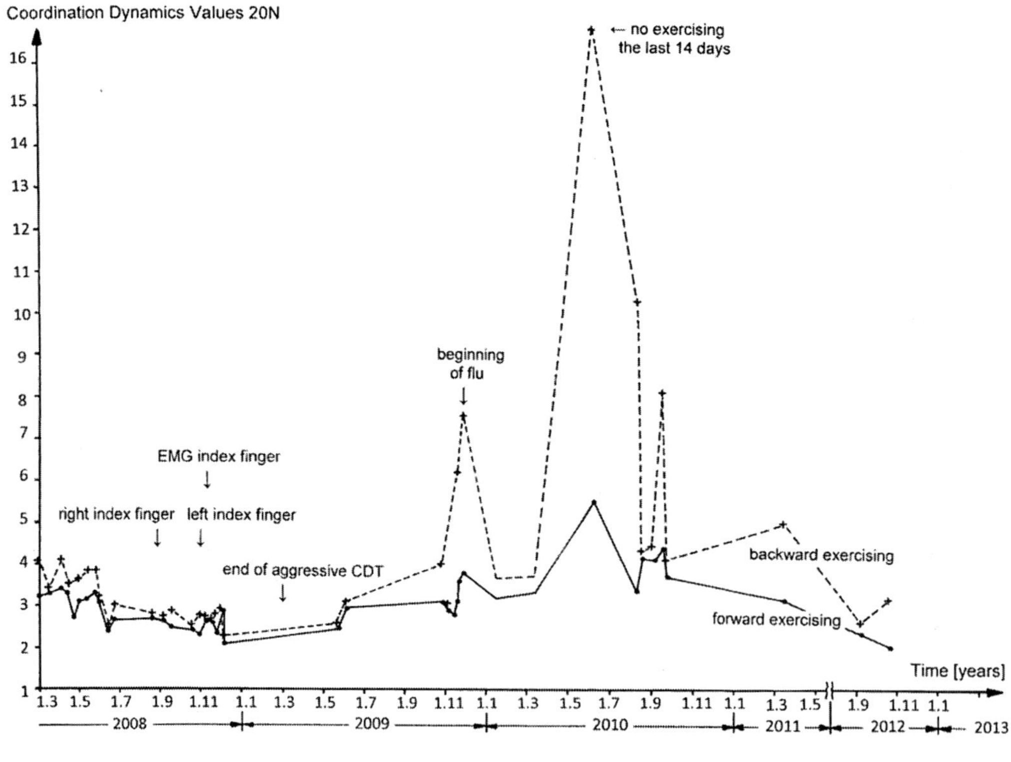

Figure 190. Relation of coordination dynamics values to therapy duration for a load of 20N and for exercising in the forward (lines and dots) and backward direction (dashed line and crosses). Note that with no intensive therapy in the beginning of 2009 the coordination dynamics values got worse (increased) and upon therapy they improved again and upon no therapy they got worse again. Flu and a break of 14 days increased the CD value strongly.

3.2. Conclusion According to Low-Load CD

1. The low-load CD values got worse with no intensive therapy.
2. Flu and a break of two weeks therapy increased (worsened) the CD values strongly.
3. It was not until four years after stopping intensive therapy that she was able to reach the low-load CD values achieved during intensive therapy.
4. The recovery time from the two week-break in therapy lies in the range between three and ten months. Such long recovery time indicates that with no therapy the molecular machinery seems to stop building new neurons and connections. To start the structural repair mechanisms again requires something in the range of six months to a year, most likely depending also on the intensity of the therapy. Therefore, it is important to tell patients to never stop the therapy for longer times than three days.

3.3. Improvement of High-Load (50N) Coordination Dynamics Values

High-load CD assessments were of course always performed with this patient with a motoric complete SCI C5/6. Only in the year 2008 Kadri became able to exercise in the forward and backward direction at a load of 50N (Figure 191). The standard assessment was started with 20N, and then continued with 50N (in forward and backward direction) and at a load of 100N the Author had to help. Then the load was reduced again. The dream that Kadri could turn by herself at a load of 100N was not achieved.

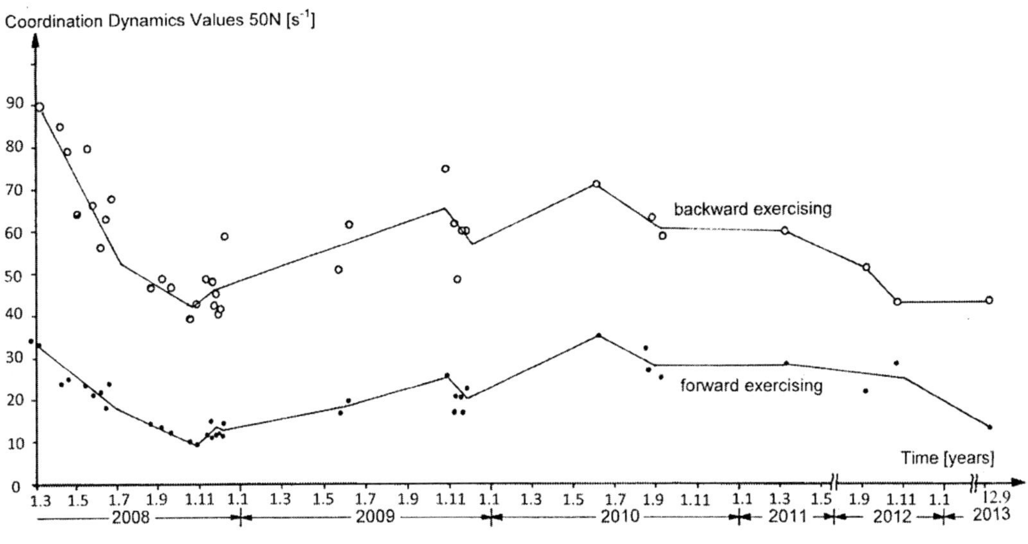

Figure 191. Relation of coordination dynamics values to therapy duration for a load of 50N and for exercising in the forward and backward direction. Note that upon no intensive therapy any more in the beginning of 2009 the 50N CD values got worse.

From Figure 191 it can be seen that the 50N-CD values improved with intensive CDT. But with no intensive therapy the values got worse (larger) again. The CD values for backward exercising were always larger (worse) than those for exercising in the forward direction. The reason is that the patient exercised first in the forward direction and then in the backward direction and she was already a bit exhausted when reaching the backward exercising. Mostly patients with C5/6 SCI can turn better in the backward direction because the biceps muscle often remains innervated.

A real high-load test, including loads up to 200N, to estimate rates of repair could not be achieved, because the patient with originally a motoric complete SCI C5/6 was not able to manage the higher loads. What healthy persons learn in one or two months (Figs.36,37), this patient was not able to learn in 8 years. But the patient with an incomplete SCI C5/6 could re-learn to exercise against loads up to 200N and could get the high-load-test value under 100 within 3 years of therapy (Figs.50,51).

3.4. Conclusions from the 50N Assessment

1. Following a period of no intensive therapy, the 50N CD values worsened.
2. Following flu and a break in therapy of 2 weeks the 50N CD values got further worsened transiently.
3. The 50N CD values for forward exercising with no intensive therapy did not reach the good values ($10s^{-1}$) of four years previously when intensive therapy was terminated. The patient had therefore lost some of the improved CNS function.
4. Even though the 50N CD values of those of 4 years ago could not be reached again, functions such as continence, physiologic breathing, cardio-vascular performance, trunk stability, some leg functions and finger functions were not lost.
5. The recovery time necessary following a therapy break of two weeks is longer than three months.

3.5. Ethics in Spinal Cord Injury Research and Clinical Trials

“Transplantation of both embryonic stem cells and embryonic stem cell-derived neural (neuronal or glial) progenitors is able to efficiently promote CNS regeneration in preclinical models of stroke, myelin deficiency, acute spinal cord injury [191,194] and Parkinson’s disease [186]”. This sentence makes the reader believe that research made already the transition from animal research to human research and can already be performed on humans.

Looking up the references one finds the title: ‘Human embryonic stem cell-derived oligodendrocyte progenitor cell transplants remyelinate and restore locomotion after spinal cord injury’. Patients with SCI or journalists would be forgiven for thinking that the SCI problem is solved. This research was performed on rats. Animal researchers often deliberately present their research in such a way as to make the reader believe that this research has already solved the SCI problem in humans. A well-known journal in medicine is even redefining medicine. It makes the reader believe that measurements in mouse are a part of medicine [192]. I have learned in my medical study that medicine has to be relevant to the cure of diseases in human.

If neurobiologists, molecular biologists and neurologists, especially from USA and Switzerland, have developed their research far enough so that potential therapies can safely be applied to humans, why then did Christopher Reeve die? Even a high tetra can live for a long time if given proper treatment (care). We will only know how Christopher Reeve died if a forensic examination is performed, for which it is probably too late now.

It would have been an honor for Christopher Reeve to push the development of proper treatment further even in death. Also, it would be of importance to get morphologic data from the human spinal cord after SCI. Anatomists made great efforts in that direction to bring the discipline of anatomy forward. On-going bladder infections and pressure ulcers do not normally overloading the kidneys due to the administration of antibiotics. But if Christopher Reeve was given some additional toxic drug therapy, then his kidneys could have become overloaded. It is astonishing how readily the highly toxic botulin toxin is used in patients to reduce muscle spasticity.

Proper research, including qualified human research, has to be performed before patients die and not only afterwards. The world seems to have forgotten the Contergan disaster. Members of this population (with no arms) are sometimes coming to rehabilitation equipment exhibitions to remind visitors: “don’t forget us that this will never happen again”. If Christopher Reeve would have had exercised on the special CDT device in the lying position (Figure 150), most likely with manual support, he could have improved at least his autonomic functions, especially the blood supply (reduction of pressure ulcers) and the breathing, since autonomic functions are partly bypassing the spinal cord via different plexuses (Figure 153). Further, the breathing automatism has a strong potency for self-repair. In dogs, the injured breathing center repairs itself within a few hours [193]. The tetra of this report regained a very good blood supply and rather normal breathing after one year of therapy. But Christopher Reeve seemed to have been isolated from the scientific community. When a patient with a spinal cord injury from USA, informing other patients about possible treatments, wanted to tell Christopher Reeve in a meeting that he should promote human and clinical research, he was held down in his wheelchair with a hand over his mouth until Christopher Reeve had left the meeting. Many SCI patients are missing Christopher Reeve, because he represented hope for them.

Specialist and general media need to set reasonable expectations of the safety and efficacy of potential therapies to avoid raising and then dashing the hopes of those living with SCI or those in government, those carrying out research, or the general public [114].

High ethical standards are required of researchers, clinicians and journalists to ensure that results are communicated to the general public in a manner that reflects honestly the safety and efficacy of a potential therapy [114]. It would be important to know to how many patients exogenous stem cell therapy was administered, how many really benefitted from it, how many suffered complications and how many died following this therapy.

Western scientist and clinicians would argue that controls are necessary to identify unambiguously whether a therapy is safe and effective, some clinicians have claimed that withholding a potential therapy from a patient with SCI is in itself unethical [114]. Many SCI patients know about CDT. But the patients are not informed that such a therapy exists! A real change from care to cure is not in the interests of the rehabilitation centers, according to my experience. For a change from care to cure a re-education of the stuff would be needed. Further, under the name of therapy trials clinicians sometimes perform research on humans often without sufficient education for that research.

There is too much false hope generated by animal physiologists, neurobiologists and researchers working in the field of genetics. Without detailed knowledge of the human anatomy, neurophysiology and pathophysiology, the animal knowledge is only of limited help to cure diseases in human. False hope also stops the patient to fight for improvements which are needed for everyday life. Why should a patient fight for two years or more with a movement-based learning therapy to get bladder and some motor functions back, if in a few years' time the walking can be gained back easily with miracle cells or pills? Such false hope is coming from qualified researchers, when they make people believe that the animal data can be easily transposed to humans and used for repair of the human CNS. Qualified human research is needed and has to be organized. Only if there is overlap between animal and human research, is there the possibility of rather safely transfer knowledge from animals to humans. As long as human research is not properly organized patients have to suffer or even to die.

Authors of recent review articles presented approximately 1000 citations on the repair of the human spinal cord following injury [106,107,186,114]. Less than 10% of the citations were from human research. Interestingly, the Author was not cited, even though students, young researchers, some physicians and even many journals find the Author easily on the internet. I learned a lot from these brilliant research articles with respect to animal research. But with respect to human research and applicability they were twenty years out of date.

'Miracle' operations are not the only danger for patients and the freedom in research. It is the worldwide research, treatment, and teaching system, that does not allow qualified human research, which is urgently needed to cure diseases and allow people to live longer with a higher quality of life. And if there are really operations with unbelievable success, then the treatment before those operations was wrong. For example, a paraplegic patient came by wheelchair from a well-known German rehabilitation center to a neurosurgery department, was operated on and walked out of the neurosurgery department two weeks later. A second patient was cured in the same way. The reason for this progress was caused by a mistake of the rehabilitation center. Bones from the spinal canal caused pressure on the spinal cord. The spinal cord stopped working but was damaged only little. A laminectomy freed the spinal cord and the spinal cord started to work again. Follow up MRI's are needed.

Chapter IX

Repair of Brain-Injured Patients in the Vigilant Coma

1. Repair Strategies at the Neuron Level to Implement Repair in Coma Patients

1. Repair depends on learning and memory formation, mediated or supported by epigenetic mechanisms. The epigenetic is the interplay between genes and environment resulting in phenotype and epigenetic landscape.
2. Epigenetic mechanisms like DNA methylation are probably sensors for movement-based learning and memory formation and fine modulators of neurogenesis in the adult CNS with CDT.
3. The epigenome consists of non-coding RNA and chromatin, a proteinaceous matrix surrounding DNA. The dynamic interactions of post-translationally modified chromatin proteins, covalently modified cytosine's inside DNA and non-coding RNA define the complex pattern of gene expression beyond the four bases of DNA.
4. The hippocampus plays an essential role in learning and memory. In the hippocampus there exists a specialized form of neural plasticity, that is, the generation of new functional neurons from stem cells occurring throughout life. Adult hippocampal neurogenesis contributes to learning and memory formation.
5. New neurons are important for learning and memory formation (besides functional reorganization), i.e., for increasing the rate of repair, for the following reasons:
    a. The insertion of new neurons helps to store the memory of the same activity that led to the creation of the neuron.
    b. Activity-dependent neurogenesis enhances the learning of new memories and degradation and clearance of previously stored unwanted memories like spasticity, because the synapses, dendrites and axons can devoted more fully to the newer memories. The old neurons with large and complex axon and dendritic trees are difficult to change. They can only be changed with sustained effort.

c. New neurons seem to improve the accuracy of relearned patterns (from model study [24]) this means that new neurons help to improve phase and frequency coordination of neuron firing and pattern stability.
d. The advantage of new neurons seems to be dramatically greater in networks that had been more active and had been required to store more memories [24]. The advantage of neurogenesis for memory storage in heavily active networks is that it provides an increased rate of repair if movement-based learning is administered aggressively and if different movements are trained.

6. Specific natural network activity is required for multiple aspects of repair. Specific activity is essential for correct migration of interneurons and controls the development and repair of their axons and dendrites. During repair there is a specific requirement of network activity in shaping the cortical integration of specific neuronal subtypes. Newly build neurons are likely electrically active shortly after their birth and participate in the early network activity that contribute to circuit maturation during repair during CDT.
7. Specific activity is required for migration and maturation at several stages of repair. A break in CDT may invalidate the whole chain of repair events. Specific interneuron subtypes require activity for migration and morphological maturation at two distinct stages of development [24]. Newly build neurons may even require specific activity for migration and maturation at several distinct stages of repair. During a break in CDT, the specific activity, required for neuron migration, maturation and network integration may not be supplied at one of these stages so that the chain of repair events is severed and the whole repair chain has to be started anew.
8. For optimal repair specific downstream and upstream activity is needed. Since coma patients cannot move by themselves, only the specific movement induced afferent input is available for repair. But since the movement induced afferent input during support by a therapist improved motor programs (Figure 22), it is likely that the specific activity induced by the movement induced afferent input can to a certain extent also offer the specific activation necessary for proliferation and migration of neurons and network maturation.
9. Drug application may undermine repair.

Altering the level of neuronal excitability within genetically targeted neurons from drug application, for example antiepileptic drugs may have profound consequences on multiple aspects of the repair of select types of neurons within a population of neurons, as well as their associated gene expression. The pain-killer ‘Contergan’, taken during pregnancy, changed gene expression and the babies were born without arms.

1. Excitation-neurogenesis coupling [24].

   a. Excitation increases or decreases neuron production directly by excitation-neurogenesis coupling.
   b. The excitation acts indirectly on the surrounding mature (hippocampal) cells through depolarization-induced release of grows factors.

c. Adult neurogenesis is enhanced by excitatory stimuli and involves Ca^{2+} channels and NMDA receptors.
d. The Ca^{2+} influx pathways are located on the proliferating stem/progenitor cells (NPCs), allowing them to directly sense and process excitatory stimuli. The Ca^{2+} signal in NPCs leads to rapid induction of a proneural gene expression pattern.

11. Integrative coordinated movements have to be trained to allow functional reorganization and new nerve cell integration across very large distances. CDT has to activate injured and uninjured networks to enhance physiologic CNS functioning and learning transfer.
12. Conclusion for optimal therapy according to the present stage of knowledge. If there is similarity between development and repair, animal (mice) data also hold in humans and the principles of neurogenesis of the hippocampus also hold in other parts of the brain, albeit to a much lesser extent, then the patient has to be trained at his limits (1) to induce substantial building of new nerve cells. The treatment has be continuously administered (2) to support all stages of repair at the progenitor level as migration, maturation and integration and the networks requiring repair have to be activated specifically (3) to generate repair-friendly micro-environmental properties in the networks. No drugs should be administered that change neuron excitability (4).

With the above repair strategies in mind, we shall now examine brain repair in coma patients.

2. The Duration of the Coma Depends on the Severity of the Brain Injury and How Early Efficient Treatment Is Started

In even more severe brain injuries than the above cases the limit of repair possible through CDT is reached. However, improvements from CDT which are possible in patients who have for example lost approximately 30% of brain tissue are of interest for medicine in general.

The patients of the Author who suffered a severe brain injury were all in the vigilant coma first. The severity of the brain injury seemed to be partly reflected in the time period necessary to get them out of the coma spontaneously. The therapy should be started in the coma stage to reduce pathologic reorganization and to shorten the coma period.

Benjamin, who suffered a severe brain injury at the age of nine when he was hit by a car, came out of the coma after seven weeks. The treatment was started as early as possible in the vigilant coma stage (Figure 118A,B). Two other boys with similarly severe brain injuries (Mario and Andrej) needed up to six months to emerge from coma mainly because their treatment started either late in the coma phase or shortly after it had ended. On the other hand,

Benjamin's brain injury, even though severe, was not as severe as in a patient Manolis of which will be reported below. In an eighteen year-old patient (Georg) with a very severe brain injury, CDT was unable to get him out of the vigilant coma. A few years after the accident, he died from a lung infection in hospital. However, a fifty year-old man who suffered a very severe brain injury following a bicycle accident came out of the coma. For years he appeared to be in a non-vigilant coma. It was only later discovered that he was able to communicate through blinking. Objective judgment is necessary to clarify whether a patient is in the coma or not.

3. Case Report: Optimal CDT in a Coma Patient

A 21-year-old male patient suffered a very severe brain injury in a car accident. A metal shard went through his skull into the brain. Parts of the brain were removed and a shunt was installed to control brain pressure. An attempt to reconstruct the skull was unsuccessful. After several operations in a university hospital he was placed in a rehabilitation center and provided with twenty minutes physiotherapy per day. His relatives felt that this was insufficient and paid for an additional hour of therapy per day. This was still insufficient to even maintain his level of CNS function. Prior to the three months of conventional therapy in a rehabilitation center the patient was able to respond 'yes' and 'no' by closing and opening his eyes respectively. After this conventional therapy he was no longer able to respond with yes or no. The ability to communicate with the patient was lost. The injury was most likely so severe that CNS functioning got continuously worse.

After the three month rehabilitation period the patient was taken home and CDT was commenced one year after the accident. There was always the fear that if the treatment is not sufficient efficient the patient may die.

Prince Friso of the Dutch royal family suffered a hypoxic brain injury and died in 2013 after being in a coma for eighteen months.

4. Visible Anatomical Damage due to a Car Accident

An MRI was performed one year after the accident at the same time that CDT was started. It showed that many parts of the brain were injured. (1) Figure 192A,B,C seem to show a hydrocephalus, but the ventricle enlargements were secondary to the loss of nervous tissue. If one compares Figure 192 with Figure 45, also of a patient with severe brain injury following a car accident but no enlarged ventricles, then it is obvious that the ventricles of the patient of Figure 192 were significantly enlarged. The first, second and third ventricles were enlarged in the coma patient. The fourth ventricle was normal. A picture of the four ventricles of normal size in relation to the brain size is given in Figure 192E. MRI pictures with the ventricles are obtained by virtually 'slicing' the brain (Figure 192Ea) and thus getting slides. The shape and size of the ventricles depends on the height of the sections. It was obvious that the patient had lost very much brain tissue. (2) The corpus callosum was very thin. Impaired communication between the two hemispheres has to be expected. (3) The frontal lobe was injured (Figure 192C) on both sides. (4) The left thalamus was slightly injured. (5) The frontal and parietal lobe showed atrophy. (6) There was great loss of parenchyma of the right hemisphere where the metal shard had entered. (7) Valerian degeneration of the pyramidal tract took place probably due to the injury of the sensory-motor cortex. (8) The cervical spinal cord seemed to be undamaged, but bones seem to touch the spinal cord and may exert some pressure with the consequence of loss of spinal cord functions (Figure 193A). The Author's spinal cord shown in Figure 193B for comparison is free; bones do not touch the cord. (9) No sign of acute ischemia can be seen in the patient's brain. (10) The valve artifact from the shunt can be seen in Figure 192C and D. There may have been no part of the brain left undamaged.

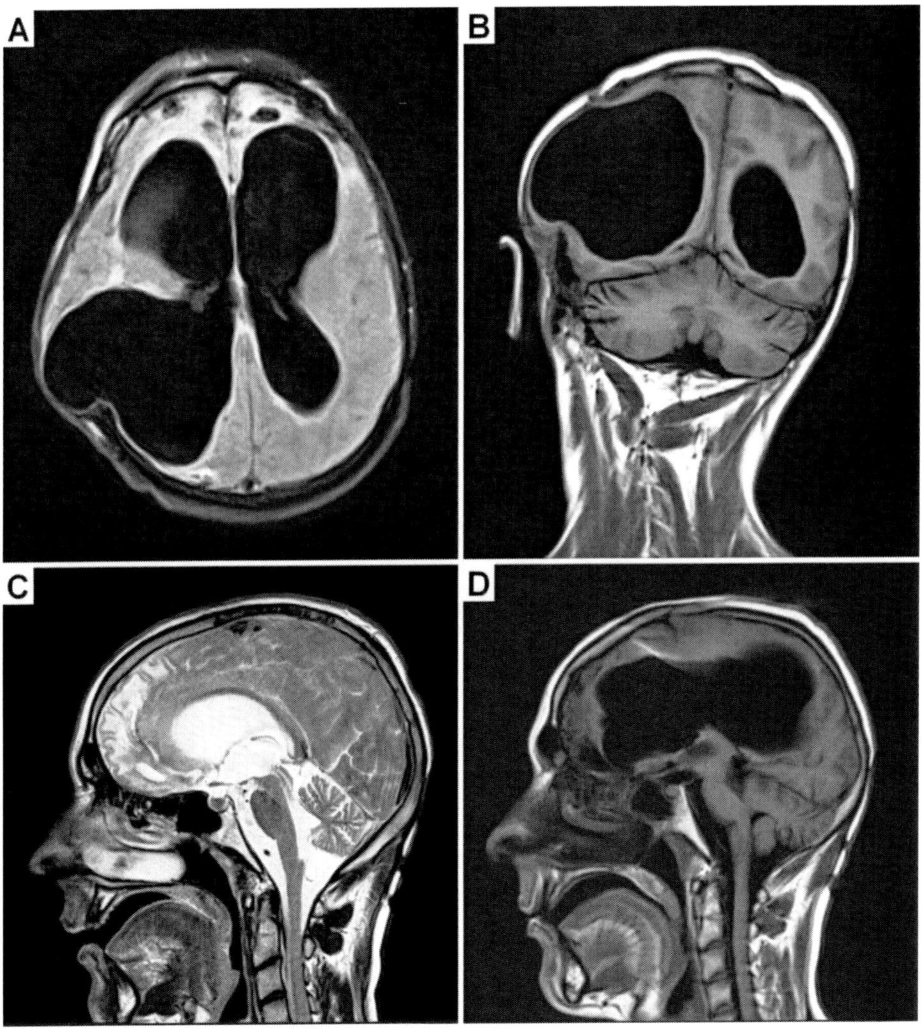

Figure 192. MRI of a 22-year-old male patient with a very severe traumatic brain injury. The pictures were taken one year after the car accident. Since a metal piece went into the brain during the accident, brain tissue had to be removed. A shunt was installed to regulate brain pressure. Because of the loss of nervous tissue the ventricles became enlarged looking like a hydrocephalus. Many parts of the brain were damaged during the accident. Both frontal lobes were damaged (A, C). The artifact from the shunt can be seen in 'C'.

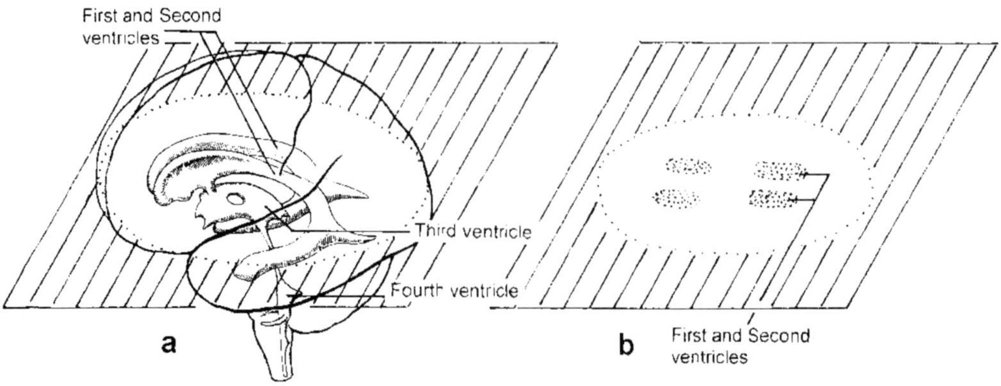

Figure 192E. Picture of the human brain with its ventricles. The ventricles have normal size. A comparison with the enlarged ventricles of Figure 192 show that the ventricles of the brain-injured patient are very much enlarged and that the patient lost may be 30% of the brain tissue.

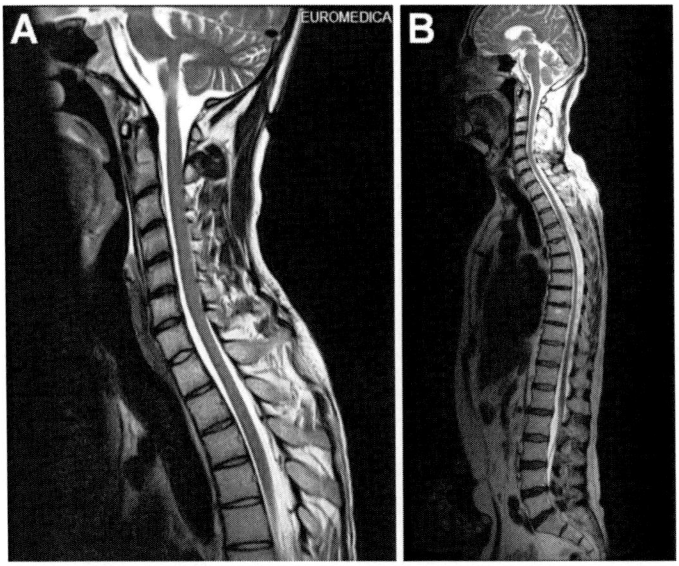

Figure 193. A. MRI of the spinal cord of a 22-year-old patient with traumatic brain injury. Vertebra may touch the spinal cord and may give rise to spinal cord pressure and impaired spinal cord functioning. B. The spinal cord of the Author. The spinal cord is 'free'.

5. Pathologic CNS Organization

The left leg of the coma patient showed extensor spasticity and the right leg flexor spasticity. The right arm showed extensor spasticity and the left arm flexor spasticity. Strong rigor was present, but no tremor at the beginning of therapy. The joints were not working properly due to having lain in bed for one year with insufficient mobilization. A bit of clonus developed later on in the left leg and also a bit of tremor developed in the hands. There was some dysregulation of the sympathetic division. One foot was warm, the other one was cold. During exercising the cold leg was sweating.

When exercising on the special device in the lying position (Figure 194A) the brain pressure was higher, because the liquor transport via the valve was not as good. Upon exercising in the sitting position on a special CDT device or during sky-walking (Figure 194B,C,D) the liquor transport was better. The brain pressure could be felt by touching the skin above the open skull.

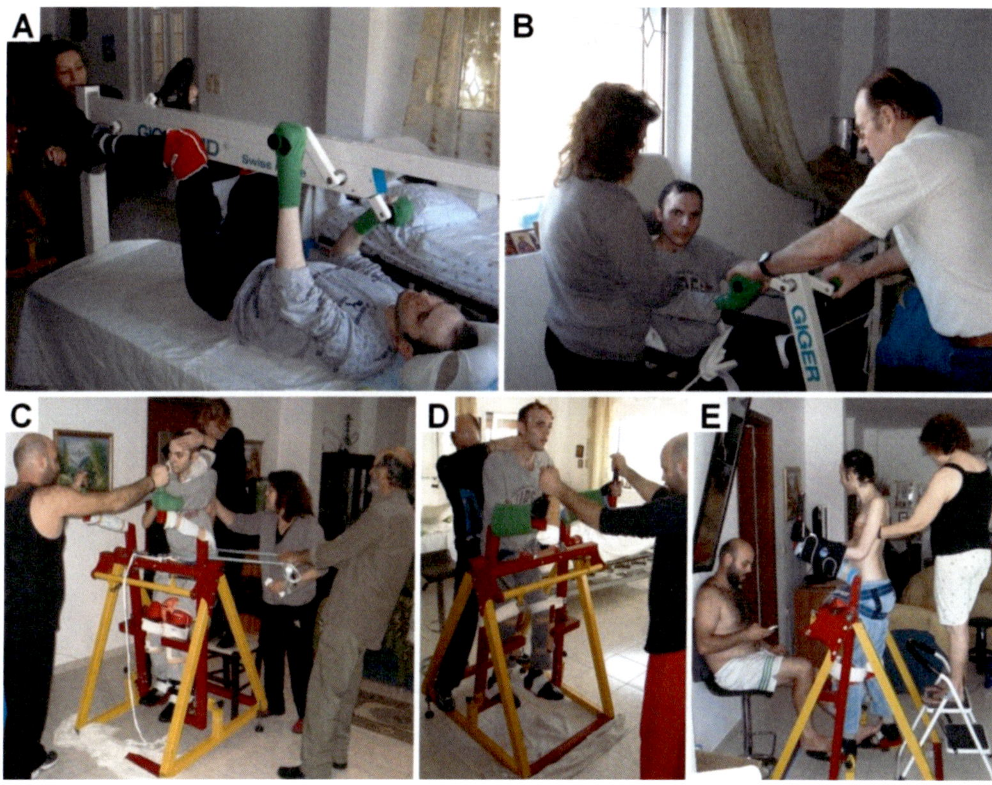

Figure 194. Movements performed with a 22-year-old male patient in the vigilant coma. All movements are passively performed with the patient, because the patient is in the coma and does not move by himself. A,B. Exercising on the special CDT device in the lying position (both hands are fixed in the prone position) and sitting position. C,D,E. Exercising on the sky-walker. Note that the trunk and head control improved from 'C' to 'E'; in 'E' no support of the head was needed any more after 5 months of CDT.

6. The Pain and Consciousness Problem

Ideally, treatment would have been started slowly and then reach a treatment time of twenty hours per week in approximately three months. However, it was not possible to increase of treatment intensity slowly in this patient because the benefit from the treatment may have been insufficient to counter the ongoing pathologic processes, putting the patient's life at risk.

The joints were a significant obstacle to therapy. Because they were not really used for one year, it was difficult to mobilize them. For movement-based learning therapy the joints obviously need to work.

Pain and other receptors will have informed the damaged CNS about problems in joints, shortened tendons and stiff muscles. In spite of the damaged thalamus, the thalamus will have attempted to inform the cerebrum of this pain. Whether the hemispheres got the pain messages (natural impulse patterns coding pain) and processed them in the right way is unclear. The natural impulse patterns coding pain (for natural impulse patterns coding touch and pain see pages 292 to 297 of [14]) (Figure 5) could have been changed by the damaged thalamus, so that the cerebrum could not recognize these patterns properly. The other possibility is that the impulse patterns were physiologic, but the hemispheres could not recognize them properly because of their damage. Due to the thalamus damage and severe damage of the hemispheres (both frontal lobes were injured and also one part of the sensory-motor cortex was severely injured), the pain coding impulse patterns will have been altered and the processing in the hemispheres will have been also changed. Still it seemed that the patient's CNS recognized pain and his expression hardened and sometimes even tears were in his eyes when performing certain movements. It is unclear whether the patient really felt pain in the functional deepness of CNS organization where the consciousness is also sited or whether the pain input patterns changed the impression in the face more in the periphery of CNS organization (a vegetative organizational state) in an automatism or reflex like manner.

Pain, or the perception thereof, and the changing impression of the face was posed a wider problem. His mother, brothers and sisters thought that Manolis was communicating with them. The Author could not safely prove that the patient was transiently out of coma. The possibility cannot be excluded that sometimes, when the CNS of the patient was in a good organizational state, achieved by exercising on the special CDT device, the patient responded consciously to his mother.

7. Difficulties in to Applying the Necessary Efficient Treatment to Coma Patients

In mild brain injuries, the patient's CNS can largely repair itself without treatment. In severe CNS injuries, the patient's nervous system needs treatment so that its functioning does not become too pathologic. In very severe brain injuries, a patient's CNS may destroy itself if the treatment is inefficient; the repair functions are in a sense overtaken by the ongoing pathologic processes. As was shown in Chapters III, IV and V, when comparing the rate of learning in a patient with a severe CNS injury with subjects with a mild or no injury, it turned out that the rate of learning was reduced by a factor of 20 to 50 in patients with severe injuries. A patient with a severe CNS injury may need in a first approximation 20 to 50 years to learn what a healthy person can in one year. It must be borne in mind that, in the severely injured nervous system, change takes a long time and the rate of learning depends on the kind of injury.

When the treatment with the coma patient Manolis was started a year after the car accident, the CNS injury was so severe that the nervous system seemed to be destroying itself because the pathologic processes such as high brain and blood pressure, bleeding, toxins from destroyed nerve cells seemed to be stronger than the repair processes. There was therefore the danger that the patient could die with or without treatment. Another obstacle was his mother. She devoted herself to her son. She did a very thorough job with his care and contributed to the treatment. But she was over-protecting her son. When the treatment increased in intensity it could most likely be seen in his face (see below), then she mother wanted to stop the treatment.

The patient would probably die if he would not get sufficient treatment. With some treatment the patient may not die and some functions in the functional periphery of the patient's CNS would get better or recover. However, the patient may stay permanently in the vigilant coma. This may not be a meaningful life and in the long-term the patient may die of a complication like a lung infection. The neurosurgeon, who operated on the patient stated that, according to his experience, his relatives should be happy if they could maintain the patient's current level. With optimal treatment as far as that is possible in such a patient however, he might be brought out of the coma, whatever the mental and motoric conditions of the patient would be afterwards.

8. Performed Movements

The performed coordinated movements with the coma patient were exercising on the special CDT device in the lying and sitting position and the sky-walking (Figure 194). Additionally the patient obtained eight hours conventional physiotherapy per week. Mobilization and stretching are important, but they contribute only little to repair. All movements had to be performed passively. Due to the flexor and extensor spasticity, rigor and stiff joints with a small angle of motion, it was difficult for the therapists to enable the patient to move his arms, legs and trunk. The patient altogether received twenty hours movement therapy plus five hours speech therapy per week.

9. Progress and Obstacles to Repair after Five Months of CDT

Even though the patient could only move passively, some progress in repair was achieved. At the beginning of therapy several persons were needed to keep the patient in the walking position (Figure 194C), then only two (Figure 194D) and after four months of therapy the patient could sometimes keep his head in the upright position (Figure 194E). The trunk and head control had improved in some similarity to development. The improvement of the trunk control and reduction of the scoliosis like bending of the trunk is astonishing because of the extent of the damage to parts of the sensory-motor cortex. Swallowing also improved and the blood pressure improved (reduced). It seemed that the biggest improvement was with respect to automatisms and vegetative functions.

10. Regression of the Coma

The improvement of automatisms and vegetative functions indicates some repair. Can the specific neural network activation with its functional and structural repair also reach the consciousness? Since in coma patients the coma terminates earlier upon CDT (refer to Figure 115A), it is likely that the movement-induced coordinated afferent input from the passive movements helps to get the patient out of the coma. The quality and intensity of the treatment in combination with the severity of the injury will dictate responsible whether consciousness will be regained.

11. Repair after Nine Months of Therapy – First Signs of Regression of Coma

The protection reflexes of the head improved further, including the blink reflex. The blood pressure, a major problem at the beginning, became healthy (after sleep for example, 90/66 mmHg, pp = 77). The patient learned to swallow: he could eat and drink a bit, but it was reflex activated. The most apparent progress was that the patient seemed to come out of the coma transiently while exercising on the special CDT device in the lying position. After 500 turns on the lying machine, the patient appeared to be more alert and he seemed to be able to communicate with his relatives by opening and closing his eyes. Whether he was in fact indicating saying "yes" and "no" by eye movements was unclear because of the brain damage and especially the damage to the sensory-motor cortex which seem to have also affected the motor functions of volitional eye movements. In the early stages of repair, early functions are inconsistent.

In patients with somewhat less severe brain injury, there was a clear change from the vigilant coma to the conscious state. In this very severe brain injury there does not seem to be any clear transition from the coma to the non-coma state. All movements and functions seem to be disturbed or mixed up. Only the protection responses of the eyes seem to be good and precise. Swallowing was also trained and improved as a consequence.

12. Repair after 16 Months of Non-Optimal Therapy – Limited Improvements

After 16 months of non-optimal CDT, the condition of Manolis had improved only a little. The trunk stability improved further. Consciousness however, did not. A follow-up MRI showed the same anatomical condition as 16 months before. Figure 195 shows for example the enlarged first and second ventricles, which have roughly the same shape as 16 months before (Figure 192A).

There may still have been some pressure from the vertebra onto the spinal cord as in Figure 193A, to be one cause for the high spasticity in the patient. It is not clear whether a further operation would be justified. The anesthetic would put the patient's severely injured brain under severe strain. The display of the blood vessels of the brain via additional contrast medium showed some ischemia. Optimal CDT under this condition could also improve the blood supply of the brain further.

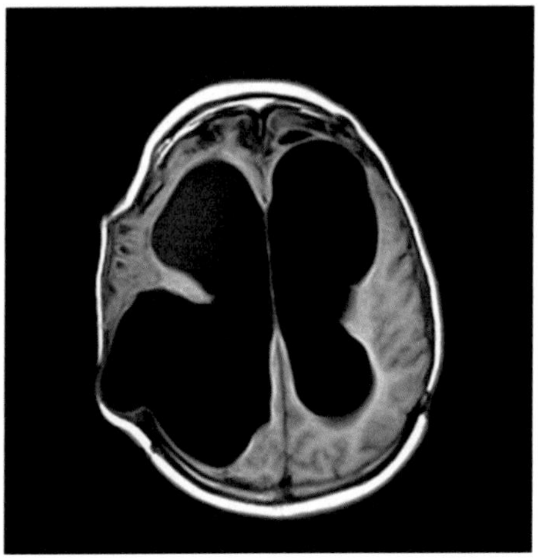

Figure 195. First and second ventricles 16 months after the very severe brain injury of a now 23 years old patient.

13. Control of Functional and Structural Repair

Probably the principle that local cellular environments, at the network place of injury and distantly for organization of patterns, are important in controlling functional reorganization and neurogenesis. The repair potential may depend on the kind of local activity (1), access to local activity (2), ability of the local environment to induce activity-sensing competence for repair (3) and factors intrinsic to the resident neurons themselves. Further, functional reorganization and neurogenesis might exist in modes that are switched on and off by the inputs from the movement-induced afferent input. Progenitor cells could be able to distinguish local activity related to new memory storage and repair of (movement) patterns (which should trigger neurogenesis) from activity relating to patterns with no repair load placed on the remaining circuits.

The neuromodulatory inputs available so far are the movement induced coordinated afferent input from the passively performed sky-walking and exercising on the special CDT devices. It is unclear whether this level of therapy will ever be sufficient for the recovery of full-consciousness in an injury of such severity and given the extent of the loss of nervous tissue.

14. Judging CNS Organization by the Impression of the Face

Figure 196 shows that the improvement of CNS repair can be seen also in the face of a patient with severe cerebrum and cerebellum injury. The appearance of the face improved over ten years with the CNS repair achieved through CDT (Figure 196).

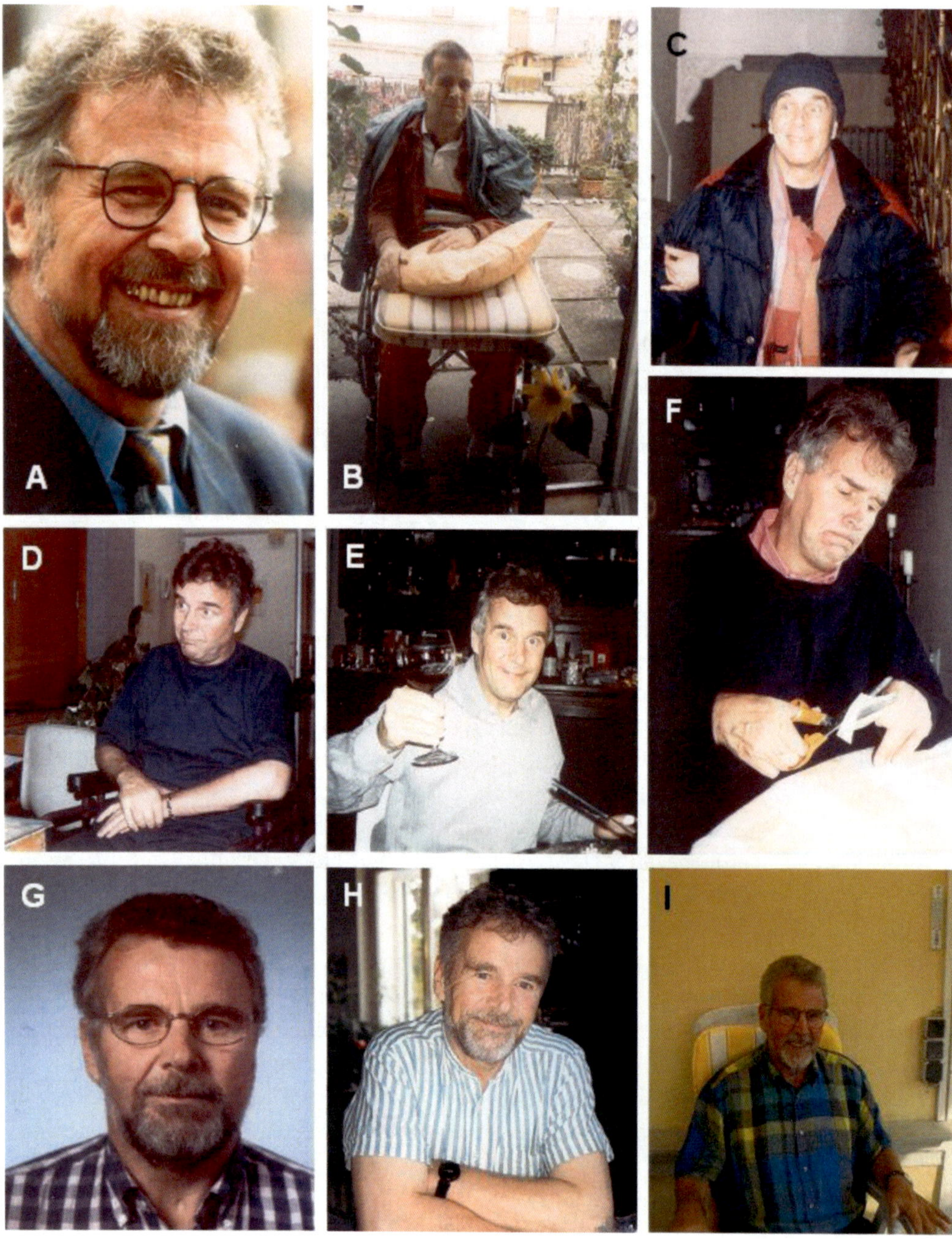

Figure 196. Therapy-related improvement of the impression of the face of the patient who suffered severe cerebellar injury: before the accident (a), after the accident (1995) till 2006 (b-i).

Despite suffering a severe brain injury and being in a coma, the face of the patient Benjamin looks quite healthy (Figure 197). The impression of the face of this patient Manolis, however, seems to indicate a more severe injury (Figure 198). The eyes were veering in one direction, but would react to noise. After exercising on the special CDT device, after approximately 1000 revolutions, the eyes were not veering so much anymore, but were moving more from right to left and backwards.

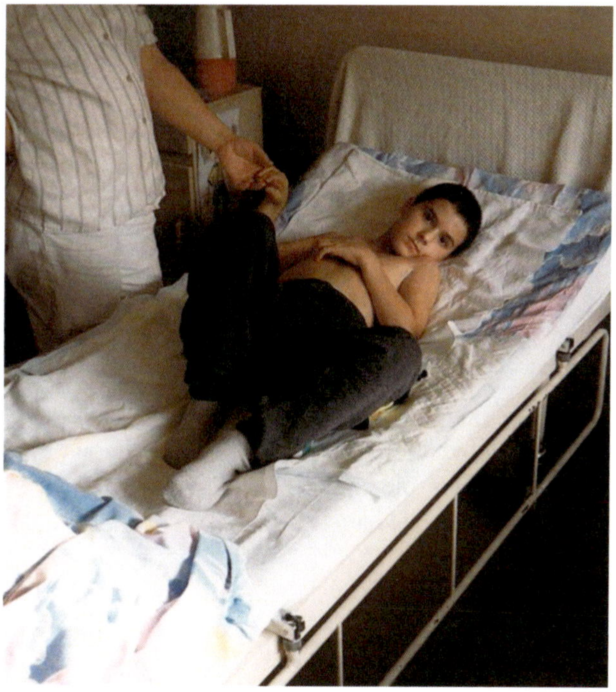

Figure 197. Picture of a 9-year-old patient Benjamin in the vigilant coma following a car accident. The picture was made shortly before he came out of the 7 weeks lasting coma. Note, the impression of the face looks healthy.

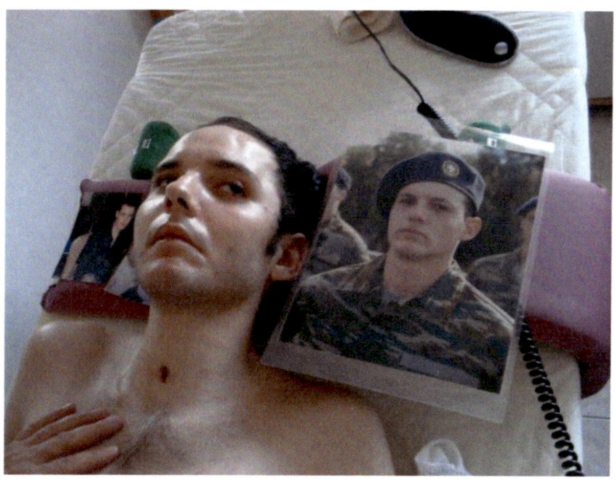

Figure 198. Picture of a 22-year-old patient being 1.5 years in the vigilant coma following a car accident. The impression in his face does not look as healthy as the one of the picture made 2 years before when the patient was in the army.

The eyes became livelier, more excited, as if the CNS of Manolis organized itself in a more differentiated manner. The integrative coordinated afferent input improved CNS functioning in the short-term memory. But whether such repeated improvement of CNS functioning is sufficient to get the patient out of the coma was unclear after five months of CDT and subsequently seemed to be more plausible after nine months of CDT. Similar experience was made with another young man who was in a coma for nearly two years due to a very severe brain and spinal cord injury. The more arms and legs were moved, the healthier and livelier the impression of his face became.

15. Brain Pressure Control

The brain pressure of the patient could be felt when touching the head where the skull was missing. The neurosurgeon checked that the shunt and the valve were working well. According to the neurosurgeon, high brain pressure should be avoided. The brain pressure rose in the lying position. This is likely due to the extended periods spent lying in bed or when exercising on the special device in the lying position (Figure 194A). However, too low brain pressure should also be avoided because of the risk of bleeding in the brain. The brain pressure reduced when exercising in the sitting position on the special CDT device (Figure 194B) or exercising on the sky-walker (Figure 194CDE).

16. Normalization of Blood Pressure upon CDT

The blood pressure was very high at the beginning of therapy. The diastolic pressure was often over 90mm Hg. First the blood pressure increased upon exercising and then the pressure maintained its level during therapy and after five months of therapy, the blood pressure reduced in general and transiently while exercising on the special CDT device in the lying position. The high blood pressure at the beginning of therapy was probably mainly caused by the malfunctioning of the severely damaged CNS. With the improvement in CNS function, the blood pressure became normal, indicating an improvement in the vegetative nervous system. The cardio-vascular performance had improved after five months of therapy and was nearly normal after nine months of therapy.

An average change of the cardio-vascular performance upon therapy in the short-term memory was the following. When lying relaxed in bed, the blood pressure was (110/68; 81) (systolic pressure/diastolic pressure; pulse rate). The impression of the face was similar to that in Figure 198. After being exercised on the special CDT device in the lying position (Figure 194A) for 200 turns (turning frequency below 1 Hz), the pressure was (125/80; 93). The nervous system was more activated by the movement induced afferent input. Some pain receptors in the joints may have been activated. After 900 turns, the blood pressure decreased (123/75; 81). Probably nervous system functioning had improved in the short-term memory with the consequence that the cardio-vascular performance became better. With slightly faster turning, the pressure and the heart rate increased (126/83; 87). Velocity-dependent spasticity may have been activated more and may have induced more pain. Upon reducing the speed of turning, the heart rate reduced again (126/84: 82).

It was shown that in severe cervical spinal cord injury (Kadri) the cardio-vascular performance could be improved to such an extent that within six months the nutrition of the skin reached the point that below the injury level the skin would no longer bleed when touched firmly and pressure ulcers no longer occurred. Here it is shown that even in very severe brain injury leading to a coma of more than one year's duration, the cardio-vascular performance could also be repaired.

It is likely that in aging (see below) the cardio-vascular performance and microcirculation can also be improved through CDT, even though arteriosclerosis will hinder the improvement. The effects of CNS degeneration as nerve cell death and in consequence impaired CNS organization can be partly compensated for.

17. Non-Optimal CDT

Even though the brain pressure issue was seemingly solved, the cardio-vascular performance was nearly normal, and the administered therapy time was approximately twenty hours per week, the number of performed coordinated movements was still too low. The number of administered turns on the special CDT devices reduced from 23,000 per week at the beginning to fewer than 10,000 after five months to nine months. When pushing the patient closer to his limits and administer more exercising, the blood pressure went up and the impression in the face of the patient became harder as if he was suffering more pain. His mother stopped the therapy to protect his son. It is very important to push the patient to his limits to enhance neurogenesis to get more urgently needed nerve cells for repair. At critical sites of CNS organization new nerve cells are needed certainly for rebuilding the networks. It is unlikely that with a mere functional re-organization the patients CNS can be improved that much that he comes fully out of the coma. It is not just the loss of perhaps 30% of the brain matter causing the severe functional deficits; in this case that too many parts of the brain were damaged. A neurosurgeon told the Author that he removed nearly one hemisphere of her secretary due to a malign astrocytoma, but after the operation she could go on with her work in spite of the loss of brain matter.

18. A Lack of a Satisfactory Outcome for Patients Who Have Lost Significant Brain Matter

The relatives of the patient were doing a very good job with respect to care, but they were reluctant to push the patient to his limits so as not to cause him discomfort. For these extremely severe CNS injuries, there seems to be no meaningful solution. It is important to discover just how much repair is possible in such severe injuries.

With even much more loss of brain matter, brain death can be reached.

Chapter X

Brain Death

1. The Rationale of Recording Single-Nerve Fiber Action Potentials from Brain-Dead Humans

The measurements performed on brain-dead humans with the single-nerve fiber action potential recording method gave new understanding on the functioning of the neuronal networks of the somewhat 'healthy' human CNS. It was found that one principle of human CNS organization is the phase and frequency coordination among neuronal and sub-neuronal network firing.

Measurements with the same equipment from patient with spinal cord injury (offering the surgeon some diagnostic opportunities during an operation to implant a bladder stimulator) showed that this phase and frequency coordination among neuron firing became substantially impaired and lead to pathologic CNS self-organization. The recorded natural firing patterns from sets of single-neurons were supported by surface EMG measurements from single-motor units.

The difference between the functioning of the healthy neuronal networks with those from patients with spinal cord injury, lead to the treatment called Coordination Dynamics Therapy (CDT). Part of the repair strategy is to improve this impaired phase and frequency coordination by exercising on the special CDT device. The exercising on the special CDT device improves this phase and frequency coordination and consequently improves CNS functioning.

The measurements taken from brain-dead humans, including recordings of natural impulse patterns from muscle spindles, urinary bladder afferents and skin afferents and their effects on the human neuronal networks organization, quantified by the natural impulse pattern from extrafusal and intrafusal motoneurons, helped a few hundred patients to improve the functioning of their nervous systems without risks [14].

Explantation of organs from a brain-dead human may help a few patients, who need replacement organs.

However, recordings of single-nerve fiber action potentials from a group of neurons help to develop and improve causal repair treatments for the injured, malfunctioning or degenerating human CNS in millions of patients.

2. Repair of Remained Neuronal Networks in the Brain-Dead Human

To better understand repair in very severe brain injuries, we have to ask the question whether the remaining neuronal networks in certain brain-dead humans can also be improved in their functioning and is there any sense of doing it?

In the former GDR (German Democratic Republic) where there were only limited diagnostic devices to safely diagnose brain death; to provide a greater safety margin, brain dead humans were sometimes kept over longer periods of time. In ambiguous cases the respirator was often not switched off to allow the severely damaged CNS more time for a possible recovery. One criterion to diagnose brain death was a steadily decreasing body temperature. But sometimes it happened that after temperature decease the temperature increased again as if the patient's CNS were recovering. What had happened? The surviving spinal cord networks reorganized themselves and took temperature regulation over. When brain death is diagnosed, the remaining networks, including the spinal cord and parts of the caudal brain stem normally die quickly. Also, single-nerve fiber action potentials had to be recorded quickly before remaining CNS functions are lost. But in some cases of brain death, when the respirator was not switched off, the remaining networks seem to be able to reorganize and reestablish or take over certain functions again, such as body temperature regulation or even motor function. For example, it once happened that the leg of a brain-dead human leg was hanging out of bed. The patient's leg moved itself back into bed after being touched.

There is therefore evidence that the remaining neuronal network parts in the CNS of a brain-dead human, as the spinal cord with parts of the brain stem, can reorganize themselves and may learn to generate certain functions. Also in complete spinal cord injury, the completely disconnected caudal spinal cord section can be trained to generate a bladder reflex, for example. The difficulty lies in getting the disconnected spinal cord under supra-spinal control.

This poses the question of how is brain death defined. If the whole brain and brain stem died, then the patient has suffered brain death. However, if parts of the brain stem remain functional and 20% to 50% of the brain remains, is such a patient dead or not? Taking a patient, like the coma patient of above, but with a loss of even more nervous tissue. Should great energy be spent in attempting to repair CNS function including movement-based learning? The possibility cannot be excluded that there could exist brain-injured patients who have lost a lot of nervous tissue or increasing brain pressure is in progress, who would recover through CDT but not without therapy.

Given financial pressures, it is regrettably sometimes thought desirable to let the patient die and explant the organs for other patients. Sometimes not much resources or effort is spent on severely injured patients. At the intensive care unit, movement-based learning therapy could be administered.

The patient could exercise or can be made to exercise on the special CDT device in the lying position as in the case of the coma patient (Figure 194A). There are even motivated hospital staff who would be interested in applying movement-based learning (own experience).

3. Learning from Coma Patients

To keep coma patients alive for a few years may cost a lot of money which may be needed urgently for other patients. But if qualified research is performed with severely injured patients, at least some patients would have a better outcome and the society would learn more about the repair possibilities of the human CNS. Patients with mild or medium severe brain injury would in turn benefit from the progress of repair in severely brain-injured patients.

4. Measuring Human CNS Organization at the Neuron Level

As has been shown [14], a method has been developed (single-nerve fiber action potential recording method) to explore human CNS functioning at the neuron level. By being able to analyze natural firing patterns of receptors and the natural impulse patterns (Figure 199) of sets of single neurons in the human CNS, the functioning of human neuronal networks can be analyzed (Figure 25). Convergence and divergence of oscillatory firing single neurons or sub-networks can be studied. Synchronization and de-synchronization has already been measured [14]. Information flow and pattern organization can be studied via multi-frequency coupling. By combining the single-nerve fiber action potential recording method with the single-motor unit surface EMG, functional integration can be measured across very large distances.

5. Progress in Repairing the Human CNS by the Measurements from Brain-Dead Humans and Patients

The possibility of learning about the self-organization of the human CNS was achieved by the recording of single-nerve fiber action potentials from brain-dead humans. Only then was the recording conditions quality sufficiently high to get good enough single AP wave forms for safe identification of impulse patterns. But for the recording of single APs and for the identification of the nerve roots the human neurophysiology and human anatomy has to be known in detail.

Figure 199A shows in a cadaver dissection the opened spinal canal. The dura mater is fixed to have excess to the cauda equine nerve roots. When the arachnoid is not opened one can see the nerve roots swimming in the cerebrospinal fluid. The very thin arachnoid cannot be seen in Figure 199A. After cutting the arachnoid one has excess to the nerve roots for measuring. The nerve roots of the cauda equine are not fixed together and are 'swimming' in the cerebrospinal fluid. The recording layout in brain-dead humans is similar to the cadaver dissection, only the erector spinae is not removed. The root from which should be recorded is lifted out of the fluid and placed on the two pairs of wire electrodes not to shunt the recorded action potentials. A photo of a real recording layout in a patient is shown in Figure 4B. In Figure 199A the recording is shown schematically. The most important contribution of the scientific basis for repairing the human CNS is coming from the usage of this recording method, because it can be measured on the single-neuron level how the human CNS communicates with the outside world via impulse patterns in afferent and efferent nerve fibers.

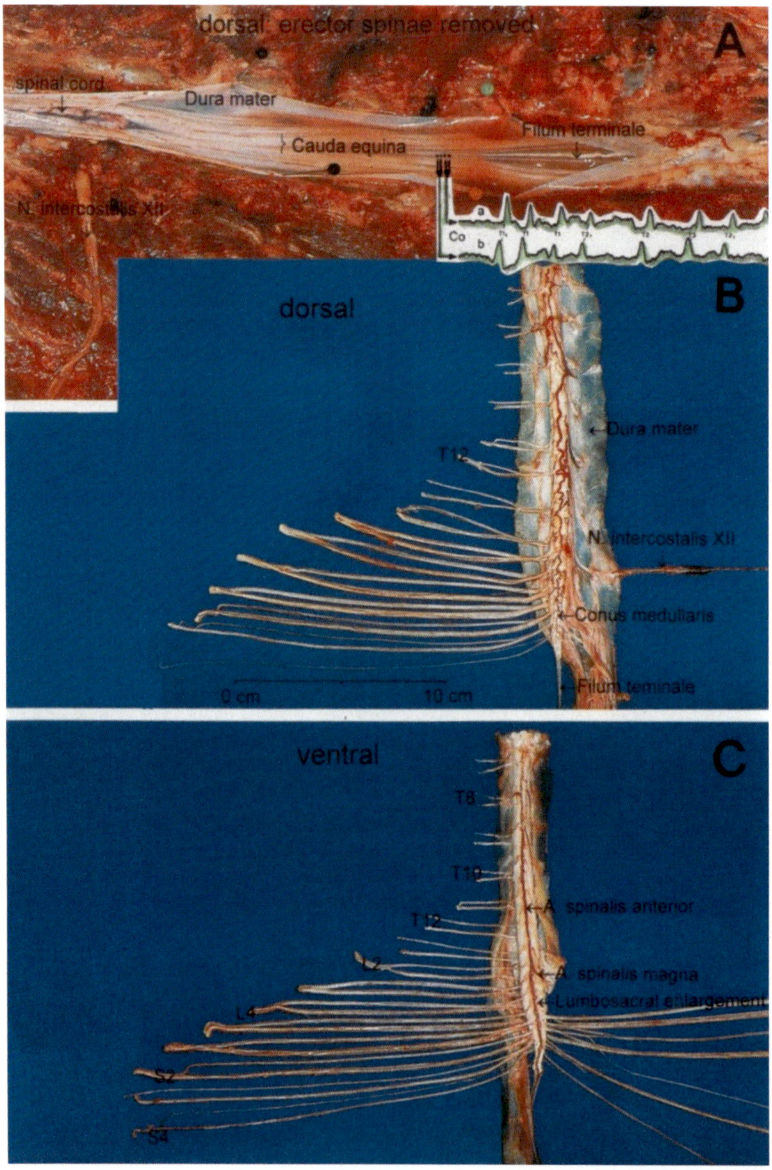

Figure 199. A. Cauda equine and intercostal XII (subcostales). Laminectomy, the spinal dura mater opened and erector spinae removed. Nerve roots lie in a position similar to a horse's tail (cauda equine. Cadaver dissection. The recording of single-nerve fiber action potentials from a coccygeal root is schematically introduced (compare with Figure 1). B. The dorsal spinal cord, the cauda equine nerve roots, the dura mater and the intercostal nerve XII removed and split up. The roots are cut at the dura mater. Note that the ventral roots are thinner than the dorsal roots. C. The ventral spinal cord and the cauda equina nerve roots. The passage of the spinal magna artery indicated. Note the root interconnections of the left side (right side of the Figure).

It has been nearly always so in science that new measuring technics brought progress in research or treatment. It should be emphasized that this recording technic is unique, because it can measure simultaneously the afferent and efferent impulse traffic under natural conditions on the single-neuron level in human and can be used to measure partly the organization of the human neuronal networks under physiologic and pathologic conditions.

From peripheral nerves single-nerve fiber APs cannot be recorded, because the epineurium shunts the AP currents. In frogs, on the other hand, single-nerve APs can be recorded because the nerves have nearly no epineurium, which shunts the currents during the recording [14]. Only, more understanding of the functioning of the frog brain will only of little help for repairing the human CNS and continence is completely different organized.

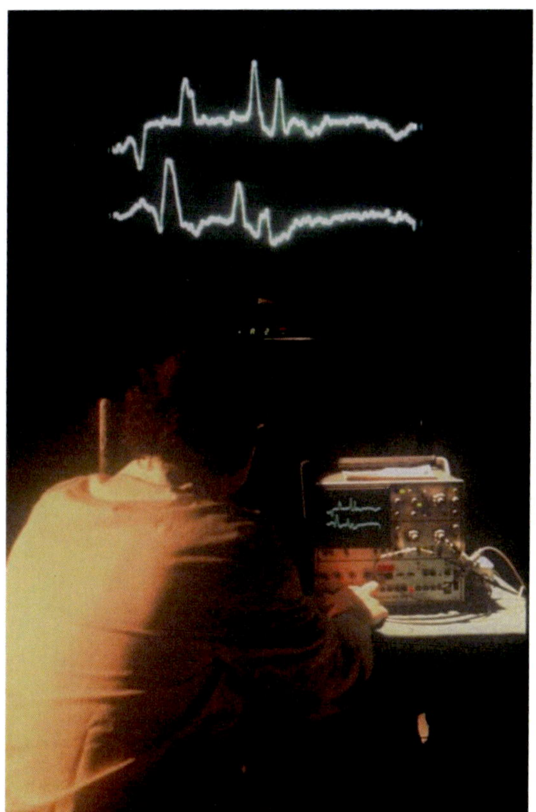

Figure 199D. Analyzing of impulse patterns by the Author from the stored recorded single-nerve fiber action potentials.

Figures 199BC show removed caudal spinal cords. The nerve roots are shifted and placed in a way to see the dorsal and ventral nerve roots. It can be seen that the sacral ventral roots are much thinner than the dorsal roots. Ideal roots for recording single-nerve fiber action potentials are the ventral S4 root (Figure 199C), a cross section is shown in Figure 1, the S5 root and the Co root, because they are thin and contain not too many fibers to allow the separation of impulse patterns.

The anatomical knowledge from brain-dead humans and approximately 100 cadaver dissections is partly used in operations to identify nerve roots during surgery, when the spinal canal is opened only little (Figure 4B).

After finding a good recording position and condition with the two wire electrodes (Figure 199A) slight impulse patterns of single-nerve fibers are obtained (and stored) somewhat in the middle of the human body and the organization of the human CNS can be analyzed (Figure 199D) at the single-neuron level under natural conditions.

Sometimes even thin nerve root interconnections do exist, so that it becomes possible to record from thin nerve fiber bundles to obtain simultaneously high quality recordings of impulse patterns of afferent and efferent single-nerve fibers. Such nerve root interconnections can be seen in Figure 199E on the lower right side of the picture. The picture documents original anatomical research performed by the Author.

One way to explore the human CNS is by starting with animal research only to reach the human level later on. But often the research goal of the human CNS is not reached. A second way to analyze the human CNS on the single-neuron level is to start at the most simplest part of the human CNS and this is the spinal cord and especially its conus medullaris (Figure 199E), where also recordings from single neurons can be taken. In the conus medullaris of humans mainly the urinary bladder and sexual functions are located. In animals, on the other hand, excessive tale functions are organized in the conus medullaris additionally so that the CNS neuronal network structure is more complex and more difficult to analyze. As this research shows, the simplest part of the human CNS is already tremendously complicated. In animal research with respect to clinical application it is normally directly switched from the animal research to the clinical trial, leaving the human research stage out with the consequence that many treatments developed in animals do not work in human patients.

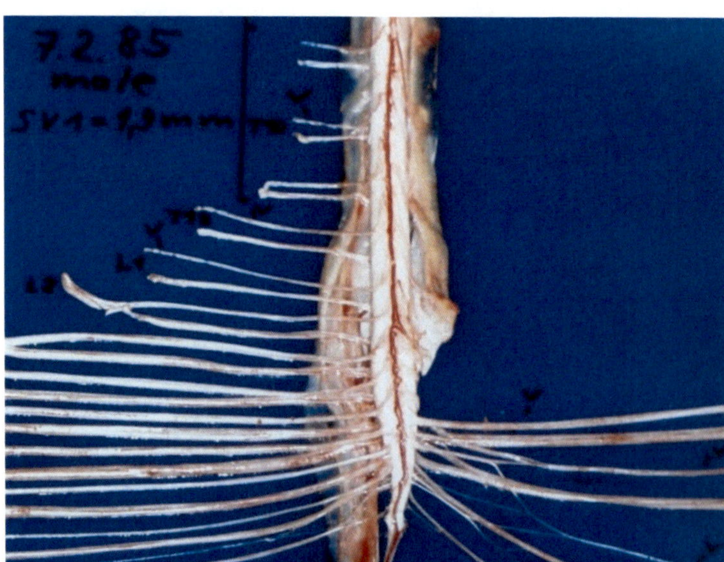

Figure 199E. Original view of the ventral caudal human spinal cord. Dissection performed by the Author. Note the root interconnections at the right lower part of the picture. V= ventral, S = sacral, L = lumbar.

Chapter XI

Improvement of CNS Functioning in Brain-Injured Athletes

1. Paralympic Games

Since the functioning of the human CNS can be improved or repaired by movement-based learning, it is a good idea to organize competitions for certain movements among different kinds of disabled patients. For patients with nervous system injuries, running, jumping and exercising on the special CDT device would be good. Since different kinds of movements are needed for nervous system repair, it would be even better to have competitions in which different kinds of movements to be performed as in triathlon or pentathlon.

2. Training among Disabled Athletes

Since Benjamin could not obtain optimal treatment at all times, he decided to try to repair his CNS further with disability sport for. He decided to train 100m and 200m sprint among healthy athletes.

According to the brain injury he had something like a hemiparesis on the left side. Since the legs are generated by simpler neural networks than hands and arms and therefore are easier to repair (like in stroke), the legs and feet could perform better than arms and hands. During forward walking the left leg was swinging laterally; the walking performance was poor. During backward walking the positioning of the left leg was better. During running in the forward or backward direction the performance was best. With the high activation of the neuronal networks during running, probably the mainly healthy networks influenced (majored) the pathologic ones and forced them to adapt.

Benjamin could run well, in spite of the severe brain injury, because the injury was such that running was preserved. Walking was trained in the vigilant coma (Figure 118B). Shortly after the regression of the coma Benjamin could run quite well (Figs.115B (right), 118C). It is disappointing that Benjamin could move the left arm and hand better (Figure 118C) before the long break in therapy in 2004 (Figure 128) than in 2013 (Figs.200B,201C). It seems as if it is very difficult to compensate later on for a missed treatment during development. Therefore a brain injury or malformation has to be treated as early as possible. Benjamin is now 24 years-old and was then 15-years-old.

Figure 200. Training of two patients with severe brain injury (24 (Benjamin, left) and 28 (Sotiris, right) years old) in interpersonal coordination. A. Patients are walking in interpersonal coordination. Both benefitted from it. Sotiris was contributing the coordinated arm and leg movements and Benjamin the speed and the motivation. B. The additional walking of the Author in interpersonal coordination with the patients did not improve the walking performance of the two patients. C. Bear walking. D. Spider walking. Spider walking is especially beneficial for repair because the palms of the hands and the soles of the feet are touching the ground (inducing important movement-induced afferent input to the CNS for better organization) even if the Achilles tendons are too short and some spasticity is present.

To improve Benjamin's running and CNS functioning further, an interval training of running was trained. But instead of having a rest between two sprints, he exercised on a special CDT device in between the sprints. The sprint was performed below the limit not to exhaust him. The reason to alternate running and exercising on the special CDT device was as follows. With the improvement of running in the short-term memory, the phase and frequency coordination of neuron firing would get worse in the short-term memory. To compensate for that, he improved the phase and frequency coordination by exercising on the special CDT device. After interval running Benjamin trained other movement on the sports field, like bear walking (Figure 200C) and spider walking (Figure 200D). Spider walking is especially beneficial because the sole of the foot is getting good natural afferent input at the same time as the palm of the hand. Simultaneous coordinated input to the palm of the hand and the sole of the foot is a developmentally important input for repair. The Author experienced earlier that when a brain-injured patient is walking on a treadmill and has a lot of spasticity, then, touching the hand reduced spasticity and improved the walking pattern. First the Author trained with the patient Benjamin alone and then together with another patient (Sotiris) with a similar injury and age.

Sotiris had a balanced severe brain injury (Figure 45) and re-learned to crawl, jump and walk. He had nearly no spasticity. He had poor balance, was slow moving and had cognitive difficulties.

Benjamin and Sotiris could train very well together. They both benefitted from the training. Sotiris dictated the coordination by interpersonal coordination and Benjamin speeded up the walking in the forward and backward direction (Figure 200A).

Benjamin and Sotiris trained walking in interpersonal coordination to improve the performance and they trained running. During running there was no interpersonal coordination, but Benjamin could push Sotiris to enlarge the session of interval training of running.

Also, Sotiris had no real argument to complain about the high intensity of training because both, Benjamin and Sotiris, were fit and Benjamin had the mental power to train at the limit into the range of overreaching. There was competition between them, even though Benjamin could run much faster than Sotiris (Figure 201A,B,C).

Figure 201. High-intensity interval training (HIT) of two brain-injured patients. A,B,C. Patients are running together. Sotiris was running only slowly because of balance problems but arms and legs moved coordinated. Benjamin was running much faster but had spasticity in the left hand and the neck (head pointed to the left side) (C). D. During the interval both patients trained alternately on the special CDT device. In the shown situation Sotiris was pushing Benjamin to turn faster. E. After the HIT the patients trained with the Author on the special CDT devices and the sky-walker. Benjamin is exercising on the sky-walker (left) to reduce the spasticity in the left hand and Sotiris exercises on the special CDT device when sitting on the ball to train simultaneously the balance.

The times for running with intervals of special CDT training are plotted in Figure 202. As can be seen from Figure 202, Benjamin was running approximately 100m in 15 seconds. Twice as fast as Sotiris (30s/100m). Both still benefitted from it. When exercising on the special CDT device in between the sprints they were pushing each other to exercise more precisely (Figure 201D). This was also the case during bear (Figure 200C) and spider walking (Figure 200D).

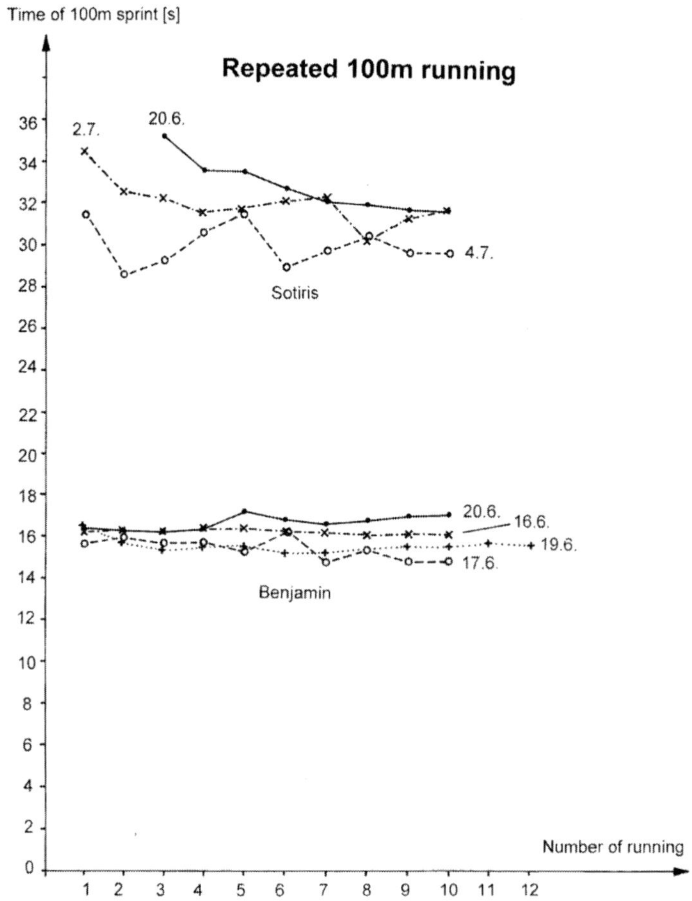

Figure 202. Hundred meter running times of the high-intensity interval training of two brain-injured patients. Numbers at curves are the dates of the running days in 2013.

3. Running and Special CDT Device Interval Training in Disabled Athletes

The running times in sequence of Benjamin and Sotiris with exercising on the special CDT device during the intervals are drawn in Figure 202. As can be seen, Benjamin was running much faster than Sotiris. The stability of the running pattern in Benjamin was higher because the running times varied less. In Sotiris the curves of the repeated running times declined more than in Benjamin indicating that his CNS functioning had a higher potential to be improved further, or put it differently, it should be easier to further improve the running times in Sotiris than in Benjamin.

4. Running and Special CDT Device Interval Training in Healthy Athletes

Running and special CDT device interval training can also be performed in healthy athletes. No CNS functions optimally with respect to phase and frequency coordination among the neurons, also not that of healthy athletes. In some athletes the coordinated firing of neurons, quantified by the coordination dynamics of coordinated arm and leg movements, is very good and poor in others. Improving CNS function via improving the coordinated firing of neurons and in consequence the coordinated movements, the performance and stability of the athlete's movements can be improved. Because the coordinated firing of the motor units also improves, muscle power will also improve. The benefits of this are high, especially in sport disciplines in which coordination is essential.

5. Ethics in Sport of Disabled Athletes

Benjamin could run 100m in the range of 15s. One obstacle to even faster running was the spasticity of the left side caused by the hemiparesis due to the brain injury. Since arm and hand movements are generated by more complex neural networks than the legs and feet, the injury affected most the movements of the left arm and hand. To make Benjamin run faster, there are two possibilities. First, to give up the left arm and optimize the running with one arm only moving and second, to try to reduce the spasticity in the left arm and hand and to improve their functioning.

The first solution is more successful to run faster at least in the near future, but it is unethical because it indulges the disability. The second possibility is ethical but will only yield results over a much longer period of time. Benjamin's coach, preferred to forsake the left arm and optimize the remaining functions.

The Author, on the other hand, wanted to include the left arm in the running performance to give priority to the repair of Benjamin's CNS. The choice between the two possibilities is not easy. If Benjamin gives up the left arm, he will be placed in a lower category in competitions. Therefore, he will be more likely to win and to earn money with which he could buy devices to improve his CNS function.

Since the CNS can be repaired better by training different movements, there should also be competitions with different movements, something like triathlon or pentathlon. The movements should be chosen with sense.

To have competitions for patients with spinal cord injury in the lying position is unethical; they have to be made to train in the upright position and not in the sitting or lying position and should use arms and legs where possible.

6. Supercompensation

In healthy humans or humans with mild CNS organization impairment there occur phases of transient better performance when performing the high-load test repeatedly over several months (Figures 39, 43, 44).

When the Author is exercising on the special CDT device and is increasing the load continuously after approximately 2000 to 3000 revolutions, he feels that he can train with the same effort at higher loads (Newton's) as if the CNS and the microbiologic machinery were working better. After a break of approximately half an hour this easy turning at high loads (improved performance) ends. Is this supercompensation in the short-term memory? Such supercompensation has to be explored further, because super performance is of interest in sport and probably is important to improve the rate of repair in patients with CNS injury. The supercompensation is probably more than just an improvement of performance during an interval training as jumping on springboard or running. In severe brain injury also the performance improves during interval training before the performance gets worse because of exhaustion. But is this supercompensation? With more understanding a good definition of supercompensation is needed.

Chapter XII

Rate of Repair and the Severity of CNS Injury

1. The Rate of Repair Declines with the Severity of the CNS Injury

The high-load coordination dynamics test is used for quantifying the rate of CNS repair. The time required for learning or repair differs strongly between an uninjured and injured nervous system. A healthy athlete learned to get the high-load test value under $100s^{-1}$ in approximately half a month (Figure 40) and a healthy person needed approximately one or two month. Even among athletes the values can vary quite significantly. Persons with a mild malfunctioning of the CNS needed three (Figs.43,204, mild cerebral palsy) and 4 months (Figure 44, mild epilepsy) to get the high-load value under $100s^{-1}$. The young man with mild cerebral palsy was probably comparable good in movement performance because he walked every day 7km to school, which was healthy in general and improved CNS functioning.

Patients with severe traumatic brain injury could reach the high-load test value of $100s^{-1}$ in approximately 32 (Figure 46b, Sotiris) and 48 months (Figure 128, Benjamin). A cerebral palsy patient with high intelligence but poor walking performance needed 19 months of therapy of varying intensity to reach the high-load value of 100, whereas a cerebral palsy girl with poor intelligence and moderate walking performance could not manage the high-load test after eight years. Over the years she turned only with 20 and 50N (Figure 113, Popi) and never learned to exercise at a load of 150 and 200N.

A patient with an incomplete C5/6 cervical spinal cord injury (approximately 50% injury) reached the high-load test value of $100s^{-1}$ in approximately 13 months (Figure 51; Sten). A patient with a motoric complete C5/6 cervical spinal cord injury could not reach the value of 100 within eight years. She never could exercise by herself at a load of 100N. Exercising at a load 50N was just possible for her (Figure 191, Kadri).

An athlete (Figure 57A) who suffered a thoracic complete spinal cord injury, succeeded in reaching a value of 100 in a few months (Figure 59, Avo, 200N values have similar dependence as the high-load test value). But because of his strong upper-body strength and the ability to go over limits, he could manage to reach that value by turning only with the arms, which is not achieving any repair, only an optimization of what is left.

Two patients with severe traumatic brain injury to whom no CDT was administered for longer times could not reach the high-load test value of 100 within fourteen years (168 months (Figure 128, Mario and Andrej)) and will probably never reach such quality of CNS organization.

Patients with very severe brain injury and being longer than one year in the vigilant coma state never managed the high-load test so far. Also brain dead humans cannot perform the high-load test, even though some reflexes can re-appear in some of them if kept ventilated for longer times.

These data of movement-based learning (for repair or improving CNS functioning) show that the times for repair strongly increase with the severity of the injury (Figure 203). What a healthy person learns in half or a month a patient with a severe brain injury needs approximately three years. Or, what a healthy person learns in one year, patients with severe brain injury cannot learn at all.

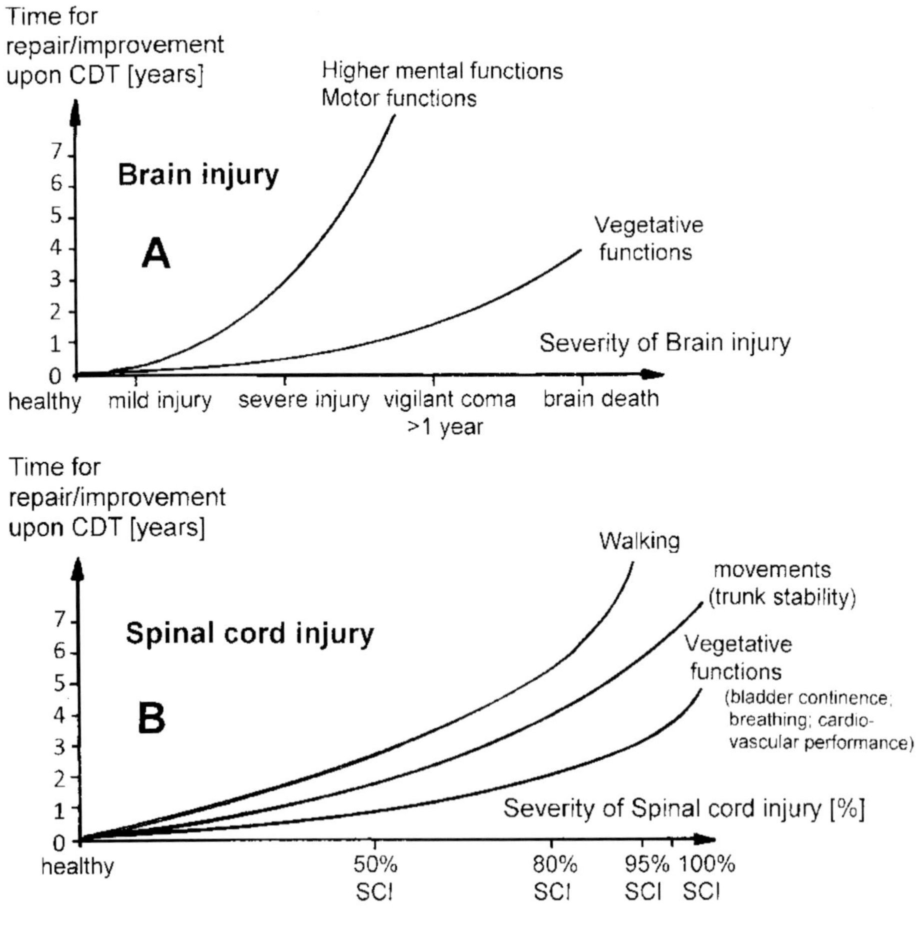

Figure 203. Approximate repair times in dependence on the severity of the brain (A) and spinal cord injury (B).

2. Repair Depends on the Kind of Injury

As was shown above, the repair of movements, quantified by the high-load coordination dynamics values, depends also on the kind of CNS injury. Patients with a complete spinal cord injury (SCI) C5/6 cannot relearn walking, but the higher mental functions are not impaired, whereas patients with a severe brain injury can relearn walking, but their higher mental functions may still be impaired. In SCI the outcome of learning is in part determined by the site of the injury (Figure 29) and injury severity. In brain injury it is more difficult to relate the rate of repair to the site and the severity of the injury. Injury of brain stem nuclei is often not taken into account. Thalamus injury has a strongly negative effect on the rate of repair. In some injuries, for example in hypoxic brain injury, it is difficult to locate the site of injury. Even though it is difficult to clearly relate the rate of repair to the site of brain injury, some important conclusions can be made about the quality of repair.

3. Comparisons of the Improvements of High-Load Cds for the Healthy and Injured CNS

Figure 204 shows the high-load values in sequence of a very well-functioning CNS (A, the Author), a healthy young man or athlete (B), a female patient with mild epilepsy (C) and a young man with a very severe brain injury. The measured values varied between 30 and approximately 300.

Even though the time periods of training vary between one month and a year, it is obvious that the high-load values, characterizing the functioning of the CNS, vary dramatically. That the patient with the severe brain injury and the biggest impairment of phase and frequency coordination had the highest (worst) values is understandable, because his CNS was severely damaged. But that the top athlete Andrus Värnik, World-Championship javelin gold medal winner, does not have similar values to the Author is astonishing. This is because one would think that a top athlete would have a very well-functioning nervous system.

This means firstly that sportsmen can improve their performance further, especially in terms of coordination, by integrating CDT into their training program. Secondly, the CD values characterize only one, albeit an important, aspect of CNS functioning. Football or tennis players could certainly improve their ball skills when including CDT in their training program. By improving the coordination dynamics, there is learning transfer to vegetative and higher mental functions. It is also possible to reduce stress and improve mental disorders, but further research is necessary.

4. High-Load Training Has a High Impact on CNS Repair

The rationale for the Author to perform many high-load tests in sequence was firstly to gain an insight into how hard it is to do them. A physician should feel what he expects from his patient. Secondly, it needed to be ascertained whether the test can be done every day for a sustained period. Thirdly, it is of interest to determine just how good (low) high-load values can be achieved in general. As Figure 204 shows, high-load tests can be done every day without overloading the body, as demonstrated by the fact that the Author's values did not get worse. Performing the high-load test is hard and requires mental fortitude. As expected the values improved continuously with successive test performances.

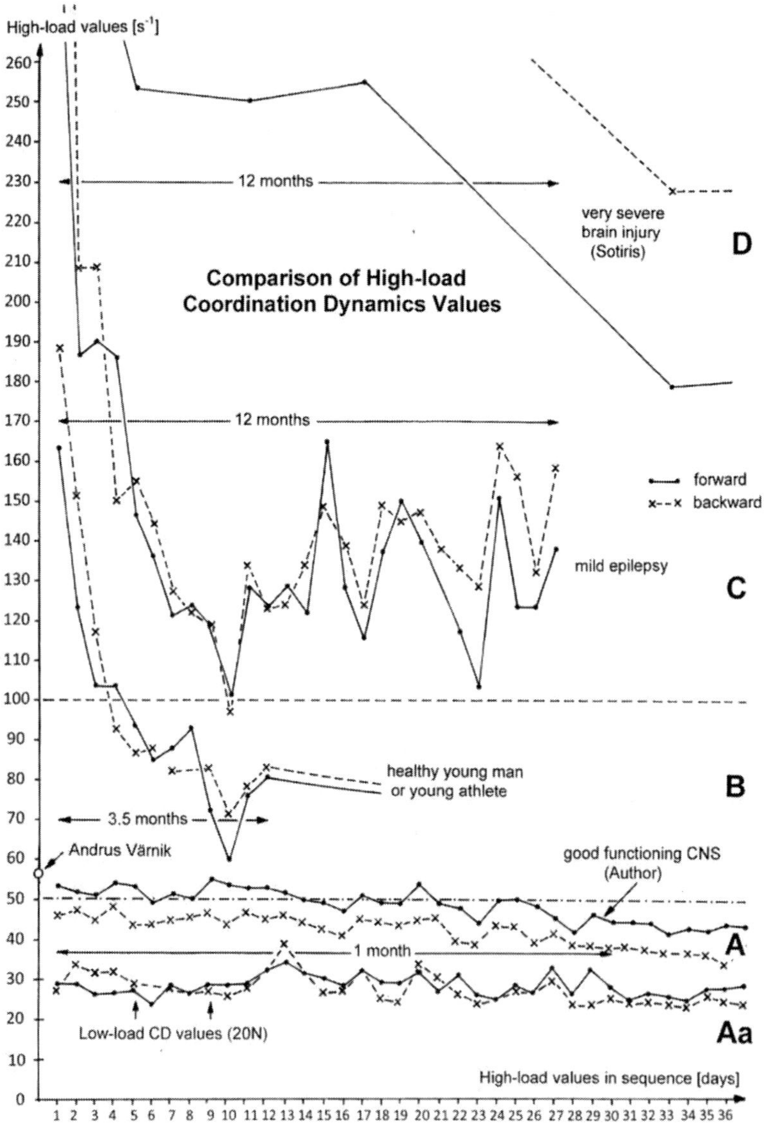

Figure 204. Comparison of the high-load values in sequence of a very well-functioning CNS (A, the Author), a healthy young man or athlete (B), a female patient with mild epilepsy (C) and a young man with a very severe brain injury. "Aa" shows the low-load values in sequence for the Author. The scale for low-load values is that for high-load divided by 10 (for example, 30 to 3). The high-load (HL) CD values are obtained by summing up the single CD values, Δ (HL CD value = Δ20N + Δ50N + 100N + Δ150N + 200N + Δ150N + Δ100N + Δ100N + Δ50N + Δ20N + Δ20N).

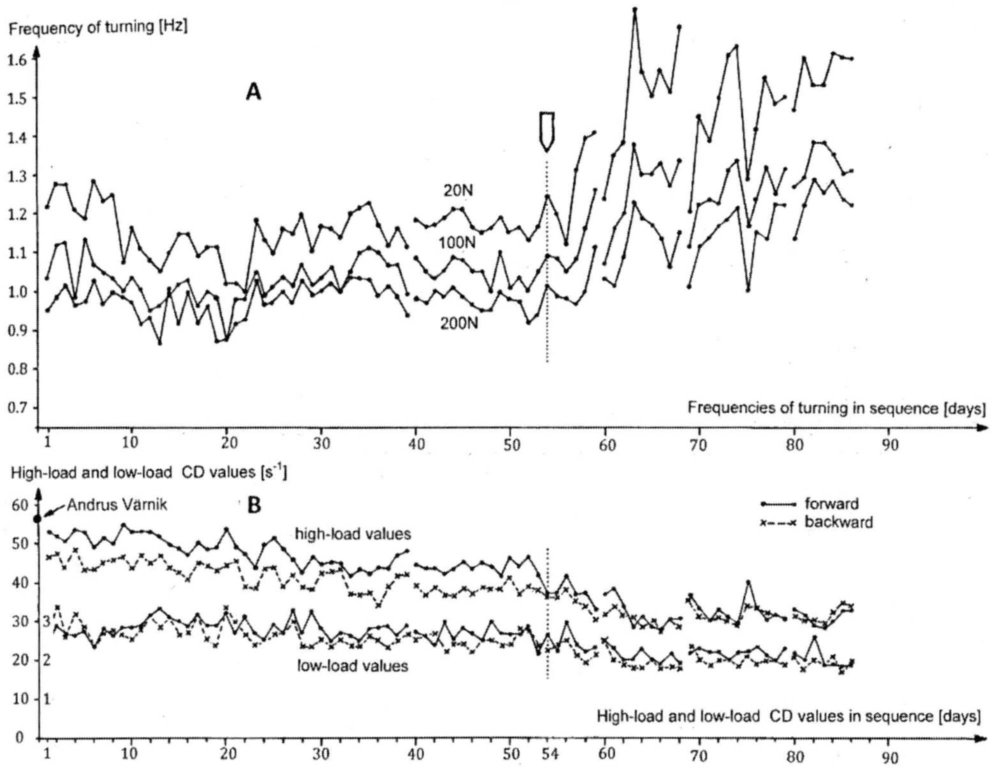

Figure 205. Continuous high-load tests (every day apart from a few breaks) of the Author with a load increase from 20 up to 200N and back to 20N. In "A" the turning frequencies are given for 20, 100 and 200N for exercising in the forward direction. In "B" the high-load test values and the CDs for 20N (lower traces) are given for exercising in the forward and backward direction. Note the strong improvement of the CD values at the 54's day of exercising in sequence.

Unexpectedly, after 54 days of continuous high-load tests (Figure 205) the high-load test values got significantly better and the turning frequency got increased, as if some specific changes had taken place in the CNS and improved its functioning. The improvement of the high-load CD values does not originate in the increased turning frequency f (CD values = df/dt:f), because a person's performance is best if the person turns at his natural frequency. Varying the turning frequency from the inner frequency helps to repair the CNS by increasing motor pattern variability, but the best value for turning is achieved at the own natural frequency and the best values are used to characterize CNS function. That the Author could turn faster with better rhythmicity was therefore due to an improvement in CNS function. The inner frequency increased as a consequence of better CNS function. High-load testing in sequence every day up to 200N seems to improve CNS function and helps in repair, provided of course that the patient is able to do it.

Chapter XIII

CDT on Aging, After Operations and After Cancer Treatment

1. Aging

1.1. To Live Longer with a Better Quality of Life

If only single cells or small groups of cells die in a somewhat distributed manner through parts or all of the CNS, then a repair-friendly microenvironment can be generated for structural and functional repair through CDT. However, if a severe injury is confined to one area, it is then very difficult to create such an environment. Due to the loss of nerve cells, the coordinated firing of neurons becomes impaired and CNS functions become impaired. With CDT the coordination dynamics can be improved and CNS functions become more physiologic again. In the elderly it is unlikely that any significant number of new nerve cells can be built throughout the CNS apart from within some structures like the hippocampus. However, through intensive movement therapy some new nerve cells can probably be built even later in life [21] and structural repair can contribute to functional repair, which is the substantial part. Even if new nerve cells cannot be built, functional repair can still optimize what is left of the degenerating CNS.

Many generations dream of a "Jungbrunnen" (fountain of youth) to become young and healthy again. The famous painter Lucas Cranach had pictured such a "Jungbrunnen". In his painting the old people enter the "Jungbrunnen" one side and leave it young and healthy on the other side (Figure 206). Such a dream cannot become reality by using CDT. However, it is possible to reduce biological age through repair.

In patients with spinal cord (Kadri) and brain injury (Popi) bladder functions could be repaired. Therefore bladder continence can probably be improved in the elderly through exercising on the special CDT device and (if possible) jumping on a springboard. Repair is however compromised in that the main feeder artery of the conus medullaris is often very arteriosclerotic. The blood supply to the sacral micturition and sexual function center is then strongly limited and the functions become impaired. The cardio-vascular performance can be improved upon exercising on the special CDT device, because CDT can improve the functioning of the vegetative nervous system including the micro-circulation of the blood and lymph vessel system. The functioning of the arterioles is controlled by the vegetative nervous system division, which can be improved in its functions upon CDT. An old saying in medicine is "you are as old as your blood vessels are". In an elderly patient with lung fibrosis (the oxygen is hindered to reach the blood vessels because of the fibrosis due to smoking) the oxygen saturation of the blood could be improved upon exercising on the special CDT device. The patient needed less additional oxygen. The patient could breathe better. The progress was probably achieved upon improved blood supply in the lung. The lung functions were optimized. Therefore also the breathing can be improved in elderly. If the functioning of the blood vessels can be improved and new cells can be built upon CDT, then also eye functions should become better upon CDT. Operations may be needed only later. Movements can anyway be improved in elderly. By sitting on a chair or lying during exercising, the elderly can even exercise on special CDT devices if they have prostate problems.

Figure 206. The fountain of youth painted by Lukas Cranach (Der Jungbrunnen von Lukas Cranach).

A patient with amyotrophic lateral sclerosis could maintain his level of health from four hours of CDT per day over the course of a year. When the patient took a long break in therapy and his doctor changed his medication, the patient died. If challenging patients can no longer move by themselves, supported exercising on the special CDT devices can allow them to live longer.

It is also probable that CDT can improve or delay dementia.

1.2. Exogenous Stem Cell Therapy Is Unlikely to Work

Stem cell therapy to improve CNS functioning has been performed in more than fifty clinics around the world. There are no statistics published on how many stem cell therapies were successful and how many patients suffered complications from the therapy as uncontrolled CNS organization, for example epilepsy or uncontrolled growing, for example cancer. Measurements of any improvements are often insufficient and unscientific. If repair is not measured properly, improvement of CNS functioning cannot be judged. If stem cell therapy is accompanied by a movement-based therapy, some progress in CNS functioning can probably be achieved from the movement therapy alone. It is likely that the stem cell treatment is clinically irrelevant. If a patient has had no therapy before, even movement therapy of low efficiency can bring some progress.

From the theoretical point of view it is unlikely that stem cell therapy will be successful. A comparably mild, in this case the spinal cord injury in Figure 29 will be taken as an example. If stem/progenitor cells are injected into the cavity of the injury site, the cells cannot be stimulated to proliferate, migrate and become integrated, because the necessary microenvironment is missing. It is mentioned above that for excitation-neurogenesis coupling Ca^{++} channels have to be opened by membrane depolarization. But the membranes of the injected cells cannot be depolarized, because of missing surrounding activating neuronal networks. If in aging single cells die, then native stem cells can likely be stimulated to divide and proliferate because the necessary microenvironment is present.

A combination of coordination dynamics therapy (CDT) and exogenous stem cell therapy is unlikely to work. As shown in Figure 29, CDT will induce repair around the injury site because of the natural integrative activation of the neural networks. However, this repair-friendly microenvironment will probably not reach the injected cells. Even though new nerve cells are urgently needed for repair (see above), one cannot simply acquire and use them by adding them exogenously. Much more qualified research on adult born and embryonic stem cells is needed, including its relationship to structural repair in humans, before they should be employed in a clinical setting.

Figure 207. Funnel from Nürnberg (Nürnberger Trichter). Fehlt dir‘s an Weisheit in manchen Dingen, Lass dir von Nürnberg den Trichter bringen. ("If you miss wisdom in some fields, let you bring the Funnel from Nuremberg").

The dream of the Early Modern period, namely that it is possible to administer intelligence and wisdom via the funnel of Nürnberg (Nürnberger Trichter, Figure 207), that means improving the complexity and connectivity of CNS functioning, is only a dream. If it were easily possible to change the neuronal networks in human for improved functioning than humans would have no stable character, because the character is manifested in the complex structure and functioning of the neuronal networks. It is not simply a matter of adding new cells. But disregarding all the above arguments and assuming that the injected stem/progenitor cells could be integrated into the existing adult networks, than the repair time would have to be considered. It was shown here that the repair times of the brain and spinal cord are in the range of months to several years (Figure 203). If stem cell therapy would work, then one has to expect that the repair would need months till years and not just a few days to a few weeks as was told to patients to whom stem cell therapy was administered.

2. Epigenetic Modification for Anticancer Treatment by Movement-Based Learning

2.1. Activation of Tumor-Suppressor Gene by Exercise

If movement-based learning can suppress genes by modifying the epigenome, it may also be able to activate genes with an anti-carcinogenic effect.

Long term CDT performed at the limit always moved CNS functioning towards more physiologic and more precise functioning. Unspecific pathophysiologic CNS functions like spasticity reduce. In the framework of the system theory of pattern formation, the stability of physiologic functions increased (the potential well became deeper) and the stability of pathologic states (like spasticity) decreased (the potential well became less deep) (Figures 18,19). Since the nervous system is involved in nearly all body functions, it could well be that CDT enhances specific physiologic functions and inhibits non-specific pathologic functions. Cancer is the uncontrolled unspecific growth of cells. It may therefore be that movement-based learning has an anticancer effect.

It was reported that exercise might exert an anticancer effect by switching on a tumor-suppressor gene [41]. Significant changes in the DNA methylation status of 43 genes occurred in women who engaged, after having completed treatment for breast cancer, in a six month moderate-intensity aerobic exercise program. An association between gene expression level and overall breast cancer survival was found in three of the women. One of these three was L3MBTL1, a putative tumor-suppressor gene. In the exercising patients, L3MBTL1 was de-methylated, indicating an increase in gene expression. Elevated expression of this gene may be associated with a lower risk of recurrence and improved survival [41].

2.2. Authors Own Experience with Anticancer Effect and Body Function Repair upon Coordination Dynamics Therapy (CDT)

The Author suffered a 'squamous cell carcinoma (epithelioma)' (a malign tumor) in the maxilla (stage between 1 and 2). The tumor was removed by surgery and a neck dissection was performed: Two lymph nodes with formation of metastases were removed and two stages of further lymph nodes and lymph vessels were removed. Radiation therapy and chemotherapy were administered to the tumor area to reduce the risk of tumor recurrence from 30% to 15%. To mitigate the side effects of the anti-cancer treatment, the Author performed CDT for fifteen hours per week, increasing to twenty-five hours per week. Three biopsies of strange growing in the following five years and other diagnostics, including two PET, showed no sign of malign growth. The exercising on the special CDT device involved coordinated arm, leg and trunk movements and additionally simultaneously coordinated neck, tongue and lip movements.

One year after the tumor extirpation and neck dissection, a reconstruction of the maxilla was performed. The Author was fit by the time he went for the fibula-transplantation operation. He even exercised one hour before the operation on the special CDT device. During the nine hour operation, pieces of the fibula and the skin with its supplying artery and vein were removed from the right leg. Bone and skin were placed to the former site of the tumor and artery and vein were connected with the temporalis blood vessels. The transplant had to receive blood supply within seven hours in order to avoid rejection. The Author was informed that during the long operation he will get edema and breathing difficulties because of the removed lymph nodes and vessels following a previous neck dissection. However, only minor neck edema occurred during the operation because the CDT had partly repaired the lymph system. One day after the reconstructive operation parts of the transplant were blue, indicating poor blood supply, even though the transplanted artery was open and was supplying blood. The surgeon informed the Author that he would lose some tissue because of poor blood supply.

The Author, having brought the special CDT device into the hospital, exercised on the special device, in spite of the many tubes he was connected to, to improve the blood supply to the head including the transplant site. As the result of the enhancement of blood circulation, the transplant area received enough blood and healed fully. No transplanted tissue was lost. The surgeon was amazed that his complicated work was fully successful. Therefore CDT had helped to reduce the risk severe neck edema occurred during the operation and a tracheotomy could be avoided and had improved the blood supply also in the head so that the transplant was fully accepted. The Author's hour training before the operation may have helped in this long-lasting and difficult operation and may have helped to improve the lymph circulation in the short-term memory to reduce edema during the operation. Later on in several small operations performed under local anesthesia parts of the fibula bone had to be removed, because the fibula bone piece was too large for the implantation of teeth, a vestibule had to be reconstructed and implants for the teeth made.

In conclusion, all surgery went well, aided by the continuously performed CDT. CDT helped to repair the body after cancer treatment and may have helped so far to exert anticancer effect. CDT includes not only the exercising on the special device but also fast walking and jogging. Since the Author had a hip necrosis on the right side and could only walk with pain, he decided to get an artificial hip to be able to walk well again and to be able to jog a little. The hip operation was successful (see below), so that the Author could again perform the whole movement program of CDT.

Interestingly, the skin from the leg changed into mucosa in the mouth after transplantation and the hair of the skin did not grow any more in the mouth, even though one hair follicle seem to be still there as the Author can feel with the tongue. The dentist argued that nearby mucosa cells migrate into the transplanted skin area, because the changes from skin to mucosa started from the sites of healthy mucosa. But it may also be that the change from skin to mucosa cells was induced by gene expression changes induced by function and/or surrounding changes.

It is very important to note that following the neck dissection (and tumor resection, radiation and chemo therapy) lost or impaired automatisms such as taste, swallowing, coughing and others which partly reappeared spontaneously were enhanced by CDT. I believe that the exercising of the complicated coordination's between pace and trot gait when exercising on the special CDT device, which stimulate the autonomic nervous system, helped to achieve better repair. Even though the eyes are a part of the nervous system and should be reached by CDT, the improvement of the eye functions (wetting of the cornea, opacity of the vitreous body) seems to be most difficult to repair by CDT. Five years after the cancer treatment the wetting of the cornea and the opacity of the vitreous body had improved. Based upon CDT most likely new cells of the eyes were built to improve eye functions. The further repair of mouth, throat and neck functions and automatisms improved quality of life dramatically. Even though, after five years of CDT not all side effects of the chemo and radiation therapy have disappeared. Sleep is still more disturbed than before the occurrence of the cancer. The chemotherapy probably impaired nervous system functioning and in consequence many body regulations. It is known that the drug Vinblastine, which was administered to the Author, is nervous system toxic. The patients may get polyneuropathy.

2.3. Difference in the Power of Regeneration

The reconstructive surgery performed one year after the cancer operation was a big load to the Author and was important for the separation of the nasal and oral cavities. It is also interesting with respect to the power of regeneration of peripheral nerve fibers. With the extirpation of the epithelioma and the surrounding maxilla and teeth, one to two cm of the upper lip became denervated. Since there was no regeneration over the following five years, there was no feeling in the denervated part of the lip and the corresponding lip muscles atrophied. There seem to have been no sprouting of nerve fibers from the healthy lip parts to the denervated ones. On the other hand, the mucosa of the mouth quickly re-innervated, regardless of whether skin from the leg, mucosa from another part of the mouth or pig skin was used for implantation and covering of the transplanted bone.

The area where a transplant was made changed into mucosa and quickly re-innervated. If the healing in the mouth had not been that strong, the Author would have had difficulty with eating and speech. Every change in the oral cavity changes the sounds of speech. The healthy speech centers of the brain have such plasticity that the speech is repaired within two to three weeks. If, on the other hand, the speech centers are impaired by a traumatic brain injury or malformation, then it is very difficult to improve speech. It needs years. It is the CNS which is the genius and can compensate for many problems in the periphery.

It is believed that the power of regeneration of nerve fibers is bigger the closer the injury is to the brain (more rostral). As discussed above, lips and oral cavity mucosa are both close to the brain, but their power of regeneration was very different. High quality reconstructive surgery in combination with movement medicine can show what repair and healing is possible. In cervical C3/C4 spinal cord injuries for example breathing is impaired in addition to the loss of muscle functions. Volitional movement-based learning therapy is not possible any more. But possibly the phrenic nerve can be repaired by passive movement therapy. Only qualified treatment can show which functions can be repaired and which ones cannot.

2.4. CDT after Hip Replacement

The Author developed a necrotic hip joint on the right side. Exercising on special CDT devices improves normal joint function. However, this hip joint could not be improved in its functioning. There was probably a genetic defect. To reduce the side effects of the cancer treatment the Author performed CDT (see above). However, it is also necessary to perform walking and running. The Author therefore decided to get an artificial hip joint in order to walk again without pain. One day before entering the hospital the Author brought his special CDT device to the hospital. I exercised before and after the hip operation, apart from the one day after the operation when resting in bed was mandatory. When leaving the hospital after a few days, the ward physician said that the healing process was fast.

The surgeon, who operated on the Author, advised me that I should not walk too much in order not to overload the hip. Otherwise the healing process, including the building of a new joint capsule would take longer. Two orthopedic physicians upon whom he operated trained too much and the healing process took six months longer. At home the Author exercised on the special device and walked and ran a bit. When he was walking too much the pain increased. The overloading of the transplanted hip set the Author back approximately two weeks in the recovery process. Moving improves the healing process, but the patient has to be careful not to over-train during the healing process. The exercising on the special CDT device was very good for improving the healing process and because very little load is exerted on the hip joint.

Generally exercising on the special CDT device in the sitting or lying position improves joint functioning because the movement stimulus for repair is given; the joint should get enough time to heal and overloading should be avoided.

2.5. CDT after Breast Cancer Treatment to Reduce Edema

As reported above, in order to reduce the risk of a tumor recurrence after the removal of the epithelioma radiation and chemotherapy (Vinblastine) was administered. Lymph nodes and lymph vessels were removed from the neck during the neck dissection. The lymph drainage became impaired and edema occurred in the face and the neck. The Author could successfully reduce the edema within one year when exercising on the special CDT device for at least one hour per day. With the coordinated arm and leg movements, head, lips and tongue were moved in coordination with the arm movements. During the nine hour reconstruction of the maxilla, only minimal edema occurred so that a tracheotomy could be avoided. This repair of the lymph system means that the improvement of the cardio-vascular performance upon CDT means that not only the blood supply becomes improved but also the lymph vessel system for the whole body is repaired. Such a lymph vessel repair will not only be possible for the neck and face by CDT. The lymphatic system inside the body will also be repaired.

Further, it is likely that, with a repaired lymphatic system, the immune system is in a better position to fight against remaining cancer cells.

A gynecologist colleague of the Author suffered breast cancer at the age of thirty-four. The cancer and the lymph nodes of the axilla were removed. As usual, edema appeared in the axilla. With exercising on the special CDT device the edema reduced because of a reconstruction of the lymphatic system in the axilla and a better drainage of the lymph due to the movement. The treating physician who was draining the lymph and was astonished that only so little edema was generated. When the gynecologist explained him that she reduced the edema by exercising on the special CDT device, he did want to listen to her, even though she was a colleague.

2.6. Prolonged Fasting and CDT May Trigger Stem Cell-Based Regeneration of Damaged, Old Immune or Other Systems

It was reported that exercise might exert an anticancer effect by switching on a tumor-suppressor gene [41]. Therefore the Author performed CDT for 15 hours per week to lower the risk of a recurrence of cancer (see above). Vision was also impaired with a delay of one year (cloudiness). While chemotherapy, in this case Vinblastine, and radiation save lives, it causes significant collateral damage to the immune and other systems. In the Author at least the immune system (occurrence of herpes), the nervous system (poorer sleep, frequency sound in the ears) and the eye functions were damaged. The toxic damage of the nervous system can hopefully be repaired by CDT over several years. But can the immune system be regenerated efficiently?

It was recently reported that cycles of prolonged fasting not only protect against immune system damage—a major side effect of chemotherapy—but also induce immune system regeneration, shifting stem cells from a dormant state to a state of self-renewal [195].

Because during prolonged fasting organisms minimize energy expenditure in part by rapidly reducing the size of a wide range of tissues, organs, and cellular populations including blood cells, the reversal of this effect during a resumption of eating represents one effect of the most potent strategies to regenerate and/or rejuvenate damaged systems and organs in a coordinated manner. With respect to the immune system that means that when you starve, the system tries to save energy, and one of the things it can do to save energy is to recycle a lot of the immune cells that are not needed, especially those that may be damaged. "PKA is the key gene that needs to shut down in order for these stem cells to switch into regenerative mode. It gives the 'okay' for stem cells to go ahead and begin proliferating and rebuild the entire system" [195].

The body got rid of the parts of the system that might be damaged or old, the inefficient parts, during the fasting. This is the assumption. Now, if you start with a system heavily damaged by chemotherapy or aging, fasting cycles can generate, literally, a new immune system. Reported results indicate that cycles of an extreme dietary intervention represent a powerful means to modulate key regulators of cellular protection and tissue regeneration but also provide a potential therapy to reverse or alleviate the immunosuppression or immunosenescence caused by chemotherapy treatment and aging, respectively, and possibly by a variety of diseases affecting the hematopoietic and immune systems and other systems and organs [195].

The possibility cannot be excluded that prolonged fasting, in combination with CDT may improve CNS functioning after hypoxic or traumatic brain injury. The genetic code is probably not damaged. The state of the epigenome is unclear.

The feasibility and effect of fasting in cancer patients undergoing chemotherapy is quite complicated [199,200]. Practice guidelines for patients undergoing/underwent chemotherapy need to be established. Still, calorie restriction or short term starvation may be an effective and reproducible intervention for increasing life span, reducing oxidative damage, enhancing stress resistance and delaying/preventing aging and age-associated diseases.

2.7. Prolonged Fasting, CDT and Training at Power Limit to Rejuvenate the Body

The Author decided to fast in addition to performing CDT to regenerate and/or rejuvenate his whole body. By training, best at limits, and resuming eating after fasting one may get new cells and may go in the direction of fountain of youth. For regeneration and/or rejuvenation three simultaneous therapies were used by the Author: Firstly, CDT was performed, that means movement-based learning, to repair the nervous system and may switch on tumor-suppressor genes. In the morning, when exercising up to 3000 turns and 130N, the super-compensation was reached in the short-term memory at approximately 2500 turns. This was felt by the Author that he could turn easily and fast for approximately 10 minutes. Secondly, by performing the high-load test and going to the limits by exercising at 200N, new nerve cells and other cells may be building. Thirdly, prolonged fasting should replace damaged cells due to prolonged anesthesia, chemotherapy, radiotherapy and aging by newly build healthy cells. The fasten periods were three to five days. The macroscopic movements performed during CDT may give the micro-physiologic environment and processes the information needed to achieve the physiologic task of regeneration and/or rejuvenation of the body.

It seems also logic to fast and move strongly at the same time to rejuvenate the body. Our genetics were made or adapted in the last five million years. When prehistoric men were hungry, they had to run first with an empty stomach to catch animals to get food. Food restriction and running and fighting (coordinated movements) therefore also formed genetics by selection. Also nowadays, after or during wars, when people did not have sufficient food and had to walk long distances, they seem to have been healthier than when there is plenty of food and no need to move much.

2.8. Reduction of the Blood Pressure by Prolonged Fasting and CDT

Figure 208 shows the improvements in health, quantified by high-load (HL) CD values and blood pressure, when performing CDT, HL-exercising and repeated prolonged fasting. In "A" the high-load (a) and the low-load (b) CD values are shown, which increase slightly with the seven periods of fasting. After termination of prolonged fasting, the CD values recovered to values of before fasting. The increase and decrease of the HL CD values can be best seen with the continuous five days fasting. The low-load CD values changed only little.

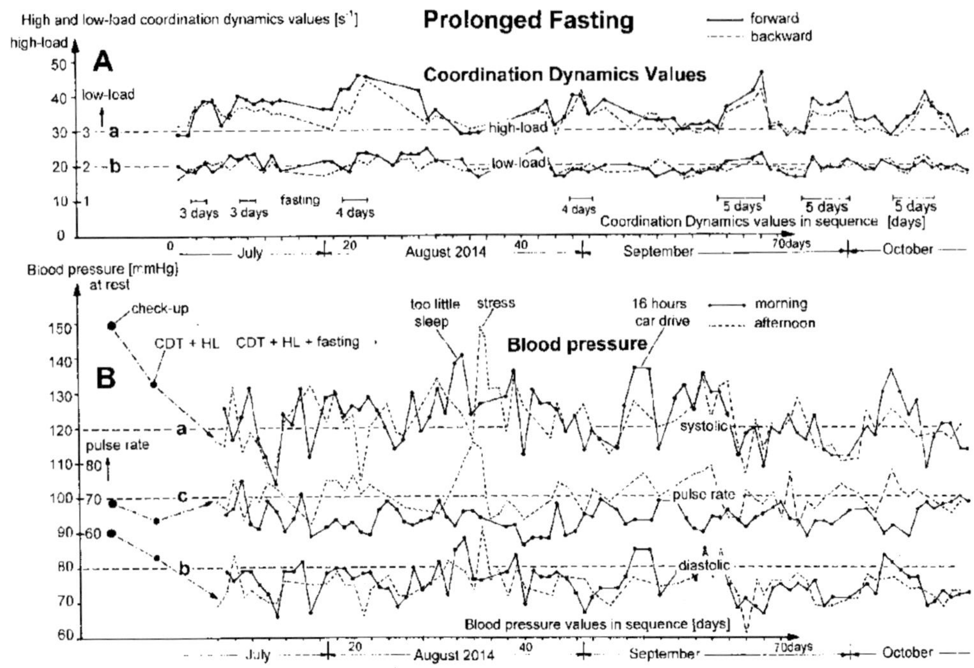

Figure 208. High (a) and low-load (b) coordination dynamics (CD) values (A) and blood pressure and pulse rate (B) in dependence of prolonged fasting during CDT. Note that the high-load CD values increase transiently following fasting. Note further, the systolic (Ba) and diastolic blood pressure (Bb) decrease transiently with prolonged fasting especially with 5 days fasting.

It is important to lower blood pressure. In Figure 208B a blood pressure comparison is given with CDT (check-up) alone, with CDT and HL exercising, and with CDT, HL exercising and repeated fasting together. The blood pressure reduced from performing only CDT (150/90) through performing CDT and high-load exercising (~133/84) to performing CDT, HL exercising and fasting (~120/75). The pulse rate did not change very much. The reduction in blood pressure was unexpected and was therefore measured only occasionally at the beginning. In the morning the blood pressure was mostly measured at rest approximately 30 minutes after exercising on the special CDT device when going with the load up to 130N. In the afternoon the pressure was measured after exercising up to 200N. It seems that the blood pressure was a bit lower after exercising up to 200N. Anyhow, it seemed that the prolonged fasting and CDT, including high-load exercising, have a natural impact on lowering of the blood pressure. When the Author visited a practitioner, she said that I may have to use β-blocker in the future because of the high diastolic blood pressure. But the blood pressures reduced with fasting and CDT and were after treatment in the range of 120/75 which is healthy for men. It has to be seen how much the blood pressure will rise after terminating repeated prolonged fasting and high-load exercising.

With respect to the lowering of the diastolic blood pressure it seems that the repeated prolonged fasting for five days and the eating again for five days was most efficient to reduce the blood pressure. But it can also be seen from Figure 208B that too little sleep, long car journeys (16 hours) and especially unpleasant noises (stress) increased the blood pressure strongly. The reduction of the blood pressure by CDT and fasting only makes sense if dangerous blood pressure increases can be avoided.

2.9. Can the Administration of CDT, HL Exercising and Prolonged Fasting Replace in Some Cases B-Blocker Administration?

Perhaps in some cases, the performed simultaneous therapies (prolonged fasting and CDT) are an alternative to the usage of β-blocker. But to exercise on the special CDT device during fasting up to a load of 200N is hard and needs a lot of mental discipline and the cardio-vascular performance has also to be good to not endanger the heart.

2.10. First Results of Reduction of the Biological Age by Prolonged Fasting and Exercising

The Author (180cm) feels good after the repeated prolonged fasting, probably mainly because of the reduction of the body weight from 91kg (slightly overweight) to 80kg (normal weight). The weight reduction by 11kg for altogether 29 days of fasting results in approximately 0.4kg per fasting day. The Author ate approximately 15 cranberries each day in order to take necessary vitamins. To digest all the vitamins he was eating a bit, something like 0.1 liter milk with 2% fat. Traditional knowledge in Finland is that five cranberries were sufficient to get all vitamins per day if they are not frozen and were kept in cold water.

Whether the blood pressure stays down after prolonged fasting and HL exercising remains to be seen. Even during 5 days fasting and exercising at high load (200N) on the special CDT device no heart problems could be felt. Stress on the other hand the Author feels immediately on the heart. Obviously the stress is the risk factor for a heart attack and not the exercising. It even seems that the exercising on the special CDT device at low to medium loads seem to calm the nervous system down. Future research has to show what exercise conditions on the special CDT device can reduce the over-excitation of the sympathetic nervous system division.

In patients with brain or spinal cord injury we expected structural repair of the neuronal networks after one year or more due to the building of new cells. Therefore we can expect rejuvenation due to exercising and prolonged fasting up to one year or more. Such long times are supported by the recovery time from a hip transplantation. Even though the Author is exercising quite a lot, he has still not fully recovered from the implantation of an artificial hip on the right side of two years ago. The gluteus maximus and medius are still smaller than on the healthy side as the Author can see easily in the mirror. But he does not feel any impairment of moving due to the reduced muscles. Jumping on springboard may help the partly atrophied muscles recover more quickly in that it provides a means of non-weight bearing exercise. But vigorous jumping may overload the surrounding tissue of the artificial hip. It is interesting and disappointing that through walking with the majority of the weight on the left healthy side to avoid pain for more than five years has damaged nervous system a little. Even though the Author is very good at exercising for high loads on the special CDT device, still his CNS has difficulty in generating the pace and trot gate coordination patterns

(Figure 208Cb, lower trace, 137N). The pathologic walking due to pain trained the CNS to generate pathologic movement patterns, which takes more than two years to get rid of. Therefore for a hip transplantation one may not wait too long for the operation.

In patients with CNS injury structural repair was best when treatment was applied continuously for longer periods than one year. However exercising and repeated prolonged fasting are demanding and takes a lot of time. The Author has done treatment in the range of 6 to 8 months. This could be worthwhile to go on for some time.

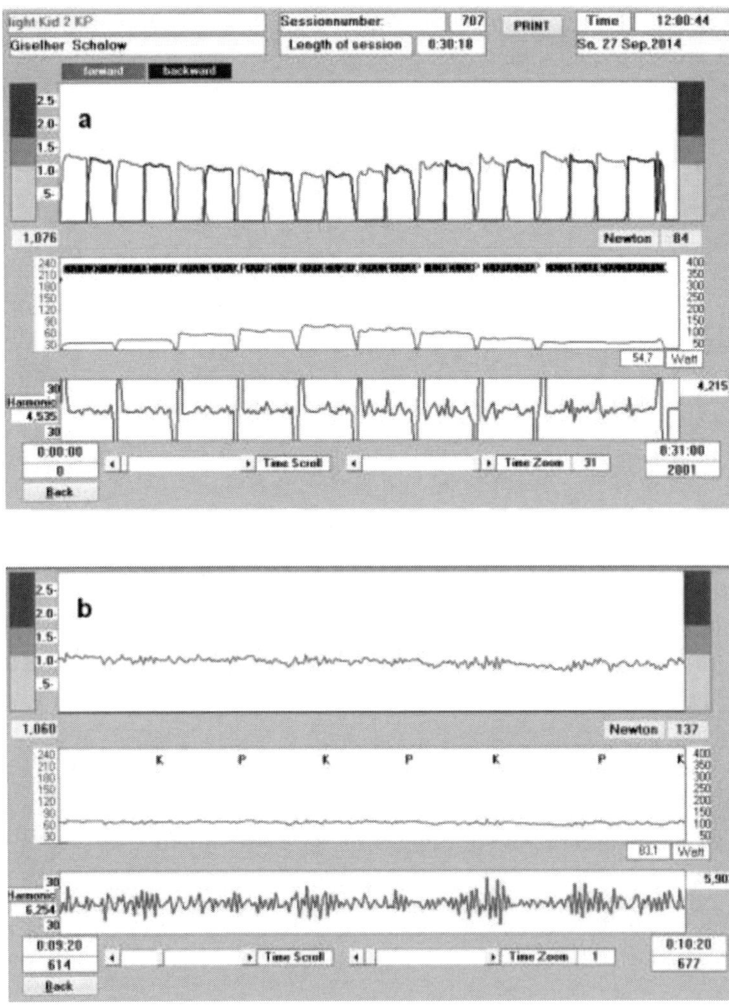

Figure 208C. High-load (HL) exercising of the Author. a. Overall test. Note the nice stepwise function of the 'Watt' trace. The interruptions are present, because the Author had to leave the device for changing the load. b. One minute exercising. Note the repeated increase of the arrhythmicity of exercising (lower trace) for the pace and trot gait coordination.

Is the nervous system really significantly involved in the process of rejuvenation of the body? It seems so. First, the nervous system is involved in nearly all functions of the body. Second, when the Author felt uncomfortable during fasting, he trained on the special CDT device for a while and felt better again. Third, it was analyzed above (Chapter II,6) that in membranes of stem/progenitor cells the Ca^{2+} channels are only opened to modulate gene expression and stimulate excitation-neurogenesis coupling if they are depolarized which means that activity is needed for rejuvenation. It is certain the rejuvenation needs more activity than pictured by Lucas Cranach (Figure 206).

The increase of the high-load CD values in Figure 208A can be understood that with fasting the power reduced and the Author turned slower to escape the high load (load escape). The CD value equals df/dt:f. When exercising at a lower frequency f, the CD value will increase.

But the low-load CD values also increased a bit (got worse) which cannot be understood with a transient weakness of the body. It seems therefore that the fasting impaired also the coordination dynamics. The Author exercised therefore more often on the special CDT device to compensate for the disturbed CNS organization. The impaired CNS functioning was also felt by the Author when he was driving a car on a narrow road. He felt a bit unsafe and had to concentrate more to stay safely on the road. When eating again this mild coordination problem terminated and automatically he could easily drive safely on the road again. There seems to be a correlation between fasting and coordinated firing of the nerve cells in the nervous system.

Improvement of the Immune System

As a first feeling of the Author, with repeated prolonged fasting the immune system got weaker, however, with eating again the immune system recovered and worked even better than before fasting. Especially the repeated five day fasting seemed to improve the functioning of the immune system. A skin disease, appearing approximately two years after radiation and chemo therapies, disappeared.

Repair of the Eye Functions

The impairment of vision, caused by chemo and radiation therapy, reduced. The drying of the eyes at night improved only little. When the Author wakes up at night, it is not clear whether he wakes up because of the impaired sleep pattern or because the eyes got too dry. By being awake, the eyes become quickly wetted again. As if the eye wettability automatism did not work sufficiently during sleep. It still has to be seen how the long-term outcome is.

Hip Joint Repair

Most functions of the human body can be improved by CDT including some diseases caused by genetics.

Probably some genetically caused diseases can partly be repaired by changing gene expression via the epigenome, but others cannot. For example, the genetic disease Down syndrome can be improved by CDT. An overused or mal-used hip joint can be improved in its functioning by CDT. A genetically determined hip joint necrosis cannot be repaired by CDT. The Author suffered a one-sided hip necrosis with aging which he could not repair with CDT by more than five year treatments. The not affected healthy hip of the other side still functions well.

In conclusion, repeated prolonged fasting in combination with CDT seems to rejuvenate the whole human body to a certain extent and not only the immune system. After reconstructive surgery fasting and CDT may help to improve the outcome. The question remains, to what extent can gene expression change in the transplanted cells. Further research has to be done to show how much rejuvenation is possible and what program is optimal to live longer with a better quality of life.

Figure 208D shows an 82-year old patient who re-learned running after severe stroke. Only CDT was administered.

Figure 208. 82-year old patient after severe stroke. A. Patient is using the stick during walking. B,C. Patient is running over Minoan culture (Knossos, Greece).

Chapter XIV

CDT in Premature Born Babies, Babies and Children

1. Treatment Has to Start as Early as Possible

Coordination dynamics therapy (CDT) has to be started as early as possible. That means the therapy has to be started in babies (Figure 209) if development is not progressing correctly. Since the therapy is healthy on a wider level, no harm can result from early intervention. In the case of pre-natal or perinatal brain damage, therapy should be started within days. Otherwise in cerebral palsy treatment should be started around an age of one or two years when pathologic development first manifests.

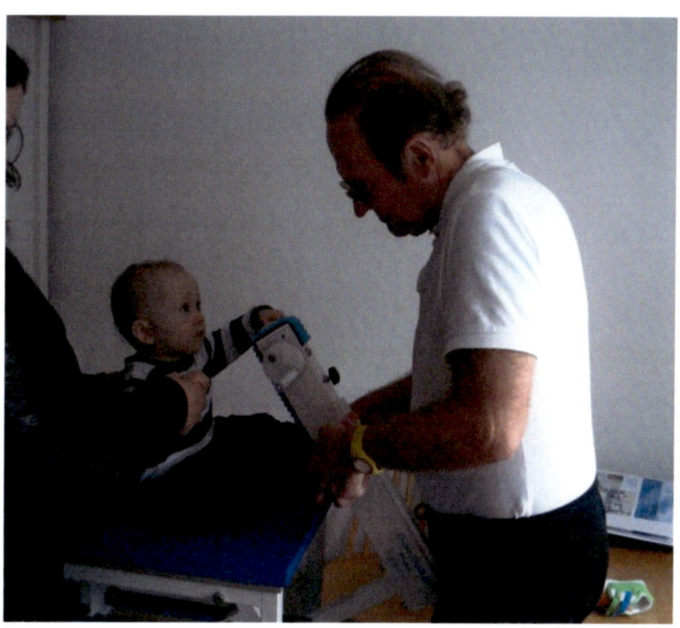

Figure 209. Training a baby on a special CDT device. Support is given by the Author. Even though the baby cannot turn by itself such exercising improves its CNS functioning.

2. Treatment in Premature Born Babies

In premature babies or when hypoxic brain injury may have taken place, therapy could be started in an incubator. Even if only few coordinated movements can be administered, the baby will benefit from it because the rate of learning is extremely fast, especially because plenty of new nerve cells can be built for repair before normal delivery at nine months. The correction "en route" with ongoing early development should be efficient. In aging, on the other hand, the problem was that nearly no new nerve cells can be built to support the functional repair by structural repair. The younger the patient, the more nerve cells are built. Especially in premature born babies plenty of new nerve cells are built anew due to the unfinished formation of the brain.

3. Improvement of Healthy Development

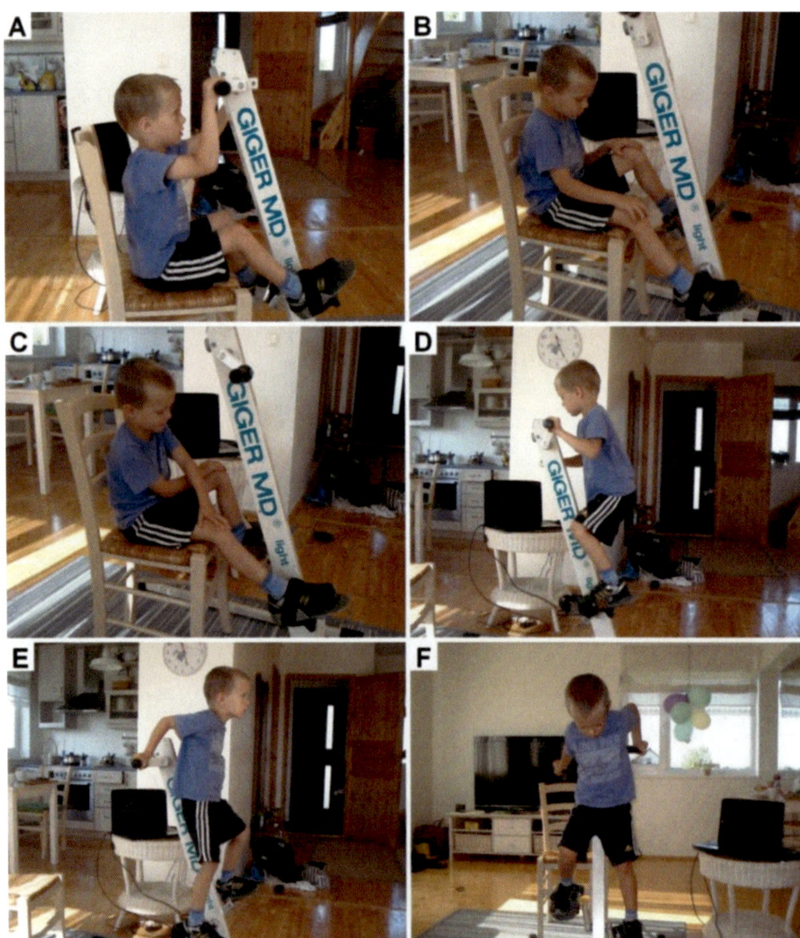

Figure 210. Different training patterns realized when exercising on the special CDT device. The healthy 4.5-year-old boy Jonas exercises in the normal sitting position with pattern change (A), the pace gait (B) and trot gait pattern (C) with no pattern change and two standing positions with pattern change from pace to trot gait (D,E,F).

If healthy young children are early acquainted with the special CDT device then they like to train with it and even experiment with the different ways one can use the device. For them, the special device is a toy to play with. The healthy boy Jürgen was introduced early to the device (Figure 33C,D,E) and also wanted to exercise if an instrument was available. He tried what could be done with the device and was open to exercise on the device in different ways (Figure 210). He could easily train in the standing position, which is difficult (Figure 210D,E,F). From the standing position (Figure 210E,F) he could easily jump off without support, which is quite a difficult task. It seems that the development of the healthy CNS can be improved if children train on a special CDT device in different ways. The improvement of the natural development could be of especial help if a child wants to become an athlete later on. But also depression and anxiety can also be improved in that way manner (see below).

In Figure 210A the four and a half year-old healthy Jonas is exercising in the normal sitting position. The exercising with crossed arms to train the corpus callosum is not shown. In 'B' he is exercising the pace gate coordination. Co-movement is induced between the hands and legs (knees) because of the simultaneous afferent input from the skins. In 'C' Jonas is exercising the trot gait coordination with additional induction of co-movement between hands and legs. These two movements are also useful to train in stroke patients. In Figure 210D,E,F Jonas is training in the standing position. Such training position is beneficial to improve the coordination between arms and legs for higher loads. Football and tennis players could improve the performance of ball kicking or hitting by training that pattern.

4. Myelomeningocele (Spina Bifida)

In Figure 211, the three year-old patient Kiki with a malformation of the caudal spinal cord (spina bifida) is exercising in the sitting (A) and lying position (B). Her higher mental functions are normal. But she is incontinent and cannot move her legs. Therefore, her mother is supporting the legs in Figure 2011A and the legs are fixed in 211B. When the Author supported her legs she liked to crawl. If knees and pelvis would be supported, she could train sky-walking without weight support like in similarity to Figure 104I-L.

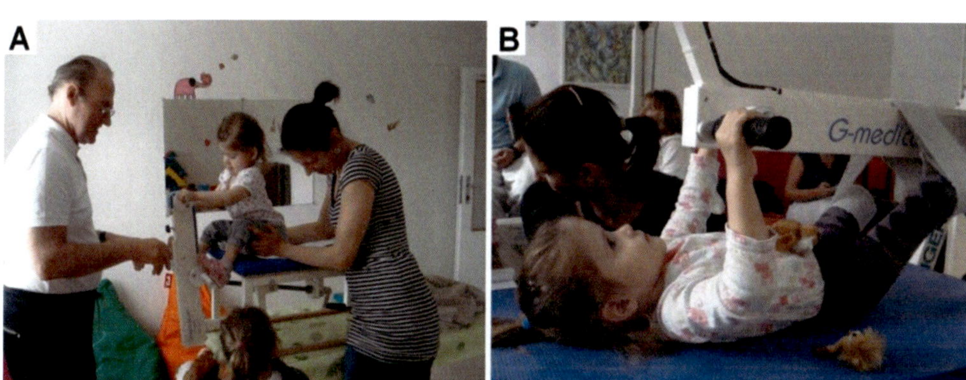

Figure 211. Three-year-old girl Kiki with a myelomeningocele during exercising on a special CDT device in the sitting (A) and lying position (B). For training in the sitting position the mother is needed for support; for training in the lying position only bandages are needed for support.

It was shown that in a child with spina bifida the CNS could partly be repaired when CDT was administered intensively for a few years [40]. It is therefore likely that the CNS in Kiki could partly be repaired, if the mother could apply CDT to her daughter. However, they do not have the money to pay for devices and mainstream medicine is offering only one to two hours per week instead of the necessary twenty. In twenty years Kiki will be a young lady who is incontinent and cannot walk. Very often society fails to properly care for its children.

5. War Children (Kriegskinder)

Figure 212. War children (Kriegskinder, Königsberg).

The CNS of children cannot only suffer damage from a traumatic brain injury, infection or malformation, but also from the fallout of conflict zones, especially in the absence of parents. They can suffer nightmares which may last for the rest of their lives. Figure 212 shows an example of such children. The older sister is trying to protect the younger brother as she is taking him and some things in a carriage to a safer place for winter. The boy is crying because he cannot cope with the situation. In Figure 213 a girl is shown among the ruins of central Berlin.

Figure 213. War children (Kriegskinder, Berlin).

The Author experienced such situations as a child. To get rid of historic trauma can be difficult. CDT can help to get rid of nightmares. Exercising on the special CDT device is especially helpful in getting rid of nightmares because the complicated coordination's which are trained go deeply into the complexity of CNS organization where also the higher mental functions are probably located. The higher mental functions can be reached and improved in their organization. Such therapy may need years. Running, on the other hand, is a very good movement in 'movement medicine', but the complexity of CNS functioning is not efficiently reached.

6. Depression and Repair of Higher Mental Functions

Depression can be counteracted by antidepressant drugs. If depression is inherited, in that it is genetically determined, drug therapy will also work. However, drugs will only mask the symptoms not improve the functioning of the neuronal networks. The genetically determined disease may progress in spite of the drugs. This pathologic CNS organization may generalize over a period of years so that other higher mental functions also become impaired or unstable. The drugs can be changed to cover different symptoms, but the disease is still progressing. But applying CDT additionally, the progress of the disease may be stopped or the disease can even be cured. I expect that the repair of the higher mental functions will take years. In similarity to motor functions, being euphoric or depressive are organizational states whose stability can be changed and more balanced with respect to each other and the stability of pathologic emotional states can be reduced. In contrast to the impairment of motor functions, it seems that the low-load and high-load coordination dynamics values do not reflect changes in the higher mental functions, even though the exercising on the special CDT devices seem to improve the higher mental functions since the higher mental functions improve in children with cerebral palsy. The babies or young infants cannot exercise by themselves on the special

CDT device but the functioning of their CNS improved following supported exercising (Figure 209). Therefore, when exercising on the special CDT device, CNS functioning improves, even though it cannot always be measured or the patient cannot exercise independently.

Epilogue

The Adventure of Performing Medical (Human) Research

1. Funding Problem Published in 1993

This research has primarily been undertaken using personally saved money, including the equipment (300 000 DM over six years, the entirety of my savings) to bring this clinical research up to the level presented in this book. In Germany, every second day a paraplegic dies because of bladder infections or by suicide. The Author qualified in technical, natural and medical sciences and worked together with R. Miledi at the institute of Biophysics of Sir Bernard Katz. Instead of continuing to work on the trophic effect or on Ca channels, he changed to basic clinical research. He was advised from well-known basic research workers and neurosurgeons that clinical research is worthy work if something can be achieved. At the beginning, reviewers argued it were not possible to record single-nerve fiber action potentials in humans with wire electrodes. Then the argument was that it should only be possible to record from thick nerve fibers, which are of no interest any more. The author had to use a computer; he has to use average values. A basic animal reviewer could not see the clinical implications and asked then for the ethics of this research, without giving reasons for his ethical considerations. Some American reviewers argued also that if it is possible to record afferents and efferents simultaneously in humans, then this will have consequences. Initially the Deutsche Forschungsgemeinschaft (DFG) supported the research, but refused to accept the application for further money because of organizational reasons. After an intervention by R. v. Weizsäcker (president of Germany) the DFG accepted the application, but refused to allocate money because this research project has been judged unqualified and ethically unjustified; they did not want to change the decision even after more than ten papers had appeared in different journals. Now the DFG raised the organizational argument again and argued that the author cannot apply for money for the same research project again. The Max Planck Institution for Clinical Research was not interested in the research project, arguing that it exceeded the scope of the field covered by them. Approximately twenty other institutions refused financial or other assistance for various reasons. The " Ministerium für Forschung und Technologie, Forschung im Dienste der Gesundheit" promised financial aid for several years by establishing in the future a Neurotraumatology Research Project, but money was refused for the time being because it was needed for the reunification of West- and Middle-Germany. An institution from Switzerland and from the USA refused money with the argument that it is better to keep the funds in their own countries, even though paraplegics would be able to receive treatment in other countries, if available. The trial to allocate funding from private persons by an advertisement in the "Züricher Zeitung" in Switzerland was stopped by an ethical committee. At the rehabilitation conference IRMA VI in Madrid it was reported that in Amsterdam, it is practically impossible to get funding for research in rehabilitation. The questions remains, how many paraplegics have still to die until the world's sense of moral responsibility is piqued so that the necessary research can be organized, which has to cover everything from the basic research up to the patient. What is the justification for killing animals in basic animal research if no clinical application is ultimately intended? This funding problem of human and clinical research was published in 1993 [58].

2. History of Tte Human Research Project Published in 2010 and 2013 [14]

Based on the Author's background of being an engineer in electronics, a theoretical physicist (Dr.rer.nat.), a physician (Dr.med.), and post-doc with Sir Bernard Katz, this research project to cure CNS injuries was started in Finland in 1983. It was first supported for two years by the 'Deutsche Forschungsgemeinschaft' (DFG). Further funding was refused. Ethical and quality objections were raised. The support of the former Federal President Richard von Weizsäcker could not overturn the decision of the DFG. The DFG did not want to confirm that the reviewers of the application for funding had a comparable background as the Author. The Author (from West-Berlin) continued the research on personally saved money. He worked together with Departments of Neurosurgery, Neurology, Pathology, and Forensic Medicine of the University of Greifswald in the former German Democratic Republic (GDR), even though this was officially not possible. Since measurements were also performed on brain-dead humans, an ethics committee was established in the former GDR, which decided that the research project was ethically justified. The Author received partial funding from the GDR, even though this was officially not possible for a subject from West-Berlin (Figure 214). Before the re-unification of the GFR and GDR, he could not get funding anymore because of shortage of money. The habilitation procedure (Dr.med.habil.) based on the development of the single-nerve fiber action potential recording method in humans was started before the fall of the Berlin wall and was completed thereafter. When working for six years in a rehabilitation center in Switzerland, the Author applied twice for support from the Swiss National Fund (SNF). Financial support for the research was refused. In recent times, this research project of CNS injury repair has been ignored or there were attempts to destroy it in several ways, including exercising pressure on editors to prevent the publication of papers, exercising pressure on organizers of conferences to prevent presentations, and writing letters to patients and institutions informing them that the method has no merit. No real theoretical or clinical scientific discussion could ever be achieved by the Author on the basis of human neurophysiology, even though the results of this research have tremendous clinical consequences. The health care system in Europe and North America prefers to let patients with very severe CNS injuries die without any qualified treatment rather than working on and proving new treatments. The Author has been working for 25 years as a human neurophysiologist and clinical research worker in several countries of Europe, mainly on personally saved money to find a way to bring qualified movement treatments for CNS repair to patients. Animal physiologists and clinicians have not accepted human neurophysiology and the Author has been excluded from the research community.

ERNST-MORITZ-ARNDT-UNIVERSITÄT

DDR 22 GREIFSWALD, DOMSTRASSE 11

AM: 19.05.1987 AZ:

B e s c h e i n i g u n g

Herr Dr. Dr. S c h a l o w, Giselher
ist in der Zeit vom 01.03.1987 bis 01.03.1988
Stipendiat an der Ernst- Moritz- Arndt- Universität Greifswald
und erhält ein monatliches Stipendium in Höhe von
900,-- Mark.

Ernst-Moritz-Arndt-Universität
22 Greifswald
Direktorat für
Internationale Beziehungen

Dr. paed. A. Kloß
Direktor f. Internat. Beziehungen

Figure 215. Funding approval of the Ernst-Moritz-Arndt University of Greifswald at times of the former German Democratic Republic (GDR) (Deutsche Demokratische Republik) for the Author from Berlin-West before the fall of the Berlin wall. Such a funding was for political reasons not possible.

3. Research across the Berlin Wall

At the Neuropathology Department of the Turku University the anatomy of the spinal cord was clarified within two years. When there was nobody who could be interested in measuring single-nerve fibre action potentials in Finland, the Author moved country in order to continue the research project of repairing the human CNS. I thought neurosurgeons of the "Deutsche Demokratische Republik" (GDR) could be interested in such research, because researchers and clinicians from countries across Eastern Europe wanted to have contact with the West. Living in West-Berlin, I had to pass via certain roots (transit ways) through the GDR to reach West-Germany or other European countries. Once I left the transit route to the North (forbidden area for transit) and went to Rostock University and asked for the Chief of the Neurosurgery Department. I was told that there was only a Neurosurgery Department at Greifswald University situated in the North of West-Berlin. I wrote to the Chief of the Neurosurgery Department (Prof. Lang), and he was interested. Subsequently, research cooperation was started to repair human CNS injury between a subject from West-Berlin and a university in the GDR.

On the way from West-Berlin to Greifswald I had to cross the border (the Berlin Wall) for leaving West-Berlin. I used the border crossing at Stolpe. Especially the wall at the 'Brandenburger Tor' (Figure 215) was difficult for me to see because before the building of the wall I used to walk through it to reach the 'Staatsoper unter den Linden' or the 'Komische Oper'. Because of lack of money I changed 1 DM (Westmark) into approximately 4 Mark of the DDR (Ostmark) and could then afford to visit the opera quite often.

Figure 215. Brandenburg Gate in Berlin.

On the way to the North I passed two Russian barracks and a Stalin monument. Sometimes I was stopped by Russian military police and I was asked whether I had seen any escaped Russian soldiers. However, I was not inspected. There was supposed to be also a concentration camp close to Neubrandenburg where members of the "Hitler Jugend" were put to. Nobody had told me more details because of fear. If denunciated, one was also put to that concentration camp from which there was no return. In West-Berlin I lived close to an American army barrack (Finkensteinallee) and sometimes I saw American military police. I felt safe when I saw American soldiers because when I was in an orphanage, the American soldiers helped the orphanage, especially at Christmas we got chocolate and other nice things to eat. My first ice cream I ever tasted was on my fourteenth birthday and was invited to the American barracks.

At the beginning of this cooperation between West-Berlin and the GDR, I had to pay 25 "Mark der DDR" per day for using the research facilities. Then later on I had to change 25DM in 25 Mark der DDR (Zwangsumtausch). Even later on I received one year of research funding from Ernst-Moritz-Arndt University of Greifswald (900Mark der DDR) (Figure 215).

The cooperation and the funding was not only helpful in order to progress the research project but was also politically interesting, since the GDR wanted that West-Berlin was seen as an independent country (which the administration of West-Berlin and West-Germany did not want). The Germans did not want to have their country split into different parts and have their culture partly destroyed. Therefore, it was officially not possible to have research cooperation between a researcher from West-Berlin and a professor from the GDR. Such cooperation was not possible, therefore, and the minister did not officially know anything about it, even though he knew about me. Because of this situation some persons I worked together with in Greifswald thought that I was an unofficial member of the secret police (inoffizieller Mitarbeiter der Staatssicherheit (Stasi)), which they told me after the fall of the Berlin wall. There was not any political pressure exerted on me, even though I felt some pressure from my surroundings. A very interesting fact was that I had been told that there were reports written about me and for me it was clear that not only my research work was judged.

Since at the beginning I had to pay every day for being in the GDR, I had travelled quite often from West-Berlin to Greifswald and was of course inspected every time, which was unpleasant. With time the border police had got to know me quite well and instead of inspecting me I explained to them what research I had been doing in Greifswald University. Most impressive for them was to look at cross-sections of nerve roots of about $1m^2$ size (for example Figure 155) to see the great many axons of different sizes are contained in nerve roots. But still the crossing of the boarder was stressful, because sometimes I was searched to see whether I was transporting forbidden things. And of course there were small necessary things I brought to Greifswald. Since certain things had been missing in the former socialistic countries I took small things with me to help my friends who helped me to organize my research project, as for example a pocket calculator, glutaraldehyd and epon for working with the electron microscope. Because some administration of the GDR thought of this, I was sometimes inspected thoroughly and this was nasty. I had the feeling of losing all human dignity. But to have friends who helped me was important to realise the research project. Already for measuring single-nerve fibre action potentials from brain-dead human cadavers (HTs), I needed an anaesthetist for sufficient ventilation and medication, then I needed a neurosurgeon to do the laminectomy, which is not simple in HTs and I needed a physician or medical student who helped me with the measuring. When working on HTs, the mental burden is high and one needs a lot of motivation: one's mind tells you that you are working on a cadaver, but one's instinct tells you that you may kill a human being because the heart is beating. When the artificial ventilation is terminated of an HT, the heart may beat for another one or two hours (the heart is not only uses only ATP; it can also work with ADP in an emergency). A pathologist would not start an autopsy on a cadaver in which the heart was still beating. I got a lot of help in Greifswald to realise the research project when no politics were coming in. The measurements in HTs were an essential step to make repair of the injured human CNS possible. Even though many other scientists and clinicians expected that the success could come only over the long term, I got a lot of support. The only country in the world which truly supported this research project was the former GDR (Deutsche Demokratische Republik). All decisions were free from the politics of the time. The DFG from West-Germany decided that this research project had not been ethically justified, most likely without using any ethical committee. In Greifswald an ethical committee was formed to decide whether the measuring in HTs was justified. The committee decided that such measuring was ethically justified. Since, for financial reasons, there was a lack of dialyse filters, kidneys had been explanted excessively from HTs in the GDR with permission of relatives. I got those HTs which could not be used for kidney explantation as for example that the former patient had an infection. Further, an explanted kidney can help one patient. Qualified medical research on HTs may help millions of patients. Also, when I tried to further clarify the innervation of the human urinary bladder everything was on legal terms. For further clarifying the anatomy (under the microscope) I needed fresh cadavers from young persons and those ones one can get in the forensic medicine. A forensic autopsy was accompanied by a lawyer to see whether the autopsy was legally performed. After the termination of the autopsy the professor of the forensic medicine explained the lawyer my research project and asked whether it was allowed that I could do a further dissection. The lawyers gave the permission. Also, the urinary bladder was not involved in the death of the persons. However, as soon as politics came in, the situation changed. A friend of mine from the Radiology department showed me a Roentgen picture of a gunshot wound into the pelvis. I had difficulty in identifying the seemingly exploded pelvis. What had happened with that patient? He was an inhabitant of the GDR and wanted to leave the country. He went with the cars from the West to the border and he tried to run across the last checkpoint. A border policeman was pursued him, reached him, and shot him with machine-gun.

4. Back to Research

The scientific problem in this research project was now, if it was possible to measure human CNS function at the neuron level? Before the Author left the Physiology department of the University of Saarland, he experimented on different animals to see whether the cauda equina nerve roots might have been be suitable for measuring single-nerve fibre action potentials in humans. But to be able to record single-nerve fibre action potentials one needed equipment. Since the DFG no longer wanted to support the research project (see above), equipment was acquired on a private basis. Then the success came, namely that it was possible to measure single-nerve fibre action potentials in human during an operation. It had mostly been so in the history of research that if a new measuring method was developed progress in science was achieved. And it was clear to me that a substantial step forward in medicine would be possible. However, the journal *Nature* was not interested in publishing about this new electrophysiological recording method in human in spite of the fact that there had been not so many methods in human to measure CNS functioning at the neuron level. Journals from the GDR published about the recording technique. With the fall of the Berlin Wall, the journals were taken over by the West. It was sad when an editor said goodbye because we had been fighting for progress under very difficult conditions.

The measurements from HTs had been broad necessary knowledge for judging the self-organisation of human CNS neural networks: phase and frequency coordination of neuron firing and the stability of networks pattern. Desmeth and Eccles probably liked that research and the Editor and founder of the Journal Electromyography and Clinical Neurophysiology (Roselle) had given plenty of space for publishing.

The clinical application was not reached in the GDR. When I applied for further funding, I was told that this was not possible, because there was no money. The system was going to end because of lack of money. I was invited to the Town hall. The administration probably wanted to see a private researcher first-hand; no political pressure was put on me. I was reported to the secret police on a few occasions. Because I was living in student accommodation, the students wanted to discuss the political situation with me. Some of the medical students were former soldiers. One student was genuinely convinced that it was justified to shoot somebody attempting to climb over the Berlin Wall. When I said there was no justification for killing any Germans who wanted to go from one part to the other, I was reported for that. I was not afraid to be put into prison, but I was afraid to lose my research conditions. I was convinced that I would not find such research conditions again and so it was.

The research project ended somewhat before the fall of the Berlin Wall because of divisions among professors and not due to political reasons. I habilitated at Greifswald University in medicine (human neurophysiology), which is the highest qualification in medicine in Germany. I started the habilitation with professors from the GDR and ended it with professors from the West, which took nearly all professorships over after the fall of the Berlin Wall. Not all professors whom came from the West were talented. But the GDR was destroyed ('wurde abgewickelt'). Some of the medical students became aggressive towards some professors from the West as I experienced when I was doing some teaching during my habilitation.

5. Switzerland

The research project was continued in Switzerland in a neuro-rehabilitation centre and the clinical application was achieved. The practised physiotherapy methods in the clinic were mainly Bobath and Vojta. These methods were good some 30 years ago, but are now out-of-date because of low rate of repair. After six years I was thrown out of the rehabilitation centre by mobbing. The reputation and appearance of the clinic seemed to be more important than the progress of the patient, as it is in most rehabilitation centres. When I criticised the low level of neuro-rehabilitation at Zürich University, the professors shouted so loudly that one could not hear my arguments. I had known that students sometimes had argued that loud to drown out the argument of their professors (in the 1960's at the Free University of Berlin).

6. Back to Finland

I went to small rehabilitation conference in Helsinki. When I realized that the level was too low, I became angry and went to a neurology conference held simultaneously in Turku, Finland. There, a chief of neurology and neuro-rehabilitation departments became interested in coordination dynamics therapy (CDT). The building up of an intensive care room for treating stroke and other patients with the special CDT device in the lying position was not successful in the Neurology Department, because the staff's main objection was that they were already too busy with their present workload. There was no money for additional staff. But CDT was used in the Neuro-rehabilitation Department. The physiotherapists liked to do CDT. However, the patients could only get two hours therapy per day (10 hours per week) for six weeks. Such a therapy time is insufficient for the repair of a severely injured CNS. Patients were not permitted to pay for additional treatment. Only 10 to 20 years later it became possible to pay privately for additional treatment. I left the university hospital because the repair of the CNS did not seem to be possible to me under these conditions. Two physicians in Finland became interested in CDT, built up a private therapy place, employed physiotherapists and I supervised the CDT. Patients with CNS injury improved with CDT for three months and at least 20 hours per week. TV broadcasts supported this new treatment. But when patients went to the state rehabilitation center close to Helsinki and argued that they preferred to get therapy in the private place (and that the state is paying for it), because there they would get better and longer treatment for the same amount of money, the administration withdraw the clinic's permit. To put pressure on the physicians, the letters from the administration were brought by the police in a police car. The argument was somehow that CDT is not a physiotherapy method and also not a movement therapy. These arguments were wrong, but the administration was and is always in the dominant position. They even wanted to cancel the practising certificate of one physician. The physicians went to court and lost at the Highest Court, even though the expert witness, gave favourable evidence to the Court about CDT. The Author was not asked to give evidence to the Court about CDT. On the documentation CD of the Court the private communication between the administration and the Court was deleted, as the physicians told me. It seemed that high positions in the rehabilitation were given to members of political parties. By trying to compete with the state rehabilitation system, one was attacking members of political parties and they found ways to stop new treatment. If courts are not independent, then the progress in medicine is on the losing side. Before the therapy place in Finland was closed down, the physicians had built up a small therapy place in Spain. Even though there were a lot of small administrative problems, there were no problems with the administration. But Finish patients could not come from Finland, unless they paid privately. According to EU regulations, a patient has the right to go to another EU country for a certain treatment if he cannot get it in his own country. The Finish administration ignored the EU regulation. The physicians are now taking their case before an EU court. The EU seems to be on their side, because the EU administration wants competition. Even if the physicians were to win after a number of years, the private therapy clinics had been destroyed and patients lost the opportunity of better treatment. One patient, who wanted CDT died waiting for the therapy clinic to reopen.

7. Slovakia

At the age of nine, the son of a friend of mine (Benjamin, Figure 118) was hit by a car while playing in the street. He suffered a severe brain injury. Since my friend had been doing the English correction of my papers for several years, he knew about CDT and asked me to come to treat his son. In a rehabilitation clinic CDT was administered to him under my supervision. The boy Benjamin got better and the parents of other patients also wanted CDT for their children. The neuro-paediatrician, the physiotherapist and of course the children with severe brain injuries and scoliosis liked CDT. The members of the rehabilitation clinic wanted me there. Suddenly the Chief of the Clinic did not want me there anymore and pressure was put on the neuro-paediatrician to stop cooperating with me. A reason was not given. Then parents and relations of children wanted to build up a small private therapy place for children. My medical certificates were checked thoroughly, but still we did not get the permission to found the therapy place for children. Even though my friend had several friends in high positions in universities from his medical studies, we could not find out who was preventing us from bringing help to the children.

8. USA

There are patients in the USA who would like to come for treatment to Europe. But because of the inadequacy of public healthcare, they have no money to come and care and treatment of CNS injuries becomes expensive over the long-term. Christopher Reeve could have afforded to come to Europe. Friends of mine pushed me to apply for money from the Christopher Reeve Foundation and to contact Christopher Reeve. But the application for funding never reached the 25 reviewers; it had been stopped before. A patient with a high cervical spinal cord injury (range C3 or C4) has nearly no movement functions left, is dependent on those around him and can be easily manipulated. It seems that Christopher Reeve had been used to get founding for biological research and not for medical research.

9. Italy

A rehabilitation physician from North Italy performed something like oscillator formation training and became interested in CDT. He invited me to give a three day course, including theory and practice. After the course he wanted to treat his patients with CDT. He applied for funding to buy instruments. The hospital administration promised that he would get them, but he did not get them. Then the motivated rehabilitation physician bought a special CDT device by himself and started to treat his patients with it. But then the administration forbade him to administer CDT in order to protect the patients from becoming upset by a treatment being different than the norm, namely passive movement therapy. Physiotherapists are used to do these movements and many patients like it because they do not need to work hard. The only disadvantage of this treatment is that the patients practically do not get better.

10. Estonia: The Golden Years of CDT

A friend of mine from Finland suggested many times to try out Tartu University in Estonia whether progress in neuro-rehabilitation would be possible there. Since there existed a faculty for physiotherapy and sport sciences from socialistic times, there was really interest in movement sciences and research. Further, the university tried to help small businesses. So it was possible to establish a small therapy place in Tallinn which worked nicely for a few years. Patients and research benefited from it. When the Author left the country, the system switched mainly back to the out-of-date situation as in other countries. The unit of theory and practice, which was very beneficial for medicine, lasted only transiently. Especially since neurosurgeons argued that there was nothing to CDT, the effect on the whole system was minimal. Also, some Master Degree physiotherapists argued, what was the sense of learning CDT if I can only treat a patient for twenty minutes one or two times per week? I believe in the units of theory and practice and teaching and research. But there must be support from those around it to allow it to function.

13. Conclusion

There are more examples to show that it is difficult to bring progress in neuro-rehabilitation. I have been attacked personally many times but I have never been attacked scientifically. It would be very interesting and of great help for medicine if we could get more knowledge concerning the self-organisation of the human CNS. The human genetics (epigenome) was strongly influenced in the last 5 million years and movements were an important part of behaviour. Changes of behaviour in the last ten thousand years will have not affected the genetics that much. Movements and movement-based learning may therefore be an important part of healthcare and to live longer with a higher quality of life. For the time being, progress in medicine is mainly coming from devices and diagnostics and not so much from new methods of repair or healing. It is important, but difficult, to organise qualified medical research. To replace medical research by biological research is not the solution.

In Neurotherapy we need a human neurophysiologic basis, ethics as given by Hippocrates and an efficient therapy, including devices from which the injured CNS can learn to improve its functioning (Figure 216).

Figure 216. For neurotherapy human neurophysiology, ethics (Hippocrates) and an efficient therapy is needed.

Repair is Cheaper than Care

For ethical reasons a patient should have the right to efficient therapy and should be informed of its existence. Unfortunately, it seems that the short-term cost of treatment is the most important consideration. Since CDT has to be applied at least for twenty hours per week for up to three years, it seems that the therapy costs are too high for the society to offer such therapy to the patient. It will be shown that the care is more expensive in the long-term.

1. Cost Evaluation of Spinal Cord Injury C5/6 Patients

For cost evaluation a patient with a complete cervical spinal cord injury C5/6 is used. Such a patient is transferred in Switzerland, if he has no further injuries, to a rehabilitation center for nine months of rehabilitation. One day costs 1,500 Euro plus the cost of operations and equipment such as wheelchairs, catheters and other things. Because of inefficient treatment, the bladder is not repaired and the blood supply below the injury level is not restored. In the patient of Figure 144 it took three years to repair the bladder and six months to repair the blood supply below the injury level. Spasticity reduced. Because of some regeneration over the injured segments (Figure 143), the patient's trunk stability improved (Figure 144) and she could sit if necessary in a normal chair and could turn over in bed at night. This patient suffered the spinal cord injury at the age of 16 and could finish high-school at an age of 20 and studied thereafter. With slight improvement in finger functions, regained trunk stability, full blood supply of the skin and no continence problems any longer and her privacy restored. She no longer needed care nor hospitalization. She can have a somewhat normal life for the next 50 years including a career and is no burden to the social or health care system.

It is now tried to estimate the life-time costs, if bladder and skin blood supply are not repaired and spasticity is not reduced by movement-based learning. The nine months cost of rehabilitation of approximately 270days x1500Euro = 405,000Euro. Including operation and facilities, the 9 months cost rises to approximately 0.5 million Euro. Now the costs for the next 50 years are evaluated. On average every year such a patient gets a pressure ulcer, because of insufficient blood supply to the skin and of the poor healing process because of the missing blood supply. Care has to be taken in a rehabilitation center for approximately 30 days to heal the ulcer. If necessary in operations necrotic tissue is taken away and replaced by healthy tissue. The 30 days in the rehabilitation center cost 30 x 1500 = 45000Euro. If we include the cost for operations, the treatment perhaps of six bladder infections per year and the problems to reduce and handle the spasticity, the patient costs approximately 60000Euro per year. Obviously, the rehabilitation centers raison d'etre is the treatment of complications. Often such patients need helpers at home to manage with everyday life. A complete tetra cannot catheterize himself to empty the urinary bladder, whereas a para can. Often patients with a cervical spinal cord injury cannot have a qualified full-time job. The costs for the lack of a proper career are difficult to estimate. Altogether the patient may cost 0.1million Euro per year. For the next 50 years such a patient costs at least 5million Euro.

Now it will be evaluated what a patient costs when bladder and blood supply are repaired, spasticity reduced and trunk stability regained by some regeneration of the spinal cord.

The patient has to be one or two months in a neurosurgery or orthopedic department of a hospital and then transferred home. Since 'exercise medicine' is still foreign for hospitals, the patient should stay as short as possible in the hospital. Still the treatment has to start in the hospital. The hospital costs are 50 x 1500 = 75000 Euro. Afterwards the patient has to be transferred home and is getting 15 hours therapy per week in an up-to-date physiotherapy place which has to be controlled by an expert witness of the insurance company. At home the patient has to exercise additionally on the special CDT device and sky-walker. The 60 hours therapy per month cost 60 x 50Euro = 3000Euro/ months. For care by a helper or the parents and the transport from home to the therapy place 4000Euro may be needed. Including some medical goods, we end up with 8000Euro per months that means 100.000Euro per year and 300.000Euro per 3 years. With 50.000Euro for home medical devices and 150.000 for other costs, we end up with a therapy cost of 0.5million Euro per three years. Later on little costs occur.

According to this simple and very approximate evaluation the life-care is 10 times more expensive than the limited repair. Patients with a spinal cord injury want to walk again. But more important is the repair of the vegetative functions like urinary bladder repair (and sexual function), improvement of the cardio-vascular performance, including the microcirculation of the skin, and the improvement of the trunk stability including the improvement of breathing. What actually makes on the long-term the limited repair cheaper than the care? By improving the general health of the patient and avoiding thus the complications, the rehabilitation center or hospital is not needed any more. With the repair of the skin nutrition no pressure ulcers occur any more. With the repair of bladder functions, no catheterization is needed any more and bladder infections do not occur any more. With the improvement of the trunk muscles the breathing improves because the intercostal muscles start to work again. Because the lung is ventilated better, lung infections do not occur any more. The improved trunk stability with some control of the pelvis allows the patient to turn in the bed at night and sit on a normal chair. With the reduction of the spasticity, spasmolytic drugs with strong side effects are not needed any more. The repair of the skin nutrition and urinary bladder functions and the improvement of the trunk stability make the limited repair cheaper than the life-time care. Without a substantial repair of the spinal cord this repair is possible because the different plexus and the sympathetic trunk are bypassing the spinal cord (Figure 153) and can be reorganized by CDT. Therefore the first goal is not to make patients with a spinal cord injury walk again, but to repair the vegetative functions. Without efficient treatment pressure ulcers and urinary bladder care increase the health care costs tremendously. The largest contribution of the high costs for care results from the stay of the patient in hospitals or rehabilitation centers. The hospitals can mainly be avoided with an up-to-date repair treatment. Also, often patients catch in rehabilitation centers antibiotic resistant bacteria.

Improved costs evaluations of spinal cord injury patients should be given by health insurance companies from patients who obtained care for the 50 years following the injury.

2. Costs in Cerebral Palsy

In cerebral palsy the situation is quite different. There are no problems with cardiovascular performance and therefore no pressure ulcer problems. Continence problems occur mainly in severe malformation or meningitis. Mostly there are problems with walking.

Different kinds of supports are needed, which often cause skin problems due to exerted pressure to push the foot or leg into the right position. The main costs will come for the care they need for 60 to 70 years especially when the higher mental functions are strongly impaired. Some mothers do a very good job in looking after their child. But first, also the mothers cannot contribute care for 24 hours per day and second an institution has to look after the grown up cerebral palsy children when the parents are old or have died. To care for a patient with severe brain injury is often mentally draining. The Author experienced it with the patient of Figure 112. It is unbelievable how patient mothers can be when they try to educate their child. Special schools and job centers are needed. There is little doubt that the costs can be reduced strongly if the cerebral palsy children get early efficient treatment, which improves CNS functioning. Again cost evaluations are needed and how much can the costs reduced if efficient treatment is applied.

References

[1] Schalow, G., (2002). Stroke recovery induced by coordination dynamic therapy and quantified by the coordination dynamic recording method. *Electromyogr. Clin. Neurophysiol.* 42, 85-104.

[2] Schalow, G., (2002). Improvement after traumatic brain injury achieved by coordination dynamic therapy. *Electromyogr. Clin. Neurophysiol.* 42, 195-203.

[3] Schalow, G., (2002). Recovery from spinal cord injury achieved by 3 months of coordination dynamic therapy. *Electromyogr. Clin. Neurophysiol.* 42, 367-376.

[4] Schalow, G., (2003). Partial cure of spinal cord injury achieved by 6 to 13 months of coordination dynamic therapy. *Electromyogr. Clin. Neurophysiol.* 43, 281-292.

[5] Schalow, G., Pääsuke, M., Ereline, J. and Gapeyeva, H., (2004). Improvement in Parkinson's disease patients achieved by coordination dynamics therapy. *Electromyogr. Clin. Neurophysiol.* 44, 67-73.

[6] Schalow, G., Jaigma, P., (2005). Cerebral palsy improvement achieved by coordination dynamics therapy. *Electromyogr. Clin. Neurophysiol.* 45, 433-445.

[7] Schalow, G. and Jaigma, P., (2006). Improvement in severe traumatic brain injury induced by coordination dynamics therapy in comparison to physiologic CNS development. Electromyogr. *Clin. Neurophysiol.* 46, 195-209.

[8] Schalow, G., (2006). Hypoxic brain injury improvement induced by coordination dynamics therapy in comparison to CNS development. *Electromyogr. Clin. Neurophysiol.* 46, 171-183.

[9] Schalow, G., (2006). Cerebellar injury improvement achieved by coordination dynamics therapy. *Electromyogr. Clin. Neurophysiol.* 46, 433-439.

[10] Schalow, G., Jaigma, P. and Belle, V.K., (2009). Near-total functional recovery achieved in partial spinal cord injury (50% injury) after 3 years of coordination dynamics therapy. *Electromyogr. Clin. Neurophysiol.* 49, 67-91.

[11] Schalow, G., (2009). Building of New Motoneurons in the Human Spinal Cord upon Coordination Dynamics Therapy to Improve Finger Functions in Motoric Complete Cervical Spinal Cord Injury. In: Berkovsky, T.C. (Ed.), Handbook of Spinal Cord Injuries, Chapter 4. pp. 231-264, Nova Science Publishers.

[12] Schalow, G., (2010). Scientific basis for learning transfer from movements to urinary bladder functions for bladder repair in patients with spinal cord injury. *Electromyogr. Clin. Neurophysiol.* 50, 339-395.

[13] Schalow, G., (2010). Cure of urinary bladder functions in severe (95%) motoric complete cervical spinal cord injury in human. *Electromyogr. Clin. Neurophysiol.* 50, 155-179.
[14] Schalow, G., (2013). Human Neurophysiology: Development and repair of the human central nervous system. Nova Science Publishers, New York.
[15] Schalow, G., (2005). Phase and frequency coordination between neuron firing as an integrative mechanism of human CNS self-organization. *Electromyogr. Clin. Neurophysiol.* 45, 369-383.
[16] Schalow, G., (2009). Relative Coordination between Neuron firing and Generation of a Motor Program in the human CNS. In: Berkovsky, T.C. (Ed.), Handbook of Spinal Cord Injuries, Chapter 5. pp 265-295. Nova Science Publishers.
[17] Schalow, G., (2009). Impaired coordination between oscillatory firing FF and FR-type motor units in Parkinson's disease and patients with spinal cord injury. In: Berkovsky, T.C. (Ed.), Handbook of Spinal Cord Injuries, Chapter 15. pp 501-517. Nova Science Publishers.
[18] Kelso, J.A.S., (1995). Dynamic Patterns. The Self-Organization of Brain and Behavior. MIT Press, Cambridge.
[19] Alvarez-Buylla, A. and Lim, D.A., (2004). For the long run maintaining germinal niches in the adult brain. *Neuron* 41, 683-686.
[20] Cao, L., Jiao, X., Zuzga, D.S. et al, (2004). VEGF links hippocampal activity with neurogenesis, learning and memory. *Nat. Genet.* 36, 827-835.
[21] Covic, M., Karaca, E. and Lie, D.C., (2010). Epigenetic regulation of neurogenesis in the adult hippocampus. *Heredity* 105, 122-134.
[22] Colvis, C.M., Pollok, J.D., Goodman, R.H. et al, (2005). Epigenetic mechanisms and gene networks in the nervous system. *J. Neurscience* 25(45), 10379-10379.
[23] Ge, S., Yang, C., Hsu, K., Ming, G. and Song, H., (2007) A critical period for enhanced synaptic plasticity in newly generated neurons of the adult brain. Neuron 54: 559-566.
[24] Deisseroth, K., Singla, S., Toda, H. et al, (2004). Excitation-neurogenesis coupling in adult neural stem/progenitor cells. *Neuron* 42, 535-552.
[25] Drapeau, E., Montaron, M.F., Aguerre, S. et al, (2007). Learning-induced survival of new neurons depends on the cognitive status of aged rats. *J. Neurophysiol.* 27, 6037-6044.
[26] Katnelson, A., (2011). How 'sham' brain surgery could be killing off valuable therapies for Parkinson's disease. *Nature* 476, 143-144.
[27] Palmer, T.D., Willhoite, A.R. and Gage, F.H., (2000). Vascular niche for adult hippocampal neurogenesis. *J. Comp. Neurol.* 425. 479-494.
[28] Schalow, G. and Schmidt, H., (1979). Local development of action potentials in slow muscle fibres after complete or partial denervation. *Proc. R. Soc. B* 203, 445-457.
[29] Schalow, G. and Schmidt, H., (1977). Effect of length and temperature on the induction of action potentials in denervated slow muscle fibres of the frog. *Pflügers Arch.* 372, 17-22.
[30] Schalow, G. and Lang, G., (1989). Elektrodiagnosis of human dorsal sacral nerve roots by recording afferent and efferent extracellular action potentials. *Neurosurg. Rev.* 12, 223-232.
[31] Schalow, G., (2005). Tremor in Parkinson's disease patients can be induced by uncontrolled activation and uninhibited synchronization of α_2-motoneuron firing to

which α_1-motoneuron firing synchronizes. *Electromyogr. Clin. Neurophysiol.* 45, 393-406.

[32] Zhao, C., Deng, W. and Gage, F.H., (2008) Mechanisms and functional implications of adult neurogenesis. *Cell* 132, 645-660.

[33] Miledi, R., Parker, I. and Schalow, G., (1981) Calcium transients in normal and denervated slow muscle fibres of the frog. *J. Physiol.* 318, 191-206.

[34] Song, H., Stevens, C.F., Gage, F.H., (2002) Astroglia induce neurogenesis from adult neural stem cells. *Nature* 417, 39-44.

[35] Marco Garcia, N.V., Karayannis, T. & Fishell, G., (2011) Neuronal activity is required for the development of specific cortical interneuron subtypes. Nature 472, 351-355.

[36] Batista-Brito, R. & Fishell, G., (2009). The developmental integration of cortical interneurons into a functional network. *Curr. Top. Dev. Biol.* 87, 81-118.

[37] Blankenship, A.G. & Feller, M.B., (2010). Mechanisms underlying spontaneous patterned activity in developing neural circuits. *Nature Rev. Neurosci.* 11, 18-29.

[38] Wang, C.L. et al., (2007). Activity-dependent development of callosal projections in the somatosensory cortex. *J. Neurosci.* 27, 11334-11342.

[39] Batista-Brito, R. and Fishell, G., (2009). The developmental integration of cortical interneurons into a functional network. *Current. Top. Dev. Biol.* 87, 87-118.

[40] Schalow, G. and Nyffeler, T., (2001). Koordinationsdynamik-Therapie: Myelomeningozele (Spina bifida). Physiotherapie.

[41] Corless, J., (2012). Powering up. Nature 485, S62-S63.

[42] Popper, K.R., (1972). Objective knowledge: An evolutionary approach. London, O.U.P..

[43] Piper, M.C. and Darrah, J., (1994). Motor Assessment of the Developing Infant. W.B. Saunders Company.

[44] Bareyre, F.M. et al., (2004). The injured spinal cord spontaneously forms a new intraspinal circuit in adult rats. *Nature Neurosci.* 7, 269-277.

[45] Raineteau, O. and Schwab, M.E., (2001). Plasticity of motor systems after incomplete spinal cord injury. *Nature Rev.* Neurosci. 2, 263-273.

[46] Grillner, S., (2003). The motor infrastructure: From ion channels to neuronal networks. *Nature Reviews Neuroscience* 4, 573-586.

[47] Jeka, J.J. and Kelso, J.A.S., (1995). Manipulating symmetry in the coordination dynamics of human movements. *Journal of Experimental Psychology* 21 (2), 360-374.

[48] Schalow, G., (2006). Symmetry diagnosis and treatment in coordination dynamics therapy. Electromyogr. *Clin. Neurophysiol.* 46, 421-431.

[49] Schalow, G., (1991). Conduction velocities and nerve fibre diameters of touch, pain, urinary bladder and anal canal afferents and α and γ-motoneurons in human dorsal sacral roots. Electromyogr. *Clin. Neurophysiol.* 31, 265-296.

[50] Schalow, G. and Pääsuke, M., (2003). Low-load coordination dynamics in athletes, physiotherapists, gymnasts, musicians and patients with spinal cord injury, after stroke, traumatic brain lesion and with cerebral palsy. *Electromyogr. Clin. Neurophysiol.* 43, 195-201.

[51] Schalow, G., Pääsuke, M and Kolts, I, (2003). High-load coordination dynamics in athletes, physiotherapists, gymnasts, musicians and patients with CNS injury. *Electromyogr. Clin. Neurophysiol.* 43, 353-365.

[52] Schalow, G., Vaher, I. and Jaigma, P., (2008). Overreaching in coordination dynamics therapy in an athlete with a spinal cord injury. *Electromyogr. Clin. Neurophysiol.* 48, 83-95.

[53] Schalow, G., (2008). Stem cell therapy and Coordination dynamics therapy improve Spinal cord injury. *Electromyogr. Clin. Neurophysiol.* 48, 233-253.

[54] Schalow, G., (1991). Oscillatory firing of single human sphincteric α_2 and α_3-motoneurons reflexly activated for the continence of urinary bladder and rectum - restoration of urinary bladder function in paraplegia. *Electromyogr. Clin. Neurophysiol.* 31, 323-355.

[55] Schalow, G., (1993). Spinal oscillators in man under normal and pathologic conditions. *Electromyogr. Clin. Neurophysiol.* 33, 409-426.

[56] Miledi, R., Parker, I. and Schalow, G., (1981). Calcium transients in normal and denervated slow muscle fibres of the frog. *J. Physiol.* 318, 191-206.

[57] Schalow, G. and Wattig, B., (1993). Recruitment of α and γ-motoneurons in rats, dogs and humans. Electromyogr. *Clin. Neurophysiol.* 33, 387-400.

[58] Schalow, G., (1993). Action potential patterns of intrafusal γ and parasympathetic motoneurons, secondary muscle spindle afferents and an oscillatory firing α_2-motoneuron, and the phase relations among them in humans. *Electromyogr. Clin. Neurophysiol.* 33, 477-503.

[59] Schalow, G., (2003). Pääsuke, M., Ereline, J. and Gapeyeva, H. Coordination dynamics in Parkinson's disease patients and healthy subjects quantified by the coordination dynamics recording method and sEMG. *Electromyogr. Clin. Neurophysiol.* 43, 473-485.

[60] Schalow, G., Pääsuke, M., Jaigma, P., (2005). Integrative re-organization mechanism for reducing tremor in Parkinson's disease patients. *Electromyogr. Clin. Neurophysiol.* 45, 407-415.

[61] Schalow, G., Bersch, U., Göcking, K. and Zäch, G.A., (1994). Interspike intervals of secondary muscle spindle and urinary bladder afferents in relation to the oscillation periods of sacral spinal oscillators for continence in man. *Physiological Research* 43, 7-18.

[62] Schalow, G. and Zäch, G.A., (2000). Reorganization of the Human CNS, Neurophysiologic measurements on the coordination dynamics of the lesioned human brain and spinal cord. Theory for modern neurorehabilitation (31 case reports). *Gen. Physiol. Biophys.* 19, Suppl. 1, 1-244.

[63] Schalow, G. and Zäch, G.A., (1996). External loops of human premotor spinal oscillators identified by simultaneous measurements of interspike intervals and phase relations. *Gen. Physiol. Biophys.* 15, Suppl.1, 95-119.

[64] Schalow, G., Pääsuke, M., Jaigma, P., (2005). Integrative re-organization mechanism for reducing tremor in Parkinson's disease patients. *Electromyogr. Clin. Neurophysiol.* 45, 407-415.

[65] Brown, P., (2000). Cortical drives to human muscle: the Piper and related rhythms. *Progress in Neurobiology* 60, 97-108.

[66] McAuley, J.H., (2003). The physiological basis of clinical deficits in Parkinson's disease. *Progress in Neurobiology* 69, 27-48.

[67] Windhorst, U., (1988). How Brain-Like is the Spinal Cord? Springer Verlag, Berlin.

[68] Eriksson, P.S., Perfilieva, E., Björk-Eriksson, T., Alborn, A.M., Nordborg, C., Peterson D.A. and Gage, F.H., (1998). Neurogenesis in the adult human hippocampus. *Nature* Medicine 4, 1313-1317.

[69] Praag, van H., Kempermann, G., Gage H., (1999). Running increases cell proliferation and neurogenesis in the adult mouse dentate gyrus. *Nature Neurosci.* 2(3), 266-270.

[70] Latash, M.L., (1998). Neurophysiologic Basis of Movement. Human Kinetics.

[71] Jankovic, J., (1987). Pathophysiology and clinical assessment of motor symptoms in Parkinson's disease. In: Koller, W.C. (ed.), Handbook of Parkinson's Disease. Marcel Dekker, New York, 99-126.

[72] Jellinger, K., (1987). The pathology of parkinsonism. In: Marsden, C.D. and Fahn, S. (eds.), Movement Disorders 2, Betterworths, London, 124-165.

[73] Makabe, H. and Sakamoto, K., (2002). Evaluation of postural tremor of finger for neuromuscular diseases and its application to the classification. *Electromyogr. Clin.* Neurophysiol. 42, 205-218.

[74] Pääsuke, M., Mottus, K., Ereline, J., Gapeyeva, H. and Taba, P., (2002). Lower limb performance in older female patients with Parkinson's disease. *Aging Clin. Exp. Res.* 14, 185-191.

[75] Carlson, A., (2002). Treatment of Parkinson's with L-DOPA. The early discovery phase, and a comment on current problems. *J. Neural. Transm.* 109, 777-787.

[76] Kandel, E.R., (1984). Steps Towards a Molecular Grammar For Learning: Explorations Into The Nature of Memory. Wiley, New York.

[77] Schalow, G., (2002). Non-drug induced spasticity reduction achieved by coordination dynamic therapy in CNS injury. *Electromyogr. Clin. Neurophysiol.* 42, 281-293.

[78] Jaigma, P., Schalow, G. and Pääsuke, M., (2004). The effect of 3-month Schalow coordination dynamic therapy on movement coordination characteristics of the limbs in subjects with cerebral palsy. *Acta Kinesiologiae Universitatis Tartuensis* 9, 66-78.

[79] Schalow, G., (2006). Functional development of the CNS in pupils between 7 and 19 years. Electromyogr. *Clin. Neurophysiol.* 46, 159-169.

[80] Hebb, D.O., (1949). The organization of behaviour. New York, Wiley.

[81] Feldenkrais, G., (1972). Awareness through movement: health exercises for personal growth. *Harper and Row*, London.

[82] Gelfan, S. and Tarlov, I.M., (1955). Differential vulnerability of spinal cord structures to anoxia. *J. Neurophysiol.* 18, 170-188.

[83] Gelfan, S. and Tarlov, I.M., (1959). Interneurons and rigidity of spinal origin. *J. Physiol.* 146, 594-617.

[84] Krogh, E., (1950). The effect of acute hypoxia on the motor cells of the spinal cord. *Acta Physiol. Scand.* 20, 263-292.

[85] Gerstenbrand, F., (1998). Wir müssen alles daran setzen eine Euthanasie apallischer Patienten zu verhindern. Wachkoma 2.

[86] Kim, K.H.S., Relkin, N.R., Lee, K.-M. and Hirsch, J., (1997). Distinct cortical areas associated with native and second languages. *Nature* 388, 171-174.

[87] De Marco Garcia, N.V., Karayannis, T. and Fishell, G., (2011). Neuronal activity is required for the development of specific cortical interneuron subtypes. *Nature* 472, 352-356.

[88] Schalow, G., (2001). Time axis calibration in human CNS organization for judging dysfunction. *Electromyogr. Clin. Neurophysiol.* 41, 485-505.

[89] Stahl, B., Hensele, I., Turner,R., Geyer S. and A. Kotz, S.A., (2013). How to engage the right brain hemisphere in aphasics without even singing: evidence for two paths of speech recovery. *Front. Hum. Neurosci.* 7, 35.

[90] Varela, F., Lachaux, J.-P., Rodriguez, E. and Martinerie, J., (2001). The brainweb: Phase synchronization and large-scale integration. *Nature Reviews/Neuroscience* 2, 229-239.

[91] Schalow, G. and Zäch, G.A., (1999). Koordinationsdynamik-Therapie als Lernprozess (Teil 2). *Physiotherapie.* (Poliomyelitis)

[92] Schalow, G. and Zäch, G.A., (1996). Mono- and polysynaptic drive of oscillatory firing α_1 (FF) and α_2-motoneurons (FR) in a patient with spinal cord lesion. *Gen. Physiol. Biophys.* 15, Suppl.1, 57-74.

[93] Praag van H., Kempermann G., Gage, H., (1999). Running increases cell proliferation and neurogenesis in the adult mouse dentate gyrus. *Nature Neurosci.* 2(3), 266-270.

[94] Gould E., Beylin A., Tanapat P., Reeves A., Shors T.J., (1999). Learning enhances adult neurogenesis in the hippocampal formation. *Nature Neurosci.* 2(3), 260-265.

[95] Greenough, W.T., Cohen, N.J., Jurasko, J.M., (1999). New neurons in old brains: learning to survive? *Nature Neurosci.* 2(3), 203-205.

[96] Kandel, E.R., Schwartz, J.H. and Jessell, T.M., (1991). *Principles of neural Science. Appleton & Lange,* East Norwalk.

[97] Ge, S., Yang, C., Hsu, K., Ming, G. and Song, H., (2007). A critical period for enhanced synaptic plasticity in newly generated neurons of the adult brain. *Neuron*, 54, 559-566.

[98] Thelen E., Smith L.B., (1994): A Dynamic Approach to the Development of Cognition and Action. MIT Press, Cambridge.

[99] Freund H.-J., (1996). Remapping the brain. *Science* 272, 1754.

[100] Kempermann, G., Kuhn, H.G., Gage, F.H., (1997). More hippocampal neurons in adult mice living in an enriched environment. *Nature* 386, 493-495.

[101] Bobath, B., (1978). Adult Hemiplegia: Evaluation and Treatment. William Heinemann Medical Books Ltd, London.

[102] Koepchen, H.P., Lazar, H., Klüßendorf, D., Hukuhara, T., (1986). 'Medullary apneusis' by lesions and cooling in the ventrolateral solitary tract and genesis of respiratory rhythm. *J. Autonom. Nerv. Syst. Suppl.*, 63-69.

[103] Schalow, G., Blanc, Y., Jeltsch, W. and Zäch, G.A., (1996). Electromyographic identification of spinal oscillator patterns and recouplings in a patient with incomplete spinal cord lesion: Oscillator formation training as a method to improve motor activities. *Gen. Physiol. Biophys.* 15, Suppl. 1, 121-220.

[104] Schalow, G., (1993). Recruitment of motoneurons in the occasional firing mode in paraplegics. Electromyogr. *Clin. Neurophysiol.* 33, 401-408.

[105] Windle, W.F., Smart, J.O. and Beers, J.J., (1958). Residual function after subtotal spinal cord transection in adult cats. *Neurology* 8, 518-521.

[106] Bradbury, E.J. and McMahon, S.B., (2006). Spinal cord repair strategies: why do they work? *Nat. Rev. Neurosci.* 7, 644-653.

[107] Harel, N.Y. and Strittmatter, S.M., (2006). Can regenerating axons recapitulate developmental guidance during recovery from spinal cord injury. *Nat. Rev. Neurosci.* 7, 603-616.

[108] Haken, H., Kelso, J.A., & Bunz, H., (1985). A theoretical model of phase transitions in human hand movements. *Biological Cybernetics*, 39, 139-156.

[109] Fritsch B., (1995). Evolution of the ancestral vertebrate brain. In: The Handbook of Brain Theory and Neural Networks (M.A. Arbib, Ed.) pp. 373-377, MIT Press.

[110] Kaufmann S.A., (1993). The Origin of Order: Self-Organization and Selection in Evolution. Oxford University Press, New York.

[111] Parisi D., Cecconi F., Nolfi S., (1990). Econets: Neuronal networks that learn in the environment. *Network* 1, 149-168.

[112] Miller G., Todd P.M. and Hedge S.U., (1989). Designing neural networks using genetic algorithms. In: Proceedings of the Third International Conference on Genetic Algorithms (L. Nadel and D. Stein, Eds.) pp 379-384, San Mateo, CA: Morgan Kaufmann.

[113] Belew R.K., McInerney J., Schraudolph N.N., (1992). Evolving networks: Using the genetic algorithm with connectionist learning. In: Artificial Life II (C.G. Langton, Ed.), pp 511-547 Reading, MA: Addison-Wesley.

[114] Thuret, S., Moon, L.D.F. and Gage, Gage, F.H., (2006). Therapeutic interventions after spinal cord injury. *Nat. Rev. Neurosci.* 7, 628-643.

[115] Nolfi S., Parisi D., (1995). "Genotypes" for neural networks. In: The Handbook of Brain Theory and Neural Networks (M.A. Arbib, Ed.) pp 431-434, MIT Press.

[116] Nolfi S., Parisi D., (1997). Learning to adapt to changing environments in evolving neuronal networks. Adaptive Behav. 1(5), 75-98.

[117] Schalow, G., *Aho, A. and Lang*, G., (1992). Microanatomy and number of nerve fibres of the lower intercostal nerves with respect to a nerve anastomosis. Donor nerve analysis. I (IV). *Electromyogr. Clin. Neurophysiol.* 32, 171-185.

[118] Schalow, G., (1992). Number of fibres and fibre diameter distributions of nerves innervating the urinary bladder in humans - Acceptor nerve analysis. II (IV). *Electromyogr. Clin. Neurophysiol.* 32, 187-196.

[119] Schalow, G., (2009). The classification and identification of human somatic and parasympathetic nerve fibres including urinary bladder afferents and efferents is preserved following spinal cord injury. *Electromyogr. Clin. Neurophysiol.* 49, 263-286.

[120] Schalow, G., (2009). Coordination impairment between the somatic and parasympathetic nervous system divisions in the human sacral micturition centre following spinal cord injury. *Electromyogr. Clin. Neurophysiol.* 49, 337-367.

[121] Schalow, G., (2010). Phase relation changes between the firings of α and γ-motoneurons and muscle spindle afferents in the sacral micturition centre during continence functions in brain-dead human and patients with spinal cord injury. *Electromyogr. Clin. Neurophysiol.* 50, 3-27.

[122] Schöner, G., Zanone, P.G. and Kelso, J.A.S., (1992). Learning as a change of coordination dynamics: Theory and Experiment. *J. Motor Behavior* 24, 29-48.

[123] Haken, H., Kelso, J.A., & Bunz, H., (1985). A theoretical model of phase transitions in human hand movements. Biological Cybernetics 39, 139-156.

[124] Schalow, G. and Zäch, G.A., (1994). Nerve compound action potentials analysed with the simultaneously measured single fibre action potentials in humans. *Electromyogr. clin. Neurophysiol.* 34, 451-465.

[125] Schalow, G., (2005). Tapering of human nerve *fibres. Gen. Physiol. Biophys.* 24, 427-448.

[126] Passatore G.M., Fillipi G.M., Grassi G., (1985). Cervical sympathetic nerve stimulation can induce an intrafusal muscle fibre contraction in the rabbit. In: The Muscle Spindle (I.A. Boyd and M.H. Gladden, Eds.) pp 221-226, Stockton Press, New York.

[127] Pavlidis, T., (1973). Biological Oscillators: Their Mathematical Analysis. Academic Press, New York.

[128] Hess, W.R., (1981). Biological order and human society. In: Akert, K. (Ed.), Biological Order and Brain Organization - Selected Works of W.R. Hess, pp 3-15, Springer-Verlag, Berlin.

[129] Schalow, G., (1992). Recruitment within the groups of γ_1, α_2 and α_3-motoneurons in dogs and humans following bladder and anal catheter pulling. *Gen. Physiol. Biophys.* 11, 101-121.

[130] Chklovskii, D.B., Mel, B.W. and Svoboda, K., (2004). Cortical rewiring and information storage. *Nature* 431, 782-787.

[131] Nottebohm, F., (2002). Why are some neurons replaced in adult brain? *The Journal of Neuroscience* 22, 624-628.

[132] Ferretti, P., Zhang, F. and O'Neill, P., (2003). Changes in spinal cord regenerative ability through phylogenesis and development: lessons to be learned. *Dev. Dyn.* 226, 245-256.

[133] Sperry, R.W., (1945). The problem of central nervous reorganization after nerve regeneration and muscle transposition. *Quart. Rev. Biol.* 20, 311-369.

[134] Sperry, R.W., (1947). Effect of crossing nerves to antagonistic limb muscles in the monkey. *Arch. Neurol. Psychiat. Chicago* 58, 452-473.

[135] Weiss, P. and Brown, P.F., (1941). Electromyographic study on recoordination of leg movements in poliomyelitis patients with transposed tendons. Proceedings of the Society for Exper. *Biology and Medicine* 48, 384-387.

[136] Bhardwj, R.D. et al, (2006). Neocortical neurogenesis in humans is restricted to development. *PNAS* 103 (33), 12564-12568.

[137] Guttmann, L., (1973). Spinal cord injuries. Comprehensive management and research. Blackwell, Oxford.

[138] Nash, M.S., (2005). Exercise as a health-promoting activity following spinal cord injury. *J.Neurol. Phys. Ther.* 29, 87-103.

[139] Schalow, G., (1990). Feeder arteries, longitudinal arterial trunks and arterial anastomoses of the lower human spinal cord. *Zbl. Neurochir.* 51, 181-184.

[140] Li, Y., Field, P.M. and Raisman, G., (1997). Repair of adult rat corticospinal tract by transplants of olfactory ensheathing cells. *Science* 277, 2000-2002.

[141] Li, Y., Field, P.M. and Raisman, G., (1998). Regeneration of adult rat corticospinal axons induced by transplanted olfactory ensheathing cells. *J. Neurosci.* 18, 10514-10524.

[142] Windle, W.F., Smart. J.O. and Beers, J.J., (1958). Residual function after subtotal spinal cord transaction in adult cats. *Neurology* 8, 518-52.

[143] Nathan, P.W., (1994). Effects on movement of surgical incisions into the human spinal cord. Brain 117,: 337-346.

[144] Schalow, G., (1983). Elektrophysiologische und morphologische Untersuchungen zur Differenzierung der Tonusfasern des Frosches. Inauguraldissertation, Universität des Saarlandes.

[145] Miledi, R., Parker, I. and Schalow, G., (1977). Calcium transients in frog slow muscle fibres, in frog skeletal muscle fibres using arsenazo III. *Nature*, 268 No. 5622, 750-752.

[146] Miledi, R., Parker, I. and Schalow, G., (1981). Calcium transients in normal and denervated slow muscle fibres of the frog. *J. Physiol.* 318, 191-206.

[147] Wong, R.,O., Chernjavsky, A., Smith, S.J. and Shatz, C.J., (1995). Early functional neural networks in the developing retina. *Nature* 374, 716-718.

[148] Penn, A.A., Riquelme, P.A., Feller M.B. and Shatz, C.J., (1998). Competition in retinogeniculate patterning driven by spontaneous activity. *Science* 279, 2108-2112.

[149] Hubermann, A.D. et al, (2008). Architecture and activity-mediated refinement of axonal projections from mosaic of genetically identified ganglion cells. *Neuron* 59, 425-438.

[150] Spitzer, N.C., (2006). Electrical activity in early neuronal development. *Nature* 444, 707-712.

[151] Root, C.M., Velazquez-Ulloa, N.A., Monsalve, G.C., Minakova, E. and Spitzer, N.C., (2008). Embryonically expressed GABA and glutamate drive electrical activity regulating neurotransmitter specification. *J. Neurosc.* 28, 4777-4784.

[152] Cancedda, L., Fiumelli, H., Chen, K. and Poo, M.M., (2007). Excitatory GABA action is essential for morphological maturation of cortical neurons in vivo. *J. Neurosc.* 27, 5224-5235.

[153] Wang, C.L. et al, (2007). Activity-dependent development of callosal projections in the somatosensory cortex. *J. Neurosci.* 27, 11334-11342.

[154] Knudsen, E.I., (2002). Instructed learning in the auditory localization pathway of the barn owl. *Nature* 417, 322-328.

[155] Knudson. E.I., (1998). Capacity for plasticity in the adult owl auditory system expanded by juvenile experience. *Science* 279, 1531-1533.

[156] Walton, K.D. et al, (1992). Identification of a critical period for motor development in neonatal rats. Neuroscience 51, 763-767.

[157] Kudo, N., Nishimaru, H. and Nakayama, K., (2004). Developmental changes in rhythmic spinal neuronal activity in the rat fetus. *Prog. Brain Res.* 143, 49-55.

[158] Allein, A.E., Meyrand, P. and Branchereau, P., (2005). Ontogentic changes of the spinal GABAergic cell population are controlled by the serotonin (5-HT) system: implication of 5-HT , receptor family. *J. Neurosci.* 25, 8714-8724.

[159] Lander, C. et al, (1997). A family of activity-dependent neuronal cell-surface chondroitin sulphate proteoglycans in cat visual cortex. *J. Neurosc.* 17, 1928-1939.

[160] McGee, A.W. et al, (2005). Experience-driven plasticity of visual cortex limited by myelin and Nogo receptor. *Science* 309: 2222-2226.

[161] Pizzorusso, T. et al, (2002). Reactivation of ocular dominance plasticity in the adult visual cortex. *Science* 298, 1248-125.

[162] Gasser H.S., Grundfest H., (1939). Axon diameter in relation to the spike dimension and the conduction velocity in mammalian A fibres. *Amer. J. Physiol.* 127, 393-414.

[163] Hunt C.C., (1954). Relation of function to diameter in afferent fibres of muscle nerve. J. Gen. Physiol. 38, 117-131.

[164] Hursh J.B. (1939): Conduction velocity and diameter of nerve fibres. *Amer. J. Physiol.* 127, 131-139

[165] Lloyd D.P.C., Chang H.T., (1948). Afferent fibres in muscle nerves. *J. Neurophysiol.* 11, 199-207.

[166] de Groat, W.C., (1975). Nervous control of the urinary bladder of the cat. *Brain. Res.* 87, 201-211.

[167] Torrens, M. and Morrison, J.F.B., (1987). The Physiology of the Lower Urinary Tract, Springer Verlag.

[168] Gossard, J.-P., Brownstone, R.M., Barajon, I. and Hultborn, H., (1994). Transmission in a locomotor-related group Ib pathway from hindlimb extensor muscles in the cat. Exp. Brain Res. 98, 213-228.

[169] Lawson, J., (1974). Pelvic anatomy, I. Pelvic floor muscles. *Ann. R. Coll. Surg. Engl.* 54, 244-252.

[170] Percy, J.P., Swash, M., Neill, M.E. and Parks, A.G., (1980). Electrophysiological study of motor nerve supply of pelvic floor. *Lancet* 3,16-17.

[171] Garry, R.C. and Garven, H.S.D., (1957). The ganglia, afferent nerve-endings and musculature of the urethra in the cat. *J. Physiol.* 139, 1P-2P.

[172] Kugelberg, E. and Hagbarth, K.E., (1958). Spinal mechanisms of the abdominal and erector spinae skin reflexes. *Brain* 81, 290-304.

[173] Nadelhaft, I., Ropollo, J., Morgan, C. and De Groat, W.C., (1983). Parasympathetic preganglionic neurons and visceral primary afferents in monkey sacral spinal cord revealed following application of horseradish peroxydase to pelvic nerve. *J. Compar. Neurol.* 216, 36-52.

[174] Lawson, J., (1974). Pelvic anatomy, I. Pelvic floor muscles. *Ann. R. Coll. Surg. Engl.* 54, 244-252.

[175] Percy, J.P., Swash, M., Neill, M.E. and Parks, A.G., (1980). Electrophysiological study of motor nerve supply of pelvic floor. *Lancet* 3,16-17.

[176] Bors, E. and Porter, R.W., (1970). Neurosurgical consideration in bladder dysfunction. Urol. Int. 25, 114-133.

[177] Gosling, J.A., Dixon, J.S., Humpherson, J.R., (1983). Functional anatomy of the urinary tract. Churchill Livingston distributed, Gower Medical Publishing, London, pp 5-18.

[178] Evans, J.P., (1936). Observation on the nerves of supply to the bladder and urethra of the cat, with a study of their action potentials. *J. Physiol.* 86, 396-413.

[179] Schalow, G. and Barth, H., (1992). Group conduction velocities and nerve fibre diameters of α and γ-motoneurons from lower sacral nerve roots of the dog and humans. *Gen. Physiol. Biophys.* 11, 85-99.

[180] Schalow, G., (1992). Recruitment within the groups of γ_1, α_2 and α_3-motoneurons in dogs and humans following bladder and anal catheter pulling. *Gen. Physiol. Biophys.* 11, 101-121.

[181] Schalow, G., (1992). Ventral root afferent and dorsal root efferent fibres in dog and human lower sacral nerve roots. *Gen. Physiol. Biophys.* 11, 123-131.

[182] Schalow, G., (2005). Tapering of human nerve fibres. *Gen. Physiol. Biophys.* 24, 427-448.

[183] Bhardwj, R.D. et al., (2006). Neocortical neurogenesis in humans is restricted to development. *PNAS*, 103 (33), 12564-12568.

[184] Spalding, K.L., Bhardwaj, R.D., Buchholz, B.A., Druid, H. & Frisen J., (2005). Retrospective birth dating of cells in humans. *Cell* 122, 133-143.

[185] Ferretti, P., Zhang, F. & O'Neill, P., (2003). Changes in spinal cord regenerative ability through phylogenies and development: lessons to be learned. *Dev. Dyn.* 226, 245-256.

[186] 86. Martino, G. and Pluchino, S., (2006). The therapeutic potential of neural stem cells. *Nature Reviews / Neuroscience* 7, 395 -406.

[187] Gage, F.H., (2002). Neurogenesis in the adult brain. *The Journal of Neuroscience* 22(3), 612-613.

[188] Gould, E., and Gross, C.G., (2002). Neurogenesis in Adult Mammals: Some progress and problems. *The Journal of Neuroscience* 22 (3), 619-623.

[189] Praag, H. van, Shubert, T., Zhao, C. and Gage, F.H., (2005). Exercise enhances learning and hippocampal neurogenesis in aged mice. *The Journal of Neuroscience* 25(38), 8680-8685.

[190] Leuner, B., Gould, E. and Shors, T.J., (2006). Is there a link between adult neurogenesis and learning? *Hippocampus* 16, 216-224.

[191] Keirstead, H.S. et al, (2005). Human embryonic stem cell derived oligodendrocyte progenitor cell transplants remyelinate and restore locomotion after spinal cord injury. *J. Neurosci.* 25, 4694-4705 (rat).

[192] Kerschensteiner, M. et al, (2005). In vivo imaging of axonal degeneration and regeneration in the injured spinal cord. *Nature Med.* 11, 572-577 (rat).

[193] Koepchen, H.P. et al, (1986). 'Medullary apneusis' by lesions and cooling in the ventrolateral solitary tract and genesis of respiratory rhythm. *J. Autonom. Nerv. Syst.* Suppl., 63-69.

[194] McDonald, J.W. et al, (1999). Transplanted embryonic stem cells survive, differentiate and promote recovery in injured rat spinal cord. *Nature Med.* 5, 1410-1412.

[195] Chia-Wei Cheng et al, (2014). Prolonged fasting reduces IGF-1/PKA to promote hematopoietic-stem-cell-based regeneration and reverse Immunosuppression. *Cell Stem Cell* 14, 810-823.

[196] Koepchen, H.P., (1990). Physiology of rhythms and control systems: An integrative approach. In: Haken H, Koepchen HP editors. *Rhythms in Physiological Systems*. Berlin, pp 3-20.

[197] Lambertz, M. & Langhorst, P., (1998). Simultaneous changes of rhythmic organization in brainstem neurons, respiration, cardiovascular system and EEG between 0.05 Hz and 0.5 Hz. *J. Auton. Nerv. Syst.* 68, 58-77.

[198] Thelen, E. and Smith, L.B., (1994). A dynamic approach to the development of cognition and action. MIT Press, Cambridge.

[199] Safdie, F.M et al. (2009). Fasting and Cancer treatment in Humans: A Case series report. *Aging* 1, No 12, 1-20.

[200] Raffaghello, L. et al. (2010). Fasting and differential chemotherapy protection in patients. *Cell Cycle* 9:22, 4474-4476.

Index

D

E

F

G

H

I

J

L

M

N

O

P

Q

R

S

T

U

V

W